T0190075

CLASSICAL DYNAMICS

Recent advances in the study of dynamical systems have revolutionized the way that classical mechanics is taught and understood. *Classical Dynamics: A Contemporary Approach* is a new and comprehensive textbook that provides a complete description of this fundamental branch of physics.

The authors cover all the material that one would expect to find in a standard graduate course: Lagrangian and Hamiltonian dynamics, canonical transformations, the Hamilton–Jacobi equation, perturbation methods, and rigid bodies. They also deal with more advanced topics such as the relativistic Kepler problem, Liouville and Darboux theorems, and inverse and chaotic scattering. Key features of the book are the early gradual introduction of geometric (differential manifold) ideas, and detailed treatment of topics in nonlinear dynamics (such as the KAM theorem) and continuum dynamics (including solitons).

The book contains over 200 homework exercises. A solutions manual is available exclusively for instructors. Additional unique features of the book include the many solved exercises (Worked Examples) as well as other illustrative examples designed to expand on the material of the text.

It will be an ideal textbook for graduate students of physics, applied mathematics, theoretical chemistry, and engineering, and will also serve as a useful reference for researchers in these fields.

Jorge José studied at the National University of Mexico. He has held positions at Brown University, University of Chicago, University of Mexico, the Theoretical Physics Institute of Utrecht University, the Netherlands, and the Van der Waals Institute at Universiteit van Amsterdam. He is a fellow of the American Physical Society. Currently he is the Matthews Distinguished University Professor and the Director of the Center for Interdisciplinary Research on Complex Systems at Northeastern University.

Eugene Saletan, an emeritus professor of physics at Northeastern University, earned his Ph.D. in physics from Princeton. He has written two other books: *Theoretical Mechanics* with Alan Cromer and *Dynamical Systems* with Marmo, Simoni, and Vitale.

CLASSICAL DYNAMICS:

A CONTEMPORARY APPROACH

JORGE V. JOSÉ

and

EUGENE J. SALETAN

CAMBRIDGE
UNIVERSITY PRESS

CAMBRIDGE
UNIVERSITY PRESS

University Printing House, Cambridge CB2 8BS, United Kingdom

Published in the United States of America by Cambridge University Press, New York

Cambridge University Press is part of the University of Cambridge.

It furthers the University's mission by disseminating knowledge in the pursuit of education, learning and research at the highest international levels of excellence.

www.cambridge.org
Information on this title: www.cambridge.org/9780521636360

© Cambridge University Press 1998

First published 1998
8th printing 2012

A catalogue record for this publication is available from the British Library

Library of Congress Cataloguing in Publication data

José, Jorge V. (Jorge Valenzuela) 1949–
Classical dynamics : a contemporary approach / Jorge V. José,
Eugene J. Saletan
    p.   cm.
1. Mechanics, Analytic. I. Saletan, Eugene J. (Eugene Jerome),
1924– . II. Title.
QA805.J73   1998   97-43733
531'.11'01515 – dc21   CIP

ISBN 978-0-521-63176-1 Hardback
ISBN 978-0-521-63636-0 Paperback

. . . Y lo dedico, con profundo cariño, a todos mis seres
queridos que habitan en ambos lados del río Bravo.

JVJ

Ellen – *Nel mezzo del cammin di nostra vita.*

EJS

CONTENTS

LIST OF WORKED EXAMPLES

PREFACE

Among the first courses taken by graduate students in physics in North America is Classical Mechanics. This book is a contemporary text for such a course, containing material traditionally found in the classical textbooks written through the early 1970s as well as recent developments that have transformed classical mechanics to a subject of significant contemporary research. It is an attempt to merge the traditional and the modern in one coherent presentation.

When we started writing the book we planned merely to update the classical book by Saletan and Cromer (1971) (SC) by adding more modern topics, mostly by emphasizing differential geometric and nonlinear dynamical methods. But that book was written when the frontier was largely quantum field theory, and the frontier has changed and is now moving in many different directions. Moreover, classical mechanics occupies a different position in contemporary physics than it did when SC was written. Thus this book is not merely an update of SC. Every page has been written anew and the book now includes many new topics that were not even in existence when SC was written. (Nevertheless, traces of SC remain and are evident in the frequent references to it.)

From the late seventeenth century well into the nineteenth, classical mechanics was one of the main driving forces in the development of physics, interacting strongly with developments in mathematics, both by borrowing and lending. The topics developed by its main protagonists, Newton, Lagrange, Euler, Hamilton, and Jacobi among others, form the basis of the traditional material.

In the first few decades following World War II, the graduate Classical Mechanics course, although still recognized as fundamental, was barely considered important in its own right to the education of a physicist: it was thought of mostly as a peg on which to hang quantum physics, field theory, and many-body theory, areas in which the budding physicist was expected to be working. Textbooks, including SC, concentrated on problems, mostly linear, both old and new, whose solutions could be obtained by reduction to quadrature, even though, as is now apparent, such systems form an exceptional subset of all classical dynamical systems.

In those same decades the subject itself was undergoing a rebirth and expanding, again in strong interaction with developments in mathematics. There has been an explosion in the study of nonlinear classical dynamical systems, centering in part around the discovery of novel phenomena such as chaos. (In its new incarnation the subject is often also called Dynamical Systems, particularly in its mathematical manifestations.)

What made substantive new advances possible in a subject as old as classical mechanics are two complementary developments. The first consists of qualitative but powerful geometric ideas through which the general nature of nonlinear systems can be studied (including global, rather than local analysis). The second, building upon the first, is the modern computer, which allows quantitative analysis of nonlinear systems that had not been amenable to full study by traditional analytic methods.

Unfortunately the new developments seldom found their way into Classical Mechanics courses and textbooks. There was one set of books for the traditional topics and another for the modern ones, and when we tried to teach a course that includes both the old and the new, we had to jump from one set to the other (and often to use original papers and reviews). In this book we attempt to bridge the gap: our main purpose is not only to bring the new developments to the fore, but to interweave them with more traditional topics, all under one umbrella.

That is the reason it became necessary to do more than simply update SC. In trying to mesh the modern developments with traditional subjects like the Lagrangian and Hamiltonian formulations and Hamilton–Jacobi theory we found that we needed to write an entirely new book and to add strong emphasis on nonlinear dynamics. As a result the book differs significantly not only from SC, but also from other classical textbooks such as Goldstein's.

The language of modern differential geometry is now used extensively in the literature, both physical and mathematical, in the same way that vector and matrix notation is used in place of writing out equations for each component and even of indicial notation. We therefore introduce geometric ideas early in the book and use them throughout, principally in the chapters on Hamiltonian dynamics, chaos, and Hamiltonian field theory.

Although we often present the results of computer calculations, we do not actually deal with programming as such. Nowadays that is usually treated in separate Computational Physics courses.

Because of the strong interaction between classical mechanics and mathematics, any modern book on classical mechanics must emphasize mathematics. In this book we do not shy away from that necessity. We try not to get too formal, however, in explaining the mathematics. For a rigorous treatment the reader will have to consult the mathematical literature, much of which we cite.

We have tried to start most chapters with the traditional subjects presented in a conventional way. The material then becomes more mathematically sophisticated and quantitative. Detailed applications are included both in the body of the text and as Worked Examples, whose purpose is to demonstrate to the student how the core material can be used in attacking problems.

The problems at the end of each chapter are meant to be an integral part of the course. They vary from simple extensions and mathematical exercises to more elaborate applications and include some material deliberately left for the student to discover.

An extensive bibliography is provided to further avenues of inquiry for the motivated student as well as to give credit to the proper authors of most of the ideas and developments in the book. (We have tried to be inclusive but cannot claim to be exhaustive; we apologize for works that we have failed to include and would be happy to add others that may be suggested for a possible later edition.)

Topics that are out of the mainstream of the presentation or that seem to us overly technical or represent descriptions of further developments (many with references to the literature) are set in smaller type and are bounded by vertical rules. Worked Examples are also set in smaller type and have a shaded background.

The book is undoubtedly too inclusive to be covered in a one-semester course, but it can be covered in a full year. It does not have to be studied from start to finish, and an instructor should be able to find several different fully coherent syllabus paths that include the basics of various aspects of the subject with choices from the large number of applications and extensions. We present two suggested paths at the end of this preface.

Chapter 1 is a brief review and some expansion of Newton's laws that the student is expected to bring from the undergraduate mechanics courses. It is in this chapter that velocity phase space is first introduced.

Chapter 2 and 3 are devoted to the Lagrangian formulation. In them geometric ideas are first introduced and the tangent manifold is described.

Chapter 4 covers scattering and linear oscillators. Chaos is first encountered in the context of scattering. Wave motion is introduced in the context of chains of coupled oscillators, to be used again later in connection with classical field theory.

Chapter 5 and 6 are devoted to the Hamiltonian formulation. They discuss symplectic geometry, completely integrable systems, the Hamilton–Jacobi method, perturbation theory, adiabatic invariance, and the theory of canonical transformations.

Chapter 7 is devoted to the important topic of nonlinearity. It treats nonlinear dynamical systems and maps, both continuous and discrete, as well as chaos in Hamiltonian systems and the essence of the KAM theorem.

Rigid-body motion is discussed in Chapter 8.

Chapter 9 is devoted to continuum dynamics (i.e., to classical field theory). It deals with wave equations, both linear and nonlinear, relativistic fields, and fluid dynamics. The nonlinear fields include sine–Gordon, nonlinear Klein–Gordon, as well as the Burgers and Korteweg–de Vries equations.

In the years that we have been writing this book, we have been aided by the direct help and comments of many people. In particular we want to thank Graham Farmelo for the many detailed suggestions he made concerning both substance and style. Jean Bellisard was very kind in explaining to us his version and understanding of the famous KAM theorem, which forms the basis of Section 7.5.4. Robert Dorfman made many useful suggestions after he and John Maddocks had used a preliminary version of the book in their Classical Mechanics course at the University of Maryland in 1994–5. Alan Cromer, Theo Ruijgrok, and Jeff Sokoloff also helped by reading and commenting on parts of the book. Colleagues from the Theoretical Physics Institute of Naples helped us understand the geometry that lies at the basis of much of this book. Special thanks should also go to Martin Schwarz for clarifying many subtle mathematical points to Eduardo Piña, and to Alain Chenciner. We are particularly grateful to Professor D. Schlüter for the very many corrections that he offered. We also want to thank the many anonymous referees for their constructive criticisms and suggestions, many of which we have tried to incorporate. We should also thank the many students who have used early versions of the book. The questions they raised and their suggestions have been particularly helpful. In addition, innumerable discussions with our colleagues, both present and past, have contributed to the project.

Last, but not least, JVJ wants to thank the Physics Institute of the National University of Mexico and the Theoretical Physics Institute of the University of Utrecht for their kind hospitality while part of the book was being written. The continuous support by the National Science Foundation and the Office of Naval Research has also been important in the completion of this project.

The software used in writing this book was Scientific Workplace. Most of the figures were produced using Corel Draw.

TWO PATHS THROUGH THE BOOK

In conclusion we present two suggested paths through the book for students with different undergraduate backgrounds. Both of these paths are for one-semester courses. We leave to the individual instructors the choice of material for students with really strong undergraduate backgrounds, for a second-semester graduate course, and for a one-year graduate course, all of which would treat in more detail the more advanced topics.

Path 1. For the "traditional" graduate course.

Comment: This is a suggested path through the book that comes as close as possible to the traditional course. On this path the geometry and nonlinear dynamics are minimized, though not excluded. The instructor might need to add material to this path. Other suggestions can be culled from path two.

Chapter 1. Quick review.

Chapter 2. Sections 2.1, 2.2, 2.3.1, 2.3.2, and 2.4.1.

Chapter 3. Sections 3.1.1, 3.2.1 (first and third subsections), and 3.3.1.

Chapter 4. Sections 4.1.1, 4.1.2, 4.2.1, 4.2.3, and 4.2.4.

Chapter 5. Sections 5.1.1, 5.1.3, 5.3.1, and 5.3.3.

Chapter 6. Sections 6.1.1, 6.1.2, 6.2.1, 6.2.2 (first subsection), 6.3.1, 6.3.2 (first three subsections), and 6.4.1.

Chapter 7. Sections 7.1.1, 7.1.2, 7.4.2, and 7.5.1.

Chapter 8. Sections 8.1, 8.2.1 (first two subsections), 8.3.1., and 8.3.3 (first three subsections).

Path 2. For students who have had a good undergraduate course, but one without Hamiltonian dynamics.

Comment: A lot depends on the students's background. Therefore some sections are labeled IN, for "If New." If, in addition, the students' background includes Hamiltonian dynamics, much of the first few chapters can be skimmed and the emphasis placed on later material. At the end of this path we indicate some sections that might be added for optional enrichment or substituted for skipped material.

Chapter 1. Quick review.

Chapter 2. Sections 2.1.3, 2.2.2–2.2.4, and 2.4.

Chapter 3. Sections 3.1.1 (IN), 3.2, 3.3.1, 3.3.2 (IN), and 3.4.1.

Chapter 4. Sections 4.1.1 (IN), 4.1.2, 4.1.3, 4.2.1 (IN), 4.2.2, 4.2.3, and 4.2.4 (IN).

Chapter 5. Sections 5.1.1, 5.1.3, 5.2, 5.3.1, 5.3.3, 5.3.4 (first two subsections), and 5.4.1 (first two subsections).

Chapter 6. Sections 6.1.1, 6.1.2, 6.2.1, 6.2.2 (first and fourth subsections), 6.3.1, 6.3.2 (first four subsections), 6.4.1, and 6.4.4.

Chapter 7. Sections 7.1.1, 7.1.2, 7.2, 7.4, and 7.5.1–7.5.3.

Chapter 8. Sections 8.1, 8.2.1, 8.3.1, and 8.3.3.

Chapter 9. Section 9.1.

Suggested material for optional enrichment:

Chapter 2. Section 2.3.3.

Chapter 3. Section 3.1.2.

Chapter 4. Section 4.1.4.

Chapter 5. Sections 5.1.2, 5.3.4 (third subsection), and 5.4.1 (third and fourth subsections).

Chapter 6. Sections 6.2.3, 6.3.2 (fifth and sixth subsections), 6.4.2, and 6.4.3.

Chapter 7. Sections 7.1.3, 7.3, 7.5.4, and the appendix.

Chapter 8. Sections 8.2.2 and 8.2.3.

Chapter 9. Section 9.2.1.

CHAPTER 1

FUNDAMENTALS OF MECHANICS

CHAPTER OVERVIEW

This chapter discusses some aspects of elementary mechanics. It is assumed that the reader has worked out many problems using the basic techniques of Newtonian mechanics. The brief review presented here thus emphasizes the underlying ideas and introduces the notation and the geometrical approach to mechanics that will be used throughout the book.

1.1 ELEMENTARY KINEMATICS

1.1.1 TRAJECTORIES OF POINT PARTICLES

The main goal of classical mechanics is to describe and explain the motion of macroscopic objects acted upon by external forces. Because the position of a moving object is specified by the location of every point composing it, we must start by considering how to specify the location of an arbitrary point. This is done by giving the coordinates of the point in a coordinate system, called the *reference system* or *reference frame*. Each point in space is associated with a set of three real numbers, the *coordinates* of the point, and this association is *unique*, which means that any given set of coordinates is associated with only one point or that two different points have different sets of coordinates. Probably the most familiar example of this geometric construction is a *Cartesian* coordinate system; other examples are *spherical polar* and *cylindrical polar* coordinates.

Given a reference frame, the position of a point can be specified by giving the radius vector $\mathbf{x}$ which goes from the origin of the frame to the point. In a Cartesian frame, the components of $\mathbf{x}$ with respect to the three axes X_1, X_2, X_3 of the frame, are the coordinates x_1, x_2, x_3 of the point. One writes

$$\mathbf{x} = \sum_{i=1}^{3} x_i \mathbf{e}_i \equiv x_1 \mathbf{e}_1 + x_2 \mathbf{e}_2 + x_3 \mathbf{e}_3, \tag{1.1}$$

where $\mathbf{e}_i$ is the unit vector in the ith direction. We are assuming here that space is three dimensional and Euclidean and that it has the usual properties one associates with such a space (Shilov, 1974; Doubrovine et al., 1982). This assumption is necessary for much of what we will say, becoming more explicit in the next section, on dynamics.

To simplify many of the equations, we will now start using the *summation convention*, according to which the summation sign Σ is omitted together with its limits in equations such as (1.1). An index that appears twice in a mathematical term is summed over the entire range of that index, which should be specified the first time it appears. The occasional exceptions to this convention will always be explained. If an index appears singly, the equation applies to its entire range, even in isolated mathematical expressions. Thus, "The $\alpha_{ij}\xi_j$ are real" means that the expression $\alpha_{ij}\xi_j$, summed over the range of j, is real for each value of i in the range of i. If an index appears more than twice, its use will be explained.

Accordingly, Eq. (1.1) can be written in the form

$$\mathbf{x} = x_i \mathbf{e}_i. \tag{1.2}$$

The points one deals with in mechanics are generally the locations of material particles: what are being discussed here are not simply geometric, mathematical points, but *point particles*, and $\mathbf{x}$ then labels the position of a particle. When such a particle moves, its position vector changes, and thus a parameter t is needed to label the different space points that the particle occupies. Hence $\mathbf{x}$ becomes a function of t, or

$$\mathbf{x} = \mathbf{x}(t). \tag{1.3}$$

We require the parameter t to have the property of increasing monotonically as $\mathbf{x}(t)$ runs through *successively later positions*. This concept of successively later positions is an intuitive one depending on the ability to distinguish the order in which events take place, that is, between *before* and *after*. For any two positions of the particle we assume that there is no question as to which is the earlier one: given two values t_1 and t_2 of t such that $t_1 < t_2$, the point occupies the position $\mathbf{x}(t_2)$ after it occupies $\mathbf{x}(t_1)$. Clearly t is a quantification of the intuitive idea of time and we will call it "time" without discussing the details at this point. Later, in the section on dynamics, the elusive concept of time will be quantified somewhat more precisely. For the kinematic statements of this section, t need have only two properties: (a) it must increase monotonically with successive positions and (b) the first and second derivatives of $\mathbf{x}$ with respect to t must exist and be continuous.

For a given coordinate system, Eq. (1.3) can be represented by the set of three equations

$$x_i = x_i(t), \quad (i = 1, 2, 3). \tag{1.4}$$

It should be borne in mind that the three functions $x_i(t)$ appearing in (1.4) depend on the particular coordinate system chosen. Once the frame is chosen, however, and the $x_i(t)$ are given, Eqs. (1.4) becomes a set of three parametric equations for the *trajectory* of the particle (the curve it sweeps out in its motion) in which t appears as the parameter. In these terms, the main goal of classical mechanics as applied to point particles (i.e., to describe and explain their motion) reduces to finding the three functions $x_i(t)$ or to finding the vector function $\mathbf{x}(t)$.

1.1.2 POSITION, VELOCITY, AND ACCELERATION

The reason physicists are interested in making quantitative statements about the properties of a system is to compare theoretical predictions with experimental measurements. Among the most commonly measured quantities, particularly relevant to geometry and to the kinematic discussion of classical mechanics, is *distance*. The definition of distance is known as the *metric* of the space for which it is being defined. The Euclidean metric (i.e., in Euclidean space) defines the distance D between two points x and y, with coordinates x_i and y_i, as

$$D = \sqrt{\sum_i (x_i - y_i)^2}. \tag{1.5}$$

We will use this definition of distance also to discuss velocity.

The reason for bringing in the velocity at this point is that trajectories are usually found from other properties of the motion, of which velocity is an example. Another reason is that one is often interested not only in the trajectory itself, but also in other properties of the motion such as velocity.

If t is the time, the velocity $\mathbf{v}$ is defined as

$$\mathbf{v}(t) = \dot{\mathbf{x}}(t) \equiv \frac{d\mathbf{x}}{dt}. \tag{1.6}$$

(Here and in what follows, the dot ($\cdot$) over a symbol denotes differentiation with respect to t.) It is convenient to write $\mathbf{v}$ in terms of distance l along the trajectory. Let s be any parameter that increases smoothly and monotonically along the trajectory, and let $\mathbf{x}(s_0)$ and $\mathbf{x}(s_1)$ be any two points on the trajectory. Then the definition of Eq. (1.5) is used to define the distance along the trajectory between the two points as

$$l(s_0, s_1) = \int_{s_0}^{s_1} \left(\frac{dx_i}{ds} \frac{dx_i}{ds} \right)^{1/2} ds. \tag{1.7}$$

Note the use of the summation convention here: there is a sum over i. Although this definition of l seems to depend on the parameter s, it actually does not (see Problem 10). The trajectory can be parameterized by the time t or even by l itself, and the result would be the same.

If l is taken as the parameter, $\mathbf{v}$ can be written in terms of l:

$$\mathbf{v} = \frac{d\mathbf{x}}{dl} \frac{dl}{dt}. \tag{1.8}$$

But $d\mathbf{x}/dl$ is just the unit vector $\boldsymbol{\tau}$ tangent to the trajectory at time t. To see this, consider Fig. 1.1, which shows a section of a space curve. The tangent vector at the point $\mathbf{x}$ on the curve is in the direction of $\mathbf{T}$. The chord vector $\Delta\mathbf{x}$ between the points $\mathbf{x}$ and $\mathbf{x} + \Delta\mathbf{x}$ approaches parallelism to $\mathbf{T}$ in the limit as $\Delta\mathbf{x} \to 0$. In this limit the vector $\boldsymbol{\tau}$ can be expressed as

$$\boldsymbol{\tau} = \lim_{l \to 0} \frac{\Delta\mathbf{x}}{\Delta l} \equiv \frac{d\mathbf{x}}{dl}, \tag{1.9}$$

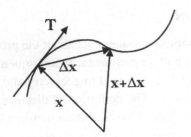

FIGURE 1.1
A vector **T** tangent to a space curve.

which is of unit length and parallel to **T**. Then (1.8) becomes

$$\mathbf{v} = \tau \frac{dl}{dt} = \tau v, \tag{1.10}$$

which says that **v** is everywhere tangent to the trajectory and equal in magnitude to the *speed* $v = \dot{l}$ along the trajectory.

Another important property of the motion is the acceleration, which is defined as the time derivative of the velocity, or

$$\mathbf{a} = \dot{\mathbf{v}} \equiv \ddot{\mathbf{x}} \equiv \frac{d\mathbf{v}}{dt}. \tag{1.11}$$

Then Eq. (1.10) implies that

$$\mathbf{a} = \frac{d\tau}{dt} v + \tau \frac{dv}{dt}. \tag{1.12}$$

The acceleration is related to the bending or curvature of the trajectory. To see this note first that $\dot{\tau}$ is perpendicular to the trajectory. Indeed, τ is a unit vector, so that $\tau \cdot \tau = 1$, and therefore

$$\frac{d}{dt}(\tau \cdot \tau) = 0 = 2\tau \cdot \frac{d\tau}{dt} :$$

$d\tau/dt$ is perpendicular to τ and hence to the curve. Let **n** be the unit vector in this perpendicular direction, called the *principal normal vector*: $\mathbf{n} = \dot{\tau}/|\dot{\tau}|$. The *curvature* κ of the trajectory, the inverse of the *radius of curvature* ρ (see Problem 11) is defined by

$$\lim_{t_1 \to t_2} \frac{|\tau(t_1) - \tau(t_2)|}{|l(t_1) - l(t_2)|} \equiv \left| \frac{d\tau}{dl} \right| = \kappa. \tag{1.13}$$

Thus $\kappa v = |\dot{\tau}|$, or $\dot{\tau} = \kappa v \mathbf{n}$. When these expressions for $\dot{l}$ and τ are inserted into (1.12), one obtains

$$\mathbf{a} = \kappa v^2 \mathbf{n} + \ddot{l}\tau. \tag{1.14}$$

In this expression the second term is the tangential acceleration and the first is the centripetal acceleration (recall that $\kappa = 1/\rho$).

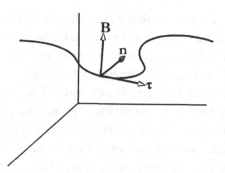

FIGURE 1.2
The tangent, normal, and binormal vectors to a space curve.

The acceleration lies in the plane formed by τ and $\mathbf{n}$, called the *osculating plane*. This can be thought of as the instantaneous plane of the curve; it is the plane that at each instant is being swept out by the tangent vector to the curve. The unit vector $\mathbf{B}$ normal to the osculating plane (Fig. 1.2), called the *binormal vector*, is given by (here the *wedge* $\wedge$ stands for the cross product, often denoted by $\times$)

$$\mathbf{B} = \tau \wedge \mathbf{n}.$$

Since $\dot{\tau}$ is parallel to $\mathbf{n}$ and $\mathbf{n}$ is a unit vector, the rate of change of $\mathbf{B}$ is parallel to $\mathbf{n}$. The torsion $\theta(t)$ of the curve is defined by writing $\dot{\mathbf{B}}$ in the form $\dot{\mathbf{B}} = -\theta \dot{l}\mathbf{n}$, or

$$\frac{d\mathbf{B}}{dl} = -\theta\mathbf{n}.$$

It can be shown (see Problem 11) that

$$\dot{\tau} = \kappa \dot{l}\mathbf{n},$$
$$\dot{\mathbf{n}} = -\kappa \dot{l}\tau + \theta \dot{l}\mathbf{B}.$$

These equations, known as the Frenet–Serret formulas, play an important role in the differential geometry of space curves and trajectories.

Of course, the first and the second time derivatives of $\mathbf{x}$ do not exhaust all possible properties of the motion; for instance one could ask about derivatives of higher order. From the purely mathematical point of view the study of these higher derivatives might be interesting, but physics essentially stops at the second derivative. This is because of Newton's second law $\mathbf{F} = m\mathbf{a}$, which connects the motion of a particle with the external forces acting on it in the real physical world. We therefore now turn from the mathematical treatment of the motion of point particles to a discussion of the physical principles that determine that motion. This section on kinematics has established the language in which we now state the axioms lying at the basis of classical mechanics, axioms that are justified ultimately by experiment.

1.2 PRINCIPLES OF DYNAMICS

1.2.1 NEWTON'S LAWS

Real objects are not point particles, and yet we shall continue to use point kinematics to describe their motion. This is largely because small enough objects approximate points, and their motion can be described with some accuracy in this way. Moreover, as we will show later, results about the motion of extended objects can be derived from axioms about idealized point particles. These axioms, as we have said, are ultimately justified by experiment, so we will try to state them in terms of idealized "thought experiments."

In general the objects and the motions we treat are restricted in two ways. The first is that the small objects we speak of, even though they are to approximate points, may not get too small. As is well known, at atomic dimensions classical mechanics breaks down and quantum mechanics takes over. That breakdown prescribes a lower limit on the size of the objects admitted in our discussions. The other restriction is on the magnitudes of the velocities involved: we exclude speeds so high that relativistic effects become important. Unless otherwise stated, these restrictions apply to all the following discussion.

Our treatment starts with the notion of an *isolated particle*. An isolated particle is a sufficiently small physical object removed sufficiently far from all other matter. What is meant here by "sufficiently" depends on the precision of measurements, and the statements we are about to make are statements about limits determined by such precision. They are true for measurements of distance whose uncertainties are large compared to characteristic dimensions of the object ("in the limit" of large distances or of small objects) and for measurements of time whose uncertainties are large compared to characteristic times of changes within the object ("in the limit" of long times). These distances and times, though long compared with the characteristics of the object, should nevertheless be short compared to the distance from the nearest object and the time to reach it. When isolation is understood in this way, the accuracy of the principles stated below will increase with the degree of isolation of the object.

It follows that there are two ways to test the axioms: the first is by choosing smaller objects and removing them further from other matter, and the second is by making cruder measurements. We say this not to advocate cruder measurements, but to indicate that the relation between theory and experiment has many facets. Not only are the axioms actually statements about the result of experiment, but the very terms in which the axioms are stated must involve detailed experimental considerations, even in an "axiomatized" field like classical mechanics. Similar considerations apply to the two restrictions described above. The detection of quantum or relativistic effects depends also on the accuracy of measurement.

1.2.2 THE TWO PRINCIPLES

The laws of mechanics can be formulated in two principles. The first of these quantifies the notion of time [see the discussion after Eq. (1.3)] for the purposes of classical mechanics and states Newton's first law in terms of this quantification. The second states conservation of momentum for two-particle interactions, which is equivalent to Newton's third law. Together they pave the way for a statement of Newton's second law. Both principles must be understood as statements about idealized experiments. They are based on years, even centuries, of constantly refined experiments. We continue to assume that physical space is three dimensional, Euclidean, and endowed with the usual Euclidean metric of Eq. (1.5).

PRINCIPLE 1

There exist certain frames of reference, called *inertial*, with the following two properties.

Property A) Every isolated particle moves in a straight line in such a frame.

Property B) If the notion of time is quantified by defining the unit of time so that one particular isolated particle moves at constant velocity in this frame, then every other isolated particle moves at constant velocity in this frame.

This defines the parameter t to be linear with respect to length along the trajectory of the chosen isolated particle, and then Property B states that any other parameter t, defined in the same way by using some other isolated particle, will be linearly related to the first one. (Buried in this statement is the nonrelativistic idea of simultaneity.) Although it is not practical to measure time in terms of an isolated (*free*) particle, it will be seen eventually that the laws of motion derived from the two principles imply that the rotation of an isolated rigid body about a symmetry axis also takes place at a constant rate and can thus be used as a measure of time. In practice the Earth is usually used for this purpose, but corrections have to be introduced to take account of the fact that the Earth is not really isolated (in terms of the measuring instruments being used) or completely rigid. The most accurate modern timing devices are atomic and are based on a long chain of reasoning stretching from the two principles and involving quantum as well as classical concepts.

The existence of one inertial frame, as postulated in Principle 1, implies the existence of many more, all moving at constant velocity with respect to each other. Two such inertial frames cannot rotate with respect to each other, for then a particle moving in a straight line in one of the frames would not move in a straight line in the other. The transformations that connect such inertial frames are called Galileian; more about this is discussed in later chapters.

> REMARK: Since inertial frames are defined in terms of isolated bodies, they cannot in general be extended indefinitely. In other words, they have a local character, for extending them would change the degree of isolation. Suppose, for instance, that there exist two inertial frames very far apart. If they are extended until they intersect, it may turn out that a particle that is free in the first frame is not free in the second. Considerations such as these play a role in physics, but are not important for the purposes of this book (Taylor, 1964). □

PRINCIPLE 2

Conservation of Momentum. Consider two particles 1 and 2 isolated from all other matter, but not from each other, and observed from an inertial frame. In general they will not move at constant velocities: their proximity leads to accelerations (the particles are said to *interact*). Let $\mathbf{v}_j(t)$ be the velocity of particle j at time t, where $j = 1, 2$. Then there exists a constant $\mu_{12} > 0$ and a constant vector $\mathbf{K}$ independent of the time such that

$$\mathbf{v}_1(t) + \mu_{12}\mathbf{v}_2(t) = \mathbf{K} \tag{1.15}$$

for all time t. Moreover, although $\mathbf{K}$ depends both on the inertial frame in which the motion is being observed and on the particular motion, μ_{12} does not: μ_{12} is always the same number for particles 1 and 2. Even if the motion is interrupted and a new experiment is performed, even if the interaction between the particles is somehow changed (say from gravitational to magnetic, or from rods of negligible weight to rubber bands), the same number μ_{12} will be found provided the experiment involves the same two particles 1 and 2. If a similar set of experiments is performed with particle 1 and a third particle 3, a similar result will be obtained, and the same is true for experiments involving particles 2 and 3. In this way, one arrives at

$$\mathbf{v}_2(t) + \mu_{23}\mathbf{v}_3(t) = \mathbf{L},$$
$$\mathbf{v}_3(t) + \mu_{31}\mathbf{v}_1(t) = \mathbf{M}. \tag{1.16}$$

As before, $\mathbf{L}$ and $\mathbf{M}$ depend on the particular experiment, but the $\mu_{ij} > 0$ do not.

Existence of Mass. The μ_{ij} are related according to

$$\mu_{12}\mu_{23}\mu_{31} = 1. \tag{1.17}$$

That completes the statement of Principle 2.

It follows from (1.17) (see Problem 13) that there exist positive constants m_i, $i = 1, 2, 3$, such that Eqs. (1.15) and (1.16) can be put in the form

$$m_1\mathbf{v}_1 + m_2\mathbf{v}_2 = \mathbf{P}_{12},$$
$$m_2\mathbf{v}_2 + m_3\mathbf{v}_3 = \mathbf{P}_{23}, \tag{1.18}$$
$$m_3\mathbf{v}_3 + m_1\mathbf{v}_1 = \mathbf{P}_{31},$$

where the $\mathbf{P}_{ij}$, called *momenta*, are constant vectors that depend on the experiment. The m_i are, of course, the masses of the particles. It should now be clear that Principle 2 states the law of conservation of momentum in two-particle interactions.

The masses of the particles are not unique: it is the μ_{ij} that are determined by experiment, and the μ_{ij} are only the ratios of the masses. But given any set of m_j, any other set is obtained from the first by multiplying all of the m_j by the same constant. What is done in practice is that some body is chosen as a standard (say, 1 cm^3 of water at 4°C and atmospheric pressure) and the masses of all other bodies are related to it. The important thing is that once such a standard has been chosen, there is just one number m_i associated with each body, independent of any object with which it is interacting.

The vectors on the right-hand sides of Eqs. (1.18) are constant, so the time derivatives of these equations yield

$$m_1\mathbf{a}_1 + m_2\mathbf{a}_2 = \mathbf{0},$$
$$m_2\mathbf{a}_2 + m_3\mathbf{a}_3 = \mathbf{0}, \tag{1.19}$$
$$m_3\mathbf{a}_3 + m_1\mathbf{a}_1 = \mathbf{0}.$$

These equations are equivalent to (1.18): their integrals with respect to time yield (1.18). They [or rather their analogs obtained by differentiating (1.15) and (1.16)] could have been used in place of (1.15) and (1.16) to state Principle 2. In fact Eqs. (1.19) are perhaps a more familiar starting point for stating the principles of classical mechanics, for what they assert is Newton's third law. Indeed, let the force **F** acting on a particle be defined by the equation

$$\mathbf{F} = m\mathbf{a} = m\frac{d^2\mathbf{x}}{dt^2}. \tag{1.20}$$

This is Newton's second law (but see Discussion below). Now turn to the first of Eqs. (1.19). The experiment it describes involves the interaction of particles 1 and 2: the acceleration $\mathbf{a}_1$ of particle 1 arises as a result of that interaction. One says there is a force $\mathbf{F}_{12} = m_1\mathbf{a}_1$ on particle 1 due to the presence of particle 2 (or *by* particle 2). Similarly, there is a force $\mathbf{F}_{21} = m_2\mathbf{a}_2$ on particle 2 by particle 1. Then the first of Eqs. (1.19) becomes

$$\mathbf{F}_{12} + \mathbf{F}_{21} = \mathbf{0}, \tag{1.21}$$

which is Newton's third law.

DISCUSSION

The two principles are equivalent to Newton's three laws of motion, which in their usual formulation involve a number of logical difficulties that the present treatment tries to avoid. For instance, Newton's first law, which states that a particle moves with constant velocity in the absence of an applied force, is incomplete without a definition of force and might at first seem to be but a special case of the second law. Actually it is an implicit statement of Principle 1 and was logically necessary for Newton's complete formulation of mechanics. Although such questions would seem to arise in understanding Newton's laws, we should remember that it is his formulation that lies at the basis of classical mechanics as we know it today. The two principles as we have given them are, in fact, an interpretation of Newton's laws. This interpretation is due originally to Mach (1942). Our development is closely related to work by Eisenbud (1958).

It is interesting that Eq. (1.20), Newton's second law, is now a definition of force, rather than a fundamental law. Why, one may ask, has this definition been so important in classical mechanics? As may have been expected, the answer lies in its physical content. It is found empirically that in many physical situations it is $m\mathbf{a}$ that is known a priori rather than some other dynamical property. That is, the force is what is specified independent of the mass, acceleration, or many other properties of the particle whose motion is being studied. Moreover, forces satisfy a *superposition principle*: the total force on a particle can be found by adding contributions from different agents. It is such properties that elevate Eq. (1.20) from a rather empty definition to a dynamical relationship. For an interesting discussion of this point see Feynman et al. (1963).

> **REMARK:** Strangely enough, in one of the most familiar cases of motion, that of a particle in a gravitational field, it is not the force $m\mathbf{a}$ that is known a priori, but the acceleration **a**. ☐

1.2.3 CONSEQUENCES OF NEWTON'S EQUATIONS

INTRODUCTION

The general problem in the mechanics of a single particle is to solve Eq. (1.20) for the function $\mathbf{x}(t)$ when the force $\mathbf{F}$ is a given function of $\mathbf{x}$, $\dot{\mathbf{x}}$, and t. Then (1.20) becomes the differential equation

$$\mathbf{F}(\mathbf{x}, \dot{\mathbf{x}}, t) = m\frac{d^2\mathbf{x}}{dt^2}. \tag{1.22}$$

A mathematical solution of this second-order differential equation [i.e., a vector function $\mathbf{x}(t)$ that satisfies (1.22)] requires a set of initial conditions [e.g., the values $\mathbf{x}(t_0)$ and $\dot{\mathbf{x}}(t_0)$ of $\mathbf{x}$ and $\dot{\mathbf{x}}$ at some initial time t_0]. There are theorems stating that, with certain unusual exceptions (see, e.g., Arnol'd, 1990, p. 31; Dhar, 1993), once such initial conditions are given, the solution exists and is unique. In this book we will deal only in situations for which solutions exist and are unique.

For many years physicists took comfort in such theorems and concentrated on trying to find solutions for a given force with various different initial conditions. Recently, however, it has become increasingly clear that there remain many subtle and fundamental questions, largely having to do with the *stability of solutions*. To see what this means, consider the behavior of two solutions of Eq. (1.22) whose initial positions $\mathbf{x}_1(t_0)$ and $\mathbf{x}_2(t_0) = \mathbf{x}_1(t_0) + \delta\mathbf{x}(t_0)$ are infinitesimally close [assume for the purposes of this example that $\dot{\mathbf{x}}_1(t_0) = \dot{\mathbf{x}}_2(t_0)$]. One then asks how the separation $\delta\mathbf{x}(t)$ between the solutions behaves as t increases: will the trajectories remain infinitesimally close, will their separation approach some nonzero limit, will it oscillate, or will it grow without bound? Let $D\mathbf{x}(t) = |\delta\mathbf{x}(t)|$ be the distance between the two trajectories at time t. Then the solutions of the differential equation are called *stable* if $D\mathbf{x}(t)$ approaches either zero or a constant of the order of $\delta\mathbf{x}(t_0)$ and *unstable* if it grows without bound as t increases.

Stability in this sense is highly significant because initial conditions cannot be established with absolute precision in any experiment on a classical mechanical system, and thus there will always be some uncertainty in their values. Suppose, for example, that a certain dynamical system has the property that it invariably tends to one of two regions A and B that are separated by a finite distance. In general the final state of such a system can in principle be calculated from a knowledge of the initial conditions, and therefore each initial condition can be labeled a or b, belonging to final state A or B. If there exist small regions that contain initial conditions with both labels, then in those regions the system is unstable in the above sense. In fact there exist dynamical systems in which the two types of initial conditions are mixed together so tightly that it is impossible to separate them: in every neighborhood, no matter how small, there are conditions labeled both a and b. Then even though such a system is entirely deterministic, predicting its end state would require knowing its initial conditions with infinite precision. Such a system is called *chaotic* or said to exhibit *chaos*. An everyday example of chaos, somewhat outside the realm of classical mechanics, is the weather: very similar weather conditions one day can be followed by drastically different ones the next.

This kind of instability was not of major concern to the founders of classical mechanics, so one might guess that it is a rare phenomenon in classical dynamical systems. It turns out that exactly the opposite is true: most classical mechanical systems are unstable in the sense defined above. The most common exceptions are systems that can be reduced to a collection of one-dimensional ones. (One-dimensional systems are discussed in Section 1.5.1, but reduction to one-dimensional systems will come much later, in the discussion of Hamiltonian systems.) For many reasons, the leading early contributors to classical mechanics were concerned mainly with stable systems, and for a long time their concerns continued to dominate the study of dynamics. In recent years, however, significant progress has been made in understanding instability, and later in this book, especially in Chapter 7, we will discuss specific systems that possess instabilities and in particular those solutions that manifest the instabilities.

FORCE IS A VECTOR

Equation (1.22) has an important property related to the acceleration of a particle as measured by observers in different inertial frames. Suppose that according to one observer a particle has position vector $\mathbf{x}$, with components x_i, $i = 1, 2, 3$ in her Cartesian coordinate system. Now consider another observer looking at the same particle at the same time, and suppose that in the second observer's frame the position vector of the particle is $\mathbf{y}$, with coordinates y_i. It is clear that there must be some *transformation law*, a sort of dictionary, that translates one observer's coordinates into the other's, or they could not communicate with each other and could not tell whether they are looking at the same particle. This dictionary must give the x_i in terms of the y_i and vice versa: it should consist of equations of the form

$$y_i = f_i(x, t), \quad x_i = g_i(y, t), \tag{1.23}$$

where the x in the argument of the three functions f_i denotes the collection of the three components of $\mathbf{x}$ and similarly for y in the arguments of the g_i. Note that the transformation law in general depends on the time t.

If the f_i functions are known, Eq. (1.23) can be used to calculate the velocities and accelerations as seen by the second observer in terms of the observations of the first. Indeed, one obtains (don't forget the summation convention)

$$\dot{y}_i = \frac{\partial f_i}{\partial x_j}\dot{x}_j + \frac{\partial f_i}{\partial t} \tag{1.24}$$

and

$$\ddot{y}_i = \frac{\partial f_i}{\partial x_j}\ddot{x}_j + \frac{\partial^2 f_i}{\partial x_j \partial x_k}\dot{x}_j\dot{x}_k + 2\frac{\partial^2 f_i}{\partial x_j \partial t}\dot{x}_j + \frac{\partial^2 f_i}{\partial t^2}. \tag{1.25}$$

So far we have not assumed any particular properties of the two coordinate systems, but if they are both Cartesian, the transformation between them must be linear, that is, the f_i

functions, which give the y_j in terms of the x_i, must be of the form

$$y_i = f_i(x, t) = f_{ik}(t)x_k + b_i(t). \qquad (1.26)$$

If, in addition, both frames of reference are inertial, they are moving at constant velocity and are not rotating with respect to each other, and then the $b_i(t) = \beta_i t$ must be linear in t and the $f_{ik}(t) = \phi_{ik}$ must be time independent (i.e., constants). Indeed, if $\mathbf{x} = 0$, then $y_i = b_i$, so the b_i term alone gives the y coordinates of the x origin, and the $f_{ik}(t)$ determine the linear combinations of the x_k that go into making up the y_i. In that case the last three terms of (1.25) vanish, and thus

$$\ddot{y}_i = \phi_{ik}\ddot{x}_k. \qquad (1.27)$$

This shows that the acceleration as measured by the second observer is a linear homogeneous function of the acceleration as measured by the first.

It is important that the same ϕ_{ij} coefficients appear in the expressions for the acceleration and for the position [in Eq. (1.26) the $f_{ik}(t)$ are now the constants ϕ_{ik}]. This is interpreted (see the Remark later in this section) to mean that the acceleration is a vector, that the transformation law for its components is the same as that for the components of the position vector. That, together with the $\mathbf{F} = m\mathbf{a}$ equation, guarantees that force is also a vector. What is more, as we shall now show, the converse of this is also true: if force is to be a vector, then acceleration must be a vector, and then $f_i(x, t) = f_{ik}(t)x_k + b_i(t)$. (This is also true if one demands that the acceleration seen by one observer should be determined entirely by the acceleration (i.e., be independent of the velocity) seen by the other.)

Before we prove the converse, we make a comment. It is actually *relative* position vectors that transform like accelerations and forces. A relative position is the difference of two position vectors, the vector that goes from one particle to another. Let there be two particles, with position vectors $\mathbf{x}$ and $\mathbf{x}'$ in one frame and $\mathbf{y}$ and $\mathbf{y}'$ in the other. Then if $\mathbf{x} - \mathbf{x}' = \Delta\mathbf{x}$ and $\mathbf{y} - \mathbf{y}' = \Delta\mathbf{y}$, Eq. (1.26) yields

$$\Delta y_i = f_{ik}(t)\Delta x_k. \qquad (1.28)$$

It is seen that in the transformation law for relative positions the b_i of (1.26) drop out and thus that the equations are homogeneous, even before restrictions are placed on the transformation properties of the components and before the frames are assumed to be inertial. But it is also seen that when the frames are inertial, so that the f_{ij} are constants, the transformation law for the relative positions is of exactly the same form as for the acceleration. Incidentally, this is true also for relative velocities, which can be defined either as the time derivatives of relative positions or, equivalently, as the difference of the velocities of two particles.

> **REMARK**: In elementary mechanics, vectors (e.g., displacement, velocity, acceleration, force) are required to have three (Cartesian) components that combine properly under the vector addition law when the vectors are added. Furthermore, there is a transformation law by which the components of a vector in one frame can be found from

its components in another. Moreover, for vector equations to be frame independent (i.e., to be true in all frames if they are true in one) *the components of all vectors must transform according to the same law.* This means that the ϕ_{ij} appearing in the transformation law for the (relative) position vectors must be the same as those appearing in the law for other vectors, a principle that we have used in the above discussion.   □

We now return to the proof of the converse. If the acceleration is a vector, its transformation law must be linear and homogeneous, which means that each of the last three terms of (1.25) must vanish. But that implies that all the second derivatives of the $f_i(x, t)$ must vanish and hence that the $f_i(x, t)$ must be linear in the x_k and in t, that is, of the form

$$f_i(x, t) = \phi_{ik}x_k + \beta_i t, \tag{1.29}$$

where the ϕ_{ij} and β_i are constants, and then the two frames are *relatively* inertial and Eq. (1.27) follows. This proves the converse: it establishes the transformation law of the position if acceleration is a vector. This also exhibits the intimate connection between Newton's laws and inertial frames, a connection manifested here in terms of transformation properties. The result may be summarized as follows:

If the acceleration is a vector in one frame, then it is a vector in and only in relatively inertial frames. If force is a vector and Newton's laws are valid in one inertial frame, then all of the frames in which they are valid are inertial.

Stated in terms of the observers (for after all, it is the observers who determine the frames, not the other way around), if Newton's laws are to be valid for different observers, those observers can be moving with respect to each other only at constant velocity. The transformation defined by the f_i functions of (1.29) is called a Galileian transformation.

1.3   ONE-PARTICLE DYNAMICAL VARIABLES

We have said that the general problem of single-particle mechanics is to solve Eq. (1.22), $\mathbf{F} = m\mathbf{a}$, for $\mathbf{x}(t)$ when the force $\mathbf{F}$ is given. We now extend this statement of the general problem: to find not only $\mathbf{x}(t)$ but also $\dot{\mathbf{x}}(t)$. Once $\mathbf{x}(t)$ is known, of course, $\dot{\mathbf{x}}(t)$ is easily found by simple differentiation; so this extension of the problem may seem redundant. Let it be so; in due course the reason for it will become clear.

Solving (1.22) is often easier said than done, for in many specific cases it is hard to find practical ways to solve the equation. Often, however, one is not interested in all of the details of the motion, but in some properties of it, like the energy, or the stable points, or, for periodic motion, the frequency. Then it becomes superfluous actually to find $\mathbf{x}(t)$ and $\dot{\mathbf{x}}(t)$. For this reason other useful objects are defined, which are generally easier to solve for and sufficient for most purposes. These objects, functions of $\mathbf{x}$ and $\dot{\mathbf{x}}$, are called *dynamical variables.* Many dynamical variables are quite important. Their study leads to a deeper understanding of dynamics, they are the terms in which properties of dynamical systems are usually stated, and they are of considerable help in obtaining detailed solutions of Eq. (1.22).

In this section we briefly discuss momentum, angular momentum, and energy in single-particle motion. In Section 1.4 we extend the discussion to systems of several particles.

1.3.1 MOMENTUM

We have already mentioned the first of these dynamical variables in Eq. (1.18). The *momentum* **p** of a particle of mass m moving at velocity **v** is defined by

$$\mathbf{p} = m\mathbf{v}. \tag{1.30}$$

With this definition, (1.22) becomes

$$\mathbf{F} = \frac{d\mathbf{p}}{dt}. \tag{1.31}$$

A glance at Eq. (1.18) shows that the total momentum of two particles is just the sum of the individual momenta of both particles. This will be discussed more fully in the next section, where the definition of momentum will be extended to systems of particles (even if the mass is not constant; see the discussion of variable mass in Section 1.4.1), and it will be seen that Eq. (1.31) is in a form more suitable to generalization than is (1.22). An important property of the extended definition of momentum is evident from Eq. (1.18): the total momentum of two particles is constant if the particles are interacting with each other and with nothing else. In terms of forces, this means that the total momentum of the two-particle system is constant if the sum of the *external forces* acting on it is zero. Similarly, if the total force on one particle is zero, then according to (1.31) $d\mathbf{p}/dt = 0$ and **p** is constant. In fact it is the constancy of momentum, its conservation in this kind of situation, that singles it out for definition. All this assumes, as usual, that all measurements are performed in inertial systems. Otherwise, Eq. (1.31) is no longer true and momentum is no longer conserved even in the absence of external forces.

1.3.2 ANGULAR MOMENTUM

The *angular momentum* **L** *of a particle about some point* **S** is defined in terms of its (linear) momentum **p** as

$$\mathbf{L} = \mathbf{x} \wedge \mathbf{p}, \tag{1.32}$$

where **x** is the position vector of the particle measured from the point **S**. In general **S** may be any moving point, but we shall restrict it to be an *inertial point*, that is, a point moving at constant velocity in any inertial frame (more about this in Section 1.4). The time derivative of (1.32) is

$$\dot{\mathbf{L}} = \frac{d}{dt}(\mathbf{x} \wedge \mathbf{p}) = \mathbf{v} \wedge \mathbf{p} + \mathbf{x} \wedge \dot{\mathbf{p}}.$$

But $\mathbf{v} \wedge \mathbf{p} = m(\mathbf{v} \wedge \mathbf{v}) = \mathbf{0}$, and then with (1.31) one arrives at

$$\dot{\mathbf{L}} = \mathbf{x} \wedge \dot{\mathbf{p}} = \mathbf{x} \wedge \mathbf{F} \equiv \mathbf{N}, \tag{1.33}$$

which defines the *torque* $\mathbf{N}$ about the point S. This equation shows that the relation between torque and angular momentum is similar to that between force and momentum: torque is the analog of force and angular momentum is the analog of momentum in problems involving only rotation (about inertial points). In particular, if the torque is zero, the angular momentum is conserved.

1.3.3 ENERGY AND WORK

IN THREE DIMENSIONS

If the force $\mathbf{F}$ of Eq. (1.22) is a function of t alone, it is easy to find $\mathbf{x}(t)$ by integrating twice with respect to time:

$$\mathbf{x}(t) = \mathbf{x}(t_0) + (t - t_0)\mathbf{v}(t_0) + \frac{1}{m} \int_{t_0}^{t} dt' \int_{t_0}^{t'} \mathbf{F}(t'') \, dt'', \tag{1.34}$$

where $\mathbf{x}(t_0)$ and $\mathbf{v}(t_0)$ are position and velocity at the initial time t_0. Unfortunately, however, $\mathbf{F}$ is almost never a function of t alone but a function of $\mathbf{x}$, $\mathbf{v}$, and t. It is only when $\mathbf{x}$ and $\mathbf{v}$ are known as functions of t that $\mathbf{F}$ can be written as a function of t alone and Eq. (1.34) prove useful. But $\mathbf{x}$ and $\mathbf{v}$ are not known as functions of the time until the problem is solved, and when the problem is solved there is nothing to be gained from (1.34). Thus Eq. (1.34) is almost never of any help.

Often, however, $\mathbf{F}$ is a function of $\mathbf{x}$ alone, with no $\mathbf{v}$ or t dependence. Then one can take the line integral with respect to $\mathbf{x}$ along the trajectory, obtaining

$$\int_{\mathbf{x}(t_0)}^{\mathbf{x}(t)} \mathbf{F}(\mathbf{x}) \cdot d\mathbf{x} = \int_{t_0}^{t} \mathbf{F}(\mathbf{x}) \cdot \dot{\mathbf{x}} \, dt = m \int_{t_0}^{t} \frac{d^2\mathbf{x}}{dt^2} \cdot \frac{d\mathbf{x}}{dt} \, dt$$

$$= \frac{1}{2} m \int_{t_0}^{t} \frac{d}{dt} (\dot{x}^2) \, dt \tag{1.35}$$

$$= \frac{1}{2} m v^2(t) - \frac{1}{2} m v^2(t_0), \tag{1.36}$$

where $v^2 = \mathbf{v} \cdot \mathbf{v}$. This equation doesn't give $\mathbf{x}(t)$ in terms of the initial conditions, but it does give the magnitude $v(t)$ of the velocity, provided the integral on the left-hand side can be performed. But there are two problems with performing the integral. The first involves the upper limit $\mathbf{x}(t)$. Recall that the problem is to find $\mathbf{x}(t)$, so $\mathbf{x}(t)$ is hardly useful as an upper limit. What Eq. (1.35) actually describes is how $v(t)$ depends on the upper limit: it gives v as a function of $\mathbf{x}$ rather than of t. To find v as a function of t requires finding $\mathbf{x}$ as a function of t some other way. Since the time t actually plays no role in Eq. (1.35), t and t_0 can be dropped from the limits of the integral and the argument of v can be changed to $\mathbf{x}$. What the equation really says is that no matter how long it takes for the trajectory to be covered, the line integral on the left-hand side is equal to $\frac{1}{2} m v^2(\mathbf{x}) - \frac{1}{2} m v^2(\mathbf{x}_0)$.

The second problem with the integral on the left-hand side of (1.35) is that, being a line integral, it depends on the path of integration, in this case on the entire trajectory: in general, even if its end point $\mathbf{x}$ is known, $v(\mathbf{x})$ is not. But in spite of these difficulties, Eq. (1.35) is quite useful.

The usefulness of Eq. (1.35) involves the dynamical variable $\frac{1}{2}mv^2$ appearing in it, which is the *kinetic energy T*:

$$T = \frac{1}{2}mv^2. \tag{1.37}$$

Let the trajectory (more generally, the path) of the integral on the left-hand side of (1.35) be called C; then the integral itself is defined as the *work W_C* done by the force $\mathbf{F}$ along C:

$$W_C \equiv \int_C \mathbf{F} \cdot d\mathbf{x}. \tag{1.38}$$

Equation (1.35) can now be written in the form $W_C = [T(t) - T(t_0)]_C$, or, in view of the fact that $\mathbf{x}$, not t, is the argument of $\mathbf{v}$,

$$W_C = [T(\mathbf{x}) - T(\mathbf{x}_0)]_C. \tag{1.39}$$

In words, the work done by the force $\mathbf{F}$ along the trajectory is equal to the change in kinetic energy between the initial point $\mathbf{x}_0$ and the final point $\mathbf{x}$ of the trajectory.

Equation (1.39) is known as the *work–energy theorem*. Both sides of it, the work W and the change ΔT of the kinetic energy, depend on the entire path of integration, not just on its end points. This path dependence reduces the practical value of the theorem, because often only some of the points of a trajectory are known without the trajectory being known in its entirety. Fortunately, however, it turns out that for many forces that occur in nature the integral that gives W is not path dependent; it depends on nothing but the end points. Forces for which this is true are called *conservative*.

Conservative forces are common in physical systems, so we now study some of their properties. Let $\mathbf{F}$ be a conservative force (it should be borne in mind that in all of these equations $\mathbf{F}$ is a function of $\mathbf{x}$), and let C_1 and C_2 be any two paths connecting two points $\mathbf{x}_1$ and $\mathbf{x}_2$, as shown in Fig. 1.3. Then, by the definition of a conservative force,

$$\int_{C_1} \mathbf{F} \cdot d\mathbf{x} = \int_{C_2} \mathbf{F} \cdot d\mathbf{x},$$

or

$$\int_{C_1} \mathbf{F} \cdot d\mathbf{x} - \int_{C_2} \mathbf{F} \cdot d\mathbf{x} = \oint \mathbf{F} \cdot d\mathbf{x} = 0,$$

where the third integration is around the closed path from $\mathbf{x}_1$ to $\mathbf{x}_2$ and back again. Thus if $\mathbf{F}$ is conservative, $\oint \mathbf{F} \cdot d\mathbf{x} = 0$ around *any* closed path, for $\mathbf{x}_1$ and $\mathbf{x}_2$ are arbitrary. Stokes's theorem can now be used to transform the closed line integral to a surface integral, giving

$$\oint_C \mathbf{F} \cdot d\mathbf{x} = \int_\Sigma (\nabla \wedge \mathbf{F}) \cdot d\mathbf{S}, \tag{1.40}$$

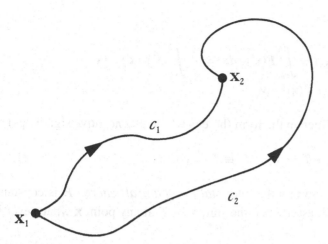

FIGURE 1.3
Two paths connecting the points x_1 and x_2.

where Σ is any smooth surface bounded by the closed path C of the line integral, and $d\mathbf{S}$ is the vector element of area on Σ. Since the left-hand side of (1.40) is zero for every closed contour C, the right-hand side is zero for every surface Σ, and hence

$$\nabla \wedge \mathbf{F} = 0 \tag{1.41}$$

is a necessary condition for $\mathbf{F}$ to be conservative. The argument can be reversed to show that Eq. (1.41) is also a sufficient condition. Thus Eq. (1.41) characterizes conservative forces.

Equation (1.41) may be recognized as the integrability condition for the existence of a single function that is a solution $V(\mathbf{x})$ of the three partial differential equations (one for each component)

$$\mathbf{F}(\mathbf{x}) = -\nabla V(\mathbf{x}) \tag{1.42}$$

(the negative sign is chosen for later convenience). In other words, if $\mathbf{F}$ is conservative, there exists a function V satisfying Eq. (1.42) and defined by it up to an additive constant. This important function V is called the *potential energy* of the dynamical system whose force is $\mathbf{F}$.

Up to now, a one-particle dynamical system has been characterized by its vector force function $\mathbf{F}$. That is, given $\mathbf{F}$, the dynamical problem is defined: solve (1.22) with that given $\mathbf{F}$. From now on, if the force is conservative, the characterization can be simplified: all that need be given is the single scalar function V, and $\mathbf{F}$ is easily obtained from V via (1.42). This reduction from the three components of a vector function to a single function simplifies the problem considerably. Later, when we deal with systems more complicated than one-particle systems, it will be seen that the simplification obtained from a potential energy function can be even greater.

Let us return to Eq. (1.38), defining the work along a path C, and now let C be a contour going from some point $\mathbf{x}_0$ to some arbitrary point $\mathbf{x}$. Then (1.38) may be written in the form

$$W_C \equiv W(\mathbf{x}, \mathbf{x}_0) = \int_{\mathbf{x}_0}^{\mathbf{x}} \mathbf{F}(\mathbf{x}') \cdot d\mathbf{x}' = -\int_{\mathbf{x}_0}^{\mathbf{x}} \nabla V(\mathbf{x}') \cdot d\mathbf{x}'$$
$$= V(\mathbf{x}_0) - V(\mathbf{x}) \equiv V_0 - V.$$

Equation (1.39) can now be written in the form (here is where that negative sign helps)

$$V + T = V_0 + T_0 \equiv E = \text{const.} \tag{1.43}$$

The last equality, namely the assertion that the *total (mechanical) energy* E is constant, comes from the first one, which asserts that the sum $V + T$ at any point $\mathbf{x}$ whatsoever is the same as it is at $\mathbf{x}_0$.

We have derived *conservation of energy* for any dynamical system consisting of a single particle acted upon by a conservative force. Because the total energy E is a sum of T and V, and because V is defined only up to an additive constant, E is also defined only up to an additive constant. This constant is usually chosen by specifying V to be zero at some arbitrarily designated point. It is clear that conservation of energy does not solve the problem of finding $\mathbf{x}(t)$ in conservative systems, but it is a significant step. We will see in what follows that in one-dimensional systems it leads directly to a solution.

It is important to note that the three dynamical variables $\mathbf{p}$, $\mathbf{L}$, and E that have been defined are associated with conservation laws. Historically, it is their conservation in many physical systems that has led these functions of $\mathbf{x}$ and $\dot{\mathbf{x}}$ to be singled out for special consideration and definition.

APPLICATION TO ONE-DIMENSIONAL MOTION

The properties of dynamical systems depend crucially on the "arena" in which the motion takes place, and one important aspect of this arena is its dimension. For example, the arena for the single-particle dynamical systems we have been discussing is three-dimensional Euclidean space. Later we will be discussing systems with $N > 1$ particles, and then the arena consists of N copies of three-space (one copy for the position of each particle), or a $3N$-dimensional space. In another way, the arena for a single particle constrained to remain on the surface of a sphere is the two-dimensional spherical surface. The dimension of the arena is called the number of *degrees of freedom* of the system or simply the number of its *freedoms*. It is the number of functions of the time needed to describe fully the motion of the system.

The simplest of all arenas is that for a particle constrained to move along a straight line. This is clearly a one-freedom system (there are other one-freedom systems; e.g., a particle constrained to a circle). One-freedom systems are of particular interest not only because they are the simplest, but also because often a higher-freedom problem is solved by breaking it up into a set of independent one-freedom problems and then solving each of those by the technique we will now describe. This technique reduces the problem to performing a one-dimensional integration or, as is said, to *quadrature*.

Let distance along the line be x and let the force $\mathbf{F}$ (also along the line) acting on the particle be a function only of x. The equation of motion is then

$$m\ddot{x} = F(x). \tag{1.44}$$

In one dimension all forces that depend only on x are conservative. Thus F is the negative derivative of a potential energy function $V(x)$ defined up to an arbitrary additive constant, and Eq. (1.43) can be written in the form

$$\frac{1}{2}m\dot{x}^2 + V(x) = E, \tag{1.45}$$

where E is the total energy of the system and is constant. This equation is obtained by integrating (1.44) once (E is the constant of integration): the dynamical variable on the left-hand side of (1.45) is called the *energy first-integral*.

Equation (1.45) can be solved for $1/\dot{x} = dt/dx$ and integrated to yield

$$t - t_0 = \sqrt{\frac{m}{2}} \int_{x_0}^{x} \frac{dx}{\sqrt{E - V(x)}}, \tag{1.46}$$

where x_0 is the position of the particle at time t_0, and x is its position at time t. If this equation is inverted, it yields the solution of the problem in the form

$$x = x(t - t_0, E). \tag{1.47}$$

Thus if E and $x_0 = x(t_0)$ are known, the problem is solved in principle. In practice t_0 is usually taken to be zero without loss of generality, so that the only constants that appear in the final expression are x_0 and E.

Most of the time the integral on the right-hand side of (1.46) cannot be performed in terms of elementary functions. Essentially Eq. (1.46) is just (1.45) rewritten. Thus we turn to (1.45) to see what information it provides before integration. It turns out that it contains much information about the various kinds of motion that belong to different values of E. For example, since the kinetic energy $\frac{1}{2}m\dot{x}^2$ is always positive, the particle cannot enter a region where $E - V(x)$ is negative. Starting in a region where $E - V(x)$ is positive, the particle can move toward higher values of the potential V until it gets to a point where $V(x) = E$ and then can go no further. This is a *turning point* for the motion, and in this way E determines the turning points.

In analyzing the possible motions, it helps to plot a graph of V as a function of x with horizontal lines representing the various constant values of E. Figure 1.4 is an example of such a graph. If the particle starts out to the left of x_1 moving to the right with total energy $E = E_0$, it cannot pass x_1; as it approaches x_1 it slows down (its kinetic energy $\frac{1}{2}m\dot{x}^2 = E - V$ decreases) and comes to rest at x_1. Then because $F = -dV/dx$ is negative, it accelerates to the left. It makes no physical sense to say that the particle with energy

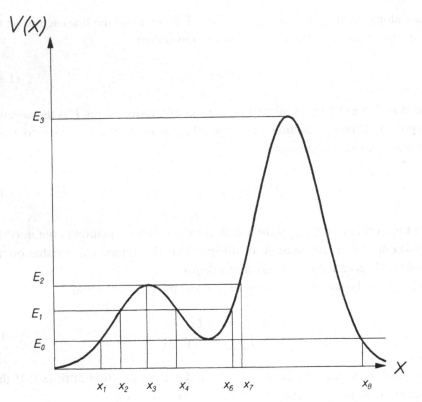

FIGURE 1.4
A one-dimensional potential energy diagram. The curve indicates the potential energy $V(x)$.

E_0 starts at some point like x_3. However, with that energy it could start somewhere to the right of x_8. For energies between E_0 and E_2 the region from x_3 to x_7 becomes physically accessible to the particle. For example, if the energy is E_1, the particle can be trapped between x_4 and x_6. For energies between E_2 and E_3, motion starting from the left of x_7 will be bounded on the right but not on the left. Finally, if $E \geq E_3$, the motion is unbounded in both directions.

Consider the bounded motion at energy E_1. The time to complete one round trip from x_4 to x_6 and back again is called the *period P*. According to Eq. (1.46) the period is given by

$$P = \sqrt{2m} \int_{x_4}^{x_6} \frac{dx}{\sqrt{E - V(x)}}, \qquad (1.48)$$

where we have used the fact that the time to go from x_4 to x_6 is the same as the time to go back. This is a general formula for the period of bounded motion between any two turning points, with the limits of the integral being the two roots of the radical in the denominator.

We turn to an explicit example, related to the simple pendulum. Let

$$V(x) = -A \cos x, \quad -\infty < x < \infty, A > 0. \qquad (1.49)$$

This potential is plotted in Fig. 1.5(a); it consists of a series of identical potential wells

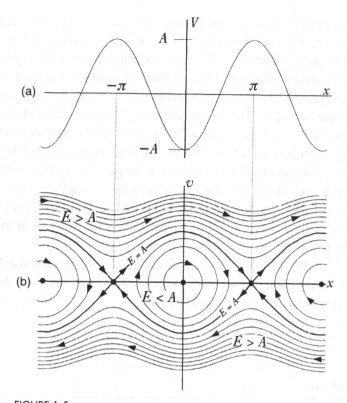

FIGURE 1.5

(a) The potential energy function $V(x) = -A \cos x$. (b) The phase portrait for the system whose potential is $V(x) = -A \cos x$ (see Section 1.5.1 for further explanation).

separated by identical potential barriers. The total energy is given by

$$E = \frac{1}{2}m\dot{x}^2 - A \cos x \equiv \frac{1}{2}mv^2 - A \cos x. \tag{1.50}$$

We start by considering motion with $E < A$, which lies entirely within one of the wells. The initial value of x determines which of the wells traps the particle, but since the motion within the well is the same no matter which well is the trap, there is no loss of generality in considering bounded motion restricted to the well at $-\pi \le x \le \pi$. Since the kinetic energy $\frac{1}{2}m\dot{x}^2$ is nonnegative, the lowest possible value for E is $-A$. The period of bounded motion is given in general by Eq. (1.48), in this case by

$$P = \sqrt{2m} \int_{x_1}^{x_2} \frac{dx}{\sqrt{E + A \cos x}}, \tag{1.51}$$

where $x_1 < 0$ and $x_2 > 0$ (as follows also from the condition that $E < A$) are the roots of the radical in the denominator of the integrand, or

$$x_{1,2} = \pm \cos^{-1}(E/A).$$

The integral cannot be evaluated in general in terms of elementary functions, but it is related to the *complete elliptic integral K* (Abramowitz and Stegun, 1976); this is a tabulated function, and hence the period can be calculated numerically. (Figure 1.5(b) is discussed in Section 1.5.1.) A special value of E is its minimum $E = -A$. At this energy, the two turning points merge at $x = 0$, and the particle is forced to remain stationary there, and therefore the point $x = 0$ is an *equilibrium point*. This is a point of *stable* equilibrium: if the mass is located near it, the force on the mass is directed toward the equilibrium point, causing the particle to accelerate toward it. For greater values of E, approaching A, the period of oscillation is longer. At $E = A$ the integral diverges and the period becomes infinite. For this value of E the turning points are at $x = \pm\pi$, but only nominally, for they are not reached in any finite time. However, with this value of E the particle could actually start out at $x = \pm\pi$, but then its velocity would have to be zero and it would remain at that value of x. Thus $x = \pm\pi$ are also equilibrium points, but these are *unstable*. That is, if the mass is located near one of them, the force on the mass is directed away from the equilibrium point, causing the particle to accelerate away from it.

If E is allowed to increase further, so that $E > A$, the motion is no longer bounded. For each such value of E there is no value of x where the kinetic energy vanishes, and the particle moves either to the right or the left, never turning back.

We will return to this system in Section 1.5.1.

1.4   MANY-PARTICLE SYSTEMS

1.4.1   MOMENTUM AND CENTER OF MASS

CENTER OF MASS

We now turn to dynamical systems composed of $N > 1$ particles interacting with each other and acted on also by external forces. Such a many-particle problem can in general be very complicated, and even the solution of the three-body gravitational-interaction problem cannot be reduced to simple integrations. Fortunately, important general results can be obtained that allow significant simplification in the formulation of such problems and lead to at least partial solutions. In a way these results beg the question: they avoid solving the general problem (to find the motion of all N particles), replacing it by other, less detailed questions. Although these other questions have important and useful answers that provide considerable insight, their chief virtue may be that they have any answers at all.

Suppose that in a system consisting of N particles a known external force is applied to each particle, but that the details of their mutual interactions are unknown. What can be said about their motion? To begin to answer this question, Principles 1 and 2 must be extended to interactions of several particles. As was mentioned in the discussion of Eq. (1.20), forces satisfy a superposition principle: the total force on a particle can be found by adding the contributions from different agents. It follows that the total force acting on a particle is the vector sum of all the forces exerted on it by each of the other particles in the system (the *internal* forces) plus all of the external forces.

The superposition principle implies that all forces can be analyzed in terms of two-particle interactions. Consider, for example, a system of three particles, A, B, and C.

Particle A, in interacting with B and C, feels no additional force having to do with the interaction of B with C: the force on A is exactly the sum of the force it would feel if only B were present and the force it would feel if only C were present. The superposition principle is an assumption additional to Newton's laws (or to the two Principles of Mechanics), a result of delicate experimental observation. One could imagine a different universe, in which the force on A depended on B, on C, and also on the interaction between B and C. (In fact something of this sort often happens when the objects that are acted on are not exactly particles, say small charged conducting spheres. Then the charge on each sphere redistributes in a way that depends on the other charged spheres in the vicinity, and that alters the forces. This shows why the experiments must be delicate.) The present treatment is valid only if all interactions between particles arise from such two-particle interactions.

> **REMARK:** We could have started with a more general statement, similar to (1.15), but involving $N > 2$ particles and leading directly to the conservation of momentum for N-particle systems, just as (1.15) led directly to (1.18). $\qquad\square$

Let $\mathbf{F}_{ij}$ be the force exerted on the ith particle by the jth. Then $\mathbf{F}_{ii} = \mathbf{0}$, and Eq. (1.21) implies that

$$\mathbf{F}_{ij} = -\mathbf{F}_{ji}. \tag{1.52}$$

The equation of motion of the ith particle is then (we abandon the summation convention here – all sums are indicated by summation signs)

$$m_i \ddot{\mathbf{x}}_i = \mathbf{F}_i + \sum_{j=1}^{N} \mathbf{F}_{ij}, \tag{1.53}$$

where $\mathbf{F}_i$ is the total external force on the ith particle, m_i is its mass, and $\mathbf{x}_i$ is its position. Clearly this equation cannot be solved unless all of the $\mathbf{F}_{ij}$ are known, and they will be changing as the particle moves. But a certain amount of information is obtained if this equation is summed over i. This sum yields

$$\sum_i m_i \ddot{\mathbf{x}}_i = \sum_i \mathbf{F}_i \equiv \mathbf{F}, \tag{1.54}$$

where the *total external force* $\mathbf{F}$ is the sum of only the external forces, since according to Eq. (1.52)

$$\sum_{i,j} \mathbf{F}_{ij} = \sum_{i,j} \mathbf{F}_{ji} = -\sum_{i,j} \mathbf{F}_{ij} = \mathbf{0}. \tag{1.55}$$

The first equality here is obtained by interchanging the roles of i and j in the sum. It is convenient in Eq. (1.54) to factor out the total mass $M = \sum_i m_i$, so that

$$\mathbf{F} = M\left(\frac{1}{M}\sum_i m_i \ddot{\mathbf{x}}_i\right) = M\ddot{\mathbf{X}}, \tag{1.56}$$

where

$$\mathbf{X} \equiv \frac{1}{M} \sum_i m_i \mathbf{x}_i \tag{1.57}$$

is defined as the *center of mass* of the system of particles. Equation (1.56) is a useful result, for what it tells us is that the center of mass moves as though the total mass were concentrated there and were acted upon by the total external force. This result is known as the *center-of-mass theorem*.

MOMENTUM

The center of mass is a useful concept for many reasons. For instance, the total momentum of the system of particles, the sum of the individual momenta, is seen to be

$$\mathbf{P} = \sum_i m_i \dot{\mathbf{x}}_i = M\dot{\mathbf{X}}, \tag{1.58}$$

as if the total mass of the system of particles were concentrated at $\mathbf{X}$. In terms of $\mathbf{P}$, Eq. (1.56) can be rewritten

$$\mathbf{F} = \dot{\mathbf{P}},$$

and therefore if the total external force $\mathbf{F} = 0$ the total momentum of the system of particles is constant and hence conserved.

Equation (1.56) is an important result. For example, it can be used to show (see Problem 16) that a large system can be made up of smaller ones in almost exactly the same way that any system is made up of its particles: the center of mass of the large system and its motion can be found from the centers of mass of its subsystems and their motion. Just as small systems can be combined in this way to make up larger ones, any system can be broken up into smaller parts: from the motion of the center of mass of a system of particles inferences can be derived about the motion of its parts. Equation (1.56) tells us in what sense the laws of motion apply to systems and subsystems and the extent to which we can disregard their internal structure.

VARIABLE MASS

In this subsection we show the usefulness of restating Newton's second law in the form $\mathbf{F} = \dot{\mathbf{P}}$. This will be done on a many-particle system in the limit as the number of particles $N \to \infty$ and the masses of the point particles become infinitesimal, which means that the sums over particles will become integrals over a *continuous* mass distribution. Not all of the mass of the system will be acted upon by the external forces, and the part that is acted upon will be allowed to vary. That will model a system of varying mass. An example is a rocket, from which mass is ejected as it moves. There are other systems to which mass is added (see Problem 25). Thus dynamical systems whose mass varies in this way arise strictly within the bounds of classical mechanics.

REMARK: The restriction to low velocities made in Section 1.2 allows us to avoid the question of relativistically varying mass. □

We will discuss only a simple system of this type, a continuous mass distribution that can be described by a single parameter λ. Each infinitesimal particle of the system is labeled by a value of λ, and that value stays with the particle as it moves. Suppose, for instance, that the distribution is one dimensional along some curve (see Problem 25); then λ could be the original length along the curve. Let the mass density ρ within the distribution be written as a function of this single parameter: $\rho = \rho(\lambda)$. The parametrization is chosen so that λ varies from 0 to some maximum value Λ, and then the total mass of the system is

$$M = \int_0^\Lambda \rho(\lambda) \, d\lambda. \tag{1.59}$$

Let the system be acted on by a total external force **F**. It then follows from the center-of-mass theorem that

$$\mathbf{F} = \frac{d^2}{dt^2} \int_0^\Lambda \mathbf{x}(t, \lambda) \rho(\lambda) \, d\lambda, \tag{1.60}$$

where $\mathbf{x}(t, \lambda)$ is the position at time t of the particle whose parameter is λ.

Assume now that the force acts only on a constantly changing part of the total mass, in particular that part of it for which $\lambda \leq \mu(t)$, where $\mu(t)$ is some differentiable monotonically increasing function. This means that as time increases, more and more of the mass is acted upon by the force. For simplicity, we assume further that for $\lambda > \mu(t)$ the velocity $\partial \mathbf{x}(t, \lambda)/\partial t = 0$, which means that before the force acts on any part of the mass, that part is stationary. Then the integrals in (1.59) and (1.60) can each be broken up into two separate integrals, one from $\lambda = 0$ to $\lambda = \mu(t)$, and the other from $\lambda = \mu(t)$ to $\lambda = \Lambda$. Thus Eq. (1.59) may be written in the form

$$M = \int_0^{\mu(t)} \rho(\lambda) \, d\lambda + \int_{\mu(t)}^\Lambda \rho(\lambda) \, d\lambda \equiv m(\mu) + \int_{\mu(t)}^\Lambda \rho(\lambda) \, d\lambda, \tag{1.61}$$

defining $m(\mu)$, which is the part of the mass acted on by the force. We will use the name *small system* for that part of the total system whose mass is $m(\mu)$. Now take one of the derivatives with respect to t in Eq. (1.60) into the integral sign, so that the integral will contain $\partial \mathbf{x}/\partial t$, whose vanishing for $\lambda > \mu$ will cause the second integral from μ to Λ to vanish. Thus (1.60) becomes

$$\mathbf{F} = \frac{d}{dt} \int_0^\Lambda \frac{\partial \mathbf{x}}{\partial t} \rho(\lambda) \, d\lambda = \frac{d}{dt} \int_0^{\mu(t)} \frac{\partial \mathbf{x}}{\partial t} \rho(\lambda) \, d\lambda$$

$$= \left[\frac{\partial \mathbf{x}}{\partial t} \rho(\lambda) \right]_{\lambda = \mu(t)} \frac{d\mu}{dt} + \int_0^{\mu(t)} \frac{\partial^2 \mathbf{x}}{\partial t^2} \rho(\lambda) \, d\lambda, \tag{1.62}$$

where we have used Leibnitz's theorem for taking the derivative of an integral with variable limits. The value of $\partial \mathbf{x}/\partial t$ in the first term in (1.62) is taken in the limit as λ approaches μ from below:

it is the velocity $\mathbf{v}$ at which mass is moving as it joins the small system at $\lambda = \mu$. Moreover, $\rho(\lambda)|_{\lambda=\mu} \cdot d\mu/dt$ is the rate dm/dt at which mass is entering the small system, or the rate at which the mass of the small system is growing. Indeed, it is evident from Eq. (1.61) that $\rho(\lambda)|_{\lambda=\mu} = dm(\mu)/d\mu$. Thus the first term on the last line of (1.62) is just $\mathbf{v}\, dm/dt$. The next term is by definition equal to $m(\mu)\ddot{\mathbf{X}}(\mu)$, where $\mathbf{X}$ is the center of mass of the small system [see Eq. (1.57)]. Then if the argument μ is dropped from the variables, Eq. (1.62) becomes

$$\mathbf{F} = \mathbf{v}\frac{dm}{dt} + m\frac{d\mathbf{V}}{dt},\tag{1.63}$$

where $\mathbf{V} = \dot{\mathbf{X}}$. In general $\mathbf{v} \neq \mathbf{V}$, but if the small system is moving as a whole (i.e., if all of its parts are moving at the same velocity) then $\mathbf{v} = \mathbf{V}$. In that case, it turns out that

$$\mathbf{F} = \frac{d\mathbf{P}}{dt},\tag{1.64}$$

where $\mathbf{P} = m(\mu)\mathbf{V}$ is the momentum of the small system. With this understanding, Eq. (1.64) can be understood as a generalization of Eq. (1.22).

How does one deal with Eq. (1.63)? That is, how and for what does one solve it? An example of a system to which it can be applied is a rocket. In a rocket, $\mathbf{F} = 0$ and $\mathbf{v}$ is the velocity at which the rocket fuel is ejected. The rate at which mass is ejected is dm/dt, and (1.63) can then be solved for $\mathbf{V}$, the velocity of the rocket (whose mass m is constantly decreasing). Both dm/dt and $\mathbf{v}$ are also given by some equations, differential equations that govern the chemical or other reactions that control the rate at which fuel is burned. Problem 25 presents another example. It involves a chain piled up next to a hole in a table, with some initial length hanging down. The problem is to find how much hangs down at time t later. This problem can be handled by applying the above discussion, with the role of the small system being played by the hanging part of the chain. In this case $\mathbf{v} = \mathbf{V}$. In general there must be additional relations between the variables m, $\mathbf{V}$, and $\mathbf{v}$ in Eq. (1.63) if the equation is to be solvable, or there must be other equations for some of these variables.

1.4.2 ENERGY

Unlike the momentum, the kinetic energy of a system of particles, defined as the sum of the individual kinetic energies, is not the kinetic energy of a particle of mass M located at the center of mass. It is found (see Problem 17) that

$$T = \frac{1}{2}M\dot{X}^2 + \frac{1}{2}\sum_i m_i \dot{y}_i^2,\tag{1.65}$$

where

$$\mathbf{y}_i = \mathbf{x}_i - \mathbf{X}\tag{1.66}$$

is the position of the ith particle relative to the center of mass. Therefore the kinetic energy of a system of particles is equal to the kinetic energy of the center of mass plus the sum of kinetic energies of the particles with respect to the center of mass.

In analogy with the one-particle problem, conservation of energy can be derived by calculating the total work that must be done on all the particles in carrying the system from an initial configuration 0 to a final configuration f:

$$\sum_i \int_{x_{i0}}^{x_{if}} \left(\mathbf{F}_i + \sum_j \mathbf{F}_{ij} \right) \cdot d\mathbf{x}_i = \sum_i \int_{x_{i0}}^{x_{if}} m_i \ddot{\mathbf{x}}_i \cdot d\mathbf{x}_i. \tag{1.67}$$

It is clear that the right-hand side of this equation is just the difference $T_f - T_0$ between the initial and final total kinetic energies. As for the left-hand side, if each external force $\mathbf{F}_i$ is derivable from a potential V_i, the first term becomes

$$\sum_i \int_{x_{i0}}^{x_{if}} \mathbf{F}_i \cdot d\mathbf{x}_i = \sum_i \left(V_i^0 - V_i^f \right) = V_{\text{ext}}^0 - V_{\text{ext}}^f. \tag{1.68}$$

If the order of summation and integration over i is changed, the second term on the left-hand side of Eq. (1.67) becomes [use Eq. (1.55)]

$$\int_{x_{i0}}^{x_{if}} \sum_{i,j} \mathbf{F}_{ij} \cdot d\mathbf{x}_i = - \int_{x_{j0}}^{x_{jf}} \sum_{i,j} \mathbf{F}_{ij} \cdot d\mathbf{x}_j$$

$$= \frac{1}{2} \int_{x_{i0}}^{x_{if}} \sum_{i,j} \mathbf{F}_{ij} \cdot d(\mathbf{x}_i - \mathbf{x}_j) = \frac{1}{2} \int_{x_{i0}}^{x_{if}} \sum_{i,j} \mathbf{F}_{ij} \cdot d\mathbf{x}_{ij}.$$

Assume now that $\mathbf{F}_{ij}$ is a function only of the relative position $\mathbf{x}_{ij} = \mathbf{x}_i - \mathbf{x}_j$ and that it is derivable from a potential function V_{ij}, also a function only of $\mathbf{x}_{ij}$. Then by changing back the order of summation and integration, one obtains

$$\int_{x_{i0}}^{x_{if}} \sum_{i,j} \mathbf{F}_{ij} \cdot d\mathbf{x}_l = \frac{1}{2} \sum_{i,i} \left(V_{ij}^0 - V_{ij}^f \right) = V_{\text{int}}^0 - V_{\text{int}}^f. \tag{1.69}$$

As shown in elementary texts (Reitz & Milford, 1964, p. 106), the expression in the center of this equation is in fact the amount of work it takes to move all of the particles of the system from their initial configuration to the final one.

The results of Eqs. (1.67), (1.41), and (1.69) are summarized by the equation

$$(V_{\text{ext}} + V_{\text{int}} + T)^0 = (V_{\text{ext}} + V_{\text{int}} + T)^f, \tag{1.70}$$

which is the statement of conservation of energy that includes the internal as well as external potential energies.

REMARK: It is interesting that $\mathbf{F}_{ij} = -\mathbf{F}_{ji}$, and yet $V_{ij} = V_{ji}$. Verify this; it's instructive.

□

1.4.3 ANGULAR MOMENTUM

Like the total kinetic energy, the total angular momentum of a system of N particles is not just the angular momentum of a particle of mass M located at the center of mass.

We calculate the total angular momentum of such a system about the origin of an inertial reference system. The total, obtained by summing the angular momenta of all the particles, is

$$L = \sum_i m_i \mathbf{x}_i \wedge \dot{\mathbf{x}}_i.$$

The total angular momentum about the center of mass is

$$\mathbf{L}_C = \sum_i m_i \mathbf{y}_i \wedge \dot{\mathbf{y}}_i,$$

where y_i is defined in Eq. (1.66). With that definition the expression for $\mathbf{L}$ becomes

$$L = \sum_i m_i (\mathbf{X} + \mathbf{y}_i) \wedge (\dot{\mathbf{X}} + \dot{\mathbf{y}}_i)$$
$$= \sum_i m_i \mathbf{X} \wedge \dot{\mathbf{X}} + \sum_i m_i \mathbf{y}_i \wedge \dot{\mathbf{y}}_i + \sum_i m_i \mathbf{y}_i \wedge \dot{\mathbf{X}} + \sum_i m_i \mathbf{X} \wedge \dot{\mathbf{y}}_i.$$

The last two terms vanish because $\sum m_i \mathbf{y}_i$ is the position of the center of mass relative to the center of mass and is hence zero. Thus the total angular momentum is

$$L = \mathbf{L}_C + M\mathbf{X} \wedge \dot{\mathbf{X}}. \tag{1.71}$$

This result states that the total angular momentum about the origin of an inertial system (and hence about any point moving at constant speed in an inertial frame) is the sum of the angular momentum of the total mass as though all the mass were concentrated at the center of mass plus the *internal angular momentum* – the angular momentum of the particles about the center of mass.

In the previous section it was seen that for a single particle the applied torque about the origin is equal to the rate of change of the angular momentum about the origin. Is this true also for a system of particles? If it is assumed that the internal torques do not contribute to the change in the total angular momentum, just as the internal forces do not contribute to the change in the total linear momentum (see Problem 18), one may consider only the externally applied torque. The external forces apply a total torque about an arbitrary (moving) point $\mathbf{Z}$ equal to

$$N_Z = \sum_i \mathbf{z}_i \wedge \mathbf{F}_i = \sum_i m_i \mathbf{z}_i \wedge \ddot{\mathbf{x}}_i,$$

where $\mathbf{x}_i$ is the position of the ith particle in an inertial frame and $\mathbf{z}_i = \mathbf{x}_i - \mathbf{Z}$ is its position relative to $\mathbf{Z}$, as shown in Fig. 1.6. It then follows that

$$N_Z = \sum_i m_i \mathbf{z}_i \wedge (\ddot{\mathbf{z}}_i + \ddot{\mathbf{Z}}) = \frac{d}{dt} \sum_i m_i \mathbf{z}_i \wedge \dot{\mathbf{z}}_i + M\zeta \wedge \ddot{\mathbf{Z}}, \tag{1.72}$$

where $\zeta = \mathbf{X} - \mathbf{Z}$ is the position of the center of mass relative to $\mathbf{Z}$. The first term on the right-hand side of this equation is just $\dot{\mathbf{L}}_Z$, and therefore $N_Z = \dot{\mathbf{L}}_Z$ if and only if $\zeta \wedge \ddot{\mathbf{Z}} = 0$.

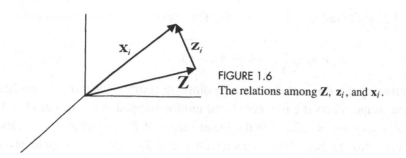

FIGURE 1.6
The relations among $\mathbf{Z}$, $\mathbf{z}_i$, and $\mathbf{x}_i$.

Therefore $\ddot{\mathbf{Z}}$ must be parallel to the line connecting $\mathbf{Z}$ and the center of mass or, as special cases, $\ddot{\mathbf{Z}} = 0$ or $\zeta = 0$. When $\ddot{\mathbf{Z}} = \mathbf{0}$, the point $\mathbf{Z}$ is moving at constant velocity relative to the origin of the inertial frame. When $\zeta = \mathbf{0}$, the point $\mathbf{Z}$ is the center of mass of the system. Thus

$$\mathbf{N}_Z = \dot{\mathbf{L}}_Z \qquad (1.73)$$

if (but not only if) $\mathbf{Z}$ is an inertial (nonaccelerating) point or the center of mass, even if the center of mass is accelerating. In the one-particle case, torques only about fixed (or inertial) points were considered, so this problem did not occur.

1.5 EXAMPLES

1.5.1 VELOCITY PHASE SPACE AND PHASE PORTRAITS

The state of a dynamical system can be characterized by its velocity and position. For a particle in ordinary three-dimensional *configuration space*, the velocity and the position are both vectors, with three components each, so even in this simple case a complete description of the state of the system requires six variables. As the system evolves, these six variables all change, and a calculation of the system's time development requires calculating six functions of the time. Fortunately, one is often interested in less than such complete solutions, for instance in just the energy or, for the case of bounded motion, in just the minimum and maximum magnitudes of the position vector. For many such purposes it is enough to know the relation between the velocity and position without knowing in detail how these dynamical variables develop in time.

THE COSINE POTENTIAL

Of great help in understanding, even in analyzing, this relation is to study graphs of velocity as a function of position. In this subsection we discuss this procedure, starting with the example of the system whose potential is given in Eq. (1.49).

The velocity and position of this system are related through Eq. (1.50), which is illustrated in Fig. 1.5(b). Each value of E yields a different set of curves in the v–x plane. To see the relation between these curves and the values of E, consider first the bounded motions (i.e., those values of E in the range $-A \le E < A$). The curves, whose equations are

$$v = \pm \sqrt{\frac{2}{m}(E + A \cos x)},$$

are closed curves crossing the x axis at the turning points (remember, the radical vanishes at these points). The graph shows the v–x curves for motion trapped in the potential well at $x = 0$ and partially also at $x = \pm 2\pi$. For the lower values of E, very close to $-A$, the turning points are very close to the bottom of each well at $x = 2n\pi$, and the particle never gets far from these equilibrium points. The potential can be approximated there by just the first two terms of Taylor's series (we consider only the well at $x = 0$),

$$V(x) = V(0) + x V'(0) + \frac{1}{2} x^2 V''(0) + \cdots \approx A\left(-1 + \frac{1}{2} x^2\right),$$

and then the v–x relation is given by

$$\frac{m v^2}{2(E + A)} + \frac{A x^2}{2(E + A)} = 1.$$

This is the equation of an ellipse centered at $x = 0$, with major and minor axes depending on m, A, and the energy E and lying along the x and y axes. For the lowest possible energy $E = -A$ the ellipse degenerates to the point at the origin (more generally, at $v = 0$, $x = 2n\pi$). Recall from Section 1.3.3 that this is a stable equilibrium point. As the energy increases toward A, the two-term Taylor expansion becomes less valid, and the graphs are less accurately described as ellipses. They remain closed curves (at least for $E < A$), getting larger as E increases, and hence are nested one within the other.

The case of $E = A$ corresponds to the transition from bounded to unbounded motion and is represented in the v–x graph by the curves that cross at $v = 0$, $x = (2n + 1)\pi$. As was mentioned in Section 1.3.1, the period of this motion becomes infinite. When $E > A$, the motion is unbounded; the curves representing it are the continuous ones running from left to right at the top and from right to left at the bottom of Fig. 1.5(b).

Such a graph, containing such a set of curves, is called a *phase portrait* of the dynamical system, and the v–x plane is called *velocity phase space* (the term "phase space" is reserved for something else: see Chapter 5). Each point in velocity phase space defines a *state* of the system, that is, gives its position and velocity. As the system evolves in time, the *system point* moves in velocity phase space. Suppose that in the example we have been describing certain initial values $x(0)$ and $v(0)$ are given, determining the initial system point. Through that point passes only one curve of the phase portrait, the one corresponding to the energy determined by $x(0)$ and $v(0)$ (see the next remark). The rest of the motion satisfies Eq. (1.50), so that as time progresses the system point remains on the initial curve, moving to the right (positive v) if it is above the x axis and to the left (negative v) if below. If the curve in question is one of the closed curves ($E < A$), the system point crosses the x axis at the turning point and begins to move in the opposite direction. If $E > A$, the point continues to the right or left in unbounded motion without ever turning back. Different initial points on the same curve lead to similar motions in velocity phase space: a curve of the phase portrait contains a set of motions that differ only in that each one starts at a different point on the curve. Each such curve is called an *orbit* in velocity phase space (not to be confused with orbits in configuration space).

> REMARK: One way to see that there is only one curve through each point of velocity phase space is to recall that the equation of motion $\mathbf{F} = m\mathbf{a}$ is of second order, requiring two initial conditions for a particular solution. The initial point $x(0)$, $v(0)$ in velocity phase space provides the two conditions, and therefore when it is given, the solution (i.e., the motion) is unique. See the discussion following Eq. (1.22). □

A special orbit is the degenerate ellipse at the origin, a one-point orbit. There the velocity is zero: the system remains stationary, corresponding to the stable equilibrium point. The curve corresponding to $E = A$ is called a *separatrix*, for it separates the bounded from the unbounded motions. In the region from $x = -\pi$ to $x = \pi$ the separatrix consists of four orbits: the one above the x axis, the one below it (it takes an infinite time to cover each), and the two unstable equilibrium points at $v = 0$, $x = \pm\pi$. Equilibrium points are also called *singular* points, with the stable ones being *elliptic* and the unstable ones *hyperbolic* (the orbits look like ellipses in the neighborhood of an elliptic point and like hyperbolas in the neighborhood of a hyperbolic one). This dynamical system contains an infinite number of singular points: the elliptic ones at $v = 0$, $x = 2n\pi$ and the hyperbolic ones at $v = 0$, $x = (2n + 1)\pi$. All of these are degenerate one-point orbits.

Although we have discussed only an example, it should be clear that the same kind of graphical analysis can be applied to any one-freedom system, even if it involves energy dissipation. For instance, if dissipation were added to the example, the total energy would decrease as the system point moved in velocity phase space, and all of the orbits (except the hyperbolic singular points) would spiral down to the stable equilibrium points at $x = 2n\pi$. Such a system is discussed in some detail in the next subsection.

THE KEPLER PROBLEM

This kind of graphical analysis in two dimensions can be complete only for a system with just one-freedom, because velocity phase space has dimension $2n$ for systems with n freedoms. Nevertheless, for larger systems this analysis can be applied to each freedom that can be separated out. A good example of this is the Kepler problem, which concerns the motion of a particle in a central gravitational field (e.g., a planet in its orbit around the Sun). This dynamical system will be discussed again in more detail in the next chapter, but at this point we turn our attention just to the phase portrait of its radial motion.

In the Kepler problem the radial velocity $\dot{r}$ is related to the radius r by

$$\frac{1}{2}m\dot{r}^2 + \frac{l^2}{2mr^2} - \frac{k}{r} = E; \tag{1.74}$$

$k = GmM$ where G is the universal gravitational constant and M is the mass of the attracting center. This is a different differential equation for each value l of the angular momentum, but for the rest of our discussion we restrict l to one fixed value. For fixed l the equation looks like the energy conservation equation for a one-freedom system whose potential energy is $l^2/(2mr^2) - k/r$. The term $\frac{1}{2}m\dot{r}^2$ represents the kinetic energy, and the sum of kinetic and potential energy is the constant total energy E. Figure 1.7(a) is a diagram of

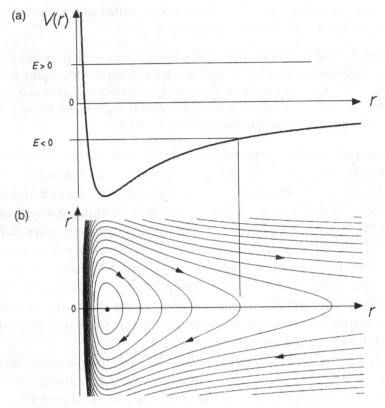

FIGURE 1.7
(a) The equivalent potential for the radial motion in the Kepler problem. (b) The phase portrait for the radial motion in the Kepler problem.

the potential, and below it, just as in the previous example, Fig. 1.7(b) is the phase portrait of this system. The closed curves of this phase portrait, as in the previous example, correspond to bound states, in which the particle has a minimum and maximum distance from the force center; but recall that these diagrams describe only the radial motion, so that while r is moving between its minimum and maximum values, in configuration space the particle is rotating about the force center. Thus these bound states represent orbits in configuration space like those of a planet orbiting the sun. These states all have negative energies E and the velocity phase space orbits cross the r axis at points that depend on E. The higher E (the closer it is to zero), the closer the inner crossing point (minimum r) is to the origin and the farther the outer one (maximum r) is from it. The negative energies end at $E = 0$, when the outer crossing point moves out to $r = \infty$. This is the separatrix for this problem, and it corresponds to the parabolic configuration-space orbit. The lower E (the more negative it gets), the further out the inner crossing point and the closer in the outer one, until they meet at a common point. This gives the lower bound of the energy; the velocity phase-space orbit degenerates to a single point, corresponding to fixed r or to the circular configuration-space orbit with the given value of l. This is the unique elliptic singular point of this system. The negative-E velocity phase-space orbits between $E = 0$ and this lower bound correspond to elliptical orbits in configuration space. The velocity phase-space orbits for $E > 0$

cross the r axis only once: each such orbit has an asymptote, for large r, that is parallel to the r axis in velocity phase space. The distance of this orbit from the r axis is the minimum speed $|\dot{r}|$ at that energy. These positive-E configuration-space orbits correspond to hyperbolic orbits in configuration space. Remember that l has been taken as fixed.

This analysis is possible for the radial motion, but it is much more complicated for the angular motion about the force center, because the equation for the angular motion involves not only the angle θ itself but also the radius r. We point out again that in systems with several freedoms this graphical analysis, essentially for one-dimensional motion, can be applied only to those freedoms that can be separated out.

This kind of treatment in velocity phase space will become more meaningful in the next chapter, which treats the Lagrangian formulation of dynamics. One of the results of the Lagrangian formulation is that the arena in which the motion takes place, the "carrier space" of the motion, is seen not as configuration space, but as velocity phase space, which will there be called the *tangent bundle*.

WORKED EXAMPLE 1.1

(a) Consider a particle of mass m in the inverse harmonic oscillator potential $V = -\frac{1}{2}kx^2$ (with one freedom). Write down the $F = ma$ equation and solve for the motion in terms of initial conditions x_0 and $\dot{x}_0$ at $t = 0$. (b) Draw the phase portrait in velocity phase space.

Solution. (a) The force is given by $F = -\partial(-\frac{1}{2}kx^2)/\partial x = kx$, so the equation of motion is

$$\ddot{x} = \omega^2 x,$$

where $\omega = \sqrt{k/m}$. This equation differs from the harmonic oscillator equation only in sign. The usual procedure for solving that equation (or, equivalently, replacing the harmonic oscillator ω by $i\omega$) yields

$$x(t) = Ae^{\omega t} + Be^{-\omega t} = a\cosh(\omega t) + b\sinh(\omega t),$$

where $A = \frac{1}{2}(a + b)$ and $B = \frac{1}{2}(a - b)$. By setting $t = 0$ one finds that $A = \frac{1}{2}(x_0 + \dot{x}_0/\omega)$ and $B = \frac{1}{2}(x_0 - \dot{x}_0/\omega)$ or that $a = x_0$ and $b = \dot{x}_0/\omega$. Thus

$$x(t) = \frac{1}{2}(x_0 + \dot{x}_0/\omega)e^{\omega t} + \frac{1}{2}(x_0 - \dot{x}_0/\omega)e^{-\omega t}$$
$$= x_0\cosh(\omega t) + (\dot{x}_0/\omega)\sinh(\omega t).$$

(b) The easiest way to draw the phase portrait is from the equation for the energy. The total energy is $E = T + V$ is given by

$$2E = m\dot{x}^2 - m\omega^2 x^2.$$

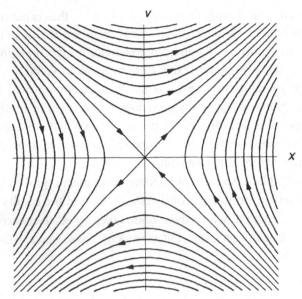

FIGURE 1.8
The phase portrait of the dynamical system of Worked Example 1.1.

For each value of E this is the equation of a hyperbola; if E is positive, the hyperbola opens vertically, and if E is negative, horizontally (Fig. 1.8). When E is zero, the hyperbolas degenerate to the two straight lines $\dot{x} = \pm\omega x$, which are also asymptotes of all of the hyperbolas. There is an unstable equilibrium point at the origin.

1.5.2 A SYSTEM WITH ENERGY LOSS

Energy is not conserved in all dynamical systems: in general it is given up in many possible ways, for example, directly to other dynamical systems, as heat, or as radiation. Systems whose energy is conserved are called *conservative*, and those that give up energy are called *dissipative*. Typical dissipative systems the reader may have come across previously are those with frictional forces. So far in this chapter we have discussed only conservative systems, though some of the problems refer to dissipative ones.

In this section we present an example of a system with one degree of freedom in which energy is dissipated by friction. The object will be to map out its phase portrait. The phase portraits of the conservative systems that have been presented so far have orbits that are closed or extend to infinity. It will be seen that the phase portrait of this system is different in an essential way.

The system consists of a mass m sliding down an inclined plane, over a hump, and up a plane on the other side, as shown in Fig. 1.9. Friction on the two planes, to the left of A and to the right of B, causes the mass to lose energy in proportion to the distance traversed, but no energy is lost on the (symmetric) hump between A and B. Eventually the mass will

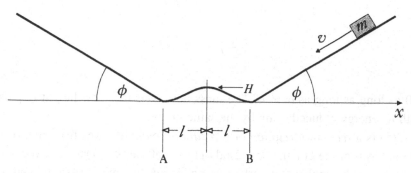

FIGURE 1.9
A simple system with energy dissipation.

lose enough energy to be trapped either to the left or right of the hump, near A or near B, depending on the initial conditions. Let the plane be tilted at an angle ϕ and choose a horizontal coordinate x in which A is at $x = -l$ and B at $x = +l$, so that the origin is right under the highest point of the hump.

Because each initial condition is represented by a point in velocity phase space, to understand the phase portrait, we will first try to find those regions where the initial conditions lead to entrapment at A or at B. We choose coordinates in velocity phase space to be the vertical height h and the velocity v along the plane.

Let E_k be the energy of the mass the kth time it arrives at the hump; it will have the same energy E_k at A, at B, and on the hump. Let ϵ be the threshold energy required to get over the hump (then $\epsilon = mgH$, where H is the height of the hump at $x = 0$). If the mass is moving to the left and has energy E_k when it reaches the hump, it first must have passed B; it then passes the hump without losing any energy, arrives at A still with energy E_k, and rises to a height h_k on the left-hand side, where its energy drops to a lower value $E_{\text{TOP},k}$ given by

$$E_{\text{TOP},k} = mgh_k = E_k - h_k\alpha;$$

here $\alpha = dE/dh$ is the rate of energy loss on the plane. If the mass is moving to the right when it reaches the hump with energy E_k, it arrives at B with energy E_k and rises the same height h_k on the right-hand side. In both cases,

$$h_k = \frac{E_k}{mg + \alpha},$$

so that

$$E_{\text{TOP},k} = E_k \left[1 - \frac{\alpha}{mg + \alpha}\right] = E_k \frac{mg}{mg + \alpha}.$$

The mass then slides down on the left-hand side to A (or on the right-hand side to B), losing an amount of energy equal to $h_k\alpha = E_k\alpha/(mg + \alpha)$. If we write $\alpha = mg\beta$, we find that the mass will come back to B in both cases (remember that it loses no energy in going from A to B) with energy

$$E_{k+1} = E_k\gamma, \tag{1.75}$$

where

$$\gamma = \frac{1-\beta}{1+\beta}. \tag{1.76}$$

Note that $\gamma > 0$ as long as long as $\alpha \neq 0$. The mass now rises to a new height h_{k+1} and returns to B with its energy reduced again by the same factor.

Equation (1.75) is a *recursion relation* or a *recursion map*: it gives the value of the $(k+1)$st energy in terms of the kth, the $(k+2)$nd in terms of the $(k+1)$st, etc. Recursion maps are very useful in the analysis of certain complicated dynamical systems and will play a role later in the book. The recursion map of Eq. (1.75) is particularly easy to solve (a solution gives all of the energies in terms of the initial one). The solution is

$$E_{k+1} = E_1 \gamma^k.$$

REMARK: If the energy loss is due to friction on the plane, α depends on both the coefficient of friction μ and the angle ϕ. Since β and γ are defined in terms of α, they also depend on μ and ϕ. In the limit $\alpha = 0$ the definitions imply that $\beta = 0$ and $\gamma = 1$.   □

Assume for the time being that the mass starts at some height h_0 on the plane on the right-hand side with a velocity v_0 to the left. It has an initial energy E_0, arrives at B with lower energy E_1, comes back to B with energy E_2, and so on, until it gets trapped near A or near B. When E_{k+1} becomes less than ϵ, that is, when $E_1 \gamma^k < \epsilon$, the mass will no longer cross the hump. Let K be the lowest integer k for which this happens; clearly K depends on E_1. Indeed, if $E_1 < \epsilon$, then the first time the mass arrives at B it fails to cross the hump and becomes trapped near B; in this case $K = 0$. If E_1 is a little larger, the mass will cross the hump the first time, rise on the plane on the left of the hump, and arrive back down at the hump with energy E_2 insufficient to cross it, becoming trapped near A; then $K = 1$. If E_1 is a little larger than that, so that E_2 is sufficient for the mass to cross the hump, but E_3 is not, it is trapped near B and $K = 2$; etc. The boundaries of the regions we are looking for in phase space correspond to those values of E_1 for which $E_1 \gamma^K = \epsilon$ for some integer K. Those boundaries are given by

$$E_1 = \epsilon \gamma^{-K} = E_0 - \alpha h_0 = mgh_0 - \alpha h_0 + \frac{1}{2}mv_0^2; \tag{1.77}$$

so far we are assuming that v_0 is to the left: $v_0 < 0$. This yields

$$h_0 = \frac{\epsilon(1/\gamma)^K}{mg(1-\beta)} - \frac{1}{2g(1-\beta)}v_0^2.$$

We now change coordinates in velocity phase space from h, v to x, v. This is useful because h is not a unique coordinate for the position: for $h \leq H$ four values of h correspond to one position, and for $h > H$ two values. The change is a simple one because $h = (x - l)\tan\phi$. Let us drop the subscripts 0 and subsume some of the constants in two new ones, P and Q. Then the boundaries between the regions of velocity phase space are given by

$$x = l + P\frac{(1/\gamma)^K}{1-\beta} - Q\frac{v^2}{1-\beta} \quad \text{for } x > l, v < 0. \tag{1.78}$$

If v_0 is to the right, that is, if $v_0 > 0$ (still for $x_0 > l$), Eq. (1.77) no longer describes the situation. The mass rises from h_0 to a new height h_0' and arrives back at h_0 again with a new energy $E_0' = E_0 - 2\alpha(h_0' - h_0)$. This new energy can then be used in the same way that E_0 was for $v_0 < 0$. A similar calculation to the one above leads to

$$x = l + P\frac{(1/\gamma)^K}{1-\beta} - Q\frac{v^2}{1+\beta} \quad \text{for } x > l, v > 0. \tag{1.79}$$

The only difference is in the factor multiplying v^2.

The calculation for $x < -l$ is quite similar to the one for $x > l$. Its result is

$$x = \begin{cases} -l - P\dfrac{(1/\gamma)^K}{1-\beta} + Q\dfrac{v^2}{1-\beta} & \text{for } x < l, v > 0, \tag{1.80} \\[3mm] -l - P\dfrac{(1/\gamma)^K}{1-\beta} + Q\dfrac{v^2}{1+\beta} & \text{for } x < l, v < 0. \tag{1.81} \end{cases}$$

So far we have calculated the boundaries of the regions to the left of A and to the right of B. In between, on the hump between A and B, the phase portrait resembles that of the cosine potential of Section 1.5.1, or at least a part of it between two neighboring elliptic equilibrium points. For the purposes of this discussion, however, the details of this part of the phase portrait are immaterial. The important question is how this part connects with the part of the phase portrait representing the motion on the two planes.

Figure 1.10 is a map of velocity phase space showing the boundaries we have just calculated. These boundaries are parabolic to the right of the line $x = l$ and to the left of the line $x = -l$, but the parabolas are different above and below the x axis, reflecting the different coefficients of v^2. On the right, the parabolas below the x axis approach the v axis more rapidly, and on the left, those above approach it more rapidly. The intercepts on the x axis are at $x = l + p\gamma^{-K}$ on the right and $x = -l - p\gamma^{-K}$ on the left, where $p = P/(1-\beta)$. The intercepts on the lines $x = \pm l$ are at

$$v = \pm\sqrt{\frac{P}{Q}}\gamma^{-K/2} \quad \text{for } v \text{ and } l \text{ both positive or both negative}$$

and at

$$v = \pm\sqrt{\frac{P}{Q}}\gamma^{-(K+1)/2} \quad \text{for } v \text{ negative and } l \text{ positive and vice versa.}$$

Here we have used the definition (1.76) of γ. These equations show that the Kth intercept on one side lines up horizontally with the $(K + 1)$st on the other, so that the symmetric phase portrait between $x = -l$ and $x = +l$ has orbits that connect the boundaries we have calculated in the rest of velocity phase space.

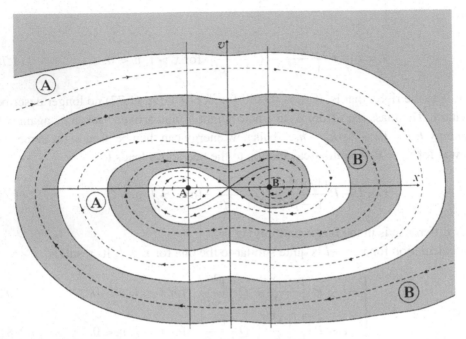

FIGURE 1.10
The phase portrait for the system with energy dissipation.

Because they separate the orbits into two classes, those that converge to A and those that converge to B, the calculated boundaries must lie on separatrices. In this system there is just one separatrix, which is made up of the calculated boundaries and of the curves connecting them between the lines $x = \pm l$: it is the long winding line that spirals in, crosses the origin, and spirals out again, separating the two snakelike regions we have labeled A and B. No phase orbit can cross the separatrix; each orbit is forced to remain in one or the other of the regions, spiraling in closer and closer to one of the two foci (A and B) at $v = 0$, $x = \pm l$ (this explains why the regions are labeled A and B). A pair of orbits is drawn in on Fig. 1.10. An interesting feature of this dynamical system is that two initial conditions on opposite sides of the separatrix, even if they are infinitesimally close to each other, give rise to motions that end up in finitely separated regions of velocity phase space.

We leave to Problem 23 the question of what happens if the initial condition lies precisely on the separatrix or on the strange short line passing through the origin (which would have been part of a separatrix if there had been no energy dissipation).

1.5.3 NONINERTIAL FRAMES AND THE EQUIVALENCE PRINCIPLE

EQUIVALENCE PRINCIPLE

Although our discussion so far has been restricted to dynamics in inertial systems, most physical experiments are actually performed in frames that are not inertial. A frame attached to the surface of the Earth, for example, is obviously not inertial, because it rotates about the Earth's axis and moves with the Earth around the Sun. It is therefore important to state the laws of mechanics in a way that can be of use also when observations are made in noninertial systems.

Suppose first that an observer moving with constant acceleration $\mathbf{A}$ relative to an inertial frame is making measurements on a particle that has no forces acting on it. Since the particle is not accelerating with respect to the inertial frame, the observer has an acceleration $\mathbf{A}$ with respect to the particle, and the particle therefore has acceleration $-\mathbf{A}$ with respect to the observer.

A stubborn observer, one who insists that the equation $\mathbf{F} = m\mathbf{a}$ applies in every frame (e.g., relative to himself) as surely as in an inertial one, would then be forced by consistency to interpret the cause of the acceleration $-\mathbf{A}$ to be an additional force $\mathbf{F}_A = -m\mathbf{A}$. A similar force, sometimes called *fictitious* or *inertial*, would seem to be applied to every other particle in the noninertial frame of reference, no matter what other "real" forces were applied to it. Here real forces are understood to be forces that can be assigned to external agents, to interactions with other particles. If such a real force $\mathbf{F}$ were applied to a particle, the observer would interpret the total force on it as $\mathbf{F}_{tot} = \mathbf{F} - m\mathbf{A}$. Indeed, in order to keep a noninteracting particle stationary in the noninertial frame, the observer would have to arrange to have a real force $\mathbf{F} = m\mathbf{A}$ applied to it.

The resulting situation looks almost exactly like a gravitational field $-\mathbf{A}$ in an inertial frame; that is, in addition to anything else, there is a gravitational force equal to $-m\mathbf{A}$ acting on each particle of mass m. This is sometimes called the Equivalence Principle, establishing the equivalence of acceleration and gravitation. This view depends on the equality between the *inertial mass*, the m that appears in Newton's second law $\mathbf{F} = m\mathbf{a}$, and the *gravitational mass*, the m that appears in the universal gravitational law $\mathbf{F} = m\boldsymbol{\Gamma}$ (here $\boldsymbol{\Gamma}$ is the gravitational field, which in the case of an attractive center of mass M has magnitude GM/r^2, where G is the universal gravitational constant). It is ordinarily assumed that these two masses are in fact equal, and for the moment let us assume it without question.

The Equivalence Principle sometimes provides a shortcut for solving problems, but it should be used with caution, for not every point in a noninertial frame accelerates at the same rate – the fictitious gravitational force may be different at different points. For instance, in a rotating frame the acceleration depends on the distance from the center of rotation. Detailed analysis (see Problem 14) shows also that in general the fictitious force can depend on the velocity of the particle.

The following simple example illustrates this shortcut. Consider a rectangular box of uniform density resting on an accelerating platform in a gravitational field, say a crate in an accelerating truck. Assuming that there is enough friction to keep the crate from slipping, at what acceleration A will it begin to tip over? The pedestrian way to solve this problem is to take moments about the center of mass (or about any point accelerating directly toward or away from the center of mass; see Section 1.4.3). The frictional force $f = mA$, which causes the acceleration, has torque of magnitude $\frac{1}{2} f h$ [Fig. 1.11(a)], and the supporting normal force that the truck applies to the crate has torque of magnitude mgd in the opposite sense, where d is the distance between the vertical through the center of mass and the point at which the normal force mg is applied. From the condition that the crate does not tilt (i.e., that the total torque is zero) one arrives at $d = \frac{1}{2} hA/g$. The maximum value of d is $\frac{1}{2} w$, and therefore the maximum acceleration is given by $A/g = w/h$. The shortcut involves recognizing that the truck bed is a noninertial frame and using the Equivalence Principle. The problem is then to find the total equivalent gravitational field $\mathbf{G}$, composed of the real field and the fictitious one: $\mathbf{G} = \mathbf{g} - \mathbf{A}$. The crate will begin to tilt when the gravitational field vector passing through the crate's center of mass fails to intersect its base [see Fig. 1.11(b), which is drawn rotated so that $\mathbf{G}$ points downward], that is, when $A/g > w/h$.

A comparison of gravitational and electrostatic forces is of some help in understanding the significance of the Equivalence Principle. In the electrostatic case, the force $\mathbf{F} = q\mathbf{E}$ on a

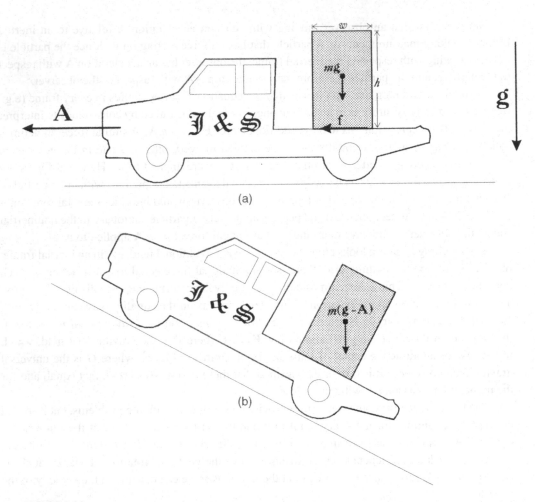

FIGURE 1.11
(a) A crate on an accelerating truck. (b) The same crate, drawn tilted so that the fictitious gravitational field **G** points straight down.

particle is obtained by multiplying its electric charge q by the electric field **E**, and the acceleration is then given by $\mathbf{a} = \mathbf{F}/m = (q/m)\mathbf{E}$. The response (acceleration) of a particle to a given field **E** can thus be modified by changing either its electric charge or its mass; two particles with the same electric charge but different masses (or different electric charges, but the same mass) will respond differently. In the gravitational case q must be replaced by the gravitational mass m_g, also known as the *gravitational charge*, and **E** by $\boldsymbol{\Gamma}$, so the acceleration is $\mathbf{a} = (m_g/m)\boldsymbol{\Gamma}$. If the gravitational mass is the same as the inertial mass m, this equation becomes $\mathbf{a} = \boldsymbol{\Gamma}$, and all particles respond in the same way. If the two masses are different, two particles with the same inertial mass but different gravitational masses will respond differently.

   The Equivalence Principle is an important part of Einstein's formulation of the general theory of relativity. It is therefore not surprising, for instance, that in the general theory gravitational forces depend in general on velocities, as do inertial forces (e.g., Coriolis forces). On the other hand, if inertial and gravitational masses are not equal, much of not only Einstein's gravitational theory, but also of Newton's, is called into doubt. Recently some researchers (Goldman et al., 1988) have investigated the possible violation of this principle.

WORKED EXAMPLE 1.2

A simple pendulum (a string of length l and negligible mass, with a small mass m attached at the end) hangs in equilibrium in a uniform gravitational field g. Its point of suspension is suddenly given a sustained acceleration **a** at an angle α with respect to the upward vertical. Describe the resulting motion of the pendulum for small α.

Solution. Use the Equivalence Principle. The acceleration **a** of the point of suspension is equivalent to a gravitational field $-\mathbf{a}$ in a frame moving with acceleration **a**. The *total gravitational field* **G** in that frame is the sum of $-\mathbf{a}$ and the uniform field **g** (of magnitude g and pointing downward). Thus $\mathbf{G} = \mathbf{g} - \mathbf{a}$. A simple vector diagram shows that the magnitude of **G** is

$$G^2 = g^2 + a^2 + 2ag \cos \alpha$$

and that the angle θ that **G** makes with the vertical is

$$\cos \theta = \frac{g + a \cos \alpha}{G}.$$

Since the pendulum is initially hanging vertically, θ is also the angle between the initial position of the pendulum and the total gravitational field **G**. Note that θ is less than α. The initial velocity of the pendulum is zero, even in the accelerating frame. The rest of the problem is solved in the standard way for a pendulum suspended in a gravitational field with given initial conditions.

ROTATING FRAMES

Another familiar type of noninertial frame is one that is rotating at a constant rate, such as the surface of the Earth. We will not treat the case in which the rate and/or axis of rotation varies. Consider two coordinate systems, one inertial and the other rotating at a constant rate about the inertial 3 axis. Let the Cartesian coordinates in the inertial system be x_j and those of the rotating one be y_j. Then the transformation equations corresponding to (1.23) are

$$
\begin{aligned}
y_1 &= x_1 \cos \omega t - x_2 \sin \omega t, \\
y_2 &= x_1 \sin \omega t + x_2 \cos \omega t, \\
y_3 &= x_3;
\end{aligned}
\tag{1.82}
$$

$$
\begin{aligned}
x_1 &= y_1 \cos \omega t + y_2 \sin \omega t, \\
x_2 &= -y_1 \sin \omega t + y_2 \cos \omega t, \\
x_3 &= y_3.
\end{aligned}
\tag{1.83}
$$

A free particle moves with $\ddot{\mathbf{x}} = \mathbf{0}$ in the inertial frame. By taking derivatives with respect to t, we get

$$\dot{y}_1 = \dot{x}_1 \cos \omega t - \dot{x}_2 \sin \omega t - \omega y_2,$$
$$\dot{y}_2 = \dot{x}_1 \sin \omega t + \dot{x}_2 \cos \omega t + \omega y_1,$$
$$\dot{y}_3 = \dot{x}_3.$$

The second derivative with respect to t then yields (use $\ddot{x}_k = 0$)

$$
\begin{aligned}
\ddot{y}_1 &= \omega(-\dot{x}_1 \sin \omega t - \dot{x}_2 \cos \omega t) - \omega \dot{y}_2 \\
&= -\omega(\dot{y}_2 - \omega y_1) - \omega \dot{y}_2 = \omega^2 y_1 - 2\omega \dot{y}_2, \\
\ddot{y}_2 &= \omega(\dot{x}_1 \cos \omega t - \dot{x}_2 \sin \omega t) + \omega \dot{y}_1 \\
&= \omega(\dot{y}_1 + \omega y_2) + \omega \dot{y}_1 = \omega^2 y_2 + 2\omega \dot{y}_1. \\
\ddot{y}_3 &= 0;
\end{aligned}
\tag{1.84}
$$

a free particle in a rotating frame looks as though it is accelerating. These equations arise because of choosing the 3 axis along the axis of rotation. It is left to the reader to show that in general

$$\ddot{\mathbf{y}} = \boldsymbol{\omega} \wedge (\boldsymbol{\omega} \wedge \mathbf{y}) + 2\boldsymbol{\omega} \wedge \dot{\mathbf{y}}, \tag{1.85}$$

where $\boldsymbol{\omega}$ is the angular velocity vector, of magnitude ω and pointing along the axis of rotation (in accordance with the right-hand-rule; see Chapter 8).

An observer in a rotating frame, seeing the particle accelerating, tends to assign a force to this acceleration. That is, in order to keep the particle from accelerating in that frame, the observer must apply a force equal to $-m\ddot{\mathbf{y}}$. This is essentially the same phenomenon as the fictitious force of the previous subsection, except that now it depends on the position $\mathbf{y}$ and the velocity $\dot{\mathbf{y}}$.

The first term of (1.85), namely $\boldsymbol{\omega} \wedge (\boldsymbol{\omega} \wedge \mathbf{y})$, is the familiar centrifugal acceleration. The second term, namely $2\boldsymbol{\omega} \wedge \dot{\mathbf{y}}$, is the Coriolis acceleration (also called the Buys Ballot acceleration). The Coriolis force is felt only by objects that are moving in the rotating frame, and then only if they are not moving parallel to the axis of rotation.

PROBLEMS

1. A gun is mounted on a hill of height h above a level plane. Neglecting air resistance, find the angle of elevation α for the greatest horizontal range at a given muzzle speed v. Find this range.

2. A mass m slides without friction on a plane tilted at an angle θ in a vertical uniform gravitational field g. The plane itself is on rollers and is free to move horizontally, also without friction; it has mass M. Find the acceleration A of the plane and the acceleration a of the mass m.

3. Figures 1.12(a)–(e) show a hand pulling a circular cylindrical object (whose mass is distributed with cylindrical symmetry). The cylinder has radius R, mass M, and moment of inertia I about its symmetry axis. The hand applies a force F by means of a weightless, flexible string. In all five cases find the acceleration A of the center of mass and the angular

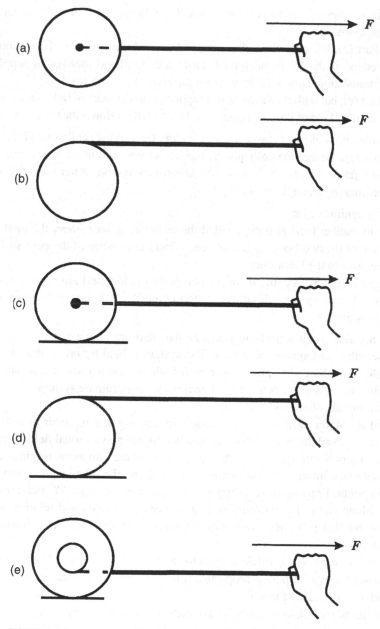

FIGURE 1.12

acceleration α of the cylindrical object; show explicitly that the work–energy theorem is satisfied.

(a) Empty space, no gravity, the string passes through the center of mass of the cylinder.

(b) Empty space, no gravity, the string is wrapped around the cylinder. [Question: How can the hand, applying the same force as in Part (a), supply the (hint) same translational kinetic energy as in Part (a) plus the extra rotational kinetic energy?]

(c) Uniform vertical gravitation sufficient, together with friction, to constrain the cylinder to roll without slipping on the surface shown. The string passes through the center of mass of the cylinder.

(d) Same as Part (c), but with the string wrapped around the cylinder. [Questions: In which direction is the frictional force? How does the hand manage to supply the necessary translational and kinetic energies different from Part (c)?]

(e) Same as Part (c), but with the string now wrapped around a shaft of radius $r < R$ within the cylinder (it's a kind of yo-yo). [Question: In which direction is the frictional force?]

4. A particle of mass m_1 makes an elastic (kinetic-energy conserving) collision with another particle of mass m_2. Before the collision m_1 has velocity v_1 and m_2 is at rest relative to a certain inertial frame which we shall call the *laboratory system*. After the collision m_1 has velocity u_1 making an angle θ with v_1.

   (a) Find the magnitude of u_1.

   (b) Relative to another inertial frame, called the *center-of-mass system*, the total linear momentum of the two-body system is zero. Find the velocity of the center-of-mass system relative to the laboratory system.

   (c) Find the velocities v_1', v_2', u_1', u_2' of the two bodies before and after the collision in the center-of-mass system. Find the scattering angle θ' (the angle between v_1' and u_1') in terms of θ.

5. Two masses m_1 and m_2 in a uniform gravitational field are connected by a spring of unstretched length h and spring constant k. The system is held by m_1 so that m_2 hangs down vertically, stretching the spring. At $t = 0$ both m_1 and m_2 are at rest, and m_1 is released, so that the system starts to fall. Set up a suitable coordinate system and describe the subsequent motion of m_1 and m_2.

6. The Earth and the Moon form a two-body system interacting through their mutual gravitational attraction. In addition, each body is attracted by the gravitational field of the Sun, which in the sense of Section 1.3 is an external force. Take the Sun as the origin and write down the equations of motion for the center of mass X and the relative position x of the Earth–Moon system. Expand the resulting expressions in powers of x/X, the ratio of the magnitudes. Show that to lowest order in x/X the center of mass and relative position are uncoupled, but that in higher orders they are coupled because the Sun's gravitational force is not constant.

7. Show that a one-dimensional particle subject to the force $F = -kx^{2n+1}$, where n is an integer, will oscillate with a period proportional to A^{-n}, where A is the amplitude. Pay special attention to the case of $n \leq 0$.

8. A yo-yo consists of two disks of mass M and radius R connected by a shaft of mass m and radius r; a weightless string is wrapped around the shaft.

   (a) The free end of the string is held stationary in the Earth's gravitational field. Assuming that the string starts out vertical, find the motion of the yo-yo's center of mass.

   (b) The free end is moved so as to keep the yo-yo's center of mass stationary. Describe the motion of the free end of the string and the rotation of the yo-yo.

   (c) The yo-yo is transported to empty space, where there is no gravitational field, and a force F is applied to the free end of the string. Describe the motion of the center of mass of the yo-yo, the yo-yo's rotation, and the motion of the free end of the string.

9. A particle in a uniform gravitational field experiences an additional retarding force $\mathbf{F} = -\alpha\mathbf{v}$, where $\mathbf{v}$ is its velocity. Find the general solution to the equations of motion and show that the velocity has an asymptotic value (called the *terminal velocity*). Find the terminal velocity.

10. Change the variable of integration in Eq. (1.7) from s to any other parameter in order to show that the distance between two points on the trajectory, as defined by (1.7), is indeed independent of the parameter.

11. (a) The concept of curvature and radius of curvature are defined by extending those concepts from circles to curves in general. The curvature κ is defined, as in Eq. (1.13), as the rate (with respect to length along the curve) of rotation of the tangent vector. Show that what Eq. (1.13) defines is in fact the rate of rotation of τ (i.e., that it gives the rate of change of the angle τ makes with a fixed direction). Show also that for a circle in the plane $\kappa = 1/R$, where R is the radius of the circle.

    (b) Derive the second of the Frenet formulas from the fact that τ, $\mathbf{n}$, and $\mathbf{B}$ are a set of orthogonal unit vectors and from the definition of θ.

12. A particle is constrained to move at constant speed on the ellipse $a_{ij}x^ix^j = 1(i, j = 1, 2)$. Find the Cartesian components of its acceleration as a function of position on the ellipse.

13. Show that if Eq. (1.17) is satisfied, there exist constants m_1, m_2, and m_3 such that Eqs. (1.15) and (1.16) can be put in the form of (1.18).

14. Consider Eq. (1.23) in two rather than three dimensions, and assume that the $\mathbf{x}$ and $\mathbf{y}$ coordinates are not both inertial, but rotating with respect to each other: $y_1 = x_1 \cos\omega t - x_2 \sin\omega t$, $y_2 = x_1 \sin\omega t + x_2 \cos\omega t$. Show that in general even if the $\mathbf{x}$ acceleration vanishes, the $\mathbf{y}$ acceleration does not. Find $\ddot{y}$ for $\ddot{\mathbf{x}} = \mathbf{0}$, but $\dot{\mathbf{x}} \neq \mathbf{0}$ and $\mathbf{x} \neq \mathbf{0}$. Give the physical significance of the terms you obtain.

15. A particle of mass m moves in one dimension under the influence of the force

$$F = -kx + \frac{a}{x^3}.$$

Find the equilibrium points, show that they are stable, and calculate the frequencies of oscillation about them. Show that the frequencies are independent of the energy.

16. Consider a system of particles made up of K subsystems, each itself a system of particles. Let M_I be the mass and $\mathbf{X}_I$ the center of mass of the Ith subsystem. Show that the center of mass of the entire system is given by an equation similar to (1.57), but with m_i and $\mathbf{x}_i$ replaced by M_I and $\mathbf{X}_I$ and the sum taken from $I = 1$ to $I = K$.

17. Express the total kinetic energy of a system of N particles in terms of their center of mass and the relative positions of the particles [i.e., derive Eq. (1.65)]. Extend the result to a continuous distribution of particles with mass density $\rho(\mathbf{x})$. (Hint: Replace the sum by an integral.)

18. In deriving Eq. (1.73) we assumed that the internal forces do not contribute to the total torque on a system of particles. Show explicitly that if for each i and j the internal force $\mathbf{F}_{ij}$ lies along the line connecting the ith and jth particles, then the internal forces indeed do not contribute to the total torque.

19. Draw the phase portrait for a particle in a uniform gravitational field. Make this a system of one-freedom by considering motion only in the vertical direction.

20. A particle of mass m moves along the x axis under the influence of the potential

$$V(x) = V_0 x^2 e^{-ax^2},$$

where V_0 and $a > 0$ are constants. Find the equilibrium points of the motion, draw a rough graph of the potential, and draw the phase portrait of the system. On these graphs indicate the relation between the energy and geometry of the orbits in velocity phase space.

21. Draw the phase portrait for the system of Problem 15.

22. Draw the phase portrait for a particle in a uniform gravitational field with a velocity-dependent retarding force the same as the one in Problem 9. Consider motion only in the vertical direction.

23. Consider the dynamical system of Section 1.5.2. Let t_k be the time it takes the mass to rise from B to the turning point at h_k and to descend again (assume it was moving to the right at B).

    (a) Find t_k in terms of t_1.

    (b) Suppose there is no hump, but that the left- and right-hand planes are connected, so that A and B are the same point (assume that the mass loses no energy in passing through this point). Show that the mass will come to rest at the bottom in a finite time, and calculate that time in terms of t_1.

    (c) Now suppose that the hump is there, and consider motion already trapped in the well at B. Show that the mass may never come to rest and that whether it does or not depends on the shape of the hump. Give one example for which it comes to rest in a finite time and one for which it does not.

    (d) Describe the motion if the initial condition lies on the short line of the phase portrait that passes through the origin and lies entirely in the region containing B. (Suppose it lies on a more general point of the separatrix. To what extent is the situation different?)

24. (a) Draw the phase portrait of a dynamical system just like that of Section 1.5.2, but with no energy dissipation on the two planes. Calculate the shapes of the phase orbits exactly to the right of B and to the left of A. (The exact shapes of the phase orbits cannot be calculated between A and B unless the exact shape of the hump is known.)

    (b) Show that in the limit as $\alpha \to 0$, the phase portrait of Section 1.5.2 approaches the one you have drawn. (Remark: This may help in understanding the phase portrait of Fig. 1.10.)

25. (a) Consider an idealized chain piled up next to a hole on a table. (By "idealized" we mean that the chain has infinitesimal links and is capable of being piled up exactly at the very edge of the hole in the table, so that no part of the chain must move to reach the hole.) Let the chain have linear mass density ρ, and assume that a finite length l of the chain is initially hanging through the hole at rest. As this length starts to fall, it pulls the rest of the chain through. Let $x(t)$ be the length of chain hanging through the hole at time t. Reduce the problem of finding $x(t)$ to quadrature [i.e., write down an integral whose solution would give the desired function, something like Eq. (1.46)]. Find $x(t)$ itself in the limit $l = 0$.

    (b) Now consider a chain of finite individual links, each of length ϵ and mass $m = \mu\epsilon$. Although the links are not infinitesimal, assume that the chain can be piled up ideally at the edge of the hole. This chain is piled up next to the hole with n of its links hanging through and initially at rest. These start to fall and pull the rest of the chain through, each link starting the next one in a completely inelastic collision (i.e., after each collision the entire falling part of the chain, including the newly added link, is falling at the same speed). Find the velocity of the chain after k links have fallen through the hole.

(c) Compare the result of part (a) with that of (b) in the limit as $\epsilon \to 0$.

(d) In both cases, check and comment upon conservation of energy.

26. Derive Eq. (1.85).

27. In the film "2001: A Space Odyssey" there is a toroidal space station rotating about a fixed axis that provides a centrifugal acceleration equal to the Earth's gravitational acceleration $g = 10 \text{ m/s}^2$ on a stationary object.

(a) Find the needed ω, assuming that the radius of the space station is 150 m.

(b) Find the (fictitious) gravitational acceleration that would be felt by a person walking at 1.3 m/s in two directions along the inner tube of the torus and across it.

(c) Find the (fictitious) acceleration that would be felt by a person sitting down or rising from a chair at 1.3 m/s.

CHAPTER 2

LAGRANGIAN FORMULATION OF MECHANICS

CHAPTER OVERVIEW

Chapter 1 set the stage for the rest of the book: it reviewed Newton's equations and the basic concepts of Newton's formulation of mechanics. The discussion in that chapter was applied mostly to dynamical systems whose arena of motion is Euclidean three-dimensional space, in which it is natural to use Cartesian coordinates. However, we referred on occasion to other situations, such as one-dimensional systems in which a particle is not free to move in Euclidean 3-space but only in a restricted region of it. Such a system is said to be *constrained*: its arena of motion, or, as we shall define below, its *configuration manifold*, turns out in general to be neither Euclidean nor three dimensional (nor $3N$-dimensional, if there are N particles involved). In such cases the equations of motion must include information about the forces that give rise to the constraints.

In this chapter we show how the equations of motion can be rewritten in the appropriate configuration manifold in such a way that the constraints are taken into account from the outset. The result is the *Lagrangian formulation* of dynamics (the equations of motion are then called *Lagrange's equations*). We should emphasize that the physical content of Lagrange's equations is the same as that of Newton's. But in addition to being logically more appealing, Lagrange's formulation has several important advantages.

Perhaps the first evident advantage is that the Lagrangian formulation is easier to apply to dynamical systems other than the simplest. Moreover, it brings out the connection between conservation laws and important symmetry properties of dynamical systems. Of great significance is that Lagrange's equations can be derived from a variational principle, a method that turns out to be extremely general and applicable in many branches of physics. One of the reasons for studying classical mechanics is to understand the Lagrangian formulation, for many equations of physics are conventionally formulated in Lagrangian terms and many conservation laws are understood also in Lagrangian terms, through their connection with symmetries. Some of the topics we mention here will be put off until Chapter 3.

2.1 CONSTRAINTS AND CONFIGURATION MANIFOLDS

In this section we change from Cartesian coordinates to others, which are more useful for dealing with dynamical systems. The new coordinates are chosen in a way that depends on the particular dynamical system for which they will be used (but they are nevertheless called *generalized coordinates*); they are adapted to that system and are more or less natural coordinates for it. The properties of the system that determine the choice are geometric: they are the number of freedoms and the shape, or topology, of the region in which the system is free to move (e.g., whether it is a sphere or an inclined plane). This region is determined by the constraints placed upon the system; it is called the *configuration manifold* $\mathbb{Q}$. The new coordinates, called the q^α, will lie on $\mathbb{Q}$, and their number will be the number of freedoms, which is also the dimension of $\mathbb{Q}$. In this section we do two things: 1. explain the idea of the configuration manifold and 2. describe the change from the Cartesian coordinates to the q^α.

2.1.1 CONSTRAINTS

We start with an example. Think of a sphere rolling on a curved surface under the action of gravity. The sphere consists of many particles whose motion is correlated so that they always form a rigid sphere and so that there is always one of them in contact with the surface and, as the body is rolling, instantaneously at rest. The forces on the sphere are far from simple. They are composed of the forces internal to the sphere (which keep it rigid), the forces applied to it by the surface on which it is rolling (which keep it in contact with that surface and prevent it from sliding), and the force of gravity. The force of gravity is known a priori, but the others, the constraining forces, are not. What is known is that under the action of gravity and the forces of constraint the body remains on the surface and continues to roll. It might seem that to describe the motion completely one would have to find the constraining forces, but it will be shown that the opposite is true, that the motion can be obtained from the gravitational force and from knowing the geometric constraints (i.e., of the shape of the surface and of the fact of rigidity); the forces of constraint, if needed, are easier to find later. This seemingly simple example of a sphere rolling on a curved surface is actually quite complicated. Most of the time we will be dealing with much simpler constraints. We now proceed to generalize this example.

CONSTRAINT EQUATIONS

The motion of a dynamical system is often constrained by external agents applying forces that are initially unknown. What is known is the geometric effect of such agents, or rather their effect combined with those applied forces that are known. Suppose one is dealing with a system of N particles and that the constraints are given by a set of K *constraint equations* of the form

$$f_I(\mathbf{x}_1, \ldots, \mathbf{x}_N, t) = 0, \quad I = 1, \ldots, K < 3N, \tag{2.1}$$

where the $\mathbf{x}_i$ are the position vectors of the N particles. The f_I are assumed to be differentiable functions of their arguments, and the t dependence describes the known way in which the constraints vary with time, independent of the motion of the particles (for instance, in the example of the rolling sphere with which we started this chapter, the surface on which it is rolling could be waving).

Constraints given by equations like (2.1) are called *holonomic* (meaning essentially integrable, from the Greek). More general constraints depend also on the velocities (rolling constraints are among them); they are given by equations of the form

$$f_I(\mathbf{x}_1, \ldots, \mathbf{x}_N; \dot{\mathbf{x}}_1, \ldots, \dot{\mathbf{x}}_N, t) = 0, \quad I = 1, \ldots, K < 3N. \tag{2.2}$$

There exist constraints that appear to be velocity dependent but are actually differential equations that can be integrated to give simply holonomic constraints. When this is not the case, velocity-dependent constraints are *nonholonomic*. In any case, it should be clear that holonomic constraints are a special case of this more general type. Finally, there are other types of constraints entirely that are not even given by equations, for example, those given by expressions of the form

$$f_I(\mathbf{x}_1, \ldots, \mathbf{x}_N, t) < 0, \quad I = 1, \ldots, K < 3N, \tag{2.3}$$

as in the case of particles restricted to a certain region of space. An example is a particle constrained to remain within a container of some given shape. Although this kind of constraint will be mentioned on occasion in the book, it will not be treated in any generality. In this chapter we will deal only with holonomic constraints.

CONSTRAINTS AND WORK

How one deals with constraints can be illustrated (Fig. 2.1) by the relatively simple example of a point particle in 3-space restricted to a surface whose equation is

$$f(\mathbf{x}, t) = 0. \tag{2.4}$$

In this example $N = 1$ (a single particle) and $K = 1$ (a single constraint equation). The Newtonian equation of motion of the particle is

$$m\ddot{\mathbf{x}} = \mathbf{F} + \mathbf{C}, \tag{2.5}$$

where $\mathbf{F}(\mathbf{x}, \dot{\mathbf{x}}, t)$ is the known external force and $\mathbf{C}$ is the unknown force of constraint that the surface exerts on the particle. We now have four equations, namely (2.4) and the three components of (2.5), for six unknown functions of the time, the three components of $\mathbf{x}$ and the three components of $\mathbf{C}$. Clearly this is not enough to determine the motion. The problem arises from the physical fact that there are many possible constraint forces $\mathbf{C}$ that will keep the particle on the surface [i.e., will lead to motions that satisfy (2.4)]. For example, suppose that the surface is a stationary plane and that a constraint force $\mathbf{C}$ has

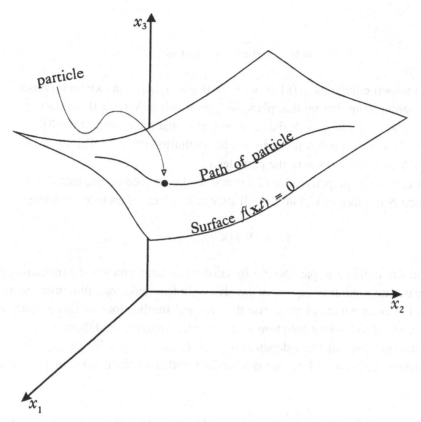

FIGURE 2.1
A particle in 3-space constrained to a two-dimensional surface given by an equation of the form
$f(\mathbf{x}, t) = 0$.

been found that will keep the particle on the plane. Now add to **C** another force that is
parallel to the plane (friction is an example of such a force) and call the sum **C'**. It is clear
then that **C'** will also keep the particle on the plane but accelerating at some different rate,
for the only difference between the two is a force along the plane. For a curved surface the
argument is similar: to a constraining force **C** can be added a force that at each point is
parallel to the surface, and the resulting force will still constrain the particle to the surface.
What is needed here is some physical input that will allow us to choose among the different
possibilities for **C**.

This input will be obtained by what seems at first an arbitrary choice (as will be
seen later, what this does physically is to place restrictions on the work done by the
constraint forces): the forces parallel to the surface will be eliminated by choosing **C** to
be perpendicular (normal) to the surface. The way to obtain a vector perpendicular to a
surface is the following:

Let $f(\mathbf{x}, t) = $ const. be the equation of any surface; then $\nabla f(\mathbf{x}, t)$ is a vector per-
pendicular to the surface at position **x** and time t, provided that $\nabla f \neq 0$ on the surface.
If $\nabla f \equiv 0$ on the surface, of course, the procedure we are outlining here will not work.
To avoid this difficulty, we shall require that the constraint has been written in such a

way that

$$\nabla f \neq 0 \quad \text{on the } f = 0 \text{ surface.} \tag{2.6}$$

For example, the two equations $f_a(\mathbf{x}) = \mathbf{s} \cdot \mathbf{x} = 0$ and $f_b(\mathbf{x}) = (\mathbf{s} \cdot \mathbf{x})^2 = 0$ constrain a particle to the same plane, but on that plane $\nabla f_a = \mathbf{s}$, while $\nabla f_b = 0$, so only $f_a = 0$ is acceptable. (If $K > 1$ and $N > 1$, the requirement is that the matrix of the $\partial f_I / \partial x^\alpha$ be at least of rank K; see the book's appendix for the definition of rank. The x^α are the $3N$ components of N position vectors of the particles.)

The constraint force perpendicular or normal to the surface (often called a *normal force* and written $\mathbf{N}$ in place of $\mathbf{C}$, but we will stick with $\mathbf{C}$) can therefore be written

$$\mathbf{C} = \lambda \nabla f(\mathbf{x}, t), \tag{2.7}$$

where λ can be any number, in particular a function of t. This removes the mathematical difficulty, because now the four equations involve only four unknown functions, namely $\lambda(t)$ and the three components of $\mathbf{x}(t)$. But the physical implications of this assumption have yet to be understood, so we now turn aside in order to understand them.

Assume that the external force depends on a potential: $\mathbf{F} = -\nabla V(\mathbf{x}, t)$. Then expressing $\mathbf{C}$ through (2.7) and taking the dot product with $\dot{\mathbf{x}}$ on both sides of (2.5) leads to

$$m\ddot{\mathbf{x}} \cdot \dot{\mathbf{x}} \equiv \frac{d}{dt}\left(\frac{1}{2}m\dot{x}^2\right) = -\nabla V \cdot \dot{\mathbf{x}} + \lambda \nabla f \cdot \dot{\mathbf{x}}. \tag{2.8}$$

Now suppose that $\mathbf{x}(t)$ is a solution of the equations of motion. Then since the particle remains on the surface, $f(\mathbf{x}(t), t) = 0$, and therefore $df/dt = 0$. But

$$\frac{df}{dt} = \nabla f \cdot \dot{\mathbf{x}} + \frac{\partial f}{\partial t},$$

and similarly

$$\frac{dV}{dt} = \nabla V \cdot \dot{\mathbf{x}} + \frac{\partial V}{\partial t}.$$

From these equations and (2.8) it follows that

$$\frac{dE}{dt} = \frac{d}{dt}\left[\frac{1}{2}m\dot{x}^2 + V\right] = \frac{\partial V}{\partial t} - \lambda \frac{\partial f}{\partial t}. \tag{2.9}$$

This means that the total energy E of the particle changes if V or f are explicit functions of the time (i.e., if the potential depends on the time or if the constraint surface is moving). We will deal at this point with time-dependent potential energy functions, and therefore if (but not only if) the surface moves, the total energy changes.

The relation between movement of the surface and energy change can be understood physically. If the energy is changing, that is, if $dE/dt \neq 0$, the work–energy theorem [Eq. (1.39)] implies that there is work being done on the system; as we are assuming that $\partial V/\partial t = 0$, Eq. (2.9) implies that the work is performed by the surface. To see how the surface does this work, suppose first that it is not moving. Since **C** is normal to the surface, it is always perpendicular to the velocity $\dot{\mathbf{x}}$, and thus $\mathbf{C} \cdot \dot{\mathbf{x}} = 0$: the rate at which work is done by the constraint force is zero. If the surface moves, however, the particle velocity need not be tangent to the surface, as shown in Fig. 2.2, and even if **C** is perpendicular to the surface $\mathbf{C} \cdot \dot{\mathbf{x}} \neq 0$: the surface through **C** can do work at a non-zero rate.

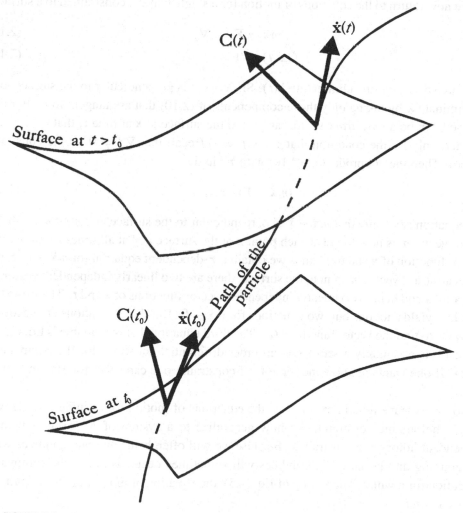

FIGURE 2.2

A two-dimensional constraint surface that depends on time. Although the constraint force **C** is always normal to the surface, the angle between **C** and the particle's velocity vector $\dot{\mathbf{x}}$ is not a right angle. Therefore the constraint force can do work on the particle.

The physical content of Eq. (2.7), i.e., the assumption that $\mathbf{C}$ is normal to the surface, should now be clear: it is that the forces of constraint do no work. On the other hand, in reality, most surfaces exert forces which have tangential components such as friction and therefore do work. Thus for the time being we are excluding frictional forces. But there can be other constraint forces, nondissipative ones, that have components parallel to the surface of constraint yet do no work (they need only be perpendicular to the velocity, like magnetic forces on charged particles). We are excluding those also. When those conditions are satisfied, the surface is called *smooth*.

2.1.2 GENERALIZED COORDINATES

We now return to the equations of motion for a single particle constrained to a surface:

$$m\ddot{\mathbf{x}} = \mathbf{F} + \lambda \nabla f, \tag{2.10}$$

$$f(\mathbf{x}, t) = 0. \tag{2.11}$$

These are solved by first eliminating $\lambda(t)$. Since $\lambda \nabla f$ is perpendicular to the surface, one can eliminate λ by taking only those components of (2.10) that are tangent to it. For this purpose let τ be an *arbitrary* vector tangent to the surface at $\mathbf{x}$ at time t, that is, a vector restricted only by the condition that $\tau \cdot \nabla f = 0$ (recall that $\nabla f \neq 0$ on the constraint surface). Then the dot product of (2.10) with τ yields

$$(m\ddot{\mathbf{x}} - \mathbf{F}) \cdot \tau = 0. \tag{2.12}$$

This equation says only that $m\ddot{\mathbf{x}} - \mathbf{F}$ is perpendicular to the surface at $\mathbf{x}$ at time t. Such a tangent vector τ is now found at each point $\mathbf{x}$ of the surface and at all times t (that is, τ is a vector function of $\mathbf{x}$ and of t); thus we obtain a τ-dependent equation for $m\ddot{\mathbf{x}} - \mathbf{F}$. Since τ is an arbitrary vector tangent to the surface, there are two linearly independent vectors at each point $\mathbf{x}$ and hence two linearly independent vector functions of $\mathbf{x}$ and t. Therefore this procedure yields not one but two equations for $m\ddot{\mathbf{x}} - \mathbf{F}$. But three equations are needed if one wants to find the vector function $\mathbf{x}(t)$. The third equation that is available is Eq. (2.11). The result is essentially a set of second-order differential equations for the components of $\mathbf{x}(t)$. If one wants to know the force $\mathbf{C}$ of constraint, one can solve for $\mathbf{x}(t)$ and return to (2.5).

So far we have found how to write the equations of motion for a single particle with a single holonomic constraint. We now generalize to a system of N particles with K independent holonomic constraints. Because we will often be using double indices without summing and because triple indices will sometimes occur, we drop the summation convention for a while. The analog of Eq. (2.5), the equation of motion of the ith particle, is (no sum on i)

$$m_i \ddot{\mathbf{x}}_i = \mathbf{F}_i + \mathbf{C}_i, \tag{2.13}$$

and the constraints are given by (2.1). As before, the constraints fail to determine the $\mathbf{C}_i$ completely, and we add the assumption of smoothness by writing the analog of (2.7), namely

$$\mathbf{C}_i = \sum_{I=1}^{K} \lambda_I \nabla_i f_I, \tag{2.14}$$

where ∇_i is the gradient with respect to the position vector $\mathbf{x}_i$ of the ith particle and the $\lambda_I(t)$ are K functions, which are as yet unknown. Like λ in the one-particle case, the λ_I will be eliminated in solving the problem. We leave to the reader (see Problem 1) the task of proving that, in analogy with the one-particle case, if the potential V satisfies $\partial V/\partial t = 0$ the total change in energy is given by

$$\frac{dE}{dt} = -\sum_I \lambda_I \frac{\partial f_I}{\partial t}, \tag{2.15}$$

so that the forces of constraint do work only if the constraint functions depend on t.

Now let the $\boldsymbol{\tau}_i$ be N arbitrary vectors "tangent to the surface," that is, vectors restricted only by the condition that

$$\sum_{i=1}^{N} \boldsymbol{\tau}_i \cdot \nabla_i f_I = 0, \quad I = 1, \dots, K. \tag{2.16}$$

If, as required in the discussion around Eq. (2.6), the matrix of the $\partial f_I/\partial x^\alpha$ is of rank K, this equation gives K independent relations among the $3N$ components of the N vectors $\boldsymbol{\tau}_i$, so that only $3N - K$ of the components are independent. Then the dot product of (2.13) with $\boldsymbol{\tau}_i$, summed over i, yields [use (2.14) and (2.16)]

$$\sum_i (m_i \ddot{\mathbf{x}}_i - \mathbf{F}_i) \cdot \boldsymbol{\tau}_i = 0. \tag{2.17}$$

This equation, the analog of (2.12), is sometimes called D'Alembert's principle. Through it, the $3N - K$ independent components of the $\boldsymbol{\tau}_i$ lead to $3N - K$ independent relations. Equations (2.1) provide K other relations, so that there are $3N$ in all from which the $3N$ components of the $\mathbf{x}_i$ can be obtained.

The problem now is to find a suitable algorithm for picking vectors $\boldsymbol{\tau}_i$ that satisfy (2.16). We will do this by sharpening the analogy to the one-particle case. In the one-particle case $\boldsymbol{\tau}$ was an arbitrary vector tangent to the surface of constraint. In the N-particle case there is no surface of constraint, so the $\boldsymbol{\tau}_i$ are not readily visualized. But Eq. (2.1) defines a $(3N - K)$-dimensional hypersurface in the $3N$-dimensional Euclidean space $\mathbb{E}^{3N}$ of the components of the $\mathbf{x}_i$, and the dynamical system is constrained to this hypersurface. That is, as the system moves and the $\mathbf{x}_i$ keep changing, the point in $3N$-space described by the collection of all the components of the $\mathbf{x}_i$ remains always on this hypersurface. We could therefore call it the *configuration hypersurface* of the dynamical system, but for reasons that will be explained in Section 2.4, we will call it its *configuration manifold* $\mathbb{Q}$. Start with N tangent vectors $\boldsymbol{\tau}_i$ that satisfy (2.16). Their $3N$ components define a ($3N$-component) vector in $\mathbb{E}^{3N}$ that is a kind of generalized tangent vector to the configuration manifold, for in Eq. (2.16) the sum is not only over i, as indicated by the summation sign, but also over the three components of each τ_i, as indicated by the dot product. Thus just as $\tau \cdot \nabla f = 0$ defines a 3-vector tangent to the $f = 0$ surface, Eq. (2.16) defines a $3N$-vector tangent to the $f_I = 0$ hypersurface $\mathbb{Q}$ (see Problem 2).

In these terms picking the τ_i to satisfy (2.16) means picking the generalized tangent vector in $3N$ dimensions. This vector will be found in several steps. The first will be to define what are called *generalized coordinates* q^α in the $3N$-space (superscripts rather than subscripts are generally used for these coordinates) for which $\mathbb{Q}$ is a coordinate hypersurface. Consider a *region* of $\mathbb{E}^{3N}$ that contains a point $\mathbf{x}_i$ of $\mathbb{Q}$, and let q^α, $\alpha = 1, \ldots, 3N$, be new coordinates in that region, a set of invertible functions of the $\mathbf{x}_i$:

$$
\begin{aligned}
q^\alpha &= q^\alpha(\mathbf{x}_1, \ldots, \mathbf{x}_N, t), \\
\mathbf{x}_i &= \mathbf{x}_i(q^1, \ldots, q^{3N}, t)
\end{aligned} \tag{2.18}
$$

for $\mathbf{x}_i$ in that region. Equations (2.18) define a *transformation* between the $\mathbf{x}_i$ and the q^α. Invertibility means that the Jacobian of the transformation is nonsingular. (The Jacobian of the transformation is the matrix whose elements are the $\partial q^\alpha / \partial x^\beta$, where the x^β are the $3N$ components of the N vectors $\mathbf{x}_i$.)

Assume further that the q^α are continuous and, because accelerations will lead to second derivatives, twice continuously differentiable functions. The first object will be to pick the q^α so that the equations of constraint become trivial (i.e., reduce to the statement that some of the q^α are constant). Then if the equations of motion are written in terms of the q^α (invertibility guarantees that they can be), those q^α that are constant will drop out. This is done by choosing the q^α so that K of them (we choose the last K) depend on the $\mathbf{x}_i$ through the functions appearing in the constraint equations. Suppressing any t dependence, we write

$$
q^{n+I}(\mathbf{x}) = R_I(f_1(\mathbf{x}), \ldots, f_K(\mathbf{x})), \quad I = 1, \ldots, K, \tag{2.19}
$$

where $\mathbf{x}$ stands for the collection of the $\mathbf{x}_i$ and $n = 3N - K$ is the dimension of the configuration manifold $\mathbb{Q}$; n is also the number of freedoms. Equations (2.19) are the last K of the $3N$ equations that give the q^α in terms of the $\mathbf{x}_i$, and as such they too must be invertible, which means that it must be possible to solve them for the $f_I = f_I(q^{n+1}, \ldots, q^{n+K})$. When the constraint conditions are imposed, they force the last K of the q^α to be constants independent of the time:

$$
q^{n+I} = R(0, \ldots, 0). \tag{2.20}
$$

This is what we mean by the constraint equations becoming trivial in these coordinates. Since the last K of the q^α remain fixed as the motion proceeds, the problem reduces to finding how the rest of the q^α, the first n, depend on the time.

The full set of q^α can be used as well as the $\mathbf{x}_i$ to define a point in $\mathbb{E}^{3N}$. That is what is meant by invertibility. Equation (2.20) restricts the point to lie in $\mathbb{Q}$: it makes the same statement as does (2.18) but in terms of different coordinates. As the first n of the q^α vary in time, the point described by the full set moves about in $\mathbb{E}^{3N}$, but it remains on the configuration manifold, and therefore the first n of the q^α form a coordinate system on $\mathbb{Q}$. The first n of the q^α are called *generalized coordinates* of the dynamical system. Hence when the equations of motion are written in terms of the generalized coordinates, they will describe the way the system moves within the configuration manifold.

From now on, Greek indices run from 1 to $n = 3N - K$, rather than from 1 to $3N$.

2.1.3 EXAMPLES OF CONFIGURATION MANIFOLDS

In this subsection we give examples of configuration manifolds and generalized coordinates for some particular dynamical systems. In these examples and in most of what follows, Greek indices will run from 1 to n.

THE FINITE LINE

The finite line, which may be curved, applies to the motion of a bead along a wire of length l (Fig. 2.3a). In this case $N = 1$ and $K = 2$ (see Problem 2), so that $n = 1$ and α takes on only the single value 1 and may be dropped altogether. Here $\mathbb{Q}$ is of dimension 1 (the dimension being essentially the number of coordinates, the number of values that α takes on), and the coordinate system on it may be chosen so that the values of q range from $-l/2$ to $l/2$.

THE CIRCLE

The circle applies to the motion of a plane pendulum (Fig. 2.3b). Denote the circle by $\mathbb{S}^1$. Again $\mathbb{Q}$ is one dimensional, and the single generalized coordinate is usually taken to be the angle and is called θ rather than q. Typically the coordinates on the circle are chosen so that θ varies from $-\pi$ to π, or from 0 to 2π. But note that both of these choices have a problem: in each of them there is one point with two coordinate values. In the first choice the point with coordinate π is the same as the one with coordinate $-\pi$, and in the second, this is true for the coordinates 0 and 2π. This lack of a unique relationship between the points of $\mathbb{Q}$ and a coordinate system is an important property of manifolds and will be treated in some detail later (see Section 2.4).

THE PLANE

The plane applies to the motion of a particle on a table (Fig. 2.3c). As before, $N = 1$, but now $K = 1$, so that $n = 2$. The coordinates are conveniently chosen to be the usual plane Cartesian, plane polar, or other familiar coordinates.

THE TWO-SPHERE $\mathbb{S}^2$

The surface of the sphere applies to the motion of a spherical pendulum, which consists of a point mass attached to a weightless rigid rod that is free to rotate about a fixed point in a uniform gravitational field (Fig. 2.3d). The coordinates usually chosen on $\mathbb{S}^2$ are the azimuth angle ϕ (corresponding to the longitude on the globe of the Earth), which varies

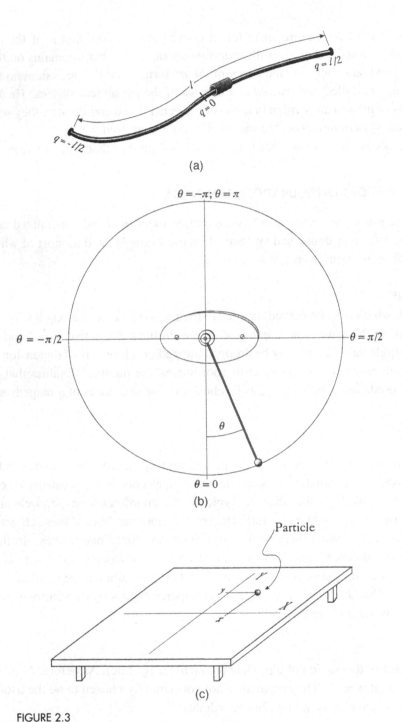

(a)

(b)

Particle

(c)

FIGURE 2.3

Examples of configuration manifolds. (a) A bead on a finite wire. (b) The configuration manifold $\mathbb{S}^1$ of the plane pendulum. The point at the top has coordinate $\theta = \pi$ as well as $\theta = -\pi$. (c) A particle on a table. The X and Y axes are shown, as well as the coordinates (x, y) of the particle. (d) The configuration manifold $\mathbb{S}^2$ of the spherical pendulum. The pendulum itself is not shown; its point of suspension is at the center of the sphere. The coordinates of the North Pole are $\theta = \pi/2$ and arbitrary ϕ. The South Pole ($\theta = -\pi/2$) can not be seen. (e) A schematic diagram of the double spherical pendulum. No attempt is made to show the 4-dimensional configuration manifold.

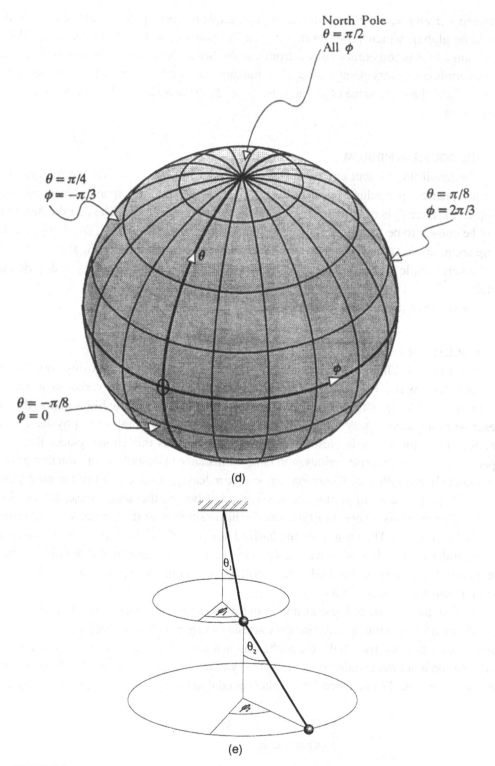

(d)

(e)

FIGURE 2.3
(*Continued*)

between $-\pi$ and π, and the colatitude or polar angle θ (corresponding to the latitude on the same globe), which varies from $-\pi/2$ at the South Pole to $\pi/2$ at the North Pole. (The range of θ is sometimes chosen from 0 at the South Pole to π at the North Pole.) In these coordinates every point with $\phi = \pi$ and any value of θ is the same as the one with $\phi = -\pi$ and the same value of θ, and at $\theta = -\pi/2$ or $\theta = \pi/2$ all values of ϕ refer to the same point.

THE DOUBLE PENDULUM

The manifold, without any particular name, but written $\mathbb{S}^2 \times \mathbb{S}^2$, for the motion of a double spherical pendulum, consists of one spherical pendulum suspended from another (Fig. 2.3e). Here $\mathbb{Q}$ is of dimension four and quite hard to visualize. The coordinates on it may be chosen to be the two azimuth angles and the two colatitudes ϕ_1, ϕ_2, θ_1, θ_2. Again there are multiple valued points on each of the two spheres. It is interesting that even such a relatively simple system has a configuration manifold that is not very easy to describe in detail.

These are of course only examples.

DISCUSSION

One may ask whether multiple-valued points can be avoided by choosing coordinates in some clever ways. For instance suppose for the plane pendulum the variation of θ were taken as $-\pi \leq \theta < \pi$. This would lead to difficulties because it would be impossible to describe in a coherent way any small interval containing the point $-\pi$. Physically, for small ϵ the point $\pi - \epsilon$ is close to the point $-\pi$, whereas this choice places them far apart, and that would cause serious problems, for instance in describing motion that passes continuously through $-\pi$. Choosing $-\pi < \theta < \pi$ leaves out one point of the circle, and $-\pi \leq \theta \leq \pi$ returns to giving one physical point two mathematical descriptions. The brief and ambiguous mention of this issue in this example was deliberate. We will return to it in Section 2.4. The reason we emphasized a *region* of $\mathbb{E}^{3N}$ in starting the description of generalized coordinates above Eq. (2.18) is that it is in general not possible to find generalized coordinates that make the constraint equations trivial and cover all of $\mathbb{E}^{3N}$ without multiple-valued points.

Notice that in none of these examples did we write out the constraint equations or the transformation equations connecting the Cartesian and generalized coordinates. In fact they are rarely written out, for usually the configuration manifold, its dimensions, and natural coordinates on it are made rather obvious by the symmetries of the system. But the equations are easily written. For instance for the plane pendulum the constraint equations could be

$$f_1(\mathbf{x}) \equiv x^2 + y^2 + z^2 - R^2 = 0,$$
$$f_2(\mathbf{x}) \equiv z = 0.$$

There is one particle and two constraint equations, so that $N = 1$ and $K = 2$, and hence $n = 1$, as is obvious from the start. The transformation equations might be (the superscripts on the qs are indices, not powers)

$$q^1 \equiv \theta = \arctan(y/x),$$
$$q^2 \equiv r = \sqrt{x^2 + y^2},$$
$$q^3 = z.$$

The last two can be written in terms of the constraint functions: $q^2 = \sqrt{f_1 - f_2^2 + R^2}$, $q^3 = f_2$. In terms of the full set of q^α (i.e., with α running from 1 to $3N$) the constraint equations become $q^2 = R, q^3 = 0$.

In the next section we use generalized coordinates to rewrite Newton's equations.

WORKED EXAMPLE 2.1

This example, an anisotropic harmonic oscillator constrained to the surface of a sphere, was first treated by Neumann (1859). We will return to it in Section 6.2.2 and again in Section 8.3.3. [Here we use only lower indices and indicate all sums by summation signs.] Consider the equations (no sum on α)

$$\ddot{x}_\alpha + \omega_\alpha^2 x_\alpha = 0, \quad \alpha = 1, \ldots, n,$$

where $\omega_\alpha^2 = k_\alpha/m$, $k_\alpha \neq k_\beta$ for $\alpha \neq \beta$. In the absence of constraints this system can be thought of either as n uncoupled one-freedom oscillators or as an anisotropic oscillator in n freedoms with n unequal frequencies ω_α. Take the second view.

(a) Find the constraint force $\mathbf{C}(\mathbf{x})$ that will keep this oscillator on the sphere $\mathbb{S}^{n-1}$ of radius 1 in $\mathbb{E}^n$, whose equation is $|\mathbf{x}|^2 \equiv \sum x_\alpha^2 = 1$. Here $\mathbf{x}$ is the vector with components x_α in $\mathbb{E}^n$. Assume that $\mathbf{C}$ is normal to $\mathbb{S}^{n-1}$ (i.e., parallel to $\mathbf{x}$). Write down the equations of motion for the constrained oscillator.

(b) Show that the n functions

$$F_\alpha = x_\alpha^2 + \sum_{\beta \neq \alpha} \frac{(x_\beta \dot{x}_\alpha - \dot{x}_\beta x_\alpha)^2}{\omega_\alpha^2 - \omega_\beta^2}$$

are constants of the motion (their importance will become evident in Section 6.2.2).

Solution. (a) Since $\mathbf{C}$ is parallel to $\mathbf{x}$, we may write $\mathbf{C}(x, \dot{x}) = \nu(x, \dot{x})\mathbf{x}$, where ν is a scalar function. Then the equations of motion are

$$\ddot{x}_\alpha + \omega_\alpha^2 x_\alpha = \nu(x, \dot{x})x_\alpha.$$

Multiply by x_α and sum over α:

$$\sum \left(x_\alpha \ddot{x}_\alpha + \omega_\alpha^2 x_\alpha^2 \right) = \nu(x, \dot{x}) \sum x_\alpha^2 = \nu(x, \dot{x}).$$

But

$$0 = \frac{d^2}{dt^2} \sum x_\alpha^2 = 2 \sum x_\alpha \ddot{x}_\alpha + 2 \sum \dot{x}_\alpha^2, \quad \text{or} \quad \sum x_\alpha \ddot{x}_\alpha = -\sum \dot{x}_\alpha^2,$$

so that the scalar function is $v(x, \dot{x}) = \sum(\omega_\alpha^2 x_\alpha^2 - \dot{x}_\alpha^2)$ and the constraint force itself is

$$\mathbf{C}(x, \dot{x}) = \mathbf{x} \sum (\omega_\alpha^2 x_\alpha^2 - \dot{x}_\alpha^2).$$

The equations of motion become (note that some indices must be changed)

$$\ddot{x}_\alpha + \omega_\alpha^2 x_\alpha = x_\alpha \sum_\beta (\omega_\beta^2 x_\beta^2 - \dot{x}_\beta^2).$$

It is seen that the equations of motion are nonlinear, which makes the problem nontrivial.

(b) Take the time derivative of F_α:

$$\frac{dF_\alpha}{dt} = 2x_\alpha \dot{x}_\alpha + \sum_{\beta \neq \alpha} \frac{2(x_\beta \dot{x}_\alpha - \dot{x}_\beta x_\alpha)(\dot{x}_\beta \dot{x}_\alpha + x_\beta \ddot{x}_\alpha - \dot{x}_\alpha \dot{x}_\beta - \ddot{x}_\beta x_\alpha)}{\omega_\alpha^2 - \omega_\beta^2}$$

$$= 2x_\alpha \dot{x}_\alpha + 2 \sum_{\beta \neq \alpha} \frac{x_\beta(v - \omega_\alpha^2)x_\alpha - x_\alpha(v - \omega_\beta^2)x_\beta}{\omega_\alpha^2 - \omega_\beta^2} (x_\beta \dot{x}_\alpha - \dot{x}_\beta x_\alpha)$$

$$= 2x_\alpha \dot{x}_\alpha - 2 \sum_{\beta \neq \alpha} x_\alpha x_\beta (x_\beta \dot{x}_\alpha - \dot{x}_\beta x_\alpha)$$

$$= 2x_\alpha \dot{x}_\alpha + 2x_\alpha^2 \sum_{\beta \neq \alpha} x_\beta \dot{x}_\beta - 2x_\alpha \dot{x}_\alpha \sum_{\beta \neq \alpha} x_\beta^2$$

$$= 2x_\alpha \dot{x}_\alpha + 2x_\alpha^2(-x_\alpha \dot{x}_\alpha) - 2x_\alpha \dot{x}_\alpha (1 - x_\alpha^2) = 0.$$

To get the second line, use the equations of motion. To get the last line, use the constraint equation.

2.2 LAGRANGE'S EQUATIONS

2.2.1 DERIVATION OF LAGRANGE'S EQUATIONS

We now turn to writing the equations of motion [i.e., Eq. (2.17)] in terms of the generalized coordinates q^α on the configuration manifold $\mathbb{Q}$. That means that $\ddot{\mathbf{x}}_i$, $\mathbf{F}_i$, and τ_i of Eq. (2.17) have to be written in terms of the q^α.

We start with the τ_i. These vectors are to be tangent to the configuration manifold $\mathbb{Q}$, which we may now think of as criss-crossed by a network of coordinate curves, as in Fig. 2.4. Any vector that is tangent to one of these curves is tangent to the manifold,

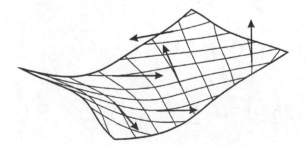

FIGURE 2.4
A surface criss-crossed by coordinate curves, with tangents to them.

and since n curves meet at each point of the manifold, n tangents to the manifold can be obtained from the n coordinate curves. Each curve is parametrized by one of the first n of the q^α (i.e., in its expression $\mathbf{x}_i(q)$ only one of the qs varies, say the γth) and, as was seen in Chapter 1, this means that the tangent to this curve is given by the derivative of $\mathbf{x}_i$ with respect to q^γ. Any linear combination of such tangent vectors is also a tangent vector.

We now make this statement more rigorous and quantitative. We prove below that the tangent vectors can be put in the form

$$\tau_i = \epsilon^\alpha \frac{\partial \mathbf{x}_i}{\partial q^\alpha},$$

where the ϵ^α are a set of n arbitrary constants. Here and in what follows we use the summation convention for the Greek indices, which run from 1 to n, but we continue to use summation signs for i. Because the $\mathbf{x}_i$ are functions of the q^α, the expressions on the right-hand side of this equation are also functions of the q^α, so the tangency point of the manifold for each τ_i is specified not in terms of the $\mathbf{x}_i$, but in terms of the q^α. The proof of tangency is obtained by inserting this expression for τ_i into Eq. (2.16) (again, there is a sum on α):

$$\sum_{i=1}^N \tau_i \cdot \nabla_i f_I = \epsilon^\alpha \sum_{i=1}^N \nabla_i f_I \cdot \frac{\partial \mathbf{x}_i}{\partial q^\alpha} = \epsilon^\alpha \frac{\partial f_I}{\partial q^\alpha} = 0. \tag{2.21}$$

The last equality follows from the fact that the f_I depend only on the last K of the qs, whereas α runs only from 1 to n. Equation (2.21) proves *generalized tangency* in $\mathbb{E}^{3N}$, in the sense described between Eqs. (2.17) and (2.18).

Now these tangent vectors can be inserted into (2.17). Since the ϵ^α are arbitrary (take n independent sets of ϵ^α, the first with only ϵ^1 nonzero, the second with only ϵ^2 nonzero, etc.), Eq. (2.17) now becomes n independent equations of the form

$$\sum_{i=1}^N (m_i \ddot{\mathbf{x}}_i - \mathbf{F}_i) \cdot \frac{\partial \mathbf{x}_i}{\partial q^\alpha} = 0, \quad \alpha = 1, \dots, n. \tag{2.22}$$

This takes care of the τ_i of Eq. (2.17): they are now written as functions of the q^α.

Next, we turn to the $\mathbf{F}_i$. Assume that the forces are conservative, so that $\mathbf{F}_i = -\nabla_i V(\mathbf{x}_1, \dots, \mathbf{x}_N)$. Then

$$\sum_{i=1}^N \mathbf{F}_i \cdot \frac{\partial \mathbf{x}_i}{\partial q^\alpha} = -\sum_{i=1}^N \nabla_i V \cdot \frac{\partial \mathbf{x}_i}{\partial q^\alpha} = -\frac{\partial V}{\partial q^\alpha}. \tag{2.23}$$

This expresses another part of the equations of motion in generalized coordinates if V is written as a function of the q^α.

Finally, we deal with the $\ddot{\mathbf{x}}_i$:

$$\ddot{\mathbf{x}}_i \cdot \frac{\partial \mathbf{x}_i}{\partial q^\alpha} = \frac{d}{dt}\left[\dot{\mathbf{x}}_i \cdot \frac{\partial \mathbf{x}_i}{\partial q^\alpha}\right] - \dot{\mathbf{x}}_i \cdot \frac{d}{dt}\frac{\partial \mathbf{x}_i}{\partial q^\alpha}. \tag{2.24}$$

But

$$\mathbf{v}_i \equiv \dot{\mathbf{x}}_i = \frac{d\mathbf{x}_i}{dt} = \frac{\partial \mathbf{x}_i}{\partial q^\alpha}\dot{q}^\alpha + \frac{\partial \mathbf{x}_i}{\partial t},$$

so that

$$\frac{\partial \mathbf{v}_i}{\partial \dot{q}^\alpha} \equiv \frac{\partial \dot{\mathbf{x}}_i}{\partial \dot{q}^\alpha} = \frac{\partial \mathbf{x}_i}{\partial q^\alpha} \tag{2.25}$$

(a sort of "law of cancellation of dots"). In the last term on the right-hand side of Eq. (2.24)

$$\frac{d}{dt}\frac{\partial \mathbf{x}_i}{\partial q^\alpha} = \frac{\partial^2 \mathbf{x}_i}{\partial q^\alpha \partial q^\beta}\dot{q}^\beta + \frac{\partial}{\partial t}\left(\frac{\partial \mathbf{x}_i}{\partial q^\alpha}\right)$$

$$= \frac{\partial}{\partial q^\alpha}\left(\frac{\partial \mathbf{x}_i}{\partial q^\beta}\dot{q}^\beta + \frac{\partial \mathbf{x}_i}{\partial t}\right) = \frac{\partial \mathbf{v}_i}{\partial q^\alpha}.$$

Inserting the last two equations into (2.24), multiplying by m_i, and summing over i, yields

$$\sum_{i=1}^{N} m_i \ddot{\mathbf{x}}_i \cdot \frac{\partial \mathbf{x}_i}{\partial q^\alpha} = \sum_{i=1}^{N}\left[\frac{d}{dt}\left(m_i \mathbf{v}_i \cdot \frac{\partial \mathbf{v}_i}{\partial \dot{q}^\alpha}\right) - m_i \mathbf{v}_i \cdot \frac{\partial \mathbf{v}_i}{\partial q^\alpha}\right]$$

$$= \frac{d}{dt}\frac{\partial T}{\partial \dot{q}^\alpha} - \frac{\partial T}{\partial q^\alpha},$$

where $T = \frac{1}{2}\sum m_i v_i^2$ is the total kinetic energy of the system of particles. Then Eq. (2.17) becomes

$$\frac{d}{dt}\frac{\partial T}{\partial \dot{q}^\alpha} - \frac{\partial T}{\partial q^\alpha} + \frac{\partial V}{\partial q^\alpha} = 0. \tag{2.26}$$

If T is written out in terms of the generalized coordinates q^α, these equations are the equations of motion in terms of the q^α. Since $\partial V/\partial \dot{q}^\alpha = 0$ (for V is a function only of the q^α, not the $\dot{q}^\alpha$), one can define a new function

$$L = T - V \tag{2.27}$$

called the *Lagrangian function* or simply the *Lagrangian*.

In terms of the Lagrangian function, the equations of motion become

$$\frac{d}{dt}\frac{\partial L}{\partial \dot{q}^\alpha} - \frac{\partial L}{\partial q^\alpha} = 0. \tag{2.28}$$

These are called *Lagrange's equations*. They are the equations we have been looking for.

In obtaining Lagrange's equations, T and V, and consequently L, were written entirely in terms of the q^α, which were defined to accommodate the constraints. But Lagrange's equations can also be used without constraints (i.e., if $K = 0$), and then the q^α can be Cartesian or any other coordinates.

Lagrange's equations have been derived from Newton's laws. They are in fact a restatement of Newton's laws written out in terms of appropriate variables that allow constraint forces to be eliminated from consideration. In Section 1.3 it was pointed out that when the force on a single particle is conservative, a dynamical system need not be defined by three functions, the components of the force, but by just the single potential energy function. Now it is seen that if in a many-particle system the forces are conservative and the constraints are smooth, the dynamical system can be defined by the single Lagrangian function. Equations (2.28) are simpler than the $\mathbf{F} = m\mathbf{a}$ equations for all of the particles plus the constraints, so Lagrange's equations are a simplification of Newton's. A full description of a system's motion now is obtained by finding its Lagrangian, specifying the initial conditions, writing out Lagrange's equations, and solving them for the $q^\alpha(t)$. The simplest way to find the Lagrangian function $L(q, \dot{q}, t)$ is to write out the kinetic and potential energies in terms of the q^α and insert them into (2.27) (we write q and $\dot{q}$ for the collection of the q^α and $\dot{q}^\alpha$). In Section 2.2.4 we will show, in contrast, that one of the most important of Lagrangians – for a charged particle in an electromagnetic field – is not just the difference of kinetic and potential energies. Indeed, the force in that case depends on the velocity and is therefore not simply the gradient of a potential. Like Newton's equations, Lagrange's are a set of second-order differential equations, but now for the $q^\alpha(t)$ rather than the $\mathbf{x}_i(t)$. To exhibit their second-order nature, write them out explicitly by expanding the time derivatives:

$$\frac{\partial^2 L}{\partial \dot{q}^\beta \partial \dot{q}^\alpha} \ddot{q}^\beta + \frac{\partial^2 L}{\partial q^\beta \partial \dot{q}^\alpha} \dot{q}^\beta + \frac{\partial^2 L}{\partial t \partial \dot{q}^\alpha} - \frac{\partial L}{\partial q^\alpha} = 0. \qquad (2.29)$$

The $\ddot{q}^\alpha$ that appear here come from the second time derivatives $\mathbf{a}$ of $\mathbf{x}$ in Newton's equations.

WORKED EXAMPLE 2.2.1

A bead of mass m slides without friction in a uniform gravitational field on a vertical circular hoop of radius R. The hoop is constrained to rotate at a fixed angular velocity Ω about its vertical diameter. Let θ be the position of the bead on the hoop measured from the lowest point (Fig. 2.5). **(a)** Write down the Lagrangian $L(\theta, \dot{\theta})$. **(b)** Find how the equilibrium values of θ depend on Ω. Which are stable, which unstable? **(c)** Find the frequencies of small vibrations about the stable equilibrium positions. Say something concerning the motion about the only stable equilibrium point when $\Omega = \sqrt{g/R}$.

Solution. **(a)** The constraint equations in spherical polar coordinates are $r = R$ and $\varphi = \Omega t$, where φ is the azimuth angle, but we do not use them explicitly. The Lagrangian is

$$L = \frac{1}{2}m(R^2\dot{\theta}^2 + R^2\Omega^2 \sin^2\theta) + mgR\cos\theta.$$

The first term is the kinetic energy T and the second is the negative potential $-V$ relative to $\theta = \pi/2$. In T the first term comes from the velocity along the hoop, and

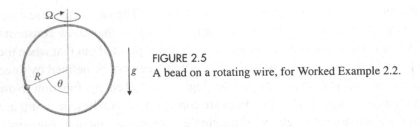

FIGURE 2.5
A bead on a rotating wire, for Worked Example 2.2.

the second from the rotation of the hoop. The constraints are built into T, for the r is constrained to be R, $\dot{r}$ is constrained to be zero (it doesn't appear), and $\dot{\varphi}$ is constrained to be Ω. This is a one-freedom system.

(b) Lagrange's equation (divided by mR^2) is

$$\ddot{\theta} = \Omega^2 \sin\theta \cos\theta - \frac{g}{R}\sin\theta \equiv F(\theta).$$

Equilibrium occurs where $\ddot{\theta} = 0$ (i.e., at $\theta = 0, \pi$) and at θ_0 given by

$$\cos\theta_0 = \frac{g}{R\Omega^2}.$$

This last equation can hold only if $\Omega \geq \sqrt{g/R}$. As Ω approaches $\sqrt{g/R}$ from above, θ_0 approaches zero, and it merges with the equilibrium point at $\theta = 0$ when $\Omega = \sqrt{g/R}$. As Ω increases, θ_0 approaches $\pi/2$. To check on the stability, think of $F(\theta)$ as a force, the negative derivative of a potential. To see whether that potential is at a minimum or a maximum at equilibrium, take the second derivative of the potential, namely $-dF/d\theta$. It is found that θ_0 is stable for $0 < \theta_0 < \pi/2$ (i.e., for all possible θ_0). The equilibrium point at $\theta = 0$ is stable for $\Omega < \sqrt{g/R}$ and unstable otherwise. (We ignore the obviously unstable equilibrium point at $\theta = \pi$.) Thus, $\theta = 0$ is the only stable point at $\Omega = 0$. As Ω increases from zero, $\theta = 0$ remains at first the only (stable) equilibrium point, until Ω gets to $\sqrt{g/R}$. At that point the equilibrium point at $\theta = 0$ bifurcates into two new stable points at $\theta_0 = \pm\arccos(g/R\Omega^2)$ (and $\theta = 0$ becomes unstable). As Ω increases, the two points at θ_0 move apart, approaching $\pm\pi/2$ in the limit of large Ω.

(c) For $\Omega > \sqrt{g/R}$, the first term in a Taylor series expansion about θ_0 yields

$$\Delta\ddot{\theta} = \left[\Omega^2(\cos^2\theta_0 - \sin^2\theta_0) - \frac{g}{R}\cos\theta_0\right]\Delta\theta = -\Delta\theta\,\Omega^2\sin^2\theta_0$$

so the (circular) frequency of small vibrations is $\Omega\sin\theta_0$. For $\Omega < \sqrt{g/R}$, a Taylor expansion about $\theta = 0$ yields

$$\ddot{\theta} = -(g/R - \Omega^2)\theta,$$

so the (circular) frequency is $\sqrt{g/R - \Omega^2}$. For $\Omega = \sqrt{g/R}$, the linear term in the Taylor series vanishes, so in this special case the vibration isn't harmonic.

2.2.2 TRANSFORMATIONS OF LAGRANGIANS

EQUIVALENT LAGRANGIANS

Although the Lagrangian function determines the equations of motion uniquely, the equations of motion do not determine the Lagrangian uniquely. That is, two Lagrangians that are different can lead to equations of motion that are the same. For example, let L_1 and L_2 be two Lagrangian functions such that the equations of motion obtained from them are *exactly the same*. Then it can be shown that there exists a function Φ on $\mathbb{Q}$ such that $L_1 - L_2 = d\Phi/dt$.

To prove this assertion, we must explain what is meant by "exactly the same." For this purpose, consider the $2n$ functions $\Lambda_{j\alpha}$, where $j = 1, 2$, and $\alpha = 1, \ldots, n$, defined by

$$\frac{d}{dt}\frac{\partial L_j}{\partial \dot{q}^\alpha} - \frac{\partial L_j}{\partial q^\alpha} \equiv \Lambda_{j\alpha}(q, \dot{q}, \ddot{q}, t),$$

so that Lagrange's equations can be written

$$\Lambda_{j\alpha}(q, \dot{q}, \ddot{q}, t) = 0.$$

Then to say that the equations of motion are exactly the same is to say that the two functions are equal. In other words

$$\Lambda_{1\alpha} = \Lambda_{2\alpha} \tag{2.30}$$

identically, for each α. We now go on to prove the assertion, but only for one-freedom (i.e., for $n = 1$), leaving the general case for Problem 4. For $n = 1$ the index α may be dropped from the equations. Write $\psi = L_1 - L_2$. Then Eq. (2.29) can be used to write (2.30) in the form

$$\Lambda_1 - \Lambda_2 \equiv \frac{\partial^2 \psi}{\partial \dot{q}^2}\ddot{q} + \frac{\partial^2 \psi}{\partial q \partial \dot{q}}\dot{q} + \frac{\partial^2 \psi}{\partial \dot{q}\partial t} - \frac{\partial \psi}{\partial q} = 0. \tag{2.31}$$

Because L_1 and L_2 are functions only of q, $\dot{q}$, and t, so is ψ, and hence the only place $\ddot{q}$ appears in (2.31) is where it is seen multiplying the first term. But (2.31) must be true for all values of $q, \dot{q}, \ddot{q}, t$. Hence the coefficient of $\ddot{q}$ must therefore vanish: $\partial^2 \psi/\partial \dot{q}^2 = 0$. This means that ψ is linear in $\dot{q}$:

$$\psi = \dot{q} F(q, t) + G(q, t). \tag{2.32}$$

This and some algebra can be used to transform (2.31) to

$$\frac{\partial F}{\partial t} - \frac{\partial G}{\partial q} = 0. \tag{2.33}$$

This is just the *integrability condition* for the pair of equations

$$F = \partial\Phi/\partial q \quad \text{and} \quad G = \partial\Phi/\partial t. \tag{2.34}$$

That is, (2.33) is the condition that there exist a *local* function $\Phi(q, t)$ that makes it possible to write F and G *locally* in the form of (2.34). When they are written that way, (2.32) becomes

$$\psi = \dot{q}\frac{\partial \Phi}{\partial q} + \frac{\partial \Phi}{\partial t} \equiv \frac{d\Phi}{dt},$$

as asserted. The emphasis on locality here is to warn the reader that it may not be possible to find a single-valued function $\Phi(q, t)$ on all of $\mathbb{Q}$. Such matters will be discussed more fully later.

What we have just proven (or rather what is proven in Problem 4) means that if a Lagrangian is changed by adding the time derivative of a function, the equations of motion will not be changed. It does not mean, however, that Lagrangians that yield the *same dynamics* must necessarily differ by a total time derivative. For example $L_a = \dot{q}^1\dot{q}^2 - q^1q^2$ and $L_b = \frac{1}{2}[(\dot{q}^1)^2 + (\dot{q}^2)^2 - (q^1)^2 - (q^2)^2]$ yield the same dynamics, but they do not differ by a total time derivative (we return to this example in Problem 3.12). See Morandi et al. (1990) and references therein.

COORDINATE INDEPENDENCE

Lagrange's equations were derived without specifying in any way the particular generalized coordinates used on $\mathbb{Q}$, and thus they are as valid in one coordinate system as in another. This is an important and useful property of the Lagrangian formulation (see Section 2.2.4 for an example of its usefulness). To observe it in detail, let $q'^\alpha(q, t)$ be a new set of generalized coordinates, in which case the functions must be invertible: the $q^\alpha(q', t)$ exist. What this means physically is that it is possible to write down the particle trajectories in terms of the q'^α as well as in terms of the q^α, and when they are known in terms of one set they can be calculated in terms of the other. When the $q^\alpha(q', t)$ are known, so are the $\dot{q}^\alpha(q', \dot{q}', t)$:

$$\dot{q}^\alpha(q', \dot{q}', t) = \frac{\partial q^\alpha(q', t)}{\partial q'^\beta}\dot{q}'^\beta + \frac{\partial q^\alpha(q', t)}{\partial t}. \tag{2.35}$$

The Lagrangian is a function of $2n + 1$ variables: the n generalized coordinates, their n time derivatives, and the time. This means that L assigns a real number (the value of the function) to each set of $2n + 1$ numbers (the values of the variables). But the set of $2n + 1$ numbers describes the physical state of the dynamical system. Therefore the Lagrangian function assigns a real number not actually to each set of $2n + 1$ numbers but to the physical state itself. When a coordinate transformation is performed, the same physical state of the system is described by a different set of $2n + 1$ numbers. Since the state doesn't change, the Lagrangian function must assign the same real value to this transformed set of $2n + 1$ numbers.

We now make this explicit mathematically. Suppose that a Lagrangian function $L(q, \dot{q}, t)$ is given in the original, unprimed coordinates. It can be written in terms of the primed ones simply by substituting the known expressions for the q^α and $\dot{q}^\alpha$ in terms of the q'^β and the $\dot{q}'^\beta$. The result will be a new (transformed) Lagrangian function L' of the primed coordinates, different from L. That is, $L'(q', \dot{q}', t)$ is defined by

$$L'(q', \dot{q}', t) \equiv L(q(q', t), \dot{q}(q', \dot{q}', t), t) \equiv L(q, \dot{q}, t), \tag{2.36}$$

the equality arising from what has just been said: the primed and unprimed variables describe the same physical state of the system. Then, as shown in Problem 3, Lagrange's equations (2.28) imply

$$\frac{d}{dt} \frac{\partial L'}{\partial \dot{q}'^\alpha} - \frac{dL'}{dq'^\alpha} = 0. \tag{2.37}$$

Because this equation and (2.28) are equally valid, Lagrange's equations are said to be coordinate independent (or *covariant* under coordinate transformations). But then these equations are somehow making a statement that does not depend on the coordinates, and the specific q^α or q'^α that appear in them play a nonessential, entirely auxiliary role. There should then be a way to write down these differential equations on the configuration manifold in such a way that the coordinates do not appear. We will do so in Chapter 3.

HESSIAN CONDITION

The equations of motion, Lagrange's and Newton's, are of second order. They are a prescription for calculating the acceleration from the state of the system (from the initial position and velocity). The initial state yields the initial acceleration, and then integration leads from the initial acceleration to later states. Lagrange's equations in the form of (2.29) may be rewritten as

$$\frac{\partial^2 L}{\partial \dot{q}^\beta \partial \dot{q}^\alpha} \ddot{q}^\beta = G_\alpha(q, \dot{q}, t),$$

where the G_α and the second partial derivatives on the left-hand side are functions of $(q, \dot{q}, t)$ that can be calculated once the Lagrangian is known. The $\ddot{q}^\beta(t_0)$ can be found by inserting the initial $q(t_0)$ and $\dot{q}(t_0)$ into the expressions for the G_α and $\partial^2 L/\partial \dot{q}^\alpha \partial \dot{q}^\beta$, which then form a numerical column vector and a numerical matrix, and by then applying the inverse of the matrix to the vector. For that to work, the *Hessian* matrix $\partial^2 L/\partial \dot{q}^\alpha \partial \dot{q}^\beta$ must be nonsingular. Hence we will always assume (unless explicitly stated otherwise) that

$$\det(\partial^2 L/\partial \dot{q}^\alpha \partial \dot{q}^\beta) \neq 0.$$

This is called the *Hessian condition*.

> **REMARK:** See Dirac (1966) for ways of dealing with systems for which this condition does not hold (in the Hamiltonian formalism). □

The rest of the procedure in solving the equations of motion involves integration. Integration can be thought of as proceeding in small time increments Δt (infinitesimal, in the limit) with the new $\dot{q}$ and q calculated from the $\ddot{q}$ obtained as described above. The later time is then treated as a new initial time, and as the procedure is iterated the orbit of the phase portrait is "unrolled" on $\mathbb{Q}$.

2.2.3 CONSERVATION OF ENERGY

The difference between the usual Lagrangian function $L = T - V$ and the energy $E = T + V$ is just in the sign of V. Is there some general way to calculate E from a knowledge of L? For a single particle in Cartesian coordinates, so long as V is independent of the velocity $\dot{x}$ (i.e., so long as $\partial V/\partial \dot{x}^\alpha = 0$), this can be done easily. Write $T = \frac{1}{2}m\dot{x}^2$; then

$$\dot{x}^\alpha \frac{\partial L}{\partial \dot{x}^\alpha} - L = \dot{x}^\alpha \frac{\partial T}{\partial \dot{x}^\alpha} - T + V = T + V = E. \tag{2.38}$$

We now take this approach to generalize from one particle to several generalized coordinates.

The analog of E as defined by (2.38) is the dynamical variable

$$E(q, \dot{q}) \equiv \dot{q}^\alpha \frac{\partial L}{\partial \dot{q}^\alpha} - L. \tag{2.39}$$

This generalized E is a candidate for the total energy in general. We first check on its conservation:

$$\dot{E} = \frac{d}{dt}\left[\dot{q}^\alpha \frac{\partial L}{\partial \dot{q}^\alpha} - L \right] = \ddot{q}^\alpha \frac{\partial L}{\partial \dot{q}^\alpha} + \dot{q}^\alpha \frac{d}{dt}\frac{\partial L}{\partial \dot{q}^\alpha} - \ddot{q}^\alpha \frac{\partial L}{\partial \dot{q}^\alpha} - \dot{q}^\alpha \frac{\partial L}{\partial q^\alpha} - \frac{\partial L}{\partial t}.$$

The first and third terms on the right-hand side cancel, and Lagrange's equations imply that so do the second and fourth terms; thus

$$\frac{dE}{dt} = -\frac{\partial L}{\partial t}. \tag{2.40}$$

This means that if L is time independent E is conserved, that is,

$$\frac{\partial L}{\partial t} = 0 \Rightarrow \frac{dE}{dt} = 0. \tag{2.41}$$

In other words, if $\partial L/\partial t = 0$, then E is a constant of the motion, like the energy.

But we have not yet established whether E is in fact the energy $T + V$. To do so we now calculate T explicitly in generalized coordinates:

$$T = \frac{1}{2}\sum_i m_i v_i^2 = \frac{1}{2}\sum_i m_i \left[\frac{\partial \mathbf{x}_i}{\partial q^\alpha}\dot{q}^\alpha + \frac{\partial \mathbf{x}_i}{\partial t} \right] \cdot \left[\frac{\partial \mathbf{x}_i}{\partial q^\gamma}\dot{q}^\gamma + \frac{\partial \mathbf{x}_i}{\partial t} \right]$$

$$= a(q, t) + b_\alpha(q, t)\dot{q}^\alpha + g_{\alpha\beta}(q, t)\dot{q}^\alpha \dot{q}^\beta,$$

where the coefficients

$$a = \frac{1}{2} \sum_i m_i \frac{\partial \mathbf{x}_i}{\partial t} \cdot \frac{\partial \mathbf{x}_i}{\partial t},$$

$$b_\alpha = \sum_i m_i \frac{\partial \mathbf{x}_i}{\partial t} \cdot \frac{\partial \mathbf{x}_i}{\partial q^\alpha}, \tag{2.42}$$

$$g_{\alpha\beta} = \frac{1}{2} \sum_i m_i \frac{\partial \mathbf{x}_i}{\partial q^\alpha} \cdot \frac{\partial \mathbf{x}_i}{\partial q^\beta}$$

are functions of q and t, but not of $\dot{q}$. Thus T is a quadratic function of the $\dot{q}^\alpha$, but it is not in general a *homogeneous* quadratic function.

We turn aside to explain homogeneity. A function f of k variables $z_1, \ldots, z_k$ is *homogeneous of degree* λ iff (if and only if) for all $a > 0$

$$f(az_1, \ldots, az_k) = a^\lambda f(z_1, \ldots, z_k).$$

For instance $z_1 + 7z_1z_2 + 3(z_2)^2$ is quadratic, and $7z_1z_2 + 3(z_2)^2$ is homogeneous quadratic. We will frequently use *Euler's theorem on homogeneous functions*: if f is homogeneous of degree λ, then

$$z_i \frac{\partial f}{\partial z_i} = \lambda f.$$

The reason that the homogeneity of T is important is that (we assume that $\partial V/\partial \dot{q}^\alpha = 0$ and $\partial V/\partial t = 0$) then Euler's theorem implies that

$$E = \dot{q}^\alpha \frac{\partial T}{\partial \dot{q}^\alpha} - L = 2T - T + V = T + V.$$

Thus homogeneity guarantees that E is the energy. So under what conditions is T homogeneous quadratic in $\dot{q}$? According to Eq. (2.42) if the $\mathbf{x}_i$ are time independent functions of q, then $a = 0$ and $b_\alpha = 0$, and T is homogeneous quadratic in the $\dot{q}^\alpha$ (it may still be a function of the q^α but not of t).

Hence if $L = T - V$ is time independent, the potential V is $\dot{q}$ independent, and the transformation from Cartesian to generalized coordinates is time independent, then E is the energy. Although this sounds like a lot of arbitrary assumptions, it is one of the most common situations. Thus Eq. (2.41) states that under these conditions energy is conserved. Note that the dynamical variable E is conserved under the single condition that L is time independent, but this one condition will not guarantee E to be the energy. To see an instance of this, we return to Worked Example 2.2.1.

WORKED EXAMPLE 2.2.2

In Worked Example 2.2.1, is $E = \dot{\theta}\partial L/\partial \dot{\theta} - L$ conserved? Compare E to the energy.

Solution. Recall that the Lagrangian

$$L = \frac{1}{2}m(R^2\dot{\theta}^2 + R^2\Omega^2 \sin^2\theta) + mgR\cos\theta$$

is time independent. Hence E is conserved. Direct calculation yields

$$E(\theta, \dot{\theta}) = \frac{1}{2}m(R^2\dot{\theta}^2 - R^2\Omega^2 \sin^2\theta) - mgR\cos\theta.$$

This is not the energy. The energy H is

$$H \equiv T + V = \frac{1}{2}m(R^2\dot{\theta}^2 + R^2\Omega^2 \sin^2\theta) - mgR\cos\theta = E + mR^2\Omega^2 \sin^2\theta.$$

Since E is constant in time but θ may vary, H is not constant in time. Energy must enter and leave the system to keep the hoop rotating at constant speed, and the term $mR^2\Omega^2 \sin^2\theta$ represents this varying amount of energy.

2.2.4 CHARGED PARTICLE IN AN ELECTROMAGNETIC FIELD

THE LAGRANGIAN

Not all Lagrangians are of the form $L = T - V$, and because not all forces are conservative a potential function V may not even exist. Nevertheless, many systems whose forces do not arise from potentials that are functions of the form $V(q)$ on $\mathbb{Q}$ can be handled by the Lagrangian formalism, but with more general Lagrangian functions. A particularly important example is that of a charged particle in an electromagnetic field, which we discuss in this subsection.

The Lorentz force on such a particle is a function of $\mathbf{v}$ as well as $\mathbf{x}$ (it is defined on what will soon be called $\mathbf{T}\mathbb{Q}$). It is given by

$$\mathbf{F} = e\left(\mathbf{E} + \frac{1}{c}\mathbf{v} \wedge \mathbf{B}\right), \tag{2.43}$$

where e is the charge on the particle (not necessarily the electron charge), $\mathbf{E}(\mathbf{x}, t)$ and $\mathbf{B}(\mathbf{x}, t)$ are the electric and magnetic fields (in this subsection we will use W for the energy to distinguish it from the electric field), and c is the speed of light in a vacuum. For the purposes of this discussion we choose the scales of space and time to be such that $c = 1$. In indicial (tensor) notation the αth component of the Lorentz force is

$$F_\alpha = e(E_\alpha + \epsilon_{\alpha\beta\gamma} v^\beta B_\gamma). \tag{2.44}$$

Here $\epsilon_{\alpha\beta\gamma}$ is the *Levi–Civita antisymmetric tensor density* defined as equal to 1 if $\alpha\beta\gamma$ is an *even* permutation of 123, equal to -1 if it is an *odd* permutation, and equal to zero if any index is repeated. (The even permutations, 123, 231, and 312, are also called *cyclic* permutations. The odd ones are 132, 321, and 213.) This is a common way of writing the cross product, and we will constantly make use of it, for the properties of $\epsilon_{\alpha\beta\gamma}$ are helpful in making calculations. (In this subsection we do not distinguish carefully between superscript and subscript indices.)

The electric and magnetic fields can be written in the form

$$
\begin{aligned}
E_\alpha &= -\partial_\alpha \varphi - \partial_t A_\alpha, \\
B_\alpha &= \epsilon_{\alpha\beta\gamma} \partial_\beta A_\gamma,
\end{aligned}
\tag{2.45}
$$

where $\varphi(\mathbf{x}, t)$ is the *scalar potential*, $\mathbf{A}(\mathbf{x}, t)$ is the *vector potential*, $\partial_\alpha \equiv \partial/\partial x^\alpha$, and $\partial_t \equiv \partial/\partial t$. These equations are usually written in the form $\mathbf{E} = -\nabla\varphi - \partial\mathbf{A}/\partial t$ and $\mathbf{B} = \nabla \wedge \mathbf{A}$. With these expressions for the fields, Newton's equations for the acceleration become

$$m\ddot{x}^\alpha = -e\partial_\alpha\varphi - e\partial_t A_\alpha + e\epsilon_{\alpha\beta\gamma}\dot{x}^\beta\epsilon_{\gamma\mu\nu}\partial_\mu A_\nu.$$

Notice that the x^α themselves do not appear in this equation, only their first and second time derivatives, so it is convenient to write these as equations for $v^\alpha \equiv \dot{x}^\alpha$:

$$m\dot{v}^\alpha = -e\partial_\alpha\varphi - e\partial_t A_\alpha + e\epsilon_{\alpha\beta\gamma}v^\beta\epsilon_{\gamma\mu\nu}\partial_\mu A_\nu. \tag{2.46}$$

We now use one of the calculational advantages of the $\epsilon_{\alpha\beta\gamma}$. It can be shown from their definition (this is an exercise on manipulating indices, left for the reader) that

$$\epsilon_{\alpha\beta\gamma}\epsilon_{\gamma\mu\nu} = \delta_{\alpha\mu}\delta_{\beta\nu} - \delta_{\alpha\nu}\delta_{\beta\mu}. \tag{2.47}$$

The order of the indices is crucial here: it determines the sign. When (2.47) is inserted into (2.46) and sums are taken over μ and ν, that equation becomes

$$m\dot{v}^\alpha = -e\partial_\alpha\varphi - e\partial_t A_\alpha + e(v^\nu\partial_\alpha A_\nu - v^\mu\partial_\mu A_\alpha). \tag{2.48}$$

The second and last terms on the right-hand side add up to $-dA_\alpha/dt$, so that (2.48) can be written in the form

$$\frac{d}{dt}(mv^\alpha + eA_\alpha) + \partial_\alpha(e\varphi - ev^\nu A_\nu) = 0.$$

This can be made to look like Lagrange's equations if L can be found such that the expression in the first parentheses is $\partial L/\partial v^\alpha$ and the second term is $\partial_\alpha L$. Since neither φ nor the A_α depend on the v^α, the Lagrangian can be chosen as (we return to writing $\dot{x}$ for v)

$$L = \frac{1}{2}m\dot{x}^2 - e\varphi(\mathbf{x}, t) + e\dot{x}^\beta A_\beta(\mathbf{x}, t)$$

$$= \frac{1}{2}m\dot{x}^2 - e\varphi(\mathbf{x}, t) + e\dot{\mathbf{x}} \cdot \mathbf{A}(\mathbf{x}, t). \tag{2.49}$$

For this system the dynamical variable W (previously E) defined in Eq. (2.39) is

$$W = \frac{1}{2}m\dot{x}^2 + e\varphi, \tag{2.50}$$

which is the sum of the kinetic energy and the charge times the scalar potential. In other words, the vector potential does not enter into the expression for W. Equation (2.41) then implies that *if both φ and $\mathbf{A}$ are t independent*, W is conserved. This is understood to mean that the sum of the particle's kinetic plus electrostatic (for φ is now time independent) potential energy remains constant, and it is explained by the fact that the magnetic force $\dot{\mathbf{x}} \wedge \mathbf{B}$ does no work on the particle because it is always perpendicular to the velocity. Nevertheless, the $\mathbf{A}$-dependent term from which the magnetic field is obtained is part of the energy of interaction between the charged particle and the electromagnetic field.

The quantity $U(\mathbf{x}, \dot{\mathbf{x}}, t) = e\varphi(\mathbf{x}, t) - e\dot{\mathbf{x}} \cdot \mathbf{A}$, called the total *interaction energy*, allows the Lagrangian to be written in the form $L = T - U$. Nevertheless the conserved quantity W is not equal to $T + U$, for U depends on the velocity.

In electrodynamics, a *gauge transformation* is a change in φ and $\mathbf{A}$ that does not change the electric and magnetic fields obtained through Eq. (2.45). It is seen from (2.45) that the fields are not changed when A_α is replaced by A'_α and φ by φ' in accordance with

$$A_\alpha \to A'_\alpha = A_\alpha + \partial_\alpha \Lambda,$$
$$\varphi \to \varphi' = \varphi - \partial_t \Lambda. \tag{2.51}$$

Since the force on a charged particle depends on $\mathbf{E}$ and $\mathbf{B}$, not on φ and $\mathbf{A}$, it should not change under a gauge transformation. Yet the Lagrangian depends on φ and $\mathbf{A}$, so one may ask how it happens that although the Lagrangian changes under a gauge transformation, the force does not.

The change in the Lagrangian under a gauge transformation can be calculated. It is given by

$$L \to L' = L + e\partial_t \Lambda + e\dot{x}^\alpha \partial_\alpha \Lambda = L + e\frac{d\Lambda}{dt}.$$

We have seen that if L and L' differ by a total time derivative, they yield exactly the same equations of motion. Thus in spite of the change in the Lagrangian, the equations of motion, and hence the force, remain the same. The Lagrangian of Eq. (2.49) is then said to be *gauge invariant*.

In classical mechanics $\mathbf{E}$ and $\mathbf{B}$ are the physically relevant aspects of the electromagnetic field; it is $\mathbf{E}$ and $\mathbf{B}$ that are "physically real." In other words, not the vector potential $\mathbf{A}$ and φ, but only their derivatives are physically relevant. In quantum mechanics, as was shown by Aharonov and Bohm (1965), physical effects depend not only on the derivatives, but also on $\mathbf{A}$ and φ themselves, through certain path integrals. For this reason gauge invariance does not play a large role in classical mechanics, whereas it becomes of paramount importance in quantum mechanics.

A TIME-DEPENDENT COORDINATE TRANSFORMATION

As is well known, a charged particle in a uniform magnetic field tends to move in a circular orbit. In this subsection we show how that is related to viewing a free particle from a rotating frame.

The Lagrangian of a free particle ($V = 0$) is simply the kinetic energy. In an inertial Cartesian frame

$$L(x, \dot{x}) = T(\dot{x}) = \frac{1}{2}m\dot{x}^\alpha \dot{x}^\alpha. \tag{2.52}$$

Moreover, the total energy $W_x = \dot{x}^\alpha \partial L / \partial \dot{x}^\alpha - L$ is conserved and equal to T.

Now consider an observer viewing a free particle from a noninertial coordinate system attached to and rotating with the Earth. Choose the coordinate axes in both the inertial and noninertial system so that the two 3 axes (the z axes) lie along the axis of rotation. Then the transformation equations from the noninertial to the inertial coordinates are

$$x^1 = y^1 \cos \omega t - y^2 \sin \omega t,$$
$$x^2 = y^1 \sin \omega t + y^2 \cos \omega t, \tag{2.53}$$
$$x^3 = y^3,$$

where ω is the angular rate of rotation and the y^α are the rotating coordinates. A straight-forward calculation shows that in the rotating frame the Lagrangian (and the kinetic energy) is given by

$$L(y, \dot{y}) = T = \frac{1}{2} m [\dot{y}^\alpha \dot{y}^\alpha + \omega^2 (\{y^1\}^2 + \{y^2\}^2) + 2\omega(\dot{y}^2 y^1 - \dot{y}^1 y^2)]. \tag{2.54}$$

The second term here looks like a repulsive potential and provides what the observer in the rotating system sees as a *centrifugal force* acting on the particle. The third term is essentially the *Coriolis force* (see Section 1.5.3). Equation (2.54) is the kinetic energy written out in the noninertial system; it is nonzero even if the y velocity is zero, for the x velocity is nonzero if the particle is off the 3 axis.

It is clear that in the y system T is not homogeneous quadratic in the $\dot{y}^\alpha$. In these coordinates the conserved dynamical variable W_y is

$$W_y \equiv \dot{y}^\alpha \frac{\partial L}{\partial \dot{y}^\alpha} - L = \frac{1}{2} m [\dot{y}^\alpha \dot{y}^\alpha + \omega^2 (\{y^1\}^2 + \{y^2\}^2)].$$

In the rotating coordinate system the kinetic energy is not conserved. The quantity conserved does not even look like the kinetic energy (i.e., it is not just $\frac{1}{2} m \dot{y}^\alpha \dot{y}^\alpha$). This exemplifies the kind of thing that can happen if the transformation from Cartesian to generalized coordinates depends on time.

Note, incidentally, that in the Lagrangian of Eq. (2.54) the third coordinate y^3 plays almost no role: the particle is free to move at constant velocity in the 3 direction. The 3 direction is special because it is the axis of the rotation (2.53). In other words, for any other rotation transformation the result would be essentially the same: the motion along the axis of the rotation would be free, and in the plane perpendicular to the axis it would be given by terms similar to the y^1 and y^2 terms in (2.54).

The transformation of (2.53) is usefully applied to the dynamical system consisting of a charged particle in a uniform magnetic field $\mathbf{B}$. The vector potential $\mathbf{A}(\mathbf{x})$ for a uniform $\mathbf{B}$ is $\mathbf{A}(x) = -\frac{1}{2} \mathbf{x} \wedge \mathbf{B}$ (this is not a gauge invariant statement: we are choosing what is called the *Landau gauge*). In tensor notation

$$A_\alpha = -\frac{1}{2} \epsilon_{\alpha\beta\gamma} x^\beta B^\gamma. \tag{2.55}$$

As is customary, we choose a coordinate system in which **B** points along the 3 axis. The 3 axis is now special in that it is parallel to **B**. Then **A** is everywhere parallel to the $(1, 2)$ plane, with components $A_1 = -\frac{1}{2}x^2 B$ and $A_2 = \frac{1}{2}x^1 B$. With this vector potential and no electrostatic potential, Eq. (2.49) gives

$$L = \frac{1}{2}m(\dot{x}^\alpha \dot{x}^\alpha) + \frac{1}{2}eB(x^1\dot{x}^2 - x^2\dot{x}^1). \qquad (2.56)$$

Note the similarity of the last terms of (2.56) and (2.54). Is it possible to go to a system rotating about the 3 axis (i.e., about **B**), as in (2.53), choosing ω so as to eliminate the term involving **B**?

We have already seen how the kinetic energy term, that is, the Lagrangian of (2.52), transforms under (2.53). A brief calculation shows that in the y coordinates the term involving **B** in (2.56) is

$$\frac{1}{2}eB(\omega[(y^1)^2 + (y^2)^2] + y^1\dot{y}^2 - y^2\dot{y}^1).$$

When this is combined with Eq. (2.54), the charged-particle Lagrangian takes on the form

$$L = \frac{1}{2}m(\dot{y}^\alpha \dot{y}^\alpha) + \frac{1}{2}\omega(m\omega + eB)(\{y^1\}^2 + \{y^2\}^2) + \left(m\omega + \frac{1}{2}eB\right)(\dot{y}^2 y^1 - \dot{y}^1 y^2).$$

We are looking for an ω that will make the last term vanish. Clearly, ω must then be chosen to be

$$\omega_L = -\frac{eB}{2m}, \qquad (2.57)$$

which is known as the *Larmor frequency* (the negative sign determines the direction of rotation; see Problem 8). With this choice, the Lagrangian becomes

$$L = \frac{1}{2}m(\dot{y}^\alpha \dot{y}^\alpha) - \frac{1}{2}m\omega_L^2(\{y^1\}^2 + \{y^2\}^2). \qquad (2.58)$$

This equation tells us that in a coordinate system rotating at the Larmor frequency, the particle in the magnetic field looks as though it is in a two-dimensional harmonic-oscillator potential whose frequency is precisely the Larmor frequency. This is a remarkable fact. Recall that the orbits of a charged particle in a magnetic field (ignoring the motion in the 3 direction) are circles of all possible radii centered everywhere in the plane. The result we have just obtained states that when viewed from a frame rotating about any arbitrary point P in the plane, every one of these circles will look like an ellipse centered at P, provided the rate of rotation is at the Larmor frequency and in the correct direction.

This procedure is a well-known and useful technique. It is particularly important in quantum mechanics, for it can be used to predict the energy spectrum of a particle in a magnetic field.

2.3 CENTRAL FORCE MOTION

2.3.1 THE GENERAL CENTRAL FORCE PROBLEM

STATEMENT OF THE PROBLEM; REDUCED MASS

The dynamical system with which Newton began his development of classical mechanics, usually called *the Kepler problem*, consists of two mass points interacting through a force that is directed along the line joining them and varies inversely with the square of the distance between them. In a generalization of this system, the force may vary in an arbitrary way, but it remains directed along the line joining the two bodies. This general system, called the *central force problem*, continues to play an important role in physics, both because there are central forces other than the inverse square force (e.g., exponentially decaying Debye screening in a plasma and the Yukawa potential) and because forces that are not central are often in some sense close to central ones and can be treated as perturbations on them.

We will assume that the magnitude of the interaction force depends only on the distance between the mass points. Let the masses of the two particles be m_1 and m_2. Then their equations of motion are

$$m_1 \ddot{\mathbf{x}}_1 = \mathbf{F}(\mathbf{x}), \quad m_2 \ddot{\mathbf{x}}_2 = -\mathbf{F}(\mathbf{x}),$$

where $\mathbf{F}$ is the force that particle 2 exerts on particle 1, and $\mathbf{x} = \mathbf{x}_1 - \mathbf{x}_2$ is the separation between the particles. Multiply the first of these equations by m_2, the second by m_1, and subtract. This yields

$$\mu \ddot{\mathbf{x}} = \mathbf{F}, \tag{2.59}$$

where $\mu = m_1 m_2 / M$ is the *reduced mass* ($M = m_1 + m_2$ is the total mass). The number of variables of the dynamical system (the dimension of $\mathbb{Q}$) has been reduced from the six components of $\mathbf{x}_1$ and $\mathbf{x}_2$ to the three components of the *relative position* $\mathbf{x}$. Equation (2.59) shows that the relative position satisfies Newton's equation for a particle of reduced mass μ acted upon by the given force $\mathbf{F}(\mathbf{x})$ directed along the position vector in $\mathbf{x}$ space (the fixed origin of $\mathbf{x}$ space is called the *force center*). The solution of this equivalent reduced-mass problem can then be combined with uniform motion of the center of mass (when the external force is zero) to provide a complete solution of the problem. Transforming to the equivalent reduced-mass problem eliminates the center-of-mass motion from consideration.

In many systems (e.g., the Sun and the Earth, the Earth and a satellite, or the classical treatment of the proton and electron in the hydrogen atom) $m_2 \gg m_1$, and then $\mu \approx m_1$. Moreover, because the center of mass then lies very close to m_2, the separation $\mathbf{x}$ between the particles is very nearly the distance from the center of mass to m_1, so the reduced-mass problem is very nearly the same as the actual one. In the limit of infinite m_2 the two problems coalesce.

We now proceed with the reduced-mass central force problem. Assume that $\mathbf{F}$ is derivable from a potential V. Because $\mathbf{F}$ is directed along $\mathbf{x}$, the potential V depends only on

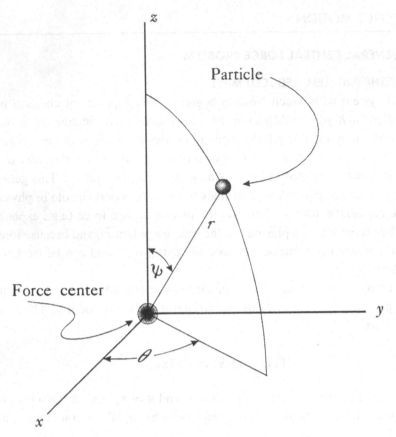

FIGURE 2.6
Spherical polar coordinates for the central force problem. (We choose the unusual notation of ψ and θ for the polar and azimuth angles in order to use r and θ when we finally get down to calculation.)

the magnitude r of $\mathbf{x}$. Two ways to see this are: 1. by calculating the gradient operator in polar coordinates or 2. by recalling that the force is everywhere perpendicular to the equipotentials; since the force is radial, the equipotential surfaces are spheres about the force center, so that $V = V(r)$. It is then natural to write the equations of motion, and therefore the Lagrangian, in spherical coordinates (Fig. 2.6). Although the angles do not appear in the potential energy, they do appear in the kinetic energy, for

$$\dot{x}^2 = \dot{r}^2 + r^2\dot{\psi}^2 + (r^2 \sin^2 \psi)\dot{\theta}^2,$$

where ψ is the polar angle (colatitude) and θ is the azimuth. The Lagrangian is therefore

$$L = \frac{1}{2}\mu[\dot{r}^2 + r^2\dot{\psi}^2 + (r^2 \sin^2 \psi)\dot{\theta}^2] - V(r).$$

REDUCTION TO TWO FREEDOMS

The number of variables can now be lowered further by reducing their number in the kinetic energy term. Since the force is central, the angular momentum vector $\mathbf{J} = \mu\mathbf{x} \wedge \dot{\mathbf{x}}$ is constant (we write $\mathbf{J}$ rather than $\mathbf{L}$ to avoid confusion with the Lagrangian). Since $\mathbf{J}$

is always perpendicular to both $\mathbf{x}$ and $\dot{\mathbf{x}}$, the fixed direction of $\mathbf{J}$ alone implies that $\mathbf{x}$ lies always in the fixed plane perpendicular to this direction. The initial conditions determine the initial direction of $\mathbf{J}$ and hence also the fixed plane in which the motion will evolve. Choose the z axis perpendicular to this plane, that is, parallel to the angular momentum vector. Then $\psi = \pi/2$ and $\dot{\psi} = 0$, so that $\dot{x}^2 = \dot{r}^2 + r^2\dot{\theta}^2$.

In these coordinates, chosen after the direction of the angular momentum has been set, the Lagrangian becomes simply

$$L = \frac{1}{2}\mu(\dot{r}^2 + r^2\dot{\theta}^2) - V(r). \tag{2.60}$$

The dynamical system described by this Lagrangian has two freedoms, represented here by the variables r and θ.

The initial configuration manifold $\mathbb{Q}$ of this system has dimension six, and hence the velocity phase space, which we will call $\mathbf{T}\mathbb{Q}$, has dimension twelve ($\mathbf{T}\mathbb{Q}$ will be discussed more fully in Sections 2.4 and 2.5). This has been reduced to four by (i) eliminating the center-of-mass motion, which reduces $\mathbb{Q}$ to three dimensions and $\mathbf{T}\mathbb{Q}$ to six, and (ii) using the fixed direction of $\mathbf{J}$ to reduce the number of freedoms by one more, leaving $\mathbb{Q}$ with two dimensions and $\mathbf{T}\mathbb{Q}$ four. Each one of these two steps seems to simplify the problem, but in actual fact part of it is swept under the rug. In step (i) the center-of-mass motion is not eliminated, but set aside. In step (ii) the final solution still requires knowing the direction of the angular momentum. There is yet a third step to come.

There are now two Lagrange equations, one for θ and one for r. They are

$$\left.\begin{array}{c} \dfrac{d}{dt}(\mu r^2\dot{\theta}) = 0, \\[2mm] \mu\ddot{r} - \mu r\dot{\theta}^2 + \dfrac{dV}{dr} = 0. \end{array}\right\} \tag{2.61}$$

The first of these is immediately integrated (it will be seen in Chapter 3 that θ is what is called an *ignorable* or *cyclic coordinate*):

$$\mu r^2\dot{\theta} = \text{const.} \equiv l. \tag{2.62}$$

The constant l turns out to be the magnitude of the angular momentum. Indeed, the magnitude of the angular momentum is just $\mu r v_\perp$, where $v_\perp$ is the component of $\mathbf{v}$ perpendicular to the radius vector, and this component is just $r\dot{\theta}$. Now solve (2.62) for $\dot{\theta}$ and insert the result into the second of (2.61), obtaining

$$\mu\ddot{r} - \frac{l^2}{\mu r^3} + \frac{dV}{dr} = 0. \tag{2.63}$$

THE EQUIVALENT ONE-DIMENSIONAL PROBLEM

Equation (2.63) looks just like an equation for the one-dimensional motion of a single particle of mass μ under the influence of the force $l^2/(\mu r^3) - dV/dr$, a force that can be obtained in the usual way from the *effective potential*

$$\mathcal{V}(r) = \frac{l^2}{2\mu r^2} + V(r). \tag{2.64}$$

In other words, the equation of motion for r can be obtained from the Lagrangian $\mathcal{L}$ for what may be called the *equivalent one-dimensional problem* (actually a set of problems, one for each l)

$$\mathcal{L}(r) = \frac{1}{2}\mu\dot{r}^2 - \mathcal{V}(r). \tag{2.65}$$

We have now taken the third step, reducing the original six dimensions of $\mathbb{Q}$ to one (parametrized by r) and the original twelve dimensions of $\mathbb{TQ}$ to two. The new $\mathbb{Q}$ is not simply $\mathbb{R}$, however, for r takes on only positive values; it is sometimes denoted as $\mathbb{R}^+$.

The way to solve the initial problem is first to solve (2.65) for $r(t)$ and then to go backward through the three reduction steps. (iii′) Insert $r(t)$ into (2.62) and solve for $\theta(t)$. This yields the solution of the reduced-mass problem in the plane of the motion. (ii′) Orient the plane (actually $\mathbf{J}$) in space in accordance with the initial conditions. (i′) Fold in the center-of-mass motion, also obtained from the initial conditions, to complete the solution.

Because $\mathcal{L}$ is of the form $\mathcal{L} = T - \mathcal{V}$, where T is the kinetic energy of the equivalent one-dimensional problem, the total energy of the equivalent one-dimensional system, namely

$$E = T + \mathcal{V} = \frac{1}{2}\mu\dot{r}^2 + \frac{l^2}{2\mu r^2} + V(r),$$

is conserved. Using (2.62) to eliminate l from $\mathcal{V}$ leads to

$$E = \frac{1}{2}\mu\dot{r}^2 + \frac{1}{2}\mu r^2\dot{\theta}^2 + V(r), \tag{2.66}$$

which is also the total energy of the reduced-mass problem. It can be shown (see Problem 7) that the total energy of the original two-particle system is (2.66) plus the energy of the center-of-mass motion.

The equivalent one-dimensional system can be handled by the methods discussed in Section 1.3.1. For instance, whether the motion in r is bounded can be studied by looking at a graph of $\mathcal{V}$.

WORKED EXAMPLE 2.3

It has been found that the experimentally determined interaction between the atoms of diatomic molecules can be described quite well by the *Morse potential* (Karplus and Porter, 1970)

$$V(r) = D(e^{-2\alpha r} - 2e^{-\alpha r}), \tag{2.67}$$

with $D, \alpha > 0$. By expanding the exponentials in a Taylor's series one can see that for $\alpha r \ll 1$ the potential is approximately harmonic (see Fig. 2.7); in general, however, the oscillation is nonlinear and the period depends on the energy. Although

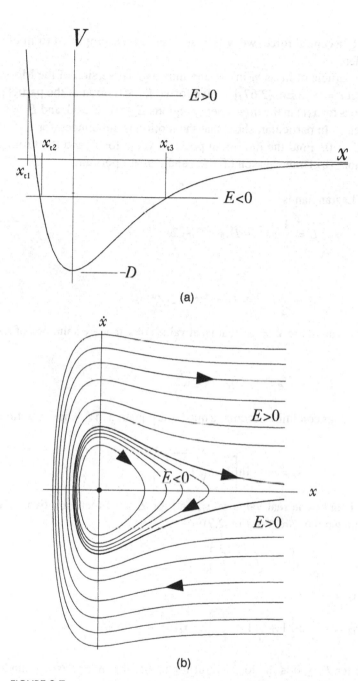

(a)

(b)

FIGURE 2.7

(a) The Morse potential $V(x) = D(e^{-2\alpha x} - 2e^{-\alpha x})$. x_{t1} is the turning point for $E > 0$. x_{t2} and x_{t3} are the turning points of a bound orbit with $E < 0$. The minimum of the potential is $V_{\min} = -D$, located at $x = 0$. (b) The phase portrait for the Morse potential. The closed curves represent bound orbits, with $E < 0$. The rest are unbounded orbits. The heavy curve is the $E = 0$ orbit, the separatrix between the unbounded and bound ones. There is an elliptic stable fixed point at $(x, \dot{x}) = (0, 0)$, represented by the heavy dot.

this is an example of a central force, we will treat it here in the simplified form of a one-freedom problem.

(a) Consider a particle of mass m in the one-dimensional version of the Morse potential [that is, set $r = x$ in Eq. (2.67)]. Solve for $x(t)$, the motion of the particle, obtaining expressions for $x(t)$ in the three energy regions $E < 0$, $E > 0$, and $E = 0$, where E is the energy. In particular, show that the motion is unbounded for $E \geq 0$ but bounded for $E < 0$. Find the minimum possible value for E and the turning points for the motion. (b) Draw a graph of $V(x)$ and a phase portrait.

Solution. (a) The Lagrangian is

$$L = \frac{1}{2}m\dot{x}^2 - D(e^{-2\alpha x} - 2e^{-\alpha x}).$$

and the energy is

$$E = \frac{1}{2}m\dot{x}^2 + D(e^{-2\alpha x} - 2e^{-\alpha x}).$$

The turning points occur where $V = E$, that is, at values of x that are solutions of the equation

$$e^{-2\alpha x} - 2e^{-\alpha x} = \frac{E}{D}.$$

By writing $e^{-\alpha x} = z$, this condition becomes a quadratic equation in z, whose solutions are

$$x_t = -\frac{1}{\alpha}\ln\left[1 \pm \sqrt{1 + \frac{E}{D}}\right].$$

For $E \geq 0$ there is only one real value of x_t, and for $E < 0$ there are two. The corresponding turning points are (see Fig. 2.7)

$$E > 0, \quad x_t = -\frac{1}{\alpha}\ln\left[1 + \sqrt{1 + \frac{E}{D}}\right],$$

$$E = 0, \quad x_t = -\frac{1}{\alpha}\ln 2,$$

$$E < 0, \quad x_{t2} = -\frac{1}{\alpha}\ln\left[1 + \sqrt{1 + \frac{E}{D}}\right], \quad x_{t3} = -\frac{1}{\alpha}\ln\left[1 - \sqrt{1 + \frac{E}{D}}\right].$$

The turning point for $E \geq 0$ is the lower limit for x, and the motion is unbounded for $x > 0$. For $E < 0$, x_{t2} is negative and x_{t3} is positive, so the motion is bounded between these two points (see Fig. 2.7). The potential has a minimum at $x = 0$, where $V(0) = -D$, so the minimum value of E is $-D$.

We use Eq. (1.46) to obtain the motion:

$$t - t_0 = \sqrt{\frac{m}{2}} \int_{x_0}^{x(t)} \frac{dx}{\sqrt{E - D(e^{-2\alpha x} - 2e^{-\alpha x})}} \equiv \sqrt{\frac{m}{2}} I, \tag{2.68}$$

where x_0 is the value of x at time $t = t_0$. To perform the integration, multiply the numerator and denominator in the integrand by $e^{\alpha x}$ and change variables to $e^{\alpha x} = u$. Then I becomes

$$I = \frac{1}{\alpha} \int_{u_0}^{u} \frac{du}{\sqrt{Eu^2 + 2Du - D}}.$$

This is a tabulated integral. For $E < 0$ the indefinite integral is given as

$$I = -\frac{1}{\alpha\sqrt{|E|}} \sin^{-1}\left[\frac{2Eu + 2D}{\sqrt{4D(D - |E|)}}\right].$$

When this is substituted into Eq. (2.68), the solution for $x(t)$ becomes

$$E < 0, \quad x(t) = \frac{1}{\alpha}\ln\left\{\frac{D - \sqrt{D(D - |E|)}\sin(\alpha t\sqrt{2|E|/m} + C)}{|E|}\right\},$$

where C is a constant of integration, which can be given in terms of E and x_0. In this $E < 0$ case it is

$$C = \sin^{-1}\left[\frac{D - |E|e^{\alpha x_0}}{\sqrt{D(D - |E|)}}\right].$$

Observe that the motion is periodic. As we mentioned above, however, unlike the case of the harmonic oscillator, the period P in the Morse potential depends on the energy:

$$P = \frac{\pi}{\alpha}\sqrt{\frac{2m}{|E|}}.$$

The solutions for $E \geq 0$ are obtained similarly. They are

$$E > 0, \quad x(t) = \frac{1}{\alpha}\ln\left\{\frac{\sqrt{D(D + E)}\cosh(\alpha t\sqrt{2E/m} + C) - D}{E}\right\}$$

and

$$E = 0, \quad x(t) = \frac{1}{\alpha}\ln\left\{\frac{1}{2} + \frac{D\alpha^2}{m}(t + C)^2\right\}.$$

In both solutions C is again a constant of integration, but its expression in terms of E and x_0 is different in the three energy regions.

It is interesting to note that for large times $x(t)$ grows exponentially in t if $E > 0$ but only logarithmically (i.e., more slowly) if $E = 0$.

For Part (b) see Fig. 2.7.

2.3.2 THE KEPLER PROBLEM

We now return to the Kepler problem, already mentioned in Section 1.5.1, for which $V(r) = -\alpha/r$, with $\alpha = G\mu M$. The problem is named for the three laws of planetary motion formulated by Johannes Kepler early in the seventeenth century on the basis of careful observations by Tycho Brahe. These laws are: (K1) the orbit of each planet is an ellipse with the Sun at one of its foci, (K2) the position vector from the Sun to the planet sweeps out equal areas in equal times, and (K3) the period T of each planet's orbit is related to its semimajor axis R so that T^2/R^3 is the same for all of the planets. In reality these three laws are only approximately true, for each planet's orbit is perturbed by the presence of the other planets. The same potential, now with $\alpha = e_1 e_2$, is called the Coulomb potential, for it is the potential energy of two opposite charges e_1 and e_2 attracting through the Coulomb force.

Although we described the orbits of this dynamical system in Chapter 1, we did not actually derive them. Here we show how to derive them, that is, how to find the function $r(\theta)$. Let the potential be $V(r) = -\alpha/r$ with arbitrary $\alpha > 0$. The needed equation for $r(\theta)$ can be obtained by rewriting (2.63) in terms of θ rather than t. To do this, (2.62) is used to replace the derivatives with respect to t by derivatives with respect to θ:

$$\frac{d}{dt} = \dot{\theta}\frac{d}{d\theta} = \frac{l}{\mu r^2}\frac{d}{d\theta} \tag{2.69}$$

and

$$\frac{d^2}{dt^2} = \frac{l^2}{\mu^2 r^2}\frac{d}{d\theta}\left(\frac{1}{r^2}\frac{d}{d\theta}\right).$$

Things become simpler if $u = 1/r$ is taken as the dependent variable instead of r itself. Then $dV/dr \equiv \alpha/r^2$ becomes αu^2 and Eq. (2.63) is

$$\frac{d^2u}{d\theta^2} + u = \frac{\alpha\mu}{l^2}, \tag{2.70}$$

which has the well-known form of the harmonic oscillator equation and can be solved at once. The solution is

$$u \equiv \frac{1}{r} = \frac{\alpha\mu}{l^2}[\epsilon\cos(\theta - \theta_0) + 1], \tag{2.71}$$

where ϵ (called the eccentricity) and θ_0 are constants of integration.

We consider only $\epsilon \geq 0$. This involves no loss of generality, since the sign of ϵ can be changed by replacing θ by $\theta + \pi$ (i.e., by reorienting the coordinate system).

First consider $\epsilon = 0$. Then r is fixed as θ varies, so the orbit is a circle.

Next, consider $0 < \epsilon < 1$. Then the expression in square brackets in (2.71) is always positive, maximal when $\theta = \theta_0$, and minimal when $\theta - \theta_0 = \pi$. That means that $r_{\min}$ (*perihelion*) is

$$r(\theta_0) = r_{\min} = \frac{l^2}{\alpha\mu(1 + \epsilon)} \tag{2.72}$$

and $r_{\max}$ (*aphelion*) is

$$r(\pi + \theta_0) = r_{\max} = \frac{l^2}{\alpha\mu(1 - \epsilon)}.$$

It is seen that θ_0 is the angle at perihelion. Reorient the coordinate system so that $\theta_0 = \pi$, and then $r_{\min} = r(\pi)$ and $r_{\max} = r(0)$. The orbit is an ellipse. Note that we have not yet found how ϵ is related to physical properties that determine the orbit (e.g., the initial conditions).

Suppose now that $\epsilon > 1$. Then there are two values of θ for which $\epsilon \cos(\theta - \theta_0) = -1$ and therefore for which $1/r = 0$. The orbit is defined only if r is finite and nonnegative, that is, for those values of θ for which $\epsilon \cos(\theta - \theta_0) \geq -1$. Again, choose $\theta_0 = \pi$, and then $\epsilon \cos \theta \leq 1$. Let Θ be one limiting angle of the orbit: $\epsilon \cos \Theta = 1$, $\Theta < \pi$. Then the orbit is restricted to $\Theta < \theta < 2\pi - \Theta$. This orbit is one branch of a hyperbola. As in the case of the elliptic orbit, perihelion occurs at $\theta = \pi$ and is again given by (2.72). Now, however, there is no aphelion: the orbit fails to close (Fig. 2.8).

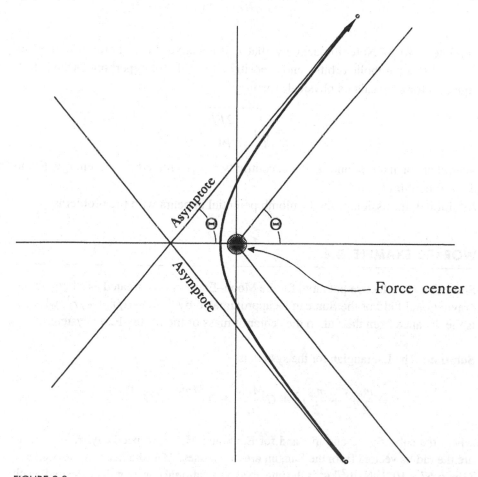

FIGURE 2.8
A hyperbolic Kepler orbit ($\epsilon > 1$). Any line through the force center at an angle greater than Θ (see the text) will intersect the orbit at some finite distance r. The half-angle between the asymptotes is also Θ.

Finally, suppose that $\epsilon = 1$. Then r is finite for all $\theta \neq \pi$, growing without bound as θ approaches π. The orbit in this case is a parabola. Again, choose $\theta_0 = \pi$, and then perihelion is as usual at $\theta = \pi$ and given by (2.72); the parabola opens in the direction of $\theta = 0$.

All of the orbits are conic sections. If one ignores perturbations, the orbits of the planets all close: they are ellipses. It is left to the problems to show that in all cases, even for parabolic and hyperbolic orbits, the Sun is at a focus of the conic section.

What remains is to relate the eccentricity ϵ to physical properties of the orbit. For this purpose consider the orbit at perihelion. When $r = r_{min}$ the planet is moving with some velocity v perpendicular to its position vector relative to the Sun. Its angular momentum is therefore $l = r_{min}\,\mu v$, and with the aid of (2.72) its kinetic energy $T = \frac{1}{2}\mu v^2$ is found to be

$$T = \frac{\mu \alpha^2 (\epsilon + 1)^2}{2l^2}.$$

Since $V = -\alpha / r_{min}$, the total energy $E = T + V$ can be written

$$E = \frac{\mu \alpha^2 (\epsilon^2 - 1)}{2l^2},$$

which relates ϵ to E. Note, incidentally, that E is negative if $\epsilon < 1$ (for closed orbits), is zero if $\epsilon = 1$ (for parabolic orbits), and is positive if $\epsilon > 1$ (for hyperbolic orbits). Solving this equation for ϵ reveals its physical meaning:

$$\epsilon = \sqrt{1 + \frac{2El^2}{\mu \alpha^2}}. \tag{2.73}$$

It is seen that for fixed μ and α, the eccentricity is determined by the energy E and the angular momentum l.

Additional discussion of the Coulomb potential appears with the problems.

WORKED EXAMPLE 2.4

Show that the effective potential for the Moon–Earth system treated as a body in the gravitational field of the Sun can be approximated by $V = -\alpha/r - \epsilon/r^3$, where r is the distance from the Sun to the center of mass of the Moon–Earth system.

Solution. The Lagrangian for the system is

$$L = \frac{m_e}{2}\dot{r}_e^2 + \frac{m_m}{2}\dot{r}_m^2 + G\frac{Mm_e}{r_e} + G\frac{Mm_m}{r_m} + G\frac{m_e m_m}{|r_e - r_m|},$$

where the subscripts e and m stand for Earth and Moon, respectively, r_j, j = e, m, are the radius vectors from the Sun, m_j are the masses, M is the mass of the Sun, and $G = 6.67 \times 10^{-11}$ N·m²/kg² is the universal gravitational constant. The desired result will be obtained by writing an approximate form of this Lagrangian.

The center of mass of the Moon–Earth system and the separation between the Earth and the Moon are

$$\mathbf{r} = \frac{m_e \mathbf{r}_e + m_m \mathbf{r}_m}{m_e + m_m}, \quad \mathbf{R} = \mathbf{r}_e - \mathbf{r}_m,$$

so that $\mathbf{r}_j = \mathbf{r} + \mathbf{R}_j$, where $R_j = \mu R / m_j$ is the distance from the Moon–Earth center of mass to m_j (here $\mu \equiv m_e m_m / (m_e + m_m) \approx m_m$, is the reduced mass). Then the kinetic energy can be put in the form (this is largely the point of introducing the reduced mass)

$$\frac{m_e}{2} \dot{r}_e^2 + \frac{m_m}{2} \dot{r}_m^2 = \frac{1}{2} m \dot{r}^2 + \frac{1}{2} \mu \dot{R}^2,$$

where $m = m_e + m_m$.

The Lagrangian is now

$$L = \frac{1}{2} m \dot{r}^2 + \frac{1}{2} \mu \dot{R}^2 + G \frac{M m_e}{r_e} + G \frac{M m_m}{r_m} + G \frac{m_e m_m}{R},$$

and what remains is to write it entirely in terms of r and R, that is to alter the first two terms of the potential energy

$$G \frac{M m_e}{r_e} + G \frac{M m_m}{r_m}.$$

This is done by approximating the $1/r_j$. From the definition $\mathbf{r}_j = \mathbf{r} + \mathbf{R}_j$ we have

$$r_j = \sqrt{r^2 + 2\mathbf{r} \cdot \mathbf{R}_j + R_j^2}.$$

Now expand $1/r_j$ in powers of R_j/r (note that $r \gg R_j$). Keeping terms only up to the quadratic (this involves the first three terms in the binomial expansion) yields

$$\frac{1}{r_j} = \frac{1}{r} - \frac{1}{2r^3} \left[2\mathbf{r} \cdot \mathbf{R}_j + R_j^2 - \frac{3}{r^2} (\mathbf{r} \cdot \mathbf{R}_j)^2 \right] + \cdots.$$

Then

$$\frac{m_e}{r_e} + \frac{m_m}{r_m} \approx \frac{m}{r} - \frac{1}{2r^3} \left[2\mathbf{r} \cdot (m_e \mathbf{R}_e + m_m \mathbf{R}_m) + m_m R_m^2 + m_e R_e^2 \right]$$

$$+ \frac{3}{2r^5} [m_e (\mathbf{r} \cdot \mathbf{R}_e)^2 + m_m (\mathbf{r} \cdot \mathbf{R}_m)^2].$$

Now use $m_e \mathbf{R}_e + m_m \mathbf{R}_m = 0$ and $m_j R_j^2 = \mu^2 R^2 / m_j$, and $m_e R_e^2 \ll m_m R_m^2 \approx m_m R^2 \approx \mu R^2$ to obtain

$$\frac{m_e}{r_e} + \frac{m_m}{r_m} \approx \frac{m}{r} - \frac{\mu}{2r^3} \left[\frac{3(\mathbf{r} \cdot \mathbf{R})^2}{r^2} - R^2 \right].$$

Finally, the approximate Lagrangian becomes (we write μ in place of $m_{\rm m}$)

$$L = \frac{1}{2}m\dot{r}^2 + G\frac{Mm}{r} + \frac{1}{2}\mu\dot{R}^2 + G\frac{m_{\rm e}\mu}{R} + \delta L,$$

where

$$\delta L = -G\frac{M\mu}{2r^3}\left[\frac{3(\mathbf{r}\cdot\mathbf{R})^2}{r^2} - R^2\right].$$

If δL were zero, the r and R parts of the system would be independent, and the r part would simply be the Kepler problem for the Moon–Earth system in the field of the Sun; δL is a perturbation. To see how small it is, estimate it as $\delta L \approx -GM\mu R^2/r^3$ and compare it with GMm/r. The relevant numbers are $m \approx 5.98 \times 10^{24}$ kg (the mass of the Earth, approximately that of Earth plus Moon), $\mu \approx 7.35 \times 10^{22}$ kg (the reduced mass approximately the mass of the Moon), $r \approx 150 \times 10^6$ km, $R \approx 3.84 \times 10^5$ km. Then

$$\frac{\delta L r}{GMm} \approx 8 \times 10^{-8},$$

so that δL is indeed a small perturbation. If it is assumed that the Moon's orbit is circular of radius R and in the plane of the Earth's orbit, then δL can be expressed as

$$\delta L = -\frac{2\epsilon}{r^3}(3\cos^2\vartheta - 1),$$

where $\epsilon = \frac{1}{4}M\mu R^2$ and ϑ is the angle between $\mathbf{r}$ and $\mathbf{R}$ (ϑ runs through 2π in 29.5 days). The average value of $\cos^2\vartheta$ over one period is $1/2$, and this can be used to obtain an average $\delta L = -\epsilon/r^3$, which is the desired result.

2.3.3 BERTRAND'S THEOREM

We now return to the general central potential problem. There are several ways in which the general case differs from the Coulomb (or Kepler) problem. One difference is that in the Coulomb problem every bounded orbit is *closed*, that is, it is retraced after some integer number n of revolutions (in Coulomb motion, in particular, $n = 1$). In terms of the velocity phase space, a closed orbit is one that returns to its initial point in $\mathbf{TQ}$ in some finite time. For the central potential this means that after some integer number n of complete rotations (i.e., when θ changes by $2\pi n$) both r and $\dot{r}$ return to their initial values. Figure 2.9(a) shows an example (in $\mathbb{Q}$) of an orbit that is closed, and Fig. 2.9(b) shows one that is not.

Are there other central potentials all of whose orbits are closed (or at least all of whose bounded orbits are closed), or is the Coulomb potential unique? The question was answered in 1873 by Bertrand in a theorem that asserts that the only central power-law potentials all of

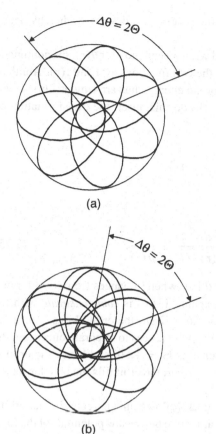

(a)

(b)

FIGURE 2.9

(a) An orbit that closes after seven periods of the r motion. In this example $\Theta = \frac{7}{2}\pi$. The radii of the inner and outer circles are $r_{\min}$ and $r_{\max}$. The orbit lies entirely within the annular region between them. (b) An orbit that fails to close. We do not try to show a complete orbit that fails to close, for it is dense in the annular region.

whose bounded orbits are closed are $V(r) = -\alpha/r$, $\alpha > 0$ (the Coulomb potential) and $V(r) = \beta r^2$, $\beta > 0$ (the isotropic harmonic oscillator). The Coulomb potential is, therefore, somewhat special in this sense. Bertrand's theorem does not state that no other potentials have closed orbits (see, for example, Problem 14), only that these are the only ones *all* of whose bounded orbits are closed. We sketch a proof below, but first we need some preliminaries.

The proof will be based on comparing the period in r (since we are interested only in bounded orbits, r must vary periodically) with the time it takes θ to change by 2π. If the orbit is closed, there must be some integer number of revolutions after which both r and $\dot{r}$ return to their initial values. It follows from conservation of angular momentum that $\dot{\theta}$ does as well, so the system returns to the initial point in $\mathbf{TQ}$. At a maximum or minimum value of r (at aphelion or perihelion) $\dot{r}$ vanishes, so if counting starts there, all that need be verified for closure is that r returns to its initial maximum or minimum value after θ has changed by an integer multiple of 2π: there must be some integer number m of periods of r in which θ changes by $2\pi n$.

Now, in analogy with Eq. (1.48) for one-dimensional motion, the time to go from some r_0 to r is given by

$$t(r) = \sqrt{\frac{\mu}{2}} \int_{r_0}^{r} \frac{dr}{\sqrt{E - V(r)}}. \tag{2.74}$$

From this, one could calculate the period and use it in the scheme just described, thereby determining whether the orbit is closed.

An easier way, however, is to eliminate the time and write an equation for $\theta(r)$, the configuration-space orbit itself. This can be done by composing the two functions $\theta(t)$ and $t(r)$. Another, simpler approach, is to replace t by θ in the equation for the energy, thus arriving at a first-order differential equation for $r(\theta)$ or, equivalently, for $\theta(r)$. To do this, insert the transformation of Eq. (2.69) into (2.66), which becomes

$$E = \frac{l^2}{2\mu r^4}\left(\frac{dr}{d\theta}\right)^2 + \frac{l^2}{2\mu r^2} + V,$$

and solve for θ:

$$\theta(r) = \sqrt{\frac{l^2}{2\mu}} \int_{r_0}^{r} \frac{dr}{r^2\sqrt{E - V(r) - l^2/(2\mu r^2)}}. \tag{2.75}$$

This integral can now be used to find the change $\Delta\theta$ in θ when r changes from its minimum value $r_{\min}$ to its maximum $r_{\max}$ and then back to $r_{\min}$ again. If this is *commensurate* with 2π (i.e., if there exist positive integers n and m such that $m\Delta\theta = 2\pi n$) the orbit closes. Since the orbit is symmetric about $r_{\min}$ (or, for that matter, about $r_{\max}$), it suffices to integrate once from the minimum to the maximum. The apsidal angle so obtained is $\Theta = \frac{1}{2}\Delta\theta$, and hence the orbit is closed if $\Theta = \pi n/m$ for some integers n and m. If $\Theta \neq \pi n/m$, the orbit fills the annular region of Fig. 2.9 densely.

We now turn to Bertrand's theorem. The proof will go as follows: first it treats circular orbits, of fixed radius r_c. These are easily shown to exist for all attractive power-law potentials of the form $V = \alpha r^k$ (Problem 14). Next, it turns to small deviations Δr from r_c and expands to first order in Δr. That will show that the perturbed orbits close only for $k \geq -2$. (The condition $k \geq -2$ is also the condition for stability of the circular orbit, treated in Problem 14.) Finally, calculations for other limiting orbits will show that only for $k = -1$ and $k = 2$ are all of the orbits closed. The detailed proof, though algebraically straightforward, is lengthy, so we will just give an outline, based on Arnol'd's treatment (Arnol'd, 1988, Section 2.8D; see also Goldstein, 1981, p. 601).

In the first step the integral of (2.75) is taken from $r_{\min}$ to $r_{\max}$ with $V = \alpha r^k$; the variable of integration is changed from r to $u \equiv 1/r$. Note that the potential is attractive only if $\alpha k > 0$, so we treat only α and k of the same sign. The integral now becomes ($u_{\min}$ is associated with $r_{\max}$ and vice versa):

$$\Theta = \sqrt{\frac{l^2}{2\mu}} \int_{u_{\min}}^{u_{\max}} \frac{du}{\sqrt{E - \mathcal{U}(u)}}, \tag{2.76}$$

where

$$\mathcal{U}(u) = V(1/u) = \frac{l^2}{2\mu}u^2 + \alpha u^{-k}$$

is V written as a function of u. It follows from Problem 14 that $u_c = 1/r_c$ is

$$u_c = \left(\frac{\alpha k \mu}{l^2}\right)^{1/(k+2)}$$

and that $\mathcal{U}'(u_c) = 0$. Now expand $\mathcal{U}(u)$ in a Taylor's series about $\mathcal{U}(u_c)$ keeping only the (quadratic) term of lowest order in $\Delta u = u - u_c$. The roots of the radical in (2.76), and hence the limits of the integral, can then be found and the integral in Δu performed by a trigonometric substitution. The result is

$$\Theta = \frac{\pi}{\sqrt{2+k}}. \tag{2.77}$$

It was seen above that if the orbit closes, Θ is a rational multiple of π, and hence so is the expression on the right-hand side of (2.77). Now, as the energy and/or momentum change, Θ changes, if at all, continuously with respect to both E and l. But if Θ changes continuously, it must pass through values that are irrational multiples of π, corresponding to orbits that do not close. We are looking for values of k all of whose orbits are closed, that is, for *all* values of E and l, so Θ cannot change with changes in E and l. Because all orbits can be reached from any given one by changing the values of E and l, it follows that if for a given value of k all orbits close, *they all have the same value of* Θ, satisfying Eq. (2.77). (We thank Alain Chenciner for explaining this to us in a private communication.) Finally, $\Theta = \pi n/m$ and Eq. (2.77) together imply that k is of the form $k = -2 + m^2/n^2$ for integer n and m, so $k \geq -2$.

REMARK: When $k = -1$, that is, for the Coulomb potential, $\Theta = \pi$. When $k = 2$, that is, for the harmonic oscillator, $\Theta = \frac{1}{2}\pi$. Their orbits close also when they are far from the circular ones. Interestingly enough, in both cases the general closed orbits are elliptical. □

The situation is more complicated for orbits that are not nearly circular. Positive and negative k must be treated separately. For positive k consider the limit of high energy E. In that limit $\Theta = \pi/2$ for the general orbit. Indeed, if one writes $u = yu_{max}$, the integral in (2.76) becomes

$$\Theta = \sqrt{\frac{l^2}{2\mu}} \int_{y_{min}}^{1} \frac{dy}{\sqrt{\mathcal{Y}(1) - \mathcal{Y}(y)}}, \tag{2.78}$$

where

$$\mathcal{Y}(y) = y^2 \left\{ \frac{l^2}{2\mu} + \alpha(yu_{max})^{-(k+2)} \right\}.$$

In the limit of high E the second term can be ignored, and then the integral reduces to $\pi/2$. But according to Eq. (2.77) this must be equal to $\pi/\sqrt{2+k}$. Thus $k = 2$.

For negative k consider the limit as E approaches zero from below. Again, set $u = yu_{max}$, where now

$$(u_{max})^{(k+2)} = \frac{2\mu\alpha}{l^2}.$$

The integral now becomes

$$\Theta = \int_{0}^{1} \frac{dy}{\sqrt{y^{-k} - y^2}}.$$

The substitution $z = y^{(2+k)/2}$ makes it easy to obtain

$$\Theta = \frac{\pi}{2+k}.$$

But according to Eq. (2.77) this must be $\pi/\sqrt{2+k}$. Thus $k = -1$.

This completes the outline of the proof of Bertrand's theorem.

2.4   THE TANGENT BUNDLE T$\mathbb{Q}$

2.4.1   DYNAMICS ON T$\mathbb{Q}$

VELOCITIES DO NOT LIE IN $\mathbb{Q}$

We now return to a discussion of Lagrange's equations in the form of Eqs. (2.28) or
(2.29), namely

$$\frac{d}{dt}\frac{\partial L}{\partial \dot{q}^{\alpha}} - \frac{\partial L}{\partial q^{\alpha}} = 0 \tag{2.28}$$

or

$$\frac{\partial^2 L}{\partial \dot{q}^{\beta}\partial \dot{q}^{\alpha}}\ddot{q}^{\beta} + \frac{\partial^2 L}{\partial q^{\beta}\partial \dot{q}^{\alpha}}\dot{q}^{\beta} + \frac{\partial^2 L}{\partial t\partial \dot{q}^{\alpha}} - \frac{\partial L}{\partial q^{\alpha}} = 0. \tag{2.29}$$

So far, although we have paid lip service to velocity phase space, we have treated these
equations as a set of second-order differential equations on the configuration manifold $\mathbb{Q}$.
In this section we extend the idea of velocity phase space to the velocity phase *manifold* T$\mathbb{Q}$.
In the process, we will exhibit the importance of manifolds by showing how the properties
of general dynamical systems are reflected in the manifold structure of T$\mathbb{Q}$.

Some of these properties are illustrated by comparing two different systems, each
consisting of two material particles. In the first system the two particles move in ordinary
Euclidean 3-space $\mathbb{E}^3$. The difference $\mathbf{v}_1 - \mathbf{v}_2$ between their velocities is a vector in $\mathbb{E}^3$
(which can be moved to any point in $\mathbb{E}^3$), and it has a clear physical meaning: it is the
relative velocity of the two points. In the second system the two particles move on the
two-dimensional surface of a sphere $\mathbb{S}^2$. The velocity vector of each particle is tangent to
the sphere: it leaves the sphere and reaches out into the $\mathbb{E}^3$ in which the sphere is imbedded.
All possible velocity vectors at any point of the sphere lie in the plane tangent to the sphere
at that point: they span that plane (the plane is a two-dimensional vector space). Much
as we might like to discuss motion on $\mathbb{S}^2$ entirely in terms of objects on $\mathbb{S}^2$ itself, we are
forced off $\mathbb{S}^2$ onto the tangent plane, in fact onto the set of all tangent planes at all points
of the sphere. And even that is not enough, for the difference between the velocity vectors
at two different points does not lie in either tangent plane, and although this difference
is the relative velocity of the two points in $\mathbb{E}^3$, it represents nothing physical that can be
described simply in terms of $\mathbb{S}^2$ itself. This is certainly different from the first two-particle
system on $\mathbb{E}^3$.

In this example it is not very important to treat the motion entirely in terms of $\mathbb{S}^2$; it can
be treated in the $\mathbb{E}^3$ in which $\mathbb{S}^2$ is imbedded. In many other dynamical systems, however,

no such higher-dimensional space is readily available. For instance the double pendulum of Fig. 2.3(e) has no obvious physical space in which to imbed the four-dimensional configuration manifold. Even though it can be shown that $\mathbb{S}^2 \times \mathbb{S}^2$, like all the manifolds we deal with in this book, can be imbedded in some $\mathbb{E}^n$, this $\mathbb{E}^n$ is not obvious and has no particular physical meaning. Even its dimension n is not obvious.

For reasons like this it is important to keep the general discussion intrinsic to the configuration manifold itself, without bringing in spaces of higher dimension except in ways that grow out of $\mathbb{Q}$ itself. It will be seen that $\mathbf{TQ}$ is constructed in this way, from $\mathbb{Q}$ itself.

On a vector space, dynamics is relatively easy to deal with, mainly because velocity vectors are similar to position vectors. On manifolds, however, problems are encountered of the kind that arise in the example of $\mathbb{S}^2$, and vectors have to be discussed in terms of tangent spaces, the analogs of the tangent planes of $\mathbb{S}^2$. As on $\mathbb{S}^2$, there is no immediately evident way to compare vectors at different points, so relative velocities present problems. One of the first hurdles in dealing with dynamics on manifolds is therefore finding a consistent way to treat tangent vectors. Another difficulty will be finding a consistent way to deal with a question we already encountered in some of the configuration manifold examples of Section 2.1: it is in general impossible to find a coordinate system that will cover a manifold, unlike a vector space, without multiple-valued points.

We now proceed to discuss such topics.

TANGENT SPACES AND THE TANGENT BUNDLE

Even while we were thinking of Lagrange's equations as second-order differential equations on $\mathbb{Q}$, the principal function we were using is a function not on $\mathbb{Q}$. The Lagrangian $L(q, \dot{q}, t)$ depends not only on the q^α, but also on the $\dot{q}^\alpha$. It is a function on a larger manifold that is called $\mathbf{TQ}$. If t is thought of as a parameter rather than another variable, the number of coordinates on $\mathbf{TQ}$ is $2n$, the q^α and $\dot{q}^\alpha$. Because it involves both the generalized velocities and the generalized coordinates, $\mathbf{TQ}$ is the analog of velocity phase space. It is not a vector space, and we will call it the *velocity phase manifold*. It is generally called the *tangent bundle* or *tangent manifold* of $\mathbb{Q}$, and thus the Lagrangian is a function on the tangent bundle $\mathbf{TQ}$.

The reason $\mathbf{TQ}$ is called the *tangent* bundle is that the generalized velocities $\dot{q}^\alpha$ lie along the tangents (recall the discussion of space curves at the end of Section 1.1) to all possible trajectories, as in the $\mathbb{S}^2$ example. The tangent bundle $\mathbf{TQ}$ is obtained from $\mathbb{Q}$ by adjoining to each point $q \in \mathbb{Q}$ the vector space, called the *tangent space* $\mathbf{T}_q\mathbb{Q}$, of all possible velocities at q, all tangent to $\mathbb{Q}$ at that point. The components of the vectors in $\mathbf{T}_q\mathbb{Q}$ are all possible values of the $\dot{q}^\alpha$. Then $\mathbf{TQ}$ is made up of $\mathbb{Q}$ plus all the $\mathbf{T}_q\mathbb{Q}$. This will be discussed in more detail later, and it will be shown how this can be done without introducing a higher-dimensional space in which to imbed $\mathbb{Q}$, in a way that is intrinsic to $\mathbb{Q}$ itself.

Since $\mathbf{TQ}$ is the manifold analog of velocity phase space, it has many of its properties. Indeed, the two-dimensional velocity phase space in the examples of Chapter 1 is actually the tangent bundle of the line. In $\mathbf{TQ}$ each point is specified by the set $(q, \dot{q})$,

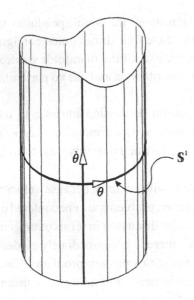

FIGURE 2.10
The tangent bundle $\mathbf{T}\mathbb{S}^1$ of the circle $\mathbb{S}^1$, constructed by attaching an infinite line to each point of $\mathbb{S}^1$. The coordinate measured along the circle is θ; the coordinate measured along the line (the fiber) is $\dot\theta$.

just as each point in those examples was specified by the set (x, v). (We will use $(q, \dot q)$ some of the time to designate the collection of coordinates $\{q^\alpha, \dot q^\alpha\}$ and some of the time abstractly to designate the point in $\mathbf{T}\mathbb{Q}$ regardless of coordinates.) Through each point $(q, \dot q)$ of $\mathbf{T}\mathbb{Q}$ passes just one solution of the equations of motion, one phase trajectory, and therefore phase portraits can be constructed on $\mathbf{T}\mathbb{Q}$ in exact analogy with velocity phase space.

Just about the only easily visualized nontrivial $\mathbf{T}\mathbb{Q}$ is the tangent bundle of the circle $\mathbb{S}^1$, the $\mathbb{Q}$ of Fig. 2.3(b). To each point θ of $\mathbb{S}^1$ adjoin all possible values of $\dot\theta$, which runs from $-\infty$ to $+\infty$. This attaches an infinite line running in both directions to each point of $\mathbb{S}^1$, generating a cylinder (Fig. 2.10). This cylinder, with coordinates $\theta, \dot\theta$, is $\mathbf{T}\mathbb{S}^1$, the $\mathbf{T}\mathbb{Q}$ manifold of the plane pendulum. The line attached to θ is $\mathbf{T}_\theta\mathbb{S}^1$ and is called the *fiber above* θ.

This example is not as simple as it looks. We have overlooked the question of how the fibers over points θ on $\mathbb{S}^1$ are related to each other away from the initial $\mathbb{S}^1$ circle. Even if two values of θ are very close, it is not immediately evident that their $\mathbf{T}_\theta\mathbb{Q}$ lines must remain close throughout their lengths. We return to this question later.

The Lagrangian can now be understood as a function on $\mathbf{T}\mathbb{Q}$: it assigns to each point $(q, \dot q)$ of $\mathbf{T}\mathbb{Q}$ a certain value. That value is independent of coordinates on $\mathbf{T}\mathbb{Q}$. Two Lagrangians L and L' that are related through a change of coordinates as in Eq. (2.35) merely express that value in different ways. In this sense the Lagrangian is itself coordinate independent (i.e., it is a *scalar function*). In an intrinsic description that does not involve coordinates, there is only the Lagrangian function L itself, not its various forms in different coordinate systems.

This can be made clearer by defining what we mean by a function. A *function* is a *map* from its *domain* onto its *range* or *codomain*. For a Lagrangian that is time independent the domain is $\mathbf{T}\mathbb{Q}$ and the range is the real numbers $\mathbb{R}$. One writes $L : \mathbf{T}\mathbb{Q} \to \mathbb{R}$. When the Lagrangian depends on time its domain is $\mathbf{TQ} \times \mathbb{R}$, where $\mathbb{R}$ is the infinite time line parametrized by the time t. The range is still $\mathbb{R}$, this $\mathbb{R}$ consisting of the real-number values taken on by L, as in the time-independent case.

The tangent bundle $\mathbf{TS}^2$ of the spherical pendulum, the sphere $\mathbb{S}^2$ with its tangent planes at all points, is a four-dimensional object which is not easily visualized.

LAGRANGE'S EQUATIONS AND TRAJECTORIES ON TQ

On $\mathbf{TQ}$ (as opposed to $\mathbb{Q}$) Lagrange's equations are a set of first-order (as opposed to second-order) differential equations: their solution in generalized coordinates are the $2n$ functions $q^\alpha(t)$, $\dot{q}^\alpha(t)$, with the $\ddot{q}^\beta$ that appear in (2.29) interpreted as the derivatives of $\dot{q}^\beta$ with respect to the time t. It may seem that there are $2n$ functions to be found from Lagrange's n equations, but there are actually n other equations, also of first order:

$$\dot{q}^\alpha = \frac{dq^\alpha}{dt}. \qquad (2.79)$$

Hence there are $2n$ equations for the $2n$ functions. The advantages of viewing the dynamical system as a set of first-order equations on $\mathbf{TQ}$ are the ones discussed in connection with velocity phase space in Chapter 1.

As a demonstration we construct the phase portrait of the plane pendulum of Fig. 2.3(b) on its tangent bundle. The phase portrait is similar to Fig. 1.5(b) except that the phase manifold is now a cylinder instead of a plane: the lines $\theta = \pi$ and $\theta = -\pi$ are identified (recall that q is called θ in this example). The potential energy is a multiple of $\cos\theta$, as it was in the example of Eq. (1.49), and the kinetic energy is a multiple of $\dot{\theta}^2$:

$$E = \frac{1}{2}ml^2\dot{\theta}^2 - mgl\cos\theta.$$

This is the same as Eq. (1.50) but in different notation. Thus wherever this expression for E is valid (i.e., for the range of θ from $-\pi$ to π) the phase portrait will be the same as the one obtained from Eq. (1.50). Figure 2.11 shows this phase portrait drawn on two views of the cylinder. Elliptic and hyperbolic points of equilibrium and a separatrix are recognized by comparing this figure with that of Chapter 1. The elliptic point corresponds to the pendulum hanging straight down, and the hyperbolic one to it balancing vertically. The separatrix consists of three distinct orbits.

Consider a fixed value of θ. As would be true if $\mathbb{Q}$ were an infinite line, on the circular configuration manifold $\mathbb{Q} = \mathbb{S}^1$ there are many trajectories passing through θ, namely all those that have different velocities $\dot{\theta}$ at that θ. Some of these trajectories oscillate back and forth on $\mathbb{S}^1$ (if the $\dot{\theta}$ is small enough), and others, for large enough $\dot{\theta}$, go completely around $\mathbb{S}^1$. But when $\dot{\theta}$ is also given, the trajectory is unique and the rest of the motion is determined. Each pair $(\theta, \dot{\theta})$ on $\mathbf{TQ}$ lies on and determines a unique trajectory, and no two different trajectories pass through the same point of $\mathbf{TQ}$.

This general geometric property of $\mathbf{TQ}$ is what makes it so useful: it separates the trajectories from each other and is a reflection of the first-order nature of the differential

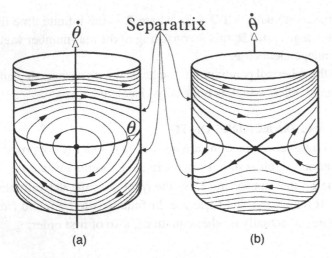

FIGURE 2.11
The phase portrait of the pendulum drawn on its tangent manifold $\mathbf{TQ}$. The point at the origin in (a) is the stable equilibrium, or elliptic point, representing the pendulum hanging motionless straight down, i.e., $(\theta, \dot{\theta}) = (0, 0)$. The two separatrices cross in (b) at the unstable equilibrium, or hyperbolic point, representing the pendulum pointing straight up, i.e., $(\theta, \dot{\theta}) = (\pi, 0)$ or, equivalently, $(\theta, \dot{\theta}) = (-\pi, 0)$ (they are the same point).

equations on $\mathbf{TQ}$. The initial conditions for a solution of $2n$ first-order differential equations are the $2n$ initial values $(q_0, \dot{q}_0) \equiv \{q^\alpha(0), \dot{q}^\alpha(0)\}$. Thus, given the initial point in $\mathbf{TQ}$, the rest of the trajectory is uniquely determined analytically by the equations of motion or graphically by the phase portrait. The solutions can therefore be written $q^\alpha = q^\alpha(t; q_0, \dot{q}_0)$, $\dot{q}^\alpha = \dot{q}^\alpha(t; q_0, \dot{q}_0)$, where $(q_0, \dot{q}_0)$ is the set of initial values of the generalized coordinates and velocities.

In practice a trajectory is not always specified by naming its initial conditions but by other properties, such as a function $\Gamma(q, \dot{q}, t)$ (e.g., the energy). Giving the initial value of Γ determines a relation among the q_0^α and $\dot{q}_0^\alpha$, and then $2n - 1$ of the initial coordinates on $\mathbf{TQ}$ determine the $2n$th, so not all $2n$ of the initial conditions need be given. A function on $\mathbf{TQ}$ is called a *dynamical variable* (this generalizes the definition of Chapter 1). The initial value of any dynamical variable, each equation of the form $\Gamma(q_0, \dot{q}_0, 0) = C$ with C a constant, obviates the need to know one of the initial coordinates. A trajectory can be specified also by other means, such as the position in $\mathbb{Q}$ at two different times (see Section 3.1). In any way it is done, what must be given are $2n$ independent pieces of information.

There is a geometric way to understand how specifying a dynamical variable reduces by one the number of needed initial conditions. Suppose that $\Gamma(q, \dot{q})$ is a time-independent dynamical variable. Then the equation $\Gamma(q, \dot{q}) = C$ defines a submanifold $\mathbb{M}_{2n-1}$ of dimension $2n - 1$ in $\mathbf{TQ}$, and if C is the initial value of Γ, the motion starts on $\mathbb{M}_{2n-1}$. The only additional needed information is where the initial point lies on this manifold, and that can be specified by giving its coordinates in some $(2n - 1)$-dimensional coordinate system on $\mathbb{M}_{2n-1}$. If Γ is time dependent the only difference is that $\mathbb{M}_{2n-1}$ changes with time.

In conclusion, Lagrange's equations can be transformed to a set of first-order differential equations on $\mathbf{TQ}$ of the form (recall that $\partial^2 L / \partial \dot{q}^\alpha \partial \dot{q}^\beta \neq 0$ by assumption)

$$\frac{dq^\alpha}{dt} = \dot{q}^\alpha, \quad \frac{d\dot{q}^\alpha}{dt} = W^\alpha(q, \dot{q}, t) \tag{2.80}$$

or, by combining the q^α and $\dot{q}^\alpha$ into a single set of $2n$ variables ξ^j, in the form

$$\frac{d\xi^j}{dt} = f^j(\xi, t), \quad j = 1, \ldots 2n. \tag{2.81}$$

This last form will be discussed more fully in the next section. The motion is found by solving the equations for $\xi(t)$ with initial conditions $\xi(t_0)$.

Much of the discussion in this book will be in terms that have been defined so far in this section. In the next subsection we introduce some of the formal properties of manifolds.

2.4.2 TQ AS A DIFFERENTIAL MANIFOLD

DIFFERENTIAL MANIFOLDS

Both $\mathbb{Q}$ and $\mathbf{TQ}$ have been called manifolds without the concept of manifolds ever being defined. In this subsection we define the concept and discuss some of the geometric properties of manifolds. We aim here to make things clearer, but without introducing the rigor that would please a mathematician. For that the interested reader is referred to books about calculus on manifolds. A good place for a physicist to start is Spivak (1968), Crampin and Pirani (1985), or Schutz (1980). For a formal and more demanding treatment of classical mechanics from the geometric point of view, see Abraham & Marsden (1985). Arnol'd (1965) is highly recommended.

We introduce manifolds by analogy with mapmaking. The mapmaker represents the surface of a sphere, part or all of the Earth's surface, on the flat surface of a *map* or *chart*. He uses many techniques: there are conical, Mercator, polar, and many other projections, each in its own way distorting some aspect of the true surface. No one of these projections covers the entire Earth without missing some points, making cuts, representing some points twice, therefore deviating from faithful representation (e.g., the exaggeration of Greenland's size in the Mercator projection). Yet maps are understood for what they are: locally reasonably faithful representations. An *atlas* is a collection of overlapping maps or charts that covers the entire Earth. In an atlas each point on the Earth's surface can be specified by the chart (or charts) on which it appears and Cartesian coordinates that are usually drawn on that chart. Each point is thereby assigned coordinates that are valid locally in one of the charts.

A manifold is a similar construct. An *n-dimensional manifold* $\mathbb{M}$ can be modeled by an n-dimensional hypersurface (e.g., the Earth's surface $\mathbb{S}^2$) in a real Euclidean vector space $\mathbb{E}^m$ (e.g., $\mathbb{E}^3$) whose dimension m is greater than n. In general such a hypersurface can not be parametrized with a single nice Cartesian-like coordinate system, so it is covered with a patchwork of overlapping n-dimensional Cartesian coordinate systems, called *local charts*. (For example, longitude and latitude can be used for local charts, but they can not be used around the poles. They must be supplemented by other charts.) This eliminates the need to know m. The collection of local charts is an *atlas*. In the regions that overlap, coordinate transformations, which are generally nonlinear, carry the coordinates of one chart into those of the other. A special case of $\mathbb{M}$ is $\mathbf{TQ}$, a manifold of dimension $2n$ whose $\mathbb{E}^m$ is not specified.

To put it differently, each local chart is a *map ϕ of a connected neighborhood* $\mathbb{U}$ in $\mathbb{M}$ (written $\mathbb{U} \subset \mathbb{M}$) into $\mathbb{R}^n$. In the mapmaking analogy, for which $n = 2$, the neighborhood $\mathbb{U}$ could be the United States and ϕ could map it onto a flat page containing a Cartesian coordinate grid, which is then the chart $\phi(\mathbb{U})$. If u is a point in $\mathbb{U}$ (written $u \in \mathbb{U}$), then $\phi(u) \in \mathbb{R}^n$ and has coordinates $\phi^j(u)$, $j = 1, \ldots, n$. Every such chart must be one-to-one, mapping each $u \in \mathbb{U}$ to just one point of $\mathbb{R}^n$. That means that each point in $\phi(\mathbb{U})$ is the image of a single point in $\mathbb{U}$ and therefore there exists an inverse map ϕ^{-1} that carries $\phi(\mathbb{U})$ back to $\mathbb{U}$. The inverse ϕ^{-1} can be used to transfer the coordinate grid to $\mathbb{U}$.

In the analogy, each point u in the United States is mapped to one point $\phi(u)$ on the chart, and its coordinates $\phi^j(u)$ can be read off the Cartesian grid on the page. The inverse ϕ^{-1} maps the chart and its Cartesian grid back to the United States on the surface of the Earth.

> **REMARK**: A neighborhood is an open set, which is most easily defined in terms of a closed one. For our purposes, a closed set is one that contains its boundaries, and an open set is one that does not. For instance the interval $0 < x < 1$ is open, for it does not contain its boundaries 0 and 1. The interval $0 \le x \le 1$ is closed. $\quad\square$

In a sense the manifold itself can be identified with the atlas, yielding an alternate model for a manifold $\mathbb{M}$: a manifold is then a collection of points that is the union of a set of denumerable sets $\mathbb{U}_p$ in $\mathbb{R}^n$, each with its own local coordinate system ϕ_p. This can be written $\mathbb{M} = \cup_p \mathbb{U}_p$. In both this atlas model and the hypersurface model we write $(\mathbb{U}, \phi)$ for a chart i.e., for the set or the neighborhood with its local Cartesian coordinate system. We think of $(\mathbb{U}, \phi)$ as the coordinate patch itself, a small piece of $\mathbb{R}^n$.

The global structure of the manifold is determined by the relations between the charts in the regions where they overlap. Suppose that $(\mathbb{U}, \phi)$ and $(\mathbb{V}, \psi)$ are two charts whose neighborhoods intersect: $\mathbb{U} \cap \mathbb{V} \equiv \mathbb{W}$ is not empty. In general the coordinate grids of the two charts will not be identical in $\mathbb{W}$. That is if u is in $\mathbb{W}$ and $\phi(u) = \zeta$, $\psi(u) = \eta$, then in general $\zeta \neq \eta$. But there is a transformation Φ (remember that ζ and η are n-component vectors) that carries one to the other:

$$\zeta = \phi(u) = \phi(\psi^{-1}(\eta)) \equiv \Phi(\eta),$$

where $\Phi = \phi \circ \psi^{-1}$ (ϕ *composed with* ψ^{-1}). Hence Φ maps $(\mathbb{V}, \psi)$ onto $(\mathbb{U}, \phi)$, one piece of $\mathbb{R}^n$ into another. In coordinates this can be written $\zeta^j = \Phi^j(\eta)$.

To visualize this, think of the United States and México together making up $\mathbb{U}$, and the United States and Canada together making up $\mathbb{V}$. Then $\mathbb{W}$ is the United States. For points in the United States, Φ maps their coordinates in the US–Canada map onto its coordinates on the US–México map. We do not go further into the analogy.

Essentially all the manifolds we will deal with in this book are finite dimensional and *differential*. A manifold $\mathbb{M}$ is differential if there exists an atlas all of whose Φ^i functions (and their inverses) for all overlapping charts are differentiable.

> **REMARK**: Except when otherwise stated, by "differentiable function" we mean one possessing derivatives to all orders, what mathematicians call C^∞. $\quad\square$

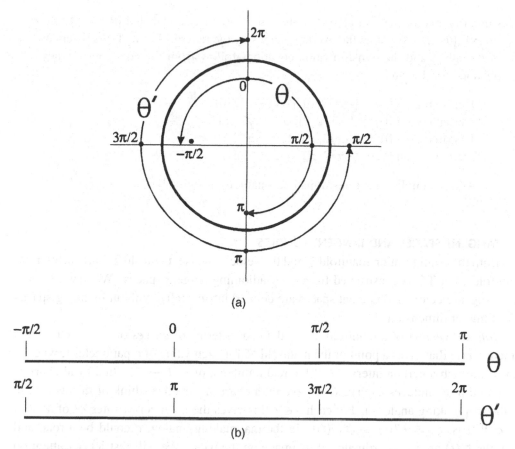

FIGURE 2.12

An atlas for the circle $\mathbb{S}^1$. (a) The circle with coordinates θ and θ' indicated. The overlap regions, namely the second and fourth quadrants, are covered by both coordinates. (b) The atlas itself: two charts on $\mathbb{E}^1$ (the line), one for θ and one for θ'.

WORKED EXAMPLE 2.5

Find an atlas for the torus $\mathbb{T}^2 = \mathbb{S}^1 \times \mathbb{S}^1$. Find transformation equations in the overlap regions.

Solution. Since $\mathbb{T}^2 = \mathbb{S}^1 \times \mathbb{S}^1$, the charts can be constructed by choosing coordinates on the two circles (θ and ϕ). At least two charts are needed for a circle. Start with the θ circle and call the coordinates in the two charts θ and θ'. An example (there are many possible atlases) is shown in Fig. 2.12, with $-\pi/2 < \theta < \pi$ and $\pi/2 < \theta' < 2\pi$ and two overlap regions. In one overlap region (the fourth quadrant) $\theta \in [-\pi/2, 0]$ and $\theta' \in [3\pi/2, 2\pi]$; the transformation equations there are $\theta' = \theta + 2\pi$. In the other overlap region (the second quadrant) $\theta \in [\pi/2, \pi]$ and $\theta' \in [\pi/2, \pi]$; the transformation equations there are $\theta' = \theta$. In this example the transformation equations are all linear. Charts for the torus can be obtained by using these and similar charts for the ϕ circle. Then the four charts may be labeled $\{\theta, \phi\}$, $\{\theta', \phi\}$, $\{\theta, \phi'\}$, $\{\theta', \phi'\}$. We label the overlap regions by the ranges of θ, and ϕ (the ranges of θ' and ϕ' and the transformation equations follow from the equations already obtained for the circle).

Overlap region I: $\theta \in [-\pi/2, 0]$, $\phi \in [-\pi/2, 0]$.
Overlap region II. $\theta \in [-\pi/2, 0]$, $\phi \in [\pi/2, \pi]$.
Overlap region III. $\theta \in [\pi/2, \pi]$, $\phi \in [-\pi/2, 0]$.
Overlap region IV. $\theta \in [\pi/2, \pi]$, $\phi \in [\pi/2, \pi]$.

As an example of the transformation equations, in region II $\theta' = \theta + 2\pi$, $\phi' = \phi$.

TANGENT SPACES AND TANGENT BUNDLES

Both the configuration manifold $\mathbb{Q}$ and the velocity phase manifold $\mathbf{TQ}$ are differential manifolds, but $\mathbf{TQ}$ is constructed from $\mathbb{Q}$ by adjoining tangent spaces. We now describe how tangent vectors and tangent spaces are defined intrinsically, without invoking spaces $\mathbb{E}^m$ of higher dimension.

Tangent vectors of a manifold $\mathbb{M}$ are defined in terms of *curves* on $\mathbb{M}$. A curve is a function $c(t)$ that gives a point of the manifold $\mathbb{M}$ for each value of a parameter t, where t varies through a certain interval I of the real numbers, or $c : I \to \mathbb{M}$. In a local chart ϕ, a curve c in $\mathbb{M}$ induces a curve $C = \phi \circ c$ on a chart of the atlas (think of this as a page in the mapmaking analogy). For each $t \in I$ this gives the local coordinates ζ^j of a point on c, that is, $\zeta^i = C^i(t) \equiv \phi^i(c(t))$. In the mapmaking analog, c could be a road and then the $\zeta^i(t)$ are the coordinates of its image on the page. We will restrict our attention to smooth curves for which all of the $C^i(t)$ are differentiable. In neighborhoods where charts overlap, the coordinates of a given curve will depend on the chart, but because our manifolds are all differentiable, smoothness is chart independent. Indeed, if $C^j(t)$ and $D^j(t)$ are the coordinates of a curve in two different charts, then

$$C^j(t) = \Phi^j(D(t)),$$

and the differentiability of Φ (and its inverse) guarantees that if D is smooth, so is C (and vice versa).

The tangent vectors to a curve C *in a chart* are defined in the usual way: they have components

$$\tau^j = \frac{dC^j}{dt}.$$

Note that we previously spoke of *unit* tangent vectors. Now the τ^j will not generally form a unit vector. Like the coordinates of curves, the τ^j depend on the chart, but the Φ maps can be used to transform among the components in overlapping regions:

$$\tau^j = \frac{\partial \Phi^j}{\partial D^k} \frac{dD^k}{dt} \equiv \frac{\partial \Phi^j}{\partial D^k} \sigma^k, \tag{2.82}$$

where the σ^k are the tangent vectors of the same curve in the other chart. At each $m \in \mathbb{M}$ the $\partial \Phi^j / \partial D^k$ form a constant matrix, so (2.82) is simply a coordinate transformation.

A tangent vector τ to a curve c at some point $m \in \mathbb{M}$ can be defined in terms of its components τ^j in a local chart ϕ in three steps. First, the curve c is used to construct its image $C = \phi(c)$ in the local chart. Second, the components of the tangent vector to the image curve at $\phi(m)$, the $\tau^j = dC^j/dt$, are found. Third, the vector τ is identified with the numbers τ^j and the chart ϕ. If another chart ψ is used, whose image curve is $D = \psi(c)$, the components of the tangent curve will be $\sigma^j = dD^j/dt$. Then τ is identified with the numbers σ^j and the chart ψ. The τ^j and σ^j are called the components, in the two charts, of the abstract vector τ on $\mathbb{M}$, and Eq. (2.82) shows how these components are related. Any local chart can be used to define τ: the components obtained in all other local charts will be independent of the defining chart.

Given a point $m \in \mathbb{M}$, a chart at m, and a set of n numbers τ^j, there is just one tangent vector τ at m whose components in that chart are the τ^j. Given two tangent vectors whose components in that chart are τ^j and σ^j, there is another whose components in that chart are $\rho^j = \alpha \tau^j + \beta \sigma^j$, where α and β are real numbers, and it defines a vector ρ at m. In this way the tangent vectors at a point form the vector space called the *tangent space* at m (Fig. 2.13). Since the tangent vectors are chart-independent objects, so is the tangent plane. Each $m \in \mathbb{M}$ has its own tangent space, denoted $\mathbf{T}_m\mathbb{M}$, the *fiber* above m. All the $\mathbf{T}_m\mathbb{M}$ are of the same dimension.

The *tangent bundle* (or *tangent manifold*) $\mathbf{TM}$ consists of $\mathbb{M}$ together with the collection of all of its tangent spaces. But this is not a complete definition, for it does not yet make $\mathbf{TM}$ a differential manifold. This is because of the problem mentioned in connection with Fig. 2.10: simply adjoining a line to each point of $\mathbb{S}^1$ does not create a differential manifold unless the lines are joined to each other, glued together, so as to make the resulting $\mathbf{TS}^1$ a cylinder in the usual sense. More generally this requires gluing together tangent spaces, which is done by defining charts on $\mathbf{TM}$ in order to distinguish local neighborhoods that tell which points in one $\mathbf{T}_m\mathbb{M}$ are close to others on a neighboring $\mathbf{T}_m\mathbb{M}$.

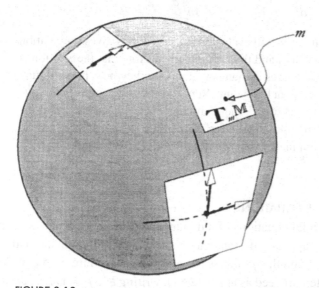

FIGURE 2.13

The manifold here is a sphere. The figure shows the manifold with three of its tangent planes. Some curves are indicated, with their tangent vectors on the planes. The heavy dot on each plane is the point of tangency, or the point on the manifold above which the plane lies.

This is done by using a local chart $(\mathbb{U}, \phi)$ on $\mathbb{M}$ to construct a local chart $(\mathbf{TU}, \mathbf{T}\phi)$ on $\mathbf{TM}$. First $\mathbf{TU}$ is taken to be the set of points in $\mathbb{U}$ and all of the tangent planes to them: a point (u, τ) in $\mathbf{TU}$ consists of a point $u \in \mathbb{U}$ and one vector τ (we drop the boldface notation) in $\mathbf{T}_u\mathbb{U}$, the tangent plane at u. The $2n$ components of (u, τ) in the chart $(\mathbf{TU}, \mathbf{T}\phi)$ are the n coordinates of u plus the n components τ^j of τ, all obtained from the chart $(\mathbb{U}, \phi)$; that is, $\mathbf{T}\phi$ maps $\mathbf{TU}$ to $\mathbb{R}^n \times \mathbb{R}^n \equiv \mathbb{R}^{2n}$, mapping (u, τ) to the point with coordinates $(\phi^j(u), \tau^k)$. If the $(\mathbb{U}, \phi)$ charts from which one starts form an atlas for $\mathbb{M}$, the $(\mathbf{TU}, \mathbf{T}\phi)$ charts defined in this way form an atlas for $\mathbf{TM}$, which then becomes a $2n$-dimensional differentiable manifold.

WORKED EXAMPLE 2.6

Let $\mathbb{U}$ and $\mathbb{V}$ be neighborhoods of a manifold $\mathbb{M}$ such that $\mathbb{U} \cap \mathbb{V}$ is not empty, and let c be a curve passing through a point $m \in \mathbb{U} \cap \mathbb{V}$ with tangent vector τ. Let Φ be the transformation in $\mathbb{U} \cap \mathbb{V}$ from the chart $(\mathbb{U}, \phi)$ to the chart $(\mathbb{V}, \psi)$.

(a) Find the components of τ in the $(\mathbb{V}, \psi)$ chart in terms of Φ and of its components in the $(\mathbb{U}, \phi)$ chart. (b) Let b be a curve tangent to c at m in terms of the $(\mathbb{U}, \phi)$ chart. Show that the two curves are tangent also in terms of the $(\mathbb{V}, \psi)$ chart, so that tangency is indeed a chart-independent concept.

Solution. (a) Write ξ^i for the coordinates in $\mathbb{V}$, and η^i for the coordinates in $\mathbb{U}$. Then $\xi = \psi(\phi^{-1}(\eta))$, or $\xi^i = \Phi^i(\eta)$, where $\Phi = \psi \circ \phi^{-1}$. Now let the equation of the curve be $\xi^i(t)$ in $(\mathbb{V}, \psi)$ and $\eta^i(t)$ in $(\mathbb{U}, \phi)$. In the two charts the components of the tangent vector are $\tau^i = d\xi^i/dt$ and $\sigma^i = d\eta^i/dt$, respectively:

$$\tau^i \equiv \frac{d\xi^i}{dt} = \frac{\partial \Phi^i}{\partial \eta^j}\frac{d\eta^j}{dt} \equiv \frac{\partial \Phi^i}{\partial \eta^j}\sigma^j.$$

When the functions and derivatives are all evaluated at $m \in \mathbb{M}$, this gives the relation between the components in one chart and those in the other. Note that the relation is a linear transformation. (b) Two curves are tangent at a point they both pass through if they have the same tangent vector at that point. Now consider two curves, with two tangent vectors, whose components are τ_1^i and τ_2^i in one chart and σ_1^i and σ_2^i in the other. Then the result just obtained shows that if $\sigma_1^i = \sigma_2^i$, then $\tau_1^i = \tau_2^i$. The converse is also true, as the transformation matrix $\partial \Phi^i/\partial \eta^j$ is nonsingular in $\mathbb{U} \cap \mathbb{V}$.

APPLICATION TO LAGRANGE'S EQUATIONS

We now return to the differential equations of a dynamical system, Lagrange's equations. The manifold on which they appear is the tangent bundle $\mathbf{TQ}$ and although it is natural to separate coordinates into the q^α and the $\dot{q}^\alpha$ this is not a natural procedure on manifolds in general. We shall therefore often proceed as at Eq. (2.81), writing ξ^j, $j = 1, \ldots, 2n$ for the coordinates on $\mathbf{TQ}$. The first n of the ξ^j are the q^α and the last n are the $\dot{q}^\alpha$, and the general point on $\mathbf{TQ}$ is called ξ (like m in the previous subsection).

The equations of motion on $\mathbf{TQ}$ are first-order differential equations (for the present we exclude equations in which the time appears explicitly) of the form

$$\frac{d\xi^j}{dt} = f^j(\xi). \tag{2.83}$$

The first n of Eqs. (2.83) read $dq^\alpha/dt = \dot{q}^\alpha$, and the last are Lagrange's equations (2.29).

The left-hand side of (2.83) is the jth component, in this chart, of the vector tangent to the orbit of the phase portrait that passes through $\xi \in \mathbf{TQ}$; the right-hand side specifies the components of that vector. Because it does so for every point $\xi \in \mathbf{TQ}$, Eq. (2.83) defines the *vector field* $f^j(\xi)$ on $\mathbf{TQ}$. The problem of solving the equations of motion is the problem of constructing the phase portrait from the vector field. Because the vector field is defined by and defines the dynamics, we will regularly refer to it as the *dynamics*. The curves of the phase portrait are called the *integral curves* of the vector field. Integrating the dynamics means finding the integral curves. Tangent vectors were defined in terms of curves: a curve determines its tangent vectors. The problem of dynamics, as we now see, is the exact opposite: to find the curves from the tangents.

We finish with two remarks about Eqs. (2.83). First, the ξ^i and the f^i in Eqs. (2.83) are defined only in a chart, so these equations have only *local* meaning. However, if they are given for each chart in an atlas, they define a *global* dynamical system on the entire manifold. Second, the equations define a vector in the tangent space at each $\xi \in \mathbf{TQ}$. But $\mathbf{TQ}$ is itself a tangent bundle, so Eqs. (2.83) involve the tangent bundle of a tangent bundle: $\mathbf{T(TQ)}$, or $\mathbf{T^2Q}$.

These topics will be discussed more fully in later chapters. In this section we have only introduced the geometric, differentiable-manifold approach to the study of dynamical systems. This introduction is meant to give a perspective, to aid in visualizing many qualitative properties of the dynamical systems we shall be discussing as we go on. More direct use of this material requires the tools of calculus on manifolds (or differential geometry), but we are putting that off until Chapter 3, when Lagrange's equations will be written out intrinsically, without reference to coordinates on $\mathbf{TQ}$.

PROBLEMS

1. Show that if all the external forces are given by a time-independent potential (i.e., if $\mathbf{F}_i = -\nabla_i V(\mathbf{x}_1, \dots, \mathbf{x}_N)$), then the rate of change of total energy is given by Eq. (2.15).

2. Let a curve in 3-space be defined by the two equations $f_I(\mathbf{x}) = 0$, $I = 1, 2$. Show that the two equations $\boldsymbol{\tau} \cdot \nabla f_I = 0$ define a tangent vector to the curve. (We write "*a* tangent vector" rather than "*the* tangent vector" because $\boldsymbol{\tau}$ is not necessarily of unit magnitude like the tangent vector defined in Eq. (1.9).)

3. Prove that if L' is defined by (2.36), then Eq. (2.28) implies Eq. (2.37).

4. Show that if two Lagrangians L_1 and L_2 in $n > 1$ freedoms give exactly the same equations of motion, then $L_1 - L_2 = d\Phi/dt$, where Φ is a function on $\mathbb{Q}$. See the proof for the case $n = 1$ in Section 2.2.2. [Hint: Just as in 3-space $\nabla \wedge \mathbf{E} = 0$ implies that there exists a

function V such that $\mathbf{E} = \nabla V$; so in n dimensions $\partial E_j/\partial x_k = \partial E_k/\partial x_j$ implies that there exists a function V such that $E_k = \partial V/\partial x_k$.]

5. Describe a possible atlas for each of the following manifolds. In each case write the transformation equations in the regions where the charts overlap. How many charts χ are there in each atlas?

   (a) The circle $\mathbb{S}^1$.

   (b) The cylinder $\mathbb{R} \times \mathbb{S}^1$.

   (c) The sphere $\mathbb{S}^2$. For this case, skip the transformation equations. [Partial answer: The lower limits for χ are 2 in all three cases.]

6. Suppose that the dynamical system represented by Eq. (2.60) has been determined; that is, that $\theta(t)$ and $r(t)$ have been found. In a Cartesian coordinate system which is not necessarily lined up either with the angular momentum $\mathbf{J}$ or with the direction of motion of the center of mass, write down the complete solution of the two-body problem in terms of the initial conditions $\mathbf{x}_1(0)$, $\dot{\mathbf{x}}_1(0)$, $\mathbf{x}_2(0)$, $\dot{\mathbf{x}}_2(0)$ and $\theta(t)$ and $r(t)$.

7. Show that the total energy of the two particles in the central force problem is the sum of $\frac{1}{2}\mu\dot{r}^2 + \mathcal{V}(r)$ and the center-of-mass energy, where μ is the reduced mass and $\mathcal{V}$ is the effective potential of the equivalent one-dimensional problem.

8. (a) It has been seen that transforming to a coordinate system rotating at the Larmor frequency changes the dynamical system of a charged particle in a uniform magnetic field to the two-dimensional isotropic harmonic oscillator. Given the sign of the charge and the direction of the magnetic field, find the direction of the rotation.

   (b) As has been pointed out, this transformation changes circular orbits centered at any point in the plane to ellipses centered at the center of rotational transformation. What is the relation among center and radius of the circular orbit, the center of the transformation, and the axes of the ellipse?

   (c) Examine and compare the two conserved quantities $L - \dot{x}^\alpha \partial L/\partial \dot{x}^\alpha$ and $L - \dot{y}^\alpha \partial L/\partial \dot{y}^\alpha$.

9. A double plane pendulum consists of a simple pendulum (mass m_1, length l_1) with another simple pendulum (mass m_2, length l_2) suspended from m_1, both constrained to move in the same vertical plane.

   (a) Describe the configuration manifold $\mathbb{Q}$ of this dynamical system. Say what you can about $T\mathbb{Q}$.

   (b) Write down the Lagrangian of this system in suitable coordinates.

   (c) Derive Lagrange's equations.

10. Consider a stretchable plane pendulum, that is, a mass m suspended from a spring of spring constant k and unstretched length l, constrained to move in a vertical plane. Write down the Lagrangian and obtain the Euler–Lagrange equations.

11. A wire is bent into the shape given by $y = A|x^n|$, $n \geq 2$ and oriented vertically, opening upward, in a uniform gravitational field g. The wire rotates at a constant angular velocity ω about the y axis, and a bead of mass m is free to slide on it without friction.

   (a) Find the equilibrium height of the bead on the wire. Consider especially the case $n = 2$.

   (b) Find the frequency of small vibrations about the equilibrium position.

12. A particle starts at rest and moves along a cycloid whose equation is

$$x = \pm \left[a \cos^{-1} \left(\frac{a-y}{a} \right) + \sqrt{2ay - y^2} \right].$$

There is a gravitational field of strength g in the negative y direction. Obtain and solve the equations of motion. Show that no matter where on the cycloid the particle starts out at time $t = 0$, it will reach the bottom at the same time. [Suggestion: Choose arclength along the cycloid as the generalized coordinate.]

13. Two masses m_1 and m_2 are connected by a massless spring of spring constant k. The spring is at its equilibrium length and the masses are both at rest; there is no gravitational field. Suddenly m_2 is given a velocity v. Assume that v is so small that the two masses never collide in their subsequent motion. Describe the motion of both masses. What are their maximum and minimum separations?

14. (a) Show that circular orbits exist for all attractive power-law central potentials. Find the radius and total energy of the circular orbit as functions of the power of r and the angular momentum.

   (b) Show that the orbits are stable only if $k \geq -2$ for all k in $V = \alpha r^k$. [Hint: It may be easier to show it separately for $V = -\alpha/r^k$ and for $V = \alpha r^k$, both with positive k.]

   (c) We mentioned, in preparation for the proof of Bertrand's theorem, that at aphelion (or perihelion) $\dot{r} = 0$. But closure of the orbit requires also that $\dot{\theta}$ at aphelion (or perihelion) be the same at all aphelion (or perihelion) points. Prove that it is.

   (d) Show that in central force motion the orbit is symmetric about r_{min} or about r_{max}.

15. Check the validity of Kepler's third law of planetary motion. Find the value of T^2/R^3.

16. Show that for all three types of Kepler orbits (ellipses, hyperbolas, parabolas) the Sun is at one of the foci.

17. Let x be the position vector, J the angular momentum vector, and p the momentum vector of a mass μ in a Coulomb potential $V = -\alpha/r$, where r is the magnitude of x.

   (a) Show that the *Runge-Lenz* vector $A = p \wedge J - \alpha\mu x/r$ is (actually each of its components is) a constant of the motion.

   (b) Show that A lies in the plane of the orbit. From the dot product of A with x obtain the equation of the orbit and show thereby that the eccentricity is given by $\epsilon = A/\mu\alpha$ and that A points along the major axis of an elliptical orbit.

18. This problem involves placing a satellite of mass m into orbit about a spherical planet (with no atmosphere) of mass $M \gg m$ and radius ρ. The mass is raised to altitude $h = R - \rho$ (i.e., to radius R) and given an initial velocity V perpendicular to the radius.

   (a) Find the eccentricity ϵ of the resulting orbit as a function of V. For what values of V is the orbit an ellipse, a circle, a parabola, or a hyperbola?

   (b) Find the distance of closest and, where appropriate, farthest approach to the planet. For this part ignore the possibility of collision with the planet.

   (c) Find the energy needed to raise the satellite into the lowest orbit that fails to collide with the planet (a surface-skimming orbit).

   (d) Now suppose V is not necessarily perpendicular to the radius. Find the escape velocity V_E as a function of the angle α between V and the radius. Be careful to avoid collisions with the planet! [Answer to part (d): Let $R/\rho = r$; $\sin\alpha_1 =$

$\sin \alpha_2 = \sqrt{1/r}$, $0 < \alpha_1 < \frac{1}{2}\pi < \alpha_2 < \pi$; $\sin \alpha_3 = 1/r$. Then $V_E = \sqrt{2GM/R}$ for $0 < \alpha < \alpha_2$; $V_E = \sqrt{(2GM/R)(r-1)/(r^2 \sin^2 \theta - 1)}$ for $\alpha_2 < \alpha < \alpha_3$; there is no escape velocity for $\alpha_3 < \alpha < \pi$.]

19. Find the general orbit (r as a function of θ) in a repulsive Coulomb potential, $V = \alpha/r$ for α positive.

20. Find the general orbit (r as a function of θ) for a particle moving in a perturbed gravitational potential

$$V = -\frac{k}{r} + \frac{\beta}{r^2}.$$

Show that if $k > 0$ and $\beta \ll k(1 - \epsilon^2)$, the orbit may be thought of as an ellipse of eccentricity ϵ precessing slowly with angular velocity

$$\omega = \frac{2\pi\beta}{\tau k a(1 - \epsilon^2)},$$

where a is the semimajor axis and τ is the orbital period.

21. A particle of mass m and electric charge e moves in a uniform electrostatic field whose potential is $\varphi = E x_1$ and a uniform magnetic field whose vector potential is $\mathbf{A} = \mathbf{e}_2 B x_1$, where E and B are constants. (We use x instead of q for Cartesian coordinates and subscripts rather than superscripts for the indices; $\mathbf{e}_k$ is the unit vector in the k direction.)

    (a) Write down the Lagrangian and describe the general motion: show that it is along cycloids, more specifically that it is uniform circular motion in the $(1, 2)$ plane superposed on a constant velocity $\mathbf{V}$ in the $(2, 3)$ plane. Write down the angular speed ω of the circular motion and the 2-component V_2 of the linear motion, both in terms of the initial conditions.

    (b) Take the 3-component of the linear motion to be zero. Find initial conditions for the following three cases: (i) the particle sometimes moves in the direction opposite to $\mathbf{V}$, (ii) the particle occasionally comes to rest, and (iii) the particle never moves in the direction opposite to $\mathbf{V}$. Draw all three types of trajectories.

    (c) Find initial conditions under which the circular part of the motion is eliminated, so the particle moves at constant velocity in the $(2, 3)$ plane.

22. As was shown in Worked Example 2.4, an effective central potential for the Moon–Earth system in the gravitational field of the Sun is $V(r) = -\alpha/r - \gamma/r^3$ (both α and γ are positive). Although this potential is clearly unrealistic close to the force center, for purposes of this discussion assume that it is valid for all r.

    (a) Discuss qualitatively how the possible trajectories depend on the energy and angular momentum.

    (b) If $\gamma = 0$, the potential is, of course, the Kepler (Coulomb) potential. For small γ the motion should look like a perturbed Kepler orbit. Show that for sufficiently small γ the orbit behaves like a Kepler ellipse precessing through an angle $\Delta\theta \approx 6\pi\alpha\gamma\mu^2/l^4$ per orbit (μ is the reduced mass and l is the angular momentum).

23. The Yukawa potential is given by $V = -(k/r)\exp(-r/\lambda)$.

    (a) Find the trajectory of a bound orbit in this potential to the first order in r/λ.

(b) For the lowest order in which a bound orbit of energy E and angular momentum l differs from a Kepler ellipse, calculate the angle $\Delta\theta$ between successive maxima of r. [Hint: To first order it looks like a Kepler orbit with a different energy. Part (b) requires going to the second order.]

24. For a particle of mass m in a Yukawa potential with given k and λ find the conditions on the angular momentum for there to be bound orbits. For each angular momentum satisfying the bound-orbit condition, show how to find the maximum energy for which the orbit will be bound. (The angular momentum of an orbit does not uniquely determine its energy.)

25. A particle of mass m moves in one-freedom in the potential

$$V(x) = -\frac{V_0}{\cosh^2 \alpha x}.$$

(a) Solve for $x(t)$, the motion of the particle, obtaining expressions for $x(t)$ in the three energy regions $E < 0$, $E > 0$, and $E = 0$, where E is the energy. In particular, show that the motion is unbounded for $E \geq 0$ but bounded for $E < 0$. Find the minimum possible value for E and the turning points for the motion.

(b) For the bound orbits, find the period as a function of E.

(c) Draw a graph of $V(x)$ and a phase diagram.

CHAPTER 3

TOPICS IN LAGRANGIAN DYNAMICS

CHAPTER OVERVIEW

One might have guessed that the Lagrangian formulation, derived simply by rewriting Newton's equations in generalized coordinates, has little new to offer, but it turns out that it provides insights that are difficult to obtain from the Newtonian point of view. This chapter starts with two topics that are intimately related to the Lagrangian point of view, the variational principle and the connection between symmetry and conservation laws (the Noether theorem). We also treat forces that do not arise from potentials and expand the application of geometric methods to Lagrangian systems.

3.1 THE VARIATIONAL PRINCIPLE AND LAGRANGE'S EQUATIONS

3.1.1 DERIVATION

THE ACTION

It is noteworthy that Lagrange's equations resemble the equations one obtains from a *variational problem*. Variational problems are classical in mathematics. An example is the *isoperimetric* problem: given a fence of a fixed length, what shape provides the largest area it can surround? The well-known answer (a circle) is obtained from equations that look very much like Lagrange's. Other variational problems arise often in physics and engineering. An example is optimization, when certain parameters have to be chosen so as to make some property, say a value or a function, an *extremum* (i.e., a maximum or minimum). Other examples are found in various branches of physics, for example, quantum mechanics, classical and quantum field theory, solid state physics, and fluid dynamics.

That Lagrange's equations look like variational equations is intriguing, for it means that a dynamical system moves in such a way as to minimize or maximize something. We will show that the dynamical system moves so as to minimize the *action*

$$S \equiv \int L(q, \dot{q}, t)\, dt,$$

a quantity that will play a larger role as we move on (e.g., in Chapter 6). Of all the possible motions the dynamical system could choose, of all the possible functions $q(t)$ that one could imagine, the actual physical motion is that for which S is minimal.

More accurately, suppose that an initial time t_0, a final time t_1, and a function $q(t)$ are given. Then the action associated with them is

$$S(q; t_0, t_1) \equiv \int_{t_0}^{t_1} L(q, \dot{q}, t)\, dt. \tag{3.1}$$

When the given $q(t)$ is inserted into the expression for $L(q, \dot{q}, t)$, the integrand becomes a function of t alone, so it can be integrated with respect to t. Clearly the value of S depends on the trajectory $q(t)$: S is a *functional* of $q(t)$, for it depends not on one value of t, but on the function q and all of t in the interval $t_0 \leq t \leq t_1$.

In this chapter we will deal only with trajectories that start and end at the same two points in $\mathbb{Q}$ (see Fig. 3.1). Two such trajectories, given by functions $q(t; a)$ and $q(t; b)$, coincide at the limits of the integral but may not in between: $q(t_0; a) = q(t_0; b) \equiv q(t_0)$ and $q(t_1; a) = q(t_1; b) \equiv q(t_1)$. If $q(t; a) \neq q(t; b)$, the corresponding actions will not be equal. There are, in fact, many possible trajectories with the same end points, and each

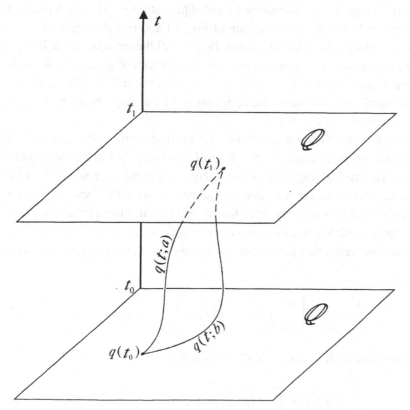

FIGURE 3.1

Two possible trajectories $q(t; a)$ and $q(t; b)$ from $q(t_0)$ to $q(t_1)$. The horizontal planes represent $\mathbb{Q}$ at the two times. A continuous family of possible trajectories would form a surface in this diagram, whose boundaries could be $q(t; a)$ and $q(t; b)$.

one yields a characteristic value of S. The physical problem is to choose among all these possibilities, to find the particular $q(t)$ that the dynamical system takes in making the trip from $q(t_0)$ to $q(t_1)$. What we are looking for is the physical trajectory that passes through the two given positions at the two distinct times t_0 and t_1. That such data determine the physical trajectory was established near the end of Section 2.4.1: the positions at two different times specifies the trajectory by giving $2n$ independent conditions.

HAMILTON'S PRINCIPLE

We now proceed to show that the physical trajectory is the one that yields the minimum value of S: *minimizing S leads to Lagrange's equations.* Consider a family of many trajectories $q(t; \epsilon)$, all starting and ending at $q(t_0)$ and $q(t_1)$, where ϵ is an index labeling each particular trajectory of the family. Just as $q(t; a)$ and $q(t; b)$ lead to different actions, each $q(t; \epsilon)$ leads to its corresponding action $S(\epsilon)$. Then the variational principle states that the physical trajectory is the one for which the action is a minimum and that this is independent of the way in which the ϵ-family of trajectories is chosen, provided it contains the physical one. This is known as *Hamilton's variational principle.*

More specifically, we require that ϵ takes its values from the real numbers and parametrizes the family of trajectories continuously and *differentiably.* This means that the family forms a connected surface in the diagram of Fig. 3.1 and that the partial derivative $\partial q(t; \epsilon)/\partial \epsilon$ exists for all values of t in the interval $[t_0, t_1]$. All the calculations will depend on ϵ only through derivatives, so ϵ can be changed without loss of generality by adding an arbitrary constant, and we choose it so that $\epsilon = 0$ labels the trajectory that gives the minimum S in that family. (Incidentally, the requirement of differentiability will be used, and is therefore needed, only at $\epsilon = 0$.)

The ϵ-families of trajectories are otherwise chosen quite arbitrarily, and it is only by chance that any one of them contains the physical trajectory. Yet in any such family, whether it includes the physical trajectory or not, there will be one trajectory for which S is a minimum. That trajectory can be included also in many other ϵ-families, but it will not in general yield the minimum S in every other family. What Hamilton's principle states is that the physical trajectory yields the minimal S of every family in which it can be included. Put mathematically, this means that for any such ϵ-family the physical trajectory satisfies the equation

$$\frac{dS}{d\epsilon}\bigg|_{\epsilon=0} \equiv \left[\frac{d}{d\epsilon} \int_{t_0}^{t_1} L(q, \dot{q}, t)\, dt \right]_{\epsilon=0} = 0. \qquad (3.2)$$

Henceforth we abbreviate $d/d\epsilon|_{\epsilon=0}$ as δ, so (3.2) becomes

$$\delta S \equiv \delta \int_{t_0}^{t_1} L(q, \dot{q}, t)\, dt = 0. \qquad (3.3)$$

It should be remembered that in all such expressions everything is evaluated at $\epsilon = 0$.

The next step is to take the derivative of the integral with respect to ϵ. The ϵ dependence arises because q and $\dot{q}$ in the Lagrangian depend on ϵ, so

$$\delta S = \int_{t_0}^{t_1} \delta L \, dt. \tag{3.4}$$

Since $\delta \equiv d/d\epsilon|_{\epsilon=0}$,

$$\delta L = \frac{\partial L}{\partial q^\alpha} \delta q^\alpha + \frac{\partial L}{\partial \dot{q}^\alpha} \delta \dot{q}^\alpha. \tag{3.5}$$

Now, $\dot{q}^\alpha \equiv dq^\alpha/dt$ is the generalized velocity along a trajectory specified by a certain value of ϵ; that is, the time derivative is taken for fixed ϵ. This should be written $\partial q^\alpha/\partial t$, rather than dq^α/dt, but in keeping with tradition and with the final resolution of the problem, we continue to use dq^α/dt. Since it is a partial derivative there is no problem about changing the order of d/dt and $\partial/\partial\epsilon$, and hence the order of d/dt and δ. Thus

$$\frac{\partial L}{\partial \dot{q}^\alpha} \delta \dot{q}^\alpha = \frac{\partial L}{\partial \dot{q}^\alpha} \frac{d}{dt} \delta q^\alpha$$

$$= \frac{d}{dt} \left[\frac{\partial L}{\partial \dot{q}^\alpha} \delta q^\alpha \right] - \left[\frac{d}{dt} \frac{\partial L}{\partial \dot{q}^\alpha} \right] \delta q^\alpha.$$

Inserting this into the expression for δL yields

$$\delta L = \left[\frac{\partial L}{\partial q^\alpha} - \frac{d}{dt} \frac{\partial L}{\partial \dot{q}^\alpha} \right] \delta q^\alpha + \frac{d}{dt} \left[\frac{\partial L}{\partial \dot{q}^\alpha} \delta q^\alpha \right]. \tag{3.6}$$

For later use we remark that this result was arrived at without any restrictions on the variation of the $q(t; \epsilon)$ at the end points.

Now insert (3.6) into (3.4). Then

$$0 = \delta S = \int_{t_0}^{t_1} \left[\frac{\partial L}{\partial q^\alpha} - \frac{d}{dt} \frac{\partial L}{\partial \dot{q}^\alpha} \right] \delta q^\alpha dt + \int_{t_0}^{t_1} \frac{d}{dt} \left[\frac{\partial L}{\partial \dot{q}^\alpha} \delta q^\alpha \right] dt. \tag{3.7}$$

[This result could also have been obtained by inserting Eq. (3.5) into Eq. (3.4) and integrating the second term by parts.] The second integral here is easily obtained: it is

$$\frac{\partial L}{\partial \dot{q}^\alpha} \delta q^\alpha \bigg|_{t_0}^{t_1} = 0,$$

which vanishes because at the end points all the trajectories come together, so that there $\delta q^\alpha = 0$. The remaining term can be written (see Section 2.2.2 for the definition of the Λ_α)

$$\int_{t_0}^{t_1} \Lambda_\alpha \delta q^\alpha \, dt = 0. \tag{3.8}$$

Now make use of the following theorem (Gel'fand and Fomin, 1963). Let f_α, $\alpha = 1, 2, \ldots, n$, be a set of n integrable functions of a real variable t on an interval I. Suppose that

$$\int_I f_\alpha h_\alpha \, dt = 0 \qquad (3.9)$$

for every arbitrary set of integrable functions h_α on the same interval, all of which vanish at its end points; then $f_\alpha = 0$ for all α. Since Hamilton's principle applies to *any* ϵ-family of trajectories, the δq^α are a set of arbitrary functions of t, like the h_α, that vanish at the end points (recall that everything is evaluated at $\epsilon = 0$). Therefore $\Lambda_\alpha = 0$, or

$$\frac{\partial L}{\partial q^\alpha} - \frac{d}{dt} \frac{\partial L}{\partial \dot{q}^\alpha} = 0. \qquad (3.10)$$

These are, of course, Lagrange's equations. What they tell us is that the $q^\alpha(t)$ functions that minimize the action, when inserted together with their derivatives into the Lagrangian function and its derivatives, are the very ones that satisfy Lagrange's equation.

In Section 3.2.2 it will be useful to view Eq. (3.9) as an inner product in a vector space $\mathbb{F}$. Think of the f_α as the components of a vector $f \in \mathbb{F}$ (and similarly h), and write

$$\int_I f_\alpha h_\alpha \, dt \equiv (f, h).$$

Then the theorem quoted above states that if f is orthogonal to all vectors (arbitrary h_α) in $\mathbb{F}$, then $f = 0$. There are some fine points that could be made more rigorous, but this will do for our purposes. The derivation of Lagrange's equations can then be put into these terms: Eq. (3.8) states that

$$(\Lambda, \delta q) = 0, \qquad (3.11)$$

and since the δq are arbitrary vectors and span $\mathbb{F}$, the vector Λ is orthogonal to all vectors in the space and therefore vanishes, which is the content of (3.10).

DISCUSSION

Similar equations were first obtained by Euler for the general mathematical variational problem, and we will henceforth refer to Eqs. (3.10) as the Euler–Lagrange (or EL) equations. An interesting discussion of the variational principle in mechanics is found in Feynman et al. (1963), Volume 2, Chapter 19. Feynman had a deep understanding of variational methods in physics and applied them, in particular, in the so-called Dirac–Feynman path integral approach to quantum mechanics (Feynman and Hibbs, 1965).

The variational principle yields the Euler–Lagrange equations not simply when S is a minimum, but more generally when its derivative vanishes. Recall from elementary calculus that a function $f(x)$ has a critical point x_c when $f'(x_c) = 0$, which can be a minimum, a maximum, or a point of inflection. The same is true of the functional S, and to determine whether the solution of the EL equations yield a maximum, a minimum, or (in this case) a saddle point requires examining the second derivative

$$\frac{d^2S}{d\epsilon^2}\bigg|_{\epsilon=0} \equiv \delta^2 S.$$

For this to be a minimum requires that $\delta^2 S \geq 0$, and it can be shown (Gel'fand and Fomin, 1963) that a necessary, but not sufficient condition for this is that

$$\frac{\partial^2 L}{\partial \dot{q}^\alpha \partial \dot{q}^\alpha} \geq 0.$$

This condition on the trace of the Hessian $\partial^2 L/\partial \dot{q}^\alpha \partial \dot{q}^\beta$ (defined at the end of Section 2.2.2) is called the *Legendre condition*. It is always satisfied by the standard Lagrangian functions of the form $T - V$ with $T = \frac{1}{2}\Sigma m_i \dot{x}_i^2$ and a velocity-independent potential (or if V depends linearly on the velocity, as for a charged particle in an electromagnetic field): for such systems the variational principle always yields a minimum for S. More generally, when L may depend on higher powers of the $\dot{q}^\alpha$, it may not yield a minimum. But in mechanics S is essentially always a minimum, and one is hardly ever concerned about this question.

Hamilton's variational principle is a global statement about the entire trajectory between $q(t_0)$ and $q(t_1)$. This is to be contrasted with the EL equations, which are local statements on $\mathbb{TQ}$ (as is the Legendre condition): they are differential equations that describe behavior in small neighborhoods. The global principle leads to local differential equations. On the other hand there is a different kind of locality in the variational principle, locality in ϵ. Even if a trajectory $q(t)$ yields a minimum S, it may be only a local minimum in ϵ. There may be other trajectories, not close to $q(t)$ in any ϵ-family, that also yield ϵ-local minima of S, perhaps even lower than that obtained from $q(t)$. In that case both trajectories satisfy the EL equations, and therefore both are physically possible (see Problem 2 for a simple example).

We finish this subsection with some physical remarks about the variational method.

The fact that the physical trajectory minimizes S gives physical significance to the action and adds it to the Lagrangian. As Feynman points out, it is as though the system feels out the action on many possible paths and chooses the one for which S is minimal. The Lagrangian, previously simply a function that emerged in rewriting Newton's equations, is now seen as the stuff of which the action is made. And in that large class of systems for which $L = T - V$, the system moves so that the integrated difference between kinetic and potential energies is as small as possible.

Variational principles have been extremely important in physics: in almost every branch the ultimate theory has been cast in this form. Usually, however, Lagrangians are not obtained from the equations of motion (the analogs of Newton's equations) but from other considerations, typically from symmetry considerations. The usual situation is the opposite of the one we have presented: before the differential equations of motion are known, the Lagrangian is obtained from other conditions. Then the Lagrangian is inserted into the variational principle, and that yields the equations of motion. In that procedure the variational principle is crucial.

There are other variational principles that are used in dynamics, but we shall not discuss them here in any detail. Later (Chapter 6) we will take up a variant in which the end points $q(t_0)$ and $q(t_1)$ are not fixed.

3.1.2 INCLUSION OF CONSTRAINTS

Constraints were discussed at the beginning of Chapter 2 in the context of Cartesian coordinates on $\mathbb{R}^3$, but constraints can arise also in non-Euclidean configuration manifolds $\mathbb{Q}$. For example, they arise and are particularly important in rigid body motion (Chapter 8). In this subsection we show, using the variational principle, how constraints can be dealt with when $\mathbb{Q}$ and the Lagrangian are already given.

The variational method of Section 3.1.1 picked out, from among *all* trajectories connecting $q(t_0)$ and $q(t_1)$, the one that minimizes S. But that method must be modified if the system is constrained, for the only trajectories available to it are those that satisfy the constraints. Suppose, therefore, that the system is subject to $K < n$ independent constraints, in general nonholonomic (velocity dependent), of the form

$$f_I(q, \dot{q}, t) = 0, \quad I = 1, 2, \ldots, K. \tag{3.12}$$

We proceed to apply the variational method, now requiring that the *comparison paths* (the trajectories from among which the physical one is to be picked out) satisfy the constraints.

We start with Eq. (3.11), except that now $\delta q \in \mathbb{F}$, whose components are $\partial q^\alpha / \partial \epsilon$, is not arbitrary, for it arises from trajectories $q(t; \epsilon)$ that are restricted by Eqs. (3.12). Equation (3.11) now says not that Λ is the null vector, but that it is orthogonal to the subspace $\mathbb{F}_q \subset \mathbb{F}$ spanned by the admissible δqs. To find the possible Λs we must find $\mathbb{F}_q$.

The subspace $\mathbb{F}_q$ can be found by establishing precisely how Eqs. (3.12) restrict the δq vectors. Start by taking the derivatives of Eqs. (3.12) with respect to ϵ:

$$\frac{\partial f_I}{\partial \epsilon} \equiv \frac{\partial f_I}{\partial q^\alpha} \frac{\partial q^\alpha}{\partial \epsilon} + \frac{\partial f_I}{\partial \dot{q}^\alpha} \frac{\partial \dot{q}^\alpha}{\partial \epsilon} = 0. \tag{3.13}$$

Now multiply each of these equations by an arbitrary sufficiently well behaved function $\mu_I(t)$ and sum over I (all sums over I are indicated by summation signs):

$$\int_{t_0}^{t_1} \sum_I \left[\mu_I \frac{\partial f_I}{\partial q^\alpha} \frac{\partial q^\alpha}{\partial \epsilon} + \mu_I \frac{\partial f_I}{\partial \dot{q}^\alpha} \frac{\partial \dot{q}^\alpha}{\partial \epsilon} \right] dt = 0.$$

Integrate by parts, as in deriving Eq. (3.7), to obtain

$$\int_{t_0}^{t_1} \sum_I \left[\mu_I \frac{\partial f_I}{\partial q^\alpha} - \frac{d}{dt} \left(\mu_I \frac{\partial f_I}{\partial \dot{q}^\alpha} \right) \right] \frac{\partial q^\alpha}{\partial \epsilon} \, dt \equiv \left(\sum_I \chi_I, \delta q \right) = 0, \tag{3.14}$$

where each χ_I is the vector whose components are (no sum on I)

$$\chi_{I\alpha} \equiv \mu_I \frac{\partial f_I}{\partial q^\alpha} - \frac{d}{dt} \left(\mu_I \frac{\partial f_I}{\partial \dot{q}^\alpha} \right).$$

Equation (3.14) gives the restrictions on the δq vectors: they are orthogonal to all possible χ_I vectors. The subspace $\mathbb{F}_q$ they span is orthogonal to the subspace $\mathbb{F}_\chi \subset \mathbb{F}$ spanned by the χ_I. (We have not proven that either $\mathbb{F}_q$ or $\mathbb{F}_\chi$ is a subspace, but let that pass.)

Since Λ is orthogonal to $\mathbb{F}_q$, which is itself orthogonal to $\mathbb{F}_\chi$, it must lie in $\mathbb{F}_\chi$. That is, there exist constants α_I such that $\Lambda = \sum \alpha_I \chi_I$. These α_I may be absorbed into the μ_I, writing $\lambda_I(t) = \alpha_I \mu_I(t)$, and then the definition of the Λ_α and Eq. (3.14) allow the result to be put in the form

$$\frac{d}{dt}\frac{\partial}{\partial \dot{q}^\alpha}\left(L + \sum \lambda_I f_I\right) - \frac{\partial}{\partial q^\alpha}\left(L + \sum \lambda_I f_I\right) = 0 \qquad (3.15)$$

(the λ_I are called *Lagrange multipliers*). This is the result of applying the variational principle with constraints: it yields a set of equations that look like the EL equations for the Lagrangian

$$\mathcal{L} = L + \sum \lambda_I f_I. \qquad (3.16)$$

We now have the n EL equations (3.15) and the K constraint equations (3.12) from which we are to find the $n + K$ functions $q^\alpha(t)$ and $\lambda_I(t)$. But there is a serious problem: Eqs. (3.15) are first-order differential equations for the λ_I, and a solution requires having initial values $\lambda_I(t_0)$. This requirement is unphysical, for it means that the initial forces of constraint (see Problem 1) or, equivalently, the initial $\ddot{q}^\alpha$ must be known. For this reason the straightforward variational approach must be rejected (for more details, see Saletan & Cromer, 1971, p. 125 ff).

If, however, the constraints are holonomic (if the f_I do not depend on the $\dot{q}^\alpha$), Eqs. (3.15) become

$$\frac{d}{dt}\frac{\partial L}{\partial \dot{q}^\alpha} - \frac{\partial}{\partial q^\alpha}\left(L + \sum \lambda_I f_I\right) = 0 \qquad (3.17)$$

or

$$\frac{d}{dt}\frac{\partial L}{\partial \dot{q}^\alpha} - \frac{\partial L}{\partial q^\alpha} - \sum \lambda_I \frac{\partial f_I}{\partial q^\alpha} = 0, \qquad (3.18)$$

which involve no time derivatives of the λ_I. The difficulty is then avoided. For holonomic constraints (3.17) are the accepted equations, and because the f_I are $\dot{q}^\alpha$ independent, the equations can be written as

$$\frac{d}{dt}\frac{\partial \mathcal{L}}{\partial \dot{q}^\alpha} - \frac{\partial \mathcal{L}}{\partial q^\alpha} = 0 \qquad (3.19)$$

with the $\mathcal{L}$ of (3.16).

What is to be done in the general case, when the f_I depend on the $\dot{q}^\alpha$? A hint is obtained by returning to holonomic constraints, taking their time derivatives, and calling these the new constraints:

$$\hat{f}_I \equiv \frac{df_I}{dt} = \frac{\partial f_I}{\partial q^\alpha}\dot{q}^\alpha = 0. \qquad (3.20)$$

The $\hat{f}_l(q, \dot{q}, t)$ are now velocity-dependent constraints, and in terms of them Eq. (3.18) becomes

$$\frac{d}{dt}\frac{\partial L}{\partial \dot{q}^\alpha} - \frac{\partial L}{\partial q^\alpha} - \sum_l \lambda_l \frac{\partial \hat{f}_l}{\partial \dot{q}^\alpha} = 0. \tag{3.21}$$

This result is generally accepted for velocity-dependent constraints even if the $\hat{f}_l(q, \dot{q}, t)$, unlike those of (3.20), are nonlinear in the $\dot{q}^\alpha$. This result can also be obtained from quasi-physical reasoning by requiring that the comparison paths have the property that *virtual displacements* (displacements between adjacent paths at a given time) do no work. We will not go into a detailed explanation here; it is discussed by Saletan & Cromer (1971).

WORKED EXAMPLE 3.1

A disk of radius R rolls along a perfectly rough horizontal plane. (This is a velocity-dependent constraint.) The disk is constrained to remain vertical (Fig. 3.2). Write down the constraint equations and solve for the motion in general.

Solution. Figure 3.2 shows generalized coordinates for this problem (the coordinates of the center of mass of the disk are x_1 and x_2). The constraint equations, stating that the disk rolls, may be written

$$f_1 = R\dot{\phi}\cos\psi - \dot{x}_1 = 0,$$
$$f_2 = R\dot{\phi}\sin\psi - \dot{x}_2 = 0.$$

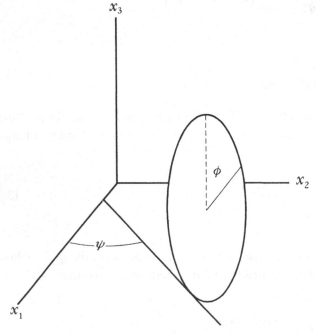

FIGURE 3.2
Generalized coordinates.

The Lagrangian is

$$L = \frac{1}{2}I_0\dot{\phi}^2 + \frac{1}{2}I_1\dot{\psi}^2 + \frac{1}{2}m(\dot{x}_1^2 + \dot{x}_2^2),$$

where I_0 is the moment of inertia of the disk about its symmetry axis and I_1 is its moment of inertia about a diameter. The equations of motion are

$$\Lambda_\alpha = \lambda_l \frac{\partial f_l}{\partial \dot{q}^\alpha},$$

which, when written out, become

$$I_0\ddot{\phi} = \lambda_1 R \cos\psi + \lambda_2 R \sin\psi, \tag{3.22}$$
$$I_1\ddot{\psi} = 0, \tag{3.23}$$
$$m\ddot{x}_1 = -\lambda_1, \tag{3.24}$$
$$m\ddot{x}_2 = -\lambda_2. \tag{3.25}$$

The constraint equations imply that $R\cos\psi = \dot{x}_1/\dot{\phi}$ and $R\sin\psi = \dot{x}_2/\dot{\phi}$, so (3.22) becomes

$$I_0\dot{\phi}\ddot{\phi} = \lambda_1\dot{x}_1 + \lambda_2\dot{x}_2 = -m(\dot{x}_1\ddot{x}_1 + \dot{x}_2\ddot{x}_2),$$

where (3.24) and (3.25) have been used. This is immediately integrated to yield

$$I_0\dot{\phi}^2 + m(\dot{x}_1^2 + \dot{x}_2^2) = \text{const.}$$

But the constraint equations imply that $\dot{x}_1^2 + \dot{x}_2^2 = R^2\dot{\phi}^2$, so $(I_0 + mR^2)\dot{\phi}^2 = \text{const.}$, or $\dot{\phi} = \text{const.}$ Thus

$$\phi = \phi_0 + \dot{\phi}_0 t.$$

From (3.23) it follows immediately that

$$\psi = \psi_0 + \dot{\psi}_0 t.$$

Inserting these results into the constraint equations and integrating gives

$$\dot{x}_1 = R\dot{\phi}_0 \cos(\psi_0 + \dot{\psi}_0 t); \quad x_1 = x_{10} + R\frac{\dot{\phi}_0}{\dot{\psi}_0}\sin(\psi_0 + \dot{\psi}_0 t);$$

$$\dot{x}_2 = R\dot{\phi}_0 \sin(\psi_0 + \dot{\psi}_0 t); \quad x_2 = x_{20} - R\frac{\dot{\phi}_0}{\dot{\psi}_0}\cos(\psi_0 + \dot{\psi}_0 t).$$

The disk rolls at a constant rate ($\dot{\phi} = \text{const.}$), changing direction also at a constant rate ($\dot{\psi} = \text{const.}$). Its center of mass moves in a circle of radius $\rho = |R\dot{\phi}_0/\dot{\psi}_0|$

centered at (x_{10}, x_{20}), for $(x_1 - x_{10})^2 + (x_2 - x_{20})^2 = \rho^2$. The disk's speed is constant, for $\dot{x}_1^2 + \dot{x}_2^2 = (R\dot{\phi}_0)^2 = \text{const}$.

It is evident from Eqs. (3.24) and (3.25) that the λ_l are the forces that cause the disk to move the way it does (see Problem 1). These are some of the forces of constraint, for without them the disk would just move in a straight line on the plane. They are only some of the constraint forces because we have completely ignored the forces that keep the disk vertical, treating the mass as though it is all concentrated at the contact point on the plane and the disk as though it has no moment of inertia about the horizontal diameter. The x_1 and x_2 components of the forces of constraint are

$$-\lambda_1 = -R\dot{\phi}_0\dot{\psi}_0 \sin(\psi_0 + \dot{\psi}_0 t), \quad -\lambda_2 = R\dot{\phi}_0\dot{\psi}_0 \cos(\psi_0 + \dot{\psi}_0 t).$$

The force of constraint is perpendicular to the velocity (a centripetal force) and is provided by friction with the plane.

For a somewhat different treatment of this problem, but with the same constraints given by equations quadratic in the velocities, see Saletan and Cromer (1971, p. 131).

3.2   SYMMETRY AND CONSERVATION

One of the most important insights gained from the Lagrangian formalism is the relation of conservation laws to *symmetries* of dynamical systems. A system is said to be symmetric if some change imposed on it does not effectively alter it. For instance, the Coulomb potential is *symmetric under rotation* about the origin: since $V = -\alpha/r$ does not depend on θ, any change in the angle does not alter V. In this section we discuss such symmetries and the way they relate to constants of the motion.

3.2.1   CYCLIC COORDINATES

INVARIANT SUBMANIFOLDS AND CONSERVATION OF MOMENTUM

An equation of the form $\Gamma(q, \dot{q}) = C$ defines a $(2n - 1)$-dimensional submanifold of $\mathbb{TQ}$. If the dynamical variable Γ is a constant of the motion and if, as was discussed in Chapter 2, the motion starts on the submanifold, it will remain on it: $\Gamma = C$ is an *invariant submanifold* for the dynamical system. This reduces the dimension of the dynamical system by one and in principle simplifies the problem. An example of how useful this can be was demonstrated in Section 2.3.1 when the two-particle central force problem was reduced from a twelve-dimensional dynamical system to one of just two dimensions (from six freedoms to one). This does not mean, however, that the entire motion is obtained from simple one-freedom dynamics [see the discussion following Eq. (2.65)].

Sometimes the EL equations provide a constant of the motion in a very simple way. If one of the generalized coordinates, say q^β, does not appear in the Lagrangian (is a *cyclic* or *ignorable* coordinate), $\partial L / \partial q^\beta = 0$ and the βth EL equation reads

$$\frac{d}{dt}\frac{\partial L}{\partial \dot{q}^\beta} = 0. \tag{3.26}$$

This obviously implies that $\partial L/\partial \dot{q}^{\beta}$ is a constant of the motion. The function $p_{\beta}(q, \dot{q}, t)$ defined by

$$p_{\beta} \equiv \frac{\partial L}{\partial \dot{q}^{\beta}} \qquad (3.27)$$

is called the *generalized momentum conjugate to* q^{β}. If q^{β} is ignorable, its conjugate momentum p_{β} provides a set of invariant submanifolds in **T**Q. That is, if the initial phase point lies on any submanifold whose equation is of the form $p_{\beta} = C$, the motion stays on that submanifold.

This is a hint of the usefulness of changing the coordinates on **T**Q from $(q, \dot{q})$ to (q, p). Later, in the Hamiltonian formulation of Chapter 5, we will do just that, but it involves going over from the tangent bundle to another manifold called the *cotangent bundle*.

Suppose, more generally, that some dynamical variable $\Gamma(q, \dot{q})$, not necessarily one of the p_{β}, is known to be a constant of the motion. Its value C is determined by the initial conditions $\Gamma(q_0, \dot{q}_0) = C$, and the equation $\Gamma(q, \dot{q}) = C$ defines a $(2n - 1)$-dimensional invariant submanifold (hypersurface) in **T**Q. Each different C defines a different invariant submanifold, and the phase orbits all lie on such invariant submanifolds. This *reduces* the $2n$-dimensional dynamical system on **T**Q to a set of new systems, each of dimension $2n - 1$ and each labeled by its particular value of Γ. In this way each constant of the motion simplifies a dynamical system by reducing its dimension by one. For this reason there are advantages to finding coordinate systems with ignorable coordinates: the momenta conjugate to the ignorable qs are immediately known to be constants of the motion.

For an example we turn to the Lagrangian (2.60) for the plane central force problem:

$$L = \frac{1}{2}\mu(\dot{r}^2 + r^2\dot{\theta}^2) - V(r).$$

The angle θ is an ignorable coordinate, so its conjugate momentum, the angular momentum $p_{\theta} \equiv \partial L/\partial \dot{\theta}$ is conserved. Each of its invariant submanifolds, given by

$$p_{\theta} \equiv \mu r^2 \dot{\theta} = l$$

with different values l, defines a new one-freedom dynamical system, as was discussed in Section 2.3.

All this can be put in terms of *symmetry*, as defined at the beginning of this section. If q^{λ} is ignorable, then $\partial L/\partial q^{\lambda} = 0$, which means that L does not change as q^{λ} varies, or L is *symmetric* (or, as we will often say, *invariant*) under translation in the q^{λ} direction in **T**Q.

TRANSFORMATIONS, PASSIVE AND ACTIVE

For the purposes of Section 3.2.2 it is useful to restate the arguments of the preceding subsection in yet another way, in terms of coordinate transformations. Consider the special family of coordinate transformations on Q that shifts just the ignorable q^{λ} by a variable amount ϵ. These transformations on $\mathbb{Q}$ induce others on $\mathbf{T}\mathbb{Q}$; if the new coordinates are called $(Q, \dot{Q})$, the transformation to these new ones from the old ones is given by

$$Q^\alpha = q^\alpha + \epsilon \delta^{\alpha\lambda}, \quad \dot{Q}^\alpha = \dot{q}^\alpha. \tag{3.28}$$

Only Q^λ actually changes; every other Q^α is equal to its q^α. Under this ϵ-family of transformation the Lagrangian becomes [L_ϵ corresponds to the L' of Eq. (2.36)]

$$L_\epsilon(Q, \dot{Q}, t) \equiv L(q(Q, t), \dot{q}(Q, \dot{Q}, t), t), \tag{3.29}$$

or, briefly, $L_\epsilon(Q, \dot{Q}, t) = L(q, \dot{q}, t)$. Notice that what enters on the right-hand side is the inverse transformation of (3.28), the one from $(Q, \dot{Q})$ to $(q, \dot{q})$. Although the ϵ that appears here plays a somewhat different role than the ϵ of the variational principle, some equations involving ϵ are the same in both situations. In particular, derivatives of L with respect to ϵ can be calculated the same way. That is,

$$\frac{\partial L_\epsilon}{\partial \epsilon} = \frac{\partial L}{\partial q^\alpha} \frac{\partial q^\alpha}{\partial \epsilon} + \frac{\partial L}{\partial \dot{q}^\alpha} \frac{\partial \dot{q}^\alpha}{\partial \epsilon} = -\frac{\partial L}{\partial q^\alpha} \delta^{\alpha\lambda} = -\frac{\partial L}{\partial q^\lambda}. \tag{3.30}$$

So far we have not used the invariance of L under this family of transformations: Eqs. (3.29) and (3.30) tell how a function L changes in general. But if L is invariant; it is the same function of the new coordinates as it was of the old ones (e.g., if it had been $\dot{q}^2 + q^2$, it would now be $\dot{Q}^2 + Q^2$). In other words,

$$L_\epsilon(q, \dot{q}, t) = L(q, \dot{q}, t) \tag{3.31}$$

[or, equivalently, $L_\epsilon(Q, \dot{Q}, t) = L(Q, \dot{Q}, t)$], and then from (3.30)

$$\frac{\partial L_\epsilon}{\partial \epsilon} \equiv -\frac{\partial L}{\partial q^\lambda} = 0. \tag{3.32}$$

Thus to say that L is invariant under translation in q^λ is the same as to say that it is invariant under this ϵ-family of transformations. [Notice the minus sign in the last part of (3.32), which arises from the fact that the inverse of (3.28) is being used.]

There are two ways to view a set of transformation equations like Eq. (3.28) and two ways to interpret Eqs. (3.29), called the *passive* and *active* views. We will use Fig. 3.3 as an example in terms of which to explain these two views. We take the transformation to be a rotation of the two-plane $\mathbf{T}\mathbb{Q}$ by $\pi/4$ and the Lagrangian to be $L(q, \dot{q}) = \exp[-(q - \sqrt{2}a)^2 - \dot{q}^2]$ on the plane. This is graphed in Fig. 3.3(a). The maximum of L occurs at $(q, \dot{q}) = (\sqrt{2}a, 0)$ and it is circularly symmetric about that point. Incidentally, this is in no sense a realistic Lagrangian.

We start with the passive view. Figures 3.3(a) and (b) show two coordinate systems on $\mathbf{T}\mathbb{Q}$ obtained from each other by a rotation of $\pi/4$. In both figures the hill illustrates a certain map $\tilde{L}$ from $\mathbf{T}\mathbb{Q}$ to $\mathbb{R}$, and L represents this map in the $(q, \dot{q})$ coordinates. In the $(Q, \dot{Q})$ coordinates the *same* $\tilde{L}$ is represented by the *different* functional form

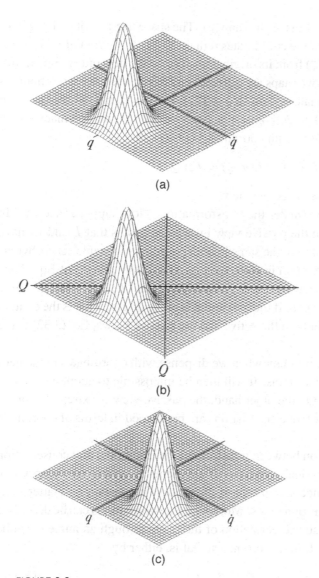

(a)

(b)

(c)

FIGURE 3.3

Active and passive views of the same transformation. (a) A graph of the (quite unrealistic) Lagrangian $L(q, \dot{q}) = \exp[-(q - \sqrt{2}a)^2 - \dot{q}^2]$. (b) Under a passive coordinate rotation by the angle $\epsilon = \pi/4$ in $\mathbf{TQ}$ the expression for L in the new coordinates is $L_\epsilon(Q, \dot{Q}) = \exp[-(Q - a)^2 - (\dot{Q} - a)^2]$. (c) Under an active transformation L changes to a new function whose expression in terms of the original coordinates is $L_\epsilon(q, \dot{q}) = \exp[-(q - a)^2 - (\dot{q} - a)^2]$.

$L_\epsilon = \exp[-(q - a)^2 - (\dot{q} - a)^2]$. In this passive view the plane and all geometric objects on it (points, vectors, maps such as $\tilde{L}$) remain fixed, but their functional forms change as the coordinates change under them. A given point $\xi \in \mathbf{TQ}$, with coordinates $(q, \dot{q})$ in one system and $(Q, \dot{Q})$ in the other, is sent by $\tilde{L}$, L, and L_ϵ to the same value in $\mathbb{R}$:

$$\tilde{L}(\xi) = L(q, \dot{q}) = L_\epsilon(Q, \dot{Q}).$$

That is the content of Eq. (3.29) in the passive view.

The active view is illustrated in Figs. 3.3(a) and (c). The space itself is rotated, together with all geometric objects on it, but the coordinates remain fixed. A rotation of $\pi/4$ (in the opposite sense) moves the map $\tilde{L}(\xi)$ from its original position to the new one in Fig. 3.3(c); the new map is called $\tilde{L}_\epsilon(\xi)$. The two maps are different and have different representations L and L_ϵ in the fixed $(q, \dot{q})$ coordinate system. If $\xi \in \mathbf{TQ}$ is held *fixed*, not allowed to move under the rotation, $\tilde{L}$ and $\tilde{L}_\epsilon$ map ξ to *different* values in $\mathbb{R}$. But if the rotation maps ξ with coordinates $(q, \dot{q})$ to the new point ξ_ϵ with coordinates $(Q, \dot{Q})$, then

$$\tilde{L}(\xi) = \tilde{L}_\epsilon(\xi_\epsilon) = L(q, \dot{q}) = L_\epsilon(Q, \dot{Q}).$$

That is the content of Eq. (3.29) in the active view.

So far L has not been invariant under the transformation. Now suppose that it is. In Fig. 3.3 that means that $a = 0$. In the passive view invariance means that L and L_ϵ have the same form – that the expressions are the same in both coordinate systems, as is obvious in the figure if $a = 0$ and the maximum is at the origin. That is the content of Eq. (3.31) in the passive view. In the active view invariance means that L and L_ϵ are in fact the same function, as is also obvious in the figure if the maximum is at the origin. That is the content of Eq. (3.31) in the active view. In both the active and the passive views, Eq. (3.32) is the differential form of (3.31).

The distinction will become important when we dispense with coordinates in the geo-metrical description of dynamical systems. It will then be impossible to adopt the passive view, for it requires coordinates. On the other hand, the passive view is extremely conve-nient, even necessary, when quantitative calculations are performed in terms of coordinate systems.

When we discussed the relation between transformations (rotations) and conservation we took the passive view. Yet physical properties of a dynamical system are independent of coordinate systems. For instance, a constant of the motion is constant no matter what coordinates are used to describe the dynamics. Moreover, transformations can be described in any coordinate system. For instance, a rotation of the plane through an angle ϵ can be described either in polar or in Cartesian coordinates, that is, either by

$$r' = r, \quad \theta' = \theta + \epsilon \tag{3.33a}$$

or by

$$\left.\begin{array}{l} x' = x \cos\epsilon - y \sin\epsilon, \\ y' = x \sin\epsilon + y \cos\epsilon. \end{array}\right\} \tag{3.33b}$$

The invariance of a Lagrangian can be expressed in either of these systems. In polar coordinates that means that θ is cyclic: $\partial L_\epsilon/\partial\epsilon = -\partial L/\partial\theta = 0$, as in Eq. (3.32), and this leads immediately to conservation of angular momentum. In Cartesian coordinates, invariance under rotation, and therefore also conservation of angular momentum, is less transparent, for there is no cyclic coordinate. The general way to connect invariance to conservation will be discussed in the next subsection, on the Noether theorem.

In conclusion, each cyclic coordinate reduces the dimension by one and yields a conserved quantity.

THREE EXAMPLES

In the free-particle Lagrangian $L = \frac{1}{2}m(\dot{x}_1^2 + \dot{x}_2^2 + \dot{x}_3^2)$ all the Cartesian coordinates are cyclic, so all three conjugate momenta $p_k \equiv \partial L / \partial \dot{x}_k = m\dot{x}_k$ are constants of the motion. They define a set of three-dimensional invariant submanifolds of the six-dimensional $\mathbf{TQ}$, one submanifold for each set of values of the p_k. Each of these submanifolds is labeled by the value of $\mathbf{p}$, and all that varies is $\mathbf{x}$. Another example is the Lagrangian of a particle in a uniform gravitational field, $L = \frac{1}{2}m(\dot{x}_1^2 + \dot{x}_2^2 + \dot{x}_3^2) - mgx_3$. Now $\partial L / \partial x_3 \neq 0$, so only p_1 and p_2 are constants of the motion: each pair of values for p_1 and p_2 defines a four-dimensional submanifold of $\mathbf{TQ}$ on which all that varies is $\mathbf{x}$ and $\dot{x}_3$.

Next, consider the plane central-force Lagrangian of Eq. (2.60), namely

$$L = \frac{1}{2}\mu(\dot{r}^2 + r^2\dot{\theta}^2) - V(r). \tag{3.34}$$

The tangent bundle $\mathbf{TQ}$ has four dimensions, with the four coordinates $r, \dot{r}, \theta, \dot{\theta}$. Equation (2.62),

$$\mu r^2 \dot{\theta} = l, \tag{3.35}$$

defines a set of three-dimensional invariant submanifolds of $\mathbf{TQ}$ labeled by the value of the conserved angular momentum l. By determining the value of l, the initial conditions determine which of these invariant submanifolds carries the dynamics, and what remain are the three coordinates $r, \dot{r}$, and θ ($\dot{\theta}$ is determined by r and l). At this point the differential equations to be solved are the second-order equation (2.63) for $r(t)$, namely

$$\mu\ddot{r} - \frac{l^2}{\mu r^3} + \frac{dV}{dr} = 0, \tag{3.36}$$

and the first-order equation (3.35) for $\theta(t)$. Hence three integrations must be performed. There are two simplifying occurrences in this example: First, the r and θ parts separate. That kind of separation will not always occur, for the remaining equations may couple the dimensions inextricably. Second, the r equation is of the Lagrangian type. That also will not always occur.

Indeed, Eq. (3.36) for $r(t)$ can be obtained from the Lagrangian $\mathcal{L}$ of Eq. (2.65). This is a separate Lagrangian dynamics that evolves on the two-dimensional $\mathbf{TQ}$ described by the coordinates r and $\dot{r}$. In this dynamics the energy E is a constant of the motion, and that defines a set of one-dimensional submanifolds, one for each value of E. The remaining first-order (and hence non-Lagrangian) dynamics is given by the differential equation

$$E = \frac{1}{2}\mu\dot{r}^2 + \frac{l^2}{2\mu r^2} + V(r). \tag{3.37}$$

The manifold on which this dynamics evolves is $\mathbb{R}^+$, described by the single coordinate r, not the tangent bundle of anything. Its solution is the single function $r(t)$. Once this function is known, Eq. (3.35) has to be solved for $\theta(t)$ on the circle $\mathbb{S}$, also a manifold of dimension one.

This shows how the two constants of the motion l and E have reduced the dimension from four to two, for there are two equations left: (3.37) and (3.35). Neither of these equations is itself of the Lagrangian type. Lagrangian dynamics always evolves on a tangent bundle, a manifold whose dimension is even. The remaining equations are of first order and evolve on one-dimensional manifolds.

A final example is that of the two-dimensional harmonic oscillator. The equations of motion are (no sum on j in this example)

$$\ddot{x}_j + \omega_j^2 x_j = 0, \quad j = 1, 2. \tag{3.38}$$

Two immediate constants of the motion are

$$E_j = \frac{1}{2}\dot{x}_j^2 + \frac{1}{2}\omega_j^2 x_j^2, \quad j = 1, 2. \tag{3.39}$$

Given a pair of values for the E_j, each defining an ellipse in the $(x_j, \dot{x}_j)$ plane, the system is reduced by two dimensions. The pair of ellipses defines a two-dimensional torus $\mathbb{T}$ in the initial four-dimensional $\mathbf{T}\mathbb{Q}$, so the reduced system lies on this torus. (The torus is generated by attaching a circle to each point of another circle, more or less the way the cylinder was generated in Section 2.4.1 by attaching a line to each point of a circle. For this construction ellipses can be used in place of circles. More about tori is presented in Chapter 6.) The reduced dynamical system on $\mathbb{T}$ is separated into two independent equations, one for x_1 and the other for x_2. As in the previous case, we are left with two separated one-dimensional systems.

3.2.2 NOETHER'S THEOREM

POINT TRANSFORMATIONS

In this subsection we show how the connection between invariance and conservation can be understood in coordinate systems in which there are no ignorable coordinates. (Later, in Section 3.4.3 we will show how it can be understood in a completely coordinate-free way.) The connection is made through Noether's theorem: if a Lagrangian is invariant under a family of transformations, its dynamical system possesses a constant of the motion, and that constant can be found from a knowledge of the Lagrangian and the transformation.

The theorem involves not single discrete transformations φ, but continuous families $\varphi(\epsilon)$ of them, as in Eqs. (3.28) and (3.33). The transformations we have dealt with in the preceding subsection were transformations on $\mathbb{Q}$ (called *point transformations*), whereas the Lagrangians, whose transformation properties we will be studying, are functions on $\mathbf{T}\mathbb{Q}$. This is possible because transformations on $\mathbb{Q}$ induce transformations on $\mathbf{T}\mathbb{Q}$. We will first explain how this comes about before going on to the theorem itself.

Equations (3.28), (3.33a), and (3.33b) are examples of ϵ-families $\varphi(\epsilon)$ of transformations on $\mathbb{Q}$. Viewed passively, each q (i.e., each set of coordinates q^α) is in general changed, for each value of ϵ, to some other $q(\epsilon)$. We assume that $\varphi(0)$ is the identity transformation, so $q(0) = q$. In this way each trajectory $q(t)$ has its coordinates changed to new coordinates

$q(t, \epsilon)$. The new coordinates satisfy

$$\dot{q}(t, \epsilon) = \frac{\partial}{\partial t} q(t, \epsilon) \tag{3.40}$$

(the partial derivative is taken with ϵ held fixed; recall the similar considerations around Eq. (3.5), where the partial with respect to t was written d/dt). This equation shows how the components of $\dot{q}$ vary with ϵ and, since $\dot{q}$ represents the local coordinate on the fiber (tangent space) in T$\mathbb{Q}$, it shows how the coordinate transformation on $\mathbb{Q}$ induces one on T$\mathbb{Q}$. That is, $\dot{q}(t) \equiv \dot{q}(t, 0)$ is changed to $\dot{q}(t, \epsilon)$. This transformation on T$\mathbb{Q}$ is called T$\varphi(\epsilon)$. To describe it geometrically, we take the active view: $\dot{q}(t, 0)$ is a vector in the tangent space at $\mathbf{T}_{q(0)}\mathbb{Q}$, and $\dot{q}(t, \epsilon)$ is a vector in the tangent space at $\mathbf{T}_{q(\epsilon)}\mathbb{Q}$. Then T$\varphi(\epsilon)$ maps each $q \equiv q(0) \in \mathbb{Q}$ to a new point $q(\epsilon) \in \mathbb{Q}$, each trajectory $q(t, 0) \equiv q(t)$ to a new trajectory $q(t, \epsilon)$, and each vector $\dot{q} \in \mathbf{T}_q\mathbb{Q}$ to a vector $\dot{q}(\epsilon) \in \mathbf{T}_{q(\epsilon)}\mathbb{Q}$.

From now on we will write φ_ϵ instead of $\varphi(\epsilon)$ because it is convenient to use the functional notation to describe where the map sends points: $q(\epsilon) = \varphi_\epsilon(q)$, and we will often suppress the ϵ dependence, as in Eq. (3.42) below, writing $q(\epsilon) = \varphi(q)$. The dependence of φ will always be restricted by requiring that the family of transformations start with the identity transformation at $\epsilon = 0$:

$$\varphi_0(q) \equiv q. \tag{3.41}$$

Under transformations of the kind being discussed, the Lagrangian changes in accordance with Eq. (3.29), which can now be written [here $\xi \equiv (q, \dot{q})$ is a point in T$\mathbb{Q}$]

$$L_\epsilon(\mathbf{T}\varphi(\xi), t) = L(\xi, t). \tag{3.42}$$

This can also be written in the form

$$L_\epsilon\left(\varphi(q), \frac{\partial}{\partial t}\varphi(q), t\right) = L(q, \dot{q}, t), \tag{3.43}$$

where $\partial\varphi(q)/\partial t$ is $\partial q(t, \epsilon)/\partial t$ [see Eq. (3.40)]. It is convenient to put these equations in terms of the inverses of φ and Tφ. Let $\psi \equiv \varphi^{-1}$ be the inverse of φ and T$\psi \equiv (\mathbf{T}\varphi)^{-1}$ be the inverse of Tφ (see Problem 7). Then the above two equations are (ψ also depends on ϵ)

$$L_\epsilon(\xi, t) = L(\mathbf{T}\psi(\xi), t) \tag{3.44}$$

and

$$L_\epsilon(q, \dot{q}, t) = L\left(\psi(q), \frac{\partial}{\partial t}\psi(q), t\right). \tag{3.45}$$

THE THEOREM
So far we have said nothing about Lagrangians that are invariant under the transformations. Assume now that L is invariant under the family of transformations, so that the

derivative of L_ϵ with respect to ϵ vanishes. The derivative can be calculated from Eq. (3.45) by using Eq. (3.6) (that was our reason for using the same parameter ϵ in both contexts). Recall that Eq. (3.6) was obtained without any restrictions on the variation with respect to ϵ, so it can be used here. What was called there the total time derivative has been here called the partial time derivative, but we now return to calling it the total time derivative. In addition, to emphasize that the derivative with respect to ϵ is also being taken, we replace the δ of Eq. (3.6) by $\partial/\partial\epsilon$. Then Eq. (3.6) becomes

$$\frac{\partial L_\epsilon}{\partial \epsilon} = \left[\frac{\partial L}{\partial q^\alpha} - \frac{d}{dt}\frac{\partial L}{\partial \dot{q}^\alpha} \right] \frac{\partial \psi(q^\alpha)}{\partial \epsilon} + \frac{d}{dt}\left[\frac{\partial L}{\partial \dot{q}^\alpha} \frac{\partial \psi(q^\alpha)}{\partial \epsilon} \right], \qquad (3.46)$$

where $\psi(q^\alpha)$ is what was previously called $q^\alpha(\epsilon)$ (recall again that ψ is ϵ dependent).

As before, $\psi_0(q(t)) \equiv q(t)$. Assume that $q(t)$ is a solution of the equations of motion. Then the EL equations imply that the first term on the right-hand side of (3.46) vanishes, so that (the derivative at $\epsilon = 0$ is written δ)

$$\delta L_\epsilon = \frac{d}{dt}\left[\frac{\partial L}{\partial \dot{q}^\alpha} \delta q^\alpha \right]. \qquad (3.47)$$

We now insert the condition that L is invariant under the transformation, that is, $\partial L_\epsilon/\partial\epsilon = 0$ (actually, it is enough to suppose that L is invariant at $\epsilon = 0$, i.e., that $\delta L_\epsilon = 0$). Then

$$\frac{d}{dt}\left[\frac{\partial L}{\partial \dot{q}^\alpha} \delta q^\alpha \right] = 0,$$

or

$$\Gamma \equiv \frac{\partial L}{\partial \dot{q}^\alpha} \delta q^\alpha \qquad (3.48)$$

is a constant of the motion.

This is Noether's theorem: if L is invariant under an ϵ family of point transformations ψ_ϵ (i.e., transformations on $\mathbb{Q}$), then Γ of Eq. (3.48) is a constant of the motion. Briefly, *to every q-symmetry of the Lagrangian corresponds a constant of the motion.* This is essentially the ignorable-coordinate result, but now stated without recourse to ignorable coordinates.

We now return to a previous example. Rewrite the plane central force problem in Cartesian coordinates:

$$L = \frac{1}{2}m(\dot{x}^2 + \dot{y}^2) + V(r),$$

where $r^2 = x^2 + y^2$ as usual. This Lagrangian is invariant under the rotations (3.33b). To make use of the theorem, calculate the δqs. They are

$$\delta x = -y, \quad \delta y = x,$$

and then

$$\Gamma = m(x\dot{y} - y\dot{x}).$$

This is the angular momentum. Invariance of L under rotation leads to conservation of angular momentum. It has been demonstrated here for only two freedoms and rotations in the plane, but it is true also in 3-space: if L is invariant under rotations in $\mathbb{R}^3$, then all three components of the angular momentum vector are conserved (see Problem 8; consider also Problem 12).

Noether's theorem can be generalized: the Lagrangian need not be strictly invariant under the transformation but can change by the addition of a total time derivative according to

$$L_\epsilon = L + \dot{\Phi}(q, t, \epsilon). \tag{3.49}$$

Recall from Chapter 2 that if two Lagrangians differ by a total time derivative of a function of q and t, they give the same equations of motion. Thus Noether's theorem says that if the Lagrangian changes by the same kind of total time derivative, it will yield a constant of the motion. Indeed, in that case

$$\delta L_\epsilon = \frac{d}{dt}\delta\Phi,$$

and when this is combined with (3.47) it leads to

$$\frac{d}{dt}\left[\frac{\partial L}{\partial \dot{q}^\alpha}\delta q^\alpha - \delta\Phi\right] = 0.$$

This means that the new function

$$\Gamma = \frac{\partial L}{\partial \dot{q}^\alpha}\delta q^\alpha - \delta\Phi \tag{3.50}$$

is a constant of the motion.

This is a more general and hence more important result. Recall that the Lagrangian of a charge in an electromagnetic field changes by a total time derivative when $\mathbf{A}$ and ϕ undergo a gauge transformation. We extend this terminology: a *Lagrangian* undergoes a *gauge transformation* whenever a total time derivative is added to it and is *gauge invariant* under a family of transformations if it satisfies (3.49). Hence Noether's theorem says that *a constant of the motion is associated with every gauge invariance of a Lagrangian under an ϵ family of transformations.*

Noether's theorem is interesting but not of great importance in particle dynamics. It becomes extremely important in field theory (Chapter 9) and in quantum mechanics. We mentioned near the end of Section 3.2.1 that Lagrangians are often chosen partly on the basis of symmetry considerations. Such physical requirements as conservation of certain dynamical variables imply, through Noether's theorem, restrictions on the possible choice of Lagrangian.

WORKED EXAMPLE 3.2

Consider the one-freedom Lagrangian

$$L = \frac{1}{2}m\dot{x}^2 - mgx$$

(a) Show that L is gauge invariant under the point transformation

$$\psi_\epsilon(x) = x + \epsilon.$$

(b) Find the associated constant of the motion.
(c) Use **(b)** to solve for the motion.

Solution. **(a)** Under this transformation $\delta x = 1$, so $\delta L = -mg \equiv -d(mgt)/dt$. Thus with $\delta\Phi = -mgt$

$$\delta L = \frac{d}{dt}\delta\Phi,$$

and L is gauge invariant.
    (b) According to (3.50)

$$m\dot{x} + mgt = \Gamma \tag{3.51}$$

is the associated constant of the motion.
    (c) Integration of (3.51) yields

$$\dot{x}(t) = \frac{\Gamma}{m} - gt,$$

and from this

$$x(t) = x_0 + \frac{\Gamma}{m}t - \frac{1}{2}gt^2.$$

Not very surprisingly, this represents the trajectory of a mass in a uniform gravitational field.

3.3 NONPOTENTIAL FORCES

Not all forces can be derived from potentials. In this section we discuss forces that depend on the velocity, particularly what are called dissipative forces. It should be mentioned that although dissipative forces must derive ultimately from microscopic forces, the connecting chain is usually complicated and their properties are often obtained directly from experiment.

3.3.1 DISSIPATIVE FORCES IN THE LAGRANGIAN FORMALISM

REWRITING THE EL EQUATIONS

In deriving the EL equations of motion for a system of particles we passed through Eq. (2.23), namely

$$\sum_{i=1}^{N} \mathbf{F}_i \cdot \frac{\partial \mathbf{x}_i}{\partial q^\alpha} = -\frac{\partial V}{\partial q^\alpha}. \tag{3.52}$$

Now, however, we no longer assume that there exists a potential energy function V that satisfies this equation. Instead we write the external forces on the system in the form

$$\mathbf{F}_i = \mathbf{F}_{Pi} + \mathbf{F}_{Di}, \tag{3.53}$$

where $\mathbf{F}_{Pi}$ is derivable from a potential $V(\mathbf{x}_1, \mathbf{x}_2, \ldots, \mathbf{x}_N)$ and $\mathbf{F}_{Di}$ is not. Then if D_α is defined by

$$\sum_{i=1}^{N} \mathbf{F}_{Di} \cdot \frac{\partial \mathbf{x}_i}{\partial q^\alpha} \equiv D_\alpha, \tag{3.54}$$

Eq. (3.52) becomes

$$\sum_{i=1}^{N} \mathbf{F}_i \cdot \frac{\partial \mathbf{x}_i}{\partial q^\alpha} = -\frac{\partial V}{\partial q^\alpha} + D_\alpha$$

and the EL equations are modified to read

$$\frac{d}{dt} \frac{\partial L}{\partial \dot{q}^\alpha} - \frac{\partial L}{\partial q^\alpha} = D_\alpha, \tag{3.55}$$

where $L = T - V$, as before.

Since the $\mathbf{F}_{Di}$ (and hence also the D_α) do not arise from a potential energy function, they do not in general lead to conservation of energy. When the energy decreases they are *dissipative* forces, and when it increases they are *driving* forces. The D_α will be called *generalized* dissipative or driving forces because they refer to generalized coordinates. Usually driving forces are time dependent, so here we concentrate mostly on dissipative ones.

THE DISSIPATIVE AND RAYLEIGH FUNCTIONS

Suppose that the nonpotential forces are functions of the velocities and that the force on each particle is directed opposite to its velocity. In particular, assume that (no sum on i in what follows unless indicated by a summation sign)

$$\mathbf{F}_{Di} = -g_i(\mathbf{v}_i)\mathbf{v}_i, \tag{3.56}$$

where g_i is a function only of $\mathbf{v}_i$. Then the dissipative force on the ith particle depends only on its own velocity, and if g_i is positive, the force is directed opposite to the direction of motion. The force then does negative work on the particle as it moves and therefore leads to energy loss. That is, if the g_i are positive, the forces are dissipative.

From Eqs. (3.54), (3.56), and (2.25) it follows that

$$D_\alpha = - \sum_{i=1}^{N} g_i(\mathbf{v}_i)\mathbf{v}_i \cdot \frac{\partial \mathbf{v}_i}{\partial \dot{q}^\alpha}.$$

But $\mathbf{v}_i \cdot \mathbf{v}_i = v_i^2$ implies that (remember: no sum on i)

$$\mathbf{v}_i \cdot \frac{\partial \mathbf{v}_i}{\partial \dot{q}^\alpha} = v_i \frac{\partial v_i}{\partial \dot{q}^\alpha},$$

so that

$$D_\alpha = - \sum_{i=1}^{N} g_i(\mathbf{v}_i) v_i \frac{\partial v_i}{\partial \dot{q}^\alpha}. \tag{3.57}$$

Assume further that the g_i depend only on the magnitudes of the velocities. This is physically quite reasonable. It implies that there is no preferred direction for the dissipative forces; the magnitude of a friction-like force on a particle, for example, depends on the speed but not on the direction in which the particle is moving. Then (3.57) becomes

$$D_\alpha = - \sum_{i=1}^{N} g_i(v_i) v_i \frac{\partial v_i}{\partial \dot{q}^\alpha} \equiv -\frac{\partial \mathcal{F}}{\partial \dot{q}^\alpha}, \tag{3.58}$$

where the last equality (the identity) defines the *dissipative function* $\mathcal{F}$. The solution of the defining differential equation for $\mathcal{F}$ is

$$\mathcal{F} = \sum_{i=1}^{N} \int_0^{v_i} g_i(z) z \, dz. \tag{3.59}$$

Here z is a variable of integration. That (3.59) is a solution of (3.58) can be verified by taking its derivative with respect to $\dot{q}^\alpha$ with the v_i treated as functions of $\dot{q}$. Although in Cartesian coordinates $\mathcal{F}$ is a function only of the velocities $\mathbf{v}_i$, when written out in terms of generalized coordinates it is not a function only of $\dot{q}$, for each Cartesian velocity $\mathbf{v}_i = \dot{q}^\alpha \partial \mathbf{x}_i / \partial q^\alpha + \partial \mathbf{x}_i / \partial t$ is a function of q as well as of $\dot{q}$. If the g_i are positive, so is $\mathcal{F}$: when the force is dissipative, $\mathcal{F}$ is a positive function.

The modified EL equations now read

$$\frac{d}{dt} \frac{\partial L}{\partial \dot{q}^\alpha}{}^\alpha - \frac{\partial L}{\partial q^\alpha} + \frac{\partial \mathcal{F}}{\partial \dot{q}^\alpha} = 0. \tag{3.60}$$

The rate of energy loss (actually the rate of change of E) is given by

$$\frac{dE}{dt} \equiv \frac{d}{dt} \left(\dot{q}^\alpha \frac{\partial L}{\partial \dot{q}^\alpha} - L \right) = \ddot{q}^\alpha \frac{\partial L}{\partial \dot{q}^\alpha} + \dot{q}^\alpha \frac{d}{dt} \frac{\partial L}{\partial \dot{q}^\alpha} - \frac{dL}{dt}$$

$$= \dot{q}^\alpha \left(\frac{d}{dt} \frac{\partial L}{\partial \dot{q}^\alpha} - \frac{\partial L}{\partial \dot{q}^\alpha} \right) = -\dot{q}^\alpha \frac{\partial \mathcal{F}}{\partial \dot{q}^\alpha}. \tag{3.61}$$

In Cartesian coordinates this is $-\sum \mathbf{v}_i \cdot \mathbf{F}_{Di}$, which is the usual expression for the rate of energy loss (power loss).

When the $g_i = b_i$ are positive constants, Eq. (3.56) implies that the dissipative forces are proportional to the speed. They are then sometimes called frictional forces, although experiments usually give a different velocity dependence for real frictional forces (Krim, 1996). The function $\mathcal{F}$ is then called the *Rayleigh function* and is given by

$$\mathcal{F} = \sum_{i=1}^{N} \frac{1}{2} b_i v_i^2. \tag{3.62}$$

3.3.2 THE DAMPED HARMONIC OSCILLATOR

In anticipation of later chapters, we turn to the best-known and simplest example of a force proportional to the velocity, namely the damped harmonic oscillator. We start with a slightly more general problem in order to demonstrate why the harmonic oscillator arises so often. Consider a system in one-freedom whose Lagrangian is

$$L = \frac{1}{2} \tau(q) \dot{q}^2 - V(q), \tag{3.63}$$

where τ is a positive definite function (i.e., greater than zero for all q) and V has a minimum at some value q_0 of q. For a small displacement x from q_0, the potential V can be expanded in a Taylor series of the form

$$V(q) = V(q_0) + V'(q_0)x + \frac{1}{2}V''(q_0)x^2 + \frac{1}{6}V'''(q_0)x^3 + \cdots.$$

Because q_0 is a minimum for V, the first derivative vanishes. Every potential energy function is defined only up to an additive constant, which can always be chosen so that $V(q_0) = 0$; then the series becomes (we change the argument of V from q to x)

$$V(x) = \frac{1}{2}kx^2 + \frac{1}{6}sx^3 + \cdots, \tag{3.64}$$

where $k \equiv V''(q_0)$ and $s \equiv V'''(q_0)$.

Now expand also the kinetic energy term, again writing everything in terms of x (note that $\dot{x} = \dot{q}$), to obtain

$$\frac{1}{2}\tau(q)\dot{q}^2 = \frac{1}{2}m\dot{x}^2 + \frac{1}{2}rx\dot{x}^2 + \cdots,$$

where $m \equiv \tau(q_0)$ is positive (by the assumption that τ is a positive definite function) and $r \equiv \tau'(q_0)$. Then if $k \neq 0$, for small enough displacements from equilibrium the Lagrangian can be approximated by

$$L = \frac{1}{2}m\dot{x}^2 - \frac{1}{2}kx^2, \tag{3.65}$$

which is the Lagrangian of the harmonic oscillator. In other words, for sufficiently small vibrations about an equilibrium point, every time-independent one-freedom dynamical system looks like a harmonic oscillator as long as $V'' \neq 0$ at the point of equilibrium.

If $V''(q_0) = 0$ but $V'''(q_0) \neq 0$, the force is quadratic in x and hence always of the same sign, always causing accelerations in the same direction. Then it is not a restoring force, and the equilibrium is not stable. More generally, for oscillations to occur, the lowest power in the Taylor expansion of the potential must be even. If $V'''(q_0)$ also vanishes, the first term in the expansion is quartic and although oscillations will occur, the system is *nonlinear*.

The Lagrangian of Eq. (3.65) is the undamped harmonic oscillator. We now add a damping force by introducing the Rayleigh function $\mathcal{F} = \hat{b}\dot{x}^2$. Then according to (3.60) the equation of motion is

$$m\ddot{x} + 2\hat{b}\dot{x} + kx = 0. \tag{3.66}$$

This is now the damped harmonic oscillator. Two forms of its solution are

$$\left. \begin{aligned} x(t) &= e^{-\beta t}(Ae^{i\omega t} + A^*e^{-i\omega t}), \\ x(t) &= e^{-\beta t}(a\cos\omega t + b\sin\omega t), \end{aligned} \right\} \tag{3.67}$$

where $\beta = \hat{b}/m$ is the *damping factor* and $\omega = \sqrt{\omega_0^2 - \beta^2}$, with $\omega_0 = \sqrt{k/m}$. The initial conditions determine the two real constants of integration (both contained in the complex constant A in the first form and in a and b in the second).

This system is still linear, because x and its derivatives appear linearly in Eq. (3.66). If $V''(q_0) = 0$ (as mentioned above) or if the dissipation is not proportional to $\dot{x}$, that is, if $\mathcal{F}$ is not quadratic in $\dot{x}$, some higher powers of x and $\dot{x}$ will appear, causing the system to become nonlinear. Nonlinear systems are considerably more complicated than linear ones. They will be discussed in some detail in Chapter 7.

The phase portrait of the undamped oscillator (i.e., with $\beta = 0$) is shown in Fig. 3.4. The phase portrait consists of nested ellipses whose semi-axes are $c \equiv \sqrt{a^2 + b^2} = |A|^2$ along the x axis and ωc along the $\dot{x}$ axis (for the undamped case $\omega = \omega_0$). Each ellipse corresponds to a given pair of values for a and b, and the energy of the motion is directly related to the area of the ellipse (Problem 15).

When $\beta \neq 0$, the phase portrait consists of curves converging on the origin of $\mathbf{TQ} = \mathbb{R}^2$, as shown in Fig. 3.5.

If $\beta < \omega_0$, so that ω is real, the curves are spirals, like the one shown in Fig. 3.5(a), and the origin of $\mathbf{TQ}$ is called the *focal point* of the dynamics. The system oscillates about $x = 0$, with the amplitude decreasing in each oscillation. This is called the *underdamped* case.

If $\beta > \omega_0$, then $\omega \equiv i\nu$ is imaginary and the system makes at most one oscillation (depending on the initial conditions) before converging to $x = 0$. This is called the *overdamped* case. Choose $\nu > 0$ and note that $\beta > \nu$. In terms of the real constant ν,

$$x(t) = e^{-\beta t}\{ae^{-\nu t} + be^{\nu t}\}$$

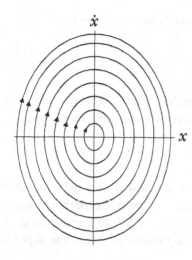

FIGURE 3.4
The phase portrait of the undamped harmonic oscillator. There is an elliptic singular point at $(x, \dot{x}) = (0, 0)$.

and

$$v(t) = -\beta x(t) + ve^{-\beta t}\{-ae^{-\nu t} + be^{\nu t}\}.$$

For large t

$$\lim_{t \to \infty} \frac{v}{x} = -\beta + \nu, \tag{3.68}$$

which means that as the system approaches the origin of $\mathbf{TQ}$ it also approaches the line that passes through the origin with (negative) slope $\nu - \beta$, as in Fig. 3.5(b). In this case the origin is called a *nodal point*.

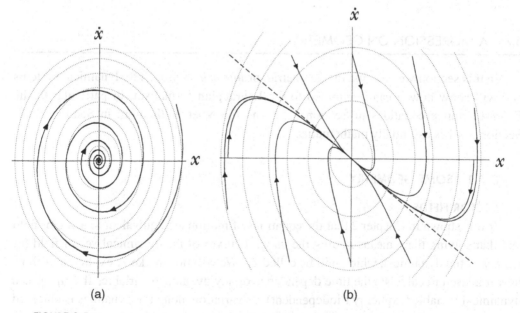

(a)   (b)

FIGURE 3.5
(a) One trajectory of the phase portrait of the underdamped harmonic oscillator. There is a focal point at $(x, \dot{x}) = (0, 0)$. The faint dotted ellipses are constant-energy curves. (b) The phase portrait of the overdamped harmonic oscillator. There is a nodal point at $(x, \dot{x}) = (0, 0)$. The faint dotted ellipses are constant-energy curves.

The case in which $\beta = \omega_0$ is called *critically damped*. The solution is then

$$x = (a + bt)e^{-\beta t}, \tag{3.69}$$

as is easily verified (Problem 16).

Incidentally, $\hat{b}$ in Eq. (3.66) can in principle (although unphysically) be taken as negative. Then β is also negative and the origin of $\mathbb{TQ}$ becomes an unstable focal point.

3.3.3 COMMENT ON TIME-DEPENDENT FORCES

We add some words on time-dependent forces. If the potential is time independent, the equations of motion derived from the usual Lagrangian $L = T - V$ are also time independent. If a system is subjected, in addition to such a generalized force $-\partial V/\partial q^\alpha$, to time-dependent but q-independent forces $F_\alpha(t)$, it will still fit into the Lagrangian formalism. The time-dependent Lagrangian is then

$$L'(q, \dot{q}, t) = L(q, \dot{q}) + q^\alpha F_\alpha(t). \tag{3.70}$$

Indeed, the equations of motion obtained from L' are

$$m_{\alpha\beta}\ddot{q}^\beta = -\frac{\partial V}{\partial q^\alpha} + F_\alpha(t).$$

We will not go more deeply into the question of time-dependent forces, but in Chapter 4 we will apply this approach to a driven harmonic oscillator.

3.4 A DIGRESSION ON GEOMETRY

In this section we describe the geometric, or *intrinsic* approach to dynamical systems and will show how it can be applied to the Lagrangian formalism, in particular to the Euler–Lagrange equations in Section 3.4.2 and the Noether theorem in Section 3.4.3. Section 3.4.1 is essentially mathematical.

3.4.1 SOME GEOMETRY

VECTOR FIELDS

It was shown in Chapter 2 that the equations of motion are equivalent to a vector field and that solving them means finding the integral curves of the dynamical vector field (or simply of the dynamics), which will be called Δ. We will now make Δ explicit and show how it is used to calculate the time dependence of any dynamical variable. If $F(q, \dot{q})$ is a dynamical variable (explicitly t independent), its variation along the dynamics is obtained by substituting solutions of the equations of motion for q and $\dot{q}$ in the argument of F. Then

$$\dot{F}(q, \dot{q}) \equiv \frac{dF}{dt} = \frac{\partial F}{\partial q^\alpha}\dot{q}^\alpha + \frac{\partial F}{\partial \dot{q}^\alpha}\ddot{q}^\alpha. \tag{3.71}$$

If the solutions are inserted, $\dot{F}$ becomes a function of t. This equation and many of those that follow are valid *locally* (i.e., in each local coordinate chart).

But Eq. (3.71) is useful even before the solutions are obtained. The equations of motion give the accelerations as functions on $\mathbf{TQ}$, and when those are inserted into the right-hand side, (3.71) tells how F varies also as a function on $\mathbf{TQ}$. Since $(\dot{q}, \ddot{q})$ are the components of Δ, the time variation of F is obtained in this way from the vector field itself.

One says that $\dot{F}$ is obtained by *applying* Δ to F. This is often written in the form

$$\dot{F} = \Delta(F), \tag{3.72}$$

where

$$\Delta = \dot{q}^{\alpha} \frac{\partial}{\partial q^{\alpha}} + \ddot{q}^{\alpha} \frac{\partial}{\partial \dot{q}^{\alpha}}. \tag{3.73}$$

Then when Δ, treated as an operator, is applied to F, the result is Eq. (3.71): the operator Δ maps functions on $\mathbf{TQ}$ to other functions on $\mathbf{TQ}$ (the $\ddot{q}^{\alpha}$ must themselves be written out as functions on $\mathbf{TQ}$). An arbitrary vector field X on $\mathbf{TQ}$, not just Δ, can be written locally in the form

$$X = X_1^{\alpha} \frac{\partial}{\partial q^{\alpha}} + X_2^{\alpha} \frac{\partial}{\partial \dot{q}^{\alpha}}, \tag{3.74}$$

where the X_1^{α} and X_2^{α} are functions on $\mathbf{TQ}$. Then every vector field, like Δ, maps functions to functions. This is written $X : \mathcal{F}(\mathbf{TQ}) \rightarrow \mathcal{F}(\mathbf{TQ})$, where $\mathcal{F}(\mathbf{TQ})$ is the space of functions on $\mathbf{TQ}$.

ONE-FORMS

There is another common way to write (3.71), which requires defining a new geometric object on $\mathbf{TQ}$ called a *one-form*. One-forms are linear functionals that map vector fields into functions: if α is a one-form on $\mathbf{TQ}$ and X is a vector field on $\mathbf{TQ}$, then $\alpha(X)$ is a function on $\mathbf{TQ}$. That one-forms are linear means that if X and Y are vector fields and f and g are functions, then

$$\alpha(fX + gY) = f\alpha(X) + g\alpha(Y).$$

The vector fields on $\mathbf{TQ}$ make up a vector space $\mathcal{X}(\mathbf{TQ})$, more or less as defined in the appendix (the difference is that unlike the vectors of the appendix, vector fields can be multiplied by functions, not just constants). The one-forms on $\mathbf{TQ}$ also make up a vector space, which we will call $\mathcal{U}(\mathbf{TQ})$, so that if α and β are one-forms, so is $\gamma \equiv f\alpha + g\beta$, defined by its action on an arbitrary vector field X:

$$\gamma(X) \equiv (f\alpha + g\beta)(X) = f\alpha(X) + g\beta(X).$$

The spaces $\mathcal{X}$ and $\mathcal{U}$ are said to be *dual* to each other. The operation of one-forms on vector fields is a generalization of the inner product on vector spaces, and for that reason we will often write the operation in *Dirac notation* as

$$\alpha(X) \equiv \langle \alpha, X \rangle.$$

In the classical literature one-forms are called *covariant vectors* and vector fields are called *contravariant vectors*.

According to Eq. (3.74) the partial derivatives $\partial/\partial q^{\alpha}$ and $\partial/\partial \dot{q}^{\alpha}$ are themselves vector fields, each with just one component equal to 1 and the rest all zero. In fact (3.74) implies that the $\partial/\partial q^{\alpha}$ and $\partial/\partial \dot{q}^{\alpha}$ form a (local) basis in $\mathcal{X}$, for the general vector field is a linear combination of them. Similarly, $\mathcal{U}$ also has a local basis, whose elements are called dq^{α} and $d\dot{q}^{\alpha}$. They are one-forms defined by their action on vector fields in $\mathcal{X}$ by

$$dq^{\beta}(X) = X_1^{\beta},$$
$$d\dot{q}^{\beta}(X) = X_2^{\beta}.$$

In particular, acting on the $\partial/\partial q^{\alpha}$ and $\partial/\partial \dot{q}^{\alpha}$ they yield local equations that look like the inner products of basis vectors in an orthogonal vector space:

$$\left. \begin{array}{l} dq^{\alpha}(\partial/\partial q^{\beta}) = \delta_{\beta}^{\alpha}, \quad d\dot{q}^{\alpha}(\partial/\partial \dot{q}^{\beta}) = \delta_{\beta}^{\alpha}, \\ dq^{\alpha}(\partial/\partial \dot{q}^{\beta}) = d\dot{q}^{\alpha}(\partial/\partial q^{\beta}) = 0. \end{array} \right\} \tag{3.75}$$

The general one-form can be written locally as

$$\omega = \eta_{\alpha} \, dq^{\alpha} + \zeta_{\alpha} \, d\dot{q}^{\alpha},$$

where the η_{α} and ζ_{α} are functions of $(q, \dot{q})$. As an example, if ω is a one-form, f a function, and X a vector field, then

$$\omega(fX) \equiv f\omega(X) \equiv (f\omega)(X), \tag{3.76}$$

where $f\omega$ is a one-form.

The differential of a function $F \in \mathcal{F}(\mathbf{TQ})$ is the one-form

$$dF \equiv \frac{\partial F}{\partial q^{\alpha}} dq^{\alpha} + \frac{\partial F}{\partial \dot{q}^{\alpha}} d\dot{q}^{\alpha}. \tag{3.77}$$

From (3.71), (3.73), (3.75), and (3.76) it follows that

$$\dot{F} = dF(\Delta) \equiv \langle dF, \Delta \rangle. \tag{3.78}$$

The definitions and notation for one-forms and vector fields are easily extended to any differentiable manifold $\mathbb{M}$, not only $\mathbf{TQ}$, by merely replacing the variables $(q, \dot{q}) \in \mathbf{TQ}$ by general variables $\xi \in \mathbb{M}$.

THE LIE DERIVATIVE

Of course $\dot{F}$ is the time derivative of F along the motion or, as we can now say, along the vector field Δ. This is often called the *Lie derivative with respect to* or *along* Δ and is written, in a notation we will often employ,

$$\Delta(F) = \dot{F} \equiv \mathbf{L}_\Delta F. \tag{3.79}$$

More generally, if $f \in \mathcal{F}$ and $X \in \mathcal{X}$, then

$$X(f) = \mathbf{L}_X(f).$$

This is a particularly useful notation, as the Lie derivative can be generalized and applied not only to functions, but to vector fields and one-forms. We start with the one-form df. The Lie derivative of the one-form df along X is defined as

$$\mathbf{L}_X \, df \equiv d(\mathbf{L}_X f); \tag{3.80}$$

that is, d and $\mathbf{L}_X$ commute when acting on functions. The rest is done by the Leibnitz rule for derivatives of products. For instance, on a general manifold $\mathbb{M}$, if ω is a one-form, it can be written locally (in a chart where the coordinates ξ^α are defined) as $\omega = \omega_\alpha \, d\xi^\alpha$, with $\omega_\alpha \in \mathcal{F}(\mathbb{M})$, and then locally

$$\mathbf{L}_X \omega = \mathbf{L}_X(\omega_\alpha \, d\xi^\alpha) = (\mathbf{L}_X \omega_\alpha) \, d\xi^\alpha + \omega_\alpha (\mathbf{L}_X \, d\xi^\alpha). \tag{3.81}$$

This defines the Lie derivative of a general one-form in each chart and hence everywhere. The Lie derivative of a function is another function, and the Lie derivative of a one-form is another one-form. In general the Lie derivative of any geometric object is another geometric object of the same kind.

The Lie derivative of a vector field is obtained by extending the Leibnitz rule to the "inner product." If Y is a vector field, then

$$\mathbf{L}_X \langle \omega, Y \rangle = \langle \mathbf{L}_X \omega, Y \rangle + \langle \omega, \mathbf{L}_X Y \rangle. \tag{3.82}$$

We state without proof (but see Problem 5) that the Lie derivative of a vector field is the vector field defined by

$$U \equiv \mathbf{L}_X Y = \mathbf{L}_X \mathbf{L}_Y - \mathbf{L}_Y \mathbf{L}_X. \tag{3.83}$$

On the right-hand side $\mathbf{L}_X \mathbf{L}_Y$ is defined by its application to functions: it is applied to a function $f \in \mathcal{F}$ by first applying $\mathbf{L}_Y$ and then applying $\mathbf{L}_X$ to the function $\mathbf{L}_Y f$. We do not prove that the right-hand side of (3.83) is in fact a vector field, but to make it plausible we point out that it has two properties of vector fields: it maps functions to functions and it satisfies the Leibnitz rule when applied to products of functions. The vector field U is called the *Lie bracket* of X and Y and is written as

$$U = [X, Y]. \tag{3.84}$$

The same bracket notation [,] is used in several other contexts. The Lie bracket, like all other brackets, is bilinear, antisymmetric,

$$[X, Y] = -[Y, X],$$

and satisfies the important *Jacobi identity*

$$[X, [Y, Z]] + [Y, [Z, X]] + [Z, [X, Y]] = 0. \tag{3.85}$$

3.4.2 THE EULER–LAGRANGE EQUATIONS

We have now covered the mathematics and now use this to write the EL equations in a coordinate-free geometric way. To do so, take θ_L to be the one-form defined locally by

$$\theta_L = \frac{\partial L}{\partial \dot{q}^\alpha} dq^\alpha. \tag{3.86}$$

It follows that

$$\mathbf{L}_\Delta \theta_L = \left(\mathbf{L}_\Delta \frac{\partial L}{\partial \dot{q}^\alpha} \right) dq^\alpha + \frac{\partial L}{\partial \dot{q}^\alpha} d(\mathbf{L}_\Delta q^\alpha).$$

The $\mathbf{L}_\Delta$ in both terms of this equation acts on functions in $\mathcal{F}(\mathbf{TQ})$, so according to Eq. (3.79) it can be replaced by d/dt, and then the equation becomes

$$\mathbf{L}_\Delta \theta_L = \left(\frac{d}{dt} \frac{\partial L}{\partial \dot{q}^\alpha} \right) dq^\alpha + \frac{\partial L}{\partial \dot{q}^\alpha} d\dot{q}^\alpha.$$

Now the EL equations can be inserted by rewriting the first term on the right-hand side of this equation. This yields

$$\mathbf{L}_\Delta \theta_L = \frac{\partial L}{\partial q^\alpha} dq^\alpha + \frac{\partial L}{\partial \dot{q}^\alpha} d\dot{q}^\alpha.$$

What appears on the right-hand side here is just dL, so the EL equations can be put in the form

$$\mathbf{L}_\Delta \theta_L - dL = 0. \tag{3.87}$$

This is the desired coordinate-free form of writing the equations of motion. The three objects, Δ, θ_L, and L that which appear in this equation are intrinsically geometric, a vector field, a one-form, and a function. They belong to the manifold $\mathbf{TQ}$ itself, rather than to any coordinate system on it. In our formulation they were all defined locally, but the definitions hold in every chart, and hence they hold globally, so that Eq. (3.87) is chart independent. In Chapter 2 we pointed out that the EL equations are coordinate independent. Now we have shown that (and how) they can be written in a consistent coordinate-independent manner.

Equation (3.87) is useful for proving theorems, for obtaining general results, and sometimes for reducing expressions in the early stages of calculations. Explicit quantitative results, however, are almost always obtained in local coordinates. The intrinsic, geometric way of writing classical equations of motion is in this sense similar to Dirac notation in quantum mechanics or to general vector-space notation (as opposed to indicial notation) in performing vector calculations.

3.4.3 NOETHER'S THEOREM

Geometric methods can be used also to prove Noether's theorem in a coordinate-free way. We will sketch the proof without presenting its details.

ONE-PARAMETER GROUPS

A construct used in the proof of the theorem, one that will be important later in the book, is the *one-parameter group* of maps. Before describing the proof itself, we explain what this means; we start with an important example. The dynamics Δ provides a family of maps of $\mathbf{TQ}$ into itself by mapping each arbitrary initial condition $\xi(0) = (q(0), \dot{q}(0))$ at time t into $\xi(t) = (q(t), \dot{q}(t))$. As time flows, each ξ is mapped further and further along the integral curve (trajectory) passing through it, and in this way the entire manifold is mapped continuously into itself by being pulled along the integral curves of Δ. We designate this continuous family of mappings by φ_t^Δ and write $\xi(t) \equiv \varphi_t^\Delta \xi(0)$. The superscript reminds us that the family comes from the vector field Δ, and the subscript names the parameter along the integral curves. The mappings satisfy the equation

$$\varphi_t^\Delta \circ \varphi_s^\Delta = \varphi_{t+s}^\Delta \tag{3.88}$$

($\circ$ is defined above Worked Example 2.5), which makes the family by definition a *one-parameter group* of transformations. Although we will not go into the technical details of what that means (see Saletan & Cromer, p. 315), we will refer to φ_t^Δ as the *dynamical* group. To prove that the dynamical group satisfies (3.88) observe where $\varphi_t^\Delta \varphi_s^\Delta$ sends the point $\xi(0)$: first φ_s^Δ maps $\xi(0)$ into $\varphi_s^\Delta \xi(0) \equiv \xi(s)$, and then φ_t^Δ maps $\xi(s)$ into $\varphi_t^\Delta \xi(s) \equiv \xi(t+s)$, which is just $\varphi_{t+s}^\Delta \xi(0)$. This is the active view.

Under the action of the dynamical group a function $F \in \mathcal{F}(\mathbf{TQ})$ is mapped from $F(\xi)$ to $F(\xi(t)) \equiv F(\varphi_t^\Delta \xi)$; since $\varphi_t^\Delta \xi$ moves along the integral curve from the initial ξ, the change of F is its change along the motion. As was shown in Section 3.4.1, its rate of change is given by the Lie derivative $\mathbf{L}_\Delta F$. If $\mathbf{L}_\Delta F = 0$, then F is a constant of the motion, invariant under φ_t^Δ, or $F(\varphi_t^\Delta \xi) = F(\xi)$.

That is an example of a one-parameter group of maps generated by a vector field. Later we will be using other transformations on $\mathbf{TQ}$, not the one generated by the dynamics, so now we generalize the example. Consider an arbitrary differentiable manifold $\mathbb{M}$ and a vector field $X \in \mathcal{X}(\mathbb{M})$. Just as $\Delta \in \mathcal{X}(\mathbf{TQ})$ generates the dynamical group φ_t^Δ, so X generates its own one-parameter group. Although X is not the dynamical vector field, it has integral curves: in local coordinates ξ^j on $\mathbb{M}$ the integral curves are, in analogy with those of Δ, solutions of the differential equations

$$\frac{d\xi^j}{d\epsilon} = X^j. \tag{3.89}$$

Here ϵ is the analog of t, measured along the integral curves as though X defined a motion on $\mathbb{M}$ and ϵ were the time. The one-parameter group that X generates is φ_ϵ^X, which acts on $\mathbb{M}$.

Under the action of φ_ϵ^X a function $f \in \mathcal{F}(\mathbb{M})$ is mapped from $f(\xi)$ to $f(\xi(\epsilon)) \equiv f(\varphi_\epsilon^X \xi)$; the change of f is its change along the integral curve. By analogy with the action of φ_t^X on F, the rate of change of f is given by the Lie derivative $\mathbf{L}_X f$. If $\mathbf{L}_X f = 0$, then f is invariant under the action of φ_ϵ^X, or $f(\varphi_\epsilon^X \xi) = f(\xi)$.

THE THEOREM

Suppose $L \in \mathcal{F}(\mathbf{TQ})$ is the Lagrangian function of a dynamical system and that $X \in \mathcal{X}(\mathbf{TQ})$ is some vector field other than Δ. What the theorem shows is that if L is invariant under φ_ϵ^X, that is, if $\delta L \equiv \mathbf{L}_X L = 0$, *for certain kinds of vector fields*, then

$$\mathbf{L}_\Delta \langle \theta_L, X \rangle = 0, \tag{3.90}$$

or the function $\Gamma \equiv \langle \theta_L, X \rangle$ is a constant of the motion, where θ_L is the one-form defined in Eq. (3.86). This is an intrinsic, coordinate-free statement of the theorem, and as such it is valid globally.

The vector fields for which the theorem holds are those whose φ_ϵ^X groups consist of point transformations (Section 3.2.2) – those that are obtained from transformations on $\mathbb{Q}$. Such vector fields are of the form

$$X_q^\alpha \frac{\partial}{\partial q^\alpha} + \frac{\partial X_q^\alpha}{\partial q^\beta} \dot{q}^\beta \frac{\partial}{\partial \dot{q}^\alpha}, \tag{3.91}$$

where the X_q^α do not depend on the $\dot{q}^\alpha$. It can be shown, by writing out $\delta q^\alpha = \mathbf{L}_X q^\alpha \equiv X(q^\alpha)$ and using the local-coordinate definition of θ_L, that locally (3.90) gives the same result as the one obtained at Eq. (3.48).

A brief sketch of the theorem's proof is the following: it is first shown that

$$\delta L = \langle \theta_L, Z \rangle + \mathbf{L}_\Delta \langle \theta_L, X \rangle, \tag{3.92}$$

where $Z = [X, \Delta]$. It is then established that $\langle \theta_L, Z \rangle = 0$ if X is of the form of (3.91). See Marmo et al. (1985) for details.

This form of Noether's theorem can be useful in that it shows that the theorem can have no simple converse. That is, given a constant of the motion Γ, there is no well-defined vector field X such that $\langle \theta_L, X \rangle = \Gamma$. For even if one could find such an X, one could add to it any vector field V such that $\langle \theta_L, V \rangle = 0$: the sum $X + V$ would do just as well as X. That there are many such Vs can be understood if one thinks of the action of a one-form as analogous to an inner product. Then the V vector fields span the analog of the orthogonal complement of a vector, which is of dimension $2n - 1$.

WORKED EXAMPLE 3.3

Return to the inverted harmonic oscillator of Worked Example 1.1, with potential $V = -\frac{1}{2}kx^2$ in one-freedom. **(a)** Its phase portrait on $\mathbf{TQ}$ is shown in Fig. 1.8. Write down the dynamical vector field Δ and discuss the relation between Δ and the phase portrait. **(b)** Obtain the equations for the integral curves, that is, solve completely for the motion on $\mathbf{TQ}$ (this is a small extension of the solution obtained in Chapter 1). **(c)** Take the Lie derivatives with respect to Δ of the energy E and the momentum $m\dot{x}$, and compare them with their time derivatives. **(d)** Show, by taking its Lie derivative with respect to Δ, that $\dot{x} - \omega x$ goes to zero (i.e., that the integral curves approach the positive-slope asymptote) exponentially in time.

Solution. **(a)** From Eq. (3.73) and the fact that $\ddot{x} = \omega^2 x$ (in this example x takes the place of q) with $\omega = \sqrt{k/m}$, the vector field is

$$\Delta = \dot{x}\frac{\partial}{\partial x} + \omega^2 x\frac{\partial}{\partial \dot{x}}.$$

This can be used to draw the vector field: the vector at each point of the $\mathbf{TQ}$ plane has horizontal (or x) component $\dot{x}$ (equal to the vertical coordinate) and vertical (or $\dot{x}$) component $\omega^2 x$ (proportional to the horizontal coordinate). In Chapter 1 the phase portrait, consisting of the integral curves of Δ, was obtained from the expression for the energy E. The integral curves are everywhere tangent to the vector field, the way "lines of force" are everywhere tangent to the magnetic field $\mathbf{B}$. If you write y for $\dot{x}$ and use standard notation for vectors in the plane, then $\Delta = y\hat{\imath} + \omega^2 x\hat{\jmath}$, whose divergence vanishes and whose curl has only a z component, like a magnetic field parallel to the (x, y) plane. Compare to Fig. 1.8.

   (b) The equations of motion on $\mathbf{TQ}$ are of first order: $dx/dt = \dot{x}$ and $d\dot{x}/dt = \omega^2 x$. The solution on $\mathbf{TQ}$ is

$$x(t) = a\cosh\omega(t - t_0) + b\sinh\omega(t - t_0),$$
$$\dot{x}(t) = a\omega\sinh\omega(t - t_0) + b\omega\cosh\omega(t - t_0).$$

This is similar to the general solution of the harmonic oscillator, except that the trigonometric functions are replaced by hyperbolic ones and some signs are different. As was found in Chapter 1, and in agreement with these solutions, the initial point in $\mathbf{TQ}$ is given by $x_0 = a$, $\dot{x}_0 = b\omega$. Then in terms of these initial conditions, the solution can be written in the form

$$\begin{vmatrix} x(\tau) \\ \dot{x}(\tau) \end{vmatrix} = \begin{vmatrix} \cosh\omega\tau & \omega^{-1}\sinh\omega\tau \\ \omega\sinh\omega\tau & \cosh\omega\tau \end{vmatrix} \begin{vmatrix} x_0 \\ \dot{x}_0 \end{vmatrix},$$

where $\tau = t - t_0$.

(c) The Lie derivative of a function with respect to a vector field is obtained by just applying the field as a derivative operator to the function. Hence

$$\mathbf{L}_\Delta E = \dot{x}\frac{\partial E}{\partial x} + \omega^2 x \frac{\partial E}{\partial \dot{x}} = -m\omega^2 x\dot{x} + m\omega^2 x\dot{x} = 0;$$
$$\mathbf{L}_\Delta(m\dot{x}) = m\omega^2 x \equiv kx.$$

In both cases this gives the time derivative: the energy is a conserved quantity and the time derivative of the momentum is the force.

(d) The Lie derivative of $\xi \equiv \dot{x} - \omega x$ is

$$\mathbf{L}_\Delta \xi = \mathbf{L}_\Delta(\dot{x} - \omega x) = -\dot{x}\omega + \omega^2 x = -\omega(\dot{x} - \omega x) = -\omega\xi \equiv \dot{\xi}.$$

Thus $\xi(t) = \xi_0 e^{-\omega t}$, as required. This shows that for large t the velocity $\dot{x}$ can be approximated by ωx.

WORKED EXAMPLE 3.4

Find the rate of energy dissipation for an underdamped harmonic oscillator.

Solution. The equation of motion (3.66) is

$$\ddot{x} + 2\beta\dot{x} + \omega_0^2 x = 0,$$

where $\beta = b/m$ and $\omega_0 = \sqrt{k/m}$. Write the solution in the first form of (3.67):

$$x = e^{-\beta t}(Ae^{i\omega t} + A^* e^{-i\omega t}),$$
$$\dot{x} = e^{-\beta t}[(-\beta + i\omega)Ae^{i\omega t} + (-\beta - i\omega)A^* e^{-i\omega t}].$$

The equation of motion gives the dynamical vector field

$$\Delta = -(2\beta\dot{x} + \omega_0^2 x)\frac{\partial}{\partial \dot{x}} + \dot{x}\frac{\partial}{\partial x}.$$

The energy is $E = \frac{1}{2}m\dot{x}^2 + \frac{1}{2}kx^2$. Consider instead

$$\varepsilon \equiv \frac{2}{m}E = \dot{x}^2 + \omega_0^2 x^2.$$

The Lie derivative of ε along Δ is

$$\mathbf{L}_\Delta\varepsilon \equiv \dot{\varepsilon} = -(2\beta\dot{x} + \omega_0^2 x)2\dot{x} + \dot{x}2\omega_0^2 x = -4\beta\dot{x}^2.$$

Now insert the expression for $\dot{x}$ into this equation. For simplicity take A to be real, which just fixes the phase of the trigonometric part: $(Ae^{i\omega t} + A^* e^{-i\omega t}) = 2A\cos\omega t$.

The expression for $\dot{\varepsilon}$ then becomes (we write $v = -\beta + i\omega$)

$$\dot{\varepsilon} = -4\beta A^2 e^{-2\beta t}(v^2 e^{2i\omega t} + v^{*2} e^{-2i\omega t} + 2|v|^2)$$
$$= 8\beta A^2 e^{-2\beta t}\{(\omega^2 - \beta^2)\cos(2\omega t) + 2\beta\omega\sin(2\omega t) - (\beta^2 + \omega^2)\}$$
$$= \frac{2}{m}\dot{E};$$

that is, the rate of energy dissipation oscillates with frequency 2ω about a steady exponential decay with a lifetime 2β.

PROBLEMS

1. Equation (3.18) together with the equations of constraint yield the equations of motion of a constrained system within the Lagrangian formalism. Take the generalized coordinates to be Cartesian; then the Euler–Lagrange equations equate $m\mathbf{a}$ to the forces derived from the potential plus other forces, which we shall call $\mathbf{N}$. Show that the $\mathbf{N}$ forces are perpendicular to the surfaces of constraint and linear in the Lagrange multipliers λ_l. Argue from the definition of the constraint forces that the $\mathbf{N}$ forces are in fact the forces of constraint. This shows the relation between the constraint forces and the λ_l.

2. A particle is free to move on the surface of a sphere $\mathbb{S}^2$ under the influence of no forces other than those that constrain it to the sphere. It starts at a point q_1 and ends at another q_2 (without loss of generality, both points may be taken to lie on a meridian of longitude).

   (a) Show that there are many physical paths (i.e., $q(t)$ functions) the particle can take in going from q_1 to q_2 in a given time τ. How many? Under what conditions are there uncountably many?

   (b) Calculate the action for each of two possible paths the particle can take and show that they are not in general equal. Now construct two new, nonphysical paths close to the original ones, going from q_1 to q_2 in the same time τ. This can be done by adding to each physical path a small distortion in the form of a function $f(t)$ such that $f(0) = f(\tau) = 0$. Show that for each of the nonphysical paths the action is greater than it is on the neighboring physical one, thus demonstrating that each physical path minimizes the action locally in ϵ, as described in the text.

3. Show explicitly that the Lagrangian $L = \frac{1}{2}m(\dot{x}^2 + \dot{y}^2) - V(r)$ is invariant under the rotation transformation of Eq. (3.33b).

4. (An exercise in geometry.) Let X be a vector field and f and g functions on a differentiable manifold $\mathbb{M}$. Show that the Lie derivative satisfies the Leibnitz rule $\mathbf{L}_X fg = f\mathbf{L}_X g + g\mathbf{L}_X f$. [It can be shown inversely (Marmo et al., 1985, p. 43) that if a map ϕ from functions to functions satisfies the Leibnitz rule $\phi(fg) = f(\phi g) + g(\phi f)$, then ϕ defines a vector field X in accordance with $\phi = \mathbf{L}_X$.]

5. (Some exercises in geometry.) Let $\{\xi^j\}$ be a local coordinate system in some neighborhood $\mathbb{U}$ of a differential manifold $\mathbb{M}$. Let f be a function and $X = X^j \partial/\partial\xi^j$ and $Y = Y^j \partial/\partial\xi^j$ be vector fields on $\mathbb{U}$.

   (a) Obtain an explicit expression in terms of local coordinates for $\mathbf{L}_X df$.

   (b) Do the same for $\mathbf{L}_X g\, df$.

(c) Do the same for $\mathbf{L}_X \mathbf{L}_Y f \equiv \mathbf{L}_X(\mathbf{L}_Y f) \equiv \mathbf{L}_X \langle df, Y \rangle$.

Those were local results. Now we find a global one.

(d) Use the Leibnitz rule (Problem 4), treating $\mathbf{L}_Y f$ or $\langle df, Y \rangle$ as a product, to show that

$$(\mathbf{L}_X Y)f \equiv \langle df, \mathbf{L}_X Y \rangle = [\mathbf{L}_X \mathbf{L}_Y - \mathbf{L}_Y \mathbf{L}_X]f,$$

with the product $\mathbf{L}_X \mathbf{L}_Y$ defined in part (c). [Hint: Start with $\mathbf{L}_X(\mathbf{L}_Y f)$.]

(e) Globally or locally, show that

$$(\mathbf{L}_X \mathbf{L}_Y)fg = f(\mathbf{L}_X \mathbf{L}_Y)g + g(\mathbf{L}_X \mathbf{L}_Y)f,$$

and then show that, in accordance with the note at the end of Problem 4, $\mathbf{L}_X \mathbf{L}_Y - \mathbf{L}_Y \mathbf{L}_X$ defines a vector field Z such that $(\mathbf{L}_X \mathbf{L}_Y - \mathbf{L}_Y \mathbf{L}_X)f \equiv \mathbf{L}_Z f$. See Eq. (3.83).

(f) Returning to local calculations, find the jth component Z^j of the vector field Z of Part (e) in terms of the X^j and the Y^j.

Answers:

(a) $\mathbf{L}_X \, df = \left(\dfrac{\partial X^j}{\partial \xi^k} \dfrac{\partial f}{\partial \xi^j} + X^j \dfrac{\partial^2 f}{\partial \xi^k \partial \xi^j} \right) d\xi^k.$

(b) $\mathbf{L}_X g \, df = \left(X^k \dfrac{\partial g}{\partial \xi^k} \dfrac{\partial f}{\partial \xi^j} + g \dfrac{\partial X^k}{\partial \xi^j} \dfrac{\partial f}{\partial \xi^k} + X^k \dfrac{\partial^2 f}{\partial \xi^j \partial \xi^k} \right) d\xi^j.$

(c) $\mathbf{L}_X \mathbf{L}_Y f = X^j \dfrac{\partial Y^k}{\partial \xi^j} \dfrac{\partial f}{\partial \xi^k} + X^j Y^k \dfrac{\partial^2 f}{\partial \xi^j \partial \xi^k}.$

(f) $Z^j = X^k \dfrac{\partial Y^j}{\partial \xi^k} - Y^k \dfrac{\partial X^j}{\partial \xi^k}.$

6. (An exercise in geometry.) Prove that the Lie bracket satisfies the Jacobi identity, Eq. (3.85).

7. Prove that if ψ is the inverse of φ, then $\mathbf{T}\psi$ is the inverse of $\mathbf{T}\varphi$ [see the discussion above Eq. (3.44)]. This is probably easiest to do in local coordinates.

8. (a) Show that the free-particle Lagrangian is invariant under all rotations in $\mathbb{R}^3$ and derive from that the conservation of angular momentum $\mathbf{J}$ by the use of the Noether theorem.

(b) Let $L = T - V$ be a single-particle Lagrangian invariant under rotations in $\mathbb{R}^3$. Show that $\mathbf{J}$ is conserved.

9. (a) Find the transformation laws for the energy function E and the generalized momentum $p_\alpha = \partial L / \partial \dot{q}^\alpha$ under a general point transformation $Q^\alpha = Q^\alpha(q, t)$. Apply the result to find the transformed energy function and generalized momenta under the following two transformations:

(b)

$$\begin{vmatrix} X \\ Y \end{vmatrix} = \begin{vmatrix} \cos \omega t & -\sin \omega t \\ \sin \omega t & \cos \omega t \end{vmatrix} \begin{vmatrix} x \\ y \end{vmatrix},$$

(c)

$$\Theta = \theta + \omega t, \quad R = r.$$

10. Given the Lagrangian $L = m / \sqrt{1 - \dot{x}^2}$, find the transformed Lagrangian function L_λ under the transformation (λ is a parameter)

$$\begin{vmatrix} \xi \\ \tau \end{vmatrix} = \begin{vmatrix} \cosh \lambda & \sinh \lambda \\ \sinh \lambda & \cosh \lambda \end{vmatrix} \begin{vmatrix} x \\ t \end{vmatrix}.$$

11. Consider a three-dimensional one-particle system whose potential energy in cylindrical polar coordinates ρ, θ, z is of the form $V(\rho, k\theta + z)$, where k is a constant.

   (a) Find a symmetry of the Lagrangian and use Noether's theorem to obtain the constant of the motion associated with it.
   (b) Write down at least one other constant of the motion.
   (c) Obtain an explicit expression for the dynamical vector field Δ and use it to verify that the functions found in (a) and (b) are indeed constants of the motion. [Note: Cylindrical polar coordinates are related to Cartesian by the transformation $x = \rho \cos \theta, y = \rho \sin \theta, z = z$.] [Partial answer: $m\dot{z} - m\rho^2\dot{\theta}/k$ is the Noether constant.]

12. (a) Describe the motion for the Lagrangian $L = \dot{q}_1\dot{q}_2 - \omega^2 q_1 q_2$. Comment on the relevance of this result to Problem 2.4.
   (b) Show that L is invariant under the family of point transformation $Q_1 = e^\epsilon q_1$, $Q_2 = e^{-\epsilon} q_2$. Find the Noether constant associated with this group of transformations. See Marmo et al. (1985), p. 128 ff.

13. Consider the continuous family of coordinate *and time* transformations

$$Q^\alpha = q^\alpha + \epsilon f^\alpha(q, t), \quad T = t + \epsilon\tau(q, t)$$

(for small ϵ). Show that if this transformation preserves the action

$$S = \int_{t_1}^{t_2} L(q, \dot{q}, t)\, dt = \int_{T_1}^{T_2} L(Q, \dot{Q}, T)\, dT,$$

then

$$\frac{\partial L}{\partial \dot{q}^\alpha}(\dot{q}^\alpha\tau - f^\alpha) - L\tau$$

is a constant of the motion. [Note: It is invariance of S that plays a role here, not the usual invariance of L. But this should not be confused with the variational derivation of the equations of motion. It is assumed here that the $q^\alpha(t)$ satisfy Lagrange's equations.]

14. Find the general orbit for a particle of mass m (in two dimensions) in a uniform gravitational field and acted on by a dissipative force according to the Rayleigh function, assuming that $k_x = k_y$. Obtain the solution in terms of the initial conditions (initial velocity and position). Find the rate of energy loss.

15. The phase portrait of the undamped harmonic oscillator consists of ellipses, with each ellipse determined by the initial conditions. It follows that the area of the ellipse is a constant of the motion. Express the area in terms of the other known constants, namely the energy E and the phase angle φ [the phase angle is defined by writing the general solution in the form $x = a \cos(\omega t + \varphi)$].

16. Verify the solution of Eq. (3.69) for the critically damped harmonic oscillator. Draw the phase portrait in $\mathbf{TQ}$.

17. Compare the time it takes the overdamped and critically damped oscillators with the same β and the same initial conditions to achieve $|x| < \epsilon$, where ϵ is some small positive constant. Show that in the limit of long times the critically damped oscillator is much closer to the x axis than the overdamped one. How much closer?

18. Find the rate of energy dissipation for the overdamped and critically damped harmonic oscillators.

19. A mass m is attached to a spring of force constant k and is free to move horizontally along a fixed line on a horizontal surface. The other end of the spring is attached to a fixed point on the line along which the mass moves. The surface on which the mass rests is a moving belt, which moves with velocity v_0 along the line of motion. The friction between the belt and the mass provides a damping factor β. Describe the motion of the mass. [Hint: It's the relative velocity that matters.] Consider all three degrees of damping (critical, overdamped, and underdamped) and treat in particular the initial condition $x = 0, \dot{x} = 0$, where distance is measured from the equilibrium length of the spring. Discuss energy conservation. [Note: In the usual undergraduate treatment, the frictional force is not proportional to the velocity as it is in the damped oscillator of this problem. It has two values, one given by the static coefficient, and the other by the lower kinetic one. Think about the subsequent motion.]

20. Consider a two-dimensional inverted oscillator with potential $V = -\frac{1}{2}(k_1 x_1^2 + k_2 x_2^2)$ and mass m. Show that unless $k_1 = k_2$ (i.e., unless the potential is circularly symmetric), asymptotically the mass moves in either the x_1 or x_2 direction.

21. (Damped simple pendulum.) Consider a simple pendulum of length R and mass m with a Rayleigh function $\mathcal{F} = \frac{1}{2}k\dot{\theta}^2$, where k is a positive constant and θ is the angle the pendulum makes with the vertical.

   (a) Write down the equation of motion.
   (b) Show that the rate at which energy is dissipated is proportional to the kinetic energy. Find the proportionality factor.
   (c) For the undamped pendulum (i.e., $k = 0$) show that the average kinetic energy $\langle T \rangle$ satisfies the equation

   $$r(\Theta) \equiv \frac{\langle T \rangle}{T_{\max}} = \frac{\int_0^\Theta (\cos\theta - \cos\Theta)^{1/2}\, d\theta}{(1 - \cos\Theta)\int_0^\Theta (\cos\theta - \cos\Theta)^{-1/2}\, d\theta},$$

   where $T_{\max}$ is the maximum kinetic energy in a period and Θ is the amplitude of the period. Numerical calculation shows that $r(\Theta)$ deviates by no more than 10% from 0.5 for Θ between 0 and $\frac{1}{2}\pi$.
   (d) Assume that the numbers of Part (c) apply for small enough k. For a clock with a damping constant k and a simple pendulum of length R, whose period is 1 sec at an amplitude of $10°$, estimate the energy that must be supplied per period to keep the clock running.

22. Use the Lagrange multiplier method to solve the following problem: A particle in a uniform gravitational field is free to move without friction on a paraboloid of revolution whose symmetry axis is vertical (opening upward). Obtain the force of constraint. Prove that for given energy and angular momentum about the symmetry axis there are a minimum and a maximum height to which the particle will go.

23. A bead is constrained to move without friction on a helix whose equation in cylindrical polar coordinates is $\rho = b$, $z = a\phi$, with the potential $V = \frac{1}{2}k(\rho^2 + z^2)$. Use the Lagrange multiplier method to find the constraint forces.

24. Derive the equations of motion for the Lagrangian $L = e^{\gamma t}[\frac{1}{2}m\dot{q}^2 - \frac{1}{2}kq^2]$, $\gamma > 0$. Compare with known systems. Rewrite the Lagrangian in the new variable $Q = e^{\gamma t/2}q$. From this obtain a constant of the motion. (It is seen that there is more than one way to deal with a dissipative system.)

CHAPTER 4

SCATTERING AND LINEAR OSCILLATORS

CHAPTER OVERVIEW

This chapter contains material that grows out of and sometimes away from Lagrangian mechanics. Some of it makes use of the Lagrangian formulation (e.g., scattering by a magnetic dipole, linear oscillations), and some does not (e.g., scattering by a central potential, chaotic scattering). Much of this material is intended to prepare the reader for topics to be discussed later: chaotic scattering off hard disks is a particularly understandable example of chaos, and linear oscillators are the starting point for the discussion of nonlinear ones, of perturbation theory, and field theory.

4.1 SCATTERING

4.1.1 SCATTERING BY CENTRAL FORCES

So far we have mostly concentrated on bounded orbits of central-force systems. Yet unbounded orbits are very common for both attractive and repulsive central forces. Indeed, for repulsive ones all orbits are unbounded. For attractive ones, the outer turning point of bounded orbits – the maximum distance of the particle from the attracting center – increases with increasing energy. If the force is weak enough at large distances, the particle may escape from the force center even at finite energies: the orbit may break open and the outer turning point may move off to infinity. Then the orbit becomes unbounded, as in the hyperbolic orbits in the Coulomb potential. The result is what is called scattering.

GENERAL CONSIDERATIONS

Consider a central-force dynamical system that possesses unbounded orbits. Let the initial conditions be that a particle is approaching the force center from far away, so far away that it may be thought of as coming from infinity, which means that it is moving on one of the unbounded orbits. When it comes close to the force center it gets deflected and goes off in a new direction. It is then said to be *scattered*, and the angle between the distant incoming velocity vector and the distant outgoing one is called the *scattering angle* (the angle ϑ in Figs. 4.1 to 4.4). Is there some information that can be obtained about the force by measuring the scattering angle?

If one particle is scattered only once by a single force center, little can be learned, because almost any angle of scattering is possible; it will depend on the details of the initial conditions. But if many particles are scattered, if many scattering experiments are performed, much can be learned by studying the distribution of scattering angles. In fact this is one of the main experimental techniques used to study interactions in physics. This section discusses some results of the classical theory of scattering.

Consider a *beam* of mutually noninteracting particles incident on a stationary *target* that consists of a collection of other particles. The target particles are the force centers that scatter the beam particles. Take J to be the *flux density*, or *intensity*, of the beam (the number of particles crossing a unit area perpendicular to the beam per unit time) and A to be its cross-sectional area. Let the target have a density of n particles per unit area. Assume for the time being that the beam particles interact with the target particles through a *hard-sphere* (or *hard-core*) interaction: both are small spheres that interact only when they collide, as in Fig. 4.1. Let the cross-sectional area of this interaction be σ: a beam particle interacts with a target particle only if its center intersects a circle of area σ around the center of the target particle. In the hard-sphere example $\sigma = \pi(r + R)^2$, where r and R are the radii of the target and beam particles, respectively. The total interaction cross section that the target particles present to the beam is

$$\Sigma = nA\sigma = N\sigma,$$

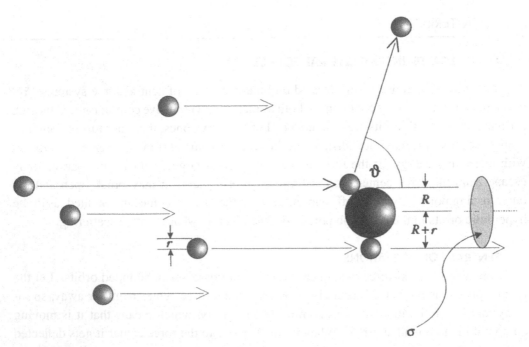

FIGURE 4.1
Hard-core scattering of spheres. The cross section is a disk of radius $R + r$.

where N is the number of target particles within the cross-sectional area of the beam. On the average the number of collisions per unit time will then be $J\Sigma$. Each collision scatters a particle, so the number of particles scattered out of the beam per unit time is, on the average, the *scattering rate*

$$S = J\Sigma = JN\sigma. \tag{4.1}$$

If a scattering experiment lasts a long enough time t, any fluctuations from the average get smoothed out, and $St = tJN\sigma$ is the total number of particles scattered.

Initially we make three assumptions. First, the force and the potential both go to zero at large distances: $\mathbf{F}(r \to \infty) = \mathbf{0}$ and $V(r \to \infty) = 0$. Second, each beam particle interacts with only one target particle at a time. This may be because the target particles are separated by distances that are large compared with the range of the interaction. If the interaction is through a potential, its range may be reduced by shielding, as when atomic electrons shield the nuclei in Coulomb scattering of heavy particles by nuclei. A beam particle has to penetrate the electron cloud before it sees a nucleus, and the electrons hardly affect the scattering because their masses are so low. Third, the target is so thin that successive interactions (multiple scattering) can be neglected.

Later, in Section 4.1.3, we will remove the last two assumptions, and it will be seen that the situation can then change radically when each beam particle interacts simultaneously with more than one target particle.

That the scattering rate of Eq. (4.1) is proportional to the flux J and to the number N of target particles is intuitively reasonable. The only factor in the equation that is intrinsic to the particular interaction is the cross section (in the hard-sphere example, this means the sizes of the spheres). For arbitrary interactions, other than hard core, Eq. (4.1) is taken as the definition of the *total cross section* σ (per particle) for the interaction.

In an actual experiment a detector is placed at a location some distance from the target and records particles coming out in a fixed direction (Fig. 4.2). Let that direction be characterized by its colatitude and azimuth angles ϑ and φ. The detector itself subtends a small solid angle $d\Omega$, and then the number dS of particles detected is some fraction of S which is proportional to $d\Omega$ and depends on ϑ and φ. The *differential cross section* $\sigma(\vartheta, \varphi)$ is then defined by the equation

$$dS = JN\sigma(\vartheta, \varphi) \, d\Omega. \tag{4.2}$$

In general σ depends also on the energy, but we do not include that dependence in the notation.

Because the total cross section σ_{tot} (we shall write σ_{tot} for σ from now on) measures the total number of particles scattered, it is related to the differential cross section by

$$\sigma_{\text{tot}} = \int \sigma(\vartheta, \varphi) \, d\Omega. \tag{4.3}$$

In this integration $d\Omega \equiv \sin\vartheta \, d\varphi \, d\vartheta$ is the element of solid angle (the element of area on the unit sphere). The integral is taken over all directions.

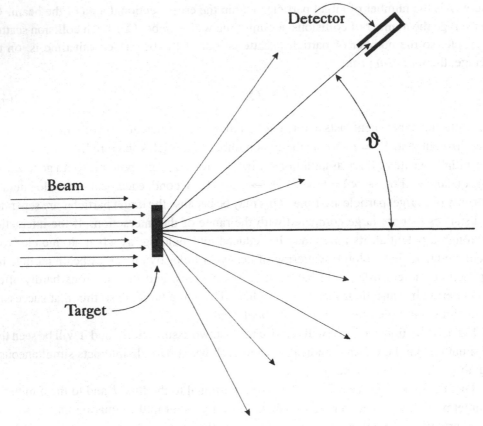

FIGURE 4.2
Experimental setup for measuring differential scattering cross sections.

Since a real detector subtends a finite solid angle, $\sigma(\vartheta, \varphi)$ is more accurately defined by a generalization of Eq. (4.3): for any finite solid angle Ω call the number of particles scattered into Ω per unit time $JN\sigma_\Omega$. Then for arbitrary Ω

$$\sigma_\Omega = \int_\Omega \sigma(\vartheta, \varphi)\, d\Omega.$$

Consider scattering by a single target particle (so that $N = 1$) with a central-force interaction. Let this target particle be in the center of the beam, so that the scattering is symmetric about the axis of the beam. For $N = 1$ Eq. (4.2) implies that

$$\sigma(\vartheta, \varphi)\, d\Omega = \frac{dS}{J}. \tag{4.4}$$

Because the scattering is symmetric about the axis of the beam, $\sigma(\vartheta, \varphi)$ is independent of φ, a function of ϑ alone. We write $\sigma(\vartheta)$ instead of $\sigma(\vartheta, \varphi)$, and then Eq. (4.3) becomes

$$\sigma_{\text{tot}} = 2\pi \int \sigma(\vartheta)\sin\vartheta\, d\vartheta, \tag{4.5}$$

and Eq. (4.4) can be written as

$$2\pi\sigma(\vartheta)\sin\vartheta\,d\vartheta = \frac{dS}{J}, \qquad (4.6)$$

where dS now represents the number of particles scattered into the solid angle between ϑ and $d\vartheta$ for all $0 \le \varphi < 2\pi$, that is, between the two cones at the origin with vertex semi-angles ϑ and $\vartheta + d\vartheta$.

This equation, like (4.4), can be used by the experimentalist to determine σ because both the number of particles dS entering the (idealized) detector per unit time and the beam intensity J are measured quantities. From one point of view, the theorist predicts σ and the experimentalist checks the prediction. From another, the experimentalist determines σ, from which the theorist tries to derive the potential. The second view is an application of *inverse scattering theory*: can $V(r)$ be determined uniquely from a knowledge of $\sigma(\vartheta)$? Inverse scattering theory will be discussed in Section 4.1.2. Here we discuss the first point of view.

Consider an incident particle aimed some distance b (called the *impact parameter*) from the target particle (Fig. 4.3). If there were no interaction, the beam particle would simply pass the target without deflection. When there is an interaction, the scattering angle ϑ depends on b, so that ϑ is a function of b or vice versa (i.e., either $\vartheta(b)$ or $b(\vartheta)$). The part of the beam that is scattered through angles between ϑ and $\vartheta + d\vartheta$ lies in a circular ring of radius $b(\vartheta)$ and of width db, as shown in Fig. 4.3. This ring has area $2\pi b\,db$, and its area times the beam intensity J is the number of particles scattered into the $d\vartheta$ interval. Then according to Eq. (4.6)

$$2\pi b(\vartheta)\,db = -2\pi\sigma(\vartheta)\sin\vartheta\,d\vartheta. \qquad (4.7)$$

If the force is a monotonic function of r, which gets stronger as r decreases, a beam particle aimed close to the target particle will encounter a stronger force and be deflected through a larger angle. Thus ϑ increases as b decreases; this is the source of the negative sign in

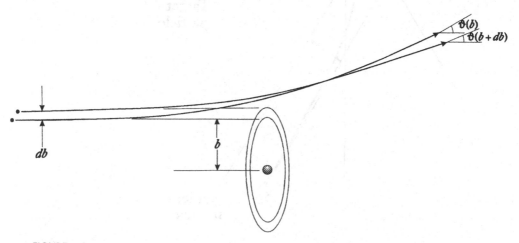

FIGURE 4.3
Illustrating the differentials $d\vartheta$ and db. $d\vartheta = \vartheta(b+db) - \vartheta(b)$. Notice that ϑ decreases as b increases. φ is not shown in the figure.

Eq. (4.7). Equation (4.7) is a differential equation for either $b(\vartheta)$ or $\vartheta(b)$. If $\sigma(\vartheta)$ is known from measurement, (4.7) can be used to find $b(\vartheta)$, and if $b(\vartheta)$ is known, (4.7) can be used to find $\sigma(\vartheta)$.

Consider the (unbounded) trajectory of one beam particle. Clearly b is related to the initial conditions, for it depends on how the particle starts its trip to the scattering center, essentially at $r = \infty$. There the total energy is just the kinetic energy $E = \frac{1}{2}mv^2$ and the magnitude of the angular momentum is simply $l = mvb$, where v is the initial speed. Suppose that the potential is known. We will now rewrite Eq. (2.75) for the particular case of scattering. When $r = \infty$ the angle θ in (2.75) becomes what is called Θ in Section 2.3.2, and then according to Fig. 4.4 the scattering angle is $\vartheta = \pi - 2\Theta$. The equations for E and l are now used to eliminate l in (2.75) in favor of b:

$$l = b\sqrt{2mE}.$$

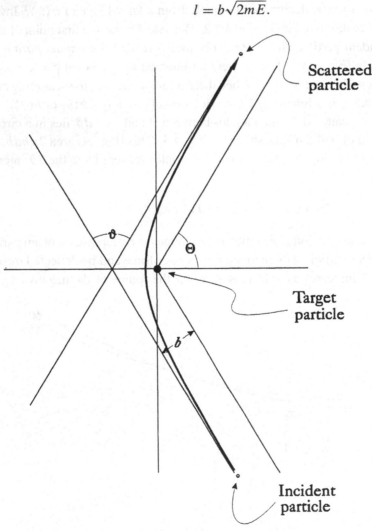

FIGURE 4.4
Asymptotes for scattering. The diagram shows that $2\Theta + \vartheta = \pi$.

Putting all this together yields

$$\vartheta = \pi - 2b \int_{r_0}^{\infty} \frac{dr}{r\{r^2[1 - V(r)/E] - b^2\}^{1/2}}, \qquad (4.8)$$

where r_0 is the minimum radius, the root of the radical. For a given potential $V(r)$ this equation gives the scattering angle as a function of b and the energy E. If E is held fixed and there is a one-to-one correspondence between ϑ and b, this equation yields $\vartheta(b)$. But, as was seen at Eq. (4.7), a knowledge of $\vartheta(b)$ is sufficient for finding the scattering cross section, so we now have a recipe for finding the cross section $\sigma(\vartheta)$ from a knowledge of $V(r)$: integrate (4.8) to obtain $\vartheta(b)$ and then solve (4.7) for $\sigma(\vartheta)$.

THE RUTHERFORD CROSS SECTION

We now go through this recipe for the Coulomb potential, obtaining the so-called Rutherford scattering formula. Doing it for the Coulomb potential spares us some work: the integration of Eq. (4.8) was essentially performed already in Chapter 2.

Suppose that an infinite-mass attractive Coulomb force center (the target particle) is approached on a hyperbolic orbit by a particle of mass m (the beam particle). If the beam particle starts out very far away it is essentially on an asymptote of the orbit, and long after it passes the target particle it will be on the other asymptote. The angle between these asymptotes is the scattering angle ϑ shown in Fig. 4.4. The relations needed to determine $b(\vartheta)$ are (some of these are repeated from above)

$$E = \tfrac{1}{2}mv^2, \quad l = mvb,$$

$$\epsilon \cos \Theta = 1, \quad \epsilon = \sqrt{1 + \frac{2El^2}{m\alpha^2}}, \quad 2\Theta + \vartheta = \pi, \qquad (4.9)$$

where α is the constant in the Coulomb potential $V = -\alpha/r$. Some algebra then leads to

$$b(\vartheta) = \frac{\alpha}{mv^2} \cot \frac{\vartheta}{2} \equiv K \cot \frac{\vartheta}{2}, \qquad (4.10)$$

where $K = \alpha/mv^2$. Now insert this equation for $b(\vartheta)$ into (4.7), which implies that

$$b \frac{db}{d\vartheta} = -\frac{1}{2}K^2 \cos \frac{\vartheta}{2} \csc^3 \frac{\vartheta}{2} = -\sigma(\vartheta)\sin \vartheta \equiv -2\sigma(\vartheta)\sin \frac{\vartheta}{2} \cos \frac{\vartheta}{2}.$$

This leads to the *Rutherford scattering cross section*

$$\sigma(\vartheta) = \frac{K^2}{4} \csc^4 \frac{\vartheta}{2} = \frac{\alpha^2}{4m^2v^4} \csc^4 \frac{\vartheta}{2}. \qquad (4.11)$$

A similar formula for Coulomb scattering is obtained if the potential is repulsive (Problem 1) and was derived and used by Rutherford in establishing the nuclear model of the atom. He was lucky in that the classical and quantum calculations give the same cross section for the Coulomb potential and for none other. When the original experiments were performed, quantum mechanics had not yet been invented.

Another property of the Coulomb cross section is that its integral σ_{tot} diverges, which means that *all* particles are scattered out of the incident beam, no matter how broad it is. The range of the Coulomb force is so long that particles get scattered no matter how far away they are from the scattering center. This long range is related to the slow rate at which the potential decreases with increasing r. For many proofs and calculations (in particular, of the total cross section) a more rapid decrease of the potential is required, often that $r^2 V(r)$ go to zero as $r \to \infty$. This is clearly not satisfied by the Coulomb potential.

The divergence of the Coulomb total cross section may seem disturbing at first sight, for there are many charges around any experiment and, to the extent that the range of the potential is infinite, the beam particles are scattered by all the charges. That would seem to contradict our assumption that the beam particles interact with just one target particle at a time. But in the experiments Rutherford analyzed, the target particles were all in atoms, so the nuclear charges of the more distant ones were completely screened by the atomic electrons. Hence on the average only one nucleus at a time contributed to the scattering.

The Rutherford scattering cross section is valid for all inverse square force laws, so it can be used also for gravitational scattering (e.g., of asteroids by planets).

4.1.2   THE INVERSE SCATTERING PROBLEM

GENERAL TREATMENT

When an experiment is performed, what is obtained is the scattering cross section or some information about it. Usually what is desired, however, is the force or the interaction potential. In this section we show how the potential can be found from information about the cross section in the central force problem. This is done essentially by inverting the procedure of Section 4.1.1. The treatment is taken from Miller (1969). We leave out some of the details; see also Keller et al. (1956).

The starting point is Eq. (4.8). In Section 4.1.1 this was used with Eq. (4.7) to obtain an expression for $\sigma(\vartheta)$, but now it will be used inversely: given $\sigma(\vartheta)$, the function $\vartheta(b)$ will be found from Eq. (4.7), and then (4.8) will be used to find an expression for $V(r)$. Assume, therefore, that $\sigma(\vartheta)$ is known from measurement and that (4.7) has been used to calculate $\vartheta(b)$. Define a new function $y(r)$ by

$$y(r) \equiv r\sqrt{1 - \frac{V(r)}{E}}. \tag{4.12}$$

The rest of the procedure involves calculating $y(r)$ from $\vartheta(b)$ and then inverting (4.12) to obtain $V(r)$.

Let the integral on the right-hand side of (4.8) be called I. It can be written in terms of $y(r)$ as

$$I = 2b \int_{r_0}^{\infty} \frac{dr}{r\sqrt{y^2 - b^2}}.$$

Assume that $y(r)$ is invertible, so that $r(y)$ exists, and change the variable of integration from r to y (this change of variables is implicit, for it depends on the potential $V(r)$, which is still to be found). Then

$$I = 2b \int_{b}^{\infty} \frac{r'(y)\,dy}{r(y)\sqrt{y^2 - b^2}} = 2b \int_{b}^{\infty} \frac{dy}{\sqrt{y^2 - b^2}} \frac{d}{dy} \ln r(y). \qquad (4.13)$$

The lower limit in the y integral is b because the lower limit in the r integral is the root r_0 of the radical in the denominator, which is given by $y(r_0) = b$. Now, according to (4.8), $\vartheta(b) = \pi - I$. But π can be written as an integral that can be combined with I, namely

$$\pi = 2b \int_{b}^{\infty} \frac{dy}{y\sqrt{y^2 - b^2}} = 2b \int_{b}^{\infty} \frac{dy}{\sqrt{y^2 - b^2}} \frac{d}{dy} \ln y,$$

so that

$$\vartheta(b) = \pi - I = 2b \int_{b}^{\infty} \frac{dy}{\sqrt{y^2 - b^2}} \frac{d}{dy} \ln \frac{y}{r(y)}. \qquad (4.14)$$

Rather than calculate this integral directly, define a new function from which $y(r)$ will be found. This new function $T(y)$ is defined in terms of $\vartheta(b)$ by

$$T(y) = \frac{1}{\pi} \int_{y}^{\infty} \frac{db}{\sqrt{b^2 - y^2}} \vartheta(b). \qquad (4.15)$$

It will be shown later that $y(r)$ can be found from $T(y)$, so the next thing to do is to perform the integration in (4.15). In principle this integration can be performed if $\vartheta(b)$ is known, but the integral is not always a simple one and may sometimes have to be calculated numerically. We assume for the sake of argument, however, that it can be performed (as is indeed the case in the example we give and in Problem 2).

Note that the integral in (4.15) diverges if $\vartheta(b)$ is a nonzero constant. But that means that the scattering angle ϑ is independent of b, which happens only if there is no scattering at all. So $\vartheta(b) = \text{constant}$ implies $\vartheta(b) = 0$, which in turn implies that $T(y) = 0$. Moreover, when scattering actually occurs, $\lim_{b\to\infty} \vartheta(b) = 0$, and if $\vartheta(b)$ approaches zero as $b^{-\lambda}$, $\lambda > 0$, the integral converges. That means that under reasonable assumptions it will always converge.

We now show the relation between $T(y)$ and $y(r)$. Eliminate $\vartheta(b)$ from (4.15) by inserting the integral expression (4.14); then (4.15) becomes

$$T(y) = \frac{1}{\pi} \int_{y}^{\infty} \frac{2b}{\sqrt{b^2 - y^2}} \left[\int_{b}^{\infty} \frac{du}{\sqrt{u^2 - b^2}} \frac{d}{du} \ln \frac{u}{r(u)} \right] db. \qquad (4.16)$$

This double integral can be performed if the order of integration is changed, but that requires care with the limits of integration. In the integral over b the limits imply that $b > y$. In the integral over u the limits imply that $u > b$. Thus $y < b < u$, so when the order is changed the double integral becomes

$$T(y) = \frac{1}{\pi} \int_y^\infty \left\{ \frac{d}{du} \ln \frac{u}{r(u)} \right\} du \int_y^u \frac{2b \, db}{\sqrt{(b^2 - y^2)(u^2 - b^2)}}. \tag{4.17}$$

The integral over b is a little complicated, but it can be performed with a trigonometric substitution and turns out to equal π. The integral over u is simpler. The result is an expression for $T(y)$ in terms of r and y, which can be immediately inverted to yield

$$r(y) = y \exp\{T(y)\}. \tag{4.18}$$

This is the general relation between $r(y)$ and $T(y)$.

> **REMARK**: When (4.18) is solved for $y/r(y)$ and the solution is inserted into (4.14), one obtains
>
> $$\vartheta(b) = -2b \int_b^\infty \frac{T'(y) \, dy}{\sqrt{y^2 - b^2}}.$$
>
> This equation and (4.15) establish that $T(y)$ and $\vartheta(b)$ are related through the Abel transform. See Tricomi (1957).   □

According to Eq. (4.12), the potential V can be written in terms of r and y:

$$V = E \frac{r^2 - y^2}{r^2}. \tag{4.19}$$

Recall that $T(y)$ is assumed to be the known function obtained from $\vartheta(b)$ by (4.15). Then when Eq. (4.18) is solved for y in terms of r and the solution inserted into (4.19), an expression is obtained for $V(r)$ and the problem is solved. In any specific case there are two tasks. The first is to calculate $T(y)$ from (4.15). That this is not always trivial will be seen in the following example of Coulomb scattering. The second is to invert (4.18). This is not always possible to do explicitly, in which case (4.18) and (4.19) constitute a pair of parametric equations with y as the parameter.

EXAMPLE: COULOMB SCATTERING

Coulomb scattering starts in principle with the Rutherford cross section and a calculation of $b(\vartheta)$. We already know from Eq. (4.10) that $b(\vartheta) = K \cot(\vartheta/2)$, so we will start with that. The function $\vartheta(b)$ is obtained by inverting $b(\vartheta)$:

$$\vartheta(b) = 2 \operatorname{arccot} \frac{b}{K}.$$

In this case Eq. (4.15) is therefore

$$T(y; K) = \frac{2}{\pi} \int_y^\infty \frac{\text{arccot}(b/K)\, db}{\sqrt{b^2 - y^2}}. \tag{4.20}$$

Since T depends on K, we have added K to its argument. This is an example of the kind of nontrivial integral to which (4.15) can give rise, and we use a clever little trick. We first take the derivative with respect to K, obtaining

$$\frac{\partial T}{\partial K} = \frac{2}{\pi} \int_y^\infty \frac{b\, db}{\{b^2 + K^2\}\sqrt{b^2 - y^2}}.$$

To integrate this, make the substitution $\zeta = \sqrt{b^2 - y^2}$. The result is

$$\frac{\partial T}{\partial K} = \frac{1}{\sqrt{K^2 + y^2}}.$$

Finally, T is obtained by integrating this with respect to K (say, by the substitution $K = y \tan\theta$) with the condition that $T(y; 0) = 0$; this yields

$$T = \ln\left\{ \frac{K}{y} + \sqrt{\left(\frac{K}{y}\right)^2 + 1} \right\}.$$

Equation (4.18) then becomes

$$r = K + \sqrt{K^2 + y^2},$$

whose solution for y^2 is

$$y^2 = r^2 - 2rK.$$

Inserting this into (4.19) finally yields

$$V = E\frac{2K}{r} = \frac{\alpha}{r}, \tag{4.21}$$

where we have used $K = \alpha/mv^2 = \alpha/2E$. Thus the Coulomb potential has been recovered from the Rutherford scattering formula [actually from Eq. (4.10) for the impact parameter, which can in turn be derived from the cross section].

4.1.3 CHAOTIC SCATTERING, CANTOR SETS, AND FRACTAL DIMENSION

So far we have been assuming that the particles in the target are so far apart that the scattered particle interacts with only one target particle at a time, a process called *one-on-one scattering*. We will now drop this assumption. For the usual kind of scattering by a continuous potential this can get quite complicated, so we start with the simplified example of a hard-core interaction. Later we will add only a few words about continuous potentials. This should not be confused with *multiple scattering*, the result of a large number of successive one-on-one scattering events in a many-particle target, which will not be treated in this book.

We will concentrate on two-dimensional scattering (i.e., in a plane) with a hard-core elastic interaction. The incoming beam is composed of noninteracting point particles (*projectiles* of radius zero), and the target is composed of infinite-mass hard disks of radius a. In the hard-core interaction the projectile interacts with only one disk at a time, so this example seems not to get away from one-on-one scattering. But the projectile can bounce back and forth between the disks, which turns out to mimic one-on-several scattering in significant ways. In particular, it leads to scattering angles that can be very irregular functions of initial data and even to orbits that are trapped and never leave the scattering region.

Because the collisions are elastic and the angle of incidence equals the angle of reflection, neither the energy nor the disk radius play any role in the scattering from each individual disk. We will therefore take $a = 1$ and will not refer to the (constant) projectile energy. The configuration manifold $\mathbb{Q}$ of the system is a plane with circular holes at the disk positions, regions that the projectile cannot enter. The four-dimensional $\mathbf{TQ}$ is obtained by attaching the two-dimensional tangent plane of the velocities above each point of $\mathbb{Q}$. In general there would be four initial conditions, but as the energy plays no role and all that matters in scattering is the incident line of the projectile (not its initial point on that line), the needed part of $\mathbf{TQ}$ is just a two-dimensional submanifold. Different numbers of target particles lead to different geometries for $\mathbf{TQ}$ and the needed submanifold.

We start the discussion with scattering by two stationary target disks. Later we will discuss three or more disks.

TWO DISKS

Assume that the centers of the two target disks are separated by a distance $D > 2$ (or $D/a > 2$ if $a \neq 1$; then what we call D is the ratio of the separation to the disk radius). Consider a projectile that collides with one of the disks at a point defined by the angle θ while moving at an angle ϕ with respect to the radial normal at that point (Fig. 4.5). Without loss of generality, consider a hit on Disk L, the disk on the left; hits on Disk R can be treated by reflection in the vertical line of symmetry between the two disks. The projectile bounces off at the same angle ϕ on the other side of the normal. Since θ ranges from $-\pi$ to π and ϕ from $-\pi/2$ to $\pi/2$, it looks like the initial conditions (θ_0, ϕ_0) can lie on half of a torus, but actually that is too big: if ϕ is small, part of Disk L is masked by Disk R (for example if $\phi_0 = 0$, a neighborhood of $\theta_0 = 0$ on Disk L cannot be reached by an incident particle). After several bounces, however, the projectile can enter some part of the (θ, ϕ) region forbidden to the initial conditions. Because in this discussion we are interested only in multiple hits, we may also restrict to the range of θ from $-\pi/2$ to $\pi/2$.

After a first collision with Disk L the projectile may hit Disk R, then Disk L again, etc. In this way, it may collide several times with the two disks before finally leaving the

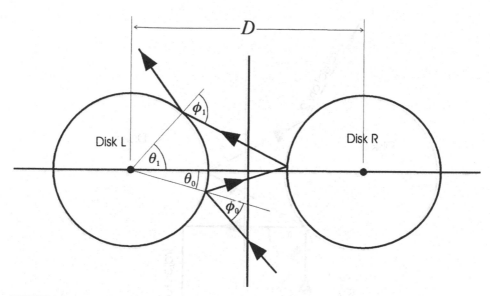

FIGURE 4.5
Illustrating the parameters in hard-core scattering from two disks. (ϕ_0, θ_0): angles for the first collision with Disk L; (ϕ_1, θ_1): angles for the second collision with Disk L.

target area. The time the projectile spends in the target area is called the *dwell* time. If it were possible for the projectile to travel exactly along the symmetry axis, it would remain trapped forever, and its dwell time would be infinite. But this *exceptional periodic orbit* can never occur, because it requires that $(\theta, \phi) = (0, 0)$, and that is forbidden by *time-reversal invariance* not only to the first hit (θ_0, ϕ_0). What this means is that given any possible orbit, another possible orbit can be constructed by replacing t by $-t$. In the scattering example, that involves simply reversing the arrows in a diagram such as Fig. 4.5, and reversing the arrows on the exceptional periodic orbit leaves it invariant. If it were reachable from any initial conditions, its time reversal, which is the orbit itself, would leave the region in a finite time. That is a contradiction, for the dwell time is infinite, so the periodic orbit is not reachable from any initial conditions.

The exceptional periodic orbit can be approached but never reached. On some orbits the (θ, ϕ) point of successive hits comes close to $(0, 0)$, but the exceptional periodic orbit is a *repeller* of almost all orbits that approach it: (θ, ϕ) eventually moves further and further from $(0, 0)$ (for this reason it is also called an *unstable* orbit). However, there exist a few *nonperiodic exceptional orbits* that approach the periodic one asymptotically. These are also infinite dwell-time orbits: on them the projectile is incident from outside the target area and, although never quite reaching the periodic orbit, it remains trapped. It never leaves.

Before discussing the analytic solution of this problem we present a brief qualitative argument for the existence of nonperiodic trapped orbits. Suppose a projectile hits Disk L at $\theta = -\pi/4$ with $\phi_1 = \pi/4$ (straight up, as in Trajectory 1 of Fig. 4.6). It will bounce off horizontally, hit Disk R, and then bounce straight down and leave the target area. Now consider Trajectory 2 of Fig. 4.6, which hits Disk L at the same θ, but at

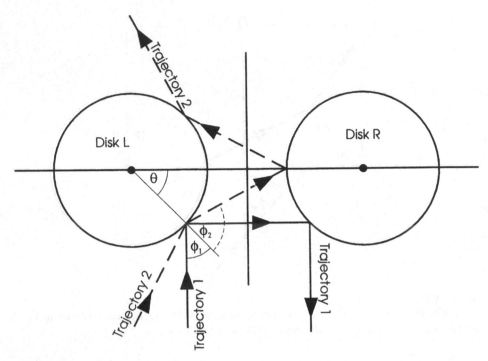

FIGURE 4.6
Two symmetric trajectories in hard-core scattering from two disks. See text for explanation.

a $\phi = \phi_2$ such that after bouncing off Disk L it hits Disk R at the point of intersection with the line joining the centers of the disks (this is possible only if $D \geq \sqrt{2} + 1$; assume it so). Symmetry then shows that the projectile will bounce off Disk R, hit Disk L directly above the first hit, and leave in the direction given by the mirror image of ϕ_2, as shown in Fig. 4.6.

Now let the incident angle vary between ϕ_1 and ϕ_2. There will be a range of ϕ values for which the projectile will leave the scattering region below the disks, as on Trajectory 1, and another range of ϕ values for which it will leave above, as on Trajectory 2. These are two open sets of trajectories, and between them is an incident ϕ whose trajectory ends up neither above nor below. That is a trapped orbit. The argument is easily extended to more general θ.

This two-disk system can be solved exactly analytically (José et al., 1992), and we now present some of the results. Assume the projectile hits Disk L initially at $(\theta, \phi) = (\theta_0, \phi_0)$, then Disk R, and then Disk L at $(\theta, \phi) = (\theta_1, \phi_1)$ (Fig. 4.5). There is a map M of the (θ, ϕ) plane that carries every possible (θ_0, ϕ_0) to its corresponding (θ_1, ϕ_1). That is, $M(\theta_0, \phi_0) = (\theta_1, \phi_1)$. The same map carries (θ_1, ϕ_1) to the subsequent hit (θ_2, ϕ_2) on Disk L (if there is one), etc., and the nth hit (if there is one) is given by $(\theta_n, \phi_n) = M^n(\theta_0, \phi_0)$. It is found that M is given by

$$\theta_{n+1} \equiv M_\theta(\theta_n, \phi_n) = \sin^{-1}[\sin\theta_n - \lambda(\theta_n, \phi_n)\sin(\phi_n - \theta_n)],$$

$$\phi_{n+1} \equiv M_\phi(\theta_n, \phi_n) = \sin^{-1}[D\sin(\phi_n - \theta_n) - \sin\phi_n],$$

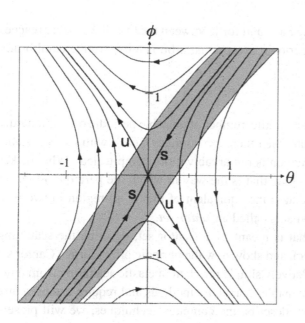

FIGURE 4.7
The region for which $-\pi/2 < \theta < \pi/2$ and $-\pi/2 < \phi < \pi/2$. The central shaded belt is the *two-hit zone* in which the projectile must hit Disk L if it is subsequently to hit Disk R. The point at $(0, 0)$ is the hyperbolic singular point, representing the exceptional periodic orbit. The curves are of the orbits under the action of the map M. The curve labeled s is the stable manifold; the curve labeled u is the unstable manifold. If the projectile is on any M-orbit other than the stable manifold, it eventually leaves the two-hit zone and escapes.

where $\lambda(\theta, \phi)$ is the solution to the equation

$$\lambda^2 + 2[\cos\phi - D\cos(\phi - \theta)]\lambda + D^2 - 2D\cos\theta = 0.$$

Successive applications of M to any initial point of the (θ, ϕ) plane carries it to a *discrete* set of points, called an *orbit under the action of M*. These points lie on curves that are invariant under such successive applications of M, some of which are shown in Fig. 4.7. The periodic exceptional orbit is at $(\theta, \phi) = (0, 0)$; the arrows show the direction in which M moves the points. It is seen that under the action of M almost all points of the plane are repelled by $(0, 0)$ (which is why it is called unstable, or a repeller); a finite number of applications will carry almost any point far from the neighborhood of $(0, 0)$ and away, for almost all orbits leave the scattering region after a finite number of hits. However, as seen from Fig. 4.7, there also exists a curve of points that approaches $(0, 0)$. This curve is called the *stable manifold* of the periodic exceptional orbit. On it lie the nonperiodic exceptional orbits, those that are trapped. There is also another curve of points, the *unstable manifold* of the periodic exceptional orbit, which is the stable manifold under the action of M^{-1} (found by reversing the direction of the arrows).

Notice how similar the structure in Fig. 4.7 is to an unstable equilibrium point in the phase portrait of a physical dynamical system (e.g., to the hyperbolic point of Fig. 2.11 in the discussion of the plane pendulum). In the present case, the (θ, ϕ) region is not the **TQ** of a dynamical system, and the flow lines of Fig. 4.7 do not map out the motion of a physical system on its velocity phase space. Nevertheless, the action of M does reproduce the state of the system every time the projectile hits Disk L, and repeated application of M forms a kind of discrete dynamical system in the (θ, ϕ) plane. In that discrete dynamics $(0, 0)$ is an equilibrium point. Its stable manifold corresponds to the line of initial conditions of the plane pendulum for which the pendulum approaches the state in which it is standing upright. The pendulum never actually reaches that state (it takes an infinite time to get there), just as the projectile, bouncing back and forth between the two disks, never reaches the periodic exceptional orbit: the periodic exceptional orbit is a hyperbolic equilibrium point of the discrete dynamics.

THREE DISKS, CANTOR SETS

In three-disk scattering there is an infinite number of unstable periodic orbits (it hardly makes sense to call them exceptional when there are so many), each with its own stable and unstable manifold. Thus the situation is remarkably more complicated. What makes the situation even more complicated than that is the way in which the periodic orbits are distributed in the parameter space that is the equivalent of the (θ, ϕ) region in two-disk scattering. They are distributed in what is called a *Cantor set*.

In what follows we explain what is meant by a Cantor set, describe the scattering parameters for the three-disk problem, and show how the periodic orbits form a Cantor set in these parameters. Even for the relatively simple problem of elastic scattering from three disks, the detailed results are not amenable to analytic methods and require extensive use of computers. Although we will not describe the computer techniques, we will present some of the results of calculations.

To explain Cantor sets we turn to the standard example of the *middle third Cantor set* (Cantor, 1915). It is constructed in the following way. From the unit closed interval $I_0 = [0, 1]$ remove the open set of its center third, leaving the closed sets $l = [0, \frac{1}{3}]$ (l for *left*) and $r = [\frac{2}{3}, 1]$ (r for *right*). Define $I_1 \equiv l \cup r$. Now remove the center thirds of l and r, leaving the sets $ll = [0, \frac{1}{9}]$, $lr = [\frac{2}{9}, \frac{1}{3}]$, $rl = [\frac{2}{3}, \frac{7}{9}]$, and $rr = [\frac{8}{9}, 1]$ (*left left*, *left right*, etc.). Define $I_2 \equiv ll \cup lr \cup rl \cup rr$. Next, remove the center thirds of ll, lr, rl, and rr to obtain lll, llr, lrl, ..., rrl, rrr, and define $I_3 \equiv lll \cup llr \cup \cdots \cup rrr$. Go on in this way, always removing the center thirds of the remaining intervals. The nth step yields the set I_n, a union of 2^n intervals designated by all possible permutations of n letters l and r, which we will call n-strings. The I_n designated by the n-strings satisfy $I_0 \supset I_1 \supset \cdots \supset I_n$. Figure 4.8 shows the result of the first three steps in this procedure. The middle third Cantor set C is then defined as what is left after an infinite number of steps, or more accurately,

$$C = \bigcap_{n=0}^{\infty} I_n. \tag{4.22}$$

Since all that is removed in any step is an open set, some points remain, and C is not empty. Indeed, it is obvious that $1, \frac{1}{9}, \frac{2}{9}, \frac{1}{3}$, and many other such points are in it; C contains an infinite number of points. It is, moreover, a pretty exotic point set, some of whose properties may at first seem contradictory. Every point in it is separated from every other. Its *measure* is zero, which means that, like the rationals, it occupies none of the length of the line. (Measure is discussed in more detail in the number-theory appendix to Chapter 7. See Halmos (1950).) Indeed, the length of I_0 is 1, that of I_1 is $\frac{2}{3}$, that of I_2 is $\frac{4}{9}$, ..., that of I_n is $(\frac{2}{3})^n$, and that of C is $\lim_{n \to \infty} (\frac{2}{3})^n = 0$. Yet C has the *power of the continuum*, which means that there are as many points in C as there are points on the line, rational and

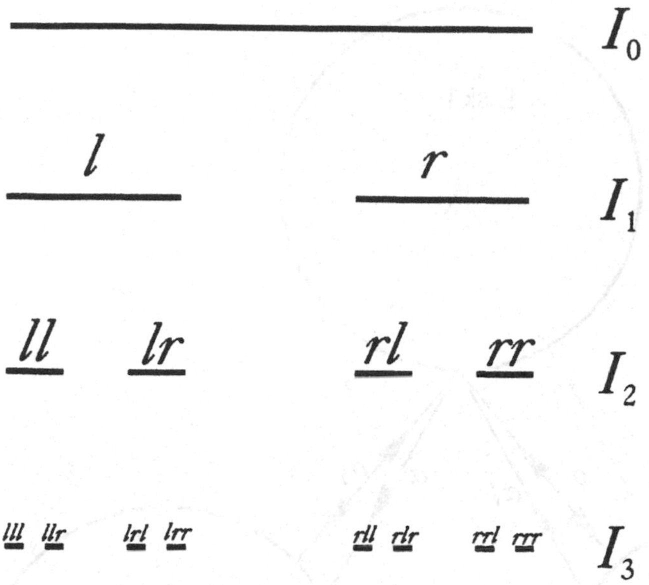

$$I_0$$

l $\qquad$ r $\qquad$ I_1

ll $\quad$ lr $\qquad$ rl $\quad$ rr $\qquad$ I_2

lll llr $\quad$ lrl lrr $\qquad$ rll rlr $\quad$ rrl rrr $\qquad$ I_3

FIGURE 4.8
The first three steps in the construction of the middle third Cantor set.

irrational. This can be demonstrated by assigning to every point in C a number in the unit interval and vice versa to every number in the unit interval a point in C. An easy way to do this is in binary notation. Every point in C can be specified by its ∞-string of ls and rs. For example the string for 0 is $lllll\ldots$, the one for $\frac{2}{9}$ is $lrlll\ldots$. Now replace the ls by 0s and the rs by 1s and place a decimal point before it all. Then the string for 0 becomes .00000..., and that for $\frac{2}{9}$ becomes .01000.... Every point in C is associated in this way with a unique binary fraction and hence with a number on the unit interval. Conversely, every number on the unit interval is associated in this way with a unique point in C. Thus C has indeed the power of the continuum.

The middle third Cantor set has another important property: it is *self-similar* or *scale invariant*. That is, after an infinite number of steps, each subinterval designated by a finite n-string looks like I_1, except that it is shorter by a factor of 3^n. For example, if lrl (from $\frac{2}{9}$ to $\frac{7}{27}$), with the center thirds taken out to all orders, is scaled up by a factor of $27 = 3^3$, it looks exactly like I_1. As will soon be seen, self-similarity is one of the features that are found in scattering from three disks arranged in an equilateral triangle.

Scattering from three or more disks with forces more general than the hard core has been treated in several papers (see Bleher et al., 1990, and references they cite). We go into some detail only for three disks with the hard-core interaction. Consider a system of three disks, labeled K, L, and M and arranged in an equilateral triangle, as illustrated in Fig. 4.9. Each disk has radius 1, and the triangle connecting their centers has sides of length $D > 2$ (or, as in the two-disk case, D is the ratio of the side of the triangle to the disk radius a). There are many periodic orbits. For example there are three that have been seen already in the two-disk case, but now between Disks K and L, between Disks L and M, and between Disks M and K. Another, going round and round from Disk K to L to M to K, etc., is

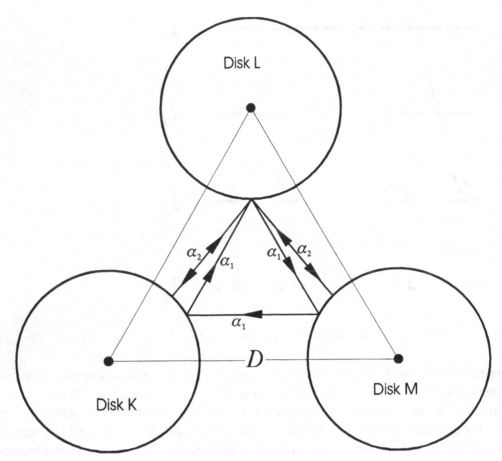

FIGURE 4.9
Hard-core scattering from three disks arranged equilaterally. Two periodic orbits are shown.

labeled α_1 in Fig. 4.9. Yet another, labeled α_2, goes from Disk K to L to M to L to K and repeats (see Problem 6). There are many others.

We will deal only with projectiles incident from below and perpendicular to the base line connecting Disks K and M. Let b (we will call it the impact parameter) be the distance from the incident trajectory to the center of the triangle. Unlike central-force scattering, this system is not rotationally invariant, so the incident angle is needed to describe the trajectory completely: the full submanifold of initial conditions is two dimensional, as expected. But for simplicity we restrict our discussion to the one-dimensional submanifold which consists only of the perpendicular trajectories. Moreover, we are interested only in collisions with $|b| < \frac{1}{2}D$, as collisions at larger values of $|b|$ will certainly leave the target area without a second hit.

Let I_0 be the b interval from $-\frac{1}{2}D$ to $\frac{1}{2}D$. If $D \leq 4$, every value of $b \in I_0$ leads to a hit on one of the disks, but if $D > 4$, there are gaps between $b = 1$ and $b = \frac{1}{2}D - 1$ and between $b = \frac{1}{2}D + 1$ and $b = D - 1$, where the projectile misses all three disks. Remove these gaps, if any, from I_0, and let I_1 be what is left (if $D \leq 4$, then $I_1 = I_0$). Figure 4.10 illustrates this procedure.

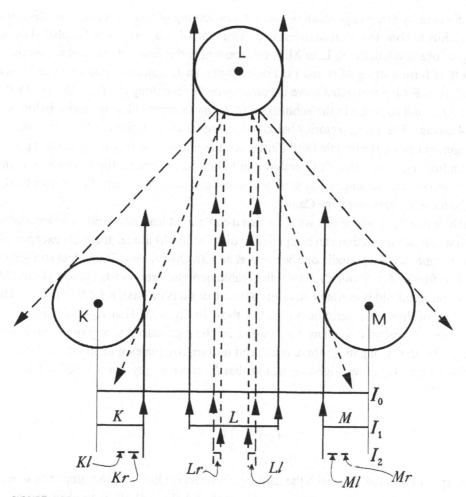

FIGURE 4.10
The first two steps in constructing the Cantor set for three-disk scattering. See text for an explanation.

Within I_1 there are subintervals for which the projectile, after colliding with whichever disk it hits first, goes on to hit a second disk, and there are other subintervals for which it leaves the target area after just one collision. Remove the one-hit subintervals, from I_1, and let I_2 be what is left, as shown in Fig. 4.10. Then I_2 represents all the trajectories that make at least two hits before leaving the target area. Now remove from I_2 those subintervals whose projectiles leave after just two hits, and let I_3 be what is left. Next, remove from I_3 those subintervals whose projectiles leave after just three hits. Go on in this way, always removing more subintervals. What is left at the nth stage is I_n, which represents all the trajectories that make at least n hits before leaving the target area, with $I_0 \supset I_1 \supset \cdots \supset I_n$.

In the limit this process yields a set of points I_∞, which will be shown in the next paragraph not to be empty. The procedure is in obvious ways similar to the one that yields the middle third Cantor set. Labeling the trajectories belonging to I_n exhibits further similarities. Each trajectory starts with a hit on one of the disks, and every subsequent hit can be denoted by l or r depending on whether, in coming off the previously hit disk, the projectile hits the one to its left or to its right. Thus each I_n trajectory can be labeled by an n-string whose first letter is K, L, or M, corresponding to the first disk hit, and whose other $n - 1$ letters form a string of ls and rs. For example, an I_4 trajectory that first hits L then M then L then K (and then must leave the target area, for it belongs to I_4) is labeled by the 4-string $Llrr$ and its b lies in the subinterval of I_4 that contains all trajectories belonging to this 4-string. The I_∞ trajectories belong to strings that are infinitely long and, by the same argument as for the middle third Cantor set, the set of these ∞-strings has the power of the continuum. (The extra K, L, and M don't alter the argument, for there are as many points on the interval of length 3 as there are on the unit interval, as was first shown by the same Cantor who discovered the Cantor set.)

To show that I_∞ is not empty, we show that it contains at least one orbit. The arguments of the first part of this section can be applied to disks K and M alone: there are exceptional orbits that approach the periodic one between K and M. One of these is incident straight up and is therefore in I_∞. Similarly, a two-disk stable periodic orbit exists between L and M, and its stable manifold is available to orbits whose first hit is on Disk K ($Kllrlrlr \ldots$). The same is true of three-disk periodic orbits, like those of Fig. 4.9. That I_∞ is not empty can be seen more formally. Given any ∞-string, it can be reproduced to any order, no matter how large, by an n-string in I_n: there is an orbit that approximates it to any finite order, no matter how high. That is enough to establish that I_∞ is not empty, for it is defined, like C, by Eq. (4.22):

$$I_\infty = \bigcap_{n=0}^{\infty} I_n.$$

The system of three disks with the hard-core interaction has been the subject of several computer and analytic investigations (see Gaspard and Rice, 1989, who also give many other references). What is often studied is the dependence of dwell time on initial data. Figure 4.11 shows a pair of graphs that Gaspard and Rice computed for the *time delay function* $T(y_0)$ (essentially the dwell time) as a function of $0 < b < 0.5$ (their y_0 is our b) for $D = 2.5$. We see that T is a highly discontinuous function. There are intervals in which the dwell time is a smooth function of b, but in others its b dependence is highly singular. The b values for which T is infinite belong to trajectories in I_∞. The self-similarity of this set of b values is made evident by the enlargement in Fig. 4.11(b). Other investigators have studied the dependence of scattering angle on initial data and have found discontinuities in it (Ott and Tél, 1993). This means that the scattering angle cannot in general be predicted with complete accuracy, as extremely small variations in the initial data may cause wild changes in the scattering angle for orbits that lie close to points in I_∞.

FRACTAL DIMENSION AND LYAPUNOV EXPONENT
So far our considerations have been almost entirely qualitative. As stated at the beginning of this subsection, quantitative analysis requires computers. It focuses on several parameters, some of which we now discuss.

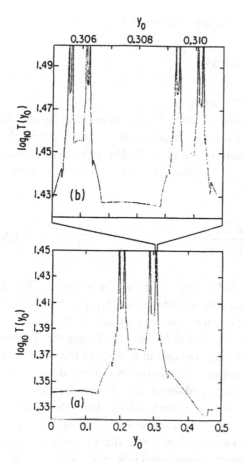

FIGURE 4.11

The logarithm of the time-delay function $T(y_0)$ in three-disk scattering with $D = 2.5$ (from Gaspard and Rice, 1989). y_0 is what we call the impact parameter b, and $T(y_0)$ is essentially what we call the dwell time (the authors define it more carefully, as the time the projectile spends inside a certain region that encloses the three disks). In Fig. (a) y_0 varies from 0 to 0.5. Figure (b) is a blow up of the y_0 interval from 0.305 to 0.311. The similarity of the two graphs reflects the self-similarity of the Cantor-like structure of the set of trapped orbits.

The first parameter is what is called the *fractal dimension* (Mandelbrot, 1982; Barnsley, 1990) of the Cantor set. There are various ways to define the dimension of an object; for the purpose of a Cantor set it is done as follows. Start with a line. To cover a line of length L with nonoverlapping intervals of length ϵ requires $N(\epsilon) \simeq L/\epsilon$ intervals. In the limit of small ϵ the approximate equality becomes exact. In an abuse of notation we write $v(\epsilon) \equiv \lim_{\epsilon \to 0} N(\epsilon) = L/\epsilon$. Next, consider a regular area A, and cover it (*tile* it) with patches of linear dimension ϵ. In the limit of small ϵ the number of patches needed is $v(\epsilon) \equiv \lim_{\epsilon \to 0} N(\epsilon) = A/\epsilon^2$. For a three-dimensional regular volume V, the equivalent formula is $v(\epsilon) = V/\epsilon^3$. The dimension, called the *Euclidean dimension,* is the exponent of ϵ in each of these expressions. The *fractal dimension* d_f is defined similarly by the exponent of ϵ, but it will be seen that for a Cantor set d_f need not be an integer. In general we have found that $v(\epsilon) = v_0 \epsilon^{-d}$ for some v_0, so d_f is defined similarly, as

$$d_f \equiv -\frac{\ln[v(\epsilon)/v_0]}{\ln \epsilon}. \tag{4.23}$$

We now show how this idea yields a noninteger fractal dimension for the middle third Cantor set C. Pick an n-string for some integer n and cover C by copies of the subinterval i_n belonging to that n-string; i_n is of length $\epsilon = 3^{-n}$. The set C consists of 2^n such subintervals, all of the same length, with gaps between them, so it can be covered by 2^n copies of i_n. Hence $v/v_0 = 2^n$, and according to (4.23) the fractal dimension of the middle third Cantor set C is

$$d_f = \frac{n \ln 2}{n \ln 3} \approx 0.631.$$

Sets of noninteger dimension are called *fractals*: C is a fractal.

A Cantor set can be formed also by removing some fraction $1/f$ other than $1/3$ of the intervals in each step. The fractal dimension d_f of such a Cantor set depends on f, and in general d_f is $\ln 2 / \ln\{2f/(f - 1)\}$ (the number of intervals in each stage is still 2^n). This result is often stated in terms of the ratio g of the length of I_{n+1} to that of I_n, which is $g = (f - 1)/f < 1$. In these terms, the fractal dimension of such a Cantor set is

$$d_f = \frac{\ln 2}{\ln 2 + |\ln g|}. \qquad (4.24)$$

These considerations can be used for the further analysis of three-disk scattering. For the Cantor set obtained in three-disk scattering it is not easy to find the ratio I_{n+1}/I_n (we use I_n for both the set and its length). It is clear that it varies with n, and one may assume that as $n \to \infty$ it approaches some limit g that depends only on D. This is because the longer the projectile bounces around among the disks, the more it loses any memory of how it got there in the first place, and g is essentially the probability that the projectile, coming off in some random direction from one of the disks, will hit one of the others (the probability enters because the length of I_n is a measure of the number of nth hits). This obviously depends on D. Thus physically g is understood as the probability that after the nth collision the particle will make an $(n + 1)$st. It follows that the probability that it will make m in a row is $g^m \equiv e^{-\gamma m}$, where $\gamma \equiv |\ln g|$. As just mentioned, the length of I_n is a measure of the number of trajectories in it. Each collision reduces the number of remaining trajectories by a factor of g, and the number in I_{n+m} is lower by a factor of $e^{-\gamma m}$ than the number in I_n. In analogy with radioactive decay, γ is a sort of lifetime for the set of trajectories. In computer simulations one can determine how the lengths of the I_n depend on n and thereby establish the lifetime and the fractal dimension of I_∞.

Let us call two trajectories *equivalent of order n* if they both belong to the same n-string. Two trajectories that are equivalent of order n will eventually belong to different k-strings, where $k > n$, and may therefore end up far apart, especially if one of them leaves the scattering region and the other goes on colliding with the disks. For this reason among others, three-disk scattering is called *chaotic* or *irregular*. What is the probability that two trajectories that are equivalent of order n are also equivalent of order $n + 1$? This depends on the length ratio of an n-string to an $(n + 1)$-string; the ratio is $g/2$ (for $I_n/I_{n+1} = g$ and there are twice as many strings in I_{n+1} as in I_n). Thus the probability that two orbits will be separated in one collision is $g/2$, and the probability that they will be separated in m collisions is $(g/2)^m = e^{-\lambda m}$, where $\lambda = \gamma + \ln 2$. The number λ, called the *Lyapunov exponent* (Liapounoff, 1949), is a measure of the instability of the scattering system, as it shows how fast two trajectories that are initially close together will separate.

It follows from (4.24) and the defining equations for γ and λ that

$$\gamma = (1 - d_f)\lambda. \qquad (4.25)$$

Bercovich et al. (1991) have studied the three-disk problem experimentally on an equivalent system of optically reflecting cylinders with a laser beam. They have successfully compared their results with computer calculations. Their paper also treats the theory, and some of our discussion is based on their work.

We emphasize that we have discussed only a one-dimensional submanifold of initial conditions. A full treatment would require dealing not only with variations in the impact parameter, but also in the incident angle. The full treatment shows that at all angles the scattering singularities (e.g., infinite dwell times) lie in a Cantor set (Eckhardt, 1988).

SOME FURTHER RESULTS

The 1990 paper of Bleher et al. treats scattering by forces other than the hard-core potential. It deals with four force centers in some detail and also discusses three-center scattering. We will not go into these matters in detail, but the results are similar to the ones we have described here. There are periodic orbits very similar to the ones in the hard-core case, also distributed in Cantor sets. The fractal dimension d_f, the "lifetime" γ, the Lyapunov exponent λ, and the relation among them play roles in many of the treatments that are similar to the ones we have discussed. A crucial difference, however, is the energy dependence: there is a critical energy above which the scattering is not chaotic. The potentials dealt with are usually in the form of several circularly symmetric hills (as in Fig. 4.12), and the critical energy for chaotic scattering is related to the maximum height of the lowest of the hills.

The treatment for three hills is similar to that for three disks. The projectile must be scattered from each of the hills at an angle large enough to interact significantly with one of the others, and if the energy of the projectile is above the maximum V_{max} of the lowest hill, the scattering angle from that hill may not be large enough.

Consider a projectile with $E > V_{max}$ incident on a *single* circularly symmetric potential hill with a finite maximum and of *finite range*, that is, one that vanishes outside some finite distance. (Strictly finite range is not necessary, but the potential must go to zero rapidly or an impact parameter cannot be defined; see Problem 3 for an example.) If the projectile is incident at a very large impact parameter b, the particle feels almost no force and the scattering angle ϑ is close to zero. If b is zero, however, the particle slows down as it passes over the force center, but ϑ is zero. If $\vartheta(b)$ is a continuous function, it follows that ϑ reaches at least one maximum for finite b as long as $E > V_{max}$. This is a generic property of potentials with finite maxima and finite

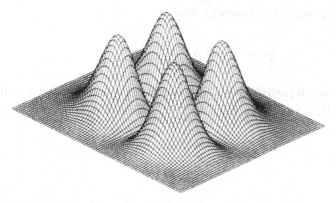

FIGURE 4.12
Four circularly symmetric potential hills in the plane.

range. It can be shown that the maximum value of ϑ is less than $\frac{1}{2}\pi$ if E is strictly greater than the maximum of the potential.

Hence if the energy of the projectile is greater than the maximum of the lowest hill, the scattering angle from that hill may not be large enough to scatter the projectile back to one of the others. It is thus seen that the maximum of the lowest hill plays a special role in scattering from several potential hills: chaos does not occur for projectile energies greater than a maximum that depends on the height of the lowest hill.

In the next section we discuss another type of scattering that exhibits similar chaos. Although there is only one scattering center, the scattering angle and the dwell time depend discontinuously on the initial data, leading to chaos and making accurate prediction impossible.

4.1.4 SCATTERING OF A CHARGE BY A MAGNETIC DIPOLE

Irregular (chaotic) scattering occurs in most scattering systems, even for some systems with a single scattering center. An example, physically more meaningful than hard-core scattering from disks, consists of an electrically charged particle in the field of a magnetic dipole. In this section we discuss this system briefly in order to demonstrate that in spite of its apparent simplicity it also exhibits irregular scattering.

THE STÖRMER PROBLEM

Scattering by a magnetic dipole has been of interest ever since the discovery of cosmic rays – charged particles impinging on the Earth from outer space. The Earth's magnetic field is taken in the first approximation to be dipolar, the field of a bar magnet. Yet although the problem has been studied since the beginning of the twentieth century, a complete analytic solution has not and cannot be found: it is what is called a nonintegrable system (Chapter 6), one whose integral curves cannot be calculated analytically. It was studied extensively by Störmer (1955) for over thirty years and is now called Störmer's problem (for a relatively early review see Sandoval-Vallarta, 1961).

As for hard-core scattering by disks (first two, then three disks), we first discuss a limit for which this system has an explicit analytic solution and will later indicate the ways in which the general problem leads to chaotic scattering.

The Lagrangian for a charge in a static magnetic field is

$$L = \frac{1}{2}mv^2 + e\mathbf{v}\cdot\mathbf{A},$$

where m is the mass of the charged particle, e is its charge, $\mathbf{v}$ is its velocity, and $\mathbf{A}$ is the vector potential [see Eq. (2.49); we take $c = 1$]. For a magnetic dipole at the origin, the vector potential is given by

$$\mathbf{A} = \frac{\mathbf{M}\wedge\mathbf{r}}{r^3},$$

where $\mathbf{r}$ is the radius vector measured from the origin and $\mathbf{M}$ is the magnetic moment of the dipole, a constant vector. Without loss of generality the z axis can be taken aligned along the dipole, so that $\mathbf{M}$ has only one component: $\mathbf{M} = M\mathbf{k}$, where $\mathbf{k}$ is the unit vector in the z direction. Because the dipole field has cylindrical symmetry, it is natural to write the Lagrangian in cylindrical polar coordinates (ρ, ϕ, z):

$$L = \frac{1}{2}m(\dot{\rho}^2 + \rho^2\dot{\phi}^2 + \dot{z}^2) + \frac{eM\rho^2\dot{\phi}}{(\rho^2 + z^2)^{3/2}}. \tag{4.26}$$

The Lorentz force $e\mathbf{v} \wedge \mathbf{B}$ on the charge does no work, so the kinetic energy $T \equiv E$ is conserved. Because ϕ is an ignorable coordinate, its conjugate generalized momentum p_ϕ is also conserved:

$$p_\phi \equiv \frac{\partial L}{\partial \dot{\phi}} = m\rho^2\dot{\phi} + \frac{eM\rho^2}{(\rho^2 + z^2)^{3/2}}. \tag{4.27}$$

In the limit of large r, when the magnetic field becomes negligibly small, this expression reverts to the usual expression for the angular momentum about the axis.

The two remaining equations of motion for ρ and z are

$$\left. \begin{aligned} m\ddot{\rho} &= m\rho\dot{\phi}^2 + \frac{2eM\rho\dot{\phi}}{(\rho^2 + z^2)^{3/2}} - \frac{3eM\rho^3\dot{\phi}}{(\rho^2 + z^2)^{5/2}}, \\ m\ddot{z} &= -\frac{3eM\rho^2z\dot{\phi}}{(\rho^2 + z^2)^{5/2}}. \end{aligned} \right\} \tag{4.28}$$

In analogy with central force motion, one can eliminate $\dot{\phi}$ from these equations by using (4.27), attempt to solve the resulting equations for $\rho(t)$ and $z(t)$, and then use (4.27) to find $\phi(t)$. Because $\ddot{z} = 0$ if $z = 0$, any motion whose initial conditions are $z(0) = \dot{z}(0) = 0$ (i.e., motion that starts in the equatorial plane with the velocity vector lying in that plane) will remain forever in the equatorial plane. We first consider this special case, the *equatorial limit*.

THE EQUATORIAL LIMIT

In the equatorial limit only the ρ equation of motion remains to be solved; it becomes

$$m\ddot{\rho} = m\rho\dot{\phi}^2 - \frac{eM\dot{\phi}}{\rho^2}, \tag{4.29}$$

and p_ϕ is given by

$$p_\phi = m\rho^2\dot{\phi} + eM/\rho. \tag{4.30}$$

Inserting (4.30) into (4.29) yields

$$m\ddot{\rho} = \frac{1}{m\rho^3}\left(p_\phi - \frac{eM}{\rho}\right)\left(p_\phi - \frac{2eM}{\rho}\right),$$

which is the equation of motion of an equivalent one-dimensional problem whose Lagrangian is

$$L_{\text{eff}} = \frac{1}{2}m\dot{\rho}^2 - V_{\text{eff}},$$

with

$$V_{\text{eff}} = \frac{1}{2m\rho^2}\left(p_\phi - \frac{eM}{\rho}\right)^2. \tag{4.31}$$

The conserved energy E for L_{eff} is

$$E = \frac{1}{2}m(\dot\rho^2 + \rho^2\dot\phi^2) = \frac{1}{2}m\dot\rho^2 + V_{\text{eff}}. \tag{4.32}$$

In fact E is the actual energy (compare to the situation in the equivalent one-dimensional problem for central force motion). We have arrived in Eq. (4.32) at a family of equivalent one-dimensional problems, one for each value of p_ϕ.

The sign of p_ϕ is important. If $p_\phi < 0$ the effective force is always positive (i.e., the effective potential is always repulsive). If $p_\phi > 0$, however, the potential is repulsive for large and small ρ but attractive for $\mu_\phi < \rho < 2\mu_\phi$, where

$$\mu_\phi = eM/p_\phi \tag{4.33}$$

(see Fig. 4.13): V_{eff} has a minimum $V_{\text{min}} = V_{\text{eff}}(\mu_\phi) = 0$ and a maximum

$$V_{\text{max}} = V_{\text{eff}}(2\mu_\phi) = \frac{1}{2m}\left(\frac{p_\phi^2}{4eM}\right)^2. \tag{4.34}$$

For $p_\phi < 0$ the incoming charged particle is repelled and scattered from the dipole. For each value of E there is a minimum ρ, which means that in the equatorial plane there is a forbidden circle about the origin that the particle cannot enter.

For $p_\phi > 0$ there are several possibilities. If the energy $E < V_{\text{max}}$, let ρ_1, ρ_2, and ρ_3 be the three positive roots of the equation $E = V_{\text{eff}}$ (Fig. 4.13). There are two forbidden regions: a circle about the origin of radius ρ_1 and a ring about the origin between the radii ρ_2 and ρ_3. The particle in the equivalent one-dimensional region is either trapped and oscillating between ρ_1 and ρ_2 or it is outside ρ_3. If it is trapped, the charge is in a bound orbit about the dipole. If it is in the outer region, the charge comes in from afar, bounces off the dipole field, and is scattered. If $E > V_{\text{max}}$, there is only one forbidden circle about the origin; the value of ρ eventually increases without bound and the charge is scattered.

From the point of view of scattering, the most interesting cases for E both greater and less than V_{max} are those in which E is close to V_{max}. As was mentioned in the second subsection of Section 1.3.3, in discussing Fig. 1.5, if $E = V_{\text{max}}$ it takes an infinite time for ρ to reach the unstable equilibrium point $\rho = 2\mu_\phi$ where $V(\rho) = V_{\text{max}}$. In the velocity phase space of the equivalent one-dimensional problem, $(\rho, \dot\rho) = (2\mu_\phi, 0)$ is a hyperbolic singular point (Fig. 4.14). The positive-ρ stable submanifold of this hyperbolic point holds the initial conditions for trapped scattering orbits (compare the two-disk hard-core problem). The incident particle, with energy $E = V_{\text{max}}$, comes in on this stable submanifold

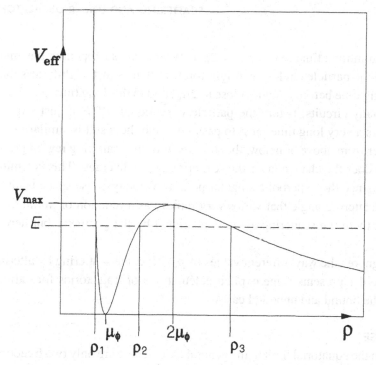

FIGURE 4.13

The potential $V_{eff}(\rho)$ of the equivalent one-dimensional system in the equatorial limit of the Störmer problem. μ_ϕ and $2\mu_\phi$ are the equilibrium values of ρ (μ_ϕ is the radius of the stable circular orbit, and $2\mu_\phi$ of the unstable one). ρ_1, ρ_2, and ρ_3 are the turning points in the equivalent one-dimensional system at energy $E < V_{max}$.

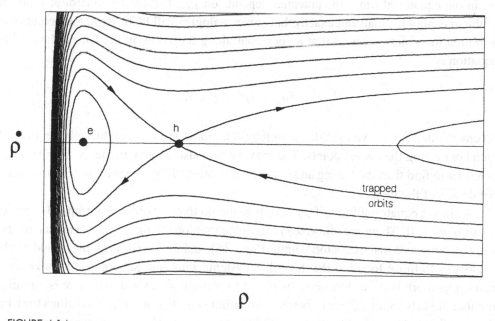

FIGURE 4.14

The phase portrait of the equivalent one-dimensional system. h is a hyperbolic singular point. e is an elliptic singular point. The stable manifold of h is the curve labeled "trapped orbits." Its mirror image in the $\dot{\rho}$ axis is the unstable manifold.

from large ρ, taking an infinite time to reach $\rho = 2\mu_\phi$. While this is happening the angle ϕ keeps changing and the particle circles the origin forever. If E is just slightly less than V_{max}, it takes a very long time before ρ comes close to $2\mu_\phi$, and in this long time ϕ changes by a large amount, many circuits, before the particle is scattered. If E is just slightly more than V_{max}, it takes a very long time for ρ to pass $2\mu_\phi$, and the result is similar. As E approaches V_{max}, either from above or below, the time spent in the scattering region grows without bound and so does the change in ϕ: the scattering angle diverges. This is similar to two-disk scattering since the hyperbolic singular point in velocity phase space leads to trapped orbits and a scattering angle that varies very rapidly as a function of initial data. The difference is that there the scattering angle is restricted by the geometry but here it isn't.

This discussion ignores the way different values of p_ϕ affect the scattering by altering V_{eff}. Worked Example 4.1 presents some explicit calculations of trajectories for various values of p_ϕ in both the bound and unbound cases.

THE GENERAL CASE

We now pass from the equatorial limit to the general case. There are only two freedoms to be considered, and the effective potential $V(\rho, z)$ is

$$V(\rho, z) = \frac{1}{2m}\left\{ \frac{p_\phi}{\rho} - \frac{eM\rho}{(\rho^2 + z^2)^{3/2}} \right\}^2. \tag{4.35}$$

As in the equatorial limit, the potential depends on p_ϕ. Unlike the equatorial limit, this problem cannot be analyzed exactly, but some analogies will help. First, the potential has a minimum, which occurs not at a point, but along the curve in the (ρ, z) plane whose equation is

$$\rho^2 \mu_\phi = (\rho^2 + z^2)^{3/2}.$$

Störmer called this curve the *talweg*, which is German for "the road through the dale," the path connecting the lowest points. The maximum is also a curve in the (ρ, z) plane, but it is harder to find than the talweg and is generally obtained by numerical calculation, as are the details of the motion.

As in the equatorial limit, if $p_\phi < 0$ there are no trapped orbits, but trapped orbits can occur if $p_\phi > 0$. There are allowed and forbidden regions in the (ρ, z) plane, just as there were on the ρ axis in the equatorial limit (Fig. 4.15), and there exist unstable periodic orbits analogous to those found in the three-disk system. Jung and Scholz (1988) have shown that trapped orbits occur whenever there is a hyperbolic point and that there is an infinite number of such points. There is therefore an infinite number of initial conditions for which the particle can be trapped, and these initial conditions have Cantor-set properties. The step from the equatorial limit to the general system is reminiscent of the step from two disks to three in hard-core scattering. As in the three-disk problem, the general solutions cannot

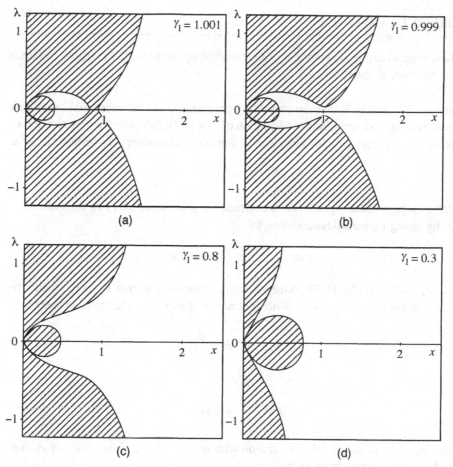

FIGURE 4.15

Allowed and forbidden regions in the meridian plane for the general Störmer problem for several values of $\gamma_1 \equiv p_\phi/(mv\mu_\phi)$, where v is the speed. λ is the latitude angle, and x is defined in Worked Example 4.1. The meridian plane is perpendicular to the equatorial plane and contains the cylindrical axis. Shaded areas are the forbidden regions. The allowed and forbidden regions in three dimensions are obtained by rotating these diagrams about the λ axis. (a) $\gamma_1 > 1$. There is an internal allowed region disconnected from the external allowed region. Particles coming from infinity cannot enter the internal allowed region, and particles in the internal one are in bounded orbits (see the next figure). (b) For $\gamma_1 < 1$ the barrier dividing the two regions breaks open into a small neck (at $\gamma_1 = 1$ the point $x = 1$ becomes a hyperbolic point). (c), (d) For smaller values of γ_1 the neck widens out and the difference between the two regions disappears, while the forbidden region near the origin gradually increases. (From Sandoval–Vallarta, 1961.)

be obtained analytically and numerical methods must be used to find the trajectories. The results are qualitatively similar to those of the three-disk problem. A more detailed and quantitative discussion will be found in the cited references. We will return to this problem in discussing adiabatic invariance in Chapter 6.

We have demonstrated that scattering, even for relatively simple systems, can be quite complex. Very small changes in the initial conditions can lead to very different scattering results. At the present writing chaotic scattering is a subject of intense research.

WORKED EXAMPLE 4.1

Obtain integral expressions for the orbit (eliminating the time) in the equatorial limit of the Störmer problem.

Solution. We follow Störmer in presenting the solution. In the equatorial limit there are three equations: apart from the equation of motion (4.29) there are conservation of generalized momentum (4.30) and conservation of kinetic energy. We write the latter as

$$v^2 = \dot{\rho}^2 + \rho^2\dot{\phi}^2,$$

where v is the constant speed. Störmer writes the three equations in a dimensionless form by changing variables according to

$$v\,dt = \mu_\phi\,d\tau, \quad \mu_\phi\rho = x,$$

where μ_ϕ, defined in Eq. (4.33), is the value of ρ at the minimum of V_{eff} and is hence the radius of the stable circular orbit. With this change of variables, the equations become

$$\ddot{x} - x\dot{\phi}^2 = -\frac{\dot{\phi}}{x^2}, \tag{4.36a}$$

$$\dot{x}^2 + x^2\dot{\phi}^2 = 1, \tag{4.36b}$$

$$x^2\dot{\phi} + \frac{1}{x} = 2\gamma_1, \tag{4.36c}$$

where the dot now denotes differentiation with respect to τ, and the constant γ_1, also introduced by Störmer, is given by

$$\gamma_1 = \frac{p_\phi}{mv\mu_\phi} = \pm\frac{p_\phi^2}{eM}\sqrt{\frac{1}{2Em}};$$

γ_1 takes on values in the range $-\infty < \gamma_1 < \infty$. Recall that in the equatorial limit we are dealing with a family of equivalent one-dimensional systems, one for each value of p_ϕ. Since γ_1 depends on both E and p_ϕ, in each equivalent one-dimensional system γ_1 depends only on E. Similarly, for a given E, it depends only on p_ϕ. In terms of the dimensionless variables, V_{eff} simplifies to

$$V_{\text{eff}}^{\text{dim}}(x) = 2E\gamma_1^2\frac{1}{x^2}\left(1 - \frac{1}{x}\right)^2.$$

As usual, to find the equation for the orbit we eliminate $d\tau$ in favor of $d\phi$. From Eq. (4.36c) we obtain

$$x^2\,d\phi = \left(2\gamma_1 - \frac{1}{x}\right)d\tau$$

Now $dx/d\phi$ can be found by substituting this into Eq. (4.36a), and finally we obtain the desired integral expression:

$$\int d\phi = \pm \int dx \, \frac{2 - 1/x}{\sqrt{x^4 - 4\gamma_1^2 x^2 + 4\gamma_1 x - 1}}.$$

Note that this is an elliptic integral whose behavior depends on whether $\gamma_1 < 1$, $\gamma_1 = 1$, or $\gamma_1 > 1$. The integral cannot be evaluated exactly for general γ_1. Curiously, it can be evaluated for $\gamma_1 = \frac{1}{2}$, yielding

$$\phi(x) = \frac{1}{2\sqrt{5}}\log\left(\frac{2x - \sqrt{5} + 1}{2x + \sqrt{5} + 1}\right) + \frac{1}{\sqrt{2}}\log\left(x^2 - x + 1\right) + \log x.$$

This analytic expression is not very transparent. When an expression suffers from such lack of transparency, as often happens, graphs can be very helpful. Some possible orbits are shown in Fig. 4.16. Figure 4.16(a) was obtained on a computer by inserting numbers into the analytic expression for $\gamma_1 = \frac{1}{2}$. The orbit is very tightly wound at the center, and numerical calculation may miss its details. This is related to the irrational number $\frac{1}{2}(\sqrt{5} + 1)$, called the *golden mean*, which is in some sense as irrational as a number can get (see the number-theory appendix to Chapter 7) and is therefore difficult to handle accurately on a computer. For more general values

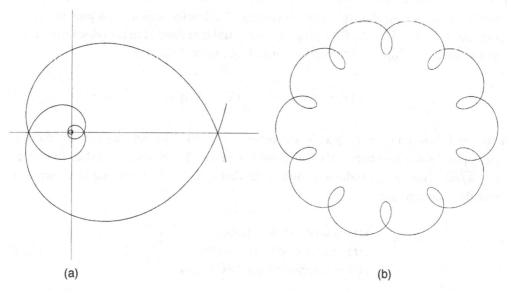

(a)   (b)

FIGURE 4.16

(a) The orbit for the exact solution at $\gamma_1 = \frac{1}{2}$. The charged particle spirals in toward the dipole and then spirals out again in the equatorial plane (see Fig. 4.15c or 4.15d). The horizontal and vertical axes are the Cartesian coordinates $x \cos\phi$ and $x \sin\phi$, respectively. In this diagram x varies from zero to 400. (b) A possible orbit for $\gamma_1 > 1$. The charged particle is trapped in the internal allowed region (see Fig. 4.15a).

of γ_1, computers are needed even earlier for the numerical evaluation of the integral. The orbit of Fig. 4.16(b) is an example.

4.2 LINEAR OSCILLATIONS

This section is devoted to free, driven, and damped linear oscillations in one and several freedoms. The material discussed here has important applications in mechanics as well as in molecular and solid state physics and field theory. It also serves as the starting point for perturbation theory of nonlinear oscillators.

An oscillating dynamical system is one whose motion is recurrent and bounded on $\mathbb{Q}$. The simplest ones are *linear* or *harmonic* oscillators. More general, *nonlinear systems* are often treated by starting with a linear approximation and trying to take account of nonlinearities by perturbative methods. Nonlinear systems will be discussed in Chapter 7.

4.2.1 LINEAR APPROXIMATION: SMALL VIBRATIONS

LINEARIZATION

If a dynamical system oscillates, it does so about one of perhaps several stable equilibrium positions. Suppose that q_0 is a stable equilibrium point of a Lagrangian dynamical system in one freedom. Section 3.3 shows that most such systems behave like harmonic oscillators for small displacements from q_0. We briefly review the argument. In the Lagrangian $L = T - V$, with $T = \frac{1}{2}m\dot{q}^2$, the potential is expanded in a Taylor series about q_0, and as long as $V''(q_0) \neq 0$ the approximate Lagrangian becomes

$$L(x, \dot{x}) \approx \frac{1}{2}m\dot{x}^2 - \frac{1}{2}kx^2 - V(q_0),$$

where $k = V''(q_0)$ and $q = q_0 + x$. The constant $V(q_0)$ does not contribute to the motion, so the (approximate) system is the one-dimensional oscillator with (circular) frequency $\omega = \sqrt{k/m}$. The second-order equation of motion is $\ddot{x} + \omega^2 x = 0$, and three ways of writing the solutions are

$$\left.\begin{aligned}
x(t) &= a\cos\omega t + b\sin\omega t, \\
x(t) &= C\cos(\omega t + \delta), \\
x(t) &= \alpha\exp(i\omega t) + \alpha^*\exp(-i\omega t),
\end{aligned}\right\} \tag{4.37}$$

where the constants a, b, C, and δ are real and α is complex, all determined by the initial conditions; α^* must be the complex conjugate of α if $x(t)$ is to be real. The energy equation

$$E = \frac{1}{2}m\dot{x}^2 + \frac{1}{2}kx^2 = \frac{1}{2}kC^2 = 2k|\alpha|^2$$

can be used to draw the phase portrait (Fig. 3.4) on $T\mathbb{Q}$, which consists of ellipses whose equations are

$$\frac{\dot{x}^2}{2E/m} + \frac{x^2}{2E/k} \equiv \frac{\dot{x}^2}{\omega^2 C^2} + \frac{x^2}{C^2} = 1$$

with all possible $E \geq 0$. The origin $x = \dot{x} = 0$ is a stable (elliptic) equilibrium point of the dynamical system on $T\mathbb{Q}$.

We now move on to n freedoms. The Lagrangian is

$$L = \frac{1}{2}m_{\alpha\beta}(q)\dot{q}^\alpha\dot{q}^\beta - V(q). \tag{4.38}$$

The first term is the kinetic energy, which in general may depend on q. The configuration manifold $\mathbb{Q}$ need not be $\mathbb{R}^n$.

Assume that the potential energy $V(q)$ has a *minimum* at some point q_0, so that $\partial V/\partial q^\beta|_{q_0} = 0 \forall \beta$. Then q_0 is a *stable* equilibrium position. This is clear physically for $n \leq 3$, for if the particle is close to q_0, the force $\mathbf{F} = -\nabla V(q)$ tends to return it to q_0.

If the second derivatives of V do not vanish, the Taylor series expansion about q_0 starts with a quadratic term:

$$V(q) = V(q_0) + \frac{1}{2}x^\alpha x^\beta \left.\frac{\partial^2 V}{\partial x^\alpha \partial x^\beta}\right|_{x=0} + x^\alpha x^\beta x^\gamma(\cdots), \tag{4.39}$$

where $x^\alpha = q^\alpha - q_0^\alpha$. The constant $V(q_0)$ can be set equal to zero. Call the (symmetric) matrix of the second derivatives K:

$$K_{\alpha\beta} = \left.\frac{\partial^2 V}{\partial x^\alpha \partial x^\beta}\right|_{x=0}. \tag{4.40}$$

Then to the lowest nonzero order the potential is

$$V(x) \approx \frac{1}{2}K_{\alpha\beta}x^\alpha x^\beta. \tag{4.41}$$

Because $x = 0$ is a minimum of V, the $K_{\alpha\beta}$ form a *positive* matrix: $V(x) \equiv K_{\alpha\beta}x^\alpha x^\beta \geq 0$, for any nonzero x ($\geq$ is used rather than $>$ because some displacements may leave the potential unchanged).

The Taylor series for the kinetic energy is

$$m_{\alpha\beta}(q) = M_{\alpha\beta} + R_{\alpha\beta\gamma}x^\gamma + \cdots.$$

The $m_{\alpha\beta}$ and $M_{\alpha\beta}$ matrices may be assumed without loss of generality to be symmetric. Since the kinetic energy is always positive, the $M_{\alpha\beta}$ form a *positive definite* matrix: $M_{\alpha\beta}\dot{x}^\alpha\dot{x}^\beta > 0$ for any nonzero $\dot{x}$. Then to lowest order in x the Lagrangian is

$$L = \frac{1}{2}M_{\alpha\beta}\dot{x}^\alpha\dot{x}^\beta - \frac{1}{2}K_{\alpha\beta}x^\alpha x^\beta. \tag{4.42}$$

The resulting equations of motion are

$$\ddot{x}^\alpha + \Lambda^\alpha_\beta x^\beta = 0, \tag{4.43}$$

where $\Lambda^\alpha_\beta = (M^{-1})_{\alpha\gamma} K_{\gamma\beta}$. The transition from q to x linearizes the problem by moving from $\mathbb{Q}$ to its (Euclidean) tangent space $\mathbf{T}_{q_0}\mathbb{Q}$ at q_0. The q coordinates on $\mathbb{Q}$ induce coordinates on $\mathbf{T}_{q_0}\mathbb{Q}$: the x^α are the components, in those coordinates, of a vector $\mathbf{x}$ in $\mathbf{T}_{q_0}\mathbb{Q}$, and the $M_{\alpha\beta}$ and $K_{\alpha\beta}$ are the matrix elements, in those coordinates, of operators $\mathbf{M}$ and $\mathbf{K}$ on $\mathbf{T}_{q_0}\mathbb{Q}$.

NORMAL MODES

In general (unless the Λ^α_β happen to form a diagonal matrix) Eqs. (4.43) are the equations of n *coupled* oscillators. The rest of the treatment involves *uncoupling* them, that is, finding directions in $\mathbf{T}_{q_0}\mathbb{Q}$ (linear combinations of the x^α, called *normal modes*) along which the oscillations are independent of each other.

In vector notation the Lagrangian on $\mathbf{T}_{q_0}\mathbb{Q}$ is

$$L = \frac{1}{2}(\dot{\mathbf{x}}, \mathbf{M}\dot{\mathbf{x}}) - \frac{1}{2}(\mathbf{x}, \mathbf{K}\mathbf{x}), \tag{4.44}$$

and the equations of motion are

$$\ddot{\mathbf{x}} + \Lambda\mathbf{x} = 0, \tag{4.45}$$

where $\Lambda \equiv \mathbf{M}^{-1}\mathbf{K}$ (since $\mathbf{M}$ is positive definite, $\mathbf{M}^{-1}$ exists). Finding the normal modes requires finding the eigenvectors of (diagonalizing) Λ: if $\mathbf{x}$ is an eigenvector belonging to eigenvalue λ, that is, if

$$\Lambda\mathbf{x} = \lambda\mathbf{x}, \tag{4.46}$$

Eq. (4.45) becomes

$$\ddot{\mathbf{x}} = -\lambda\mathbf{x} \equiv -\omega^2\mathbf{x}, \tag{4.47}$$

defining *normal frequencies* ω. The eigenvectors oscillate like independent one-freedom harmonic oscillators and thus point along the normal modes. The normal frequencies are the square roots of the eigenvalues.

Each ω must be real, so λ must be nonnegative: the eigenvalues of Λ must all be real and nonnegative. This can be proven by using the properties (symmetry and positiveness) of $\mathbf{M}$ and $\mathbf{K}$. The details are left to Problem 9, which shows that Λ has n linearly independent eigenvectors with real components belonging to nonnegative eigenvalues λ. The system can oscillate in each of the normal modes at its normal frequency $\omega = \sqrt{\lambda}$.

Let $\mathbf{a}$ be an eigenvector: $\Lambda\mathbf{a} = \omega^2\mathbf{a}$. Then

$$\mathbf{x}(t) = \mathbf{a}(\alpha e^{i\omega t} + \alpha^* e^{-i\omega t}) = 2\mathbf{a}\{\Re(\alpha)\cos\omega t + \Im(\alpha)\sin\omega t\}$$

is the general solution of Eq. (4.47), where α is a complex number and $\Re(\alpha)$ and $\Im(\alpha)$ are its real and imaginary parts. The initial conditions determine α.

The general solution is a linear combination of normal modes:

$$\mathbf{x}(t) = \sum_{\gamma=1}^{n} \mathbf{x}_\gamma(t) \equiv \sum_{\gamma=1}^{n} \mathbf{a}_\gamma [\alpha_\gamma \exp(i\omega_\gamma t) + \alpha_\gamma^* \exp(-i\omega_\gamma t)]. \tag{4.48}$$

Here the $\mathbf{a}_\gamma$, $\gamma = \{1, \ldots, n\}$, are the n normal modes, which are linearly independent and hence form a basis for $\mathbf{T}_{q_0}\mathbb{Q}$. Diagonalizing Λ is equivalent to transforming to the $\mathbf{a}_\gamma$ basis (see the appendix). The α_γ are constants determined by the initial conditions. The normal modes thus yield $2n$ linearly independent solutions of the n second-order linear differential equations of motion, and their linear combinations are the general solutions. See Problem 9.

A general solution oscillates in several different normal modes simultaneously, each at its own frequency and with its own phase. This may make the motion appear quite complicated. Sometimes the initial conditions (amplitudes and phases) happen to be such that the system moves periodically into one of the x^γ directions and stays close to it for a while, then to another, etc. To an observer it may look as though it is oscillating along each of these x^γ for a while.

Because $\mathbf{K}$ is positive, but not necessarily positive definite, Λ may have zero eigenvalues. In that case Eq. (4.45) becomes $\ddot{\mathbf{x}} = \mathbf{0}$. Thus there is no force and the potential does not change in this direction. The solution is then

$$\mathbf{x} = \mathbf{a} + \mathbf{v}t, \tag{4.49}$$

where $\mathbf{a}$ and $\mathbf{v}$ are constant vectors. These are *zero modes* in $\mathbf{T}_{q_0}\mathbb{Q}$, and the corresponding solution is the one-dimensional free particle. The displacement does not stay small and the approximation breaks down unless the force continues to vanish: the treatment works only for those zero directions in which all of the terms vanish in the Taylor expansion of V. The last sum of Eq. (4.48) changes correspondingly. This is illustrated in the following example.

WORKED EXAMPLE 4.2

An idealized *linear classical water molecule* consists of three particles in a line connected by equal springs and constrained to move along the line joining them (Fig. 4.17). The outer two particles 1 and 3 have equal masses μ and the central one has mass ν, and the spring constant is k. **(a)** Find the normal modes (describe them) and normal frequencies. **(b)** Write down the general solution. **(c)** Write down the solution with initial conditions $x_1(0) = -A$, $x_2(0) = A\mu/\nu$, $x_3(0) = 0$, and $\dot{x}_\gamma(0) = 0 \ \forall \ \gamma$.

Solution. **(a)** This is a linear problem ab initio. Choose the coordinate x_2 of mass ν to be zero at some arbitrary point along the line. The coordinates x_1 and x_3 of the

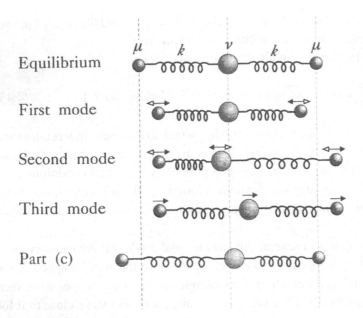

Equilibrium

First mode

Second mode

Third mode

Part (c)

FIGURE 4.17
The linear classical water molecule. The top line shows all three particles in equilibrium. The next three lines show the three normal modes described in the text. In each mode the phases are indicated by the shade of the arrows. That is, in the first mode the phase difference between the two outer particles is π, and the central particle remains at rest. In the second mode the two outer particles are in phase, while the central particle is out of phase with them. In the third mode the molecule moves as a rigid body. The last line of the diagram illustrates the example that ends the discussion of this system.

other two masses will be their deviations from equilibrium when $x_2 = 0$: If $-l$ and l are their equilibrium positions when v is at $x_2 = 0$, then x_1 and x_3 are the distances of the other two masses from $-l$ and l, respectively. (We use lower indices so as not to confuse squaring with the index 2.) The Lagrangian is

$$L = \frac{1}{2}\mu(\dot{x}_1^2 + \dot{x}_3^2) + \frac{1}{2}v\dot{x}_2^2 - \frac{1}{2}k[(x_1 - x_2)^2 + (x_2 - x_3)^2].$$

In the x coordinates, $\mathbf{M}$ and $\mathbf{K}$ are represented by the matrices

$$M = \begin{vmatrix} \mu & 0 & 0 \\ 0 & v & 0 \\ 0 & 0 & \mu \end{vmatrix} \quad \text{and} \quad K = k\begin{vmatrix} 1 & -1 & 0 \\ -1 & 2 & -1 \\ 0 & -1 & 1 \end{vmatrix},$$

and Λ by the matrix

$$\Lambda = k\begin{vmatrix} 1/\mu & -1/\mu & 0 \\ -1/v & 2/v & -1/v \\ 0 & -1/\mu & 1/\mu \end{vmatrix}.$$

The solution to the eigenvector problem gives the following normal frequencies

ω_γ and column vectors a_γ of the normal modes $\mathbf{a}_\gamma$, $\gamma = 1, 2, 3$:

$$\omega_1 = \sqrt{\frac{k}{\mu}}, a_1 = \begin{vmatrix} 1 \\ 0 \\ -1 \end{vmatrix}; \quad \omega_2 = \sqrt{k(1/\mu + 2/\nu)}, a_2 = \begin{vmatrix} 1 \\ -2\mu/\nu \\ 1 \end{vmatrix}; \quad \omega_3 = 0, a_3 = \begin{vmatrix} 1 \\ 1 \\ 1 \end{vmatrix}.$$

In the first normal mode, $x_1 = -x_3$, so the two outer masses vibrate out of phase by $180°$ at frequency ω_1 and with equal amplitudes. The central mass remains fixed (its amplitude is zero).

In the second normal mode, $x_1 = x_3$, so the two outer masses vibrate in phase at frequency ω_2 and with equal amplitudes. Because $x_2 = -2x_1\mu/\nu$, the central mass vibrates out of phase by $180°$ at the same frequency and with $2\mu/\nu$ times the amplitude. In the first and second modes the center of mass remains stationary.

In the third normal mode, $x_1 = x_2 = x_3$, so the system moves as a whole. The center of mass moves at some fixed velocity v. Clearly there is no force. This is the zero mode. See Fig. 4.17.

(b) The general solution is

$$\begin{vmatrix} x_1(t) \\ x_2(t) \\ x_3(t) \end{vmatrix} = \begin{vmatrix} 1 \\ 0 \\ -1 \end{vmatrix} \{\alpha_1 e^{i\omega_1 t} + \alpha_1^* e^{-i\omega_1 t}\} + \begin{vmatrix} 1 \\ -2\mu/\nu \\ 1 \end{vmatrix} \{\alpha_2 e^{i\omega_2 t} + \alpha_2^* e^{-i\omega_2 t}\} + \begin{vmatrix} 1 \\ 1 \\ 1 \end{vmatrix} \{a + vt\}.$$

The α_γ, a, and v are determined by the initial conditions, and in general the motion will not be in one of the normal modes.

(c) Some algebra leads to

$$\begin{vmatrix} x_1(t) \\ x_2(t) \\ x_3(t) \end{vmatrix} = -\frac{1}{2}A \begin{vmatrix} \cos\omega_1 t + \cos\omega_2 t \\ -2(\mu/\nu)\cos\omega_2 t \\ -\cos\omega_1 t + \cos\omega_2 t \end{vmatrix}.$$

The center of mass remains fixed. The central mass performs simple harmonic motion at frequency ω_2, while the two outer masses perform more complicated motion at a combination of frequencies ω_1 and ω_2. If $\omega_1 \approx \omega_2$ (i.e., if $\nu \gg \mu$), the phenomenon of beats occurs (Halliday et al., 1993), but the two outer masses beat out of phase. That is, when the amplitude of one is large, the amplitude of the other is small, and it appears that only one of the outer masses is oscillating, and at other times only the other one. Roughly speaking, energy flows back and forth between the two of them at the beat frequency $\frac{1}{2}(\omega_1 - \omega_2)$.

4.2.2 COMMENSURATE AND INCOMMENSURATE FREQUENCIES

THE INVARIANT TORUS $\mathbb{T}$

We return in this subsection to the two-freedom harmonic oscillator of Eq. (3.38), whose vector field Δ lies on the toroidal invariant submanifold $\mathbb{T}$ described after Eq. (3.39). It will be shown that the properties of Δ, and hence of the motion on $\mathbb{T}$, depend on the ratio of the two frequencies. In the process we will introduce the idea of the *Poincaré map*.

The equations of motion are $\ddot{x}_j + \omega_j^2 x_j = 0$, $j = 1, 2$; their solutions are

$$x_j(t) = A_j \cos(\omega_j t + \phi_j), \qquad (4.50)$$

where $A_j = \sqrt{2E_j}/\omega_j$ is the jth amplitude and E_j the jth energy. The torus $\mathbb{T}$ is the product of the two ellipses that make up the phase portraits of each of the oscillators (there is a different torus for each pair of E_j). At time $t = P_j \equiv 2\pi/\omega_j$ both x_j and $\dot{x}_j$ return to their original values at $t_0 = 0$, so the system point on the jth ellipse returns to its original position. The system point will return again to the same position on the jth ellipse at every time nP_j, where n is any integer. If there exist integers n_1 and n_2 such that $\omega_1/\omega_2 = n_1/n_2$, then after a time $n_1 P_1 = n_2 P_2$ the system will return to its original position on both ellipses and hence to its original position on $\mathbb{T}$. But if $\rho \equiv \omega_1/\omega_2$ is an irrational number, such integers do not exist. The system will then never return to its original position on $\mathbb{T}$. It will cover $\mathbb{T}$ densely, coming arbitrarily close to every point on $\mathbb{T}$. Each phase orbit (each integral curve of Δ) winds around and around the torus, never returning exactly to any point through which it had previously passed (it is called an *irrational winding line on the torus*). If ρ is rational, the frequencies ω_j are said to be *commensurate*. If ρ is irrational, they are said to be *incommensurate*. There is a sharp qualitative difference between two-dimensional harmonic oscillators with commensurate frequencies and those with incommensurate ones. Given any bounded region $\mathbb{U}$ on $\mathbb{T}$, the phase portrait will pass through $\mathbb{U}$ a finite number of times (if at all) if the frequencies are commensurate and an infinite number of times if they are incommensurate (Fig. 4.18).

Another way to do this is to define angle variables θ_j such that $\dot{\theta}_j = \omega_j/2\pi$ (then θ_j counts revolutions on the jth ellipse). If an incommensurate system starts at $(\theta_1, \theta_2) =$

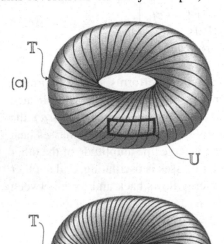

(a)

(b)

FIGURE 4.18
An attempt to illustrate the difference between the oscillator with commensurate and incommensurate frequencies. (a) The frequencies are commensurate: each trajectory passes through $\mathbb{U}$ only a finite number of times. (b) The frequencies are incommensurate: each trajectory passes through $\mathbb{U}$ an infinite number of times (of course this is impossible to illustrate accurately).

(0, 0), it will never again pass through a point whose θ_j simultaneously take on rational values.

Recall that at the beginning of Section 3.2.1 invariant submanifolds were related to constants of the motion. A particular example of an invariant submanifold is any trajectory of a dynamical system, since by definition the motion remains always on its trajectory. Can each trajectory be associated to a set of constants of the motion, $2n - 1$ dynamical variables? It is tempting to say that it can be: if you know the energy, the angular momentum, etc., you should know the trajectory. But because every neighborhood $\mathbb{U}$ on the torus, no matter how small, is traversed by every trajectory, this cannot be done on $\mathbb{T}$ for the incommensurate double harmonic oscillator. Indeed, suppose there were a function Γ on $\mathbb{T}$ that takes on different values on different trajectories but remains constant along each trajectory. Since every trajectory traverses every neighborhood $\mathbb{U} \subset \mathbb{T}$, no matter how small, an infinite number of times, Γ would have to take on all of its values in every arbitrarily small $\mathbb{U}$; it would have to be pathologically discontinuous. Any function used to label trajectories on $\mathbb{T}$ (and hence on $\mathbf{T}\mathbb{Q}$) would be everywhere discontinuous! Because we deal only with differentiable functions and because constants of the motion are functions on $\mathbf{T}\mathbb{Q}$, we must exclude such pathological functions. This means that apart from E_1 and E_2 there are no constants of the motion for this dynamical system.

However, there is a sense in which each trajectory can be labeled by $2n$ constants of the motion but only locally. Given a local neighborhood like the small rectangle in Fig. 4.18, each trajectory *segment* crossing it can be labeled uniquely, for instance by the point at which it enters the neighborhood. But segments of trajectories within the neighborhood are often parts of the same global trajectory. Thus it is possible to label each local trajectory segment by three constants of the motion (E_1, E_2, and one other); that is the sense in which there exist local constants of the motion that correspond to no global ones. In the same sense, in n freedoms there exist $2n - 1$ local constants of the motion, even though in general they do not exist globally.

THE POINCARÉ MAP

A way to describe the difference between the commensurate and incommensurate cases is by means of the *Poincaré map*, as illustrated in Fig. 4.19. To do this, one chooses a submanifold of $\mathbf{T}\mathbb{Q}$ that is crossed by the integral curves, and one studies the trace left on it by an integral curve. In the figure $\mathbb{P}$ represents such a submanifold, and a trajectory is shown intersecting $\mathbb{P}$ in two points. Write Σ for the set of intersection points (in Fig. 4.19 Σ consists of only two points). For the two-freedom oscillator $\mathbb{P}$ could be a circular cut of $\mathbb{T}$, and Σ would be the set of points where a trajectory crosses $\mathbb{P}$. Figure 4.20 shows two such possibilities for $\mathbb{P}$, the circles $\mathbb{S}_1$ and $\mathbb{S}_2$ (although both are closed curves, neither divides $\mathbb{T}$ into two separate pieces). If the frequencies are commensurate, Σ contains a finite number of points (Fig. 4.20a). If they are incommensurate, Σ contains an infinite number of points densely packed on the whole circle (Fig. 4.20b).

It is often difficult to visualize properties of complicated dynamical systems whose tangent bundles have more than two dimensions, and for such systems the Poincaré map can be quite helpful. In general $\mathbb{P}$ (a *section* of $\mathbf{T}\mathbb{Q}$ with respect to the dynamical vector

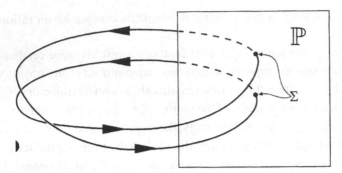

FIGURE 4.19
An illustration of the Poincaré map. $\mathbb{P}$ is a *Poincaré section*, intersected at two points by an integral curve. The two points form the set Σ. The integral curve is mapped to Σ by the Poincaré map.

field) is intersected by an integral curve (or by some set of integral curves) of the dynamics in a set of points Σ. The integral curve on $\mathbf{T}\mathbb{Q}$ is thereby mapped onto Σ on $\mathbb{P}$ (this defines the Poincaré map), the way the trajectory in Fig. 4.19 is mapped onto the two points of intersection. Properties of the dynamical system are then deduced from the structure of Σ. In the two-freedom oscillator the nature of Σ changes an infinite number of times as ρ is allowed to vary in the interval $I =]0, 1]$ (that is, open at 0 and closed at 1, hence excluding the value zero). For rational ρ it contains a finite number of points, and for irrational ρ it contains a densely packed infinite number. It is known that the *measure* of the rationals on the unit interval is zero (see the number theory appendix to Chapter 7), which implies that if ρ is chosen at random in I, it is hardly likely that a rational number will be hit. In this sense *almost all* two-dimensional harmonic oscillators are of the irrational type.

In the harmonic-oscillator example the Poincaré map is simple and probably adds little to our understanding of the motion. At least in part this is because the torus $\mathbb{T}$ and the

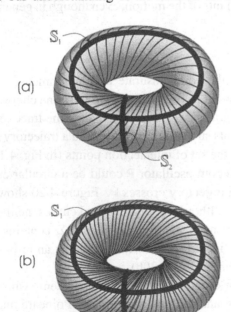

FIGURE 4.20
Another attempt to illustrate the difference between the oscillator with commensurate and incommensurate frequencies, this time with the Poincaré map. Two possible Poincaré sections are illustrated, $\mathbb{S}_1$ and $\mathbb{S}_2$. (a) Commensurate frequencies. (b) Incommensurate frequencies (again, impossible to illustrate accurately).

orbits on it are themselves so easy to visualize. For more complex dynamical systems in higher dimension, with dissipation, nonlinearity, and driving forces, a good way to visualize properties of the motion is to choose a two-dimensional submanifold for $\mathbb{P}$. Even so, the transition from simplicity (in our example, from rational ρ) to complexity (in our example, to irrational ρ) is in general more complicated. For instance, a two-dimensional section $\mathbb{P}$ might have a finite number of points in its Σ, but as the parameters change (in our example ρ is the only parameter), the finite set can become infinite and lie densely on small closed curves (as discrete points for rational ρ become infinite and dense for irrational ρ). For some values of the parameters, the closed curves may themselves smear out, and Σ may be dense in two-dimensional regions of $\mathbb{P}$. This is the transition to chaos. It turns out that the kind of result we obtained for the two-dimensional oscillator is very common: in most systems the analog of irrational ρ has higher measure than the analog of rational ρ, and in this sense most dynamical systems are chaotic. We will discuss the Poincaré map later in the book, mostly in connection with nonlinearity and chaos in Chapter 7.

4.2.3 A CHAIN OF COUPLED OSCILLATORS

In this subsection we discuss a system consisting of $2n + 1$ particles, all of the same mass m, connected to each other as in Fig. 4.21: each mass interacts with its nearest neighbors through a linear spring whose constant is K. The system we will discuss, with all masses equal and with equal coupling between all nearest neighbors, is particularly easy to handle. We will not discuss the similar system with unequal masses (even with a random distribution of masses), which has physical interest but is considerably more complicated (Mattis, 1993).

GENERAL SOLUTION

At equilibrium the separation between each nearest-neighbor pair is a. The coordinate x_ν of the νth mass is its displacement from its equilibrium position, positive to the right and negative to the left. The Lagrangian of the system is

$$L = \frac{1}{2}m \sum_{-n}^{n} \dot{x}_\nu^2 - \frac{1}{2}K \sum_{-n}^{n-1}(x_\nu - x_{\nu+1})^2. \tag{4.51}$$

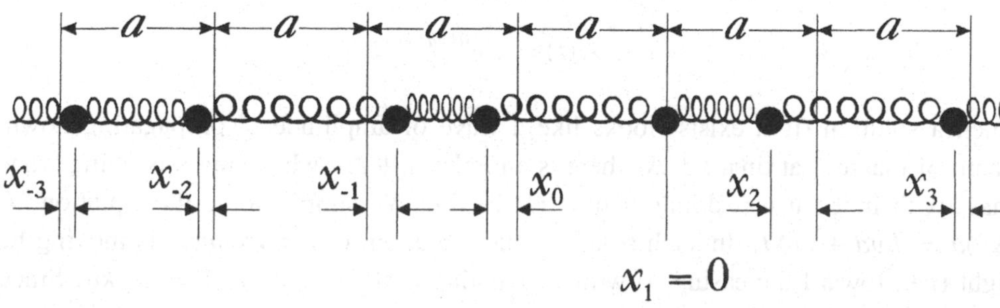

FIGURE 4.21
Part of a chain of particles, all of the same mass m, interacting through springs with force constant k. The displacement of the νth particle from its equilibrium position is x_ν, positive if to the right and negative if to the left.

The second sum, the negative of the potential energy V, goes only to $n - 1$ because it is over the springs, not the particles. Each x_κ with

$$\kappa \in \{-n + 1, -n + 2, \ldots, n - 2, n - 1\}$$

appears twice in V, both in $(x_\kappa - x_{\kappa+1})$ and in $(x_{\kappa-1} - x_\kappa)$. However, the positions x_{-n} and x_n of the end masses appear only once. The EL equations of motion (divided through by m) are

$$\begin{aligned}
\ddot{x}_{-n} + \omega_0^2(x_{-n} - x_{-n+1}) &= 0, \\
\ddot{x}_\nu - \omega_0^2(x_{\nu-1} - 2x_\nu + x_{\nu+1}) &= 0, \qquad -n + 1 \leq \nu \leq n - 1 \\
\ddot{x}_n + \omega_0^2(x_n - x_{n-1}) &= 0,
\end{aligned} \qquad (4.52)$$

where $\omega_0^2 = K/m$, as usual. For the time being we ignore the end masses and concentrate only on those whose indices are in the interval $(-n + 1, n - 1)$.

This system is of the type dealt with in Section 4.2.1, for Eqs. (4.51) and (4.52) are of the form of (4.44) and (4.45). The Λ matrix in the present case is $-\omega_0^2 \Phi$, where Φ is the $(2n + 1)$-dimensional matrix (rows and columns run from $-n$ to n)

$$\Phi = \begin{bmatrix}
\ddots & \ddots & 0 & 0 & 0 & 0 \\
\ddots & -2 & 1 & 0 & 0 & 0 \\
0 & 1 & -2 & 1 & 0 & 0 \\
0 & 0 & 1 & -2 & 1 & 0 \\
0 & 0 & 0 & 1 & -2 & \ddots \\
0 & 0 & 0 & 0 & \ddots & \ddots
\end{bmatrix}.$$

The only nonzero elements of Φ are on the main diagonal and the two adjacent ones (Φ is a *tridiagonal* matrix).

Instead of trying to diagonalize Λ analytically, we look directly for solutions of the normal mode type, but with a wavelike structure (the ν in the exponent, like those above, is an integer, not a frequency!):

$$x_\nu(t) = C e^{i(k\nu a - \omega t)}. \qquad (4.53)$$

Such a solution (if it exists) looks like a wave of amplitude C propagating down the chain of masses: at time $t + \Delta t$ there is an index $\nu + \Delta\nu$ whose mass is doing what the mass with index ν was doing at time t. To find $\Delta\nu$ simply solve the equation $k(\nu + \Delta\nu)a = k\nu a + \omega\Delta t$. In such solutions the vibration, or *disturbance*, is moving to the right (i.e., toward increasing ν) with ν changing at the rate $\Delta\nu/\Delta t = \omega/ka$. Since $a\nu$ measures the distance along the chain, the disturbance of Eq. (4.53) propagates at speed $c = a\Delta\nu/\Delta t = \omega/k$. The frequency of the disturbance is $f = \omega/2\pi$ and the wavelength is $\lambda = 2\pi/k$. Similar disturbances propagating to the left are described by (4.53) with

$kva + \omega t$ in the exponent. As always, it is the real parts of such solutions (or the solutions plus their complex conjugates) that represent the physical motion.

To find such solutions, insert (4.53) into Eq. (4.52), obtaining

$$-\omega^2 x_\nu = \omega_0^2 [e^{-ika} - 2 + e^{ika}] x_\nu.$$

This equation gives a *dispersion relation* between ω and k:

$$\omega^2 = 4\omega_0^2 \sin^2 \frac{ka}{2} \tag{4.54}$$

These results are interpreted in the following way: a wavelike solution

$$x_\nu(t, k) = C(k) \exp i\{kva - \omega(k)t\} \tag{4.55}$$

exists at every *wave vector k* whose frequency ω satisfies (4.54). The wave of (4.55) travels along the chain at a speed that depends on k (or equivalently on ω):

$$c(k) = \frac{\omega}{k} = \left| \frac{2\omega_0}{k} \sin\left(\frac{ka}{2}\right) \right|. \tag{4.56}$$

Thus far (we have not yet included the boundary conditions) this is a continuous (not countable) set of solutions, since k can take on all positive values from 0 to ∞. In the limit of long wavelength, as $k \to 0$, the speed approaches $c_0 = \omega_0 a$. For all finite values of k the wave speed is less than c_0.

The general solution $x_\nu(t)$ is a linear combination of waves with different wave vectors k, that is, an integral of the form

$$x_\nu(t) = \int_{-\infty}^{\infty} C(k) \exp i\{-\omega(k)t + kva\} \, dk. \tag{4.57}$$

The C here is not exactly the same as the one of Eq. (4.55): that one is more like the present $C(k) \, dk$. It is seen that the linear combination of solutions is a Fourier transform in which va and k are the Fourier conjugate variables. As usual, the coefficients of the linear combination, in this case the $C(k)$, are found from the initial conditions. Setting $t = 0$ in (4.57) yields

$$x_\nu(0) = \int_{-\infty}^{\infty} C(k) e^{ikva} \, dk. \tag{4.58}$$

THE FINITE CHAIN

In any finite chain the solutions depend on the boundary conditions imposed on the end-point masses. The most common boundary conditions are of four types: (A) fixed ends, (B) free ends, (C) absorbing ends, or (D) periodic boundary conditions. Condition

(A) is $x_{-n} = x_n = 0$: the chain is treated as though attached to walls at its ends. Under condition (B) the only force on the end masses comes from the springs attaching them to masses $-n + 1$ and $n - 1$. Under both (A) and (B) waves are reflected at the ends. Under condition (C) the end points are attached to devices that mimic the behavior of an infinite chain and absorb all the energy in a wave running up or down the chain: there is no reflection. Condition (D) is $x_{-n} = x_n$. This is equivalent to identifying the nth mass with the $-n$th or to looping the chain into a circle.

As an example, we treat condition (A). The equations become simpler if the masses are renumbered according to $v' = v + n$, so that v' runs from 0 to $n' = 2n$. In this example we will use only this new indexing scheme, so we drop the prime and rewrite the fixed end-point condition as

$$x_0 = x_n = 0. \tag{4.59}$$

To be completely general, we no longer restrict the new n to odd numbers. Since the masses at x_0 and x_n do not move, the dynamical system now consists of the $n - 1$ particles with indices running from $v = 1$ to $v = n - 1$.

The end-point condition can be satisfied by combining pairs of the *running-waves* of (4.55) so as to interfere destructively and cancel out at the end points, forming *standing waves*. The equation of a standing wave is

$$\tilde{x}_v(t) = C(k)e^{i\omega t}\sin(vka). \tag{4.60}$$

Such a wave vanishes at $v = 0$ and will vanish also at the other end point if $\sin(nka) = 0$, or

$$k = \frac{\pi}{na}\gamma \equiv k_\gamma, \tag{4.61}$$

where γ can be any integer, which means that only a discrete set of wave vectors k is allowed. Then Eq. (4.61) and the dispersion relation (4.54) show that there is also only a discrete set of allowed frequencies, namely

$$\omega_\gamma^2 = 4\omega_0^2\sin^2\left(\frac{k_\gamma a}{2}\right) = 4\omega_0^2\sin^2\left(\frac{\pi\gamma}{2n}\right). \tag{4.62}$$

Standing waves can be formed of two waves running in opposite directions. Indeed, the standing wave of Eq. (4.60) can be written in the form

$$\tilde{x}_v(t) = \frac{Ce^{i(kva-\omega t)} - Ce^{-i(kva+\omega t)}}{2i}.$$

This consists of two waves, $(kva - \omega t)$ running to the right and $(kva + \omega t)$ to the left. It is as though the wave running to the right is reflected at the right end, giving rise to the wave running to the left, which in turn is reflected at the left end.

When (4.61) is inserted into (4.60), the standing-wave solution is (we drop the tilde and index the wave by the value of γ to which it belongs)

$$x_{\nu,\gamma}(t) = C_\gamma e^{i\omega_\gamma t} \sin\left(\frac{\nu}{n}\gamma\pi\right). \tag{4.63}$$

It will now be shown that these solutions correspond to the normal modes of Section 4.2.1.

First we show that there is a finite number of them. According to (4.63), when the integer γ runs from 0 to n, the argument of the sine runs from 0 to $\nu\pi$. As γ continues beyond the interval $(0, n)$, both the expressions for ω_γ and $x_{\nu,\gamma}(t)$ repeat themselves [any sign change of $x_{\nu,\gamma}(t)$ can be taken up in the amplitude C_γ]. In fact $\gamma = 0$ and $\gamma = n$ give zero solutions, so distinct solutions belong only to γs in the interval $(1, n-1)$ and γ takes on only $n-1$ values: γ and ν have the same range. This is in keeping with the idea that these are normal modes, for the $(n-1)$-particle system should have $n-1$ normal modes.

To show that our solutions are the normal modes, we put them in standard form by writing the general solution in the form of Eq. (4.48). The general solution is a sum of solutions of the form of (4.63):

$$x_\nu(t) = \sum_{\gamma=1}^{n-1} x_{\nu,\gamma}(t) = \sum_{\gamma=1}^{n-1} C_\gamma e^{i\omega_\gamma t} \sin\left(\frac{\nu}{n}\gamma\pi\right). \tag{4.64}$$

This is already essentially in the desired standard form: $x_\nu(t)$ is the νth component of the vector $\mathbf{x}(t)$ on the left-hand side of (4.48). The x_ν, remember, are the displacements of the masses from their equilibrium positions; they define the initial basis. The right-hand side of (4.64) corresponds to the right-hand side of (4.48), namely

$$\mathbf{x}(t) = \sum_{\gamma=1}^{n} \mathbf{a}_\gamma \left[\alpha_\gamma \exp(i\omega_\gamma t) + \alpha_\gamma^* \exp(-i\omega_\gamma t)\right], \tag{4.65}$$

which is (4.48), except that the upper limit ν is here replaced by n. We are writing the solutions in complex form, so we needn't worry about the complex conjugate term in (4.65). In Eq. (4.64) the $\mathbf{a}_\gamma$ are also expressed in terms of their components in the initial basis. The νth component of $\mathbf{a}_\gamma$ is $a_{\gamma,\nu} = \sin(\nu\gamma\pi/n)$. All the $a_{\gamma,\nu}$ belonging to one value of γ are multiplied by the same factor $e^{i\omega_\gamma t}$ and hence so is the entire vector $\mathbf{a}_\gamma$. Thus (4.64) and (4.65) are of the same form.

The column vector representing $\mathbf{a}_\gamma$ in the initial basis, called a_γ in Section 4.2.1, is

$$a_\gamma = F_\gamma \begin{bmatrix} \sin\left(\frac{1}{n}\gamma\pi\right) \\ \sin\left(\frac{2}{n}\gamma\pi\right) \\ \sin\left(\frac{3}{n}\gamma\pi\right) \\ \vdots \\ \sin\left(\frac{n-2}{n}\gamma\pi\right) \\ \sin\left(\frac{n-1}{n}\gamma\pi\right) \end{bmatrix}, \tag{4.66}$$

where F_γ is a normalizing factor. Although the a_γ do not have to be normalized (see, e.g., Worked Example 4.2), we choose to do so here. The a_γ are orthogonal even before normalization, but the normalized ones satisfy

$$a_\gamma \cdot a_\beta \equiv \sum_\nu a_{\gamma,\nu} a_{\beta,\nu} = \delta_{\gamma\beta}.$$

It is shown in Problem 17 that $F_\gamma \equiv F = \sqrt{2/(n-1)}$, the same for all γ. The a_γ are the normal modes, a different one for each γ, each with its own normal frequency ω_γ. If the system is in a normal mode, all of the masses vibrate at the same frequency and in phase, but their amplitudes, in the ratio of the components of a_γ, are different. The general solution is made up of a linear combination of these, each with its own complex amplitude C_γ. Written out in real, as opposed to complex, notation, the general solution is

$$\mathbf{x}(t) = \sum_{\gamma=1}^{n-1} \mathbf{x}_\gamma(t) \equiv \sum_{\gamma=1}^{n-1} \mathbf{a}_\gamma A_\gamma \cos(\omega_\gamma t + \phi_\gamma), \tag{4.67}$$

where A_γ is the amplitude and ϕ_γ the phase of $\mathbf{x}_\gamma(t)$.

The Lagrangian can be written in terms of the normal modes. Let ξ_γ, $\gamma \in \{1, \ldots, n-1\}$ label the components of vectors in the $\mathbf{a}_\gamma$ basis. Since each normal mode behaves like an oscillator uncoupled from all the others, the Lagrangian must be a sum of independent Lagrangians, of the form

$$L = \sum_\gamma L_\gamma(\xi, \dot{\xi}) = \frac{m}{2} \sum_\gamma \left(\dot{\xi}_\gamma^2 - \omega_\gamma^2 \xi_\gamma^2 \right). \tag{4.68}$$

This can be verified formally by transforming the Lagrangian of (4.51) from the x_ν components to the ξ_γ components (Problem 18); Eq. (4.66) can be used to obtain the transformation matrix. The EL equations derived from (4.68) are (no sum on γ)

$$\ddot{\xi}_\gamma + \omega_\gamma^2 \xi_\gamma = 0.$$

These are the equations of $n-1$ uncoupled oscillators with frequencies ω_γ, as is to be expected if they are the normal modes.

4.2.4   FORCED AND DAMPED OSCILLATORS

FORCED UNDAMPED OSCILLATOR

We return now to the one-freedom case and consider an undamped harmonic oscillator driven by an additional, *external*, time-dependent force $F(t)$. As pointed out at Eq. (3.70), the Lagrangian is

$$L = \frac{1}{2}m\dot{q}^2 - \frac{1}{2}kq^2 + qF(t), \tag{4.69}$$

and the equation of motion is

$$\ddot{q} + \omega_0^2 q = f(t), \tag{4.70}$$

where $f = F/m$ and $\omega_0 = \sqrt{k/m}$.

This is the equation of the *forced* or *driven* oscillator. If $q_p(t)$ is any particular solution of (4.70), every other solution is of the form $q(t) = q_p(t) + q_h(t)$, where $q_h(t)$ is an arbitrary solution of the free oscillator, without a driving force (of the homogeneous equation). The solutions of the homogeneous equation have been discussed in the first part of Section 4.2.1.

Suppose that the driving force is harmonic at some frequency Ω:

$$F(t) = F_0 \cos \Omega t. \tag{4.71}$$

Physical intuition leads to the expectation that a linear oscillator driven at a frequency Ω will respond at that frequency; hence we look for a solution of the form $q_p(t) = C \cos(\Omega t + \gamma)$. When this is inserted into (4.70) it is found that $\gamma = 0$ and

$$C = \frac{f_0}{\omega_0^2 - \Omega^2},$$

where $f_0 = F_0/m$. Unless $\Omega = \omega_0$, the general solution of (4.70) is therefore

$$q(t) = A \sin(\omega_0 t + \delta) + \frac{f_0 \cos \Omega t}{\omega_0^2 - \Omega^2}, \tag{4.72}$$

where A and δ are constants of integration that depend on the initial conditions. The general solution oscillates at two frequencies, ω_0 and Ω.

The amplitude and phase of the ω_0 oscillation depend on the initial conditions, whereas those of the Ω oscillation depend on the driving frequency (and f_0). The numerator is in phase with the driving force ($\gamma = 0$), but the denominator can be positive or negative, so the Ω term is in phase with the driving force if $\Omega < \omega_0$ and is π out of phase if $\Omega > \omega_0$. Its amplitude also depends on Ω through the denominator. As Ω approaches ω_0, the amplitude of the Ω term increases without bound, and in the limit the solution is no longer valid. When $\Omega = \omega_0$ the correct solution is (see Problem 14)

$$q(t) = A \sin(\omega_0 t + \delta) + \frac{f_0 t \sin \omega_0 t}{2\omega_0} : \tag{4.73}$$

the inhomogeneous part grows linearly with time and eventually overwhelms the homogeneous part. This is unrealistic because every real system has some damping. Without damping, the system has no way to dissipate the energy that the driving force constantly pumps into it (see Problem 11).

FORCED DAMPED OSCILLATOR

With damping, the equation of motion of the oscillator becomes

$$\ddot{q} + 2\beta\dot{q} + \omega_0^2 q = f(t), \tag{4.74}$$

obtained by adding the driving term to the right-hand side of a modification of Eq. (3.66). Again, we take f to be of the form $f = f_0 \cos \Omega t$.

The easiest way to deal with this equation is to shift to the complex domain (this does not always work for nonlinear oscillators). A homogeneous solution is $q_h(t) = \alpha e^{(-\beta + i\omega)t}$, where $\omega = \sqrt{\omega_0^2 - \beta^2}$ is the *natural frequency* of the damped oscillator and α is a complex constant. The physical solution is obtained from this complex-valued function by taking its real part $e^{-\beta t}[\Re(\alpha) \cos \omega t - \Im(\alpha) \sin \omega t]$. Because this is the general form of the real solution, the complex form can be used if at the end the real part is taken ("real" often means both real as opposed to complex and physically real). If the complex number α is written in its polar form $\alpha = a e^{i\delta}$, where $a > 0$ and δ are both real, the real part of q_h is $a e^{-\beta t} \cos(\omega t + \delta)$, so that the *complex phase* δ of α is the *physical phase* of the oscillation. In complex form, the driving term is

$$f = f_0 e^{i\Omega t}, \tag{4.75}$$

where f_0 is now a complex constant. The physical driving force is the real part of f. If f_0 happens to be real, the driving force is $f_0 \cos \Omega t$. If f_0 is not real, its complex phase is its physical phase.

With the same physical insight as in the undamped case, we look for a solution the form

$$q(t) = \alpha e^{(-\beta + i\omega)t} + \Gamma e^{i\Omega t}. \tag{4.76}$$

The first term is a homogeneous solution q_h and the second is a particular solution q_p. When this $q(t)$ is inserted into (4.74) it is found that

$$\Gamma = \frac{f_0}{\omega_0^2 - \Omega^2 + 2i\beta\Omega}. \tag{4.77}$$

Again the solution is a sum of q_h, an oscillation at the natural frequency ω (not ω_0), and q_p, an oscillation at the driving frequency Ω. The natural oscillation, damped with time constant β, eventually becomes negligible; it is called the *transient response*. The amplitude of the Ω component is finite even if $\Omega = \omega_0$, for $\omega_0^2 - \Omega^2 + 2i\beta\Omega$ never vanishes provided $\beta \neq 0$. The Ω component is called the *steady state response*.

The driving force $f_0 e^{i\Omega t}$ and the response $\Gamma e^{i\Omega t}$ are not necessarily in phase (or π out of phase) as in the undamped case. The phase difference between the two complex numbers f_0 and Γ is the phase of their quotient, namely of

$$\frac{f_0}{\Gamma} = \omega_0^2 - \Omega^2 + 2i\beta\Omega.$$

Thus the phase difference γ between them is given by

$$\tan \gamma = \frac{\Im(f_0/\Gamma)}{\Re(f_0/\Gamma)} = \frac{2\beta\Omega}{\omega_0^2 - \Omega^2}.$$

The steady-state amplitude is

$$|\Gamma| \equiv C = \frac{|f_0|}{\sqrt{\left(\omega_0^2 - \Omega^2\right)^2 + 4\beta^2\Omega^2}}.$$

For a given driving amplitude $|f_0|$, the response amplitude $|\Gamma|$ is a maximum when $\Omega = \omega' \equiv \sqrt{\omega_0^2 - 2\beta^2}$.

Much of this behavior is analogous to that of AC circuits (Feynman et al., 1963, Vol. 1, Chapter 23). In circuits the driving term is the applied voltage and the response of interest is not the charge q, but the current, whose analog is $\dot{q}$. According to Eq. (4.76), the steady-state $\dot{q}$ is $i\Omega\Gamma$, so $|\dot{q}| = |\Omega\Gamma|$, which is a maximum at *resonance* (i.e., when $\Omega = \omega_0$). In mechanics also $\dot{q}$ is of more interest than q, because the kinetic energy and the rate of energy dissipation are both proportional to $\dot{q}^2$.

In the steady state, energy is dissipated at the rate it is fed in by the driving force:

$$f\dot{q} \equiv (\ddot{q} + 2\beta\dot{q} + \omega_0^2 q)\dot{q}.$$

Over a period $\ddot{q}\dot{q}$ and $q\dot{q}$ integrate out to zero, for $\ddot{q}$ and $\dot{q}$ (as well as $\dot{q}$ and q) are $\pi/2$ out of phase and hence are negative as much as they are positive. All that survives is the contribution from the term $2\beta\dot{q}^2$. The average rate at which energy is dissipated is the *intensity* I, defined by

$$\left\langle \frac{dE}{dt} \right\rangle \equiv I = \frac{\Omega}{2\pi} \int_0^{2\pi/\Omega} 2\beta[\Omega\Gamma \cos(\Omega t)]^2 \, dt = \beta(\Omega\Gamma)^2. \tag{4.78}$$

Thus $(\Omega\Gamma)^2$ or, in complex notation, $|\Omega\Gamma|^2$ is a measure of the intensity. The intensity is given by

$$I = \beta \frac{|f_0|^2}{\left(\omega_0^2/\Omega^2 - 1\right)^2 \Omega^2 + 4\beta^2}. \tag{4.79}$$

This is known as a *Lorentzian* function of Ω. The intensity I is maximal at resonance.

At resonance $\tan\gamma$ becomes infinite, so q is $\pi/2$ out of phase with the driving term; $\dot{q}$ is in phase with the driving term (which in circuit theory means that the current is in phase with the voltage). Figure 4.22 shows $|\Gamma|^2$, I, and γ as functions of Ω.

The maximum value of I is

$$I_{\text{max}} = \frac{|f_0|^2}{4\beta}.$$

A measure of the sharpness of the resonance is its *half-width* at *half maximum* $\Delta\Omega$, half the width of the I curve at half its total height. For $\beta \ll \omega_0$ it is found that $\Delta\Omega \approx \beta$ to first order in β. (The half-width of the $|\Gamma|^2$ curve is also β to first order in β.) We leave the calculation to Problem 13. Another measure of the sharpness is the *quality factor* $Q \equiv \omega'/2\beta$. To first order in β it can be shown that $Q = \omega_0/(\Delta\Omega)$. For β very small, the I curve is narrow and has a sharp maximum. Some of this is illustrated in Fig. 4.22. Many real physical systems have Q values of the order of 10^4 and higher.

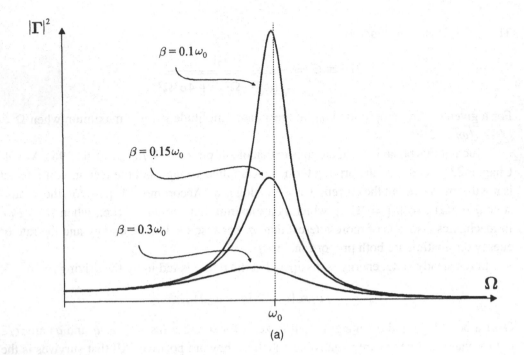

(a)

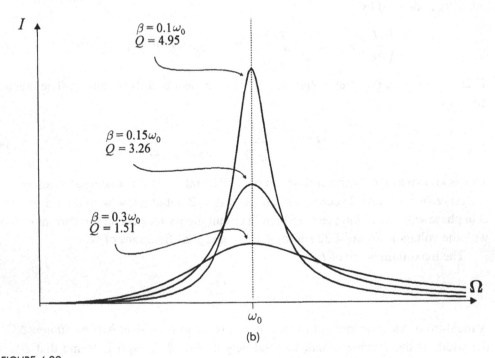

(b)

FIGURE 4.22

The response of a damped harmonic oscillator driven at different frequencies Ω. (a) The square amplitude $|\Gamma^2|$ of the response as a function of Ω for three different values of β. (b) The intensity I as a function of Ω for the same three values of β. The quality factor Q is given for each value of Ω. (c) The phase difference γ as a function of Ω for one value of β. The smaller β, the sharper the rise of γ from 0 to π, and in the undamped limit ($\beta = 0$), $\gamma = 0$ for $\Omega < \omega_0$ and $\gamma = \pi$ for $\Omega > \omega_0$.

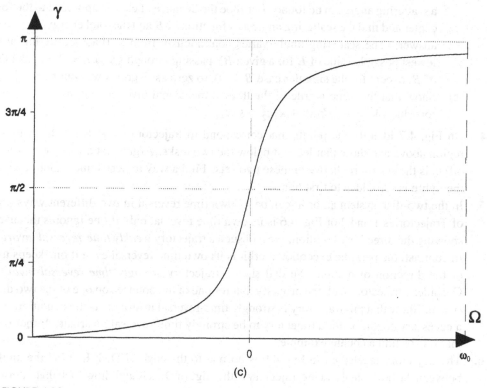

FIGURE 4.22
(*Continued*)

PROBLEMS

1. Obtain the Rutherford scattering formula for a repulsive Coulomb potential (see Problem 2.19).

2. Find the central potential whose scattering cross section is given by

$$\sigma(\vartheta) = \alpha\pi^2 \frac{\pi - \vartheta}{(2\pi - \vartheta)^2 \vartheta^2 \sin\vartheta}.$$

   [Answer: $V = K/r^2$, where the dependence of K on E and α will be found when the problem is solved.]

3. This problem involves scattering off the circularly symmetric inverted oscillator (i.e., the potential of Problem 3.20, but with $k_1 = k_2 \equiv k$).

   (a) Show that all orbits are asymptotic in the $t \to \infty$ and $t \to -\infty$ limits to straight lines passing through the origin.

   (b) It follows from Part (a) that it is impossible to define an impact parameter b in the usual sense, since b depends on the orbit's being asymptotic, in the $t \to -\infty$ limit, to a line that misses the origin by a distance b. (This has to do with the force getting stronger and stronger as the distance from the force center increases.) Nevertheless a scattering angle can be found. Let B be the distance of closest approach to the force center and find the scattering angle as a function of B and the total energy E. [Partial answer: The scattering angle varies continuously from π (back-scattering at the lowest possible value of E for a given B), passing through $\frac{1}{2}\pi$ at $E = 0$ (independent of B, except for the singular case $B = 0$), to zero as E grows without bound.]

   (c) Show that the same is true of the three-dimensional inverted symmetric oscillator, for which $V = -\frac{1}{2}(k_1 x_1^2 + k_2 x_2^2 + k_3 x_3^2)$.

4. On Fig. 4.7 identify the points that correspond to trajectories that leave the scattering region above and those that leave it below the two disks. Argue that the curve of trapped orbits is the separatrix between these two sets. Find a way to determine which region of the figure is forbidden to first hits.

5. In the two-disk system an orbit can be its own time reversal in two different ways. Each of Trajectories 1 and 2 of Fig. 4.6 is its own time reversal only if one ignores the arrows showing the direction of motion. We call such a trajectory *weakly time-reversal invariant*. In contrast, the periodic exceptional orbit is its own time reversal even if one keeps track of the direction of motion. We call such a trajectory *strongly time-reversal invariant*. Consider a trajectory with the property that for one of the bounces on one of the two disks, $\phi = 0$. Show that this trajectory is strongly time-reversal invariant. Is the condition $\phi = 0$ a necessary condition for a trajectory to be strongly time-reversal invariant? Either prove that it is or find a counterexample.

6. The trajectory labeled α_2 in Fig. 4.9 is normal to the edge of Disk K. Find the angle θ between the horizontal and the trajectory at the edge of Disk K and show such that a periodic trajectory exists even if the three disks touch each other. [Answer: $\theta = \arctan(\sqrt{3} - 2/D)$.]

7. A certain Cantor set S is constructed in the following way: starting from the interval $[0, 1]$, the middle third is removed. From the two remaining intervals the middle fourth is removed; from the four remaining intervals the middle fifth is removed; from the eight remaining intervals the middle sixth is removed; etc. Calculate the measure μ and the fractal dimension d_f of S. [Answer: $\mu = 0$, $d_f = 1$.]

8. Find the relation between the three sets of constants of Eqs. (4.37): (A, B), (C, δ), and α.

9. Let ξ and η be complex vectors in $\mathbf{T}_{q_0}\mathbb{Q}$, and define $\langle \xi, \eta \rangle = (\xi, \mathbf{M}\eta) \equiv \xi^{\alpha*} M_{\alpha\beta} \eta^\beta$, where the ξ^α and η^β may be complex.

   (a) Use the properties of the $M_{\alpha\beta}$ matrix to show that $\langle \xi, \eta \rangle$ is an inner product on $\mathbf{T}_{q_0}\mathbb{Q}$ by showing that it is linear, Hermitian, and positive definite.

   (b) Show that Λ is Hermitian with respect to this inner product and hence that Λ has n linearly independent eigenvectors (orthogonal in the inner product $\langle\ \rangle$) spanning $\mathbf{T}_{q_0}\mathbb{Q}$.

   (c) Use the properties of the $K_{\alpha\beta}$ matrix to show that $\langle \xi, \Lambda\xi \rangle \geq 0$ and hence that all of the eigenvalues of Λ are nonnegative. [Hint: $M_{\alpha\beta}$ and $K_{\alpha\beta}$ are real and symmetric. What about Λ? For help, see Saletan & Cromer (1971, p. 112). Comment: Although vectors with complex components are used to obtain the above result, the actual eigenvectors of Λ can be chosen to be real, for the $M_{\alpha\beta}$ and $K_{\alpha\beta}$ are all real.]

10. Let $\mathbf{x}(t) = \sum_\gamma \mathbf{x}_\gamma(t) = \sum_\gamma \mathbf{a}_\gamma [\alpha_\gamma \exp(i\omega_\gamma t) + \alpha_\gamma^* \exp(-i\omega_\gamma t)]$ be a solution of the free n-dimensional harmonic oscillator problem. Show that the energy $E_\gamma \equiv \frac{1}{2}(\dot{\mathbf{x}}_\gamma, \mathbf{M}\dot{\mathbf{x}}_\gamma) +$

$\frac{1}{2}(\mathbf{x}_\gamma, \mathbf{K}\mathbf{x}_\gamma)$ in each normal mode is a constant of the motion. Find an expression for E_γ in terms of α_γ, $\mathbf{a}_\gamma$, and $\mathbf{K}$. [Hint: Show that $\omega_\gamma^2 \mathbf{M}\mathbf{a}_\gamma = \mathbf{K}\mathbf{a}_\gamma$.]

11. (a) For the undamped, one-freedom, linear oscillator with natural frequency ω_0, driven at frequency $\Omega \neq \omega_0$, find the rate at which energy is pumped into the oscillator. Find the maximum energy of an oscillator starting from rest at $q = 0$.

    (b) Do the same when the driving frequency is ω_0.

12. Consider a two-freedom linear oscillating system whose Λ matrix is given by

$$\Lambda = \begin{vmatrix} 13 & -6 \\ -6 & 8 \end{vmatrix}.$$

Find its normal modes and normal frequencies. Obtain the particular motion whose initial conditions are $x_1 = 1$, $x_2 = \dot{x}_1 = \dot{x}_2 = 0$.

13. (a) Find the half-width at half maximum $\Delta\Omega_\Gamma$ of the $|\Gamma|^2$ curve in terms of β and ω_0. Show that to first order in β, $\Delta\Omega_\Gamma \approx \beta$.

    (b) Find the half-width $\Delta\Omega_I$ of the I curve in terms of β and ω_0. Show that to first order in β, $\Delta\Omega_I \approx \beta$. [Comment: $\omega_0 \approx \omega \approx \omega'$ to zeroth order in β for $\beta \ll \omega_0$, so $Q \approx \omega_0/(2\Delta\Omega)$ to first order in β, whether $\Delta\Omega_\Gamma$ or $\Delta\Omega_I$ is used for $\Delta\Omega$.]

14. For the undamped linear oscillator driven at its natural frequency, whose equation is

$$\ddot{q} + \omega_0^2 q = f_0 e^{i\omega_0 t},$$

set $q = \alpha(t)e^{i\omega_0 t}$ and find $\alpha(t)$. In this way obtain the real solution when the driving term is $f_0 \cos \omega_0 t$.

15. For the two-dimensional harmonic oscillator, suppose $\rho = \omega_1/\omega_2$ is rational.

    (a) How many points are marked on the circular section of the Poincaré map (Fig. 4.19)?

    (b) How is the Poincaré map related to Lissajous figures? Try to give a mathematical description. What happens to the Lissajous figure when ρ is irrational?

16. (This Problem also depends on real numbers that may be commensurate or incommensurate.) Consider the dynamical system consisting of a free particle in a two-dimensional square box. When it collides with the walls, the particle loses no energy (the collisions are elastic), the component of its momentum parallel to the wall is conserved, and the component of its momentum perpendicular to the wall is reversed (the angle of reflection is equal to the angle of incidence). Assume, without loss of generality, that the particle starts somewhere at the left-hand wall.

    (a) Assume that the slope of its initial trajectory is a rational number p/q, where p and q are integers. Show that the particle will return to its initial position after a finite number n of collisions. Find n. Show that the number of points at which it hits any wall is finite. Find this number for each of the four walls.

    (b) Assume that the slope is irrational. Show that the particle will never return to its initial position and that it will hit any finite interval of any of the four walls an infinite number of times.

    (c) Extend these results to a rectangular box that is not square. [Hint: Think of the walls as mirrors and consider all of the images of the box.]

17. For the chain of N masses with fixed ends show that the normal modes a_γ are orthogonal and derive the normalizing factor $F = \sqrt{2/(N-1)}$ given in the text. [Hint: Write $\cos\theta = \frac{1}{2}\{e^{i\theta} + e^{-i\theta}\}$.]

18. Show that the Lagrangian for the chain of n masses with fixed ends, treated in Section 4.2.3, transforms to the form given in Eq. (4.68).

19. A chain of $N + 1$ particles interacting through springs as in Section 4.2.3 is subject to the condition that $x_0 = x_N$; that is, we impose periodic boundary conditions, or the chain is looped on itself. This means that there are just N independent particles.

  (a) Find the normal modes. (The normal modes are running waves.)
  (b) Find the velocities of the normal-mode waves.

20. For the fixed-end chain of N particles as in Section 4.2.3, consider the initial conditions $x_1(0) = 1$, all other $x_\nu(0) = 0$, and all $\dot{x}_\nu(0) = 0$. Find the $x_\nu(t)$. [See the hint to Problem 17.]

CHAPTER 5

HAMILTONIAN FORMULATION OF MECHANICS

CHAPTER OVERVIEW

The reason it is preferable to describe the motion of a dynamical system on $\mathbf{TQ}$ rather than on $\mathbb{Q}$ is that on $\mathbf{TQ}$ the equations of motion are of first order, and hence there is just one trajectory passing through each point $(q, \dot{q})$ of the manifold: the trajectories are separated. Then $\mathbf{TQ}$ is called a *carrier manifold* for the dynamics. But $\mathbf{TQ}$ is not the only carrier manifold that separates the trajectories. There are others on which the equations of motion have a simpler form. Indeed, although the equations on $\mathbf{TQ}$ give explicit expressions for half of the variables $(q, \dot{q})$ they do not give explicit expressions for the other half. The first derivatives of the qs are simple: $dq^\alpha/dt = \dot{q}^\alpha$. But the first derivatives of the $\dot{q}$s are buried in the EL equations. It would be simpler if the equations for them were in the similarly explicit form $d\dot{q}^\alpha/dt = f^\alpha(q, \dot{q})$, where the f^α would be functions that depend on the properties of the particular dynamical system. However, there is no obvious way to do this for $(q, \dot{q})$ on the tangent bundle, for there is no general way to dig the $\ddot{q}^\alpha$ out of the EL equations.

This can be done, however, for the variables (q, p), which involve the ps (the *generalized momenta*) rather than the $\dot{q}$s. That is, if the dynamical variables and all else are written in terms of the variables (q, p) instead of $(q, \dot{q})$, the equations of motion become explicit expressions for the time derivatives of both the q^α and the p_β as functions of (q, p). This chapter is devoted to developing this idea in what is called the *Hamiltonian formalism*. The Hamiltonian formalism has other advantages as well. It yields a different perspective and leads to a deeper understanding of mechanics, allows for extensions to other fields of physics, and provides, in particular, a route to quantization (Dirac, 1958). It also allows a broader class of (*canonical*) transformations than does the Lagrangian formalism.

Since the most general goal of a mechanics problem is to find the $q^\alpha(t)$, it may be said that physics actually takes place on $\mathbb{Q}$. To the extent that this is so, $\mathbf{TQ}$ is just a mathematical construct used to simplify the description and to perform the calculations (through the Lagrangian formalism). When the calculations are complete, the physical motion is obtained by *projecting* the trajectories $(q(t), \dot{q}(t))$ from $\mathbf{TQ}$ down to the trajectories $q(t)$ on $\mathbb{Q}$, essentially by ignoring the $\dot{q}(t)$ part. Recall how $\mathbf{TQ}$ is constructed by appending its tangent plane (*fiber*) $\mathbf{T}_q\mathbb{Q}$ to each point $q \in \mathbb{Q}$. One could imagine, more generally, constructing a different manifold $\mathbb{F}$ by appending a different kind of fiber to each point of $\mathbb{Q}$ (the result would be a *fiber bundle* over $\mathbb{Q}$). If $\mathbb{F}$ could somehow be made to carry the dynamics (to separate the trajectories), and if the trajectories on $\mathbb{F}$ could be projected down to the physical trajectories on $\mathbb{Q}$, then $\mathbb{F}$ would serve our purposes as well as $\mathbf{TQ}$. In the Hamiltonian formalism this is precisely what is done: $\mathbb{F}$ is the bundle whose fibers consist of possible values of the momentum p, a manifold whose points are of the form (q, p). It is called the *cotangent bundle* or the *phase manifold* (or *phase space*) $\mathbf{T}^*\mathbb{Q}$; its fibers are not the tangent spaces to the points of $\mathbb{Q}$, but their dual spaces (defined in Section 3.4.1), called *cotangent spaces*. These statements will be clarified as we proceed.

5.1   HAMILTON'S CANONICAL EQUATIONS

We will discuss the transition to the Hamiltonian formalism from three different complementary points of view.

5.1.1   LOCAL CONSIDERATIONS

FROM THE LAGRANGIAN TO THE HAMILTONIAN

The generalized momentum p_α conjugate to q^α has been defined (Section 3.2.1) as

$$p_\alpha = \frac{\partial L}{\partial \dot{q}^\alpha}, \tag{5.1}$$

where $L : \mathbf{TQ} \times \mathbb{R} \to \mathbb{R}$ is, as always, the Lagrangian function. (The domain of L is formed by $\mathbf{TQ}$ and the time $t \in \mathbb{R}$.) The EL equations and the differential equations for the q^α can therefore be written in the form

$$\frac{dp_\alpha}{dt} = \frac{\partial L}{\partial q^\alpha}, \tag{5.2a}$$

$$\frac{dq^\alpha}{dt} = \dot{q}^\alpha. \tag{5.2b}$$

Except for the fact that L and the $\dot{q}^\alpha$ are all functions of $(q, \dot{q}, t)$ (the $\dot{q}^\alpha$ trivially so) rather than of (q, p, t), these equations are in the desired simple form. To obtain precisely the desired form requires only that the right-hand sides be written in terms of (q, p, t).

In principle this is easy to do for Eq. (5.2b): write the $\dot{q}^\alpha$ on the right-hand side as functions of (q, p, t) by inverting (5.1). Assume that this can be done (we will find necessary conditions for this later), and call the resulting functions $\dot{q}^\alpha(q, p, t)$. The right-hand side of Eq. (5.2a) can be similarly transformed: in all the $\partial L/\partial q^\beta$ replace the $\dot{q}^\alpha$ wherever they appear by $\dot{q}^\alpha(q, p, t)$. It would be easier to do the replacement in L first, turning L into a function of (q, p, t), and then to take the derivatives. This requires care, however, because the partial derivative of $L(q, \dot{q}, t)$ with respect to q^α is not the same as that of $\hat{L}(q, p, t) \equiv L(q, \dot{q}(q, p, t), t)$, for L is a function of $(q, \dot{q}, t)$, and $\hat{L}$ is a function of (q, p, t). In taking derivatives of L, the $\dot{q}$s are held fixed, whereas in $\hat{L}$ the ps are held fixed. We compare these two sets of derivatives:

$$\frac{\partial \hat{L}}{\partial q^\alpha} = \frac{\partial L}{\partial q^\alpha} + \frac{\partial L}{\partial \dot{q}^\beta}\frac{\partial \dot{q}^\beta}{\partial q^\alpha} = \frac{\partial L}{\partial q^\alpha} + p_\beta \frac{\partial \dot{q}^\beta}{\partial q^\alpha},$$

where we used (5.1). The last term on the right-hand side is the derivative of the function $\dot{q}^\beta(q, p, t)$. Now put all the functions of (q, p, t) on one side of the equation to obtain

$$\frac{\partial L}{\partial q^\alpha} = \frac{\partial}{\partial q^\alpha}[\hat{L}(q, p, t) - p_\beta \dot{q}^\beta(q, p, t)].$$

The derivative of $\hat{L}$ with respect to p_α will also be needed. It is

$$\frac{\partial \hat{L}}{\partial p_\alpha} = \frac{\partial L}{\partial \dot{q}^\beta}\frac{\partial \dot{q}^\beta}{\partial p_\alpha} = p_\beta \frac{\partial \dot{q}^\beta}{\partial p_\alpha},$$

or (again put all of the functions of (q, p, t) on one side)

$$\frac{\partial}{\partial p_\alpha}[\hat{L}(q, p, t) - p_\beta \dot{q}^\beta(q, p, t)] = -\dot{q}^\alpha. \tag{5.3}$$

This equation looks like the solution one would obtain for $\dot{q}^\alpha$ by inverting (5.1), but actually it is not, for the left-hand side itself depends on the $\dot{q}^\beta$. It is instead a differential equation for $\dot{q}^\alpha$ as a function of (q, p, t).

The function

$$H(q, p, t) \equiv p_\beta \dot{q}^\beta(q, p, t) - \hat{L}(q, p, t) \tag{5.4}$$

appears in Eqs. (5.3) and (5.4). This very important function is called the *Hamiltonian function*, or simply the *Hamiltonian*.

It is noteworthy that the Hamiltonian is the dynamical variable we have called E, written now in terms of (q, p) rather than $(q, \dot{q})$. Recall that for many dynamical systems E is the energy, and hence when $L = T - V$ the Hamiltonian is $H = T + V$ expressed in terms of (q, p). In those common cases when V is independent of the $\dot{q}^\beta$ only the kinetic energy T need be written out in terms of (q, p). It is not always that simple, however, as will be evident from Worked Example 5.1 of the Hamiltonian for a charged particle in an electromagnetic field.

With the aid of the Hamiltonian, Eqs. (5.2a, b) can now be written down entirely in terms of (q, p, t):

$$\left. \begin{array}{l} \dot{q}^\beta = \dfrac{\partial H}{\partial p_\beta}, \\[3mm] \dot{p}_\beta = -\dfrac{\partial H}{\partial q^\beta}. \end{array} \right\} \tag{5.5}$$

These are called *Hamilton's canonical equations*. They are the equations of motion in the Hamiltonian formalism. Their solution yields local expressions for the trajectories

$(q(t), p(t))$ in $\mathbf{T}^*\mathbb{Q}$, the (q, p) manifold. This manifold is called the *cotangent bundle* (which will be discussed in Section 5.2). The projections of these trajectories down to $\mathbb{Q}$ (essentially by ignoring the $p(t)$ part) are the configuration-manifold trajectories $q(t)$. We leave to Section 5.2 a more detailed discussion of $\mathbf{T}^*\mathbb{Q}$ and an explanation of the difference between it and $\mathbf{T}\mathbb{Q}$.

> REMARK: The term "canonical" means standard or conventional (according to the canons). It is used technically for dynamical systems, in particular in the Hamiltonian formalism. □

The recipe for writing down Hamilton's equations for a system whose Lagrangian is known is then the following:

  i. Use (5.1) to calculate the p_α in terms of $(q, \dot{q}, t)$.
  ii. Invert these equations to obtain the $\dot{q}^\alpha$ in terms of (q, p, t).
  iii. Insert the $\dot{q}^\alpha(q, p, t)$ into the expression for L to obtain $\hat{L}$.
  iv. Use (5.4) to obtain an explicit expression for $H(q, p, t)$.
  v. By taking the derivatives of H, write down (5.5).

The recipe works only if step *ii* is possible, that is, only if Eqs. (5.1) can be solved uniquely for the $\dot{q}^\alpha$. According to the implicit function theorem a necessary and sufficient condition for invertibility locally is that the Hessian $\partial^2 L / \partial \dot{q}^\alpha \partial \dot{q}^\beta$ be nonsingular (see Abraham and Marsden, 1985). We came across the same requirement in Section 2.2.2. As a general rule we will assume it to be fulfilled, although there will be occasions when it fails on small subsets.

Hamilton's canonical equations have been derived from the Euler–Lagrange equations, which have in turn been derived from Newton's equations of motion. Thus far the only method we have of finding the Hamiltonian function is by passing through the Lagrangian, but often it is as simple from scratch to write down one as the other. More about this will be discussed later.

WORKED EXAMPLE 5.1

(a) Obtain the Hamiltonian and **(b)** write down Hamilton's canonical equations for a charged particle in an electromagnetic field. Project them down to $\mathbb{Q}$ and show that they are the usual equations for a particle in an electromagnetic field.

Solution. **(a)** The Lagrangian (Section 2.2.4) is

$$L(q, \dot{q}, t) = \frac{1}{2}m\dot{q}^\beta \dot{q}^\beta - e\varphi(q, t) + e\dot{q}^\beta A_\beta(q, t),$$

where the index, β in this case, runs from 1 to $n = 3$. [Note the slight abuse of notation: some sums are on two superscripts and some on a superscript and a subscript. This is inevitable in the transition from $(q, \dot{q})$ to (q, p), for the $\dot{q}^\alpha$ carry superscripts and the p_α carry subscripts. Eventually all sums will be on superscript–subscript pairs.] According to Eq. (5.1)

$$p_\gamma = m\dot{q}^\gamma + eA_\gamma, \tag{5.6}$$

which is the generalized momentum canonically conjugate to q^γ. The generalized momentum is not the *dynamical momentum* $m\dot{q}^\gamma$ but has a contribution that comes from the electromagnetic field.

To obtain the Hamiltonian, solve this equation for $\dot{q}^\gamma$ and insert it into Eq. (5.4). The result is (we use $\delta^{\alpha\beta}$ to force the sums to be on superscript–subscript pairs)

$$H(q, p, t) = \frac{1}{2m} \delta^{\alpha\gamma} \{p_\alpha - eA_\alpha(q, t)\}\{p_\gamma - eA_\gamma(q, t)\} + e\varphi(q, t). \tag{5.7}$$

(b) The canonical equations obtained from this Hamiltonian are

$$\dot{q}^\beta = \frac{1}{m} \delta^{\alpha\beta} \{p_\alpha - eA_\alpha(q, t)\}, \tag{5.8a}$$

$$\dot{p}_\beta = \frac{e}{m} \delta^{\alpha\gamma} \frac{\partial A_\alpha}{\partial q^\beta} \{p_\gamma - eA_\gamma(q, t)\} - \frac{\partial \varphi}{\partial q^\beta}. \tag{5.8b}$$

The equations of motion on $\mathbb{Q}$ are obtained by eliminating the p_β from Eqs. (5.8a, b). This is done by taking the time derivatives of the $\dot{q}^\alpha$ in (5.8a) and replacing the $\dot{p}_\alpha$ that appear by their expressions in (5.8b). Equation (5.8a) yields

$$\ddot{q}^\beta = \frac{1}{m} \delta^{\alpha\beta} (\dot{p}_\alpha - e\partial_\gamma A_\alpha \dot{q}^\gamma - e\partial_t A_\alpha),$$

where $\partial_\gamma = \partial/\partial q^\gamma$ and $\partial_t = \partial/\partial t$. Now Eq. (5.8b) is used to eliminate $\dot{p}_\alpha$, but the p_γ on the right-hand side must be replaced by its expression in terms of $(q, \dot{q})$, namely by Eq. (5.6). Then the equation for $\ddot{q}^\beta$ becomes (here we no longer worry about whether the indices are superscripts or subscripts)

$$m\ddot{q}^\beta = e\{(\partial_\beta A_\gamma - \partial_\gamma A_\beta)\dot{q}^\gamma - \partial_t A_\beta - \partial_\beta \varphi\}. \tag{5.9}$$

It follows from Section 2.2.4 that $\partial_\beta A_\gamma - \partial_\gamma A_\beta = \epsilon_{\beta\gamma\alpha} B_\alpha$ and $-\partial_t A_\alpha - \partial_\alpha \varphi = E_\alpha$, where $\mathbf{B} = (B_1, B_2, B_3)$ is the magnetic field and $\mathbf{E} = (E_1, E_2, E_3)$ is the electric field (not to be confused with the energy). Thus

$$m\ddot{q}^\beta = e\{\epsilon_{\gamma\alpha\beta}\dot{q}^\gamma B_\alpha + E_\beta\},$$

which is the component form of the Lorentz force $\mathbf{F} \equiv m\mathbf{a} = e(\mathbf{v} \wedge \mathbf{B} + \mathbf{E})$, which is the required result.

WORKED EXAMPLE 5.2

(a) Obtain the Hamiltonian and the canonical equations for a particle in a central force field. (b) Take two of the initial conditions to be $p_\phi(0) = 0$ and $\phi(0) = 0$ (this is essentially the choice of a particular spherical coordinate system). Discuss the resulting simplification of the canonical equations.

Solution. (a) The Lagrangian is (Section 2.3)

$$L = \frac{1}{2}m[\dot{r}^2 + r^2\dot{\theta}^2 + (r^2\sin^2\theta)\dot{\phi}^2] - V(r), \tag{5.10}$$

where θ is the azimuth angle and ϕ is the colatitude. Denote the conjugate momenta by p_r, p_θ, and p_ϕ. The defining equations for them are

$$\left.\begin{aligned}
p_r &\equiv \frac{\partial L}{\partial \dot{r}} = m\dot{r}, \\
p_\theta &\equiv \frac{\partial L}{\partial \dot{\theta}} = mr^2\dot{\theta}, \\
p_\phi &\equiv \frac{\partial L}{\partial \dot{\phi}} = m(r^2\sin^2\theta)\dot{\phi}.
\end{aligned}\right\} \tag{5.11}$$

Inverting these equations yields

$$\left.\begin{aligned}
\dot{r} &= \frac{p_r}{m}, \\
\dot{\theta} &= \frac{p_\theta}{mr^2}, \\
\dot{\phi} &= \frac{p_\phi}{mr^2\sin^2\theta}.
\end{aligned}\right\} \tag{5.12}$$

The Hamiltonian is

$$H = p_\alpha \dot{q}^\alpha - L = \frac{p_r^2}{2m} + \frac{p_\theta^2}{2mr^2} + \frac{p_\phi^2}{2mr^2\sin^2\theta} + V(r) \equiv T + V, \tag{5.13}$$

where the kinetic energy T is the sum of the first three terms. Hamilton's canonical equations in these variables are

$$\left.\begin{aligned}
\dot{r} &\equiv \frac{\partial H}{\partial p_r} = \frac{p_r}{m}, & \dot{p}_r &\equiv -\frac{\partial H}{\partial r} = \frac{1}{mr^3}\left[p_\theta^2 + \frac{p_\phi^2}{\sin^2\theta}\right] - V'(r); \\
\dot{\theta} &\equiv \frac{\partial H}{\partial p_\theta} = \frac{p_\theta}{mr^2}, & \dot{p}_\theta &\equiv -\frac{\partial H}{\partial \theta} = \frac{p_\phi^2\cos\theta}{mr^2\sin^3\theta}; \\
\dot{\phi} &\equiv \frac{\partial H}{\partial p_\phi} = \frac{p_\phi}{mr^2\sin^2\theta}, & \dot{p}_\phi &\equiv -\frac{\partial H}{\partial \phi} = 0.
\end{aligned}\right\} \tag{5.14}$$

Comment: Notice that the three equations on the left follow immediately from the definitions of the generalized momenta. This is always the case with Hamilton's canonical equations: by studying their derivation it becomes clear that half of them are simply inversions of those definitions.

(b) If $p_\phi(0) = 0$, the last of Eqs. (5.14) implies that $p_\phi(t) = 0$ for all t. Then the canonical equations become

$$\left. \begin{array}{ll} \dot{r} = \dfrac{p_r}{m}, & \dot{p}_r = \dfrac{p_\theta^2}{mr^3} - V'(r); \\[2ex] \dot{\theta} = \dfrac{p_\theta}{mr^2}, & \dot{p}_\theta = 0; \\[2ex] \dot{\phi} = 0, & \dot{p}_\phi = 0. \end{array} \right\} \tag{5.15}$$

From $\phi(0) = 0$ and these equations it follows that $\phi(t) = 0$ for all t: the motion stays in the $\phi = 0$ plane. The result is the two-freedom system consisting of the first two lines of (5.15). The fourth equation implies that p_θ is a constant of the motion. This is the angular momentum in the $\phi = 0$ plane, called l in Section 2.3.

These examples demonstrate some properties of the Hamiltonian formulation. Like the EL equations on $\mathbf{T}\mathbb{Q}$, Hamilton's canonical equations are a set of first-order differential equations, but in a new set of $2n$ variables, the (q, p) on $\mathbf{T}^*\mathbb{Q}$. Because they are constructed from Lagrange's, Hamilton's equations contain the same information. Indeed, we see that they give the correct second-order differential equations on $\mathbb{Q}$. They can also be solved for the $q^\alpha(t)$ and the $p_\alpha(t)$, and then the $q^\alpha(t)$ are the physical trajectories on $\mathbb{Q}$. This also yields the trajectories on $\mathbf{T}\mathbb{Q}$, since the $\dot{q}^\alpha(t)$ can be obtained by differentiation. The Hamiltonian formalism can also yield the EL equations directly (Problem 5). However going from $\mathbf{T}^*\mathbb{Q}$ to $\mathbf{T}\mathbb{Q}$ gains us nothing: the Hamiltonian formalism is sufficient and autonomous.

Worked Example 5.2 dealt with the general central-force problem. We now specify to the Kepler problem, but we treat it within the special theory of relativity to show how a relativistic system can be handled within the Hamiltonian formalism. To do so requires a brief review of some aspects of special relativity. (For more details see Landau and Lifshitz, 1975, or Bergmann, 1976.)

A BRIEF REVIEW OF SPECIAL RELATIVITY

In the early part of the 20th century physics was radically altered by the special theory of relativity and by quantum mechanics. Each of these contains classical mechanics as a limit; quantum mechanics in the limit of small h (Planck's constant), and relativity in the limit of large c (the speed of light). Relativity is in some sense "more classical" than quantum mechanics, for unlike quantum mechanics it refers to particles moving along trajectories.

Einstein's special theory of relativity compares the way a physical system is described by different observers moving at constant velocity with respect to each other. Before relativity, the way to transform from one such observer's description to the other's would be to use Galilean transformations. If one observer sees a certain event taking place at position $\mathbf{x}$ and at time t, a second observer moving with velocity $\mathbf{V}$ relative to the first sees the same event taking place at position $\mathbf{x}'$ and time t' given by

$$\mathbf{x}' = \mathbf{x} - \mathbf{V}t, \quad t' = t. \tag{5.16}$$

It is assumed here that at time $t = 0$ the two observers are at the same point and that they have synchronized their clocks. It is also assumed that the two observers have lined up their axes, so that no rotation is involved. The most general Galilean transformation relinquishes all of these assumptions; in particular it involves rotations.

Experimentally it is found that the Galilean transformation is only approximately accurate: the lower the magnitude of $\mathbf{V}$, the better. The accurate relation is given by the Lorentz transformation (with the above assumptions)

$$\mathbf{x}' = \gamma(\mathbf{x} - \mathbf{V}t) + (1 - \gamma)\left\{\mathbf{x} - \frac{\mathbf{V}}{V}\left[\frac{\mathbf{x} \cdot \mathbf{V}}{V}\right]\right\}, \quad t' = \gamma\left(t - \frac{\mathbf{x} \cdot \mathbf{V}}{c^2}\right), \quad (5.17)$$

where

$$\gamma = \frac{1}{\sqrt{1 - V^2/c^2}}. \quad (5.18)$$

Equation (5.17) shows that in the limit as $c \to \infty$, or equivalently as $\gamma \to 1$, the Lorentz transformation reduces to the Galilean transformation. This is the sense in which relativity contains classical mechanics in the limit of high c.

Equation (5.17) hides some of the simplicity of the Lorentz transformation, so we write it out for the special case in which $\mathbf{V}$ is directed along the 1-axis. In that case x^2 and x^3 are unaffected, and the rest of the transformation can be written in the form

$$x^{1\prime} = \gamma\left[x^1 - \frac{Vx^0}{c}\right], \quad x^{0\prime} = \gamma\left[-\frac{Vx^1}{c} + x^0\right], \quad (5.19)$$

where $x^0 = ct$. Because of the similarity between the way x^1 and x^0 transform, it is convenient to deal with *four-vectors* $x = (x^0, x^1, x^2, x^3)$. Henceforth we will choose coordinates in space and time such that $c = 1$ (e.g., light years for distance).

The geometry of the four-dimensional space–time manifold so obtained is called *Minkowskian*. In Euclidean geometry, the square magnitude of a three-vector (or the Euclidean distance between two points) remains invariant under Galilean transformations (including rotations). The invariant may be written in the form

$$\Delta s^2 = (\Delta x^1)^2 + (\Delta x^2)^2 + (\Delta x^3)^2,$$

where $\Delta \mathbf{x}$ is the separation between two points. In Minkowskian geometry a different quadratic expression remains invariant under Lorentz transformation:

$$\Delta \sigma^2 \equiv g_{\mu\nu}\Delta x^\mu \Delta x^\nu \equiv -(\Delta x^0)^2 + (\Delta x^1)^2 + (\Delta x^2)^2 + (\Delta x^3)^2, \quad (5.20)$$

where $\Delta x \equiv (\Delta x^0, \Delta x^1, \Delta x^2, \Delta x^3)$ is the *space–time separation* between two *events*. The indices run from 0 to 3, and the *Minkowskian metric* $g_{\mu\nu}$ has elements $g_{\mu\nu} = 0$ if $\mu \neq \nu$ and $g_{11} = g_{22} = g_{33} = -g_{00} = 1$.

The equations of Newtonian physics reflect Euclidean geometry by involving only scalars and three-vectors. For instance, Newton's second law $m\ddot{\mathbf{x}} = \mathbf{F}$ contains only the scalar m and the three-vectors $\mathbf{x}$ and $\mathbf{F}$. That m is a (Euclidean) scalar means that it is invariant under rotation, and that $\mathbf{F}$ is a three-vector means that it transforms in the same way as $\mathbf{x}$. This ensures the invariance of the equation under Galilean transformations. Similarly, the equations of relativistic physics reflect Minkowskian geometry by involving only scalars and four-vectors. Therefore Newton's equations of motion must be modified.

The special relativistic generalization of Newton's equations is

$$m\frac{d^2x^\mu}{d\tau^2} \equiv ma^\mu = F^\mu\left(x, \frac{dx}{d\tau}, \tau\right). \tag{5.21}$$

Like x, both a and F are four-vectors, generalizations of the three-vector acceleration and force. The mass m in this equation is sometimes called the *rest mass*, but it is the only mass we will use here, so we call it simply the mass. The scalar τ is called the *proper time*; it is measured along the motion and defined by the analog of Eq. (5.20)

$$(d\tau)^2 = -g_{\mu\nu}\,dx^\mu\,dx^\nu. \tag{5.22}$$

The four-vector generalization of the three-vector velocity has components

$$u^\mu = \frac{dx^\mu}{d\tau}, \tag{5.23}$$

and it follows from Eq. (5.22) that

$$g_{\mu\nu}u^\mu u^\nu = -1. \tag{5.24}$$

We state without proof that $u^0 = \gamma$ and that $u^k = \gamma v^k$ for $k = 1, 2, 3$, where v^k is the nonrelativistic velocity dx^k/dt.

We now want to find the Hamiltonian form of this relativistic dynamics. The first step is to find a Lagrangian that leads to Eq. (5.21). We take as a principle that in the Newtonian (i.e., high-c) limit all results should approach the nonrelativistic ones, and therefore we write a time-independent Lagrangian modeled on a standard nonrelativistic one:

$$L_R = T(\mathbf{v}) - U(\mathbf{x}), \tag{5.25}$$

where T and U are the relativistic generalizations of the kinetic and potential energies, but $\mathbf{v}$ and $\mathbf{x}$ are the nonrelativistic velocity and position three-vectors of the particle. Since c will not appear in any of the equations, the Newtonian limit is obtained by going to speeds low with respect to the speed of light, that is, speeds much less than 1.

The three-velocity should satisfy the equation $\partial L_R/\partial v^k = p^k$ (Latin indices will be used from 1 to 3, Greek indices from 0 to 3), but the relativistic generalization of the momentum must now be used. It follows from arguments concerning conservation of both momentum and energy (Landau and Lifshitz, 1975) that the generalization involves the four-vector $p^\mu = mu^\mu$, called the relativistic momentum. The space components of p^μ are

$$p^k \equiv mu^k = m\gamma v^k \equiv \frac{mv^k}{\sqrt{1 - \mathbf{v}^2}}, \tag{5.26}$$

where $\mathbf{v}^2 = \mathbf{v} \cdot \mathbf{v} \equiv (v^1)^2 + (v^2)^2 + (v^3)^2$ and γ is redefined by generalizing Eq. (5.18). (Recall that the previous definition of γ contained the relative velocity $\mathbf{V}$ of two observers. This one contains the velocity $\mathbf{v}$ of the particle.) For $|\mathbf{v}| \ll 1$, the space components of the relativistic momentum hardly differ from the nonrelativistic ones, and the time component (i.e., $p^0 \equiv m\gamma$) is $m + \frac{1}{2}m\mathbf{v}^2$ to lowest order in the velocity, or p^0 hardly differs from the nonrelativistic kinetic energy plus a constant.

The requirement that $\partial L_R/\partial v^k = p^k$ can now be put in the form

$$\frac{\partial T}{\partial v^k} = \frac{mv^k}{\sqrt{1 - \mathbf{v}^2}},$$

whose solution is

$$T = -m\sqrt{1 - \mathbf{v}^2} + C.$$

Again, in the Newtonian limit this yields $T \approx \frac{1}{2}m\mathbf{v}^2 + \kappa$, the kinetic energy plus a constant, as it should.

Equation (5.25) can now be written out more explicitly: the relativistic Lagrangian of a conservative one-particle system is

$$L_R = -m/\gamma - U(\mathbf{x}). \tag{5.27}$$

Accordingly, the relativistic expression for the energy is

$$E_R = \frac{\partial L_R}{\partial v^k} v^k - L_R = m\gamma + U(\mathbf{x}), \tag{5.28}$$

which, in the Newtonian limit, is the usual energy (plus a constant).

This treatment of the relativistic particle exhibits some of the imperfections of the relativistic Lagrangian (and Hamiltonian) formulation of classical dynamical systems. For one thing, it uses the nonrelativistic three-vector velocity and position but uses the relativistic momentum. For another, all of the equations are written in the special coordinate system in which the potential is time independent, and this violates the relativistic principle according to which space and time are to be treated on an equal footing. Nevertheless, in the system we will treat, this approach leads to results that agree with experiment.

The Hamiltonian is then

$$H_R \equiv \frac{\partial L_R}{\partial v^k} v^k - L_R = \frac{p^k p^k}{m\gamma} + \frac{m}{\gamma} + U(\mathbf{x}).$$

Some algebra shows that $m\gamma = \sqrt{\mathbf{p}^2 + m^2}$ (where $\mathbf{p}^2 = p^k p^k$), and then

$$H_R = \sqrt{\mathbf{p}^2 + m^2} + U(\mathbf{x}). \tag{5.29}$$

Since this H_R is time independent, $dH_R/dt = 0$, and comparison with (5.28) shows that $H_R = E_R$.

THE RELATIVISTIC KEPLER PROBLEM

Now that we have established the basics of the formalism, we apply it to the *relativistic Kepler problem* or *relativistic classical hydrogen atom* (Sommerfeld, 1916). In an important step in the history of quantum mechanics, Sommerfeld used this approach in the "old quantum theory" (Born, 1960) to obtain the correct first-order expressions for the fine structure of the hydrogen spectrum. To get the Hamiltonian of the relativistic hydrogen atom, write $U = -e^2/r$ in (5.29), where $r^2 = \mathbf{x} \cdot \mathbf{x}$ and e is the electron charge in appropriate units (drop the subscript R from H):

$$H = \sqrt{\mathbf{p}^2 + m^2} - \frac{e^2}{r}. \tag{5.30}$$

As in the nonrelativistic case, the motion is confined to a plane and the number of freedoms reduce immediately to two. In plane polar coordinates the Hamiltonian becomes

$$H = \sqrt{p_r^2 + \frac{p_\theta^2}{r^2} + m^2} - \frac{e^2}{r}, \tag{5.31}$$

where (drop the subscript R also from L)

$$p_\theta = \frac{\partial L}{\partial \dot\theta} \quad \text{and} \quad p_r = \frac{\partial L}{\partial \dot r}.$$

As in the nonrelativistic case, θ is an ignorable coordinate so p_θ is conserved.

We proceed to find the equation of the orbit, that is, $r(\theta)$. The fact (see Problem 26) that

$$p_r = \frac{p_\theta}{r^2} \frac{dr}{d\theta} \tag{5.32}$$

can be used to find $r' = dr/d\theta$ from Eq. (5.31). The result (recall that $H = E$) is

$$r' = \frac{r^2}{p_\theta} \sqrt{\left(E + \frac{e^2}{r}\right)^2 - m^2 - \frac{p_\theta^2}{r^2}}.$$

Changing the dependent variable from r to $u = 1/r$ [see Eq. (2.70)] yields

$$u' = -\sqrt{(a + bu)^2 - u^2 - D}, \tag{5.33}$$

where $a = E/p_\theta$, $b = e^2/p_\theta$, and $D = m^2/p_\theta^2$. One way to handle this equation is to square both sides and take the derivative with respect to θ. The result is

$$u'' = (b^2 - 1)u + ab.$$

Like Eq. (2.70), this a harmonic-oscillator equation (if $b^2 < 1$; see Problem 27), whose solution is obtained immediately:

$$u = A \cos\{\sqrt{1 - b^2}(\theta - \theta_0)\} + \frac{ab}{1 - b^2},$$

where A and θ_0 are constants of integration (θ_0 determines the initial point on the orbit). To determine A, the solution has to be inserted back into Eq. (5.33). After some algebra, the solution for r is found to be

$$r = \frac{q}{1 + \epsilon \cos[\Gamma(\theta - \theta_0)]}, \tag{5.34}$$

where

$$\Gamma = \sqrt{1 - \frac{e^4}{c^2 p_\theta^2}}, \quad q = \frac{c^2 \Gamma^2 p_\theta^2}{e^2 E}, \quad \epsilon = \sqrt{1 + \frac{\Gamma^2(1 - m^2 c^4/E^2)}{1 - \Gamma^2}}. \tag{5.35}$$

In order to estimate the size of various terms we have reintroduced the speed of light c, which can be done entirely from dimensional considerations (e.g., m must be multiplied by the square of a velocity to have the dimensions of energy).

In spite of the similarity to the result of Section 2.3.2, the relativistic orbit as given by Eq. (5.34) is not a conic section with eccentricity ϵ. The difference lies in the factor Γ of the argument of the cosine, which almost always causes the orbits to fail to close if $\Gamma \neq 1$. From Eq. (5.35) it is seen that as $c \to \infty$ (i.e., in the nonrelativistic limit) $\Gamma \to 1$ and the orbit approaches a conic section. (It is not clear from our equations what happens to ϵ in the nonrelativistic limit, but that is treated in the Sommerfeld paper.) Before reaching this limit and under some conditions, the orbit looks like a precessing ellipse, and this precession is what gives rise to the fine structure in the hydrogen spectrum. The change $\Delta\theta$ between successive minima is calculated in Problem 11.

5.1.2 THE LEGENDRE TRANSFORM

The transition from the Lagrangian and $\mathbf{TQ}$ to the Hamiltonian and $\mathbf{T^*Q}$ is an example of what is called the *Legendre transform*. We explain this in one dimension and then indicate how to extend the explanation to more than one.

Consider a differentiable function $L(v)$ of the single variable v. For each value of v in its domain, the function gives the number $L(v)$ in its range, and its graph is the continuous curve of all the points $(v, L(v))$ (Fig. 5.1). At each point the function has a derivative (the curve has a tangent), which we write

$$p(v) \equiv dL(v)/dv. \tag{5.36}$$

The Legendre transform is a way of describing the function or reproducing the graph entirely in terms of p, with no reference to v: that is, p will become the independent variable whose values are used to construct the curve. But just as the values of v alone without the values of $L(v)$ are not enough to define the curve, the values of p alone are not enough. What is needed is a function $H(p)$ of the new variable p.

The function $H(p)$ is obtained in the following way. Start from Fig. 5.1, which shows the tangent to the curve $L(v)$ at the point $v = v_0$. The slope of this tangent is $p(v_0) \equiv p_0$, and its

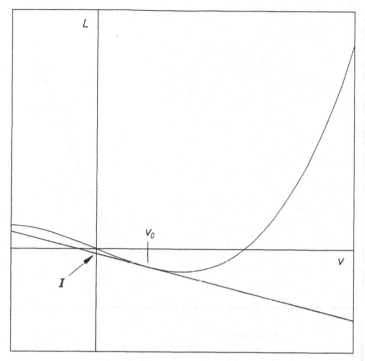

FIGURE 5.1

A function $L(v)$, which has a point of inflection at the origin. The tangent to the curve at $v = v_0$ has slope $p = dL/dv$ calculated at v_0. The intercept on the L axis is labeled I.

intercept I on the L axis is given by

$$I = L(v_0) - p_0 v_0.$$

There is a different intercept for each v on the curve (we then drop the subscript 0), so the intercept becomes a function of the point v and of the derivative p at that point:

$$I(v, p) = L(v) - pv. \qquad (5.37)$$

Suppose now that Eq. (5.36) is invertible, so that v can be obtained as a function of p. In that case each value of p uniquely determines a value of v: a given slope occurs only at one point on the curve. (The function of Fig. 5.1 does not satisfy this requirement: the same slope occurs at more than one point. We will return to this later.) When $v(p)$ is found, it can substituted for v in (5.37), which then becomes a function of p alone. Then $H(p)$ is defined by

$$H(p) \equiv -I(v(p), p) = pv(p) - L(v(p)). \qquad (5.38)$$

So far $H(p)$ has been obtained from $L(v)$. We now show that $L(v)$ can be obtained from $H(p)$, and thus that the two functions are equivalent. We will first demonstrate this graphically and then prove it analytically.

Suppose $H(p)$ is known. Each combination $(p, H(p))$ corresponds to a line of slope p and intercept $-H(p)$ on the L axis in the (L, v) plane. The set of such lines for $v > 0$ (where the

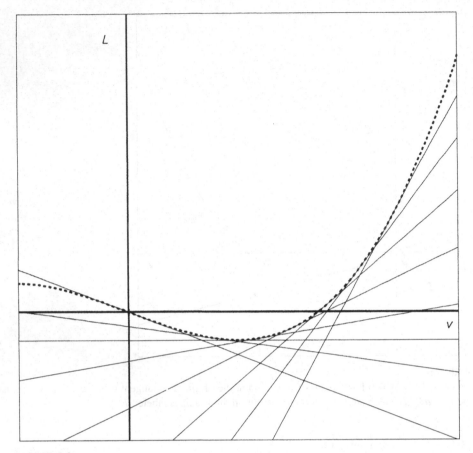

FIGURE 5.2
Tangents to the $L(v)$ curve of Fig. 5.1 for values of v above the point of inflection, obtained by calculating $(p, H(p))$ for several values of p. The dotted curve is $L(v)$, seen to be the envelope of the tangents.

relationship between v and p is unique) on Fig. 5.1 is shown in Fig. 5.2. The curve $L(v)$ is the envelope of these lines, the continuous curve that is tangent to them all. Thus the $L(v)$ curve (or part of it) has been constructed from $H(p)$.

The analytic proof requires showing that the function $L(v)$ can be obtained from $H(p)$. Assume that $H(p)$ is known, and take the derivative of (5.38) with respect to p:

$$\frac{dH}{dp} = v(p) + p\frac{dv}{dp} - \frac{dL}{dv}\frac{dv}{dp} \equiv v(p) \tag{5.39}$$

(use $dL/dv \equiv p$). The left-hand side is a function of p, so (5.39) gives v as a function of p. Now assume that this equation can be inverted to yield the function $p(v)$, which can then be used to rewrite Eq. (5.38) as a function of v:

$$L(v) = H(p(v)) - p(v)v. \tag{5.40}$$

The similarity of this equation and (5.38) is suggestive: just as H gives the slope–intercept representation of $L(v)$ in the (v, L) plane, so L gives the slope–intercept representation of $H(p)$ in the (p, H) plane.

We now turn to the conditions for invertibility. By definition any function $p(v)$ gives a unique p for each value of v; invertibility means that $p(v)$ is a one-to-one function. This means that $p(v)$ does not pass through a maximum or minimum, so $dp/dv \neq 0$. But $dp/dv = d^2L/dv^2$, so a necessary condition for invertibility is that the second derivative of L fail to vanish: L must have no point of inflection. (Similarly, a necessary condition for invertibility of $v(p)$ is that $d^2H/dp^2 \neq 0$.) This can be seen also on Figs. 5.1 and 5.2. There is a point of inflection at the origin: $d^2L/dv^2 = 0$ at $v = 0$. If tangents are drawn in Fig. 5.2 on both sides of $v = 0$, there will be more than one point v with the same slope p, the lines overlap, and the diagram becomes hopelessly confusing. Either to the right or to the left of $v = 0$, however, where $v(p)$ is single valued, the diagram is clear.

To go on to higher dimension, suppose that $L(v)$ depends on a collection of n variables $v = \{v^1, v^2, \ldots, v^n\}$. The diagrammatic demonstration is much more difficult, for the partial derivatives $\partial L(v)/\partial v^\alpha = p_\alpha(v)$ give the inclination of the tangent hyperplane at v, rather than the slope of the tangent line. The $L(v)$ hypersurface in $n + 1$ dimensions is the envelope of the tangent hyperplanes. Analytically, however, the treatment proceeds in complete analogy with the single-variable case, except that the necessary conditions for invertibility now become that the Hessians $\partial^2L/\partial v^\alpha\partial v^\beta$ and $\partial^2H/\partial p_\alpha\partial p_\beta$ be nonsingular (this is again a manifestation of the implicit function theorem).

The similarity between the Legendre transform and the treatment of Section 5.1.1 should now be clear: in the transition from the Lagrangian to the Hamiltonian formalism, the generalized velocities $\dot{q}^u$ are the v^α in the Legendre transform, and the generalized momenta are the p_α. The main difference is that the Lagrangian and Hamiltonian are functions also of the generalized coordinates q^α, so a Legendre transform is performed for each point of $\mathbb{Q}$. The Legendre transform occurs also in other branches of physics, notably in thermodynamics and field theory; its geometric content for dynamical systems is discussed in Section 5.2.

5.1.3 UNIFIED COORDINATES ON T*$\mathbb{Q}$ AND POISSON BRACKETS

THE ξ NOTATION

One of the main advantages of the Hamiltonian formalism is that the q and p variables are treated on an equal footing. In order to do this consistently we unify them into a set of $2n$ variables that will be called the ξ^j with the index j running from 1 to $2n$ (Latin indices run from 1 to $2n$ and Greek indices continue to run from 1 to n). The first n of the ξ^j are the q^β, and the second n are the p_β. More explicitly,

$$\xi^j = q^j, \quad j \in \{1, \ldots, n\},$$
$$\xi^j = p_{j-n}, \quad j \in \{n+1, \ldots, 2n\}.$$

If we now want to write Hamilton's canonical equations in the unified form $\dot{\xi}^j = f^j(\xi)$, the f^j functions must be found from the right-hand side of (5.5): the first n of the f^j are the $\partial H/\partial p_j \equiv \partial H/\partial \xi^{j+n}$, and the last n are $-\partial H/\partial q^j \equiv -\partial H/\partial \xi^j$, so the canonical equations are

$$
\left.
\begin{aligned}
\dot\xi^k &= \frac{\partial H}{\partial \xi^{k+n}}, \qquad k = 1, \ldots, n, \\
\dot\xi^k &= -\frac{\partial H}{\partial \xi^{k-n}}, \qquad k = n+1, \ldots, 2n.
\end{aligned}
\right\}
\tag{5.41}
$$

This can be written in one of the two equivalent forms

$$
\dot\xi^j = \omega^{jk} \partial_k H
\tag{5.42a}
$$

or

$$
\omega_{lj}\dot\xi^j = \partial_l H.
\tag{5.42b}
$$

Here $\partial_k \equiv \partial/\partial\xi^k$.

Equation (5.42a) introduces the $2n \times 2n$ *symplectic matrix* Ω, with matrix elements ω^{jk} given by

$$
\Omega = \begin{vmatrix} 0_n & \mathbb{I}_n \\ -\mathbb{I}_n & 0_n \end{vmatrix},
\tag{5.43}
$$

where $\mathbb{I}_n$ and 0_n are the $n \times n$ unit and null matrices, respectively. Two important properties of Ω are that $\Omega^{-1} = -\Omega$ (we use lower indices for the matrix elements of Ω^{-1}, writing ω_{jl}) and that it is antisymmetric. In other words,

$$
\Omega^2 = -\mathbb{I}, \quad \Omega^{\mathrm{T}} = -\Omega,
\tag{5.44}
$$

or, in terms of the matrix elements,

$$
\left.
\begin{aligned}
\omega^{jk}\omega_{kl} = \delta_l^j, \quad \omega_{ki}\omega_{lj}\delta^{il} = -\delta_{kj}, \quad \omega^{ki}\omega^{lj}\delta_{il} = -\delta^{kj}, \\
\omega_{ij} = -\omega_{ji}, \quad \omega^{ij} = -\omega^{ji}.
\end{aligned}
\right\}
\tag{5.45}
$$

The points of $\mathbf{T}^*\mathbb{Q}$ are labeled ξ, or equivalently (q, p), the way the points of $\mathbf{T}\mathbb{Q}$ are labeled $(q, \dot q)$.

We now argue that $\mathbf{T}^*\mathbb{Q}$ is a new and different carrier manifold from $\mathbf{T}\mathbb{Q}$. It is clearly a carrier manifold, for Eqs. (5.42a, b), Hamilton's canonical equations *in the ξ notation*, are first-order differential equations, so the trajectories are separated on $\mathbf{T}^*\mathbb{Q}$, and, as was illustrated in Worked Example 5.1, the $\mathbf{T}^*\mathbb{Q}$ trajectories project down properly to $\mathbb{Q}$. But is $\mathbf{T}^*\mathbb{Q}$ really different, not just $\mathbf{T}\mathbb{Q}$ dressed up in new coordinates? All that has been done, after all, is to replace the $\dot q^\alpha$ by p_α, and since either one of these two sets can be written unambiguously in terms of the other, this looks simply like a coordinate change; it doesn't seem very profound. We will discuss this in more detail in Section 5.2, but in the meantime we will treat $\mathbf{T}^*\mathbb{Q}$ as a manifold in its own right, different from $\mathbf{T}\mathbb{Q}$. Indeed, the

Legendre transform of Section 5.1.2 leads one to suspect that there actually is a nontrivial difference between L and $\dot{q}$ on the one hand and H and p on the other. They may contain the same information, but they do so in very different ways: to reconstruct a function from its derivatives is not the same as to have an explicit expression for it.

Equations (5.42a) are of the form of (2.83) and (3.89). They define a vector field on $\mathbf{T}^*\mathbb{Q}$ whose components are the $2n$ functions $\omega^{jk}\partial_k H$. The solutions $\xi^j(t)$ are the integral curves of the vector field. We will call this the *dynamical* vector field or the *dynamics* Δ, just as we did on $\mathbf{T}\mathbb{Q}$. As Lagrange's equations establish a vector field on $\mathbf{T}\mathbb{Q}$, the canonical equations establish a vector field on $\mathbf{T}^*\mathbb{Q}$.

Recall that in Chapter 2 dynamical variables were defined as functions of the form $f(q,\dot{q},t) \in \mathcal{F}(\mathbf{T}\mathbb{Q})$. They can also be defined as functions of the form $\hat{f}(q,p,t) \in \mathcal{F}(\mathbf{T}^*\mathbb{Q})$. If $f \in \mathcal{F}(\mathbf{T}\mathbb{Q})$ and $\hat{f} \in \mathcal{F}(\mathbf{T}^*\mathbb{Q})$ are related by $\hat{f}(q,p,t) \equiv f(q,\dot{q}(q,p,t),t)$, then $\hat{f}$ and f represent the same physical object.

Consider a dynamical variable $f(q,p,t)$ (we omit the circumflex). Just as it was possible, without solving the equations of motion, to find how a dynamical variable on $\mathbf{T}\mathbb{Q}$ varies along the motion, so it is possible to do so on $\mathbf{T}^*\mathbb{Q}$. Indeed (recall that the time derivative is also the Lie derivative along the dynamical vector field),

$$\mathbf{L}_\Delta f \equiv \frac{df}{dt} = (\partial_j f)\dot{\xi}^j + \partial_t f = (\partial_j f)\omega^{jk}\partial_k H + \partial_t f, \qquad (5.46)$$

where $\partial_t = \partial/\partial t$.

VARIATIONAL DERIVATION OF HAMILTON'S EQUATIONS

As a demonstration of the usefulness of the ξ notation, we show that Hamilton's canonical equations can be derived from a variational principle on $\mathbf{T}^*\mathbb{Q}$. It is shown in Problem 7 that Hamilton's canonical equations are the Euler–Lagrange equations for the Lagrangian $\hat{L}(q,\dot{q},p,\dot{p},t) \equiv \hat{L}(\xi,\dot{\xi},t)$, a function on the tangent bundle of $\mathbf{T}^*\mathbb{Q}$ defined in an inversion of the Legendre map by

$$\hat{L}(q,\dot{q},p,\dot{p},t) = \dot{q}^\alpha p_\alpha - H(q,p,t);$$

$\hat{L}$ is in fact $L(q,\dot{q},t)$ rewritten in terms of the ξ^k and $\dot{\xi}^k$. Since they are EL equations, the canonical equations should be derivable by a variational procedure on $\mathbf{T}^*\mathbb{Q}$ very similar to that used for the EL equations on $\mathbb{Q}$. We describe it only briefly.

The action is again defined by integrating the Lagrangian:

$$S = \int \hat{L}(\xi,\dot{\xi},t)\,dt,$$

and S is calculated now for various paths in $\mathbf{T}^*\mathbb{Q}$, rather than in $\mathbb{Q}$. That is, the two $\mathbb{Q}$ planes of Fig. 3.1 are replaced by $\mathbf{T}^*\mathbb{Q}$, and the beginning and end points $q(t_0)$ and $q(t_1)$ are replaced by $\xi(t_0)$ and $\xi(t_1)$. All this can be treated as merely a notational change, and the result will be the same EL equations, but with L replaced by $\hat{L}$ and q^α and $\dot{q}^\alpha$ replaced by ξ^k and $\dot{\xi}^k$:

$$\frac{d}{dt}\frac{\partial \hat{L}}{\partial \dot{\xi}^k} - \frac{\partial \hat{L}}{\partial \xi^k} = 0.$$

It is shown in Problem 7 that these are Hamilton's canonical equations.

The variation on $\mathbf{T}^*\mathbb{Q}$ differs, however, from that on $\mathbb{Q}$: what are held fixed at t_0 and t_1 are points in $\mathbf{T}^*\mathbb{Q}$, not in $\mathbb{Q}$. In $\mathbb{Q}$, when points q^α are held fixed, the $\dot{q}^\alpha$ are allowed to vary. In $\mathbf{T}^*\mathbb{Q}$, when points ξ^k (i.e., both the q^α and the p_α) are held fixed, the $\dot{\xi}^k$ (i.e., both the $\dot{q}^\alpha$ and $\dot{p}_\alpha$) are allowed to vary. There seems to be a contradiction. The $\dot{q}^\alpha$ are related to the p_α by the Legendre transform, so if the p_α are fixed, the $\dot{q}^\alpha$ must also be fixed; how then can they be allowed to vary? The answer is that the variational procedure is performed in $\mathbf{T}^*\mathbb{Q}$ without regard to its cotangent-bundle structure and its relation to $\mathbf{T}\mathbb{Q}$.

POISSON BRACKETS

The term involving ω^{jk} on the right-hand side of (5.46) is important; it is called the *Poisson bracket* of f with H. In general if $f, g \in \mathcal{F}(\mathbf{T}^*\mathbb{Q})$, the Poisson bracket of f with g is defined as

$$\{f, g\} \equiv (\partial_j f)\omega^{jk}(\partial_k g) \equiv \frac{\partial f}{\partial \xi^j}\omega^{jk}\frac{\partial g}{\partial \xi^k}$$

$$\equiv \frac{\partial f}{\partial q^\alpha}\frac{\partial g}{\partial p_\alpha} - \frac{\partial f}{\partial p_\alpha}\frac{\partial g}{\partial q^\alpha}. \tag{5.47}$$

The Poisson bracket (PB) has the properties of brackets that were mentioned in Section 3.4.1 in connection with the Lie bracket: it is bilinear and antisymmetric and satisfies the Jacobi identity (see Problem 20). In addition it satisfies a Leibnitz type of product rule:

$$\{f, gh\} = g\{f, h\} + \{f, g\}h \tag{5.48}$$

(the way to remember this is to think of $\{f, \bullet\}$ as a kind of derivative operator, where $\bullet$ marks the place for the function on which the operator is acting).

> **REMARK:** Bilinearity, antisymmetry, and the Jacobi identity are the defining properties of an important kind of algebraic structure called a *Lie algebra*. The function space $\mathcal{F}(\mathbf{T}^*\mathbb{Q})$ is therefore a *Lie algebra* under the PB, and Hamiltonian dynamics can be fruitfully studied from the point of view of Lie algebras. There is a lot of literature on this, too much to enumerate in detail (e.g., Arnol'd, 1988; Sudarshan & Mukunda, 1974; Marmo et al., 1987; Saletan & Cromer, 1971). The Lie algebraic point of view plays an important role also in the transition to quantum mechanics. □

In terms of the PB Eq. (5.46) becomes

$$\mathbf{L}_\Delta f \equiv \frac{df}{dt} = \{f, H\} + \partial_t f. \tag{5.49}$$

Equation (5.49) is satisfied by every dynamical variable, including the coordinates themselves. Applied to the local coordinates, it becomes

$$\dot{\xi}^j = \{\xi^j, H\}, \tag{5.50}$$

which is just another way of writing Hamilton's canonical equations. Another application of Eq. (5.49) is to the Hamiltonian. Antisymmetry of the PBs implies that $\{H, H\} = 0$, and then

$$\dot{H} = \{H, H\} + \partial_t H = \partial_t H, \tag{5.51}$$

that is, the total time derivative of the Hamiltonian is equal to its partial time derivative. This reflects the conservation of energy for time-independent Lagrangians, for it can be shown from the original definition of the Hamiltonian that $\partial_t H = -\partial_t L$.

The PBs of functions taken with the coordinates will play a special role in much of what follows, so we present them here. If $f \in \mathcal{F}(\mathbf{T}^*\mathbb{Q})$, then

$$\left.\begin{array}{c} \{\xi^j, f\} = \omega^{jk}\partial_k f, \text{ or} \\ \{q^\alpha, f\} = \partial f/\partial p_\alpha, \quad \{p_\alpha, f\} = -\partial f/\partial q^\alpha; \\ \{\xi^j, \xi^k\} = \omega^{jk}, \text{ or} \\ \{q^\alpha, p_\beta\} = -\{p_\beta, q^\alpha\} = \delta^\alpha_\beta, \quad \{q^\alpha, q^\beta\} = \{p_\alpha, p_\beta\} = 0. \end{array}\right\} \tag{5.52}$$

WORKED EXAMPLE 5.3

(a) Find the Poisson brackets of the components j_α of the angular momentum $\mathbf{j}$ with dynamical variables (we use $\mathbf{j}$ rather than $\mathbf{L}$ to avoid confusion with the Lagrangian and the Lie derivative). **(b)** Find the PBs of the j_α with the position and momentum in Euclidean 3-space. **(c)** The only way to form scalars from the momentum and position is by dot products. Find the PBs of the j_α with scalars. **(d)** The only way to form vectors is by multiplying $\mathbf{x}$ and $\mathbf{p}$ by scalars and by taking cross products. Find the PBs of the j_α with vectors. Specialize to the PBs of the j_α among themselves.

Solution. The configuration space $\mathbb{Q}$ is the vector space $\mathbb{E}^3$ with coordinates $(x^1, x^2, x^3) \equiv \mathbf{x}$. The angular momentum $\mathbf{j}$ of a single particle is defined as $\mathbf{j} = \mathbf{x} \wedge \mathbf{p}$. Here $\mathbf{p}$ is the momentum, which lies in another $\mathbb{E}^3$. The angular momentum $\mathbf{j}$, in a third $\mathbb{E}^3$, has components

$$j_\alpha = \epsilon_{\alpha\beta\gamma} x^\beta p_\gamma. \tag{5.53}$$

(As we do often when using Cartesian coordinates, we violate conventions about upper and lower indices.)

(a) From the second line of Eq. (5.52) we have

$$\{j_\alpha, f\} = \epsilon_{\alpha\beta\gamma}\{x^\beta p_\gamma, f\} = \epsilon_{\alpha\beta\gamma} x^\beta \{p_\gamma, f\} + \epsilon_{\alpha\beta\gamma}\{x^\beta, f\} p_\gamma$$

$$= \epsilon_{\alpha\beta\gamma}[-x^\beta \partial f/\partial x^\gamma + p_\gamma \partial f/\partial p_\beta].$$

(b) Use the above result with $f \equiv x^\lambda$ and $f = p_\lambda$ to get

$$\{j_\alpha, x^\lambda\} = -\epsilon_{\alpha\beta\gamma}x^\beta\delta_\gamma^\lambda = \epsilon_{\alpha\lambda\beta}x^\beta;$$

$$\{j_\alpha, p_\lambda\} = \epsilon_{\alpha\beta\gamma}p_\gamma\delta_\lambda^\beta = \epsilon_{\alpha\lambda\gamma}p_\gamma.$$

(c) Use the Leibnitz rule, Eq. (5.48), to obtain

$$\{j_\alpha, x^\lambda x^\lambda\} = 2x^\lambda\epsilon_{\alpha\lambda\beta}x^\beta = 0;$$

$$\{j_\alpha, p_\lambda p_\lambda\} = 2p_\lambda\epsilon_{\alpha\lambda\gamma}p_\gamma = 0;$$

$$\{j_\alpha, x^\lambda p_\lambda\} = \epsilon_{\alpha\lambda\gamma}x^\lambda p_\gamma + \epsilon_{\alpha\lambda\beta}x^\beta p_\lambda = 0.$$

(d) The only vectors are products of scalar functions with $\mathbf{x}$, $\mathbf{p}$, and $\mathbf{j} = \mathbf{x} \wedge \mathbf{p}$, that is, they are of the form

$$\mathbf{w} = f\,\mathbf{x} + g\,\mathbf{p} + h\,\mathbf{j},$$

where f, g, and h are scalars. By the Leibnitz rule and Part (a)

$$\{j_\alpha, fx^\lambda\} = f\epsilon_{\alpha\lambda\beta}x^\beta;$$

$$\{j_\alpha, gp_\lambda\} = g\epsilon_{\alpha\lambda\gamma}p_\gamma.$$

All that need be calculated are the PBs of the components of $\mathbf{j}$ with each other:

$$\{j_\alpha, j_\beta\} = \epsilon_{\alpha\mu\nu}\{x^\mu p_\nu, j_\beta\} = \epsilon_{\alpha\mu\nu}x^\mu\{p_\nu, j_\beta\} + \epsilon_{\alpha\mu\nu}\{x^\mu, j_\beta\}p_\nu$$

$$= \epsilon_{\alpha\mu\nu}\epsilon_{\nu\beta\gamma}x^\mu p_\gamma + \epsilon_{\alpha\mu\nu}\epsilon_{\mu\beta\gamma}x^\gamma p_\nu$$

$$= (\epsilon_{\alpha\mu\nu}\epsilon_{\nu\beta\gamma} + \epsilon_{\alpha\nu\gamma}\epsilon_{\nu\beta\mu})x^\mu p_\gamma$$

$$= [(\delta_{\alpha\beta}\delta_{\mu\gamma} - \delta_{\alpha\gamma}\delta_{\mu\beta}) + (\delta_{\gamma\beta}\delta_{\mu\alpha} - \delta_{\gamma\mu}\delta_{\alpha\beta})]x^\mu p_\gamma$$

$$= (\delta_{\gamma\beta}\delta_{\alpha\mu} - \delta_{\alpha\gamma}\delta_{\mu\beta})x^\mu p_\gamma = \epsilon_{\alpha\beta\nu}\epsilon_{\nu\mu\gamma}x^\mu p_\gamma,$$

or

$$\{j_\alpha, j_\beta\} = \epsilon_{\alpha\beta\nu}j_\nu. \tag{5.54}$$

Because h is a scalar, the Leibnitz rule implies that $\{j_\alpha, hj_\beta\} = h\epsilon_{\alpha\beta\nu}j_\nu$. Hence for any vector $\mathbf{w}$,

$$\{j_\alpha, w_\beta\} = \epsilon_{\alpha\beta\nu}w_\nu.$$

Summarizing, we have

$$\{j_1, w_1\} = \{j_2, w_2\} = \{j_3, w_3\} = 0,$$

$$\{j_1, w_2\} = w_3, \quad \{j_2, w_3\} = w_1, \quad \text{and cyclic permutations.}$$

In particular,

$$\{j_1, j_2\} = j_3 \quad \text{and cyclic permutations,}$$
$$\{j_1, j_1\} = \{j_2, j_2\} = \{j_3, j_3\} = 0.$$

The Cartesian components of the angular momentum do not *commute* with each other (the term is taken from matrix algebra). We will see later that this has important consequences.

An example will illustrate how the Poisson brackets can be used. Consider the isotropic harmonic oscillator in n degrees of freedom (for simplicity let both the mass $m = 1$ and the spring constant $k = 1$). The Lagrangian is $L = \frac{1}{2}\delta_{\alpha\beta}(\dot{q}^\alpha \dot{q}^\beta - q^\alpha q^\beta)$, which leads to

$$H = \frac{1}{2}\delta_{jk}\xi^j \xi^k.$$

Then from (5.50) with bilinearity and Leibnitz it follows that

$$\dot{\xi}^i = \frac{1}{2}\delta_{jk}\{\xi^i, \xi^j \xi^k\} = \frac{1}{2}\delta_{jk}(\xi^j \omega^{ik} + \omega^{ij}\xi^k) = \delta_{jk}\omega^{ij}\xi^k. \tag{5.55}$$

One way to solve these equations is to take their time derivatives and use the property that $\Omega^2 = -\mathbb{I}$, obtaining $\ddot{\xi}^k = -\xi^k$. These $2n$ second-order equations give rise to $4n$ constants of the motion that are connected by the first-order equations (5.55), so only $2n$ of them are independent. A different, perhaps more interesting, way to solve these equations is to write them in the form

$$\dot{\boldsymbol{\xi}} = \Lambda\boldsymbol{\xi}, \tag{5.56}$$

where $\boldsymbol{\xi}$ is the $2n$-dimensional vector with components ξ^k and Λ is the matrix with elements $\lambda_k^i = \delta_{jk}\omega^{ij}$. The solution of Eq. (5.56) is

$$\boldsymbol{\xi}(t) = e^{\Lambda t}\boldsymbol{\xi}_0, \tag{5.57}$$

where $\boldsymbol{\xi}_0$ is a constant vector whose components are the initial conditions, and

$$e^{\Lambda t} \equiv \sum_0^\infty \frac{(\Lambda t)^n}{n!}. \tag{5.58}$$

Calculating $e^{\Lambda t}$ is simplified by the fact that $\Lambda^2 = -\mathbb{I}$; indeed,

$$(\Lambda^2)_l^i \equiv \lambda_k^i \lambda_l^k = \omega^{ij}\delta_{jk}\omega^{kr}\delta_{rl} = -\delta^{ir}\delta_{rl} = -\delta_l^i.$$

This can be used to write down all powers of Λ:

$$\Lambda^3 = -\Lambda, \quad \Lambda^4 = \mathbb{I}, \quad \Lambda^5 = \Lambda, \quad \Lambda^6 = -\mathbb{I}, \ldots.$$

Then expanding the sum in (5.58) leads to

$$e^{\Lambda t} = \mathbb{I}\left(1 - \frac{t^2}{2!} + \frac{t^4}{4!} + \cdots\right) + \Lambda\left(t - \frac{t^3}{3!} + \frac{t^5}{5!} + \cdots\right)$$
$$= \mathbb{I}\cos t + \Lambda \sin t$$

(compare the similar expansion for e^{it}). The solution of the equations of motion is then

$$\boldsymbol{\xi}(t) = \boldsymbol{\xi}_0 \cos t + \Lambda\boldsymbol{\xi}_0 \sin t \qquad (5.59a)$$

or

$$\xi^k(t) = \xi_0^k \cos t + \lambda_j^k \, \xi_0^j \sin t. \qquad (5.59b)$$

Equations (5.59a, b) are just a different way of writing (5.57). Because only first-order equations were used, only the $2n$ constants of the motion ξ_0^k appear in these solutions. Finally, to complete the example, we write the (q, p) form of the solution:

$$q^\alpha(t) = q_0^\alpha \cos t + p_{0\alpha} \sin t,$$
$$p_\alpha(t) = -q_0^\alpha \sin t + p_{0\alpha} \cos t.$$

POISSON BRACKETS AND HAMILTONIAN DYNAMICS

The Poisson brackets play a special role when the motion derives from Hamilton's canonical equations. Not every conceivable motion on $\mathbf{T}^*\mathbb{Q}$ is a *Hamiltonian dynamical system*, defined in the canonical way by (5.42a) or (5.50). Any equations of the form

$$\dot{\xi}^j = X^j(\xi), \qquad (5.60)$$

with the $X^j \in \mathcal{F}(\mathbf{T}^*\mathbb{Q})$, define a dynamical system on $\mathbf{T}^*\mathbb{Q}$ (the X^j are the components of a dynamical vector field). If there is no $H \in \mathcal{F}(\mathbf{T}^*\mathbb{Q})$ such that $X^j = \omega^{jk}\partial_k H$ it may be impossible to put the equations of motion in the canonical form of (5.42a). Consider, for example, the system in two freedoms given by

$$\dot{q} = qp, \quad \dot{p} = -qp. \qquad (5.61)$$

If this were a Hamiltonian system, $\dot{q} = qp$ would equal $\partial H/\partial p$, and $\dot{p} = -qp$ would equal $-\partial H/\partial q$. But then $\partial^2 H/\partial q\, \partial p$ would not equal $\partial^2 H/\partial p\, \partial q$; thus there is no function H that is the Hamiltonian for this system. Nevertheless, this is a legitimate dynamical system whose integral curves are easily found:

$$q(t) = q_0 \frac{Ce^{Ct}}{p_0 + q_0 e^{Ct}}, \quad p(t) = p_0 \frac{C}{p_0 + q_0 e^{Ct}},$$

where $C = q_0 + p_0 = q + p$ is a constant of the motion.

One special role of the Poisson bracket is that it provides a test of whether or not a dynamical system is Hamiltonian. We now show that a dynamical system is Hamiltonian (or that the X^j are the components of a *Hamiltonian vector field*) if and only if the time derivative acts on PBs as though they were products (i.e., by the Leibnitz rule). That is, the system is Hamiltonian iff

$$\frac{d}{dt}\{f, g\} = \{\dot{f}, g\} + \{f, \dot{g}\}.$$ (5.62)

For the proof, assume that the dynamical system is Hamiltonian, and let $H(\xi, t)$ be the Hamiltonian function. Then

$$\frac{d}{dt}\{f, g\} = \{\{f, g\}, H\} + \partial_t\{f, g\}.$$

According to the Jacobi identity (and antisymmetry) the first term can be written

$$\{\{f, g\}, H\} = \{\{f, H\}, g\} + \{f, \{g, H\}\}.$$

The second term is

$$\begin{aligned}
\partial_t\{f, g\} &= \partial_t[(\partial_j f)\omega^{jk}\partial_k g] \\
&= (\partial_j \partial_t f)\omega^{jk}\partial_k g + (\partial_j f)\omega^{jk}\partial_k \partial_t g \\
&= \{\partial_t f, g\} + \{f, \partial_t g\}.
\end{aligned}$$

Combining the two results yields

$$\begin{aligned}
\frac{d}{dt}\{f, g\} &= \{\{f, H\} + \partial_t f, g\} + \{f, \{g, H\} + \partial_t g\} \\
&= \{\dot{f}, g\} + \{f, \dot{g}\}.
\end{aligned}$$

This proves that if a Hamiltonian function exists, the Leibnitz rule works for the Poisson bracket.

We now prove the converse by the following logic. If the Leibnitz rule Eq. (5.62) is satisfied for all $f, g \in \mathcal{F}(\mathbf{T^*Q})$ it is satisfied in particular by the pair ξ^l, ξ^i. The proof then shows that if (5.62) is satisfied for this pair, there exists a Hamiltonian function H such that $\omega_{jk}\dot{\xi}^k = \partial_j H$, or the existence of the Hamiltonian function follows from the Leibnitz rule.

Since $\{\xi^l, \xi^i\} = \omega^{li}$, its time derivative is zero. Hence

$$\begin{aligned}
\frac{d}{dt}\{\xi^l, \xi^i\} = 0 &= \{\dot{\xi}^l, \xi^i\} + \{\xi^l, \dot{\xi}^i\} = \{X^l, \xi^i\} + \{\xi^l, X^i\} \\
&= (\partial_j X^l)\omega^{jk}\partial_k \xi^i + (\partial_j \xi^l)\omega^{jk}\partial_k X^i \\
&= \partial_j(X^l \omega^{ji}) + \partial_k(\omega^{lk} X^i),
\end{aligned}$$

where we have used Eq. (5.60). Now multiply by $\omega_{lp}\omega_{ir}$ and sum over repeated indices to obtain

$$\partial_p Z_r - \partial_r Z_p = 0,$$

where $Z_j = \omega_{jk}X^k$. This is the local integrability condition (see Problem 2.4) for the existence of a function H satisfying the set of partial differential equations

$$\partial_j H = Z_j \equiv \omega_{jk}X^k \equiv \omega_{jk}\dot{\xi}^k.$$

This completes the proof. We have proven the result only locally, but that is the best that can be done. If the Poisson bracket behaves as a product with respect to time differentiation, there exists a *local* Hamiltonian function H, that is, the dynamical system is *locally* Hamiltonian. There may not exist a single H that holds for all of $\mathbf{T}^*\mathbb{Q}$.

We illustrate the theorem on the example of Eq. (5.61). The time derivative of $\{q, p\} \equiv 1$ is of course zero, yet

$$\{\dot{q}, p\} + \{q, \dot{p}\} = \{qp, p\} - \{q, qp\} = \{q, p\}p - \{q, p\}q$$
$$= p - q \neq 0.$$

This section has introduced the Hamiltonian formalism and the Poisson brackets on $\mathbf{T}^*\mathbb{Q}$, as well as the unified treatment of coordinates and momenta. The next section concentrates on the structure of $\mathbf{T}^*\mathbb{Q}$ and sets the stage for transformation theory, which takes advantage of the unified treatment.

5.2   SYMPLECTIC GEOMETRY

In this section we discuss the geometry of $\mathbf{T}^*\mathbb{Q}$ and the way in which this geometry contributes to its properties as a carrier manifold for the dynamics.

5.2.1   THE COTANGENT MANIFOLD

We first demonstrate the difference between $\mathbf{T}\mathbb{Q}$ and $\mathbf{T}^*\mathbb{Q}$. The tangent bundle $\mathbf{T}\mathbb{Q}$ consists of the configuration manifold $\mathbb{Q}$ and the set of tangent spaces $\mathbf{T}_q\mathbb{Q}$, each attached to a point $q \in \mathbb{Q}$. The points of $\mathbf{T}\mathbb{Q}$ are of the form $(q, \dot{q})$, where $q \in \mathbb{Q}$ and $\dot{q}$ is a vector in $\mathbf{T}_q\mathbb{Q}$. But the points (q, p) of $\mathbf{T}^*\mathbb{Q}$ are not of this form, for p is not a vector in $\mathbf{T}_q\mathbb{Q}$. We show this by turning to the one-form $\theta_L \equiv (\partial L/\partial \dot{q}^\alpha)\, dq^\alpha = p_\alpha\, dq^\alpha$ of Eq. (3.86) and comparing it to the vector field $\dot{q}^\alpha(\partial/\partial q^\alpha)$.

The $\dot{q}^\alpha$ are the local components of the vector field $\dot{q}^\alpha(\partial/\partial q^\alpha)$ on $\mathbb{Q}$: for given functions $\dot{q}^\alpha \in \mathcal{F}(\mathbb{Q})$ they specify the vector with components $\dot{q}^\alpha(q)$ in the tangent space $\mathbf{T}_q\mathbb{Q}$ at each $q \in \mathbf{T}\mathbb{Q}$. The p_α, however, are the local components of the one-form $\theta_L = p_\alpha\, dq^\alpha$. Although θ_L was introduced as a one-form on $\mathbf{T}\mathbb{Q}$, it can be viewed also as a one-form on $\mathbb{Q}$, for it has no $d\dot{q}$ part (when viewed this way, its components nevertheless depend on the $\dot{q}^\alpha$ as parameters). A one-form is not a vector field, and hence the $p_\alpha = \partial L/\partial \dot{q}^\alpha$, components of a one-form, are not the components of a vector field. In Section 3.4.1 we said that one-forms are *dual* to vectors fields, so θ_L lies in a space that is dual to $\mathbf{T}_q\mathbb{Q}$. This new space is denoted $\mathbf{T}_q^*\mathbb{Q}$ and called the *cotangent* space *at* $q \in \mathbb{Q}$. Its elements, the one-forms, map the vectors (i.e., combine with them in a kind of inner product) to functions. For example, according to Eq. (3.75) the one-form $\theta_L \equiv p_\alpha \, dq^\alpha$ combines with the vector $\dot{q}^\beta \partial/\partial q^\beta$ to form (we use Dirac notation)

$$\langle \theta_L, \dot{q}^\beta \partial/\partial q^\beta \rangle = p_\alpha \dot{q}^\beta \langle dq^\alpha, \partial/\partial q^\beta \rangle = p_\alpha \dot{q}^\beta \delta_\beta^\alpha = p_\alpha \dot{q}^\alpha.$$

As the tangent bundle $\mathbf{TQ}$ is formed of $\mathbb{Q}$ and its tangent spaces $\mathbf{T}_q\mathbb{Q}$, so the *cotangent bundle*, or *cotangent manifold*, is formed of $\mathbb{Q}$ and its cotangent spaces $\mathbf{T}_q^*\mathbb{Q}$. As a vector field consists of a set of vectors, one in each fiber $\mathbf{T}_q\mathbb{Q}$ of $\mathbf{TQ}$, so a one-form consists of a set of *covectors* (i.e., one-forms), one in each fiber $\mathbf{T}_q^*\mathbb{Q}$ of $\mathbf{T}^*\mathbb{Q}$. Therefore the Hamiltonian formalism, whose dynamical points are labeled (q, p), takes place not on the tangent bundle $\mathbf{TQ}$, but on the cotangent bundle $\mathbf{T}^*\mathbb{Q}$. Often $\mathbf{T}^*\mathbb{Q}$ is also called *phase space* or the *phase manifold*.

Hamilton's canonical equations (5.42a or b) are thus a set of differential equations on $\mathbf{T}^*\mathbb{Q}$. They set the $\dot{\xi}^j$ equal to the components of the dynamical vector field on $\mathbf{T}^*\mathbb{Q}$ (not on $\mathbb{Q}$!): locally

$$\Delta_{\mathrm{H}} = \dot{\xi}^j \partial_j, \tag{5.63}$$

or, in terms of (q, p),

$$\Delta_{\mathrm{H}} = \dot{q}^\alpha \frac{\partial}{\partial q^\alpha} + \dot{p}_\alpha \frac{\partial}{\partial p_\alpha} \tag{5.64}$$

(where we use H for Hamiltonian to distinguish Δ_{H} from the Δ_{L}, Lagrangian). Explicit expressions for the $\dot{\xi}^j$ are provided by the canonical equations (5.42a or b). The Legendre transform carries the Lagrangian to the Hamiltonian formalism: it maps $\mathbf{TQ}$ to $\mathbf{T}^*\mathbb{Q}$, sending $\mathbf{T}_q\mathbb{Q}$ at each point $q \in \mathbb{Q}$ into $\mathbf{T}_q^*\mathbb{Q}$ by mapping the vector with components $\dot{q}^\alpha$ in $\mathbf{T}_q\mathbb{Q}$ to the covector with components $p_\alpha = \partial L/\partial \dot{q}^\alpha$ in $\mathbf{T}_q^*\mathbb{Q}$. Because this covector depends on L, the Legendre transform depends on L. It sends every geometric object on $\mathbf{TQ}$ to a similar geometric object on $\mathbf{T}^*\mathbb{Q}$, in particular Δ_{L} to Δ_{H}.

Most importantly, it sends $\theta_L = (\partial L/\partial \dot{q}^\alpha) \, dq^\alpha$ (now viewed as a one-form on $\mathbf{TQ}$) to the one-form

$$\theta_0 \equiv p_\alpha \, dq^\alpha \tag{5.65a}$$

on $\mathbf{T}^*\mathbb{Q}$. The components $\partial L/\partial \dot{q}^\alpha$ of θ_L are functions of $(q, \dot{q})$ and, like the Legendre transform, they depend on the Lagrangian L. But the components of θ_0 are always the same functions of (q, p) (indeed, trivially, they are the p_α themselves). The Legendre transform and θ_L depend on L in such a way that θ_L is always sent to the same *canonical one-form* θ_0 on $\mathbf{T}^*\mathbb{Q}$.

5.2.2 TWO-FORMS

We return to the equations of motion (5.42b). The $\dot{\xi}^j$ on the left-hand side are the components of a vector field in $\mathcal{X}(\mathbf{T}^*\mathbb{Q})$ and the $\partial_j H$ on the other side are the components of the one-form dH in $\mathcal{U}(\mathbf{T}^*\mathbb{Q})$ ($\mathcal{X}$ and $\mathcal{U}$ are spaces of vector fields and one-forms, respectively). Before going on, we should explain how one-forms can be associated to vectors. The ω_{jk}, which associate the vector with the one-form, are the elements of a geometric object ω called a *two-form*. We now describe two-forms and show explicitly what it means to say that the ω_{jk} are the elements of a two-form.

Recall that in Chapter 3 a one-form α on $\mathbf{T}\mathbb{Q}$ was defined as a linear map from vector fields X to functions, that is, $\alpha : \mathcal{X} \to \mathcal{F} : X \mapsto \langle \alpha, X \rangle$. (This definition is valid for any differential manifold, for $\mathbf{T}^*\mathbb{Q}$ as well as $\mathbf{T}\mathbb{Q}$.) *Two-forms* are defined as bilinear, antisymmetric maps of *pairs* of vector fields to functions. That is, if ω is a two-form on $\mathbf{T}^*\mathbb{Q}$ and X and Y are vector fields on $\mathbf{T}^*\mathbb{Q}$, then

$$\omega(X, Y) = -\omega(Y, X) \in \mathcal{F}(\mathbf{T}^*\mathbb{Q}); \tag{5.66}$$

mathematically, $\omega : \mathcal{X} \times \mathcal{X} \to \mathcal{F}$. Since the map is bilinear, it can be represented locally by a matrix whose elements are (recall that $\partial_j \equiv \partial/\partial\xi^j$ is a vector field)

$$\omega_{jk} = \omega(\partial_j, \partial_k) = -\omega_{kj} = -\omega(\partial_k, \partial_j), \tag{5.67}$$

so that by linearity if (locally) $X = X^j\partial_j$ and $Y = Y^j\partial_j$, then

$$\omega(X, Y) = \omega_{jk}X^j Y^k. \tag{5.68}$$

What happens if a two-form is applied to only one vector field $Y^k\partial_k$, that is, what kind of object is $\omega_{jk}Y^k$? It is not a function, for its dangling subscript j implies that it has n components. A function can, however, be constructed from it and from a second vector field $X^j\partial_j$ by multiplying the components and summing over j, thus obtaining $(\omega_{jk}Y^k)X^j$, which is the right-hand side of Eq. (5.68). That means that $\omega_{jk}Y^k$ is the jth component of a one-form, for it can be used to map any vector field $X^j\partial_j$ to a function. In an obvious notation, this one-form can be called $\omega(\bullet, Y) = -\omega(Y, \bullet)$ (the $\bullet$ shows where to put the other vector field): when the one-form $\omega(\bullet, Y)$ is applied to a vector field $X = X^j\partial_j$ the result is $\omega(X, Y)$. It is convenient to introduce uniform terminology and notation for the action of both one-forms and two-forms on vector fields. Vector fields are said to be *contracted with* or *inserted into* the forms, denoted by i_X:

$$i_X\alpha \equiv \alpha(X) \equiv \langle \alpha, X \rangle \quad \text{and} \quad i_X\omega \equiv \omega(\bullet, X), \tag{5.69}$$

where X is a vector field, α is a one-form, and ω is a two-form. Then $i_X\alpha$ and $i_Y i_X\omega \equiv \omega(X, Y) = -i_X i_Y\omega$ are functions, and $i_X\omega$ is a one-form.

5.2.3   THE SYMPLECTIC FORM ω

We now proceed to see how this relates to the canonical equations. Since the ω_{jk} of Eq. (5.42b) are antisymmetric in j and k, they are the components of a two-form (which we will call ω). The left-hand side of (5.42b) is the local coordinate expression for the lth component of the one-form $i_\Delta \omega$ (we drop the subscript H on Δ, for we'll be discussing only $\mathbf{T}^*\mathbb{Q}$), and the right-hand side is the lth component of the one-form dH. The canonical equations equate these two one-forms:

$$i_\Delta \omega = dH. \tag{5.70}$$

This is the geometric form of Hamilton's canonical equations, no longer in local coordinates.

Writing this in the (q, p) notation requires an explicit expression for ω. That requires an explanation of the two ways that two-forms can be created from one-forms. The first is by combining two one-forms in the *exterior* or *wedge product*. If α and β are one-forms, their exterior product $\alpha \wedge \beta$ ("α wedge β") is defined by its action on an arbitrary vector field:

$$i_X(\alpha \wedge \beta) = (i_X\alpha)\beta - (i_X\beta)\alpha \equiv \langle \alpha, X \rangle \beta - \langle \beta, X \rangle \alpha. \tag{5.71}$$

Because $i_X\alpha$ and $i_X\beta$ are functions, $i_X(\alpha \wedge \beta)$ is a one-form. Then

$$i_Y i_X(\alpha \wedge \beta) \equiv i_Y[i_X(\alpha \wedge \beta)] = -i_X i_Y(\alpha \wedge \beta)$$
$$= \langle \alpha, X \rangle \langle \beta, Y \rangle - \langle \beta, X \rangle \langle \alpha, Y \rangle. \tag{5.72}$$

This shows that $\alpha \wedge \beta = -\beta \wedge \alpha$ maps pairs of vector fields bilinearly and antisymmetrically to functions and hence it is a two-form.

The second way to create two-forms from one-forms uses the wedge product and extends the definition of the *exterior derivative* d, which was used to construct one-forms from functions (e.g., df from f). Let $\alpha = f\,dg$ be a one-form, where $f, g \in \mathcal{F}(\mathbf{T}^*\mathbb{Q})$; then the two-form $d\alpha$ is defined as

$$d\alpha \equiv df \wedge dg. \tag{5.73}$$

In particular, if $\alpha = dg$ (that is, if $f = 1$), then $d\alpha = 0$, or $d^2 = 0$ (this reflects the fact that $\partial_j\partial_k g = \partial_k\partial_j g$). More generally (in local coordinates) if $\alpha = \alpha_k d\xi^k$ with $\alpha_k \in \mathcal{F}(\mathbf{T}^*\mathbb{Q})$, then

$$d\alpha = d\alpha_k \wedge d\xi^k = (\partial_j\alpha_k)\,d\xi^j \wedge d\xi^k. \tag{5.74}$$

Just as the $d\xi^j$ make up a local basis for one-forms, the $d\xi^j \wedge d\xi^k$ make up a local basis for two-forms, so that locally every two-form can be written as $\omega_{jk}d\xi^j \wedge d\xi^k$ with $\omega_{jk} = -\omega_{kj}$ [only the antisymmetric part of $\partial_j\alpha_k$ enters in Eq. (5.74)]. Globally $d\alpha$ is defined by its contraction with pairs of vector fields:

$$i_X i_Y d\alpha = i_X d(i_Y\alpha) - i_Y d(i_X\alpha) - i_{[X,Y]}\alpha.$$

Further application of the exterior product and derivative can be used to obtain p-forms with $p > 2$. They will be defined when needed.

The properties of two-forms can be used to obtain an explicit expression for the ω of Eq. (5.70) in the (q, p) notation. In that notation, the general two-form ω may be written (the factor $\frac{1}{2}$ is added for convenience; recall that Greek indices run from 1 to n)

$$\omega = \omega_\alpha^\beta dq^\alpha \wedge dp_\beta + \frac{1}{2}(a_{\alpha\beta}dq^\alpha \wedge dq^\beta + b^{\alpha\beta}dp_\alpha \wedge dp_\beta),$$

with $a_{\alpha\beta} = -a_{\beta\alpha}$ and $b^{\alpha\beta} = -b^{\beta\alpha}$. When this ω is contracted with the Δ of Eq. (5.64), the result must be dH. Contraction yields

$$i_\Delta \omega = \dot{q}^\mu \left[\omega_\alpha^\beta \delta_\mu^\alpha \, dp_\beta + \frac{1}{2} \left(a_{\alpha\beta} \left\{ \delta_\mu^\alpha \, dq^\beta - \delta_\mu^\beta \, dq^\alpha \right\} \right) \right]$$

$$+ \dot{p}_\mu \left[-\omega_\alpha^\beta \delta_\beta^\mu \, dq^\alpha + \frac{1}{2} \left(b^{\alpha\beta} \left\{ \delta_\alpha^\mu \, dp_\beta - \delta_\beta^\mu \, dp_\alpha \right\} \right) \right]$$

$$= \dot{q}^\mu \left[\omega_\mu^\beta \, dp_\beta + a_{\mu\beta} \, dq^\beta \right] - \dot{p}_\mu \left[\omega_\alpha^\mu \, dq^\alpha - b^{\mu\beta} \, dp_\beta \right]$$

$$= dH = -\dot{p}_\mu \, dq^\mu + \dot{q}^\mu \, dp_\mu$$

[the last equality comes from Eq. (5.5)]. Equating coefficients of $\dot{p}_\mu$ and $\dot{q}^\mu$ shows that $a_{\mu\beta} = b^{\mu\beta} = 0$ and $\omega_\alpha^\mu = \delta_\alpha^\mu$ or that

$$\omega = dq^\alpha \wedge dp_\alpha. \tag{5.75}$$

An important consequence of (5.75) is that $d\omega = 0$ (we state this without explicit proof, for $d\omega$ is a three-form); ω is then called *closed*. Note also that

$$\omega = -d\theta_0,$$

so that $d\omega = -d^2\theta_0 \equiv 0$ (again, $d^2 = 0$). Another important property of ω is that it is *nondegenerate* (the matrix of the ω_β^μ is nonsingular): $i_X \omega = 0$ iff X is the null vector field. This has physical consequences, because it implies that each Hamiltonian *uniquely* determines its dynamical vector field Δ, for if there were two fields Δ_1 and Δ_2 such that $i_{\Delta_1} \omega = i_{\Delta_2} \omega = dH$ for one H, then it would follow that

$$i_{\Delta_1} \omega - i_{\Delta_2} \omega \equiv i_{(\Delta_1 - \Delta_2)} \omega = 0.$$

Then nondegeneracy implies that $\Delta_1 = \Delta_2$. The converse is not quite true: Δ determines H only up to an additive constant: if $i_\Delta \omega = dH_1$ and $i_\Delta \omega = dH_2$, then $i_\Delta \omega - i_\Delta \omega \equiv 0 = dH_1 - dH_2 = d(H_1 - H_2)$ or $H_1 - H_2 = $ const. The physical consequence of this is that the dynamics on $\mathbf{T}^*\mathbb{Q}$ determines the Hamiltonian function only up to an additive constant, reflecting the indeterminacy of the potential energy.

A two-form which, like ω, is closed and nondegenerate is called a *symplectic form*. Recall that the canonical one-form θ_0 appears naturally in $\mathbf{T}^*\mathbb{Q}$ as a result of the Legendre transform, and hence so does $\omega \equiv -d\theta_0$. For this reason it is said that the cotangent bundle is *endowed with a natural symplectic form* (or *structure*), and $\mathbf{T}^*\mathbb{Q}$ is called a *symplectic manifold*. As will be seen, the symplectic form furnishes the cotangent bundle with a rich geometric structure.

A geometric property of the symplectic form has to do with areas in $\mathbf{T}^*\mathbb{Q}$. To see this, consider two vector fields $X = X^\alpha(\partial/\partial q^\alpha) + X_\alpha(\partial/\partial p_\alpha)$ and $Y = Y^\alpha(\partial/\partial q^\alpha) + Y_\alpha(\partial/\partial p_\alpha)$. Then

$$\omega(X, Y) = i_Y i_X \, dq^\alpha \wedge dp_\alpha = (i_X \, dq^\alpha)(i_Y \, dp_\alpha) - (i_Y \, dq^\alpha)(i_X \, dp_\alpha)$$

$$= X^\alpha Y_\alpha - Y^\alpha X_\alpha. \tag{5.76}$$

At a fixed point $\xi = (q, p) \in \mathbf{T}^*\mathbb{Q}$ these vector fields define vectors in the tangent space $\mathbf{T}_\xi(\mathbf{T}^*\mathbb{Q}) \equiv \mathbf{T}_\xi M$ (in order to avoid too many $\mathbf{T}$s, we will write $\mathbf{T}^*\mathbb{Q} = M$ for a while). If the number of freedoms $n = 1$, so that $\mathbf{T}_\xi M$ is a two-dimensional plane, Eq. (5.76) looks like the area of the parallelogram formed by the two vectors X and Y in that plane. See Fig. 5.3. In more than one freedom, (5.76) is the sum of such areas in each of the (q, p) planes of $\mathbf{T}_\xi M$, sort of a generalized area subtended by the two vector fields at that point. Thus ω provides a means of measuring areas. This will be discussed more fully in Section 5.4 in connection with the Liouville volume theorem.

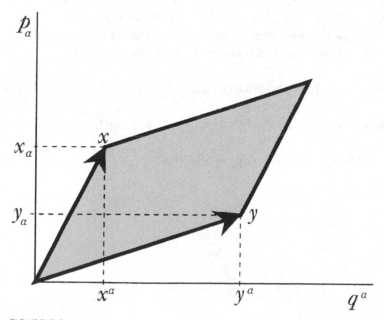

FIGURE 5.3

The (q^α, p_α) tangent plane at a point $\xi \in M$. Two vector fields X and Y project to the vectors x and y in the (q^α, p_α) plane. The components of the projected vectors are x_α and y_α on the p_α axis, and x^α and y^α on the q^α axis. The shaded parallelogram is the area subtended by x and y. The value that the function $\omega(X, Y) \in \mathcal{F}(M)$ takes on at any point $\xi \in M$ is the sum of such areas formed by the projections of X and Y into all (q, p) planes of the tangent space at ξ.

The elements ω_{jk} of the symplectic form and the ω^{jk} of the Poisson bracket, both obtained from Hamilton's canonical equations, are the matrix elements of Ω and Ω^{-1}, respectively. The symplectic form ω is therefore directly related to the Poisson bracket, and it is possible to write the relation in a coordinate-free way by using the fact that two-forms map vector fields to one-forms. Recall that H uniquely determines Δ through the one-form dH and the equation $i_\Delta \omega = dH$. In the same way any other dynamical variable f uniquely determines a vector field X_f through the one-form df and the equation

$$i_{X_f} \omega = df. \tag{5.77}$$

(Note, incidentally, that although Δ does not uniquely determine H, it does uniquely determine dH: because it is nondegenerate, ω provides a one-to-one association of vector fields to one-forms.) A vector field $X_f \in \mathcal{X}(\mathbf{T}^*\mathbb{Q})$ that is associated through Eq. (5.77) with some function in $\mathcal{F}(\mathbf{T}^*\mathbb{Q})$ is called a *Hamiltonian vector field* (X_f is *Hamiltonian with respect to* f). Not all vector fields are Hamiltonian; for example, the dynamical vector field of Eq. (5.61) is not.

WORKED EXAMPLE 5.4

Half of the components of a certain Hamiltonian vector field $X = X^\alpha (\partial/\partial q^\alpha) + X_\alpha (\partial/\partial p_\alpha)$ are $X^\alpha = \delta^{\alpha\beta} p_\beta$. Find the function f with respect to which X is Hamiltonian and the most general expression for X. Write out the differential equations for the integral curves of X. Comment on the result.

Solution. From $\omega = dq^\alpha \wedge dp_\alpha$ it follows that

$$i_X \omega = X^\alpha \, dp_\alpha - X_\alpha \, dq^\alpha = \delta^{\alpha\beta} p_\beta \, dp_\alpha - X_\alpha \, dq^\alpha \equiv df,$$

and then $df = (\partial f/\partial q^\alpha)\, dq^\alpha + (\partial f/\partial p_\alpha)\, dp_\alpha$ implies that

$$\delta^{\alpha\beta} p_\beta = \frac{\partial f}{\partial p_\alpha}$$

or that

$$f(q, p) = \frac{1}{2} \delta^{\alpha\beta} p_\alpha p_\beta + F(q),$$

with $F(q)$ arbitrary. Then $X_\alpha \equiv -\partial f/\partial q^\alpha = -\partial F/\partial q^\alpha$, and

$$X = \delta^{\alpha\beta} p_\beta \frac{\partial}{\partial q^\alpha} - \frac{\partial F}{\partial q^\alpha} \frac{\partial}{\partial p_\alpha}.$$

The integral curves (use t for the parameter) are given by

$$\frac{dq^\alpha}{dt} = \delta^{\alpha\beta} p_\beta,$$
$$\frac{dp_\alpha}{dt} = -\frac{\partial F}{\partial q^\alpha}.$$

These are the equations for a particle (of mass $m = 1$) in the potential F. The Hamiltonian f is the sum of the kinetic and potential energies.

We now use these considerations to establish the relation between ω and the Poisson bracket. Consider two dynamical variables $f, g \in \mathcal{F}(\mathbf{T}^*\mathbb{Q})$ and their Hamiltonian vector fields X_f and X_g. If one thinks of g as the Hamiltonian function, then X_g is its dynamical vector field, and according to (5.46) the time derivative of f along the motion is

$$\mathbf{L}_{X_g} f = \{f, g\} = i_{X_g} df = i_{X_g} i_{X_f} \omega = \omega(X_f, X_g). \tag{5.78}$$

This is the general result: the connection between the Poisson bracket and the symplectic form is given by

$$\{f, g\} = \omega(X_f, X_g) = -\omega(X_g, X_f). \tag{5.79}$$

Equation (5.79) is the intrinsic, coordinate-free definition of the Poisson bracket. It can then be shown that the Jacobi identity is a consequence of the closure of ω (i.e., of $d\omega = 0$).

Equations (5.78) and (5.79) are quite important. It will be seen in the next section that (5.78) is the basis for the Hamiltonian version of the Noether theorem.

5.3 CANONICAL TRANSFORMATIONS

In the Lagrangian formalism we considered transformations only on $\mathbb{Q}$. One of the advantages of the Hamiltonian formalism is that it allows transformations on $\mathbf{T}^*\mathbb{Q}$, which mix the qs and ps yet preserve the Hamiltonian nature of the equations of motion. Such *canonical transformations* are the subject of this section.

5.3.1 LOCAL CONSIDERATIONS

REDUCTION ON T*Q BY CONSTANTS OF THE MOTION

If one of the generalized coordinates q^α of a dynamical system is ignorable (or cyclic), it is absent not only from the Lagrangian L, but also from the Hamiltonian H. The equation for p_α is then $\dot{p}_\alpha = 0$, so p_α is a constant of the motion. We've seen this before: if q^α does not appear in L, the conjugate momentum is conserved. In Chapter 2 it was seen that this reduces the dimension of $\mathbf{T}\mathbb{Q}$ by one, somewhat simplifying the problem. A similar reduction takes place in the Hamiltonian formalism, but with a difference.

We demonstrate this difference on the central-force Hamiltonian of (5.13), which we repeat here:

$$H = p_\alpha \dot{q}^\alpha - L = \frac{p_r^2}{2m} + \frac{p_\theta^2}{2mr^2} + \frac{p_\phi^2}{2mr^2 \sin^2 \theta} + V(r). \tag{5.80}$$

Because

$$\dot{p}_\phi = -\frac{\partial H}{\partial \phi} = 0,$$

p_ϕ is a constant of the motion (it will be seen shortly that p_ϕ is the z component of the angular momentum $\mathbf{r} \wedge \mathbf{p}$). Because p_ϕ is conserved, the motion takes place on five-dimensional invariant submanifolds of the six-dimensional $\mathbf{T}^*\mathbb{Q}$ whose equations are $p_\phi = \mu$, where μ is any constant. On each invariant submanifold the Hamiltonian becomes

$$H = \frac{p_r^2}{2m} + \frac{p_\theta^2}{2mr^2} + \frac{\mu^2}{2mr^2 \sin^2 \theta} + V(r), \tag{5.81}$$

which is a perfectly good Hamiltonian in only two freedoms, namely r and θ. This means that the motion now takes place on a $\mathbf{T}^*\mathbb{Q}$ of four dimensions, not five: one constant of the motion has reduced the dimension by two! (This does not mean, incidentally, that ϕ is also a constant of the motion; its time variation is obtained from one of Eqs. (5.14) after $r(t)$ and $\theta(t)$ are found.) This is a general phenomenon, one further advantage of the Hamiltonian formalism. A constant of the motion reduces the dimension of $\mathbf{T}^*\mathbb{Q}$ by two. We will return to this later, in Section 5.4.2, on the Darboux theorem.

Every ignorable coordinate simplifies the problem. The best imaginable situation, in which the problem would all but solve itself, would be if *all* of the generalized coordinates were ignorable and the Hamiltonian were a function of the p_β alone. But that seems too much to wish for. If it were to happen, all of the momenta would be constant and the equations for the q^α would be of the extremely simple form $\dot{q}^\alpha = \text{const}$. It actually turns out, however, that in a certain sense this can often be arranged through a coordinate change on $\mathbf{T}^*\mathbb{Q}$, and the next chapter will show how.

DEFINITION OF CANONICAL TRANSFORMATIONS

The kind of coordinate change now to be considered, of which the one just mentioned is a special case, mixes the q^α and p_β in a way that preserves the Hamiltonian structure of the dynamical system. That is, the transformation

$$Q^\alpha(q, p, t), \quad P_\alpha(q, p, t) \tag{5.82}$$

to new variables (Q^α, P_α) has the property that if the equations of motion for (q, p) are Hamilton's canonical equations with some Hamiltonian $H(q, p, t)$, then there exists a new Hamiltonian $K(Q, P, t) \in \mathcal{F}(\mathbf{T}^*\mathbb{Q})$ such that

$$\dot{Q}^\alpha = \frac{\partial K}{\partial P_\alpha}, \quad \dot{P}_\alpha = -\frac{\partial K}{\partial Q^\alpha}; \tag{5.83}$$

in other words, K serves as a Hamiltonian function for the new variables. If such transformations exist and if a particular one can be found for which K depends only on the P_α, it will yield what we called the best imaginable situation.

Transformations (5.82) on $\mathbf{T}^*\mathbb{Q}$ that preserve the Hamiltonian formalism are called *canonical transformations*. They are very different from the point transformations of the Lagrangian formalism, because (5.82) loses track of the cotangent-bundle nature of $\mathbf{T}^*\mathbb{Q}$: the new generalized coordinates Q^α are no longer functions only of the old generalized coordinates q^α, so they are not coordinates on $\mathbb{Q}$. Canonical transformations are general in that they mix the q^α and p_β, but they are special in that they preserve the Hamiltonian formalism.

We start by redefining canonical transformations more carefully and in the ξ notation. For a while we will be working in local coordinates, so we'll restrict our considerations only to local transformations; later the definitions will be extended to global ones. Consider a dynamical system on $\mathbf{T}^*\mathbb{Q}$ with a local Hamiltonian function $H(\xi, t)$, so that Hamilton's canonical equations read $\omega_{jk}\dot{\xi}^k = \partial H/\partial \xi^j$. Let

$$\eta^k(\xi, t) \tag{5.84}$$

be an invertible local-coordinate transformation on $\mathbf{T}^*\mathbb{Q}$. If there exists a function $K(\eta, t)$ such that

$$\omega_{jk}\dot{\eta}^k = \frac{\partial K}{\partial \eta^j}, \tag{5.85}$$

the transformation Eq. (5.84) will be called locally *canonoid with respect to H*. It transforms the Hamiltonian dynamical system generated by H in the ξ variables into a Hamiltonian system generated by K in the η variables.

Here is an example in one freedom. If $H = \frac{1}{2}p^2$ (the free particle), the transformation

$$Q = q, \quad P = \sqrt{p} - \sqrt{q} \tag{5.86}$$

is canonoid with respect to H on that part of $\mathbf{T}^*\mathbb{Q} = \mathbb{R}^2$ where it is invertible (the new Hamiltonian is $K = \frac{1}{3}[P + \sqrt{Q}]^3$). The transformation of (5.86) is not canonoid, however, with respect to $H = \frac{1}{2}(p^2 + q^2)$ (the harmonic oscillator) or many other Hamiltonians (see Problem 12).

Some transformations are locally canonoid with respect to *all* Hamiltonians: they comprise the important class of locally *canonical transformations* (CTs).

To demonstrate that canonical transformations exist, we give an example. The transformation $\eta^k = \delta^{kj}\omega_{jl}\xi^l$ [in (q, p) notation, $Q = -p$, $P = q$] is canonical. To prove its *canonicity* (i.e., that it is canonical), we calculate:

$$\omega_{rk}\dot{\eta}^k = \omega_{rk}\delta^{kj}\omega_{jl}\dot{\xi}^l = \omega_{rk}\delta^{kj}\frac{\partial H}{\partial \xi^j}$$

$$= \omega_{rk}\delta^{kj}\frac{\partial H}{\partial \eta^i}\frac{\partial \eta^i}{\partial \xi^j} = \omega_{rk}\delta^{kj}\delta^{is}\omega_{sj}\frac{\partial H}{\partial \eta^i}$$

$$= \delta_{rs}\delta^{is}\frac{\partial H}{\partial \eta^i} = \frac{\partial H}{\partial \eta^r} \tag{5.87}$$

[using the defining properties of the ω_{jk} at Eq. (5.45).] Thus H is a Hamiltonian for both the old ξ and new η variables; the expression $\partial H/\partial\eta^r$ shows that H must be written in terms of the η^k or that $K(\eta) = H(\xi(\eta))$. What makes this transformation canonical is that it works for *any* Hamiltonian function H.

CHANGES INDUCED BY CANONICAL TRANSFORMATIONS

An important property of CTs (and a useful one, for it can be used to check canonicity) is that they preserve the Poisson brackets (PBs). Proving this result requires taking PBs with respect to both the new and the old variables, which will be indicated by superscripts:

$$\{f, g\}^\xi \equiv \frac{\partial f}{\partial\xi^k}\omega^{kj}\frac{\partial g}{\partial\xi^j}, \quad \{f, g\}^\eta \equiv \frac{\partial f}{\partial\eta^k}\omega^{kj}\frac{\partial g}{\partial\eta^j}.$$

The claim is that a transformation $\eta^\alpha(\xi, t)$ is canonical iff for every pair $f, g \in \mathcal{F}(\mathbf{T}^*\mathbb{Q})$ there exists a nonzero constant a such that

$$\{f, g\}^\eta = a\{f, g\}^\xi. \tag{5.88}$$

Here we prove only that if (5.88) is satisfied, the existence of a Hamiltonian H for the ξ variables implies the existence of a Hamiltonian K for the η variables. The converse is proven in the appendix to this chapter.

According to (5.88),

$$\frac{d}{dt}\{f, g\}^\eta = a\frac{d}{dt}\{f, g\}^\xi = a\{\dot{f}, g\}^\xi + a\{f, \dot{g}\}^\xi. \tag{5.89}$$

The second equality comes from Eq. (5.62) and the fact that the ξ system is Hamiltonian. Now apply Eq. (5.88) to the right-hand side of (5.89) to obtain

$$\frac{d}{dt}\{f, g\}^\eta = \{\dot{f}, g\}^\eta + \{f, \dot{g}\}^\eta.$$

According to the argument at (5.62) this means that the η system is Hamiltonian, or there exists a function $K(\eta, t)$ such that $\omega_{jk}\dot{\eta}^k = \partial K/\partial\eta^j$.

As explained in the appendix, we always take $a = 1$, so for us a CT is a transformation that preserves the PBs with $a = 1$. It follows from the appendix that preservation of the fundamental PBs is sufficient:

$$\{\xi^j, \xi^k\}^\eta = \{\xi^j, \xi^k\}^\xi, \quad \text{or equivalently,} \quad \{\eta^j, \eta^k\}^\xi = \{\eta^j, \eta^k\}^\eta. \tag{5.90}$$

Hence a fast way to check canonicity of a transformation is to calculate the PBs of the new variables. For instance in the example of Eq. (5.86) [take PBs with respect to the (q, p) variables],

$$\{Q, P\} = \{q, \sqrt{p} - \sqrt{q}\} = \{q, \sqrt{p}\} = \frac{1}{2\sqrt{p}} \neq 1;$$

$\{Q, Q\} = \{P, P\} = 0$ by antisymmetry. Because $\{Q, P\}$ is not equal to 1, the transformation is not canonical. In contrast, the example of Eq. (5.87) (in one freedom) yields

$$\{Q, P\} = \{-p, q\} = \{q, p\} = 1,$$

which proves its canonicity.

Preservation of the Poisson brackets shows that the inverse of a CT is also a CT and that the result of following one CT by another is also a CT. This is enough to establish that the CTs form a group. More about this is discussed in Section 5.3.4.

If the CT is time independent (i.e., if the $\eta^k(\xi)$ do not depend explicitly on t), invariance of the PBs can be used to find the new Hamiltonian K. Indeed,

$$\dot{\eta}^k = \{\eta^k, H\}^\xi = \{\eta^k, H\}^\eta = \omega^{kj} \frac{\partial H}{\partial \eta^j}.$$

The partial derivative with respect to η^j means that H must be written as a function of the η variables. Then $\dot{\eta}^k = \omega^{kj} \partial K / \partial \eta^j$ implies that (up to a constant)

$$K(\eta, t) = H(\xi(\eta), t): \tag{5.91}$$

the new Hamiltonian is the old one written in terms of the new coordinates. If the $\eta^k(\xi, t)$ depend explicitly on t, the situation is more complicated; we turn to that in Section 5.3.3. Note, incidentally, that the result depends not on the transformation $\eta^k(\xi)$, but on its inverse $\xi^k(\eta)$.

Under any (invertible) transformation $\eta^k(\xi)$ on the manifold $\mathbf{T}^*\mathbb{Q}$, whether canonical or not, the expressions for vector fields and differential forms also change. For example, if V is a vector field, its components X^k and Y^j in the two coordinate systems are defined by the equations

$$V = X^k \frac{\partial}{\partial \xi^k} = Y^j \frac{\partial}{\partial \eta^j}.$$

The Y^j can be found in terms of the X^k by using the relation between the derivatives:

$$X^k \frac{\partial}{\partial \xi^k} = X^k \frac{\partial \eta^j}{\partial \xi^k} \frac{\partial}{\partial \eta^j} = Y^j \frac{\partial}{\partial \eta^j},$$

so

$$Y^j = X^k \frac{\partial \eta^j}{\partial \xi^k}. \tag{5.92}$$

Usually the Y^j are needed as functions of the η variables, so the inverse transformation must be used to write the X^k and the $\partial \eta^j / \partial \xi^k$ in terms of the η^l.

Similarly, the formula $d\xi^k = (\partial \xi^k / \partial \eta^j) d\eta^j$ can be used to find the elements of a differential form in the η coordinates in terms of its elements in the ξ coordinates. For example if a two-form $\alpha = \rho_{jk} d\xi^j \wedge d\xi^k = \sigma_{il} d\eta^i \wedge d\eta^l$, then

$$\sigma_{il} = \rho_{jk} \frac{\partial \xi^j}{\partial \eta^i} \frac{\partial \xi^k}{\partial \eta^l}. \tag{5.93}$$

Equation (5.93) can be used to find the elements of the symplectic form ω in one coordinate system from its elements in the other. It turns out that the new elements are the same as the old ones. This invariance of ω under CTs can be proven by using the invariance of the PBs, for these two are intimately connected. To do so, consider the expression

$$\omega_{jk} \frac{\partial \xi^j}{\partial \eta^i} \frac{\partial \xi^k}{\partial \eta^l} \equiv \lfloor \eta^i, \eta^l \rfloor^\xi,\qquad(5.94)$$

which is called the *Lagrange bracket* of the ηs with respect to the ξs (in general the Lagrange brackets are defined for $2n$ independent functions η^j of the ξ^k). In the (q, p) notation they are given by

$$\lfloor \eta^i, \eta^l \rfloor^{(q,p)} = \frac{\partial p_\alpha}{\partial \eta^i} \frac{\partial q^\alpha}{\partial \eta^l} - \frac{\partial q^\alpha}{\partial \eta^i} \frac{\partial p_\alpha}{\partial \eta^l}.\qquad(5.95)$$

It is implied by $\omega_{jk}\omega^{kl} = \delta^l_j$ and $(\partial \xi^l/\partial \eta^k)(\partial \eta^k/\partial \xi^j) = \delta^l_j$ that the Lagrange and Poisson brackets form inverse matrices:

$$\{\eta^l, \eta^k\}^\xi \lfloor \eta^k, \eta^j \rfloor^\xi = \delta^l_j.\qquad(5.96)$$

Since the PBs are invariant under CTs, $\{\eta^l, \eta^k\}^\xi = \omega^{lk}$, and since $\lfloor \eta^i, \eta^l \rfloor^\xi$ is the inverse, $\lfloor \eta^i, \eta^l \rfloor^\xi = \omega_{il}$. Inserting this into Eq. (5.94) and comparing with Eq. (5.93) shows that the symplectic form is invariant under canonical transformations:

$$\omega_{il} = \omega_{jk} \frac{\partial \xi^j}{\partial \eta^i} \frac{\partial \xi^k}{\partial \eta^l}.\qquad(5.97)$$

In the (q, p) notation the expression for ω is therefore the same in the new and old coordinates, or

$$\omega = dQ^\alpha \wedge dP_\alpha.\qquad(5.98)$$

The generalized area defined by the symplectic form [see Section 5.2] is also invariant under CTs. This is related to the important *Liouville volume theorem*, which will be discussed in Section 5.4.1.

TWO EXAMPLES

An example of a canonical transformation that is somewhat less simple than the one of Eq. (5.87) is the following one in two freedoms:

$$Q^1 = q^1 \cos\alpha - \frac{p_2}{\beta} \sin\alpha, \quad P_1 = \beta q^2 \sin\alpha + p_1 \cos\alpha,$$

$$Q^2 = q^2 \cos\alpha - \frac{p_1}{\beta} \sin\alpha, \quad P_2 = \beta q^1 \sin\alpha + p_2 \cos\alpha.\qquad(5.99)$$

Canonicity can be checked by taking PBs. Straightforward calculation yields

$$\{Q^1, P_1\}^{(q,p)} = \cos^2\alpha + \sin^2\alpha = 1,$$
$$\{Q^1, Q^2\}^{(q,p)} = -\frac{\cos\alpha\sin\alpha}{\beta} + \frac{\sin\alpha\cos\alpha}{\beta} = 0,$$
$$\{Q^1, P_2\}^{(q,p)} = 0,$$
$$\{P_1, P_2\}^{(q,p)} = -\beta\cos\alpha\sin\alpha + \beta\sin\alpha\cos\alpha = 0.$$

All the other basic PBs vanish by antisymmetry or can be obtained from these by interchanging the indices 1 and 2. Since the PBs of the new variables are the same as those of the old, the transformation is canonical. This transformation is useful in solving dynamical systems that involve a charged particle in a constant magnetic field (Problem 16).

Another example is provided by the transition from Cartesian to polar coordinates. The transformation on $\mathbb{Q}$ (and its inverse) is given by

$$\left.\begin{array}{l} r = \sqrt{x^2 + y^2 + x^2}, \theta = \arccos\dfrac{z}{\sqrt{x^2 + y^2 + z^2}}, \phi = \arctan\dfrac{y}{x}; \\ x = r\sin\theta\cos\phi, \ y = r\sin\theta\sin\phi, \ z = r\cos\theta, \end{array}\right\} \tag{5.100}$$

where θ is the colatitude and ϕ is the azimuth angle. Since (5.100) is not a transformation on $\mathbf{T}^*\mathbb{Q}$ it cannot be canonical, but it can be the first half of a canonical transformation if appropriate equations for the momenta are added. We could find general equations for the momenta by requiring that the PBs remain invariant, but instead we will find them from the equations of Worked Example 5.2. When we have found a full set of transformation equations on $\mathbf{T}^*\mathbb{Q}$ we will discuss the canonicity of the transformation.

From $p_x = m\dot{x}$ it follows that

$$p_x = m\dot{r}\sin\theta\cos\phi + mr\dot{\theta}\cos\theta\cos\phi - mr\dot{\phi}\sin\theta\sin\phi. \tag{5.101}$$

Now insert $\dot{r}, \dot{\theta}, \dot{\phi}$ from Eqs. (5.12) into (5.101), from which we get

$$p_x = p_r\sin\theta\cos\phi + \frac{p_\theta\cos\theta\cos\phi}{r} - \frac{p_\phi\sin\phi}{r\sin\theta}. \tag{5.102}$$

It can be shown similarly that

$$\left.\begin{array}{l} p_y = p_r\sin\theta\sin\phi + \dfrac{p_\theta\cos\theta\sin\phi}{r} + \dfrac{p_\phi\cos\phi}{r\sin\theta}, \\ p_z = p_r\cos\theta - \dfrac{p_\theta\sin\theta}{r}. \end{array}\right\} \tag{5.103}$$

The momenta transform linearly, so to obtain the inverse we need only invert the matrix

$$T = \begin{bmatrix} \sin\theta\cos\phi & \dfrac{\cos\theta\cos\phi}{r} & -\dfrac{\sin\phi}{r\sin\theta} \\ \sin\theta\sin\phi & \dfrac{\cos\theta\sin\phi}{r} & \dfrac{\cos\phi}{r\sin\theta} \\ \cos\theta & -\dfrac{\sin\theta}{r} & 0 \end{bmatrix}.$$

Calculation yields

$$
T^{-1} = \begin{bmatrix} \sin\theta\cos\phi & \sin\theta\sin\phi & \cos\theta \\ r\cos\phi\cos\theta & r\sin\phi\cos\theta & -r\sin\theta \\ -r\sin\theta\sin\phi & r\sin\theta\cos\phi & 0 \end{bmatrix} \equiv \begin{bmatrix} x/r & y/r & z/r \\ xz/\rho & yz/\rho & -\rho \\ -y & x & 0 \end{bmatrix},
$$

where $\rho = \sqrt{x^2 + y^2}$. Thus the inverse is

$$
\begin{bmatrix} p_r \\ p_\theta \\ p_\phi \end{bmatrix} = \begin{bmatrix} \frac{xp_x + yp_y + zp_z}{r} \\ \frac{xzp_x + yzp_y - \rho^2 p_z}{\rho} \\ -yp_x + xp_y \end{bmatrix}. \tag{5.104}
$$

The full set of transformations on $\mathbf{T}^*\mathbb{Q}$ (and their inverses) is given by Eqs. (5.100), (5.102), (5.103), and (5.104). The last entry in (5.104) shows that p_ϕ is the angular momentum about the z axis.

The canonicity of the transformation can be demonstrated by calculating the PBs of the Cartesian coordinates with respect to the polar ones, or vice versa. We will exhibit only three of the PBs, leaving the rest to the reader. Equations (5.100) give the coordinates (as opposed to the momenta) only in terms of other coordinates, so it is relatively easy to calculate brackets like $\{x, W\}^{\text{polar}}$ or $\{r, W\}^{\text{Cartesian}}$ where W is any dynamical variable (in particular, the brackets of the coordinates among themselves vanish trivially). For example,

$$
\begin{aligned}
\{x, p_x\}^{\text{polar}} &\equiv \frac{\partial x}{\partial r}\frac{\partial p_x}{\partial p_r} + \frac{\partial x}{\partial \theta}\frac{\partial p_x}{\partial p_\theta} + \frac{\partial x}{\partial \phi}\frac{\partial p_x}{\partial p_\phi} \\
&= \sin^2\theta\cos^2\phi + \cos^2\theta\cos^2\phi + \sin^2\phi \equiv 1.
\end{aligned}
$$

The PBs of the polar variables with respect to the Cartesian ones do not involve trigonometric functions and are on the whole easier to verify. For example,

$$
\begin{aligned}
\{r, p_r\}^{\text{Cartesian}} &\equiv \frac{\partial r}{\partial x}\frac{\partial p_r}{\partial p_x} + \frac{\partial r}{\partial y}\frac{\partial p_r}{\partial p_y} + \frac{\partial r}{\partial z}\frac{\partial p_r}{\partial p_x} \\
&= \frac{x^2}{r^2} + \frac{y^2}{r^2} + \frac{z^2}{r^2} = 1.
\end{aligned}
$$

The PBs of the momenta are nevertheless complicated because they involve both coordinates and momenta. But p_ϕ does not involve z or p_z, so its PBs have four instead of six terms. For example,

$$
\{p_r, p_\phi\}^{\text{Cartesian}} \equiv \frac{\partial p_r}{\partial x}\frac{\partial p_\phi}{\partial p_x} - \frac{\partial p_r}{\partial p_x}\frac{\partial p_\phi}{\partial x} + \frac{\partial p_r}{\partial y}\frac{\partial p_\phi}{\partial p_y} - \frac{\partial p_r}{\partial p_y}\frac{\partial p_\phi}{\partial y}.
$$

Direct calculation shows that this vanishes. Thus only $\{p_r, p_\theta\}^{\text{Cartesian}}$ remains. We leave the rest to the reader.

5.3.2 INTRINSIC APPROACH

We now restate these matters briefly in the language of the intrinsic, geometric point of view. In that point of view a passive coordinate transformation is replaced by an active invertible mapping Φ of $\mathbf{T}^*\mathbb{Q}$ onto itself, carrying each point $\xi \in \mathbf{T}^*\mathbb{Q}$ to another point $\Phi\xi = \eta \in \mathbf{T}^*\mathbb{Q}$. This is written $\Phi : \mathbf{T}^*\mathbb{Q} \to \mathbf{T}^*\mathbb{Q} : \xi \mapsto \eta$.

Under such a mapping each function $f \in \mathcal{F}(\mathbf{T}^*\mathbb{Q})$ is changed as described in Chapter 3 into a new function that we will call $f' \equiv \Phi_{\mathrm{fn}} f$ (the notation Φ_{fn} is used to distinguish the way the transformation acts on the function from the way it acts on the points of $\mathbf{T}^*\mathbb{Q}$). The function f' takes on the same value at the new point η as f takes at the old point ξ:

$$f'(\eta) \equiv \Phi_{\mathrm{fn}} f(\eta) \equiv \Phi_{\mathrm{fn}} f(\Phi\xi) = f(\xi),$$

or, equivalently,

$$f'(\xi) \equiv \Phi_{\mathrm{fn}} f(\xi) = f(\Phi^{-1}\xi), \tag{5.105}$$

which is an explicit definition of $\Phi_{\mathrm{fn}} f \in \mathcal{F}(\mathbf{T}^*\mathbb{Q})$.

Vector fields and differential forms are also changed, and the criteria used to find how they change are similar to the criterion used for functions. For instance, let $X \in \mathcal{X}(\mathbf{T}^*\mathbb{Q})$ be a vector field. What is required is a new vector field X' that has the same action on the transformed function $\Phi_{\mathrm{fn}} f$ at the new point η as X has on f at ξ. That is, if $f \in \mathcal{F}(\mathbf{T}^*\mathbb{Q})$ and $g = X(f)$, then

$$X'(\Phi_{\mathrm{fn}} f)(\eta) = g(\xi) \equiv X(f)(\xi) \equiv X(f)(\Phi^{-1}\eta) = \Phi_{\mathrm{fn}}[X(f)](\eta).$$

Now write $\Phi_{\mathrm{fn}} f = h$, or $f = \Phi_{\mathrm{fn}}^{-1} h$ (Φ_{fn}^{-1} is obtained from Φ^{-1} the same way that Φ_{fn} is obtained from Φ). Then $X'(h)(\eta) = \Phi_{\mathrm{fn}}[X(\Phi_{\mathrm{fn}}^{-1} h)](\eta)$, or

$$X' = \Phi_{\mathrm{fn}} X \Phi_{\mathrm{fn}}^{-1}. \tag{5.106}$$

Linear operators on vector spaces transform in a similar way under linear transformations.

In local coordinates this leads to the same formula as (5.92): by definition,

$$X'(f)(\eta) = X'^k \frac{\partial f}{\partial \eta^k},$$

where the X'^k and f are all written as functions of the η^j. Moreover,

$$\Phi_{\mathrm{fn}} X\big(\Phi_{\mathrm{fn}}^{-1} f\big)(\eta) = \Phi_{\mathrm{fn}} X f(\xi) = \Phi_{\mathrm{fn}}\left[X^j \frac{\partial f}{\partial \xi^j} \right].$$

The Φ_{fn} in front of the bracket in the last term means that everything must be written as functions of the η^j, so

$$\Phi_{\mathrm{fn}} X\big(\Phi_{\mathrm{fn}}^{-1} f\big)(\eta) = X^j \frac{\partial \eta^k}{\partial \xi^j} \frac{\partial f}{\partial \eta^k}.$$

Equation (5.106) therefore implies that

$$X'^k = X^j \frac{\partial \eta^k}{\partial \xi^j},$$

which agrees with Eq. (5.92).

The change induced by Φ in a one-form α is found similarly: contract α with a vector field to yield a function and use the known ways in which the function and vector field change. It is convenient to denote the change of the vector field by an operator we will call Φ_{vf}:

$$X' \equiv \Phi_{vf} X = \Phi_{fn} X \Phi_{fn}^{-1}. \tag{5.107}$$

Then by a calculation similar to the one that led to Eq. (5.106), the new one-form α' is found to be

$$\alpha' = \Phi_{fn} \alpha \Phi_{vf}^{-1}. \tag{5.108}$$

This means that contracting α' with a vector field X is the same as first applying Φ_{vf}^{-1} to X, then contracting the resulting vector field with α, and then applying Φ_{fn} to the resulting function. A similar result can be obtained for two-forms, but we do not describe it in detail. The main point is that the changes induced on functions, vector fields, and differential forms of any degree can all be calculated in the intrinsic formulation. We will write $\alpha' = \Phi_{form} \alpha$, whether α be a one-form, a two-form, or even one of higher degree.

The intrinsic (active) interpretation of Eq. (5.97) is that the symplectic form ω is not changed under the mapping Φ: the new one is the same as the original one. Thus in the intrinsic formulation, a mapping $\Phi : \mathbf{T}^*\mathbb{Q} \to \mathbf{T}^*\mathbb{Q}$ is canonical iff

$$\Phi_{form} \omega = \omega. \tag{5.109}$$

This will prove important in later sections, when we'll be dealing not with single canonical transformations, but continuous families of them.

5.3.3   GENERATING FUNCTIONS OF CANONICAL TRANSFORMATIONS

GENERATING FUNCTIONS

We now discuss how canonical transformations can be generated and in part how they can be classified. So far specifying a canonical transformation requires giving the $2n$ functions $\eta^k(\xi, t)$. But the fact that the transformation is canonical means that these $2n$ functions are restricted in some way, that there must be a connection between them. What we will show is that to each local CT corresponds a single local function F on $\mathbf{T}^*\mathbb{Q}$, called the *generating function*. However, F does not specify the CT uniquely, for many different sets of the η^k functions correspond to the same F. Nevertheless, with more information about the transformation (what will be called its *type*), the F functions lead to a complete classification of the CTs, and that gives a way of specifying a local CT by naming its type and its generating function.

We first show how to obtain the generating function F in local (q, p) notation. Recall the canonical one-form $\theta_0 = p_\alpha \, dq^\alpha$ of Eq. (5.65a) whose exterior derivative satisfies $d\theta_0 = -\omega \equiv dp_\alpha \wedge dq^\alpha$. Consider also the similar one-form constructed from new coordinates obtained from a canonical transformation: $\theta_1 = P_\alpha d Q^\alpha$. The exterior derivative of θ_1 is also ω, for according to Eq. (5.98) $dP_\alpha \wedge dQ^\alpha = -\omega$, so

$$d(\theta_0 - \theta_1) \equiv d(p_\alpha \, dq^\alpha - P_\alpha \, dQ^\alpha) = 0.$$

This is, incidentally, another test for canonicity.

There is a theorem (called the Poincaré lemma; Schutz, 1980, pp.140–141) that states that if a one-form α is closed (i.e., if $d\alpha = 0$) there exists a *local* function f such that $\alpha = df$. Indeed, $d\alpha = 0$ is just the integrability condition for the existence of such an f, a condition we have invoked several times already [e.g., at Eq. (1.42), Problem 2.4, and the discussion of invertibility of Eq. (5.1) above Worked Example 5.1]. This implies that, given a CT, in each neighborhood $\mathbb{U} \subset \mathbf{T}^*\mathbb{Q}$ there exists a local function $F \in \mathcal{F}(\mathbb{U})$ such that

$$p_\alpha \, dq^\alpha - P_\alpha \, dQ^\alpha = dF. \tag{5.110}$$

Now, F can be written as a function of any set of local coordinates on $\mathbb{U}$, for instance (q, p) or (Q, P), but not in general of combinations of the new and old variables. For example, (q, P) will not do in general, because the q^α may not be independent of the P_β (although later we will consider cases in which they are). When all terms in Eq. (5.110) are written in terms of (q, p) and coefficients of the dq^α and dp_α are equated, it is found that

$$\left.\begin{aligned}
\frac{\partial F}{\partial q^\alpha} &= p_\alpha - \frac{\partial Q^\beta}{\partial q^\alpha} P_\beta, \\
\frac{\partial F}{\partial p_\alpha} &= -\frac{\partial Q^\beta}{\partial p_\alpha} P_\beta.
\end{aligned}\right\} \tag{5.111}$$

If a CT is given, Eq. (5.111) is a set of partial differential equations for $F(q, p, t)$. The integrability conditions for these equations are satisfied, for they are just $d(\theta_0 - \theta_1) = 0$.

We now restate all this in the ξ notation. To write the equations concisely, we define a new matrix Γ by

$$\Gamma = \begin{vmatrix} 0 & -\mathbb{I}_n \\ 0 & 0 \end{vmatrix} \equiv \{\gamma_{jk}\} \tag{5.112}$$

so that $\Omega^{\mathrm{T}} = \Gamma - \Gamma^{\mathrm{T}}$ or $\omega_{jk} = \gamma_{jk} - \gamma_{kj}$. Then $p_\alpha \, dq^\alpha = -\gamma_{kl}\xi^l \, d\xi^k$, and the *generating function* $F(\xi, t)$ of a canonical transformation $\eta^k(\xi, t)$ is defined by the analog of Eq. (5.110):

$$dF = \gamma_{kl}(\eta^l d\eta^k - \xi^l d\xi^k). \tag{5.113}$$

In general the transformation $\eta^k(\xi, t)$ is time dependent: the independent variables are now the ξ^k and t (and hence the ξ^k are t independent). Equations (5.111) now become

$$\frac{\partial F}{\partial \xi^k} = \gamma_{il}\frac{\partial \eta^i}{\partial \xi^k}\eta^l - \gamma_{kl}\xi^l. \tag{5.114}$$

Again, these are differential equations for $F(\xi, t)$. They give all the ξ-derivatives of F in terms of the canonical transformation $\eta^k(\xi, t)$, so the CT determines F up to an additive function of t. The integrability conditions are satisfied locally if and only if the transformation from the ξ^k to the η^k is locally canonical. It will be seen later, however, that F does not determine the CT uniquely.

THE GENERATING FUNCTION GIVES THE NEW HAMILTONIAN

Suppose that a certain dynamical system is specified by giving $H(\xi, t)$ and that a particular CT is used to transform to new variables η^k. How does one find the new Hamiltonian K? This is where the generating function comes in: F and the CT can be used to find K. To show how, we start by writing the $\dot{\eta}^l$ in two ways:

$$\dot{\eta}^l = \omega^{lj} \frac{\partial K}{\partial \eta^j}$$

$$= \frac{\partial \eta^l}{\partial t} + \{\eta^l, H\} \equiv \frac{\partial \eta^l}{\partial t} + \frac{\partial \eta^l}{\partial \xi^i} \omega^{ij} \frac{\partial H}{\partial \xi^j}$$

(the first way comes from canonicity); therefore

$$\frac{\partial \eta^l}{\partial t} = \omega^{lj} \frac{\partial K}{\partial \eta^j} - \frac{\partial \eta^l}{\partial \xi^i} \omega^{ij} \frac{\partial H}{\partial \xi^j}. \tag{5.115}$$

In order to use these equations to find K, we now obtain another equation involving $\partial \eta^l / \partial t$. Since t appears on the right-hand side of (5.114) only in the η^k functions (the ξ^k and t are independent variables), we can write

$$\frac{\partial^2 F}{\partial \xi^k \, \partial t} = \gamma_{il} \left(\frac{\partial^2 \eta^i}{\partial \xi^k \partial t} \eta^l + \frac{\partial \eta^i}{\partial \xi^k} \frac{\partial \eta^l}{\partial t} \right)$$

$$= \frac{\partial}{\partial \xi^k} \left(\gamma_{il} \frac{\partial \eta^i}{\partial t} \eta^l \right) + \gamma_{il} \left(\frac{\partial \eta^i}{\partial \xi^k} \frac{\partial \eta^l}{\partial t} - \frac{\partial \eta^i}{\partial t} \frac{\partial \eta^l}{\partial \xi^k} \right)$$

$$= \frac{\partial \psi}{\partial \xi^k} + (\gamma_{il} - \gamma_{li}) \frac{\partial \eta^i}{\partial \xi^k} \frac{\partial \eta^l}{\partial t}$$

$$= \frac{\partial \psi}{\partial \xi^k} - \frac{\partial \eta^l}{\partial t} \omega_{li} \frac{\partial \eta^i}{\partial \xi^k}, \tag{5.116}$$

where

$$\psi = \gamma_{il} \frac{\partial \eta^i}{\partial t} \eta^l.$$

Now insert (5.115) into the last term on the last line of (5.116) to get

$$\frac{\partial \eta^l}{\partial t} \omega_{li} \frac{\partial \eta^i}{\partial \xi^k} = \frac{\partial \eta^i}{\partial \xi^k} \omega_{il} \frac{\partial \eta^l}{\partial \xi^r} \omega^{rj} \frac{\partial H}{\partial \xi^j} + \omega^{lj} \omega_{li} \frac{\partial \eta^i}{\partial \xi^k} \frac{\partial K}{\partial \xi^j}$$

$$= \lfloor \xi^k, \xi^r \rfloor^\eta \omega^{rj} \frac{\partial H}{\partial \xi^j} - \delta_l^j \frac{\partial \eta^i}{\partial \xi^k} \frac{\partial K}{\partial \xi^j}$$

$$= \frac{\partial}{\partial \xi^k} (H - K).$$

Hence (5.116) becomes

$$\frac{\partial}{\partial \xi^k}\left[\frac{\partial F}{\partial t} + H - K - \psi\right] = 0,$$

which means that the sum in square brackets is a function of t alone. But F is defined only up to an additive $f(t)$, so it can always be chosen to make the sum in square brackets vanish. With such a choice,

$$\frac{\partial F}{\partial t} = \psi - H + K \equiv \gamma_{il}\frac{\partial \eta^i}{\partial t}\eta^l - H + K. \tag{5.117}$$

In (q, p) notation this becomes

$$\frac{\partial F}{\partial t} = -H + K - \frac{\partial Q^\alpha}{\partial t}P_\alpha. \tag{5.118}$$

Equation (5.117) [or (5.118)] gives K as a function of ξ^k in terms of the CT and F (the latter is itself determined by the CT). The inverse transformation must be used to write $K(\eta, t)$. Note that, in agreement with (5.91), $K = H$ if the transformation is time independent.

If Eq. (5.114) is multiplied by $\dot{\xi}^k$, summed over k, and added to $\partial F/\partial t$ from (5.117), the result is

$$\begin{aligned}\frac{dF}{dt} &= \gamma_{il}\frac{\partial \eta^i}{\partial \xi^k}\eta^l\dot{\xi}^k - \gamma_{kl}\dot{\xi}^l\dot{\xi}^k + \gamma_{il}\frac{\partial \eta^i}{\partial t}\eta^l - H + K \\ &= -\gamma_{kl}\dot{\xi}^k\xi^l - H - (-\gamma_{kl}\dot{\eta}^k\eta^l - K) \\ &= \dot{q}^\alpha p_\alpha - H - (\dot{Q}^\alpha P_\alpha - K). \end{aligned} \tag{5.119}$$

This equation is another useful relation between F, the CT, and the two Hamiltonians.

At the beginning of this section we mentioned the classification of canonical transformations. To some extent the generating functions provide such a classification, since each local CT gives rise to a generating function F unique up to an additive function of t. We have already mentioned, however, that each F does not determine a local CT. Indeed, for a given $F(\xi, t)$, Eq. (5.114) is a set of differential equations for the $\eta^k(\xi, t)$, and its solutions are not unique. For example, consider the two CTs on $\mathbf{T}^*\mathbb{Q} = \mathbb{R}^2$ given by

$$Q = p, \quad P = -q \tag{5.120a}$$

and

$$\hat{Q} = \frac{1}{2}p^2, \quad \hat{P} = -q/p. \tag{5.120b}$$

It is easily verified by checking the only nonvanishing PB in each case that these are both CTs. If either one of these is inserted into Eq. (5.111), the result is

$$\frac{\partial F}{\partial q} = p, \quad \frac{\partial F}{\partial p} = q,$$

whose solution is $F = qp$. Thus $F = qp$ is the generating function for at least two CTs. The same situation is found in more than one freedom: the set of CTs belonging to each generating function is quite large. Although up to this point our treatment does not classify the CTs, it does group them into categories, each of which belongs to a particular generating function.

GENERATING FUNCTIONS OF TYPE

There is another way, however, that generating functions can be used to classify canonical transformations. Suppose that a CT has the property that in a neighborhood $\mathbb{U} \subset \mathbf{T}^*\mathbb{Q}$ the q^α and Q^α variables are independent and can thus be used to specify points in $\mathbb{U}$. In Eq. (5.120a), for example, a property of the CT is that $Q = p$, so q and Q can be used instead of q and p to specify a point anywhere in $\mathbf{T}^*\mathbb{Q}$. A CT for which this is true will be called a transformation of Type 1. With such a transformation all functions can be written as a function of (q, Q), and then Eq. (5.110) can be put in the form

$$dF \equiv \frac{\partial F^1}{\partial q^\alpha} dq^\alpha + \frac{\partial F^1}{\partial Q^\alpha} dQ^\alpha = p_\alpha \, dq^\alpha - P_\alpha \, dQ^\alpha,$$

where $F^1(q, Q, t) \equiv F(q, p(q, Q, t), t)$. This gives the p_α and P_α immediately as functions of (q, Q, t):

$$p_\alpha = \frac{\partial F^1}{\partial q^\alpha}, \quad P_\alpha = -\frac{\partial F^1}{\partial Q^\alpha}. \tag{5.121}$$

To find the new Hamiltonian, invert to write $F(q, p, t) = F^1(q, Q(q, p, t), t)$, and then from Eq. (5.118)

$$\frac{\partial F}{\partial t} \equiv K - H - \frac{\partial Q^\alpha}{\partial t} P_\alpha = \frac{\partial F^1}{\partial t} + \frac{\partial F^1}{\partial Q^\alpha}\frac{\partial Q^\alpha}{\partial t}$$

$$= \frac{\partial F^1}{\partial t} - P_\alpha \frac{\partial Q^\alpha}{\partial t},$$

or

$$K = H + \frac{\partial F^1}{\partial t}. \tag{5.122}$$

Note the difference between $\partial F/\partial t$ and $\partial F^1/\partial t$: the former is calculated with (q, p) fixed, whereas the latter is calculated with (q, Q) fixed, which explains the term $(\partial F^1/\partial Q^\alpha)(\partial Q^\alpha/\partial t)$ in the first line.

This yields a classification of the canonical transformations of Type 1. For each generating function of Type 1, Eqs. (5.121) give unique expressions for the p_α and P_α. To obtain explicit equations for the CT itself, the first set of Eqs. (5.121) is inverted to obtain the $Q^\beta(q, p, t)$, and this is inserted into the second set to obtain the $P_\beta(q, p, t)$. But, as usual, invertibility imposes a Hessian condition:

$$\det \left| \frac{\partial^2 F^1}{\partial q^\alpha \, \partial Q^\beta} \right| \neq 0.$$

The procedure works only if the Hessian condition is satisfied, so such invertible functions of $2n + 1$ variables classify the CTs of Type 1. Moreover, Eq. (5.122) gives the new Hamiltonian K in terms of H and F^1.

Not all transformations are of Type 1. We define four types of transformations as follows:

Type 1. (q, Q) are independent.
Type 2. (q, P) are independent.
Type 3. (p, Q) are independent.
Type 4. (p, P) are independent.

For example, the transformation $Q = q$, $P = q + p$ in two freedoms is canonical and of Types 2, 3, and 4 simultaneously, but not of Type 1. The CTs of other types can also be classified by functions of $2n + 1$ variables.

As an example, consider Type 2. Write $f^2(q, P, t) = F(q, p(q, P, t), t)$. Then if everything in Eq. (5.110) is written in terms of (q, P, t) and dF is set equal to df^2, that equation becomes

$$dF = p_\alpha \, dq^\alpha - P_\beta \left(\frac{\partial Q^\beta}{\partial q^\alpha} dq^\alpha + \frac{\partial Q^\beta}{\partial P_\alpha} dP_\alpha \right)$$

$$= df^2 \equiv \frac{\partial f^2}{\partial q^\alpha} dq^\alpha + \frac{\partial f^2}{\partial P_\alpha} dP_\alpha.$$

Equating coefficients of dq^α and dP_α then yields

$$\frac{\partial f^2}{\partial q^\alpha} = p_\alpha - P_\beta \frac{\partial Q^\beta}{\partial q^\alpha} \quad \text{and} \quad \frac{\partial f^2}{\partial P_\alpha} = -P_\beta \frac{\partial Q^\beta}{\partial P_\alpha},$$

or

$$p_\alpha = \frac{\partial F^2}{\partial q^\alpha} \quad \text{and} \quad Q^\alpha = \frac{\partial F^2}{\partial P_\alpha}, \tag{5.123}$$

where

$$F^2(q, P, t) = f^2(q, P, t) + P_\alpha Q^\alpha(q, P, t)$$

$$\equiv F(q, Q(q, P, t), t) + P_\alpha Q^\alpha(q, P, t).$$

To obtain explicit equations for the CT itself, the first set of Eqs. (5.123) is inverted (again, there is an invertibility condition) to obtain the $P_\beta(q, p, t)$, and this is inserted into the second half to obtain an explicit expression for the $Q^\beta(q, p, t)$. Thus the Type 2 CTs, like those of Type 1, can also be classified by functions of $2n + 1$ variables. [Note that F^2 is not simply the generating function F written in the variables (q, P, t).] It is a simple matter to show that, in analogy with Type 1,

$$K = H + \frac{\partial F^2}{\partial t}. \tag{5.124}$$

We leave Types 3 and 4 to the problems. Of the four types, we will have most use for Type 1, in Hamilton–Jacobi theory (Section 6.1), and Type 2, in canonical perturbation theory (Section 6.3), although other types can also be used for both of these purposes. Having achieved a partial classification of the CTs in terms of generating functions of type, we now leave the question of classifying the CTs.

WORKED EXAMPLE 5.5

(a) In one freedom, consider the complex transformation

$$Q = \frac{m\omega q + ip}{\sqrt{2m\omega}}, \quad P = i\frac{m\omega q - ip}{\sqrt{2m\omega}} \equiv iQ^*, \qquad (5.125)$$

where the asterisk denotes the complex conjugate. Show that this transformation is a CT of Type 1, and find its generating function of Type 1. Apply this CT to the harmonic oscillator whose Hamiltonian is

$$H = \frac{1}{2}m\omega^2 q^2 + \frac{p^2}{2m}.$$

That is, find the new Hamiltonian, solve for the motion in terms of (Q, P), and transform back to (q, p).

(b) Do the same for the complex transformation

$$Q_t = \frac{m\omega q + ip}{\sqrt{2m\omega}}e^{i\omega t}, \quad P_t = i\frac{m\omega q - ip}{\sqrt{2m\omega}}e^{-i\omega t} \equiv iQ_t^* \qquad (5.126)$$

[the subscript t is used to avoid confusion with Eq. (5.125) and to emphasize the time dependence].

Solution. (a) Canonicity can be checked by calculating that $\{Q, P\} = 1$. (It is of some importance in quantum mechanics that $\{Q^*, Q\} = i$.) From the dependence of Q on both q and p it follows that Q and q are independent, so the CT is of Type 1 (by the same argument it is also of Types 2, 3, and 4). Equations (5.121) can be used to find the Type 1 generating function $F^1(q, Q)$: the equations are

$$p \equiv i(m\omega q - \sqrt{2m\omega}Q) = \frac{\partial F^1}{\partial q},$$

$$P \equiv i(\sqrt{2m\omega}q - Q) = -\frac{\partial F^1}{\partial Q}.$$

The solution (up to an additive constant) is

$$F^1(q, Q) = i\left(\frac{Q^2}{2} - \sqrt{2m\omega}qQ + \frac{m\omega q^2}{2}\right). \qquad (5.127)$$

Now consider the harmonic oscillator Hamiltonian. Because the CT of Eq. (5.125) is time independent, the new Hamiltonian K is just H written out in the new coordinates:

$$K = H = -i\omega Q P. \tag{5.128}$$

Hamilton's canonical equations are

$$\dot{Q} = -i\omega Q, \quad \dot{P} = i\omega P.$$

The equations for $Q(t)$ and $P(t)$ seem to be uncoupled, but since $P = iQ^*$, the P motion can be found from the Q motion. The solutions are immediate:

$$Q(t) = Q_0 e^{-i\omega t}, \quad P(t) = P_0 e^{i\omega t}, \tag{5.129}$$

where Q_0 and P_0 are complex constants determined by the initial conditions (each first-order equation needs only one initial condition). These are two complex constants, so the initial conditions seem to depend on four real constants. But $P = iQ^*$, so $P_0 = iQ_0^*$; that is, the initial conditions depend only on the two real constants of Q_0: if $Q_0 = (m\omega q_0 + ip_0)/\sqrt{2m\omega}$, then $P_0 = i(m\omega q_0 - ip_0)/\sqrt{2m\omega}$. The (q, p) motion can be obtained by inverting Eq. (5.125) or by taking real and imaginary parts:

$$\left.\begin{aligned}
q(t) &= \sqrt{\frac{2}{m\omega}}\,\Re\{Q(t)\} \equiv \sqrt{\frac{1}{2m\omega}}[Q(t) + Q^*(t)] \\
&= \sqrt{\frac{1}{2m\omega}}\left(Q_0 e^{-i\omega t} + Q_0^* e^{i\omega t}\right) \equiv q_0 \cos\omega t + \frac{p_0}{m\omega}\sin\omega t. \\
p(t) &= \sqrt{2m\omega}\,\Im\{Q(t)\} \equiv -i\sqrt{\frac{m\omega}{2}}[Q(t) - Q^*(t)] \\
&= -i\sqrt{\frac{m\omega}{2}}\left(Q_0 e^{-i\omega t} - Q_0^* e^{i\omega t}\right) \equiv -m\omega q_0 \sin\omega t + p_0 \cos\omega t,
\end{aligned}\right\} \tag{5.130}$$

where $\Re$ means "the real part of" and $\Im$ means "the imaginary part of."

Note: It is impossible to obtain a Lagrangian from the Hamiltonian of Eq. (5.128), because it does not satisfy the Hessian condition (see Problem 5; in this one-freedom case the condition is $\partial^2 H/\partial^2 P^2 \neq 0$). Hence the complex CT of this example makes it impossible to obtain the Lagrangian formalism for the generalized coordinate Q. Nevertheless this transformation is often used in quantum mechanics.

(b) We now turn to the transformation of Eq. (5.126). Again canonicity is guaranteed by the fact that $\{Q_t, P_t\} = 1$, and the CT is of Type 1 by the same argument as in part (a). Equations (5.121) now read

$$p = i(m\omega q - \sqrt{2m\omega}\,Q_t e^{-i\omega t}) = \frac{\partial F_t}{\partial q},$$

$$P_t = i(\sqrt{2m\omega}\,q e^{-i\omega t} - Q_t e^{-2i\omega t}) = -\frac{\partial F_t}{\partial Q_t}.$$

whose solution (up to an additive function of t) is

$$F_t(q, Q_t, t) = i\left(\frac{Q_t^2}{2}e^{-2i\omega t} - \sqrt{2m\omega}\,qQ_t e^{-i\omega t} + \frac{m\omega q^2}{2}\right) \qquad (5.131)$$

[the subscript t is again used to distinguish between this and Eq. (5.127)]. The similarity to Eq. (5.127) is obvious: the differences are the factors $e^{-2i\omega t}$ on the first term and $e^{-i\omega t}$ on the second.

This CT can also be inverted by calculating real and imaginary parts. The results are

$$\left.\begin{aligned} q(t) &= \sqrt{\frac{1}{2m\omega}}\{Q_t(t)e^{-i\omega t} - iP_t(t)e^{i\omega t}\}, \\[2mm] p(t) &= -i\sqrt{\frac{m\omega}{2}}\{Q_t(t)e^{-i\omega t} + iP_t(t)e^{i\omega t}\}. \end{aligned}\right\} \qquad (5.132)$$

Although the *old* Hamiltonian is given in the new coordinates by the analog of Eq. (5.128), namely $H = -i\omega Q_t P_t$, the generating function is now t dependent, so the *new* Hamiltonian is

$$\begin{aligned} K_t &= H + \frac{\partial F_t}{\partial t} = H + \omega Q_t^2 e^{-2i\omega t} - \omega\sqrt{2m\omega}\,qQ_t e^{-i\omega t} \\[2mm] &= H + \omega Q_t^2 e^{-2i\omega t} - \omega Q_t^2 e^{-2i\omega t} + i\omega P_t Q_t \\[2mm] &= 0. \end{aligned}$$

The new Hamiltonian vanishes identically! The motion is now trivial: $Q_t = \text{const.} \equiv Q_0$, $P_t = \text{const.} \equiv P_0$. A glance at Eq. (5.132) and comparison with (5.130) shows that this yields the correct (q, p) solution.

Note: If the Hamiltonian of Eq. (5.128) was pathological because no Lagrangian could be obtained from it, this one is even more so. But this pathology is useful: because $\dot{Q}_t = \dot{P}_t = 0$, both of these new variables are constants of the motion, and we are in the happy position of having no equations of motion to solve. This seems like a good way to solve dynamical systems: find a CT in which the new variables are constants of the motion. It would be useful to have a general recipe for doing this, and in the next chapter we describe such a procedure, the Hamilton–Jacobi method.

5.3.4 ONE-PARAMETER GROUPS OF CANONICAL TRANSFORMATIONS

We now extend the discussion of canonical transformations to families of CTs that depend continuously on a parameter. It will be seen that every dynamical variable on $\mathbf{T}^*\mathbb{Q}$ generates a continuous one-parameter family (in fact a group) of CTs, and this will lead to the Hamiltonian version of the Noether theorem.

INFINITESIMAL GENERATORS OF ONE-PARAMETER GROUPS; HAMILTONIAN FLOWS

An example of a one-parameter group of CTs is the physical motion itself, generated by the Hamiltonian $H \in \mathcal{F}(\mathbf{T}^*\mathbb{Q})$ with parameter t. We start from this case. The group is the dynamical group φ^Δ of Section 3.4.3 (we drop the subscript t for the group) generated by the dynamical vector field Δ. This is done on $\mathbf{T}^*\mathbb{Q}$ the same way as on $\mathbf{T}\mathbb{Q}$: each point $\xi \in \mathbf{T}^*\mathbb{Q}$ is carried along by the dynamics Δ, arriving at time t at the point $\xi(t) \in \mathbf{T}^*\mathbb{Q}$. As in Chapter 3, we write $\xi(t) = \varphi_t^\Delta \xi$, where $\varphi_t^\Delta \in \varphi^\Delta$ is now an element of the group: the collection of the φ_t^Δ for all t is the group φ^Δ. As t varies, φ_t^Δ wanders through φ^Δ, and each $\xi(t) \equiv \varphi_t^\Delta \xi$ sweeps out an integral curve of Δ, as though the points of $\mathbf{T}^*\mathbb{Q}$ were *flowing* along the trajectories.

A similar *flow* or one-parameter group φ^X can be generated in this way by any vector field X on a manifold. Now, it was shown at Eq. (5.77) that the symplectic structure of $\mathbf{T}^*\mathbb{Q}$ associates a unique *Hamiltonian vector field* $X_g \equiv G$ with each dynamical variable $g \in \mathcal{F}(\mathbf{T}^*\mathbb{Q})$. Hence each dynamical variable g on $\mathbf{T}^*\mathbb{Q}$ leads to the one-parameter group φ^G through a chain of associations: g leads to G, and G leads to $\varphi^G \equiv \varphi^g$. Since we will usually concentrate on the function g rather than its Hamiltonian vector field G we will usually write φ^g for φ^G. The function g is called the *infinitesimal generator* of the group φ^g, and the flow a *Hamiltonian flow*. Note that φ_0^g is the identity map, sending each $\xi \in \mathbf{T}^*\mathbb{Q}$ into itself.

To understand precisely how a transformation group φ^g is related to its infinitesimal generator g consider how a function $f \in \mathcal{F}(\mathbf{T}^*\mathbb{Q})$ varies under the action of φ^g. Actually, we have already discussed the variation of functions. Equation (5.78) shows that if $f, g \in \mathcal{F}(\mathbf{T}^*\mathbb{Q})$, then the Lie derivative of f with respect to G gives the variation of f along the integral curves (along the flow) of G and hence under the action of φ^g [this is a repeat of (5.78)]:

$$\mathbf{L}_G f = \{f, g\} = i_G \, df = i_G i_{X_f} \omega = \omega(G, X_f) \equiv \omega(X_g, X_f). \tag{5.133}$$

As a general example, let the f of Eq. (5.133) be one of the local coordinates ξ^j and write ϵ instead of t for the parameter of φ^g. Then (5.133) becomes

$$\mathbf{L}_G \xi^j \equiv \frac{d\xi^j}{d\epsilon} = \{\xi^j, g\} = \omega^{ji} \partial_i g. \tag{5.134}$$

This is a set of differential equations for the integral curves $\xi^j(\epsilon)$ of G. For each ϵ the solutions give an explicit expression for the transformation φ_ϵ^g, specifying that each point with coordinates ξ^j is mapped to the point $\xi^j(\epsilon) \equiv \varphi_\epsilon^g \xi^j$.

This formulation makes it possible, as long as g is independent of ϵ, to write down an expression for φ_ϵ^g as a formal power series in ϵ, one that is more than formal in those cases for which the series converges. To see how this works, start by writing $\xi(\epsilon) = \varphi_\epsilon^g \xi_0$ (we suppress the superscript j on ξ) and insert this into Eq. (5.134):

$$\frac{d\xi}{d\epsilon} = \frac{d\varphi_\epsilon^g}{d\epsilon} \xi_0 = \{\xi, g\} = \{\varphi_\epsilon^g \xi_0, g\} = \mathbf{L}_G \cdot \varphi_\epsilon^g \xi_0.$$

This may be abbreviated by the operator equation

$$\frac{d\varphi_\epsilon^g}{d\epsilon} = \mathbf{L}_G \circ \varphi_\epsilon^g \equiv \{\bullet, g\} \circ \varphi_\epsilon^g. \tag{5.135}$$

The circle ($\circ$) as usual indicates successive application or composition, and the bullet ($\bullet$) shows where to put ξ (or more generally where to put the dynamical variable on which the operator acts).

Equation (5.135) is now used to obtain the power series. At $\epsilon = 0$ the CT is the identity ($\varphi_0^g = \mathbb{I}$), so if φ_ϵ^g is to be expanded in a power series in ϵ, the zeroth-order term is just $\mathbb{I}$. Then proceed to first order. In first order one writes $\varphi_\epsilon^g \approx \mathbb{I} + \epsilon\Phi_1$, and then to lowest order in ϵ

$$\frac{d\varphi_\epsilon^g}{d\epsilon} \approx \Phi_1 \approx \mathbf{L}_G \circ (\mathbb{I} + \epsilon\Phi_1) \approx \mathbf{L}_G \equiv \{\bullet, g\},$$

where we have kept only terms of order zero and used (5.135). Thus, to first order in ϵ

$$\varphi_\epsilon^g \approx \mathbb{I} + \epsilon\{\bullet, g\}.$$

In second order write $\varphi_\epsilon^g \approx \mathbb{I} + \epsilon\Phi_1 + \frac{1}{2}\epsilon^2\Phi_2$ and take the derivative again:

$$\frac{d\varphi_\epsilon^g}{d\epsilon} \approx \Phi_1 + \epsilon\Phi_2 \approx \mathbf{L}_G\left(\mathbb{I} + \epsilon\Phi_1 + \frac{1}{2}\epsilon^2\Phi_2\right) \approx \mathbf{L}_G + \epsilon\mathbf{L}_G\Phi_1.$$

Thus $\Phi_2 = \mathbf{L}_G\Phi_1 = \{\{\bullet, g\}, g\}$, so that $\varphi_\epsilon^g \approx \mathbb{I} + \epsilon\{\bullet, g\} + \frac{1}{2}\epsilon^2\{\{\bullet, g\}, g\}$ to second order in ϵ. Since $\Phi_1 = \mathbf{L}_G$, the expression for Φ_2 may be written as $\mathbf{L}_G^2$ (and similarly for higher orders). Proceeding in this way, one obtains a power series for φ_ϵ^g that can be put in the form

$$\varphi_\epsilon^g = \mathbb{I} + \epsilon\mathbf{L}_G + \frac{1}{2!}(\epsilon\mathbf{L}_G)^2 + \frac{1}{3!}(\epsilon\mathbf{L}_G)^3 + \cdots$$

$$= \sum_0^\infty \frac{1}{n!}(\epsilon\mathbf{L}_G)^n \equiv \exp(\epsilon\mathbf{L}_G) \equiv \exp(\epsilon\{\bullet, g\}). \tag{5.136}$$

We emphasize that this expression for φ_ϵ^g is valid (when it converges) only if g is not a function of ϵ.

When applied to the Hamiltonian dynamics Δ with a t-independent Hamiltonian H, Eq. (5.136) yields a formal solution of Hamilton's canonical equations:

$$\xi(t) = \exp(t\mathbf{L}_\Delta)\xi_0 \equiv \exp(t\{\bullet, H\})\xi_0,$$

where H must be written as a function of ξ_0 and the PB is taken with respect to ξ_0. This is sometimes called *exponentiating* the Poisson bracket. For more on this group theoretical approach to Hamiltonian dynamics, see Marmo et al. (1985), Sudarshan & Mukunda (1974), and Saletan & Cromer (1971). See also Problem 32.

THE HAMILTONIAN NOETHER THEOREM

An immediate consequence of (5.133) is the Hamiltonian Noether theorem. Suppose that the Hamiltonian function H is invariant under the group infinitesimally generated by g. Since H is invariant, (5.133) implies that

$$\{H, g\} = 0. \tag{5.137}$$

The Poisson bracket here is just (the negative of) dg/dt, since $\{H, g\} = -\{g, H\}$, so g must be a constant of the motion. The converse is also true: if $dg/dt \equiv \{g, H\} = 0$, then Eq. (5.133) shows that H is invariant under φ^g. Thus we have the Hamiltonian Noether theorem and its converse: *g is a constant of the motion iff H is invariant under the group of transformations infinitesimally generated by g*. The Hamiltonian Noether theorem, unlike its Lagrangian counterpart, has a converse.

For example, let the function g be one of the local generalized coordinates q^α, say q^λ with fixed index λ. Its Hamiltonian vector field $X_\lambda \equiv X_{q_\lambda}$ is obtained from Hamilton's canonical equations for $H = q^\lambda$ (with the time t replaced by the parameter ϵ). Let the components of X_λ in the (q, p) notation be given by $X_\lambda = X_\lambda^\alpha(\partial/\partial q^\alpha) + X_{\lambda\alpha}(\partial/\partial p_\alpha)$. Then

$$\frac{dq^\alpha}{d\epsilon} \equiv X_\lambda^\alpha = \{q^\alpha, q^\lambda\} = 0, \quad \frac{dp_\alpha}{d\epsilon} \equiv X_{\lambda\alpha} = \{q^\alpha, p_\lambda\} = -\delta_\lambda^\alpha,$$

so the vector field is $X_\lambda = -\partial/\partial p_\lambda$. The integral curves of X_λ, obtained by solving the differential equations $dq^\alpha/d\epsilon = 0, dp_\alpha/d\epsilon = -\delta_\lambda^\alpha$, are

$$q^\alpha(\epsilon) = q^\alpha(0), \quad p_\alpha(\epsilon) = p_\alpha(0) - \epsilon\delta_\lambda^\alpha. \tag{5.138}$$

Thus $\varphi_\epsilon^{q_\lambda}$ moves each point $(q, p) \in \mathbf{T}^*\mathbb{Q}$ a distance ϵ in the negative p_λ direction from its initial position: the flow generated by q^λ is simply translation in the negative p_λ direction. The transformation $\varphi_\epsilon^{q_\lambda}$ of Eq. (5.138), from $(q(0), p(0))$ to $(q(\epsilon), p(\epsilon))$, is obviously canonical, and soon it will be shown that all the transformations of a Hamiltonian flow are canonical.

A similar calculation shows that the vector field $Y_\lambda \equiv Y_{p_\lambda}$, which is Hamiltonian with respect to p_λ, is $Y_\lambda = \partial/\partial q^\lambda$: the flow generated by p_λ is translation in the positive q^λ direction.

Consider a dynamical system whose Hamiltonian is H. Along the flow generated by p_λ the variation of H is given by

$$\mathbf{L}_{Y_\lambda} H = \{H, p_\lambda\} = \frac{\partial H}{\partial q^\lambda} = -\dot{p}_\lambda.$$

The Noether theorem in this case states that if H is invariant under translation in the q^λ direction (i.e., if $\mathbf{L}_{Y_\lambda} H = 0$), then $\dot{p}_\lambda = 0$ – the conjugate momentum p_λ is conserved. (This reflects a familiar Lagrangian result: the momentum p_λ is conserved iff H is independent of q^λ, i.e., iff q^λ is an ignorable coordinate.)

FLOWS AND POISSON BRACKETS

Invariance of a function under a flow can be stated in terms of Lie derivatives: $f \in \mathcal{F}(T^*\mathbb{Q})$ is invariant under the flow generated by $g \in \mathcal{F}(T^*\mathbb{Q})$ if $\varphi_{\epsilon\,\mathrm{fn}}^g f = f$ [see the beginning of Section 5.3.2. for the notation] or if $\{f, g\} \equiv \mathbf{L}_{X_g} f = 0$. Similarly any other geometric object is invariant under the flow if its Lie derivative with respect to X_g vanishes. For example, a two-form γ is invariant if $\varphi_{\epsilon\,\mathrm{form}}^g \gamma = \gamma$ or if $\mathbf{L}_{X_g}\gamma = 0$. If this is true for $\gamma = \omega$, it means that every element $\varphi_\epsilon^g \in \varphi^g$ is a canonical transformation.

To prove this, it is sufficient, according to Eq. (5.109), to show that the symplectic form ω is invariant under each of the φ_ϵ^g. Since φ_0^g is the identity, which of course leaves ω invariant, all that need be shown is that ω does not vary as ϵ changes, and thus that $\mathbf{L}_{X_g}\omega = 0$. To do so we use the *Cartan identity*,

$$\mathbf{L}_{X_g} \equiv i_{X_g} d + d\, i_{X_g}, \tag{5.139}$$

which is discussed in Problem 21. Indeed, the Cartan identity implies that

$$\mathbf{L}_{X_g}\omega = i_{X_g}\, d\omega + d\, i_{X_g}\omega = 0; \tag{5.140}$$

here the first term vanishes because ω is closed (i.e., $d\omega = 0$) and the second because it equals $-d^2 g \equiv 0$. (The same result can also be obtained in local coordinates by using Problem 23 and setting the components of X_g equal to $\omega^{kl}\partial_l g$. The vanishing of the Lie derivative of ω follows from $\partial_i\partial_j g = \partial_j\partial_i g$, which is the same as $d^2 g = 0$.) Thus each transformation $\varphi_\epsilon^g \in \varphi^g$ leaves ω invariant and hence is canonical.

For example, in two freedoms consider the infinitesimal generator

$$g(q^1, q^2, p_1, p_2) = -\left(\beta q^1 q^2 + \frac{p_1 p_2}{\beta}\right). \tag{5.141}$$

Calculation using the equation above (5.138) yields

$$\frac{dq^1}{d\epsilon} = -\frac{p_2}{\beta}, \quad \frac{dp_1}{d\epsilon} = \beta q^2,$$

$$\frac{dq^2}{d\epsilon} = -\frac{p_1}{\beta}, \quad \frac{dp_2}{d\epsilon} = \beta q^1.$$

The solution of these equations is

$$\left.\begin{aligned}
q^1(\epsilon) &= q^1(0)\cos\epsilon - \frac{p_2(0)}{\beta}\sin\epsilon, & p_1(\epsilon) &= \beta q^2(0)\sin\epsilon + p_1(0)\cos\epsilon, \\
q^2(\epsilon) &= q^2(0)\cos\epsilon - \frac{p_1(0)}{\beta}\sin\epsilon, & p_2(\epsilon) &= \beta q^1(0)\sin\epsilon + p_2(0)\cos\epsilon.
\end{aligned}\right\} \tag{5.142}$$

Compare this to Eq. (5.99).

The fact that p_α generates translation in the q^α direction depends only on the fact that $\{q^\alpha, p_\alpha\} = 1$. In fact let $f, g \in \mathcal{F}(T^*\mathbb{Q})$ be any two dynamical variables such that $\{f, g\} = 1$.

Then under the action of φ^g, the variation of f is given by

$$\frac{df}{d\epsilon} \equiv \mathbf{L}_{X_g} f = 1.$$

It follows that for finite ϵ the change in f is given by

$$\varphi^g_{\epsilon\,\text{fn}} f = f + \epsilon.$$

In other words, f varies along the Hamiltonian flow generated by g in the same way that q^α varies along the Hamiltonian flow generated by p_α. This hints that the PB relation $\{f, g\} = 1$ implies that the dynamical variables f and g can be taken as a pair of conjugate local q and p coordinates of a canonical coordinate system in a neighborhood of $\mathbf{T}^*\mathbb{Q}$. More about this is discussed when we turn to the Darboux theorem in Section 5.4.2.

5.4 TWO THEOREMS: LIOUVILLE AND DARBOUX

In this section we discuss two important theorems relating to Hamiltonian dynamics.

5.4.1 LIOUVILLE'S VOLUME THEOREM

There is more than one theorem associated with Liouville. The one we treat in this subsection discusses how regions of the phase manifold $\mathbf{T}^*\mathbb{Q}$ change under the flow of a Hamiltonian dynamical system (Liouville, 1838). We will call it *Liouville's Volume theorem*, or the LV theorem.

Suppose a dense set of initial points occupy a region $R \subset \mathbf{T}^*\mathbb{Q}$ whose dimension is $2n$; that is, dim $R = $ dim $\mathbf{T}^*\mathbb{Q}$. Then as the dynamics develops in time the points in R will move and at some later time will occupy another region $R(t)$ (Fig. 5.4). What the LV theorem says is that the *volume* of $R(t)$ is the same as the volume of $R \equiv R(0)$. With the LV theorem is associated a whole set of other assertions, which make the same sort of statement for regions whose dimensions are less than $2n$. The constants of the motion associated with those other assertions are called *Poincaré invariants*. We will also describe them in this subsection.

VOLUME

To explain and prove the LV theorem we first explain what is meant by a volume on a manifold. In a two-dimensional vector space two vectors $\mathbf{v}_1, \mathbf{v}_2$ define a parallelogram (Fig. 5.5) whose area (up to sign) is

$$A(\mathbf{v}_1, \mathbf{v}_2) = \mathbf{v}_1 \wedge \mathbf{v}_2 = \det\begin{bmatrix} v_1^1 & v_2^1 \\ v_1^2 & v_2^2 \end{bmatrix} = \epsilon_{ij} v_1^i v_2^j,$$

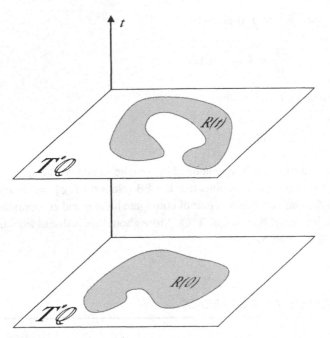

FIGURE 5.4

A three-dimensional representation of $\mathbb{W} \equiv \mathbf{T}^*\mathbb{Q} \times \{t\}$, where t is the time. The horizontal plane represents $\mathbf{T}^*\mathbb{Q}$ (the projection of $\mathbb{W}$ along the t direction). The shaded area on the $t = 0$ plane represents a region $R(0)$ at time $t = 0$. At a later time the region becomes $R(t)$. According to the LV theorem, the volumes of $R(0)$ and $R(t)$ are equal.

where v_i^k is the kth component of $\mathbf{v}_i$. Here ϵ_{ij} is the two-dimensional Levi–Civita symbol defined by $\epsilon_{12} = -\epsilon_{21} = 1$, $\epsilon_{11} = \epsilon_{22} = 0$, and the epsilon-product in the last term is one of the ways to define the determinant. In a three-dimensional space, three vectors $\mathbf{v}_1$, $\mathbf{v}_2$, $\mathbf{v}_3$ define a parallelepiped (Fig. 5.6) whose volume (up to sign) is

$$V(\mathbf{v}_1, \mathbf{v}_2, \mathbf{v}_3) = \mathbf{v}_1 \cdot (\mathbf{v}_2 \wedge \mathbf{v}_3) = \det \begin{bmatrix} v_1^1 & v_2^1 & v_3^1 \\ v_1^2 & v_2^2 & v_3^2 \\ v_1^3 & v_2^3 & v_3^3 \end{bmatrix} = \epsilon_{ijk} v_1^i v_2^j v_3^k.$$

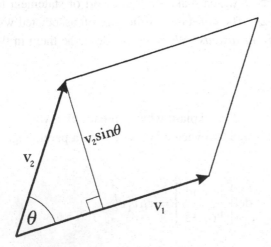

FIGURE 5.5
The area of the parallelogram subtended by the two vectors $\mathbf{v}_1$ and $\mathbf{v}_2$ is $v_1 v_2 \sin \theta \equiv |\mathbf{v}_1 \wedge \mathbf{v}_2|$.

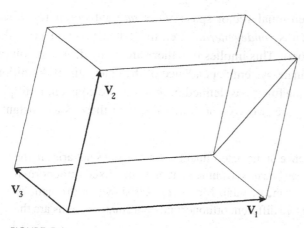

FIGURE 5.6

The volume of the parallelepiped subtended by the three vectors v_1, v_2, and v_3 is $v_1 \cdot (v_2 \wedge v_3) = \epsilon_{ijk} v_1^i v_2^j v_3^k$.

In a vector space of dimension $n = 2$ or $n = 3$, therefore, the determinant or epsilon-product can be thought of as a *volume function* on the space: it maps n vectors into the volume they subtend. This definition is in accord with several well-known properties of the volume. If the vectors are linearly dependent, the volume is zero; if a multiple of one of the v_j is added to one of the others, the volume remains the same (Fig. 5.7); the volume depends linearly on each of the vectors.

Volume defined in this way may be positive or negative, depending on the order in which the vectors are written down, but this is not a serious problem. The real problem with this definition is that it is not invariant under coordinate transformations. There are many volume functions, each defined in a different coordinate system. A more general definition (see Schutz, 1980; Crampin and Pirani, 1986; or Bishop and Goldberg, 1980) puts it more

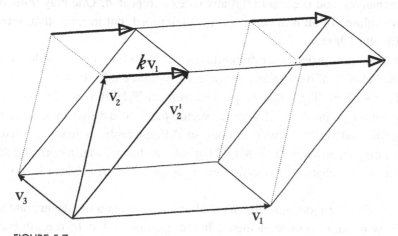

FIGURE 5.7

When any multiple of v_1 is added to v_2 the volume of the parallelepiped remains the same: The volume subtended by v_1, v_2, and v_3 is the same as that subtended by v_1, v_2', and v_3, where $v_2' = v_2 + kv_1$ for some constant k.

directly: a volume in a three-dimensional vector space V^3 is *any* antisymmetric linear function $V : V^3 \times V^3 \times V^3 \to \mathbb{R}$ that is *nondegenerate* (i.e., that takes on a nonzero value for three linearly independent vectors). This implies that there are many different volume functions, which reflects the coordinate-system dependence of the determinant definition. It is shown in Problem 29 that volume functions defined in this way are constant multiples of each other. That is, if V_1 and V_2 are two volume functions, then there is a constant c such that $V_2 = c V_1$.

> **REMARK:** The coordinate dependence of the determinant definition does not arise in the usual vector algebra on three-space because volume is defined in a fixed orthonormal coordinate system that is taken as fundamental. Moreover, it is shown in Problem 29 that volume functions defined in two different orthonormal coordinate systems are the same, i.e., that $c = 1$.   □

In the generalization to an n-dimensional vector space V^n a volume function is defined as a linear, antisymmetric, nondegenerate function

$$V : (V^n)^{\times n} \to \mathbb{R}, \tag{5.143}$$

where $(V^n)^{\times n} \equiv V^n \times V^n \times \cdots \times V^n$ (n factors). As for $n = 3$, this is equivalent to a determinant definition: given n vectors $\mathbf{v}_j \in V^n$, the volume they subtend is

$$V(\mathbf{v}_1, \ldots, \mathbf{v}_n) = \det \begin{bmatrix} v_1^1 & \cdots & v_n^1 \\ \vdots & \ddots & \vdots \\ v_1^n & \cdots & v_n^n \end{bmatrix} = \epsilon_{i_1 i_2 \cdots i_n} v_1^{i_1} v_2^{i_2} \cdots v_n^{i_n}, \tag{5.144}$$

where $\epsilon_{i_1 i_2 \cdots i_n}$ is the n-dimensional Levi–Civita symbol: it is equal to 1 if $\{i_1, i_2, \ldots, i_n\}$ is obtained by an even number of interchanges in $\{1, 2, \ldots, n\}$, is equal to -1 if by an odd number of interchanges, and is equal to 0 if any index is repeated. One may think of $V(\mathbf{v}_1, \ldots, \mathbf{v}_n)$ as the volume of an n-dimensional parallelepiped, but in more than three dimensions it is difficult to draw.

These considerations are carried to a differential manifold $\mathbb{M}$ by gluing together volume functions defined in each of its tangent spaces: at each point $\xi \in \mathbb{M}$ there is a volume function $V_\xi : (\mathbf{T}_\xi \mathbb{M})^{\times n} \to \mathbb{R}$ on $\mathbf{T}_\xi \mathbb{M}$, that maps n vectors in $\mathbf{T}_\xi \mathbb{M}$ into $\mathbb{R}$. The collection v of these V_ξ for all points $\xi \in \mathbb{M}$ maps n vector *fields* to a value in $\mathbb{R}$ at each $\xi \in \mathbb{M}$ (i.e., to a function on $\mathbb{M}$), and if it does so differentiably, it maps n vector fields linearly and antisymmetrically to $\mathcal{F}(\mathbb{M})$. That means that v is an n-form on $\mathbb{M}$, which we will call a *volume element*. A familiar example of a volume element is $d^3 x$ in three-space.

In sum, a *volume element* v on an n-dimensional manifold $\mathbb{M}$ is a nondegenerate n-form. There are many possible volume elements, but at each point $\xi \in \mathbb{M}$ (i.e., in each tangent plane $\mathbf{T}_\xi \mathbb{M}$) they are all multiples of each other (Problem 29). Because they may be different multiples at different points ξ, the ratio of any two volume elements v_1 and v_2 is a function on $\mathbb{M}$: there is an $f \in \mathcal{F}(\mathbb{M})$ such that $v_2 = f v_1$. A volume element is not quite a volume: it is a volume function in each of the tangent spaces, which are very local objects. That makes it a sort of *infinitesimal* volume function, and a volume is therefore obtained from it only by integration, as will soon be explained.

But first we turn from a general manifold $\mathbb{M}$ to $\mathbb{T}^*\mathbb{Q}$. Recall (Section 5.2.3) that the symplectic form ω is related to areas in the various (q, p) planes. It also provides a natural *canonical volume element* on $\mathbb{T}^*\mathbb{Q}$: since ω is a two-form,

$$v = \frac{1}{n!}\omega^{\wedge n} \equiv \frac{1}{n!}\omega \wedge \omega \wedge \cdots \wedge \omega \quad (n \text{ factors}) \tag{5.145}$$

is a $2n$-form and hence a volume element (it is nondegenerate because ω is nondegenerate). Equation (5.145) can be used to express v directly in the (q, p) notation. Most of the terms in the nth wedge power contain two factors of at least one of the $d\xi^k$, and antisymmetry causes them to vanish. There are $n!$ terms that fail to vanish, and these turn out to be all equal (which explains the factor $1/n!$). For example if $n = 2$,

$$\begin{aligned} v &= \frac{1}{2}[dq^1 \wedge dp_1 \wedge dq^1 \wedge dp_1 + dq^1 \wedge dp_1 \wedge dq^2 \wedge dp_2 \\ &\quad + dq^2 \wedge dp_2 \wedge dq^1 \wedge dp_1 + dq^2 \wedge dp_2 \wedge dq^2 \wedge dp_2] \\ &= dq^1 \wedge dp_1 \wedge dq^2 \wedge dp_2 \end{aligned}$$

(the first and last terms vanish by antisymmetry, and the two other terms are equal, also by antisymmetry). A similar calculation gives the general expression for higher n:

$$v = dq^1 \wedge dp_1 \wedge \cdots \wedge dq^n \wedge dp_n \equiv \bigwedge_{\alpha=1}^{n} dq^\alpha \wedge dp_\alpha \equiv \pm \bigwedge_{k=1}^{2n} d\xi^k. \tag{5.146}$$

The invariance of ω under canonical transformations leads immediately to one form of the LV theorem:

$$v \equiv \frac{1}{n!}\omega^{\wedge n} \quad \text{is invariant under CTs.}$$

This is the *differential* LV theorem: *the canonical volume element is invariant under CTs and hence under Hamiltonian flows.*

INTEGRATION ON T*Q; THE LIOUVILLE THEOREM

The *integral* LV theorem states that the (canonical) volume V of a region in $\mathbb{T}^*\mathbb{Q}$ is invariant under Hamiltonian flows. To prove the theorem, we will first show how to use v to calculate the volume V of a region through integration and then show that the volume so calculated is invariant.

We will define integration in analogy with the usual textbook definition of the Riemann integral as a limiting process applied to summation. The definition differs from the textbook definition, however, in that the volume in $\mathbb{T}^*\mathbb{Q}$ is obtained by summing small volumes which themselves are in the tangent spaces to $\mathbb{T}^*\mathbb{Q}$ rather than on $\mathbb{T}^*\mathbb{Q}$ itself.

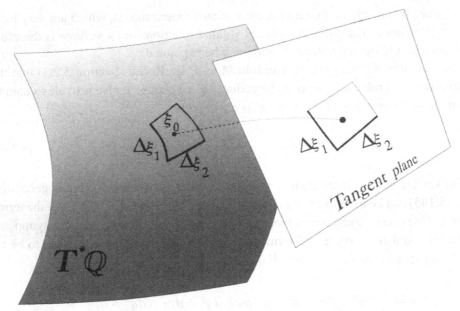

FIGURE 5.8
The cell bounded by $\Delta \xi_1$ and $\Delta \xi_2$ at a point ξ_0 on a two-dimensional $\mathbf{T}^*\mathbb{Q}$, and the parallelogram
bounded by $\Delta \xi_1$ and $\Delta \xi_2$ on the tangent plane at ξ_0. The tangent plane is drawn some distance from
$\mathbf{T}^*\mathbb{Q}$ in order to make the figure clearer.

REMARK: We use the Riemann definition of integration because it is more familiar
to most physics students than the alternative Lebesgue definition. The latter is more
general and elegant, but it requires some discussion of *measure*, which we are avoiding
at this point.   □

Start by approximating the volume ΔV of a small cell in $\mathbf{T}^*\mathbb{Q}$, a sort of parallelepiped
defined by small displacements $\Delta \xi^k$ in the coordinate directions from some point $\xi_0 \in \mathbf{T}^*\mathbb{Q}$
as in Fig. 5.8. It is only a "sort of" parallelepiped because the $\Delta \xi^k$ are not straight-line
segments, because $\mathbf{T}^*\mathbb{Q}$ is not a linear space. But for sufficiently small $\Delta \xi^k$, the region
around ξ_0 is nearly linear, so ΔV is *approximately* $\Delta \xi^1 \Delta \xi^2 \cdots \Delta \xi^{2n}$, and the smaller
the $\Delta \xi^k$, the better the approximation. What we want to show is how to obtain ΔV by
using the canonical volume function defined in the previous subsection. Thus we turn to the
tangent space $\mathbf{T}_{\xi_0}(\mathbf{T}^*\mathbb{Q})$ at ξ_0. There a volume δV can be obtained by applying the canonical
volume element v with $2n$ vectors $\Delta \xi^k$ at $\mathbf{T}_{\xi_0}(\mathbf{T}^*\mathbb{Q})$ (Fig. 5.8), and it turns out that $\delta V \approx \Delta V$.
Indeed, construct $2n$ vector fields X^k defined by (no sum on k) $X^k = \Delta \xi^k (\partial/\partial \xi_k)$. Then
contracting these with the canonical volume element at ξ_0 yields

$$\delta V = i_{X^1} i_{X^2} \cdots i_{X^{2n}} v \equiv i_{X^1} i_{X^2} \cdots i_{X^{2n}} \bigwedge_{k=1}^{2n} d\xi^k = \Delta \xi^1 \Delta \xi^2 \cdots \Delta \xi^{2n} \approx \Delta V. \qquad (5.147)$$

The volume V of a larger region $R \subset \mathbf{T}^*\mathbb{Q}$ is approximated by breaking R up into
small cells, calculating δV according to (5.147) for each cell, and summing over all the

cells in R. The smaller the $\Delta\xi^k$, the more cells there will be and the more closely each δV approximates its ΔV and hence the more accurately their sum approximates V.

The exact value of V is obtained in the limit $\Delta\xi^k \to 0$. In that limit the components of the X^k become infinitesimal. The limit expression is called the integral of v over R and is written in the following way:

$$\lim_{\Delta\xi \to 0} \sum i_{X^1} i_{X^2} \cdots i_{X^{2n}} v = V \equiv \int_R v \equiv \int_R dq^1 \wedge dp_1 \wedge \cdots \wedge dq^n \wedge dp_n. \qquad (5.148)$$

The expression $\int v$ may look a little strange, as we are accustomed to seeing differentials in integrands, but remember that v contains $2n$ differentials. The volume integral that appears here is a $2n$-fold integral.

Having defined volume, we are prepared to state the integral LV theorem. Suppose that $\eta^k(\xi)$ is a canonical transformation, viewed as a passive coordinate change. Since v is invariant under CTs, (5.146) can be used to write it in terms of the transformed variables:

$$v = \bigwedge_{\alpha=1}^{v} dQ^\alpha \wedge dP_\alpha \equiv dQ^1 \wedge dP_1 \wedge \cdots \wedge dQ^n \wedge dP_n \equiv \pm \bigwedge_{k=1}^{2n} d\eta^k, \qquad (5.149)$$

where (Q, P) are the new coordinates. Under a passive transformation, R does not change, but it has a different expression in the new coordinates. We emphasize this by writing R' instead of R. Since the volume does not change, Eq. (5.148) can also be written in the new coordinates:

$$V = \int_{R'} dQ^1 \wedge dP_1 \wedge \cdots \wedge dQ^n \wedge dP_n. \qquad (5.150)$$

Now think of $\eta^k(\xi)$ as an active transformation. In the active interpretation, (Q, P) is the transformed point, and R' is a new region, obtained by applying the CT to R. The active interpretation of Eq. (5.150) is that the *volume* of the new region R' is the same as the *volume* of R; although the region has changed, its volume has not.

It was pointed out in Section 5.3.4 that a Hamiltonian dynamical system yields a Hamiltonian flow – a continuous family of CTs (i.e., the transformation from an initial point $\xi(0) \in \mathbf{T}^*\mathbf{Q}$ to the point $\xi(t) \in \mathbf{T}^*\mathbf{Q}$ is a CT). Thus (Q, P) in Eq. (5.150) can be replaced by $(q(t), p(t))$ and R' by $R(t)$ [see Fig. 5.4; we also replace R by $R(0)$] to obtain

$$\int_{R(0)} v \equiv V = \int_{R(t)} dq^1(t) \wedge dp_1(t) \wedge \cdots \wedge dq^n(t) \wedge dp_n(t) \equiv \int_{R(t)} v(t) = V(t).$$
$$(5.151)$$

This is the *integral LV theorem*. It tells us that as a region is carried along and distorted by a Hamiltonian flow, its volume nevertheless remains fixed, that is, $dV/dt = 0$.

POINCARÉ INVARIANTS

Just as the volume element $v \equiv \omega^{\wedge n}/n!$ is invariant under CTs, and hence under any Hamiltonian dynamics, so is any other wedge power of ω. That is, $\omega^{\wedge p}$ is an invariant of the motion for all nonnegative $p \le n$ (for $p > n$ antisymmetry implies that $\omega^{\wedge p} = 0$). Indeed, by the Leibnitz rule

$$\mathbf{L}_\Delta \omega^{\wedge p} = p\omega^{\wedge(p-1)} \mathbf{L}_\Delta \omega \equiv 0.$$

In particular, $\mathbf{L}_\Delta v = 0$, which is again the differential Liouville theorem. Each $\omega^{\wedge p}$ is a $2p$-form, called the pth *differential Poincaré invariant*.

Integral Poincaré invariants are obtained by integrating the differential ones. For example, take the first differential Poincaré invariant, ω itself. Its integral over any simply connected two-dimensional surface S in $\mathbf{T}^*\mathbb{Q}$, the first integral Poincaré invariant, is

$$I_1 = \int_S \omega \equiv \int_S dq^\alpha \wedge dp_\alpha.$$

Without the details, the surface integral here is defined roughly the same way as the volume integral of Eq. (5.148): small areas ΔS on S are approximated by areas δS of small parallelograms in tangent planes to S, and these are summed. The limit of the sum for small δS is the integral. (This generalization is not immediate, for it requires finding a coordinate system tangent to S at each $\xi \in S$.) By the same argument as was used for the volume integral (invariance of ω under CTs, taking first the passive view and then the active one) one concludes that $dI_1/dt = 0$ or that

$$I_1 = \int_{S(t)} dq^\alpha(t) \wedge dp_\alpha(t)$$

is a constant of the motion. Other integral Poincaré invariants I_k, $k > 1$, are defined similarly.

Integral Poincaré invariants can also be written as integrals over closed submanifolds (boundaries of regions) in $\mathbf{T}^*\mathbb{Q}$. We start with the case of one freedom ($n = 1$). Then $\dim(\mathbf{T}^*\mathbb{Q}) = 2$, and I_1 is the volume integral of Eq. (5.151). For $n = 1$ any one-form α can be written locally as

$$\alpha = a(q, p) \, dq + b(q, p) \, dp, \tag{5.152}$$

where $a(q, p)$ and $b(q, p)$ are local functions. The exterior derivative of α is the two-form

$$d\alpha = da \wedge dq + db \wedge dp \equiv \left[\frac{\partial a}{\partial p} - \frac{\partial b}{\partial q} \right] dp \wedge dq$$

(use $dq \wedge dq = dp \wedge dp = 0$). Now let S be a connected finite area on $\mathbf{T}^*\mathbb{Q}$; the boundary ∂S of S is a closed contour, and according to Stokes's theorem (think of a and b as the components of a vector in two dimensions, and then $d\alpha$ is essentially the curl of that vector; see also Schutz, 1980),

$$\oint_{\partial S} \alpha \equiv \oint_{\partial S} (a \, dq + b \, dp) = \iint_S \left[\frac{\partial a}{\partial p} - \frac{\partial b}{\partial q} \right] dp \, dq.$$

Now write the double integral with a single integral sign, drop the circle from the contour integral (the integration is once over ∂S), and, as in the integrals discussed above, replace $dp\, dq$ by $dp \wedge dq$. Then Stokes's theorem becomes

$$\int_{\partial S} \alpha = \int_S d\alpha. \tag{5.153}$$

To apply this to I_1, set $a = p$ and $b = 0$, so that $\alpha = p\, dq$ and $d\alpha = dp \wedge dq$. Then Eq. (5.153) becomes (for $n = 1$)

$$\int_{\partial S} p\, dq = \int_S dp \wedge dq \equiv -I_1. \tag{5.154}$$

Hence I_1 can be written as a contour integral as well as a surface integral.

We have used Stokes's theorem by integrating over a two-dimensional region S in a two-dimensional $\mathbb{T}^*\mathbb{Q}$, yet it is equally valid for integrals over two-dimensional S regions in manifolds $\mathbb{M}$ of higher dimension:

$$\int_{\partial S} p_\alpha\, dq^\alpha = \int_S dp_\alpha \wedge dq^\alpha \equiv -I_1. \tag{5.155}$$

If $\dim \mathbb{M} = m > 2$, moreover, Stokes's theorem can be generalized to integrals over regions of dimension higher than two. Let $\beta^{(r)}$ be an r-form $1 \le r \le m - 1$, on $\mathbb{M}$ and let Σ, of dimension $2 \le s \le m$, be a bounded region in $\mathbb{M}$ with boundary $\partial \Sigma$. Then the *generalized Stokes's theorem* reads

$$\int_{\partial \Sigma} \beta^{(r)} = \int_\Sigma d\beta^{(r)}.$$

This makes it possible to write all of the integral Poincaré invariants I_k as integrals over closed submanifolds, but we do not go into that here.

We end with a remark about the sign in Eq. (5.155). It depends on the sense of the integral around ∂S, that is, the order in which the dp_α and dq^α are written. There are caveats about what is known technically as *orientability* of manifolds. The interested reader is referred to Schutz (1980).

DENSITY OF STATES

The LV theorem has important applications in statistical physics, having to do with the density of states in the phase manifold. We discuss this at first without any reference to statistics and at the end of this subsection we mention the application to statistical mechanics.

Suppose a finite (but large) number N of points is distributed in an arbitrary way over the phase manifold, as though someone had peppered $\mathbb{T}^*\mathbb{Q}$ (Fig. 5.9). Label the points of the distribution ξ_K, $K \in \{1, 2, \ldots, N\}$. Now break up $\mathbb{T}^*\mathbb{Q}$ into N cells, each of volume V_K and each containing just one of the points, and define a *discrete density function* $D'(\xi_K)$ on the distribution by

$$D'(\xi_K) = \frac{1}{N V_K}. \tag{5.156}$$

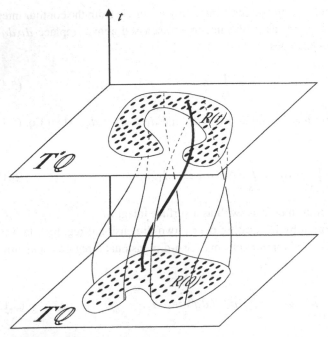

FIGURE 5.9

The same $\mathbb{W}$ as in Fig. 5.4, with the same region R. The dots on the $t = 0$ plane represent a discrete set $\xi_K(0)$ of initial conditions within $R(0)$. These become $\xi_K(t)$ at time t. Some of the integral curves that start on the boundary of $R(0)$ are drawn in; they all reach the boundary of $R(t)$. An integral curve starting at one of the interior $\xi_K(0)$ is drawn in; it reaches one of the $\xi_K(t)$. If an integral curve starting at one of the interior $\xi_K(0)$ were to leave $R(t)$ at some t, it would have to pierce the boundary and hence intersect one of the boundary's integral curves, which can not happen. Thus the number of points in $R(t)$ does not change as t increases, and hence the LV theorem implies that their density also does not change.

The explicit form of $D'(\xi_K)$ depends on the way the N points are distributed over $\mathbf{T}^*\mathbb{Q}$. For instance, in regions where the points are tightly packed, the V_K are smaller than in regions where the packing is looser, and hence $D'(\xi_K)$ takes on relatively large values in such regions. Assume for the time being that the total volume of $\mathbf{T}^*\mathbb{Q}$ is finite (that $\mathbf{T}^*\mathbb{Q}$ is *compact*). Then the discrete density function defined in this way can be *normalized* in the sense that

$$\sum_{K=1}^{N} D'(\xi_K)V_K = 1. \tag{5.157}$$

Now increase N, so the packing gets tighter and the V_K get smaller, but do this so that the products NV_K all have finite limits as N grows without bound and the separations between the ξ_K (and the V_K) shrink to zero. In the limit D' approaches a function $D(\xi)$ defined on the entire manifold, called a *normalized density distribution function*, which is assumed to be continuous and finite. The sum in Eq. (5.157) becomes an integral:

$$\int D(\xi)v = 1. \tag{5.158}$$

The v here is the same as in Eq. (5.148), and the integral is understood in the same way (sum small volumes in the tangent spaces, each volume weighted now by a value of D, and then go to the limit of small volumes). The integral in Eq. (5.158), over the entire manifold, converges because $\mathbf{T}^*\mathbb{Q}$ is assumed compact (if $\mathbf{T}^*\mathbb{Q}$ is not compact assume that $D(\xi)$ vanishes outside some finite region).

Now consider an integral like (5.158), but over a finite region $R \subset \mathbf{T}^*\mathbb{Q}$:

$$v_R \equiv \int_R D(\xi)v. \tag{5.159}$$

This integral can be obtained in much the same way as that of Eq. (5.158), by going to the limit of a finite sum like that of (5.157), but taken only over the ξ_K in R. Before reaching the limit, the finite sum is equal to

$$\sum_{K=1}^{n_R} D'(\xi_K)V_K = \frac{n_R}{N},$$

where n_R is the number of points in R. Thus v_R is a measure of the number of points in R.

As the system develops in time, the density distribution will change and hence is a function of both ξ and t. We write $D(\xi, t)$. A given region R in $\mathbf{T}^*\mathbb{Q}$ will also develop in time, and so will the volume element v. Then v_R becomes time dependent:

$$v_R(t) = \int_{R(t)} D(\xi, t)v(t), \tag{5.160}$$

where we have written $v(t)$ and $R(t)$ as in (5.151). If the dynamics Δ is Hamiltonian, the volume of $R(t)$ does not change, and we now argue that neither does $v_R(t)$, that it is in fact independent of the time. This is because v_R is a measure of the number of states in $R(t)$, and states cannot enter or leave the changing region. Indeed, in order for a state to enter or leave $R(t)$ at some time t, an integral curve of Δ from the interior of $R(t)$ must cross the boundary at time t. But as is seen in Fig. 5.9, the boundary is itself made up of integral curves, those that start on the boundary of $R(0)$. For an internal integral curve of Δ to leave (or an external one to enter) $R(t)$ it must intersect an integral curve of the boundary, and that is impossible because there is only one integral curve passing through each point of $\mathbf{T}^*\mathbb{Q}$. Hence the number of states in $R(t)$ does not change: $dv_R/dt = 0$.

Because this is true for any time t and for any region $R(0)$ and because according to the LV theorem the volume of $R(t)$ remains constant, the density $D(\xi, t)$ must also be invariant under the Hamiltonian flow. In terms of Poisson brackets, if H is the Hamiltonian of Δ,

$$\mathbf{L}_\Delta D \equiv \{D, H\} + \frac{\partial D}{\partial t} = 0. \tag{5.161}$$

In three dimensions this looks like a *continuity equation*, a conservation law for the normalized number of points v_R, which is generally written in the form

$$\frac{\partial D}{\partial t} + \nabla \cdot \mathbf{J} = 0,$$

where $\mathbf{J}$, the *number current*, measures the rate at which points leave or enter the boundary of a *fixed* region R. This equation can be understood also in higher dimensions as a continuity equation with $\nabla \cdot \mathbf{J}$ replaced by $\partial_k J^k$, where J^k is the rate of flow in the ξ^k direction. From Eq. (5.161) it is seen that $\partial_k J^k = \{D, H\} \equiv \partial_k D\omega^{kj}\partial_j H$, or

$$J^k = D\omega^{kj}\partial_j H. \tag{5.162}$$

The LV theorem plays an important role in the formulation of classical statistical mechanics, which deals with systems containing on the order of $N = 10^{23}$ particles. It is hopeless to try to solve each and every equation of motion for such a large set of interacting particles, and there are essentially two approaches to this problem. Both use the LV theorem, but in different ways. They are the Gibbs "ensemble" approach and the Boltzman "kinetic" approach. This is not the place to go into a discussion of statistical mechanics, but we indicate where the LV theorem is used.

The Gibbs approach assumes an ensemble, or set of systems, all described by the same Hamiltonian, each with its own set of positions and momenta in its own $6N$-dimensional $\mathbf{T}^*\mathbb{Q}$, yet all with identical macroscopic properties (temperature, pressure, volume). The time evolution of $\mathbf{T}^*\mathbb{Q}$ satisfies the LV theorem and the properties of the macroscopic system are then obtained by averaging the density $D(\xi)$ over the ensemble. The hypothesis that the macroscopic properties can be obtained in this way is known as the *ergodic hypothesis*. Although no general proof has yet been found for this hypothesis, the theoretical results it yields agree with experiment to a remarkable extent.

The Boltzmann approach deals instead with only one system, whose phase manifold is of dimension six. The LV theorem is used extensively in this formulation.

WORKED EXAMPLE 5.6

In one freedom consider a density distribution which at time $t = 0$ is Gaussian (we use x, rather than q)

$$D_0(x, p) \equiv D(x, p, 0) = \frac{1}{\pi \sigma_x \sigma_p} \exp\left\{ -\frac{(x - a)^2}{\sigma_x^2} - \frac{(p - b)^2}{\sigma_p^2} \right\} \tag{5.163}$$

($\sigma_x > 0$ and $\sigma_p > 0$) for a standard Hamiltonian of the form

$$H = \frac{p^2}{2m} + V(x) \tag{5.164}$$

on $\mathbf{T}^*\mathbb{Q} \equiv \mathbb{R}^2$. The maximum of this distribution is at $(x, p) = (a, b)$.

(a) Show that $D_0(x, p)$ has elliptical symmetry about its maximum. That is, show that the curves of equal values of D_0 are ellipses in (x, p). Under what condition is the symmetry circular?

(b) Show that

$$\int_{\mathbb{R}^2} D_0 v \equiv \int_{-\infty}^{\infty} dx \int_{-\infty}^{\infty} dp\, D_0 = 1. \tag{5.165}$$

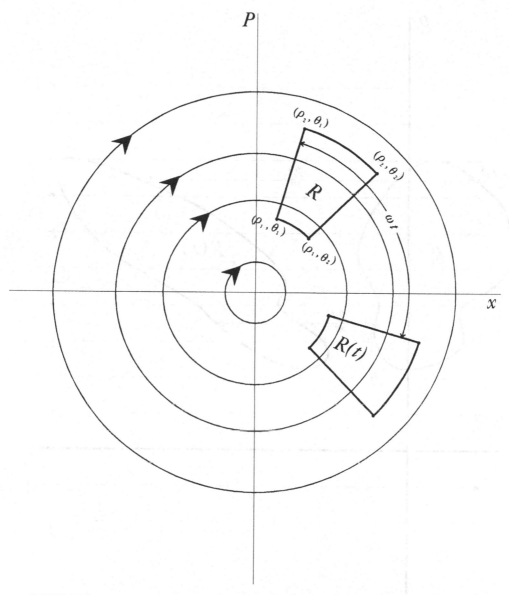

FIGURE 5.10

Phase diagram on $\mathbf{T}^*\mathbb{Q}$ for the harmonic oscillator of Worked Example 5.6(c). $P = p/m\omega$. R is an arbitrarily chosen region of $\mathbf{T}^*\mathbb{Q}$ at time $t = 0$, which moves to $R(t)$ at some time t later. (ρ, θ) are polar coordinates on $\mathbf{T}^*\mathbb{Q}$.

(c) For $V = \frac{1}{2}m\omega^2 x^2$, find the density distribution $D(x, p, t)$ at some later time.

(d) Choose a convenient area R and study the way it moves and changes shape under the motion.

(e) Repeat Part (d), but for the free-particle (i.e., for $V = 0$).

Solution. (a) D_0 is constant where the argument of the exponential is a constant, that

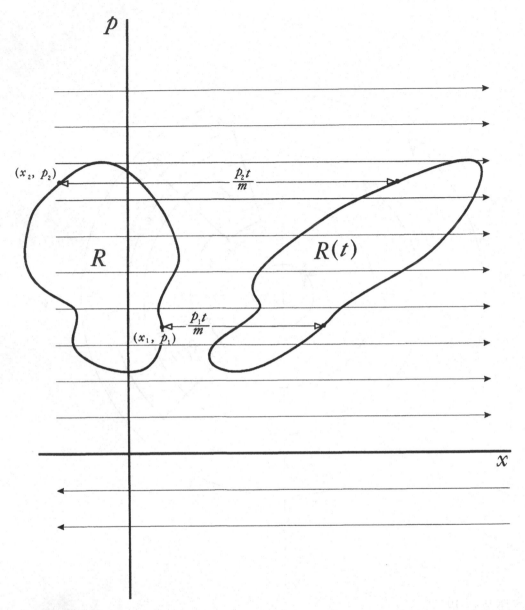

FIGURE 5.11
Phase diagram on $\mathbf{T}^*\mathbb{Q}$ for the free particle of Worked Example 5.6(e). R is an arbitrarily chosen region of $\mathbf{T}^*\mathbb{Q}$ at time $t = 0$, which moves to $R(t)$ at some time t later.

is, along the curve on which x and p satisfy the equation

$$\frac{(x-a)^2}{\sigma_x^2} + \frac{(p-b)^2}{\sigma_p^2} = K^2,$$

where $K > 0$ is a constant. This is the equation of an ellipse centered at $(x, p) = (a, b)$ and with semiaxes $\sigma_x K$ and $\sigma_p K$. The ellipse is a circle if $\sigma_x = \sigma_p$.

(b) We now use the fact that

$$\int_{-\infty}^{\infty} \exp\left(-\frac{x^2}{\alpha^2}\right) dx = \alpha\sqrt{\pi}.$$

The integral in Part (b) is therefore

$$\frac{1}{\pi\sigma_x\sigma_p} \int_{-\infty}^{\infty} \exp\left(-\frac{\xi^2}{\sigma_x^2}\right) d\xi \int_{-\infty}^{\infty} \exp\left(-\frac{\eta^2}{\sigma_p^2}\right) d\eta,$$

where $\xi = x - a$ and $\eta = y - b$. The result follows immediately.

(c) To say, as in Eq. (5.161), that $\mathbf{L}_\Delta D = 0$ is to say that the value of D remains fixed along an integral curve. In other words, if $(x(t), p(t))$ is an integral curve that starts at (x_0, p_0), then $D(x(t), p(t), t) = D(x_0, p_0, 0)$, or more generally,

$$D(x, p, t) = D(x(-t), p(-t), 0).$$

Thus we need to find the solutions of the equations of motion. As this is the harmonic oscillator, the solutions are

$$x(t) = x_0 \cos \omega t + \frac{p_0}{m\omega} \sin \omega t, \quad p(t) = -m\omega x_0 \sin \omega t + p_0 \cos \omega t,$$

where (x_0, p_0) is the initial point. Therefore, given a point (x, p),

$$x(-t) = x \cos \omega t - \frac{p}{m\omega} \sin \omega t, \quad p(-t) = m\omega x \sin \omega t + p \cos \omega t,$$

and

$$D(x, p, t) = \frac{1}{\pi\sigma_x\sigma_p} \exp\left\{-\frac{(x(-t) - a)^2}{\sigma_x^2} - \frac{(p(-t) - b)^2}{\sigma_p^2}\right\}.$$

At each time t the density function retains its elliptical symmetry, except that the ellipses are centered at the point whose coordinates are given by $x(-t) = a$ and $p(-t) = b$ and are rotated through an angle ωt. The new center is at

$$x = a \cos \omega t + \frac{b}{m\omega} \sin \omega t, \quad p = -m\omega a \sin \omega t + b \cos \omega t,$$

which is just the point to which the original center has moved under the motion. In one period the center of the distribution returns to its original point in $\mathbf{T}^*\mathbb{Q}$.

(d) We could choose R to be an ellipse, but we choose a truncated circular segment (Fig. 5.10). The p axis in Fig. 5.10 is divided by $m\omega$ to convert the elliptical integral curves into circles: we write $P = p/m\omega$. The vertices of R are labeled in polar coordinates (ρ, θ): they are at (ρ_1, θ_1), (ρ_1, θ_2), (ρ_2, θ_2), and (ρ_2, θ_1). After a time t each point (x, P) moves to the point $(x(t), P(t))$ given by

$$x(t) = x \cos \omega t + P \sin \omega t, \quad P(t) = -x \sin \omega t + P \cos \omega t.$$

This is a rotation in the (x, P) plane: each point (ρ, θ) moves to $(\rho, \theta + \omega t)$. Thus the segment R is simply rotated through the angle ωt about the origin into the new segment $R(t)$. A similar result is obtained for an area of any shape: it is simply rotated in the (x, P) plane. In the (x, p) coordinates there is some distortion as the p axis is multiplied by $m\omega$, but at the end of each period the area returns to its original position and original shape.

(e) The harmonic oscillator is peculiar in that every area R maintains its shape as the system moves. For the free particle, the motion starting at (x, p) is given by

$$x(t) = x + pt/m, \quad p(t) = p.$$

Let R be an arbitrary shape, as in Fig. 5.11. Each point (x, p) in R moves in a time t through a distance in the x direction equal to pt/m, so points with large values of p move further than points with small values of p. As a result R gets distorted, as shown in the figure.

One of the things that makes the harmonic oscillator special is that the period is independent of the energy, which makes the (angular) rate of motion the same along all of the integral curves. This means that at the end of each period, the area R returns to its original position and shape. The area keeps its shape on the way because we are using P instead of p and because the motion is uniform around the circle (x depends sinusoidally on t). If the period were different on different integral curves, as in the case of the Morse potential (Worked Example 2.3), R would become distorted and wind around the origin in a spiral whose exact shape would depend on the details of the motion. See Problem 31.

5.4.2 DARBOUX'S THEOREM

The elements of a two-form make up a matrix, and under any coordinate transformation this matrix changes (see the book's appendix), and under a general coordinate-dependent transformation the matrix changes in a coordinate-dependent way. This is true of any two-form, including the symplectic form ω, so under a general coordinate transformation on $\mathbf{T}^*\mathbb{Q}$ its elements will not necessarily be constants. Darboux's theorem says, however, that local coordinates can always be chosen so as to make the matrix elements of ω not only constants, but the very same constants as the matrix elements of Ω of Eq. (5.43). That is, in any $2n$-dimensional symplectic manifold $\mathbb{M}$, there exist local *canonical coordinates* q^α, p_α in terms of which the symplectic form takes on the *canonical form* $\omega = dq^\alpha \wedge dp_\alpha$. This means that the PBs of these coordinates satisfy Eq. (5.52). Thus locally every symplectic manifold of a given dimension looks essentially the same: they are all locally isomorphic. Recall the definition of a manifold in terms of local coordinate patches (Section 2.4.2), in each of which the manifold looks like a Euclidean space. Darboux's theorem tells us that in these local quasi-Euclidean patches coordinates can always be chosen to be canonical.

The theorem also makes a second statement: *any* function $S \in \mathcal{F}(\mathbb{M})$ (with some simple restrictions) can be chosen as one of the q^α or p_α, or any dynamical variable you choose can be one of the canonical coordinates. Once that coordinate is chosen, the others are somewhat restricted, but, as will be seen, there is still considerable latitude. An example of this is the central force problem of Section 5.3.1.

THE THEOREM

We won't actually prove the theorem, but we will describe briefly a proof that is essentially that of Arnol'd (1988). The brief description is all that is needed for applying the theorem (the reader wanting the details is referred to Arnol'd's book).

We concentrate on a neighborhood $\mathbb{U} \subset \mathbb{M}$. Proving the theorem requires finding $2n$ functions q^α, $p_\alpha \in \mathcal{F}(\mathbb{U})$ that satisfy the PB relations of Eq. (5.52), for when these $2n$ functions are chosen as the coordinates, ω is in the canonical form of Eq. (5.43). There are two steps to the proof.

We start with the second statement. First we show that given any function $S \in \mathcal{F}(\mathbb{M})$ (with the condition that $dS \neq 0$ in $\mathbb{U}$), there exists another function $R \in \mathcal{F}(\mathbb{U})$ whose PB with S is $\{R, S\} = 1$. The function S, through its vector field X_S or its one-parameter group φ^S, generates motion in $\mathbb{U}$ the same way that p_λ generates translations in the direction of increasing q^λ. Then in $\mathbb{U}$ a function R is constructed to increase along the motion so that $\{R, S\} = 1$: S generates translation in the direction of increasing R.

Rather than prove this analytically we illustrate it in Fig. 5.12. The directed curves of Fig. 5.12 represent the integral curves of X_S (or the flow of φ^S). At some point ξ_0 in $\mathbb{U}$ set $S = 0$ (by adding a constant to S) and pass a submanifold $\mathbb{N}$ through ξ_0 that crosses all the integral curves of X_S in $\mathbb{U}$. Pick a point $\xi \in \mathbb{U}$ not on $\mathbb{N}$; it necessarily lies on an integral curve of X_S and belongs to some value of the parameter λ of the one-parameter group φ^S. If we set $\lambda = 0$ on $\mathbb{N}$, then $\varphi^S_\lambda \in \varphi^S$ carries a point on $\mathbb{N}$ to ξ, and we will call λ the *distance* from $\mathbb{N}$ to ξ. This is illustrated in the figure. Varying λ then maps out the integral curve passing through the chosen ξ, just as varying t for a fixed initial point in a dynamical system maps out a dynamical integral curve. Now define the function $R(\xi)$ for each $\xi \in \mathbb{U}$ as the distance λ of ξ from $\mathbb{N}$, so that by definition the rate of change of R along the integral curves is $dR/d\lambda = 1$. But as is true for all dynamical variables, the rate of change of R along the integral curves of X_S is $\{R, S\}$. Hence $\{R, S\} = 1$. Thus the desired function R has been constructed.

If $n = 1$, this completes the proof of the theorem, for R can be taken to be q^1, and S to be p_1. The proof for higher n is by induction: the theorem is assumed proven for manifolds of lower dimension $2n - 2$ and then the $n = 1$ proof extends it to dimension $2n$. Unfortunately $\mathbb{N}$ cannot serve as the manifold of lower dimension, for its dimension is $2n - 1$, not $2n - 2$, and being of odd dimension it cannot have canonical coordinates. Instead, the lower-dimensional manifold to which the assumption hypothesis is applied is the $(2n - 2)$-dimensional submanifold $\mathbb{L}$ of $\mathbb{N}$ given by $q^1 = p_1 = 0$ (the coordinates of ξ_0 are $q^1 \equiv S = 0$ and $p_1 \equiv R = 0$, so $\mathbb{L}$ passes through ξ_0; see Fig. 5.13). The rest of the proof, which we do not describe, shows that the symplectic form ω on $\mathbb{M}$ can be broken into two parts:

$$\omega = dR \wedge dS + \omega|_{\mathbb{L}}, \tag{5.166}$$

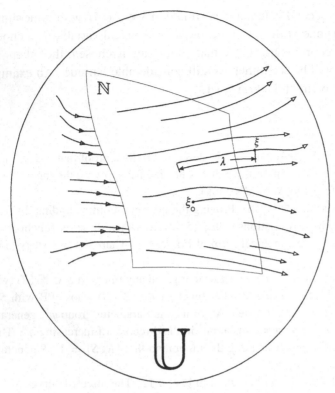

FIGURE 5.12
A neighborhood $\mathbb{U}$ of $\mathbb{M}$ (both of dimension $2n$) containing the submanifold $\mathbb{N}$. In the diagram $\mathbb{U}$ is represented by a region in $\mathbb{R}^3$, which masks the fact that its dimension $2n$ is even. (The lowest nontrivial value of n is 2, so that $\mathbb{U}$ has dimension 4, which makes it impossible to draw.) Similarly, $\mathbb{N}$ is represented by a 2-dimensional surface, which masks its odd dimension $2n - 1$. The point ξ_0 lies in $\mathbb{N}$. The integral curves of X_S (the flow lines of φ^S) are shown, passing through $\mathbb{N}$. For the point ξ shown, $R(\xi) = \lambda$. For any point in $\mathbb{N}$ (including ξ_0), $R(\xi) = 0$.

where $\omega|_\mathbb{L}$ is a symplectic form on $\mathbb{L}$. Then the inductive assumption implies that $\omega|_\mathbb{L}$ can be written in the canonical form

$$\omega|_\mathbb{L} = \sum_{\alpha=2}^{n} dq^\alpha \wedge dp_\alpha,$$

where q^α, p_α, $\alpha \in (2, \ldots, n)$, are canonical coordinates on $\mathbb{L}$. When R is taken as q^1 and S as p_1, the symplectic form ω is in canonical form. An important aspect of this is that the q^α and p_α, $\alpha \in (2, \ldots, n)$, in addition to having the correct PB commutation properties with each other, commute with q^1 and p_1. For more details, see Arnol'd (1988).

REDUCTION

Darboux's theorem is important in reducing the dimension of a dynamical system: it is what lends each constant of the motion its "double value," as mentioned at Eq. (5.81). The central

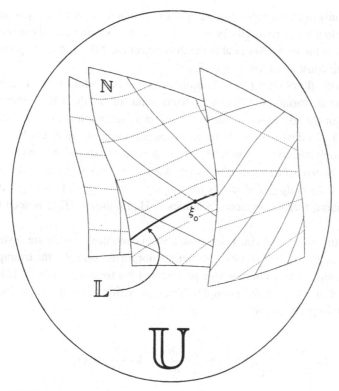

FIGURE 5.13

The neighborhood $\mathbb{U}$ of Fig. 5.12, showing three constant-R (equivalently, constant-q^1) submanifolds. The submanifold $\mathbb{N}$, on which $q^1 = 0$, contains a further submanifold on which $p_1 = 0$: this is $\mathbb{L}$, of dimension $2n - 2$, represented in the diagram by the heavy curve. (Because $\mathbb{N}$, whose dimension is $2n - 1$, is represented by a two-dimensional surface, $\mathbb{L}$ must be represented by a one-dimensional curve. This masks the fact that even if $n = 2$, the dimension of $\mathbb{L}$ is greater than 1.) The constant-p_1 submanifolds in $\mathbb{U}$ are not shown, but they intersect the constant-q^1 submanifolds in the broken curves (actually in manifolds of dimension $2n - 2$). The solid curves (these are in fact one-dimensional curves) on the constant-q^1 submanifolds are the integral curves of X_R, along which only p_1 varies. These form part of the coordinate grid.

force problem can be used to illustrate this and to point out some of the subtleties involved in reduction. Recall how the constancy of p_ϕ was used to reduce the central force problem and obtain Eq. (5.81) from (5.80). We will use that as an example from which to generalize the reduction procedure.

Suppose a constant of the motion S is known (p_ϕ in the example) so the dynamical system is known to flow on the $(2n - 1)$-dimensional invariant submanifolds whose equations are $S = $ const. ($p_\phi = \mu$ in the example). By Darboux's theorem, S can be taken as the first of the momentum coordinates in a canonical coordinate system on $\mathbf{T}^*\mathbb{Q}$, and its canonically conjugate generalized coordinate R can be found (in the example S is p_ϕ and R is ϕ). When the Hamiltonian H is written in this canonical coordinate system (in the example H is already written that way), it does not depend on R (in the example ϕ is ignorable) because S is a constant of the motion. That is, $\dot{S} = 0$, and since $\dot{S} = -\partial H/\partial R$, this means that $\partial H/\partial R = 0$. Thus on each $(2n - 1)$-dimensional submanifold the dynamics does not depend on R and involves only $2n - 2$ other variables (in the example, ϕ fails to appear not only in the Hamiltonian, but also in Hamilton's canonical equations obtained from it). This results in a *set* of reduced $(2n - 2)$-dimensional *Hamiltonian* dynamical systems, one for each constant value of S. Notice that the dimension has been reduced by two: that is the "double value" of the constant of the motion S.

Each of the reduced systems flows on a $(2n - 2)$-dimensional submanifold $\mathbb{L} \subset \mathbf{T}^*\mathbb{Q}$ on which neither S nor R appear as variables. That is, a fixed value of S labels the submanifold $\mathbb{L}$, and R is ignorable (in the example, μ labels $\mathbb{L}$, and ϕ is ignorable; the remaining variables on $\mathbb{L}$ are r, p_r, θ, p_θ). To deal with the reduced dynamics and show that it is Hamiltonian, one breaks up the symplectic form ω on $\mathbf{T}^*\mathbb{Q}$ into two parts in accordance with Eq. (5.166) by identifying $\omega|_\mathbb{L}$ (ω *restricted to* $\mathbb{L}$) as what is left when $dR \wedge dS$ is subtracted from ω (in the example $dR \wedge dS$ is $d\phi \wedge dp_\phi$, and $\omega|_\mathbb{L}$ is $dr \wedge dp_r + d\theta \wedge dp_\theta$). It is important to understand how it comes about that the reduced dynamics is Hamiltonian. This is seen by analyzing $\omega|_\mathbb{L}$.

Here is where Darboux's theorem comes in. It says that a local canonical coordinate system can be chosen of the form $\{R, S, \xi^3, \ldots, \xi^{2n}\}$ with the ξ^k being coordinates on $\mathbb{L}$ (in the example they are r, p_r, θ, p_θ) such that $\omega|_\mathbb{L}$ contains only wedge products of the form $d\xi^j \wedge d\xi^k$, $j, k \in \{3, \ldots, 2n\}$, none involving dR or dS (as in the example). Then the action of ω on a vector field X can also be broken up into two parts. Write

$$X = X_R \frac{\partial}{\partial R} + X_S \frac{\partial}{\partial S} + \sum_{k=3}^{2n} X^k \frac{\partial}{\partial \xi^k} \equiv X_R \frac{\partial}{\partial R} + X_S \frac{\partial}{\partial S} + X|_\mathbb{L},$$

so that

$$i_X \omega = i_X(dR \wedge dS) + i_X \omega|_\mathbb{L}, \tag{5.167}$$

The X^k, and hence $X|_\mathbb{L}$, may depend on R and S, but S is a constant on $\mathbb{L}$, so even if it appears in $X|_\mathbb{L}$ it won't affect the reduced dynamics. Moreover, if $X = \Delta$ is the dynamics, it does not depend on the ignorable coordinate R (in the example, the dynamical vector field is independent of ϕ; see Eq. (5.14), whose right-hand sides are the components of Δ). Then both $\omega|_\mathbb{L}$ and $X|_\mathbb{L} \equiv \Delta|_\mathbb{L}$ (ignoring its possible S dependence) depend only on the ξ^k, and $i_\Delta \omega$ can be rewritten as

$$i_X \omega = i_\Delta(dR \wedge dS) + i_{\Delta|_\mathbb{L}} \omega|_\mathbb{L}. \tag{5.168}$$

But since $\partial H / \partial R = 0$,

$$dH = \frac{\partial H}{\partial S} dS + \sum_3^{2n} \frac{\partial H}{\partial \xi^k} d\xi^k = \frac{\partial H}{\partial S} dS + dH|_\mathbb{L}. \tag{5.169}$$

When this is inserted into Eq. (5.168), it implies that $i_\Delta dS = 0$ (we already knew this, for S is a constant of the motion) and that $i_\Delta dR = \partial H / \partial S = \dot{R}$ (in the example, there is no $\partial / \partial p_\phi$

part in Δ, and the $\partial/\partial\phi$ part is $\partial H/\partial p_\phi = \dot{\phi}$). More to the point, it also implies that

$$i_{\Delta|_\mathbb{L}}\omega|_\mathbb{L} = dH|_\mathbb{L} \tag{5.170}$$

[in the example these are the canonical equations obtained from the Hamiltonian of Eq. (5.81), namely the upper four of Eqs. (5.14) with p_ϕ replaced by μ]. This shows that the reduced dynamics $\Delta|_\mathbb{L}$ on $\mathbb{L}$ is Hamiltonian, as asserted. Moreover, the Hamiltonian of the reduced system is just H restricted to $\mathbb{L}$ and the symplectic form is just ω restricted to $\mathbb{L}$.

So far we have dealt with only one constant of the motion. If more than one is used, the situation gets more complicated. Suppose that S_1 and S_2 are two constants of the motion and that S_1 has already been used to reduce the system by two dimensions. If S_2 is to reduce it further, it must be a coordinate on $\mathbb{L}$, a function only of the ξ^k. This is possible iff $\{S_1, S_2\} = 0$. Thus the double reduction will work for each of the two constants only if they are *in involution* (i.e., if they commute). Only then can S_1 and S_2 be generalized momenta in a canonical coordinate system and thus both appear in the argument of H. If both do appear, their two canonically conjugate variables R_1 and R_2 will be ignorable and the reduction will be by two additional dimensions, for a total of four. But if S_1 and S_2 are not in involution, they cannot both appear as momenta in the argument of H, and the second reduction is just by one dimension (only R_1 *or* R_2 is ignorable), to a manifold of dimension $2n - 3$. Because its dimension is odd, the manifold cannot be symplectic, and the Hamiltonian formalism can no longer be used.

This happens if one tries to use more than one component of the angular momentum vector for the canonical reduction of a rotationally invariant Hamiltonian system, for the components of the angular momentum do not commute, as was seen in Worked Example 5.3. This implies also that the three components of the angular momentum cannot be used as canonical momenta in the Hamiltonian description of a dynamical system (but see Marsden et al., 1991, and Saletan and Cromer, 1971, p. 207). We do not deal with the general case of noncommuting constants of the motion (Marmo et al., 1985).

It is therefore always helpful to find the maximum number of commuting constants of the motion (i.e., in involution). In the best of all possible cases, n of them can be found: $S_1, S_2, \ldots, S_n$. Then all of their canonically conjugate coordinates $R_1, R_2, \ldots, R_n$ are ignorable, and H can be written as a function of the S_k alone. The equations of motion are then trivial: $\dot{S}_k = 0$ and $\dot{R}_k = \text{const}$. This possibility will be discussed in some detail in Chapter 6. (Full sets of n commuting constants of the motion are reminiscent of, in fact related to, complete sets of commuting observables in quantum mechanics.)

WORKED EXAMPLE 5.7

Reduce the central force problem of Eq. (5.13) by using the constant of the motion $j^2 \equiv \mathbf{j} \cdot \mathbf{j}$.

Solution. Use Eqs. (5.102) and (5.103) to write the Cartesian components of $\mathbf{j}$ in terms of the polar coordinates and momenta (e.g., $j_x = yp_z - zp_y$):

$$j_x = -p_\theta \sin\phi - p_\phi \frac{\cos\theta\cos\phi}{\sin\theta},$$
$$j_y = p_\theta \cos\phi - p_\phi \frac{\cos\theta\sin\phi}{\sin\theta}, \tag{5.171}$$
$$j_z = p_\phi.$$

Squaring each of these and adding them up yields

$$j^2 = p_\theta^2 + \frac{p_\phi^2}{\sin^2\theta}. \tag{5.172}$$

With this, the Hamiltonian of Eq. (5.13) becomes

$$H = \frac{1}{2m}\left[p_r^2 + \frac{j^2}{r^2}\right] + V(r), \tag{5.173}$$

in which j^2 is a constant, reducing the problem to one freedom. The equations of motion on each $j^2 = $ const. submanifold are

$$\left.\begin{array}{l} \dot{r} = \dfrac{\partial H}{\partial p_r} = \dfrac{p_r}{m}, \\[2mm] \dot{p}_r = -\dfrac{\partial H}{\partial r} = \dfrac{j^2}{mr^3} - V'(r). \end{array}\right\} \tag{5.174}$$

Comments:

1. These would be the same as the first two equations of (5.15) if j^2 were equal to p_θ^2, but that is true only if $p_\phi = 0$. In the present reduction we did not need to make that assumption.
2. The reduction is immediately by four dimensions, from six to two.
3. In principle, because it commutes with both r and p_r (this follows from Worked Example 5.3) the dynamical variable j^2 can be taken as a generalized momentum in a canonical coordinate system on $\mathbf{T}^*\mathbb{Q}$. In accordance with Darboux's theorem the symplectic form can then be written

$$\omega = dr \wedge dp_r + d\psi \wedge dj^2 + \text{another term}, \tag{5.175}$$

where ψ is a dynamical variable that serves as the coordinate conjugate to j^2. The "other term" is necessary because $\mathbf{T}^*\mathbb{Q}$ is of dimension six; it and the last term, still to be found, contains a third pair of conjugate variables. That is, $(d\psi \wedge dj^2 + \text{another term})$ is the $\omega|_L$ of the theorem. But neither ψ nor the third pair of variables is easy to find, and therefore it is difficult to write ω in this way. That is, completing the coordinate system is not trivial, and we avoid it by not trying to write out ω in these terms.

PROBLEMS

1. A particle in a uniform gravitational field is constrained to the surface of a sphere, centered at the origin, with a radius $r(t)$ which is a given function of the time. Obtain the Hamiltonian and the canonical equations. Discuss energy conservation. Is the Hamiltonian the total energy?

2. Consider the one-freedom Hamiltonian

$$H(q, p) = \frac{p^2}{2m} e^{-q/a}$$

   for a particle of mass m.

   (a) Solve for $q(t)$ and $p(t)$. Show that if $\mathbf{T}^*\mathbb{Q}$ is taken to be the half-plane with $p > 0$, the system looks like a particle with a retarding force proportional to $\dot{q}^2$. Find the particular motion with the initial conditions $q(0) = 0$, $p(0) = mv$, where $v > 0$. Find the limits of q, p, and $\dot{q}$ as $t \to \infty$. What is the relation of H to the kinetic energy $T = \frac{1}{2}m\dot{q}^2$? At what rate is T dissipated?

   (b) What term has to be added to the Hamiltonian to provide an additional positive constant force?

3. Given the Hamiltonian

$$H = q^1 p_1 - q^2 p_2 - a(q^1)^2 + b(q^2)^2,$$

   where a and b are constants, show that the three functions

$$f_1 = (p_2 - bq^2)/q^1, \quad f_2 = q^1 q^2, \quad f_3 = q^1 e^{-t}$$

   are constants of the motion. Are they functionally independent? Do there exist other independent constants of the motion? If they do exist, find them until you obtain the maximum number of functionally independent ones. Show explicitly that Poisson's theorem (see Problem 19) is valid. [Note: This Hamiltonian is unusual in that it does not come from any Lagrangian (see Problem 5). Indeed, one of the equations of motion on $\mathbb{Q}$ is of first order. Hint: If f and g are functionally dependent (i.e., if one can be written as a function of the other or both can be written as functions of a third function) then $df \wedge dg = 0$. Can you see why?]

4. Consider a particle moving in the plane under the influence of the generalized potential (generalized because it is velocity dependent) $V = (1 + \dot{r}^2)/r$, where r is the distance from the origin. Write down the Hamiltonian in polar coordinates. Discuss the conservation of angular momentum (is H invariant under rotation?). Reduce the problem on $\mathbb{Q}$ to a first-order differential equation in r.

5. Given a Hamiltonian function $H(q, p, t)$, how does one obtain the corresponding Lagrangian? That is,

   (a) describe the inverse of the procedure that leads to Eq. (5.4).

   (b) Establish necessary conditions for this procedure.

(c) Show that the EL equations are derivable from Hamilton's canonical equations and the procedure of Part (a).

[Partial answers: (a) $L(q, \dot{q}, t) = \dot{q}^\beta p_\beta(q, \dot{q}, t) - H(q, p(q, \dot{q}, t), t)$. (b) $\partial^2 H / \partial p^\alpha \partial p^\beta$ is nonsingular.]

6. We say that a Lagrangian L undergoes a gauge transformation when any function of the form $d\psi(q, t)/dt$ is added to it (see Section 2.2.4.) How does the Hamiltonian change under such a gauge transformation? [Remark: The p_α change to new momenta p'_α, and the transformation from (q, p) to (q, p') is a particular kind of CT called a *gauge* CT. See Problem 8.]

7. Define a Lagrangian function $\hat{L}(q, \dot{q}, p, \dot{p}, t) \equiv \hat{L}(\xi, \dot{\xi}, t)$ on $\mathbf{T}^*\mathbf{Q}$ by

$$\hat{L}(q, \dot{q}, p, \dot{p}, t) = \dot{q}^\alpha p_\alpha - H(q, p, t).$$

Show that Hamilton's canonical equations are the EL equations for the Lagrangian $\hat{L}$, that is, show that

$$\frac{d}{dt} \frac{\partial \hat{L}}{\partial \dot{\xi}^k} - \frac{\partial \hat{L}}{\partial \xi^k} = 0.$$

[Hint: This is easier to do in the (q, p) notation. Incidentally, this result implies that Hamilton's canonical equations can be derived from a variational principle.]

8. A gauge transformation of the electromagnetic field [Eq. (2.51)] changes both the Hamiltonian of Eq. (5.7) and the momentum of Eq. (5.6).

   (a) Show that the change is a CT of Type 2, find its generating function, and show that the new Hamiltonian is given by Eq. (5.124).

   (b) Call a CT like that of Part (a) a *gauge* CT. Show in n freedoms that any CT that leaves all of the q^α invariant is a gauge CT.

9. Obtain the following equations for transformations of Types 3 and 4.

   Type 3:

   $$q^\alpha = -\partial F^3 / \partial p_\alpha, \quad P_\alpha = -\partial F^3 / \partial Q^\alpha, \quad K = H + \partial F^3 / \partial t.$$

   Type 4:

   $$q^\alpha = -\partial F^4 / \partial p_\alpha, \quad Q^\alpha = \partial F^4 / \partial P_\alpha, \quad K = H + \partial F^4 / \partial t.$$

   Express $F^3(p, Q, t)$ and $F^4(p, P, t)$ in terms of $F(q, p, t)$.

10. Of what Type or Types j is the identity transformation $Q^\alpha = q^\alpha$, $P_\alpha = p_\alpha$? Find the generating functions F^j of all possible Types for this transformation.

11. From the solution to the relativistic Kepler problem find the change $\Delta\theta$ between successive minima of the precessing elliptic orbit when Γ of Eq. (5.35) is close to 1. [Partial answer: $\Delta\theta$ is proportional to p_θ^{-2}.] [Note: This result is not the same as one of Einstein's original three tests of the general theory. Einstein showed in 1915 that general relativistic effects perturb the Kepler potential by an additive term proportional to $1/r^2$ and used that

to calculate the degree to which the orbit of Mercury fails to close. That $\Delta\theta$ is also proportional to p_θ^{-2}, but by a larger factor. See Worked Example 6.4 and Goldstein, 1981, p. 338, Problem 26.]

12. **(a)** Prove that the transformation of Eq. (5.86) is canonoid with respect to $H = \frac{1}{2}p^2$ and that the new Hamiltonian K is the one given.

    (b) Show that this transformation is not canonoid with respect to $H = \frac{1}{2}(p^2 + q^2)$. [Hint: Find the equations for $\dot{Q}$ and $\dot{P}$ and show that no new Hamiltonian exists.]

    (c) Find a region of $\mathbf{T}^*\mathbb{Q} = \mathbb{R}^2$, as large as possible, on which the transformation is invertible.

13. Consider the function $F(\xi, t) = pq - \frac{1}{2}p^2 t/m$ on $\mathbf{T}^*\mathbb{Q} = \mathbb{R}^2$.

    (a) Find all CTs for which F is the generating function.

    (b) Consider a system whose Hamiltonian is $H = T + V(q)$, where $T = \frac{1}{2}p^2/m$. From among the CTs found in Part (a) find those for which the new Hamiltonian is $K = V(q(Q, P))$. [This is reminiscent of the interaction representation in quantum mechanics.]

14. Let

$$Q^1 = (q^1)^2, \quad Q^2 = q^1 + q^2, \quad P_\alpha = P_\alpha(q, p), \quad \alpha = 1, 2$$

be a CT in two freedoms.

    (a) Complete the transformation by finding the most general expressions for the P_α.

    (b) Find a particular choice for the P_α that will reduce the Hamiltonian

$$H = \left(\frac{p_1 - p_2}{2q_1}\right)^2 + p_2 + (q^1 + q^2)^2$$

to

$$K = P_1^2 + P_2.$$

Use this to solve for the $q^\alpha(t)$.

15. Let $Q^\alpha = A^\alpha{}_\beta q^\beta$ be part of a time-independent CT from (q, p) to (Q, P), where the $A^\alpha{}_\beta$ are the elements of a nonsingular constant matrix A. Write $P_\alpha = f_\alpha(q, p)$.

    (a) Show that then $P_\alpha = B_\alpha{}^\beta p_\beta + g_\alpha(q)$, where the $B_\alpha{}^\beta A^\gamma{}_\beta = \delta^\gamma_\alpha$. That is, show that the $B_\alpha{}^\beta$ are the elements of a matrix B, the inverse transpose of A. Show that the g_α satisfy a condition that causes the P_α to be defined only up to a gauge transformation (see Problem 8). [Note that $g_\alpha = 0 \ \forall \ \alpha$ satisfies these conditions.]

    (b) Apply this to the case in which A is a rotation in $\mathbb{R}^3$. [Rotations are discussed in some detail in Chapter 8.]

16. A particle of mass m and charge e moves in the $(1, 2)$ plane under the combined influence of the harmonic oscillator potential $\frac{1}{2}m\omega_0^2[(q^1)^2 + (q^2)^2]$ and a magnetic field whose vector potential is $\mathbf{A} = (0, hq^1, 0)$, where h is a constant. Use the CT of Eq. (5.99) with $\tan 2\alpha = 2m\omega(eh)^{-1}$ and β chosen for convenience to find the motion in general and in the two limits $h = 0$ and $h \to \infty$.

17. The $\mathbb{Q}$ part of a CT in one freedom $(T^*\mathbb{Q} = \mathbb{R}^2)$ is $Q = \arctan(\lambda q/p)$, $\lambda \neq 0$. Complete it by finding $P(q, p, t)$. [Answer: $P = \frac{1}{2}(p^2/\lambda + \lambda q^2) + R(q, p, t)$, where R is an arbitrary function that is homogeneous of degree zero in the pair of variables (q, p).]

18. A particle of mass m is subjected to the force $F = -kq - \alpha/q^3$.

    (a) Show that a possible Hamiltonian for this system is $H = p^2/(2m) + \frac{1}{2}kq^2 + Ap/q$, where A is a properly chosen constant.

    (b) Use the CT of Problem 17 to solve for the motion.

    (c) Discuss the relation of H to the Hamiltonian H' that would be found by simply finding a potential $V(q)$ from the given F, constructing the Lagrangian $L' = T - V$, and then obtaining the Hamiltonian by the usual procedure.

19. Prove Poisson's theorem: if $F, G \in \mathcal{F}(T^*\mathbb{Q})$ are constants of the motion of a Hamiltonian dynamical system, so is their PB $\{F, G\}$.

20. Let $f, g, h \in \mathcal{F}(T^*\mathbb{Q})$.

    (a) Prove that the Poisson bracket satisfies the Jacobi identity, i.e., that $\{f, \{g, h\}\} + \{g, \{h, f\}\} + \{h, \{f, g\}\} = 0$.

    (b) Prove that the Poisson bracket satisfies the Leibnitz rule, i.e., that $\{f, gh\} = g\{f, h\} + \{f, g\}h$. [Both of these proofs are easier in the ξ notation than in the (q, p) notation.]

21. Demonstrate that the Cartan identity $\mathbf{L}_X = i_X d + d i_X$ holds when applied (a) to $g\,df$ and (b) to $dg \wedge df$. (This identity is true in general, but its full proof requires dealing with forms of higher degree.) [Answer to Part (b): (Note: To avoid confusion, always write expressions like $f\,dg$, never $(dg)f$, because $d(f\,dg) = df \wedge dg \neq dg \wedge df$.) The second equality of the first of the two calculations below comes from Part (a). The solution would be complete without the last two lines of the first calculation, but we carry it further to exhibit some manipulations in the exterior calculus.

$$\mathbf{L}_X(dg \wedge df) = \mathbf{L}_X d(g\,df) = d[(i_X d + d i_X)(g\,df)]$$
$$= d i_X d(g\,df) = d i_X(dg \wedge df)$$
$$= d[(i_X\,dg)\,df - (i_X\,df)\,dg]$$
$$= (d \cdot i_X\,dg) \wedge df - (d \cdot i_X\,df) \wedge dg.$$

We have used $d^2 = 0$ and we use it again below.

$$(i_X d + d i_X)(dg \wedge df) = d i_X(dg \wedge df) = \mathbf{L}_X(dg \wedge df),$$

which completes Part (b).]

22. Let $f, g \in \mathcal{F}(T^*\mathbb{Q})$, and let $F, G \in \mathcal{X}(T^*\mathbb{Q})$ be the associated Hamiltonian vector fields, that is, $i_F\omega = -df$ and $i_G\omega = -dg$. Show that the vector field $[G, F]$ is Hamiltonian with respect to the Poisson bracket $\{f, g\}$. [Hint: Use Problem 3.5(d) and (e), the Leibnitz product rule, and the Cartan identity of Problem 21.]

23. Given a one-form $\alpha = \alpha_k\,d\xi^k$, a two-form $\omega = \omega_{jk}\,d\xi^j \wedge d\xi^k$, and a vector field $X = X^k\partial_k$, calculate the components of the one-form $\mathbf{L}_X\alpha$ and of the two-form $\mathbf{L}_X\omega$.

[Hint: Remember that $\mathbf{L}_X d\xi^k = d\mathbf{L}_X\xi^k$.] [Answer: $(\mathbf{L}_X\alpha)_k = (\partial_j\alpha_k)X^j + \alpha_j(\partial_k X^j)$ and $(\mathbf{L}_X\omega)_{ij} = (\partial_k\omega_{ij})X^k + (\omega_{kj}\partial_i - \omega_{ki}\partial_j)X^k$.]

24. Show that $i_{\partial_j}(d\xi^k) \equiv \langle d\xi^k, \partial_j \rangle = \delta_j^k$.

25. Let $\eta^k(\xi, t)$ be a CT from the ξs, or from (q, p), to the η^k, or to (Q, P), and define $\hat{L}(\xi, \dot{\xi}, t)$ as in Problem 7. Define $\hat{L}'(\eta, \dot{\eta}, t)$ similarly by

$$\hat{L}'(Q, \dot{Q}, P, \dot{P}, t) = \dot{Q}^\alpha P_\alpha - K(Q, P, t).$$

Assume that

$$\frac{d}{dt}\frac{\partial\hat{L}}{\partial\dot{\xi}^k} - \frac{\partial\hat{L}}{\partial\xi^k} = 0,$$

that is, that the ξ^k satisfy Hamilton's canonical equations. Show that then

$$\frac{d}{dt}\frac{\partial\hat{L}'}{\partial\dot{\xi}^k} - \frac{\partial\hat{L}'}{\partial\xi^k} = 0.$$

[Note: Because the transformation is canonical, a result similar to that of Problem 7 holds: Hamilton's canonical equations in the η^k for the Hamiltonian K are the EL equations in the η^k for the Lagrangian $\hat{L}'$. Thus if the EL equations hold in the ξ^k, they hold in the η^k. But beware! You are asked to prove something different, for the equations that appear here are neither the ξ nor the η EL equations; not the ξ equations because $\hat{L}$ and $\hat{L}'$ are different functions on $\mathbf{T}^*\mathbb{Q}$, and not the η equations because those are

$$\frac{d}{dt}\frac{\partial\hat{L}'}{\partial\dot{\eta}^k} - \frac{\partial\hat{L}'}{\partial\eta^k} = 0,$$

and the ξ^k and η^k are different variables. This is discussed further in Chapter 6.]

26. Use the Hamiltonian of Eq. (5.31) to verify that $r^2 p_r = r' p_\theta$, where $r' = dr/d\theta$; that is, derive Eq. (5.32).

27. For the relativistic Kepler problem show (a) that if $p_\theta > e^2/c$, the orbit never passes through the center of force (as in the nonrelativistic problem), but (b) that if $p_\theta < e^2/c$, the orbit is a spiral, and argue that the radius r from the center of force reaches zero in a finite time.

28. (a) Prove that $\Delta\sigma^2$ of Eq. (5.20) is invariant under Lorentz transformations.
    (b) (This is harder!) Prove that the only linear transformation under which $\Delta\sigma^2$ is invariant is the Lorentz transformation. [Hints: Do this in two dimensions (one space dimension and the time t). Take one of the two events to be the origin, with coordinates $(0, 0)$. What you should find is that there is a parameter β that can be interpreted as V/c.]

29. Let $\mathbb{V}$ be an n-dimensional vector space, and consider a set of n vectors $\mathbf{v}_k \in \mathbb{V}$, with components v_k^j in one coordinate system and w_k^j in another, $k, j \in \{1, 2, \ldots, n\}$. Consider the two volumes

$$V_1(\mathbf{v}_1, \ldots, \mathbf{v}_n) = \det\begin{bmatrix} v_1^1 & \cdots & v_n^1 \\ \vdots & \ddots & \vdots \\ v_1^n & \cdots & v_n^n \end{bmatrix}$$

and

$$V_2(\mathbf{v}_1, \ldots, \mathbf{v}_n) = \det \begin{bmatrix} w_1^1 & \cdots & w_n^1 \\ \vdots & \ddots & \vdots \\ w_1^n & \cdots & w_n^n \end{bmatrix}.$$

Show that $V_2 = cV_1$, where c is the determinant of the transformation matrix between the two coordinate systems. Show that if the coordinate systems are both orthonormal, $c = \pm 1$.

30. Consider a free particle in one freedom confined to a box of length $2l$, with perfectly reflecting walls at l and $-l$. Study the evolution of the density distribution function which at $t = 0$ is given by $D(q, p, 0) = (ab)^{-1}$ for $0 < q < a$, $0 < p < b$, and $D(q, p, 0) = 0$ elsewhere in $\mathbf{T}^*\mathbb{Q}$, where $a < l/2$. That is, check whether D is normalized and see what happens to the region in which $D \neq 0$ as time progresses.

31. A one-freedom system consists of a bead of mass m sliding on a frictionless V-shaped wire which is standing upright in a uniform gravitational field. Assume (unrealistically) no energy loss as the bead passes the low point of the V, where the two straight portions of the wire meet. Solve for the motion and find how the period depends on the amplitude. Choose q to be the horizontal component of the bead's position vector, and draw the phase diagram for this system. Discuss the motion of a region $R \subset \mathbf{T}^*\mathbb{Q}$ and make a rough diagram (as in Worked Example 5.6). [Suggestion: Choose a narrow wedge-shaped region for R.]

32. Apply the method of Eq. (5.136) to the simple case of a harmonic oscillator in one freedom. Show explicitly that the power series in t converges to the usual result.

33. Show that if a function $f(q)$ is a constant of the motion of a Hamiltonian dynamical system, the Hamiltonian H is degenerate in the sense that $\det(\partial^2 H/\partial p_\alpha \partial p_\beta)$ vanishes. [Hint: Take the PB of f with H.]

34. Half of the components of a certain Hamiltonian vector field $X = X^\alpha(\partial/\partial q^\alpha) + X_\alpha(\partial/\partial p_\alpha)$ are $X^\alpha = q^\alpha$. Find the function f with respect to which X is Hamiltonian and the most general expression for X. Write out the differential equations for the integral curves of X. Solve the q part of the equations. Find a simple special case and solve the p part. Comment on the solutions.

35. **(a)** Half of the components of a certain Hamiltonian vector field in two freedoms ($\mathbf{T}^*\mathbb{Q}$ is of dimension four) $X = X^\alpha(\partial/\partial q^\alpha) + X_\alpha(\partial/\partial p_\alpha)$ are $X^1 = q^2$, $X^2 = -q^1$. Find the function f with respect to which X is Hamiltonian and the most general expression for X. Write out the differential equations for the integral curves of X. Solve the q part of the equations. Find a simple special case and solve the p part. Comment on the solutions.

(b) Solve the same problem as Part (a) except for $X^1 = q^2$, $X^2 = q^1$.

CANONICITY IMPLIES PB PRESERVATION

In this appendix we prove that canonicity implies preservation of the Poisson brackets (Currie and Saletan, 1972).

Let $H(\xi, t)$ be any function in $\mathcal{F}(\mathbf{T}^*\mathbb{Q})$ and write $\dot{\xi}^k = \omega^{kj} \partial_j H$. Then a transformation $\eta(\xi, t)$ is canonical if for every H there exists a $K(\eta, t)$ such that

$$\dot{\eta}^k = \omega^{kj} \frac{\partial K(\eta, t)}{\partial \eta_j}. \tag{5.176}$$

What we will prove is that canonicity of η implies that there exists a nonzero constant a such that

$$\{f, g\}^\eta = a \{f, g\}^\xi \tag{5.177}$$

for any two functions $f, g \in \mathcal{F}(\mathbf{T}^*\mathbb{Q})$.

The proof will be obtained by calculating the PB $\{\xi^j, \xi^k\}^\eta \equiv \mu^{jk}(\xi, t)$. (Note that the μ^{jk} are defined as functions of the ξ^k.) The logic of the proof is that if K exists for every Hamiltonian H, it exists for specific choices of H, in particular for simple ones. By choosing certain Hamiltonians it is first shown that the μ^{jk} are constants. Then by choosing others the constants are calculated. It should be clear from the start that the μ^{jk} are defined only by the transformation $\eta^k(\xi, t)$ and do not depend on the Hamiltonians that are used to calculate them.

If for every H there exists a K, surely K exists if $H \equiv c$ is a constant. In that case $\dot{\xi}^k = 0$. Like the time variation of any function, that of ξ can be described in the η coordinates, and because K exists, the Leibnitz rule of Eq. (5.62) for the PB yields

$$\frac{d\mu^{jk}}{dt} \equiv \frac{d}{dt} \{\xi^j, \xi^k\}^\eta = \{\dot{\xi}^j, \xi^k\}^\eta + \{\xi^j, \dot{\xi}^k\}^\eta$$
$$= \{0, \xi^k\}^\eta + \{\xi^j, 0\}^\eta \equiv 0.$$

Thus the μ^{jk} are constant for the motion generated by $H = c$, or (now in the ξ coordinates)

$$\frac{d\mu^{jk}}{dt} = \{\mu^{jk}, c\}^\xi + \frac{\partial \mu^{jk}}{\partial t} = 0,$$

and since the PB here vanishes identically,

$$\frac{\partial \mu^{jk}}{\partial t} = 0: \tag{5.178}$$

the μ^{jk} are independent of t. We emphasize the fact that the μ^{jk} do not depend on the choice of H even though a particular H was used to discover that they are time independent.

Next, if for every H there exists a K, surely K exists if $H = c_l \xi^l$ where the c_l are arbitrary constants. In that case $\dot{\xi}^k = \omega^{kj} c_j$, and again describing the motion in the η coordinates yields

$$\frac{d\mu^{jk}}{dt} = \{\dot{\xi}^j, \xi^k\}^\eta + \{\xi^j, \dot{\xi}^k\}^\eta = \{\omega^{jl} c_l, \xi^k\}^\eta + \{\xi^j, \omega^{kl} c_l\}^\eta \equiv 0,$$

so that [using (5.178)]

$$\frac{d\mu^{jk}}{dt} = \{\mu^{jk}, c_l \xi^l\}^\xi = \frac{\partial \mu^{jk}}{\partial \xi^m} \omega^{ml} c_l = 0.$$

This must be true or arbitrary c_l. Since the ω^{ml} form a nonsingular matrix,

$$\frac{\partial \mu^{jk}}{\partial \xi^m} = 0: \tag{5.179}$$

the μ^{jk} are independent of the ξ^m or are identically constants: $d\mu^{jk}/dt \equiv 0$. It remains to find what constants they are.

Finally, if for every H there exists a K, surely K exists if $H = \frac{1}{2}c_{lm}\xi^l\xi^m$, where the $c_{lm} = c_{ml}$ are otherwise arbitrary constants. Then $\dot{\xi}^k = \omega^{kj}c_{jl}\xi^l$, and

$$0 \equiv \frac{d\mu^{jk}}{dt} = \{\dot{\xi}^j, \xi^k\}^\eta + \{\xi^j, \dot{\xi}^k\}^\eta = \omega^{jm}c_{ml}\{\xi^l, \xi^k\} + \{\xi^j, \xi^l\}c_{lm}\omega^{km},$$

or

$$\Omega CM - MC\Omega = 0,$$

where Ω and M are the antisymmetric matrices of the ω^{jk} and μ^{jk}, respectively, and C is the symmetric matrix of the c_{jk}. Multiply this on the left by $-\Omega$ and on the right by Ω, and use the fact that $\Omega^2 = -\mathbb{I}$, to obtain

$$CM\Omega = \Omega MC. \tag{5.180}$$

This must be true for all possible symmetric C matrices, in particular for $C = \mathbb{I}$, so $M\Omega = \Omega M$, and (5.180) becomes

$$C(M\Omega) = (M\Omega)C.$$

Thus $M\Omega$ commutes with all symmetric matrices. The only matrices that commute with all symmetric matrices are multiples of the unit matrix (proven in the last paragraph of this appendix), so $M\Omega = k\mathbb{I}$, or $M = -k\Omega$. If $-k$ is replaced by a, the result is

$$\mu^{jk} \equiv \{\xi^j, \xi^k\}^\eta = a\omega^{jk},$$

and

$$\{f, g\}^\eta = \frac{\partial f}{\partial \xi^j}\frac{\partial \xi^j}{\partial \eta^l}\omega^{lm}\frac{\partial \xi^k}{\partial \eta^m}\frac{\partial g}{\partial \xi^k} = \frac{\partial f}{\partial \xi^j}\{\xi^j, \xi^k\}^\eta \frac{\partial g}{\partial \xi^k}$$

$$= a\frac{\partial f}{\partial \xi^j}\omega^{jk}\frac{\partial g}{\partial \xi^k} = a\{f, g\}^\xi,$$

which completes the proof.

It is convenient always to take $a = 1$. A more general canonical transformation, one for which $a \neq 1$, can be obtained, for instance, by following one with $a = 1$ by the trivial transformation $\eta^k = \sqrt{a}\xi^k$. (Questions: How would you obtain one for which $a = -1$? What can you say about $a = 0$?)

It is interesting that not all Hamiltonians are used in this proof, but only those that are quadratic in the variables. This shows that such quadratic Hamiltonians play a special role in symplectic dynamics.

We indicate how to prove that the only matrices that commute with all symmetric ones are multiples of the unit matrix. Suppose C is a symmetric matrix and that N commutes with it. Let x_λ be a column eigenvector of C belonging to eigenvalue λ, so that $Cx_\lambda = \lambda x_\lambda$. Then

$$CNx_\lambda = NCx_\lambda = \lambda Nx_\lambda,$$

and Nx_λ is also an eigenvector of C belonging to the same eigenvalue. If the eigenspace is of dimension one (i.e., if λ has multiplicity one), Nx_λ must be a multiple of x_λ, and hence x_λ is an eigenvector also of N. Now, every column vector is an eigenvector of multiplicity one of *some* symmetric matrix, so if N commutes with all symmetric matrices, every column vector is an eigenvector of N. That is possible only if N is a multiple of the unit matrix.

CHAPTER 6

TOPICS IN HAMILTONIAN DYNAMICS

CHAPTER OVERVIEW

Most dynamical systems cannot be integrated in closed form: the equations of motion do not lend themselves to explicit analytic solution. There exist strategies, however, for approximating solutions analytically and for obtaining qualitative information about even extremely complex and difficult dynamical systems. Part of this chapter is devoted to such strategies for Hamiltonian dynamics.

The first two sections describe Hamilton–Jacobi (HJ) theory and introduce action–angle (AA) variables on $\mathbf{T}^*\mathbb{Q}$. HJ theory presents a new and powerful way for integrating Hamilton's canonical equations, and motion on $\mathbf{T}^*\mathbb{Q}$ is particularly elegant and easy to visualize when it can be viewed in terms of AA variables. Both schemes are important in their own right, but they also set the stage for Section 6.3.

6.1 THE HAMILTON–JACOBI METHOD

It was mentioned at the end of Worked Example 5.5 that there would be some advantage to finding local canonical coordinates on $\mathbf{T}^*\mathbb{Q}$ in which the new Hamiltonian function vanishes identically, or more generally is identically equal to a constant. In this section we show how to do this by obtaining a Type 1 generating function for a CT from the initial local coordinates on $\mathbf{T}^*\mathbb{Q}$ to new ones, all of which are constants of the motion. The desired generating function is the solution of a nonlinear partial differential equation known as the Hamilton–Jacobi (or HJ) equation. Unfortunately, the HJ equation is in general not easy to solve. In fact it is often impossible to solve it analytically in closed form even in physically important situations. Nevertheless, the HJ equation and the procedure that evolves from it for integrating the equations of motion proves to be an extremely useful tool, in general more so than Hamilton's canonical equations (which can be just as intractable). This is because the HJ method can be used to deal with many difficult dynamical systems as distortions of other systems that are easier to solve. We will discuss this in Section 6.3.2 on Hamiltonian perturbation theory.

Because it lends itself to perturbative calculations, the HJ method is extremely important in celestial mechanics. In addition, it played a crucial role in the old quantum theory
and thereby in the early development of quantum mechanics (see Born, 1960).

6.1.1 THE HAMILTON–JACOBI EQUATION

DERIVATION

We start by using the ideas of Section 5.3.3 to obtain a differential equation for the
Type 1 generator of a local CT from the point $\xi(t) \in \mathbf{T}^*\mathbb{Q}$ to a set of $2n$ local constants of
the motion η^j. (As explained in Section 4.2.2, there do not always exist $2n$ *global* constants
of the motion, so in general the method will work only locally.) Then the new Hamiltonian
K must be independent of the new variables η^j. Indeed, the η^j are all constants of the
motion, so Hamilton's canonical equations read

$$\dot{\eta}^j \equiv 0 = \omega^{jk}\frac{\partial K}{\partial \eta^k}:$$

K depends on none of the η^k, or K is identically a constant. Conversely, if the CT give a
new Hamiltonian $K = \text{const.}$, the new variables are constants of the motion. The CT from
the ξ^j to the η^j will have to depend on t, and therefore the generating function will also
depend on t. When the generating function is found, Eqs. (5.121) can be used to find $q(t)$
and $p(t)$. Because in general the procedure works only locally, the CT obtained is a local
one, which implies that it is valid only for a finite time, the time during which the system
remains in the local region in which the $2n$ constants of the motion are valid. Such a region
always exists, for the initial point in $\mathbf{T}^*\mathbb{Q}$ defines the motion at least locally.

If it is possible to make K identically equal to a constant, that constant can be set equal
to zero without losing generality, because an arbitrary constant can always be added to
the Hamiltonian without changing the motion. We proceed, therefore, to look for a local
CT whose new Hamiltonian K is identically equal to zero. As usual, we write $\eta(\xi, t)$ or
$Q(q, p, t)$, $P(q, p, t)$ for the transformation. Which of the new variables are called Ps and
which Qs is unimportant, but it is determined by the type of the CT. We choose the CT to
be of Type 1 (though it is often chosen also to be of Type 2). It is usual to write $S(q, Q, t)$
for the generating function that we have called $F^1(q, Q, t)$ in Section 5.3.3, for it turns out
to be almost the same as the action defined in the Lagrangian context of Section 3.1 (more
about this in the last part of this section). With this in mind, Eq. (5.122) becomes

$$\frac{\partial S}{\partial t} = -H.$$

In this equation everything, including the momenta p_α in the argument of $H(q, p, t)$,
must be written as functions of (q, Q, t). This can be done through Eq. (5.121): $p_\alpha = \partial S/\partial q^\alpha$. Then

$$\frac{\partial S}{\partial t} = -H\left(q, \frac{\partial S}{\partial q}, t\right). \tag{6.1}$$

This is the celebrated *Hamilton–Jacobi (HJ) equation*, a partial differential equation for the Type 1 generating function from (q, p) to a set of $2n$ local constants of the motion (Q, P). For actual physical systems, the equation is always nonlinear, since the momenta, and hence also the $\partial S/\partial q^\alpha$, appear quadratically in the Hamiltonian.

We will discuss solving the HJ equation in Section 6.1.2. In the meantime, assume that a solution for $S(q, Q, t)$ has been found. Because $K \equiv 0$ by construction, the motion in the (Q, P) coordinates is trivial:

$$Q^\alpha(t) = \text{const.}, \quad P_\beta(t) = \text{const.}$$

To find the motion in the (q, p) coordinates, use Eqs. (5.121):

$$p_\alpha = \frac{\partial S}{\partial q^\alpha}, \quad P_\alpha = -\frac{\partial S}{\partial Q^\alpha}. \tag{6.2}$$

Invert the second of these to find the q^β in terms of (Q, P, t) and insert that into the first to find the p_β in terms of (Q, P, t). This can be done only if the Hessian satisfies the condition $|\partial^2 S/\partial q^\alpha \partial Q^\beta| \neq 0$, but recall that this is a condition already imposed on a Type 1 generator. The problem is thereby solved, since $q(t)$ and $p(t)$ have been obtained in terms of the $2n$ constants of the motion (Q, P).

In summary, the *HJ method* for integrating a dynamical system consists of solving Eq. (6.1) for $S(q, Q, t)$ with the Hessian condition $|\partial^2 S/\partial q^\alpha \partial Q^\beta| \neq 0$, and then of using Eq. (6.2) to obtain the motion.

PROPERTIES OF SOLUTIONS

Equation (6.1) is a *partial* differential equation for S involving only the variables q^α and t. Although the solutions of *ordinary* differential equations depend on constants of integration, the solution of a partial differential equation depends on functions (e.g., Cauchy data). It is therefore not clear how to obtain a solution S of the HJ equation that will depend on the n constants Q^α.

Nevertheless, there are many such solutions, called *complete solutions*, and Eq. (6.2) requires that we find not general solutions, but complete solutions. Therefore we will concentrate on complete solutions of Eq. (6.1). For each dynamical system there exist many very different complete solutions, in which the Q^α are very different constants. We exhibit two such different solutions for the simplest of systems, the free particle in one freedom, whose Hamiltonian is $H = \frac{1}{2} p^2$ (we set $m = 1$). Each of the two functions

$$S_a(q, Q_a, t) = \frac{(q - Q_a)^2}{2t}, \tag{6.3a}$$

$$S_b(q, Q_b, t) = q\sqrt{2Q_b} - Q_b t \tag{6.3b}$$

is a complete solution of the HJ equation (see Problem 1). Indeed,

$$\frac{1}{2}\left(\frac{\partial S_a}{\partial q}\right)^2 = \frac{1}{2}\left(\frac{q - Q_a}{t}\right)^2 = \frac{(q - Q_a)^2}{2t^2} = -\frac{\partial S_a}{\partial t}$$

and

$$\frac{1}{2}\left(\frac{\partial S_b}{\partial q}\right)^2 = \frac{1}{2}(\sqrt{2Q_b})^2 = Q_b = -\frac{\partial S_b}{\partial t}.$$

Also, each satisfies the Hessian condition. Yet these are very different functions. It is found in Problem 1 that Q_a is the initial position and Q_b is the energy: S_a generates a CT to constants of the motion (Q_a, P_a) of which Q_a is the initial position, and S_b generates a CT to (Q_b, P_b) of which Q_b is the energy. This reflects the fact that the generator of the transformation from (q, p) to the set of constants of the motion (Q, P) differs for different (Q, P) sets: each (Q, P) set gives rise to its own complete solution. Our example is particularly simple because it is for one freedom.

Notice that Eq. (6.1) contains only derivatives of S, so if an arbitrary constant C is added to any solution of the HJ equation, it remains a solution. But this constant cannot be one of the Q^α, since for an additive constant $\partial S/\partial C = 1$, and thus $\partial^2 S/\partial C \partial q^\alpha = 0$ for all α and the Hessian condition would not be satisfied.

For time-independent Hamiltonians there is a standard technique for determining one of the Q^α by separating out the time dependence of S. One looks for a solution in the special form

$$S(q, Q, t) = W(q, Q) + T(t). \tag{6.4}$$

Then Eq. (6.1) becomes

$$\frac{dT}{dt} = -H\left(q, \frac{\partial W}{\partial q}\right). \tag{6.5}$$

The left-hand side of (6.5) is independent of the q^α and the right-hand side is independent of t. The two sides must therefore be equal to the same constant, which we call $-Q^1$, and that leads immediately to the first step of the solution:

$$T = -Q^1 t \quad \text{or} \quad S(q, Q, t) = W(q, Q) - Q^1 t. \tag{6.6}$$

The HJ equation now becomes

$$H\left(q, \frac{\partial W}{\partial q}\right) = Q^1, \tag{6.7}$$

which is a partial differential equation for *Hamilton's characteristic function* $W(q, Q)$. In this way the problem is reduced to solving a differential equation in one less variable. We will call (6.7) the *time-independent HJ equation*. We emphasize that Eq. (6.4) is an *ansatz*: it restricts the form of S. But if it leads to a complete solution, this form will do as well as any other. We have seen before that when H is time independent, it is conserved and is usually the energy. Thus we will often write E for Q^1. An arbitrary constant can still be added to S, say by replacing t by $t - t_0$; then $C = Q^1 t_0$.

Before going on to discuss a broadly used method for finding solutions of (6.7) and giving some examples, we show the relation between the HJ equation and the action defined in Chapter 3.

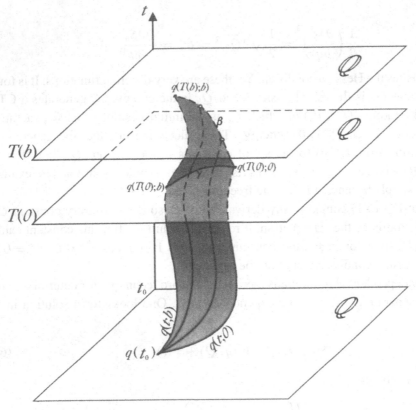

FIGURE 6.1

The two paths $q(t;0)$ and $q(t;b)$ start at the same point $q(t_0)$ in $\mathbb{Q}$ at time t_0, but they end at different points $q(T(0);0)$ and $q(T(b);b)$ at different times. These paths form two boundaries of the shaded surface swept out by an ϵ-family of trajectories, all starting at the same point $q(t_0)$. The ϵ-surface intersects the $t = T(0)$ plane in the curve γ. The third curve bounding the ϵ-surface is β. It consists of the end points of the paths that make up the ϵ-family: its equation is $q(T(\epsilon);\epsilon)$ for $0 \le \epsilon \le b$, and it connects $q(T(0);0)$ to $q(T(b);b)$. Two other curves of the ϵ-family are shown (small circles show their end points), with values of ϵ between 0 and b.

RELATION TO THE ACTION

In this subsection we show that the action $S(q;t_0, t)$ defined in Section 3.1 is among the solutions of the HJ equation.

We will be referring to Fig. 6.1, which represents $\mathbb{Q} \times \mathbb{R}$ in the same way as does Fig. 3.1. Consider two paths $q(t;0)$ and $q(t;b)$ starting at the same point $q(t_0) \equiv q(t_0;0) = q(t_0;b)$. (In Chapter 3 the analogous paths were labeled a and b rather than 0 and b.) The new paths differ from those of Chapter 3 in that they end at different points and at different times, $q(T(0);0)$ and $q(T(b);b)$, whereas those of Chapter 3 all ended at the same time t and position q. (The new paths can be chosen also to start at different points and times, but our choice involves no loss of generality.)

The action along the paths is defined exactly the same way as in Chapter 3:

$$S(q(T(\epsilon);\epsilon), T(\epsilon)) = \int_{t_0}^{T(\epsilon)} L(q(t;\epsilon), \dot{q}(t;\epsilon), t)\,dt, \qquad (6.8)$$

where ϵ is either 0 or b for the two paths illustrated in the figure (we suppress the t_0 dependence of S). As in Chapter 3, ϵ is then made a differentiable parameter which varies from 0 to b. By taking the ϵ derivative of S, we will obtain a differential equation for $S(q, T)$, and this equation will turn out to be the HJ equation (T ends up as the time variable in S because t is used for the variable of integration).

We are now dealing with an ϵ-family of paths that sweeps out the shaded surface in the space $\mathbb{Q} \times \mathbb{R}$ of Fig. 6.1. The surface has three boundaries. Two of them are $q(t; 0)$ and $q(t; b)$, which intersect the $T(0)$ hyperplane at $q(T(0); 0)$ and $q(T(0); b)$. In addition, $q(t; b)$ reaches the $T(b)$ hyperplane at $q(T(b); b)$. The third boundary of the shaded surface is the curve β of the end points of the $q(t; \epsilon)$ paths. The points of β are given by $q(T(\epsilon); \epsilon)$ with $0 \leq \epsilon \leq b$. Finally, the shaded surface intersects the $T(0)$ plane in the curve γ, whose points are given by $q(T(0); \epsilon)$.

For each path the action S is calculated by integrating from the initial point to the final point, from $q(t_0)$ to $q(T(\epsilon); \epsilon)$ (i.e., up to β). To see how S varies with ϵ we must calculate its derivative along β. Let f be any function on $\mathbb{Q} \times \mathbb{R}$ that is differentiable with respect to ϵ. We will write $\Delta f \equiv df/d\epsilon|_{\epsilon=0}$ for its derivative along β. This should not be confused with $\delta f \equiv \partial f/\partial\epsilon|_{\epsilon=0}$ of Chapter 3, which is its derivative along γ, that is, holding T fixed. Then since S depends on both T and the q^α,

$$\Delta S = \frac{\partial S}{\partial T}\Delta T + \frac{\partial S}{\partial q^\alpha}\Delta q^\alpha.$$

Here Δq^α is given by

$$\Delta q^\alpha = \dot{q}^\alpha \Delta T + \delta q^\alpha;$$

$\dot{q}^\alpha$ is the time derivative holding ϵ fixed, and δq^α is the ϵ derivative along γ holding T fixed. Hence

$$\Delta S = \left(\frac{\partial S}{\partial T} + \frac{\partial S}{\partial q^\alpha}\dot{q}^\alpha\right)\Delta T + \frac{\partial S}{\partial q^\alpha}\delta q^\alpha. \tag{6.9}$$

According to Eq. (6.8), however,

$$\Delta S = \frac{d}{d\epsilon}\int_{t_0}^{T(\epsilon)} L(q, \dot{q}, t)\, dt,$$

or, according to the Leibnitz rule for taking the derivative of an integral (remember that everything is evaluated at $\epsilon = 0$; we suppress the $|_{\epsilon=0}$ notation),

$$\Delta S = \int_{t_0}^{T(0)} \frac{\partial}{\partial\epsilon} L(q, \dot{q}, t)\, dt + L(q_0, \dot{q}_0, T(0))\frac{dT}{d\epsilon}, \tag{6.10}$$

where the subscript 0 on q and $\dot{q}$ indicates that the argument of the function is $(T(0); 0)$. The upper limit in the integral in Eq. (6.10) is the fixed value $T(0)$ of t independent of ϵ, so the integral is the same as that of Eq. (3.4), except that the t_1 there is replaced by $T(0)$ here. The integral may therefore be called δS and expanded in accordance with Eq. (3.7):

$$\delta S = \int_{t_0}^{T(0)} \left[\frac{\partial L}{\partial q^\alpha} - \frac{d}{dt} \frac{\partial L}{\partial \dot{q}^\alpha} \right] \delta q^\alpha dt + \int_{t_0}^{T(0)} \frac{d}{dt} \left[\frac{\partial L}{\partial \dot{q}^\alpha} \delta q^\alpha \right] dt,$$

where the derivatives of L are evaluated along the $\epsilon = 0$ path $q(t; 0)$. Because the upper limit is the fixed time $T(0)$, the δq^α are calculated for fixed time and hence are what we have called δq^α above: they are the derivatives of the q^α along the curve γ, which lies entirely in the $T(0)$ plane.

Assume now that $q(t; 0)$ is the physical path satisfying the EL equations. Then the first integral of the last equation vanishes, independent of the δq^α. The second term, however, does not vanish because the derivatives of the q^α along γ do not vanish (in Chapter 3 all the paths ended at the same point, so the δq^α vanished). The integration can be performed to give

$$\int_{t_0}^{T(0)} \frac{d}{dt} \left[\frac{\partial L}{\partial \dot{q}^\alpha} \delta q^\alpha \right] dt = \frac{\partial L}{\partial \dot{q}^\alpha} \delta q^\alpha,$$

where everything is evaluated at $t = T(0)$. Insertion of this into Eq. (6.10) yields

$$\Delta S = L \Delta T + \frac{\partial L}{\partial \dot{q}^\alpha} \delta q^\alpha. \tag{6.11}$$

Comparison with Eq. (6.9) shows that

$$\frac{\partial S}{\partial q^\alpha} = \frac{\partial L}{\partial \dot{q}^\alpha} \equiv p_\alpha, \quad \frac{\partial S}{\partial T} = L - p_\alpha \dot{q}^\alpha = -H.$$

Thus S satisfies the HJ equation.

If we had allowed the paths to vary also at the starting point $q(t_0)$, we would have obtained a similar equation showing that $\partial S/\partial q^\alpha(t_0) = -p_\alpha(t_0)$. Then S becomes just the kind of Type 1 generator of a canonical transformation from (q, t) to constants of the motion that are the initial positions and moments $(q(t_0), p(t_0))$. Thus the action is found to be a complete solution of the HJ equation.

The calculation we have just performed is sometimes called a variational derivation of the HJ equation. A variational derivation, however, would imply that some derivative vanishes and hence that something is minimized by the variation, yet here nothing has been minimized. Certainly ΔS does not vanish, as it does if the paths all arrive at the same point at the same time. This calculation shows only that the action satisfies the HJ equation.

6.1.2 SEPARATION OF VARIABLES

Solving the Hamilton–Jacobi equation can be difficult. As far as we know there is no general method for solving it in all situations. In this section we describe the method of *separation of variables*, a commonly used technique that works only for certain particular Hamiltonians in certain special local charts. This technique is important for two reasons. The first is that it often meets with success. But the major reason is that the dynamical systems for which it works are often the starting points for perturbative calculations, and that makes it possible to obtain accurate approximations for other systems, which would otherwise be intractable.

There also exist dynamical systems whose HJ equations cannot be separated, but which can nevertheless be integrated in other ways (see Perelomov, 1990). In this book we do not discuss those other ways.

In the following subsections sums will be indicated by summation signs.

THE METHOD OF SEPARATION

The separation method was already used in Section 6.1.1 to separate out the time for time-independent H. In a similar way, if the Hamiltonian is independent of q^α (i.e., if q^α is ignorable), q^α can be separated out by an ansatz similar to Eq. (6.4). Write $W = \bar{W} + W_\alpha(q^\alpha)$, where $\bar{W}$ depends on all the other q^β, and solve Eq. (6.7) for $\partial W / \partial q^\alpha \equiv dW_\alpha/dq^\alpha \equiv W'_\alpha$. This leads to an equation, one of whose sides depends on only q^α and the other on all the other q^β. Hence the two sides are equal to the same constant, as at Eq. (6.7).

More generally, a similar procedure can be used to separate out $m \leq n$ of the q^α in W if a solution of (6.7) exists that can be written in the form

$$W = \sum_{\alpha=1}^{m} W_\alpha(q^\alpha, Q) + \bar{W}_m, \tag{6.12}$$

where each W_α depends only on one of the q^α (though it may depend on all of the Q^β) and $\bar{W}_m$ depends on the q^β with $\beta > m$. If such a solution exists, the first m of the q^α are said to be separated out and the HJ equation is said to be *partially separable*; if this can be done with $m = n$, the equation is *(completely) separable*, in which case the $\bar{W}_m$ term drops out of Eq. (6.12), and

$$W = \sum_{\alpha=1}^{n} W_\alpha(q^\alpha, Q). \tag{6.13}$$

Each W_α is a function only of its q^α, so the partial differential equation becomes an ordinary differential equation in n variables (in m if the equation is only partially separable, remaining a partial differential equation in the other $n - m$).

In the completely separable case of Eq. (6.13) the ordinary differential equation in n variables is converted to a set of n equations in one variable by a procedure similar to the one described for an ignorable coordinate. Equation (6.13) is inserted into (6.7), which can sometimes simply be solved for one of the $W'_\alpha \equiv \partial W_\alpha/\partial q^\alpha$ ($\partial/\partial q^\alpha$ rather than d/dq^α because each W_α depends in general on several of the Q^β) isolating the terms involving q^α from all others (see the example below):

$$W'_\alpha + g_\alpha(q^\alpha) = F_\alpha(q).$$

where F_α contains no q^α dependence. Again, each side must be a constant:

$$W_\alpha' + g_\alpha(q^\alpha) = \text{const.}$$

This first-order ordinary differential for W_α involves only the single variable q^α and can always be reduced to quadrature. Then the equation $F_\alpha = \text{const.}$ is separated further in the same way, and so on.

Sometimes the situation is not so simple and more complicated procedures are needed. For example, after inserting Eq. (6.12) into (6.7), one looks for a function $f(q^1, \ldots, q^{n-1})$ (only the first $n - 1$ of the q^α appear) such that the q^n-dependence in fH will be isolated:

$$f H(q, W') = \bar{H}_{n-1}(q, W') + H_n\left(q^n, W_n'\right) = f Q^1, \tag{6.14}$$

where $\bar{H}_{n-1}$ does not depend on q^n or on W_n'. Then

$$H_n = f Q^1 - \bar{H}_{n-1},$$

and by the same reasoning again, one arrives at

$$H_n\left(q^n, W_n'\right) = \text{const.},$$

which is a first-order ordinary differential equation for W_n', again involving only the single variable q^n. What remains then is the equation

$$f Q^1 - \bar{H}_{n-1} = \text{const.},$$

which involves only the $n - 1$ variables $q^1, \ldots, q^{n-1}$. This procedure is illustrated in Worked Example 6.1 below.

If either procedure we have described (or any other) is successful, it results in n equations of the form

$$H_\alpha\left(q^\alpha, W_\alpha'\right) = Q^\alpha, \quad \alpha = 1, \ldots, n. \tag{6.15}$$

When these equations are all solved for the W_α, the solution S of the HJ equation can be written

$$S(q, Q; t) = \sum_{\alpha=1}^{n} W_\alpha(q^\alpha, Q) - Q^1 t, \tag{6.16}$$

and the equations of motion can be obtained from

$$p_\alpha = \frac{\partial W_\alpha}{\partial q^\alpha}, \quad P_\alpha = -\sum_\beta \frac{\partial W_\beta}{\partial Q^\alpha} + \delta_{1\alpha} t. \tag{6.17}$$

WORKED EXAMPLE 6.1

Write down the HJ equation for the two-dimensional central force problem in polar coordinates. Show that the equation is separable in these coordinates, and solve for Hamilton's principal function (reduce it at least to quadrature). From the result, prove that angular momentum is conserved.

Solution. The Hamiltonian for the plane central-force problem is

$$H = \frac{p_r^2}{2m} + \frac{p_\theta^2}{2mr^2} + V(r). \tag{6.18}$$

Write $W = W_r + W_\theta$; then the time-independent HJ equation is

$$\frac{1}{2m}\left[\left(\frac{\partial W_r}{\partial r}\right)^2 + \frac{1}{r^2}\left(\frac{\partial W_\theta}{\partial \theta}\right)^2\right] + V(r) = Q_1 \equiv E$$

(we use subscripts on the Qs so as not to confuse the indices with powers). This is where the procedure of (6.14) comes in. If both sides are multiplied by $f(r) = 2mr^2$, this equation becomes

$$r^2\left(\frac{\partial W_r}{\partial r}\right)^2 + 2mr^2V(r) - 2mr^2Q_1 = -\left(\frac{\partial W_\theta}{\partial \theta}\right)^2,$$

which separates the r and θ parts. Thus

$$W_r' = \sqrt{2mQ_1 - 2mV(r) - r^{-2}Q_2^2},$$
$$W_\theta' = Q_2.$$

Integration then yields

$$\left. \begin{aligned} W_r(r) &= \sqrt{2m}\int \sqrt{Q_1 - V(r) - (2mr^2)^{-1}Q_2^2}\, dr, \\ W_\theta(\theta) &= Q_2\theta. \end{aligned} \right\} \tag{6.19}$$

It follows immediately that $p_\theta = Q_2$ is a constant.

Separable systems have interesting properties. One property follows from the first set of Eqs. (6.17), in which each p_α is written as a function of the constants Q^β and of only its conjugate q^α. Since in any particular motion all of the Q^β are fixed, there is a direct relation between each p_α and its q^α as the motion proceeds, independent of how the other generalized coordinates move. That is, for fixed values of the Q^β, if at any time during the motion q^α is known, Eqs. (6.17) give p_α at that time, independent of the P_β. This means that all particular motions with a given set of Q^β, even if their sets of P_β are different, yield the same *relation* between each of the p_α and its conjugate q^α, though the *function*

$q^\alpha(t)$ depends in general on all of the P_β. The P_β also tell how the motions in the different (q^α, p_α) planes of $\mathbf{T}^*\mathbf{Q}$ are related to each other. The time plays a role also in another property of separable systems, namely the special nature of P_1. Every constant of the motion is a dynamical variable and can hence be written as a function of (q, p, t). For the constants (Q, P) in particular this is done by inverting the first set of Eqs. (6.17) to obtain expressions for $Q^\alpha(q, p, t)$ that are then inserted into the second set to obtain $P_\alpha(q, p, t)$. All of the (Q, P) functions obtained in this way are independent of t with the exception of P_1, for which $\partial P_1/\partial t = 1$. This means that P_1 is the initial time, usually written t_0.

Although the HJ equation for a given dynamical system may be separable in one set of coordinates, it is not in general separable in another: finding separable HJ equations involves not only studying the dynamical systems themselves, but also the coordinates in which they are written. The conditions for separability are not simple, and much has been written about them (e.g., see Perelomov, 1990). Symmetry considerations often help. For instance, a problem with spherical symmetry (e.g., a central force problem) is likely to be separable in spherical polar coordinates.

EXAMPLE: CHARGED PARTICLE IN A MAGNETIC FIELD

As an example consider the two-freedom system of a charged particle in the (x, y) plane subjected to a constant uniform magnetic field $\mathbf{B}$ perpendicular to the plane. Let the charge on the particle be e and choose the vector potential $\mathbf{A}$ to have only a y component (this is known as the *Landau gauge*). Then $A_y = Bx$, and the Hamiltonian of the system is

$$H(x, y, p_x, p_y) = \frac{1}{2m}p_x^2 + \frac{1}{2m}(p_y + bx)^2,$$

where $b = -eB$ (we take $c = 1$). Assume that the HJ equation is separable, writing $S = X(x) + Y(y) - Q^1t$. The time-independent equation (multiplied by $2m$) is

$$2mQ^1 = X'^2 + (Y' + bx)^2.$$

Solving for Y' separates the equation:

$$Y' = -bx + \sqrt{2mQ^1 - X'^2} \equiv Q^2$$

(the superscript 2 on Q is an index, not the square of Q). The y part can be integrated immediately:

$$Y(y) = Q^2y,$$

and for the time being the x part may be left in the form of an indefinite integral, as in Worked Example 6.1:

$$X(x) = \int \sqrt{2mQ^1 - (Q^2 + bx)^2}\, dx.$$

Thus S is

$$S = \int \sqrt{2m Q^1 - (Q^2 + bx)^2}\, dx + Q^2 y - Q^1 t. \tag{6.20}$$

The important things about S are its derivatives, so it is convenient to leave parts of S as indefinite integrals: the derivative is obtained simply by undoing the integration. For instance, the equations for the p_j are

$$p_x \equiv \frac{\partial S}{\partial x} = \sqrt{2m Q^1 - (Q^2 + bx)^2}; \quad p_y \equiv \frac{\partial S}{\partial y} = Q^2. \tag{6.21}$$

The equations for the P_α require integration, however:

$$P_1 \equiv -\frac{\partial S}{\partial Q^1} = t - m \int \frac{dx}{\sqrt{2m Q^1 - (Q^2 + bx)^2}} \equiv t - m I_1,$$

$$P_2 \equiv -\frac{\partial S}{\partial Q^2} = \int \frac{(Q^2 + bx)\, dx}{\sqrt{2m Q^1 - (Q^2 + bx)^2}} - y \equiv I_2 - y.$$

To calculate I_1 make the substitution $Q^2 + bx = \sqrt{2m Q^1} \sin\theta$. The integral evaluates to

$$I_1 = \frac{1}{b} \sin^{-1}\left\{ \frac{Q^2 + bx}{\sqrt{2m Q^1}} \right\}.$$

To evaluate I_2 without actually performing the integration, write $z = Q^2 + bx$. Then $p_x = \partial X/\partial x = b\partial X/\partial z$, and $I_2 = -\partial X/\partial Q^2 = -\partial X/\partial z$. Hence

$$I_2 = -\frac{p_x}{b} = -\frac{1}{b}\sqrt{2m Q^1 - (Q^2 + bx)^2}.$$

Collecting the results, yields

$$\left. \begin{array}{l} t - P_1 = \dfrac{m}{b} \sin^{-1}\left\{ \dfrac{Q^2 + bx}{\sqrt{2m Q^1}} \right\}, \\[3mm] P_2 + y = -\dfrac{1}{b}\sqrt{2m Q^1 - (Q^2 + bx)^2}. \end{array} \right\} \tag{6.22}$$

Equations (6.21) and (6.22) can now be used to give the motion in $T^*\mathbb{Q}$.

We will, however, write down the motion in $\mathbb{Q}$ and make only some remarks about $T^*\mathbb{Q}$. The first of Eqs. (6.22) shows that P_1 is equal to t_0 (it always is, in the time-independent case). Inverting the equation yields

$$x(t) = x_0 + r \sin\left[\frac{b}{m}(t - t_0) \right],$$

where $x_0 = -Q^2/b$ and $r = \sqrt{2m Q^1}/b$. Thus $x(t)$ oscillates about $x = x_0$ with amplitude r and frequency $\omega_C = b/m \equiv eB/m$, called the *cyclotron frequency*. After some manipulation, the second of Eqs. (6.22) yields

$$(y - y_0)^2 + (x - x_0)^2 = r^2,$$

where $y_0 = -P_2$. This shows that the trajectory is a circle centered at (x_0, y_0) with radius r. Since x_0, r, and y_0 depend on the arbitrary constants of integration Q^1, Q^2, and P_2, they are also arbitrary: the dynamical system consists of all possible circles in the plane. The only fixed quantity that has emerged is the frequency ω_C, twice the Larmor frequency ω_L of Section 2.2.4.

As for $\mathbf{T}^*\mathbb{Q}$, we look first at the (x, p_x) plane. From the first of Eqs. (6.21) it follows that

$$\frac{p_x^2}{b^2} + (x - x_0)^2 = r^2. \tag{6.23}$$

This is the equation of an ellipse centered at $(x_0, 0)$ with semiaxes in the ratio $1/b$ (see Fig. 6.5). Every trajectory in the (x, p_x) plane of $\mathbf{T}^*\mathbb{Q}$ is an ellipse on the x axis, all of the same eccentricity, but of different sizes. In the (y, p_y) plane, the trajectory is an oscillation in y for a fixed value of p_y. It may seem strange that p_y is constant even though y is not. This has to do with the choice of gauge: p_y is not $m\dot{y}$, for one of the canonical equations is $\dot{y} = (p_y + bx)/m$. In summary, the constants of integration play the following roles: Q^1 determines r; Q^2 determines the x coordinate of the center of the circle in $\mathbb{Q}$ and the center of the ellipse in $\mathbf{T}^*\mathbb{Q}$; $P_1 \equiv t_0$ determines where on the circle in $\mathbb{Q}$ (or where on the ellipse in $\mathbf{T}^*\mathbb{Q}$) the motion starts; and P_2 determines the y coordinate of the center of the circle in $\mathbb{Q}$.

The same dynamical system can be written out in other gauges as well, because a gauge transformation will not alter the motion in $\mathbb{Q}$ even though it changes the Hamiltonian. Consider in particular the gauge transformation $\mathbf{A} \to \mathbf{A}' = \mathbf{A} + \nabla\Lambda$, where $\Lambda = -Bxy/2$. The new Hamiltonian is then

$$H(x, y, p_x, p_y) = \frac{1}{2m}(p_x - by/2)^2 + \frac{1}{2m}(p_y + bx/2)^2. \tag{6.24}$$

The HJ equation for this Hamiltonian is harder to separate (see Problem 2). Separability of the HJ equation depends on many factors. Earlier we mentioned that it can depend on the coordinate system. We see here that it can depend also on the gauge in which it is written.

WORKED EXAMPLE 6.2

Treat the relativistic two-freedom Kepler problem of Section 5.1.1 by the HJ method, separating the HJ equation in polar coordinates. That is: **(a)** Find an expression for S (it may involve an integral). **(b)** Obtain an explicit expression for the orbit (no integrals) and show that for small enough angular momentum the particle will spiral down into the force center.

Solution. **(a)** The Hamiltonian of Eq. (5.31) is

$$H = \sqrt{p_r^2 + \frac{p_\theta^2}{r^2} + m^2} - \frac{e^2}{r},$$

Since H is time independent, write $S = W - Et$ and then the time-independent HJ equation becomes

$$E = \sqrt{(W_r')^2 + \frac{(W_\theta')^2}{r^2} + m^2} - \frac{e^2}{r}. \tag{6.25}$$

Equation (6.25) is easily separated to yield

$$W_\theta' = r\sqrt{\left(E + \frac{e^2}{r}\right)^2 - (W_r')^2 - m^2} \equiv p_\theta,$$

where $p_\theta \equiv Q^\theta$ is the (constant) angular momentum, and $W_\theta = p_\theta\theta$. The equation for W_r is

$$E = \sqrt{(W_r')^2 + \frac{p_\theta^2}{r^2} + m^2} - \frac{e^2}{r}, \tag{6.26}$$

whose solution leads to

$$S(r, \theta, E, p_\theta) = -Et + W_\theta + W_r$$

$$= -Et + p_\theta\theta + \int \sqrt{\left(E + \frac{e^2}{r}\right)^2 - \frac{p_\theta^2}{r^2} - m^2} \, dr. \tag{6.27}$$

(b) Recall that $\partial S/\partial Q^\alpha = P_\alpha = \text{const.}$ Now the Q^α are E and p_θ. The derivative with respect to E yields the equation for $r(t)$, but we do not calculate that. The derivative with respect to p_θ yields the equation for the orbit $r(\theta)$:

$$\theta - \theta_0 = p_\theta \int \left\{ \left(E + \frac{e^2}{r}\right)^2 - \frac{p_\theta^2}{r^2} - m^2 \right\}^{-\frac{1}{2}} \frac{dr}{r^2}$$

$$= -\int \{A + Bu - \gamma u^2\}^{-\frac{1}{2}} du, \tag{6.28}$$

where $\theta_0 = P_\theta$, $u = 1/r$, and [compare with Eq. (5.35); factors of c are made explicit there]

$$A = \frac{E^2 - m^2}{p_\theta^2}, \quad B = \frac{2e^2 E}{p_\theta^2}, \quad \gamma = 1 - \frac{e^4}{p_\theta^2}.$$

We had already obtained an expression for the orbit at Eq. (5.35), where γ was called Γ^2. It was assumed in Chapter 5 that $\gamma \equiv \Gamma^2$ is positive, but it is seen from the definition that for small enough p_θ it will become negative. To perform the integration of Eq. (6.28), therefore, we distinguish three cases: γ positive, negative, and zero.

$\gamma > 0$. The integral can be performed by completing the square and by setting

$$u - \frac{B}{2\gamma} = \sqrt{(B^2 + 4A\gamma)/4\gamma^2} \cos \phi.$$

The result obtained is that of Eq. (5.34).

$\gamma < 0$. The integral can be performed by completing the square and by setting

$$u - \frac{B}{2\gamma} = \sqrt{(4A|\gamma| - B^2)/4\gamma^2} \tan \phi.$$

The result obtained for $\theta - \theta_0$ is

$$\theta - \theta_0 = \frac{1}{\sqrt{|\gamma|}}[\ln\{2\sqrt{|\gamma|(A + Bu - \gamma u^2)} + B - 2\gamma u\} - \ln K].$$

We have included an arbitrary constant of integration here (which effectively sets θ_0) because by a judicious choice of K the solution for $r = 1/u$ becomes

$$r = \frac{\gamma}{\Delta \cosh\{\sqrt{|\gamma|}(\theta - \theta_0)\} + B/\gamma},$$

where $\Delta = \sqrt{B^2 + 4A\gamma}$. The particle spirals down to the force center. In the non-relativistic case, the particle hits the force center only if the angular momentum $p_\theta = 0$, but in the relativistic case also if $|p_\theta| < e^2$.

$\gamma = 0$. This is the nonrelativistic case, which was treated in Section 2.3.2.

WORKED EXAMPLE 6.3

Show that the HJ equation for a particle attracted to two fixed (nonrelativistic) gravitational centers in the plane is separable in *confocal elliptical coordinates*. Obtain integral expressions for Hamilton's characteristic function W.

(This problem was first shown to be integrable by Euler in 1760 and then by Lagrange in 1766. It has been treated since then by several authors, and the most recent complete analysis known to us was by Strand and Reinhardt in 1979 in connection with the ionized hydrogen molecule. It is of general interest in $\mathbb{R}^3$, but this example refers only to $\mathbb{R}^2$.)

Solution. Confocal elliptic coordinates ξ and η are related to Cartesian (x, y) coordinates by

$$x = c \cosh \xi \cos \eta,$$
$$y = c \sinh \xi \sin \eta.$$

The ranges of ξ and η are $-\infty < \xi < \infty$ and $0 \leq \eta < 2\pi$. To find the coordinate

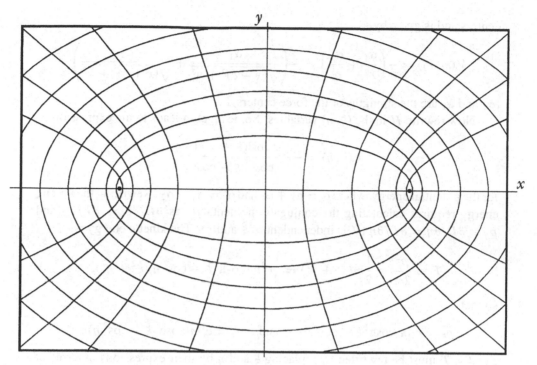

FIGURE 6.2

Confocal elliptic coordinates (ξ, η) in the plane, related to Cartesian coordinates (x, y) by $x = a \cosh \xi \cos \eta$, $y = a \sinh \xi \sin \eta$. The curves of constant ξ are ellipses and those of constant η are hyperbolas. The two dots are the common foci of all of the ellipses and hyperbolas at $x = \pm a$.

curves (Fig. 6.2) first let $\xi = $ const., and write $\cosh \xi = A$ and $\sinh \xi = B$. Then

$$\frac{x^2}{A^2} + \frac{y^2}{B^2} = c^2, \quad \xi \text{ fixed:}$$

each constant-ξ curve is an ellipse. Next, let $\eta = $ const., and write $\cos \eta = a$ and $\sin \eta = b$. Then

$$\frac{x^2}{a^2} - \frac{y^2}{b^2} = c^2, \quad \eta \text{ fixed:}$$

each constant-η curve is a hyperbola opening in the $\pm x$ directions. We state without proof that all of these curves have the same foci at $y = 0, x = \pm c$ and that the ellipses intersect the hyperbolas at right angles.

For this problem, set the foci of the elliptic coordinate system at the fixed gravitational centers $(x, y) = (\pm c, 0)$. The Hamiltonian is

$$H = \frac{p^2}{2m} + V(r_1, r_2),$$

where the potential energy V depends on the distances r_1 and r_2 from the two force centers and is given by

$$V(r_1, r_2) = -\left(\frac{\alpha_1}{r_1} + \frac{\alpha_2}{r_2}\right) = -\left(\frac{\alpha_1}{\sqrt{(x-c)^2 + y^2}} + \frac{\alpha_2}{\sqrt{(x+c)^2 + y^2}}\right)$$

(α_1 and α_2 are the strengths of the force centers).

Now rewrite H in the (ξ, η) variables. Some algebra leads to the expression

$$V(\xi, \eta) = -\frac{1}{c}\frac{\alpha \cosh \xi + \alpha' \cos \eta}{\cosh^2 \xi - \cos^2 \eta}$$

for the potential energy, where $\alpha = \alpha_1 + \alpha_2$ and $\alpha' = \alpha_1 - \alpha_2$. Rewriting the kinetic energy requires calculating the conjugate momenta $p_\xi \equiv \partial L / \partial \dot{\xi} = \partial T / \partial \dot{\xi}$ and $p_\eta \equiv \partial L / \partial \dot{\eta} = \partial T / \partial \dot{\eta}$ (V is independent of $\dot{\xi}$ and $\dot{\eta}$). The kinetic energy is

$$T \equiv \frac{p^2}{2m} \equiv \frac{1}{2}m(\dot{x}^2 + \dot{y}^2) = \frac{1}{2}c^2 m(\cosh^2 \xi - \cos^2 \eta)(\dot{\xi}^2 + \dot{\eta}^2),$$

so that

$$p_\xi = c^2 m(\cosh^2 \xi - \cos^2 \eta)\dot{\xi} \quad \text{and} \quad p_\eta = c^2 m(\cosh^2 \xi - \cos^2 \eta)\dot{\eta}.$$

Finally, T must be rewritten by replacing $\dot{\xi}$ and $\dot{\eta}$ by their expressions in terms of p_ξ and p_η.

When all this is done, the Hamiltonian in the new coordinates becomes

$$H = \frac{p_\xi^2 + p_\eta^2 - \alpha \cosh \xi + \alpha' \cos \eta}{2c^2 m(\cosh^2 \xi - \cos^2 \eta)}.$$

The time-independent HJ equation then reads

$$\left(\frac{\partial W}{\partial \xi}\right)^2 + \left(\frac{\partial W}{\partial \eta}\right)^2 = 2Ec^2 m(\cosh^2 \xi - \cos^2 \eta) + \alpha \cosh \xi - \alpha' \cos \eta.$$

The ξ and η terms are easily separated: writing $W = W_\xi + W_\eta$ leads to

$$\left(\frac{\partial W_\xi}{\partial \xi}\right)^2 = Q + 2Ec^2 m \cosh^2 \xi + \alpha \cosh \xi,$$

$$\left(\frac{\partial W_\eta}{\partial \eta}\right)^2 = -Q - 2Ec^2 m \cos^2 \eta - \alpha' \cos \eta.$$

The desired integral expressions are

$$W_\xi = \int [Q + 2Ec^2 m \cosh^2 \xi + \alpha \cosh \xi]^{1/2} \, d\xi.$$

$$W_\eta = \int [-Q - 2Ec^2 m \cos^2 \eta - \alpha' \cos \eta]^{1/2} \, d\eta.$$

These are elliptic integrals.

The problem in three freedoms, also separable and integrable, is discussed in detail by Strand and Reinhardt (1979). There are certain limits of this problem which are also of physical interest. See Problems 14 and 16.

6.1.3 GEOMETRY AND THE HJ EQUATION

In this section we describe the geometric content of the HJ equation and show how its solutions yield the phase portrait of a dynamical system Δ, that is, how it constructs the integral curves (Crampin and Pirani, 1986, pp. 160–2). In the process we clarify the nature of complete solutions and exhibit the role of the n constants Q^α. The discussion is local, restricted to a neighborhood $\mathbb{U} \subset \mathbf{T}^*\mathbb{Q}$ that has a nonempty intersection with $\mathbb{Q}$, but we will treat $\mathbb{U}$ as though it were $\mathbf{T}^*\mathbb{Q}$ itself. We will discuss only the time-independent HJ equation.

We deal first with energy $E = 0$ (recall that $Q^1 = E$). Let $W_0 \in \mathcal{F}(\mathbb{Q})$ be any solution, not necessarily a complete solution, of the zero-energy HJ equation

$$H\left(q, \frac{\partial W_0}{\partial q}\right) = 0, \tag{6.29}$$

and form the functions

$$f_\alpha = p_\alpha - \partial W_0/\partial q^\alpha \in \mathcal{F}(\mathbf{T}^*\mathbb{Q}). \tag{6.30}$$

The n equations

$$f_\alpha = 0 \tag{6.31}$$

define an n-dimensional submanifold $\mathbb{W}_0$ of the $2n$-dimensional $\mathbf{T}^*\mathbb{Q}$, and we now show that the zero-energy curves of the phase portrait lie within $\mathbb{W}_0$, that is, that $\mathbb{W}_0$ is an invariant manifold. Since $\mathbb{W}_0$ is defined by (6.31), we must show that the f_α do not change along the motion, that is, that $\mathbf{L}_\Delta f_\alpha = 0 \,\forall f_\alpha$, where

$$\Delta = \frac{\partial H}{\partial p_\beta} \frac{\partial}{\partial q^\beta} - \frac{\partial H}{\partial q^\beta} \frac{\partial}{\partial p_\beta}$$

and the f_α are given by (6.30). The Lie derivatives are indeed zero:

$$\begin{aligned}
\mathbf{L}_\Delta f_\alpha &\equiv \left\{ \frac{\partial H}{\partial p_\beta} \frac{\partial}{\partial q^\beta} - \frac{\partial H}{\partial q^\beta} \frac{\partial}{\partial p_\beta} \right\} \left(p_\alpha - \frac{\partial W_0}{\partial q^\alpha} \right) \\
&= -\frac{\partial H}{\partial p_\beta} \frac{\partial^2 W_0}{\partial q^\beta \partial q^\alpha} - \frac{\partial H}{\partial q^\alpha} = -\frac{\partial}{\partial q^\alpha} H\left(q, \frac{\partial W_0}{\partial q}\right) = 0.
\end{aligned} \tag{6.32}$$

Of course the zero-energy integral curves also lie in the $(2n - 1)$-dimensional submanifold $\mathbb{H}_0 \subset \mathbf{T}^*\mathbb{Q}$ defined by $H = 0$. In fact the entire n-dimensional submanifold $\mathbb{W}_0$ lies in $\mathbb{H}_0$, for $\mathbb{W}_0$ is defined by (6.31), so on $\mathbb{W}_0$

$$H(q, p) = H\left(q, \frac{\partial W_0}{\partial q}\right) = 0$$

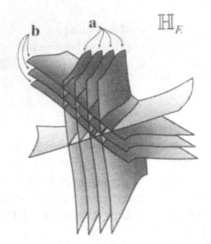

FIGURE 6.3
Diagram illustrating the interior of one of the $\mathbb{H}_E$ for $n = 2$ with $\dim(\mathbb{H}_E) \equiv 2n - 1 = 3$. The eight surfaces in the diagram are all (pieces of) $\mathbb{W}_E$ submanifolds. The four surfaces labeled **a** are in one set of nonintersecting $\mathbb{W}_E$ submanifolds, and the three surfaces labeled **b** are in another such set. The eighth lies in another such set and is drawn to illustrate the abundance of $\mathbb{W}_E$ submanifolds: there are many others, and they can intersect in complicated ways. But the integral curves of the phase portrait lie in all of the $\mathbb{W}_E$ submanifolds, so they must lie in their intersections. In this $n = 2$ case the intersections are curves, so the integral curves are the intersections themselves.

[where the last equality comes from (6.29)]. This means that all points $(q, p) \in \mathbb{W}_0$ lie in $\mathbb{H}_0$, and therefore $\mathbb{W}_0 \subset \mathbb{H}_0$.

To generalize to nonzero energy E, replace H in the preceding discussion by $H_E = H - E$ and consider the zero-energy solutions of the HJ equation for H_E, which are energy-E solutions for H. Write $\mathbb{W}_E$ and $\mathbb{H}_E$ for the analogs of $\mathbb{W}_0$ and $\mathbb{H}_0$, and then the conclusion is that the energy-E integral curves lie in $\mathbb{W}_E$ and that $\mathbb{W}_E \subset \mathbb{H}_E$.

Each $\mathbb{H}_E$ contains all points in $\mathbf{T}^*\mathbb{Q}$ that belong to energy E, and since no point $(q, p) \in \mathbf{T}^*\mathbb{Q}$ can have two energies, the $\mathbb{H}_E$ manifolds do not intersect. Moreover, since every $(q, p) \in \mathbf{T}^*\mathbb{Q}$ has some energy E, the $\mathbb{H}_E$ manifolds fill up $\mathbf{T}^*\mathbb{Q}$. But since there are many solutions of the HJ equation for each E, there are many $\mathbb{W}_E$ submanifolds within each $\mathbb{H}_E$, and these can intersect in complicated ways.

From all of the $\mathbb{W}_E$ in $\mathbb{H}_E$ choose a nonintersecting subset, one through each point of each $\mathbb{H}_E$. Because $\dim \mathbb{H}_E = 2n - 1$ and $\dim \mathbb{W}_E = n$, this yields an $(n - 1)$-dimensional set of nonintersecting $\mathbb{W}_E$, parametrized by $n - 1$ numbers: each $\mathbb{W}_E$ is labeled by a different set of $n - 1$ numbers. Figure 6.3 illustrates this for $n = 2$ (then $\dim \mathbb{H}_E = 3$). It shows a portion of one of the $\mathbb{H}_E$, in which a set of nonintersecting $\mathbb{W}_E$ form a continuous stack of two-dimensional surfaces. A finite subset of one such stack are labeled **a** in the figure, and another is labeled **b**. Each stack is parametrized by one number, which we will call Q^2: each $\mathbb{W}_E$ in the stack is labeled by a different value of Q^2. For $n > 2$ the $n - 1$ numbers are called Q^α, $\alpha = 2, 3, \ldots, n$.

Now repeat the procedure for all values of E. The result is a set of nonintersecting $\mathbb{W}_E$ submanifolds, one passing through each point of $\mathbf{T}^*\mathbb{Q}$, parametrized now by $E \equiv Q^1$ and the Q^α with $\alpha > 1$, in all by n numbers: each $\mathbb{W}_E$ is labeled by a different set of n numbers.

The $\mathbb{W}_E$ are not unique, for there are many sets of nonintersecting $\mathbb{W}_E$ submanifolds that can be chosen. All are equally valid. For example, Fig. 6.3 shows $\mathbb{W}_E$ surfaces from three different stacks.

Because the $\mathbb{W}_E$ are labeled not by $E \equiv Q^1$ alone, but by $Q \equiv \{Q^1, Q^2, \ldots, Q^n\} \equiv Q$, we will now call them $\mathbb{W}_Q$. Each $\mathbb{W}_Q$ is defined through Eq. (6.31), or rather through its generalization to arbitrary E, by a different solution W of the HJ equation, which can be labeled by Q and written $W(q, Q)$. Because it depends on the n constants Q, each one of the $W(q, Q)$ is a complete solution of the (time independent) HJ equation. Different sets of $\mathbb{W}_Q$ manifolds are defined by different W_Q functions. Equations (6.3a, b) show two such different complete solutions (for the time-dependent HJ equation).

What we set out to show is how this maps out the integral curves. But it was already seen at Eqs. (6.2) that the integral curves are given by

$$F_\alpha \equiv P_\alpha + \frac{\partial W}{\partial Q^\alpha} = 0, \quad \alpha = 2, 3, \ldots, n, \tag{6.33}$$

where the P_α are constants. Indeed, these are $n-1$ equations, so they define a curve within each n-dimensional $\mathbb{W}_Q$, and to verify that it is an integral curve (a one-dimensional invariant manifold) requires only showing that the F_α are invariant under the motion:

$$\mathbf{L}_\Delta F_\alpha = \left\{ \frac{\partial H}{\partial p_\beta} \frac{\partial}{\partial q^\beta} - \frac{\partial H}{\partial q^\beta} \frac{\partial}{\partial p_\beta} \right\} \left(P_\alpha + \frac{\partial W}{\partial Q^\alpha} \right)$$

$$= \frac{\partial H}{\partial p_\beta} \frac{\partial^2 W}{\partial q^\beta \partial Q^\alpha} = \frac{\partial}{\partial Q^\alpha} H \left(q, \frac{\partial W}{\partial q} \right) = 0,$$

vanishing because $H = E = $ const. provided the only Q^α that vary are those with $\alpha \neq 1$.

Because time dependence is left out, Eq. (6.33) yields only the orbits in $\mathbf{T}^*\mathbb{Q}$. However, time dependence can be included by using the Q^1 equation $P_1 + \partial W/\partial Q^1 = 0$, but we do not go into that.

6.1.4 THE ANALOGY BETWEEN OPTICS AND THE HJ METHOD

Hamilton came upon the HJ equation while studying optics, and it was only later that Jacobi established its relevance also to dynamics. This reflects a certain analogy between optics and mechanics, one that is connected with HJ theory. In this section we describe that analogy.

We start from Snell's law for the refraction of a light ray at a smooth surface between two media M_1 and M_2 with indices of refraction n_1 and n_2:

$$n_1 \sin \theta_1 = n_2 \sin \theta_2.$$

Here θ_k is the angle between the light ray and the normal to the surface. Compare this to the path of a particle in two dimensions that moves from a region R_1 of constant potential energy V_1 to another R_2 of a different constant potential $V_2 > V_1$. Suppose that the two regions are separated by a line parallel to the y axis (Fig. 6.4), that the particle is incident on the separating line at an angle θ_1 with respect to its normal, and that it has enough momentum to pass from R_1 to R_2. Let θ_2 be the angle at which the particle leaves the normal. Since the only force the particle feels in crossing the line is in the x direction, the y component p_y of the momentum is the same in both regions, but because p_x changes in going from one region to the other, the magnitude of the momentum p_1 in R_1 is different from p_2 in R_2. The θ_k are related to the p_k by $p_k \sin \theta_k = p_y$, or

$$p_1 \sin \theta_1 = p_2 \sin \theta_2.$$

This is clearly similar to Snell's law: all particles incident on the separating line with a given momentum of magnitude p_1 behave like light rays passing from one medium to another. (The index of refraction n always satisfies the condition $n \geq 1$, which is not true for the momentum

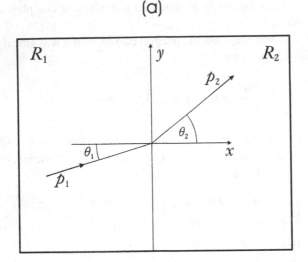

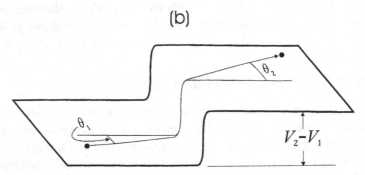

FIGURE 6.4

Two views of the dynamical system consisting of a particle of given momentum p_1 passing from a region with constant potential energy V_1 to another with potential energy V_2. (a) Path in configuration space. (b) Energy diagram.

of a particle, but if the optical and dynamical equations are divided by n_1 and p_1 respectively, the analogy is perfect.)

One way to explain Snell's law is by Huygens's principle (see, e.g., Ohanian, 1989, p. 905). It can also be derived from *Fermat's variational principle of least time* (for the derivation, see Hecht and Zajac, 1974), according to which a light ray, in moving from a point $\mathbf{x}_1$ to another point $\mathbf{x}_2$, travels so as to minimize the time. The mechanical analog of this is also a variational principle, but not one of those we have already discussed. The present one states that the integral

$$A \equiv \int p_\alpha \dot{q}^\alpha \, dt \tag{6.34}$$

is an extremum for the physical path if the comparison paths all have the same total energy. As in Section 3.1, the integral is taken between fixed end points in $\mathbb{Q}$. We do not prove this result: a proof can be found in Whittaker (1944, p. 247). We want only to exhibit the analogy between

Eq. (6.34) and Fermat's principle, which we will do by rewriting (6.34) and putting Fermat's principle in the form of an integral. Rather than treat this subject in all of its generality, we restrict the discussion to motion in $\mathbb{R}^3$.

Start with Fermat's principle. Geometrical optics deals with the trajectories of light rays as they traverse media of different indices of refraction n. The *index of refraction* of a medium is defined as

$$n(\mathbf{x}, \mathbf{x}') = \frac{c}{v} \geq 1,$$

where c is the velocity of light in vacuum and v is the velocity in the medium. In general n can depend on the point $\mathbf{x}$ in the medium and on the direction $\mathbf{x}'$ of the ray, but we assume the medium to be isotropic, so $n \equiv n(\mathbf{x})$ is a function only of position. The *optical path length l* of a light ray traveling from $\mathbf{x}_1$ to $\mathbf{x}_2$ is defined by

$$l = \int_{\mathbf{x}_1}^{\mathbf{x}_2} n \, ds,$$

where ds is the element of distance along the ray. Since $v = ds/dt$,

$$l = c \int_{\mathbf{x}_1}^{\mathbf{x}_2} \frac{1}{v} \, ds = c \int_0^T dt,$$

where T is the time it takes the ray to go from $\mathbf{x}_1$ to $\mathbf{x}_2$. Minimizing T is the same as minimizing l, and Fermat's principle may be written as

$$\frac{d}{dt} \int_{\mathbf{x}_1}^{\mathbf{x}_2} n(\mathbf{x}) \, ds \equiv \frac{dl}{dt} = 0.$$

(Actually this implies that l is an extremum, not necessarily a minimum.)

Now rewrite Eq. (6.34) for a particle of mass m in a potential $V(\mathbf{x})$. Then $p_\alpha \dot{q}^\alpha = mv^2$, and the integral from $\mathbf{x}_1$ to $\mathbf{x}_2$ becomes

$$\int_{\mathbf{x}_1}^{\mathbf{x}_2} mv^2 dt = \int_{\mathbf{x}_1}^{\mathbf{x}_2} mv \, ds = \int_{\mathbf{x}_1}^{\mathbf{x}_2} p \, ds.$$

If the physical trajectory can be found from the extremum of this integral, p is the analog of n in Fermat's principle, which is in agreement with the example of Snell's law. We write $n \propto p$.

The analogy can be carried further, and the wave properties of optical phenomena can be related to the HJ equation. The propagation of a wave can be described in terms of *wave fronts*, moving surfaces on which the phase of the wave is constant, usually chosen to be the maximum or minimum of the phase. If the medium is isotropic, as we are assuming, the rays are everywhere perpendicular to the wave fronts. Although a light wave is electromagnetic and is therefore a *vector wave* (e.g., it is described by the electric field $\mathbf{E}$), we will consider only *scalar waves* (which means essentially that we consider just one component of $\mathbf{E}$). Such a scalar wave is described by a *wave function* of the form

$$\psi(\mathbf{x}, t) = |\psi(\mathbf{x}, t)| e^{-i\phi(\mathbf{x}, t)}.$$

The amplitude $|\psi(\mathbf{x}, t)|$ gives the strength of the wave and $\phi(\mathbf{x}, t)$ gives its phase at all positions and times $(\mathbf{x}, t)$.

For dealing with the wave front it is the phase that is important, for the wave fronts are defined by $\phi(\mathbf{x}, t) = C$, where C is a constant. It will be seen that ϕ is the analog of the solution S of the HJ equation. Because ϕ is constant on the wave front,

$$\frac{d\phi}{dt} \equiv \nabla\phi \cdot \dot{\mathbf{x}} + \frac{\partial\phi}{\partial t} = 0$$

there, or

$$\nabla\phi \cdot \dot{\mathbf{x}} = -\frac{\partial\phi}{\partial t}. \tag{6.35}$$

But $\nabla\phi \cdot \dot{\mathbf{x}}$ is just $|\nabla\phi|$ times the speed of the wave along the ray, for $\nabla\phi$ is perpendicular to the wave front, so the speed of the wave is $-(\partial\phi/\partial t)/|\nabla\phi|$.

A *plane wave* is one whose fronts are planes and whose rays are parallel straight lines perpendicular to the fronts. In an ideal situation without dissipation, $|\psi|$ of a plane wave is constant and its phase is given by the equation

$$\phi(\mathbf{x}, t) = \mathbf{k} \cdot \mathbf{x} - \omega t + \alpha, \tag{6.36}$$

in which $\mathbf{k}$ is the constant *wave vector* pointing in the direction of propagation (its magnitude is $k = 2\pi/\lambda$, where λ is the wavelength), $\omega = 2\pi/\tau$ is the (circular) frequency (τ is the period), and α is a phase angle. Equation (6.35) implies that $\mathbf{k} \cdot \mathbf{x} = \omega$: a plane wave moves at a speed equal to $\dot{x} = \omega/k$.

In small enough regions even a curved wave front is approximately plane with an approximate $\mathbf{k}$ and an approximate ω. Assume that $|\psi|$ changes very little over a wavelength: more specifically, assume that $|\nabla\psi| \ll |\mathbf{k}\psi|$. This is called *the short-wavelength* (or *high-frequency*) *approximation*. In order to make sense of this assumption, we find the approximate $\mathbf{k}$ and ω by expanding ϕ in a Taylor series in a small region of space and for short times. The first few terms, the only ones we will keep, are

$$\phi = \phi_0 + \mathbf{x} \cdot \nabla\phi + t\frac{\partial\phi}{\partial t} + \cdots.$$

Comparison with (6.36) shows that in this approximation

$$\mathbf{k} \approx \nabla\phi \quad \text{and} \quad \omega \approx -\frac{\partial\phi}{\partial t}.$$

The short-wavelength approximation can now be written $|\nabla\psi| \ll |\psi\nabla\phi|$. Assume for the purposes of this discussion that the frequency remains fixed. Then in this approximation the speed of the curved wave front is $\dot{x} = \omega/|\nabla\phi|$. But according to the definition of the index of refraction, $\dot{x} = c/n$, so $|\nabla\phi| = n\omega/c$, or

$$(\nabla\phi)^2 = \frac{\omega^2}{c^2}n^2.$$

This result is known as the *eikonal equation* of geometric optics. It implies that $|\nabla\phi|$ is *inversely proportional* to the speed of the wave.

We have seen that the momentum p is the particle analog of the n in optics. Thus $|\nabla\phi|$ is the analog of $\omega p/c$. We extend the analogy from the magnitudes to the vectors themselves, writing $c\nabla\phi/\omega \propto \mathbf{p}$. Recall that if S is a solution of the HJ equation, $p_\alpha = \partial S/\partial q^\alpha$, or $\mathbf{p} = \nabla S$. Therefore $c\nabla\phi/\omega \propto \nabla S$, or

$$\frac{c}{\omega}\phi(\mathbf{x}, t) \propto S(\mathbf{x}, t).$$

The momentum is always tangent to the trajectories of the dynamical system in $\mathbb{R}^3$, and the equation $\mathbf{p} = \nabla S$ implies that the trajectories are everywhere perpendicular to the $S = $ const. surfaces (we will call these surfaces Σ). In just the same way, the light rays are everywhere perpendicular to the $\phi = $ const. surfaces, the wave fronts. Thus the dynamical system looks just like the optical system: the dynamical trajectories correspond to the light rays and the Σ surfaces correspond to the wave fronts.

In quantum mechanics dynamical systems are described by complex waves; in fact quantum mechanics is also called wave mechanics. The wave function Ψ of quantum mechanics satisfies the Schrödinger equation, and if Ψ is written in the form

$$\Psi(\mathbf{x}, t) = A(\mathbf{x}, t)\exp\left\{i\frac{S(\mathbf{x}, t)}{\hbar}\right\},$$

the function S satisfies the HJ equation in the limit as $\hbar \to 0$ (Maslov and Fedoriuk, 1981). Here $\hbar = h/2\pi$, where h is Planck's constant. In other words, the wave analogy is deeper than simply between optics and dynamics: it actually describes the semiclassical limit of quantum mechanics.

Our discussion is restricted to $\mathbb{R}^3 \equiv \mathbb{Q}$. Unlike trajectories on $\mathbf{T}\mathbb{Q}$, the trajectories in $\mathbb{Q}$ are not separated: they intersect in complicated ways. It is possible, however, to form a particular subset whose trajectories are initially separated. This can be done by giving conditions to specify an initial Σ surface in $\mathbb{R}^3$ and initial momenta $\mathbf{p}$ as a function of position on Σ. Even though they start off separated and perpendicular to Σ, these trajectories will in general intersect at some later time. In the optical analogy this means that the rays will intersect and therefore that the wave front will have singular points. It then follows that the dynamical S function will also have singular points. Such singular points, called *caustics*, play an important role in the semiclassical limit.

6.2 COMPLETELY INTEGRABLE SYSTEMS

6.2.1 ACTION–ANGLE VARIABLES

INVARIANT TORI

It was pointed out in the first part of Section 6.1.2 that in a separable system each p_α can be written as a function of its own conjugate q^α. This means that a graph connecting

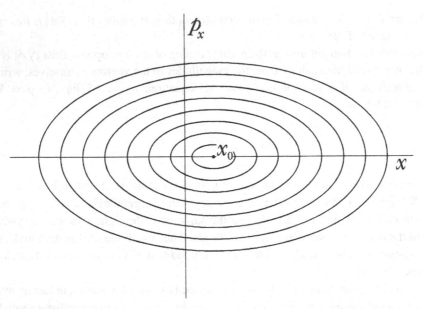

FIGURE 6.5

The relation between p_x and x as given by Eq. (6.23) for fixed x_0 and several values of r. The ratio of the minor axes to the major axes is $1/b$. The ellipses are all centered at $x_0 = -p_y/b$. In a given motion r and x_0 (and hence p_y) are fixed, and therefore as time proceeds x and p_x vary so as to remain on one of these ellipses. In a different (x, p_x) plane, in particular for a different value of p_y, the center of the ellipses is at a different point on the x axis.

p_α and q^α can be drawn in the α submanifold of $\mathbf{T}^*\mathbb{Q}$, defined by the $2n - 2$ equations $p_\beta = \text{const.}, q^\beta = \text{const.}, \beta \neq \alpha$. An example was given for a charge in a magnetic field: see Eq. (6.23) and Fig. 6.5.

For most physical systems the momenta appear quadratically in the Hamiltonian, and the resulting square root causes the dependence of each p_α on q^α to be double-valued. This reflects the physics: for a given position, the velocity, and hence also the momentum, can be positive or negative, even when many other constants are fixed, such as the energy. That is illustrated by the ellipses of Fig. 6.5.

Figure 6.5 and similar graphs of p_α as a function of q^α are restricted to fixed α submanifolds of $\mathbf{T}^*\mathbb{Q}$. They are not phase portraits, which consist of integral curves that move from one α submanifold to another as the motion proceeds, for in a motion *all* of the q^β and p_β change, even for $\beta \neq \alpha$. The curve in the α submanifold is the *projection* of the motion onto that submanifold. (In one freedom the α submanifold is $\mathbf{T}^*\mathbb{Q}$ itself, so the α projection is the full phase portrait.) We will write $\mathcal{C}_\alpha$ for each curve of the α projection.

Figure 6.5 shows a typical α projection for one kind of separable system with bounded orbits (unless explicitly stated otherwise, we will deal only with time-independent Hamiltonians): the $\mathcal{C}_\alpha$ are closed curves, though not in general ellipses. A different kind of separable system with bounded orbits is represented by the plane pendulum of Section 2.4.1. This is a one-freedom system, whose $\mathbf{T}^*\mathbb{Q}$ is the two-dimensional surface of a cylinder, so the α projection is the phase portrait of Fig. 2.11 (Fig. 2.11 is the $\mathbf{T}\mathbb{Q}$ phase portrait; the one in $\mathbf{T}^*\mathbb{Q}$ is essentially the same because $p_\theta = ml^2\dot{\theta}$). Because $\mathbf{T}^*\mathbb{Q}$ is a cylinder, two types of closed orbits are possible. One type consists of C_α curves like those of the charged particle in a magnetic field, curves that enclose the elliptic point at $(0, 0)$ and can be contracted to points. These are called *librational orbits* and correspond to oscillations of the pendulum about its equilibrium position. The other type consists of C_α curves that run completely around the cylindrical $\mathbf{T}^*\mathbb{Q}$ and cannot be contracted to points. These are called *rotational orbits* and correspond to rotations of the pendulum completely around its support.

What the pendulum and charged particle have in common is the boundedness of the orbits. Boundedness leads to periodicity in the α submanifold. That is, the system point remains on a fixed C_α, and after a certain time it returns to its original position, repeating over and over.

In a completely separable system with bounded orbits, all of the C_α look like those of Fig. 6.5 or Fig. 2.11: each trajectory in $\mathbf{T}^*\mathbb{Q}$ traces out a closed C_α in every one of the α submanifolds, and each C_α is *homotopic* to (can be continuously distorted into) a circle $\mathbb{S}^1$. Each trajectory of the phase portrait therefore lies on all of these "circles" simultaneously. In two freedoms there are two circles, and for each given point (q^1, p_1) on the first $\mathbb{S}^1$ circle there is a second complete $\mathbb{S}^1$ circle of (q^2, p_2) values: to each point of $\mathbb{S}^1$ is attached another $\mathbb{S}^1$. This describes a 2-torus $\mathbb{T}^2 = \mathbb{S}^1 \times \mathbb{S}^1$, so the motion takes place on such a $\mathbb{T}^2$. Although this is easy to visualize in two freedoms, it is much harder in n, when the motion takes place simultaneously on n circles, or on an n-torus $\mathbb{T}^n = \mathbb{S}^1 \times \cdots \times \mathbb{S}^1$ in $\mathbf{T}^*\mathbb{Q}$. The dynamics flows entirely on such tori: each $\mathbb{T}^n$ is an invariant manifold. Each $\mathbb{T}^n$ intersects the α submanifold in one of the C_α curves and each carries many motions. The initial conditions for each particular motion are given by naming all the C_α on which it lies (and thereby its torus) and by naming the initial point on each of its C_α.

There are many such tori, corresponding to the many C_α curves in each of the α submanifolds. Because integral curves can not intersect, neither can the tori, and hence the dynamics flows on a set of nonintersecting tori, as illustrated in Fig. 6.6. A system whose integral curves are organized in this way is called *completely integrable*.

Although it is often helpful to visualize tori, the visualization should not be taken too literally, especially in higher dimensions. Even the 2-torus $\mathbb{T}^2$ imbedded in $\mathbb{R}^3$, for example, can often be visualized in two ways. In Fig. 6.7, torus A can be thought of as one circle S with another s attached at each point, or as s with S attached at each point. Torus B is nested in A in one visualization and linked with A in the other. For our purposes, both pictures represent the same system of tori. When the coordinates in $\mathbb{R}^3$ are well specified, however, only one visualization of the two tori is possible.

THE ϕ^α AND J_α

In this subsection we will show that for a completely integrable system it is possible to transform canonically from (q, p) to new coordinates (ϕ, J) that directly specify both the torus $\mathbb{T}^n \subset \mathbf{T}^*\mathbb{Q}$ on which a particular motion lies and the rate at which each C_α curve (each circle $\mathbb{S}^1$) of that torus is traversed.

The new momenta J_α are used to specify the torus, and since the tori are invariant manifolds, the J_α must be constants of the motion. That implies that the new generalized

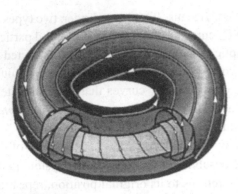

FIGURE 6.6

Illustration of flow on two-dimensional tori in $\mathbb{R}^3$ (since $\mathbf{T}^*Q$ is necessarily of even dimension, the diagram may be misleading). Four nested tori of a one-dimensional set are shown, with integral curves indicated. A typical C_α curve runs around a torus either vertically or horizontally (see Fig. 6.10). Integral curves, however, move simultaneously along both kinds of C_α ($\alpha = 1, 2$ in this diagram), and so they wind around the torus. Both the frequencies ν^α and their ratio ν^1/ν^2 are different on different tori. [Another misleading aspect of this kind of diagram: when a 2-torus is imbedded in a space of dimension higher than three, the separation of its interior from its exterior is unclear and it is possible to pass from the interior to the exterior without intersecting the torus (think about the difference between a circle – a 1-torus – in $\mathbb{R}^2$ and the same circle in $\mathbb{R}^3$).]

position coordinates ϕ^α cannot appear in the Hamiltonian, for

$$\dot{J}_\alpha = -\partial H/\partial \phi^\alpha = 0. \tag{6.37}$$

In other words, $H = H(J)$. The J_α are called *action variables*.

The coordinate ϕ^α will describe the motion on C_α, and all of the ϕ^α will be normalized to increase by 2π each time a C_α is traversed, that is,

$$\oint_{C_\alpha} d\phi^\alpha = 2\pi, \tag{6.38}$$

where the integral is taken once around one of the C_α in the α projection. This means that

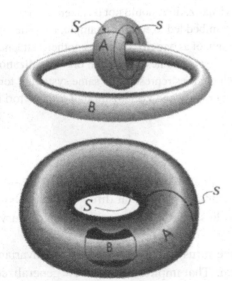

FIGURE 6.7

Two visualizations of a pair of 2-tori imbedded in $\mathbb{R}^3$. Torus A is $S \times s$. The circle s links torus B. If torus A is developed by attaching S at each point of s, the two tori are linked. If S is drawn next to torus B and torus A is developed by attaching s at each point of S, torus B lies inside of torus A.

the ϕ^α are local coordinates only: they increase by 2π in one trip around C_α and it takes two charts to cover the closed curve, like the angle around a circle. The ϕ^α are thus called *angle variables*. Because H is a function only of the J_α, the $\phi^\alpha(t)$ are linear in the time. That is,

$$\dot\phi^\alpha = \frac{\partial H}{\partial J_\alpha} = \nu^\alpha(J), \tag{6.39}$$

and since the J_α, and therefore also the ν^α, are constants of the motion, this integrates immediately to

$$\phi^\alpha(t) = \nu^\alpha t + \Theta^\alpha, \tag{6.40}$$

where the Θ^α are constants. Thus the ν^α are *frequencies* that describe the rate at which the C_α curves are traversed: on each C_α the motion is periodic with period $T_\alpha = 2\pi/\nu^\alpha$. The (ϕ, J) variables are called *action–angle* (AA) *variables*. By construction, from Eqs. (6.37) and (6.39), they form a canonical set [there is a CT, yet to be found, from (q, p) to (ϕ, J)]:

$$\left.\begin{aligned} \{J_\alpha, J_\beta\} = \{\phi^\alpha, \phi^\beta\} = 0, \quad \{\phi^\alpha, J_\beta\} = \delta^\alpha_\beta, \\ \omega = \sum_\alpha d\phi^\alpha \wedge dJ_\alpha. \end{aligned}\right\} \tag{6.41}$$

Because $\dot J_\alpha = 0$, the dynamical vector field is

$$\Delta = \sum_\alpha \nu^\alpha \frac{\partial}{\partial \phi^\alpha}.$$

The periodicity in the ϕ^α will allow us eventually to write the motion itself as a Fourier series.

THE CANONICAL TRANSFORMATION TO AA VARIABLES

We now establish more accurately the existence of the new variables and show how to obtain the CT from (q, p) to (ϕ, J). Because J_α determines the C_α in the α projection, and because q^α determines the point on C_α (up to the sign of p_α), the set (q, J) determines the point in $\mathbf{T}^*\mathbb{Q}$. It is therefore reasonable to assume that the transformation is of Type 2. Instead of $F_2(q, J)$ we will write $\tilde W(q, J)$ for the Type 2 generating function (we will see later that it is related to Hamilton's characteristic function W), so that

$$p_\alpha = \frac{\partial \tilde W}{\partial q^\alpha}, \quad \phi^\alpha = \frac{\partial \tilde W}{\partial J_\alpha}, \tag{6.42}$$

and when we have found $\tilde W$ we will have found the CT. Now rewrite Eq. (6.38) in the form (remember, we are not using the summation convention)

$$2\pi = \oint_{C_\alpha} \frac{\partial^2 \tilde W}{\partial q^\alpha \partial J_\alpha} dq^\alpha = \frac{\partial}{\partial J_\alpha} \oint_{C_\alpha} \frac{\partial \tilde W}{\partial q^\alpha} dq^\alpha = \frac{\partial}{\partial J_\alpha} \oint_{C_\alpha} p_\alpha \, dq^\alpha. \tag{6.43}$$

This suggests that J_α is

$$J_\alpha \equiv \frac{1}{2\pi} \oint_{C_\alpha} p_\alpha \, dq^\alpha, \tag{6.44}$$

which is $(2\pi)^{-1}$ times the area enclosed by C_α in the α projection [it is also the Poincaré invariant of Eq. (5.155)]. This, too, is reasonable, since the motion projects onto C_α, so the area it encloses is a constant of the motion. Equation (6.44) is therefore taken to be the defining equation for the action variables.

Since AA variables refer to systems whose HJ equation is separable, there is an intimate relation between AA variables and HJ theory. Recall that (Q, P) of HJ theory is a set of constants of the motion, and because there are $2n$ of them, any other constant of the motion can be expressed as a function of (Q, P). Moreover, the J_α are special: they are functions only of the Q^β, independent of the P_β. This is because the C_α are defined by the equations $p_\alpha = \partial W_\alpha/\partial q^\alpha$, and these are independent of the P_β. Indeed, writing out the equations yields

$$2\pi J_\alpha = \oint_{C_\alpha} \frac{\partial W_\alpha(q^\alpha, Q)}{\partial q^\alpha} \, dq^\alpha \equiv \oint_{C_\alpha} \frac{\partial W}{\partial q^\alpha} dq^\alpha \equiv 2\pi \hat{J}_\alpha(Q) \tag{6.45}$$

(where we have used $\partial W/\partial q^\alpha = \partial W_\alpha/\partial q^\alpha$). In principle this gives the J_α as functions of the Q^β; however, for the CT to (ϕ, J) from (q, p) we need the J_α as functions of (q, p). These can be obtained by inserting the $Q^\alpha(q, p)$ into the $\hat{J}_\alpha(Q)$:

$$J_\alpha(q, p) \equiv \hat{J}_\alpha(Q(q, p)). \tag{6.46}$$

Equation (6.46) is the first half of the CT from (q, p) to (ϕ, J).

To find the second half, invert Eq. (6.45) to obtain $Q^\alpha(J)$ (invertibility will be discussed later), and then the first of Eqs. (6.42) is satisfied if $W(q, Q(J))$ is taken to be $\tilde{W}(q, J)$. Indeed,

$$\frac{\partial \tilde{W}}{\partial q^\alpha} = \frac{\partial W}{\partial q^\alpha} \equiv p_\alpha.$$

We will write

$$W_\alpha(q^\alpha, Q(J)) = \tilde{W}_\alpha(q^\alpha, J) \quad \text{and} \quad \tilde{W} = \sum_\alpha \tilde{W}_\alpha, \tag{6.47}$$

so that (note the subscript α)

$$p_\alpha = \frac{\partial \tilde{W}_\alpha}{\partial q^\alpha}. \tag{6.48}$$

Thus we take $\tilde{W}$ to be the Type 2 generating $\tilde{W}$ from (q, p) to (ϕ, J); it is

$$\tilde{W}(q, J) = \sum_\alpha \tilde{W}_\alpha(q, J) \equiv \sum_\alpha W_\alpha(q, Q(J, q)).$$

Then the CT is completed by the second of Eqs. (6.42).

To check, we show that the ϕ^α obtained from this CT satisfy (6.38). The $d\phi^\alpha$ on C_α appearing in (6.38) can be calculated from (6.42). On C_α the q^β for $\beta \neq \alpha$ are constant, and all of the J_α are constant, so that on C_α (no sum on α),

$$d\phi^\alpha \equiv d\frac{\partial \tilde{W}}{\partial J_\alpha} = \frac{\partial^2 \tilde{W}_\alpha}{\partial J_\alpha \partial q^\alpha} dq^\alpha.$$

Therefore

$$\oint_{C_\alpha} d\phi^\alpha = \frac{\partial}{\partial J_\alpha} \oint_{C_\alpha} \frac{\partial \tilde{W}}{\partial q^\alpha} dq^\alpha = \frac{\partial}{\partial J_\alpha} \oint_{C_\alpha} p_\alpha \, dq^\alpha = 2\pi \frac{\partial J_\alpha}{\partial J_\alpha} = 2\pi, \qquad (6.49)$$

which satisfies Eq. (6.38).

We have now arrived at an elegant view of a completely integrable dynamical system with bounded orbits. The cotangent bundle is decomposed into an n-dimensional set of n-dimensional tori $\mathbb{T}^n$, and the integral curves lie on these tori, as illustrated in Fig. 6.6. The new canonical coordinates (ϕ, J) on $\mathbf{T}^*Q$ are perfectly adapted for describing the motion: the values of the J_α label the torus on which a particular motion is taking place, and the ϕ^α are coordinates on the torus that vary linearly with the time, as in Eq. (6.40). The equations of motion are

$$\dot{\phi}^\alpha = \frac{\partial Q^1}{\partial J_\alpha} \equiv \nu^\alpha, \quad \dot{j}^\alpha = -\frac{\partial Q^1}{\partial \phi^\alpha} \equiv 0, \qquad (6.50)$$

where Q^1 is the Hamiltonian expressed as a function of (ϕ, J). The values of the frequencies ν^α are completely determined by the torus. We have seen (Section 4.2.2) that motion on tori can have nontrivial properties, and we shall expand on this shortly.

> **REMARK:** If the original (q, p) coordinates are Cartesian, the J_α have the dimensions of angular momentum or action, which is one reason why they are taken to be the new momenta. Then if H has the dimensions of energy, the ϕ^α will be dimensionless (pure numbers). □

We return to the invertibility of Eq. (6.45), which requires that

$$D \equiv \det\left(\frac{\partial J_\alpha}{\partial Q^\beta}\right) = \det\left(\oint \frac{\partial^2 W}{\partial q^\alpha \partial Q^\beta} dq^\alpha\right) \neq 0.$$

Write D in the form

$$D = \oint \det\left(\frac{\partial^2 W}{\partial q^\alpha \partial Q^\beta}\right) dq^1 \cdots dq^n$$

and then use the fact that W is a solution of the HJ equation. The determinant in the integrand is nonzero by assumption, and it follows that the integral itself is nonzero. That Eq. (6.45) is invertible is interesting. Each C_α is determined by the Q^β, and that the Q^β

are determined by the J_α means that each C_α is determined by its enclosed area $2\pi J_\alpha$. In other words, every C_α has a unique area in the α submanifold, and this has to do with the way the tori are nested.

In summary, AA variables are obtained for separable systems in the following 4-step procedure:

Step 1. Separate and solve the HJ equation for the W_α.

Step 2. Use Eq. (6.45) to define the J_α and invert it to find the $Q^\alpha(J)$. Assuming that the integral in (6.45) can be performed, $Q^\alpha(J)$ and the known relation between each p_α and its conjugate q^α (i.e., the C_α curve) yield the J part of the CT from (q, p) to (ϕ, J).

Step 3. Define $\tilde{W}$ by Eq. (6.47).

Step 4. Use this $\tilde{W}$ in Eq. (6.42) to generate the rest of the CT from (q, p) to (ϕ, J). (The J part of the CT is found largely in Step 2.)

EXAMPLE: A PARTICLE ON A VERTICAL CYLINDER

We give an example to illustrate this procedure and to show the difference between the treatment of librational and rotational modes. This example shows also that AA variables can be used with systems whose constraints are not necessarily holonomic.

Consider a particle of mass m constrained to move on a cylinder of radius R standing vertically in a uniform gravitational field of strength g (Fig. 6.8), with the additional nonholonomic constraint that $x \geq 0$. When the particle reaches the bottom of the cylinder at $x = 0$ it bounces elastically: the vertical component of its velocity (and momentum) reverses direction, and the horizontal component does not change. Between bounces the

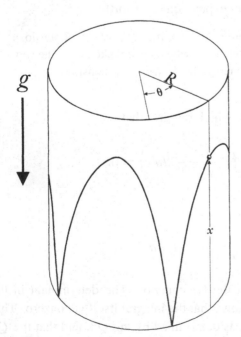

FIGURE 6.8
A particle constrained to move on the surface of a cylinder in a uniform gravitational field. At $x = 0$ the x component of the momentum changes sign.

Hamiltonian is

$$H = \frac{p_\theta^2}{2mR^2} + \frac{p_x^2}{2m} + mgx.$$

The angle θ varies from $-\pi$ to π.

Step 1. The time-independent HJ equation for $W(\theta, x) = W_\theta(\theta) + W_x(x)$ is

$$E = \frac{1}{2mR^2} \left(\frac{\partial W_\theta}{\partial \theta}\right)^2 + \frac{1}{2m} \left(\frac{\partial W_x}{\partial x}\right)^2 + mgx,$$

where $E = Q^1$. The variables are separated by setting $(W_x')^2 + 2m^2gx = Q^x$, and it is found that

$$W \equiv W_x + W_\theta = -\frac{1}{3m^2g} \left(Q^x - 2m^2gx\right)^{3/2} + R\theta\sqrt{2mE - Q^x}.$$

Step 2. According to (6.45) $2\pi J_x$ is given by $\int_{C_x} (\partial W_x/\partial x)\, dx$. But

$$\frac{\partial W_x}{\partial x} = p_x = \sqrt{Q^x - 2m^2gx}, \tag{6.51}$$

so

$$J_x = \frac{1}{2\pi} \oint_{C_x} p_x\, dx = \frac{1}{2\pi} \oint_{C_x} \sqrt{Q^x - 2m^2gx}\, dx,$$

where the C_x are shown in Fig. 6.9: they are parabolas truncated and closed at $x = 0$. The closure of the parabolas comes from the nonholonomic constraint: the momentum jumps discontinuously from its negative value $-p_x$ to $+p_x$. The integral is just the area enclosed by C_x and can be calculated in terms of the x_{max} of each curve, which is the root of the radical: $x_{max} = Q^x/(2m^2g)$. Taking the square root to be positive, J_x is given by

$$J_x = \frac{1}{\pi} \int_0^{x_{max}} \sqrt{Q^x - 2m^2gx}\, dx = \frac{1}{3\pi m^2 g}(Q^x)^{3/2}. \tag{6.52}$$

Notice that in this librational mode the x integration is from 0 to x_{max} for the positive root (positive p_x) and back from x_{max} to 0 for the negative root.

Inversion yields $Q^x(J_x) = (3\pi m^2 g J_x)^{2/3}$.

To find J_θ, use $W_\theta = \theta R\sqrt{2mE - Q^x}$, so that $p_\theta = R\sqrt{2mE - Q^x}$, which is constant (angular momentum is conserved). Hence

$$J_\theta = \frac{1}{2\pi} \oint_{C_\theta} p_\theta\, d\theta = p_\theta = R\sqrt{2mE - Q^x}.$$

The θ submanifold is a cylinder, like the $\mathbf{T}^*\mathbb{Q}$ of the plane pendulum. But in this case *all* of the C_θ curves are rotational, in opposite senses for opposite signs of p_θ. The curves for the positive and negative roots (the two signs of p_θ) are disconnected. The integration is now once around one of these curves (e.g., for positive p_θ from $\theta = 0$ to 2π). This is different from the librational mode, where the integral was up and back.

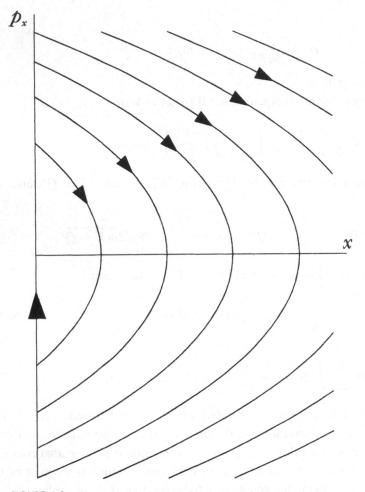

FIGURE 6.9
The x projection of the phase portrait. Each C_x is a closed curve consisting of a piece of a parabola
$p_x^2 = Q^x - 2m^2gx$ and the line closing it along the p_x axis. The arrows on the graph show the direction
of motion and of integration for $\oint p_x \, dx$. When the system point reaches $x = 0$ at a negative value of
p_x, it jumps discontinuously to the positive value of p_x at $x = 0$ on the same parabola.

Inversion now requires finding E as a function of the Js.

Step 3. To find $\tilde{W}_x$ we use the following trick. Since $p_x(x, J) = \partial \tilde{W}_x(x, J)/\partial x$,

$$\tilde{W}_x(x, J) = \int_{x_0}^{x} p_x \, dx = \int_{x_0}^{x} \sqrt{Q^x - 2m^2gx'} \, dx',$$

(where x_0 can be chosen for convenience). This is the same as the integral from which
J_x was found, except for the limits of integration. In fact it would have been convenient
to find $\tilde{W}_x$ first and then change the limits to find J_x. This trick can always be used in a
one-freedom problem (the integrand in the expression for $\tilde{W}$ is then $\sqrt{2m\{E - V(x)\}}$) and
generally for the first of several separable freedoms. In any case, setting x_0 so as to give

no contribution from the lower limit, one obtains

$$\tilde{W}_x = \int_{x_0}^x \sqrt{Q^x - 2m^2 gx'}\, dx' = -\left[(\pi J_x)^{2/3} - \left(\frac{1}{3m^2 g} \right)^{2/3} 2m^2 gx \right]^{3/2},$$

where Q^x has been written in terms of J_x.

The θ part is simpler:

$$\tilde{W}_\theta = \int^\theta R\sqrt{2mE - Q^x}\, d\theta = \theta J_\theta,$$

so that

$$\tilde{W} = \theta J_\theta - \left[(\pi J_x)^{2/3} - \left(\frac{1}{3m^2 g} \right)^{2/3} 2m^2 gx \right]^{3/2}.$$

Step 4. The full CT from $(x, \theta, p_x, p_\theta)$ to $(\phi^x, \phi^\theta, J_x, J_\theta)$ is obtained as follows: for J_x, solve Eq. (6.51) for Q^x and insert into (6.52). J_θ has already been found. ϕ^x and ϕ^θ are found by taking derivatives of $\tilde{W}$ and inserting $J_x(x, p_x)$. The result is

$$J_x = \frac{1}{3\pi m^2 g} \left(p_x^2 + 2m^2 gx \right)^{3/2}, \quad \phi^x = -\frac{\pi p_x}{\sqrt{p_x^2 + 2m^2 gx}},$$

$$J_\theta = p_\theta, \qquad\qquad\qquad \phi^\theta = \theta.$$

To obtain the motion in AA variables, first rewrite the Hamiltonian as

$$H = \frac{J_\theta^2}{2mR^2} + \frac{1}{2m} \left(3\pi m^2 g J_x \right)^{2/3}.$$

Then each $\phi^\alpha(t)$ is linear in t with frequency $\nu^\alpha \equiv \dot{\phi}^\alpha = \partial H / \partial J_\alpha$, or

$$\nu^x = \frac{1}{3m}(3\pi m^2 g)^{2/3} J_x^{-1/3}, \quad \nu^\theta = \frac{J_\theta}{mR^2}.$$

The motion is qualitatively different depending on whether ν^x and ν^θ are commensurate (see Section 4.2.2 and Problem 3).

We check that ϕ^x is normalized:

$$\oint d\phi^x = \oint \frac{\partial \phi^x}{\partial x} dx = 2\int_0^{x_{\max}} \frac{\partial \phi^x}{\partial x} dx = [2\phi^x]_0^{x_{\max}}$$

$$= -2 \left(\frac{\pi^2}{J_x} \right)^{1/3} \sqrt{(\pi J_x)^{2/3} - \left(\frac{1}{3m^2 g} \right)^{2/3} 2m^2 gx} \,\Bigg|_0^{x_{\max}}$$

$$= -2\sqrt{\pi^2 - \frac{\pi^2 x}{x_{\max}}} \,\Bigg|_0^{x_{\max}} = 2\pi$$

(this requires calculating $x_{\max}$ in terms of J_x). Since $\phi^\theta \equiv \theta$ its normalization is trivially verified.

When the AA solution is transformed back to cylindrical coordinates, it is found, not surprisingly, that $x = x_0 + \dot{x}_0 t - \frac{1}{2} g t^2$ and $\theta = \theta_0 + \dot{\theta}_0 t$.

WORKED EXAMPLE 6.4

This problem returns to the Kepler problem in two dimensions (see Worked Example 6.1). **(a)** For $V = -\alpha/r$, find the CT from the polar variables $(r, \theta, p_r, p_\theta)$ to the AA variables. Obtain explicit expressions for J_r and J_θ and integral expressions for ϕ_r and ϕ_θ. Find the Hamiltonian $H(J)$ and show that the orbits close. **(b)** Add the term ϵ/r^2 to the potential, and find the J_k and H, this time showing that the orbits do not close in general. Calculate the angular separation between successive minima in r exactly, and from that to first order in ϵ for small ϵ.

Solution. **(a)** Step 1 is taken care of by Worked Example 6.1. According to Eq. (6.19)

$$\left. \begin{aligned} W_r(r) &= \sqrt{2m} \int \sqrt{Q_1 - V(r) - (2mr^2)^{-1} Q_2^2} \, dr, \\ W_\theta(\theta) &= Q_2 \theta, \end{aligned} \right\} \tag{6.53}$$

from which $p_\theta = Q_2 = \text{const.}$ (remember that we are using subscripts on the Qs and that $Q_1 \equiv E$).

Step 2 defines the Js by

$$\left. \begin{aligned} J_r &\equiv \frac{1}{2\pi} \oint_{C_r} p_r \, dr = \frac{\sqrt{2m}}{2\pi} \oint_{C_r} \sqrt{E - V(r) - (2mr^2)^{-1} J_\theta^2} \, dr, \\ J_\theta &\equiv \frac{1}{2\pi} \oint_{C_\theta} p_\theta \, d\theta = p_\theta \equiv Q_2. \end{aligned} \right\} \tag{6.54}$$

The θ equation is inverted automatically: $Q_2 = J_\theta$. Note that the θ projection is rotational, as opposed to librational.

If the first of Eqs. (6.54) can be integrated, the equation can be inverted and the J part of the CT obtained. We are dealing now only with closed C_r curves, which means bounded (elliptical) orbits in $\mathbb{Q}$. This is a librational mode, so the integral around C_r is twice the integral between the turning points of the motion, which are the roots r_1 and r_2 of the radical. Now insert $V(r) = -\alpha/r$. The integral for J_r becomes

$$J_r = \frac{\sqrt{2m}}{\pi} \int_{r_1}^{r_2} \sqrt{E + \frac{\alpha}{r} - \frac{J_\theta^2}{2mr^2}} \, dr. \tag{6.55}$$

The integration is usually performed in the complex plane. We do not go into that here but merely state the result:

$$J_r = -|J_\theta| + \alpha \sqrt{-\frac{m}{2E}}. \tag{6.56}$$

In order to obtain one of the CT equations from this, J_θ and E must be written in polar coordinates:

$$J_r = -p_\theta + m\alpha r \left\{2m\alpha r - p_r^2 r^2 - p_\theta^2\right\}^{-1/2}. \tag{6.57}$$

Solving (6.56) for $E \equiv Q_1$ both inverts the equation and gives the Hamiltonian in terms of the action variables:

$$Q_1 \equiv E = -\frac{\alpha^2 m}{2(J_r + |J_\theta|)^2} = H(J). \tag{6.58}$$

In Step 3 $\tilde{W} \equiv \tilde{W}_r + \tilde{W}_\theta$ is obtained, in accordance with (6.47), by substituting J_θ for Q_2 and Eq. (6.58) for Q_1 in Eq. (6.53):

$$\tilde{W}_r = \sqrt{2m} \int \sqrt{E + \frac{\alpha}{r} - \frac{J_\theta^2}{2mr^2}}\, dr,$$

$$\tilde{W}_\theta = \theta J_\theta,$$

where we have not written out E in full.

For Step 4, the ϕs are obtained from the derivatives of $\tilde{W}(q, J)$:

$$\left.\begin{aligned}
\phi_r &= \frac{\partial \tilde{W}_r}{\partial J_r} + \frac{\partial \tilde{W}_\theta}{\partial J_r} = v_r \sqrt{\frac{m}{2}} \int \left\{E + \frac{\alpha}{r} - \frac{J_\theta^2}{2mr^2}\right\}^{-1/2} dr, \\
\phi_\theta &= \frac{\partial \tilde{W}_r}{\partial J_\theta} + \frac{\partial \tilde{W}_\theta}{\partial J_\theta} = \sqrt{\frac{m}{2}} \int \left\{E + \frac{\alpha}{r} - \frac{J_\theta^2}{2mr^2}\right\}^{-1/2} \left\{v_\theta - \frac{J_\theta}{mr^2}\right\} dr + \theta,
\end{aligned}\right\} \tag{6.59}$$

where $v_k = \partial H / \partial J_k$ for k either r or θ. Both of the integrals that appear here can be performed, but the results are not particularly informative and are difficult to invert. The full CT is then given by (6.57), the second of (6.54), and (6.59).

It follows from (6.58) that up to sign $v_r = v_\theta$, since $\partial H / \partial J_r = \pm \partial H / \partial J_\theta$. This means that C_r is traversed at the same rate as C_θ, or that $T_r = T_\theta$ (the sign ambiguity accounts for the two senses in which C_r can be traversed). Since the periods are the same, r and θ return to their starting points at the same time, and the $\mathbb{Q}$ orbit closes: the orbits of the Kepler problem are closed [see the discussion following Eq. (2.75)].

(b) Adding ϵ/r^2 to the potential replaces J_θ^2 in the first of Eqs. (6.54) and in some other equations by $J_\theta^2 + 2m\epsilon$. In particular, Eq. (6.55) is changed, and Eq. (6.56) then becomes

$$J_r = -\sqrt{J_\theta^2 + 2m\epsilon} + \alpha\sqrt{-\frac{m}{2E}}$$

(positive square root). This means that H is now

$$H = -\frac{\alpha^2 m}{2\left(J_r + \sqrt{J_\theta^2 + 2m\epsilon}\right)^2}. \tag{6.60}$$

The two frequencies are no longer equal. If we write D for the denominator in Eq. (6.60), we get

$$\frac{\nu_r}{\nu_\theta} = \frac{\partial D/\partial J_r}{\partial D/\partial J_\theta} = \frac{\sqrt{J_\theta^2 + 2m\epsilon}}{J_\theta}.$$

The r circuit and the θ circuit are traversed at rates given by ν_r and ν_θ, respectively. This means that when one r circuit is completed, the fraction ν_θ/ν_r of the θ circuit is covered. In one circuit θ changes by 2π, so in each r circuit the θ orbit fails to close by

$$\Delta\theta = 2\pi\left(\frac{\nu_\theta}{\nu_r} - 1\right) = 2\pi\left(\frac{J_\theta - \sqrt{J_\theta^2 + 2m\epsilon}}{\sqrt{J_\theta^2 + 2m\epsilon}}\right).$$

The sign of $\Delta\theta$ depends on the sign of ϵ. By expanding this result for $\Delta\theta$ in a Taylor series, one finds that to first order in ϵ (see the note on Problem 5.11)

$$|\Delta\theta| \approx \frac{m\epsilon}{p_\theta^2}.$$

6.2.2 LIOUVILLE'S INTEGRABILITY THEOREM

COMPLETE INTEGRABILITY

This section is devoted to another important theorem due to Liouville, one that leads to a different definition of a completely integrable system on $\mathbf{T}^*\mathbb{Q}$. The new definition is something like the inverse of the previous section's, for rather than arriving at the J_α, it starts from constants of the motion that resemble them. Loosely speaking, the theorem states that if a system has n independent constants of the motion that commute with each other, then it admits AA variables. We now present the theorem in more detail.

Suppose that a Hamiltonian dynamical system Δ in n freedoms possesses n *independent* constants of the motion I_α that are in involution (i.e., commute with each other; see Section 5.4.2):

$$\{I_\alpha, I_\beta\} = 0, \tag{6.61}$$

$\alpha, \beta \in \{1, 2, \ldots, n\}$. One of the I_α may be the Hamiltonian H itself, provided the Hamiltonian is a constant of the motion (i.e., the system is conservative).

Independence of the I_α means that (this is an extension of the hint in Problem 5.3; see also Problem 4 in this chapter)

$$dI_1 \wedge dI_2 \wedge \cdots \wedge dI_n \neq 0. \tag{6.62}$$

The following different aspect of independence is used in the theorem: Let Σ_α be the $(2n-1)$-dimensional submanifold defined by equation $I_\alpha = \sigma_\alpha = \text{const}$. Write $\mathbf{I} = \{\sigma_1, \sigma_2, \ldots, \sigma_n\}$

for a set of constant values of the I_α and define the intersection of the Σ_α as Σ_I:

$$\Sigma_I \equiv \Sigma_1 \cap \Sigma_2 \cap \cdots \cap \Sigma_n.$$

Then the I_α are independent functions iff dim $\Sigma_I = n$. [For example, in 3-space two functions $f(x, y, z)$ and $g(x, y, z)$ are independent iff the intersection of the 2-dimensional surfaces $f = $ const. and $g = $ const. is a 1-dimensional curve.] We make one further assumption: each Σ_I is *compact*, which means essentially that it is contained in a finite region of $T^*\mathbb{Q}$ (for our purposes a more technical definition is not needed).

Before proceeding, we point out that the Σ_I are invariant manifolds. This is because the I_α are constants of the motion, so each Σ_α is an invariant submanifold and hence so is Σ_I for each set I of σ_α constants. That means that Δ is tangent to all of the Σ_I; thus its integral curves (the Hamiltonian flow) lie on the Σ_I submanifolds.

A Hamiltonian dynamical system satisfying these two properties (i.e., n independent constants of the motion in involution and compact Σ_I submanifolds) is now defined as a dynamical system *completely integrable in the sense of the Liouville integrability (LI) theorem*.

The *Liouville integrability theorem* for such systems states that (a) each Σ_I is homotopic to an n-dimensional torus $\mathbb{T}^n$ and (b) there exist canonical coordinates (ϕ, J) such that the values of the J_α label the torus (the Σ_I) and are therefore constants of the motion and the $d\phi^\alpha/dt$ are constant on each torus: (ϕ, J) are a set of AA variables.

Although it starts from a different premise, the LI theorem leads to the same elegant view of integrable dynamical systems as does Section 6.2.1. The new approach is more general, however, in that it does not presuppose separability in some coordinate system on $\mathbb{Q}$. This is an important difference, for there exist dynamical systems that are not separable on $\mathbb{Q}$ but are completely integrable in the sense of the LI theorem (see Toda, 1987).

For a detailed proof of the LI theorem, see Arnol'd (1988), Section 49 (in fact the LI theorem is also known as the Liouville–Arnol'd theorem). Here we will only outline the proof and comment on some of the aspects of the theorem.

THE TORI

The theorem establishes that each Σ_I is a torus by showing that there are certain *principal directions* on it in which finite translations map *every* point back to where it started: a translation far enough in a principal direction is the identity transformation.

The principal directions are found by combining the CTs generated by the I_α. The I_α generate CTs in the same way that any other dynamical variable does (Section 5.3.4): each of the I_α generates its Hamiltonian vector field X_α and its own one-parameter group φ^α of CTs (do not confuse φ with ϕ). Involution, Eq. (6.61), implies that each of the I_β is invariant (Σ_I is an invariant manifold) under any φ^α group, so all of the X_β are tangent to the Σ_I; hence each φ^α maps each Σ_I into itself. (In this context the Hamiltonian of a conservative system is no different from any other I_α.) Recall the translations in the q^α directions of $T^*\mathbb{Q}$ generated by the p_α. The φ^α are analogously generated by the I_α. Just as involution of the p_α implies that translations in the q^α directions commute, so involution of the I_α implies that the φ^α commute: involution implies commutation.

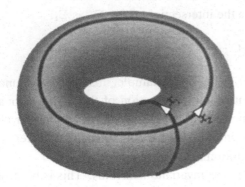

FIGURE 6.10
The torus $\mathbb{T}^2$ with two coordinates x^1 and x^2 shown. The coordinate directions also form a basis of principal directions (see the text). The two circles representing the x^1 and x^2 axes cannot be continuously deformed to points or to each other.

Commutation is important, for it makes the φ^α behave like translations in independent directions, and that will make $\Sigma_\mathbf{I}$ look like a torus. If $\varphi_\epsilon^\alpha \in \varphi^\alpha$ (ϵ is the parameter) is applied to some $\xi_0 \in \Sigma_\mathbf{I}$ for fixed $\mathbf{I}$ and ϵ is allowed to increase, ξ_0 remains on $\Sigma_\mathbf{I}$ while it gets translated farther and farther from its initial position. The compactness of $\Sigma_\mathbf{I}$ implies, however, that ξ_0 can't be translated arbitrarily far. In the proof of the LI theorem it is shown that as ϵ grows, ξ_0 is eventually brought close to its initial position again. By combining several φ_ϵ^α with different α and ϵ, it can be brought even closer, and there are combinations of the φ_ϵ^α that map every $\xi \in \Sigma_\mathbf{I}$ right back to its initial position: these are the identity transformations that lie in the principal directions. The commutation makes $\Sigma_\mathbf{I}$ look like a torus rather than a sphere. On the 2-sphere $\mathbb{S}^2$, for example, rotation by 2π about any axis brings every point back to where it started, but rotations by arbitrary angles about different axes do not commute. On $\mathbb{T}^2$, however, there are such transformations that do commute. This is illustrated in Fig. 6.10. The maps $x^1 \to x^1 + a$ and $x^2 \to x^2 + b$ commute for all a, b, and if a and b are both chosen to be 2π, each map sends every point back to where it started.

The identity transformations in the principal directions are thus combinations of the φ_ϵ^α. Every combination of the φ_ϵ^α is necessarily of the form

$$\varphi_\mathbf{e} = \varphi_{\epsilon_1}^1 \varphi_{\epsilon_2}^2 \cdots \varphi_{\epsilon_n}^n,$$

where

$$\mathbf{e} = \{\epsilon_1, \epsilon_2, \ldots, \epsilon_n\} \tag{6.63}$$

is an n-dimensional vector labeling each such combination. Transformations in the principal directions are labeled by particular $\mathbf{e}$ vectors. In the proof of the LI theorem it is shown that there are n linearly independent $\mathbf{e}$ vectors $\mathbf{e}_\alpha$ whose linear combinations with any integer coefficients label all possible identity transformations (there is an infinite number of them). The proof is then concluded by showing that this is possible only on $\mathbb{T}^n$. We will call the n linearly independent $\mathbf{e}_\alpha$ vectors a *basis of principal vectors* and the directions they define a *basis of principal directions*. On $\mathbb{T}^2$, for instance (refer again to Fig. 6.10), a basis of principal vectors is $(2\pi, 0)$ and $(0, 2\pi)$, corresponding to the translations $x^1 \to x^1 + 2\pi$ and $x^2 \to x^2 + 2\pi$. On $\mathbb{T}^n$ there are n similar closed loops (circuits) along the basis of principal directions, on which the coordinates can be chosen also to vary from 0 to 2π.

THE J_α

It is not hard to find canonical coordinates (ϕ, J) on $\mathbf{T}^*Q$ such that (a) the J_α label the $\Sigma_{\mathbf{I}}$ tori and (b) the dynamics is then given by $\dot\phi^\alpha = $ const. For instance, the I_α themselves can be chosen as the J_α, for they satisfy condition (a). But so does any independent invertible set of functions J_α formed from them (invertible in order that they label the tori). Moreover, any such set of J_α is involutive, and since they label invariant tori the J_α do not vary in time. If ϕ^α are their canonical conjugates, $\dot{J}_\alpha \equiv \partial H/\partial\phi^\alpha = 0$, so H is a function only of the J_α, and it follows that $\dot\phi^\alpha$ is a constant, just as for the AA variables of Section 6.2.1. Since there are many independent invertible J_α sets that can be formed from the I_α, one may ask whether one set is better than another. Is there some way to choose among the possibilities?

There is indeed a way to choose – some sets are better than others. It may turn out that with one choice the line traced out on the torus by the ϕ^α coordinate, the ϕ^α "axis," winds irrationally around $\Sigma_{\mathbf{I}}$, covering it completely (more accurately, coming arbitrarily close to every point on it; see Fig. 4.18). Every neighborhood of $\Sigma_{\mathbf{I}}$, no matter how small, would then contain points whose ϕ^α coordinates were arbitrarily far apart. Such a pathological ϕ^α coordinate can be avoided, however, if the ϕ^α "axis" lies in one of the principal directions. Since the J_α generate translations in the ϕ^α directions, they must hence be chosen to generate translations along independent principal directions. Thus n acceptable J_α must satisfy two criteria: they must be independent functions of the I_α and they must generate translations along a basis of principal directions. The coordinates along the principal directions can be chosen to range from 0 to 2π.

We assert that certain linear combinations of the I_α formed with the $\mathbf{e}_\alpha$ are such acceptable J_α, where the $\mathbf{e}_\alpha$ are principal basis vectors. Choose any fixed $\mathbf{e}$ vector, not necessarily in the principal basis. Then the one-parameter group generated by $J \equiv \sum_\alpha \epsilon_\alpha I_\alpha$ is $\varphi^1_{\lambda\epsilon_1} \varphi^2_{\lambda\epsilon_2} \cdots \varphi^n_{\lambda\epsilon_n}$, where the ϵ_α are the components of $\mathbf{e}$, as in Eq. (6.63) (the ϵ_α are fixed and λ is the group parameter). Indeed, take the derivative at $\lambda = 0$:

$$\frac{d\xi(\lambda)}{d\lambda} = \{\xi, J\} = \sum \epsilon_\alpha \{\xi, I_\alpha\} = \sum \epsilon_\alpha \frac{d(\varphi^\alpha_\lambda \xi)}{d\lambda}$$

$$= \sum \frac{d(\varphi^\alpha_{\lambda\epsilon_\alpha} \xi)}{d\lambda} = \frac{d}{d\lambda}\left(\varphi^1_{\lambda\epsilon_1} \varphi^2_{\lambda\epsilon_2} \cdots \varphi^n_{\lambda\epsilon_n}\right)\xi.$$

The last equality uses the fact that at $\lambda = 0$ each of the $\varphi^\alpha_{\lambda\epsilon_\alpha}$ is the identity. It follows that if $\mathbf{e}$ is one of the principal basis vectors $\mathbf{e}_\alpha$, then J generates a transformation in a principal direction. If $\mathbf{e}_\alpha = \{\epsilon_{\alpha;1}, \epsilon_{\alpha;2}, \ldots, \epsilon_{\alpha;n}\}$, the n dynamical variables

$$J_\alpha = \sum_\beta \epsilon_{\alpha;\beta} I_\beta$$

generate translations along the corresponding basis of principal directions. These J_α are independent (because the $\mathbf{e}_\alpha$ are a basis) and hence satisfy the requirements for an acceptable set, and the ϕ^α canonically conjugate to them are nonpathological coordinates on $\mathbb{T}^n$. To make the ϕ^α vary from 0 to 2π, the ϵ_α vectors must be properly normalized, but even without normalization the construction would yield an acceptable set of canonical coordinates.

The principal directions are illustrated by the circles we have called the x^1 and x^2 "axes" of Fig. 6.10. Each of these circles has the property that it cannot be continuously shrunk to a point. To explain what this means, we first turn to $\mathbb{R}^2$. The $\mathbb{R}^2$ plane has the

topological property that any circle on it can be shrunk to a point by letting the radius
go continuously to zero. Every $\mathbb{R}^n$ has the same topological property, and so does the
sphere $\mathbb{S}^n$. The torus $\mathbb{T}^n$, however, does not: the x^1 and x^2 circles are obvious coun-
terexamples. Moreover, these two circles cannot be continuously deformed into each
other: they are *inequivalent circuits*. In general $\mathbb{T}^n$ has sets of n inequivalent circuits that
cannot be shrunk to points, and the principal directions are along such circuits. This
is another way to understand the LI theorem: it shows that under the conditions of the
theorem there exist canonical coordinates (ϕ, J) on $\mathbf{T}^*\mathbb{Q}$ such that the J_α label the tori
and the ϕ^α are coordinates along n inequivalent circuits (see Ozorio de Almeida, 1990,
Section 6.1).

This completes our description of the LI theorem.

EXAMPLE: THE NEUMANN PROBLEM

This example represents an anisotropic n-freedom harmonic oscillator constrained to
a sphere $\mathbb{S}^{n-1}$ in $\mathbb{R}^n$. It was first treated by Neumann (1859) for $n = 3$; later Uhlenbeck
(1982) showed it to be completely integrable in the sense of the LI theorem for arbitrary
finite n. The same system was discussed briefly in Worked Example 2.1, but now it will be
treated more fully by the Hamiltonian formalism. (As in Worked Example 2.1, we do not
use the summation convention and write x, rather than q, for the coordinates, using only
lower indices.)

We first present the system in three, and then describe its extension to n freedoms. The
Hamiltonian H on $\mathbf{T}^*\mathbb{R}^3$ is given by

$$H = \frac{1}{2}|\mathbf{j}|^2 + \frac{1}{2}\sum_{\alpha=1}^{3} k_\alpha x_\alpha^2, \tag{6.64}$$

where the x_α, $\alpha = 1, 2, 3$, are Cartesian coordinates of a particle, and $\mathbf{j} \equiv \mathbf{x} \wedge \mathbf{p}$ is its angular
momentum. Although the second term of the Hamiltonian resembles the potential energy
of the anisotropic oscillator, the first term does not look like its kinetic energy.

The dynamical system of this H is an anisotropic harmonic oscillator constrained to
$\mathbb{S}^2$. To prove it, we show that H is the sum of the kinetic and potential energies of such an
oscillator. The second term is obviously the potential energy. To see that the first term is
the kinetic energy, notice first that $\{|\mathbf{x}|^2, H\} = 0$, for the x_α commute among themselves
and $|\mathbf{j}|^2$ commutes with $|\mathbf{x}|^2$ (it was shown in worked Example 5.3 that $|\mathbf{j}|^2$ commutes with
all scalars). Thus $|\mathbf{x}|^2$ is a constant of the motion, which implies that the motion lies on
the two-dimensional surface of the sphere defined by the initial position. (This is different
from Worked Example 2.1, where the sphere was of radius 1.) The kinetic energy of a
particle on a sphere is $|\mathbf{j}|^2$ up to a constant multiple involving its mass, so the first term of
H is the kinetic energy of the constrained particle. Thus the system is as asserted.

Because it is constrained to $\mathbb{S}^2$, the system does not have three freedoms, but two, even
though it is written in terms of Cartesian coordinates on $\mathbb{R}^3$. This is a reflection of the result
of Problem 5.33: H is a degenerate Hamiltonian and the constraint is a consequence of
H itself (no additional constraint equation is required). It is evident from the physics that

if the oscillator is isotropic (i.e., if the three k_α are equal), the constraint force cancels the spring force and there is no total force on the particle. Therefore the system is nontrivial only if the k_α are unequal.

We now generalize to n freedoms. In going from $\mathbf{T}^*\mathbb{R}^3$ to $\mathbf{T}^*\mathbb{R}^n$ the Hamiltonian will be written in terms of Cartesian coordinates on $\mathbb{R}^n$, although it will be seen that there are actually only $n - 1$ freedoms. To generalize from 3 to n, the j_α are replaced by

$$J_{\alpha\beta} = x_\alpha p_\beta - x_\beta p_\alpha \qquad (6.65)$$

(we use only lower indices) and the Hamiltonian is written

$$H = \frac{1}{4} \sum_{\alpha,\beta} J_{\alpha\beta}^2 + \frac{1}{2} \sum k_\alpha x_\alpha^2. \qquad (6.66)$$

The extra factor of $\frac{1}{2}$ on the first term of H is necessary because of the antisymmetry of the $J_{\alpha\beta}$. As in the case of $n = 3$, it turns out that $|\mathbf{x}|^2 \equiv \sum x_\alpha^2$ is a constant of the motion, so that the motion takes place on the sphere $\mathbb{S}^{n-1}$ in $\mathbb{R}^n$. We first show that $|\mathbf{x}|^2$ is a constant of the motion by calculating $\{|\mathbf{x}|^2, H\}$. Because the x_α all commute among themselves, we need only calculate the PB of $|\mathbf{x}|^2$ with the $J_{\alpha\beta}^2$ term of H. This will be done in several steps, because some of the intermediate results will be needed later (remember, there are no sums without summation signs):

$$\{x_\gamma, J_{\alpha\beta}\} = \{x_\gamma, (x_\alpha p_\beta - p_\alpha x_\beta)\} = x_\alpha \delta_{\gamma\beta} - x_\beta \delta_{\gamma\alpha}, \qquad (6.67a)$$

$$\{x_\gamma, J_{\alpha\beta}^2\} = 2J_{\alpha\beta}\{x_\gamma, J_{\alpha\beta}\} = 2J_{\alpha\beta}(x_\alpha \delta_{\gamma\beta} - x_\beta \delta_{\gamma\alpha}), \qquad (6.67b)$$

$$\{x_\gamma^2, J_{\alpha\beta}^2\} = 4x_\gamma J_{\alpha\beta}\{x_\gamma, J_{\alpha\beta}\} = 4x_\gamma J_{\alpha\beta}(x_\alpha \delta_{\gamma\beta} - x_\beta \delta_{\gamma\alpha}), \qquad (6.67c)$$

$$\sum_{\alpha,\beta} \{x_\gamma^2, J_{\alpha\beta}^2\} = 4x_\gamma \left[\sum_\alpha J_{\alpha\gamma} x_\alpha - \sum_\beta x_\beta J_{\gamma\beta} \right] = 8x_\gamma \sum_\alpha J_{\alpha\gamma} x_\alpha, \qquad (6.67d)$$

$$\sum_{\gamma,\alpha,\beta} \{x_\gamma^2, J_{\alpha\beta}^2\} = 8 \sum_{\gamma,\alpha} x_\gamma x_\alpha J_{\alpha\gamma} = 0. \qquad (6.67e)$$

Antisymmetry of the $J_{\alpha\beta}$ is used in obtaining both the last equality of Eq. (6.67d) and Eq. (6.67e). From Eq. (6.67e) it follows that $\{|\mathbf{x}|^2, H\} = 0$ and hence that $|\mathbf{x}|^2$ is a constant of the motion.

To show that this is a completely integrable system we use the LI theorem and exhibit n constants of the motion I_α in involution, which were first found by Uhlenbeck:

$$I_\alpha = x_\alpha^2 + \Phi_\alpha, \qquad (6.68)$$

where (again, see Worked Example 2.1)

$$\Phi_\alpha = \sum_{\beta \neq \alpha} \frac{J_{\alpha\beta}^2}{k_\alpha - k_\beta}.$$

In the case of $n = 3$ the Φ_α are

$$\Phi_1 = \frac{j_3^2}{k_1 - k_2} + \frac{j_2^2}{k_1 - k_3}, \quad \Phi_2 = \frac{j_1^2}{k_2 - k_3} + \frac{j_3^2}{k_2 - k_1}, \quad \Phi_3 = \frac{j_2^2}{k_3 - k_1} + \frac{j_1^2}{k_3 - k_2}.$$

We first show that the I_α are constants of the motion, that is, that $\{I_\alpha, H\} \equiv \{x_\alpha^2, H\} + \{\Phi_\alpha, H\} = 0$. In the first term, x_α^2 commutes with the potential energy term of H, and according to Eq. (6.67d)

$$\{x_\gamma^2, H\} \equiv \frac{1}{4} \sum_{\alpha,\beta} \{x_\gamma^2, J_{\alpha\beta}^2\} = 2x_\gamma \sum_\alpha J_{\alpha\gamma} x_\alpha. \tag{6.69}$$

The term $\{\Phi_\alpha, H\}$ is more complicated; it involves the PBs of powers of the $J_{\alpha\beta}$. Equation (6.65) implies that $\{J_{\alpha\beta}, J_{\lambda\mu}\} = 0$ unless exactly one of α or β equals exactly one of λ or μ. If more than one are equal (e.g., if $\alpha = \lambda$ and $\beta = \mu$) the PB also vanishes. Thus it is enough to calculate this PB for $\alpha \neq \beta \neq \mu \neq \alpha$, or

$$\{J_{\alpha\beta}, J_{\alpha\mu}\} = \{x_\alpha p_\beta, x_\alpha p_\mu\} + \{x_\beta p_\alpha, x_\mu p_\alpha\} - \{x_\alpha p_\beta, x_\mu p_\alpha\} - \{x_\beta p_\alpha, x_\alpha p_\mu\}$$
$$= -p_\beta x_\mu + x_\beta p_\mu = J_{\beta\mu}.$$

We write down the general formula without an explicit derivation and two other formulas that follow:

$$\{J_{\alpha\beta}, J_{\lambda\mu}\} = J_{\lambda\beta}\delta_{\alpha\mu} + J_{\mu\alpha}\delta_{\beta\lambda} + J_{\alpha\lambda}\delta_{\beta\mu} + J_{\beta\mu}\delta_{\alpha\lambda}, \tag{6.70a}$$

$$\{J_{\alpha\beta}^2, J_{\lambda\mu}\} \equiv 2J_{\alpha\beta}\{J_{\alpha\beta}, J_{\lambda\mu}\} = 2J_{\alpha\beta}(J_{\lambda\beta}\delta_{\alpha\mu} + J_{\mu\alpha}\delta_{\beta\lambda} + J_{\alpha\lambda}\delta_{\beta\mu} + J_{\beta\mu}\delta_{\alpha\lambda}), \tag{6.70b}$$

$$\sum_{\alpha,\beta} \{J_{\alpha\beta}^2, J_{\lambda\mu}\} = 2 \sum_{\alpha,\beta} J_{\alpha\beta}(J_{\lambda\beta}\delta_{\alpha\mu} + J_{\mu\alpha}\delta_{\beta\lambda} + J_{\alpha\lambda}\delta_{\beta\mu} + J_{\beta\mu}\delta_{\alpha\lambda}) = 0. \tag{6.70c}$$

Equation (6.70c) implies that the $J_{\alpha\beta}$ term of Φ_α commutes with the $J_{\alpha\beta}$ term of H. We are left with

$$\frac{1}{2} \sum_\gamma \{\Phi_\alpha, k_\gamma x_\gamma^2\} \equiv \frac{1}{2} \sum_\gamma k_\gamma \left\{ \sum_{\beta \neq \alpha} \frac{J_{\alpha\beta}^2}{k_\alpha - k_\beta}, x_\gamma^2 \right\} \equiv B.$$

A way to eliminate the $\beta = \alpha$ term in the sum is to add a small quantity ϵ to the denominator and take the sum over all values of β. The antisymmetry of the $J_{\alpha\beta}$ eliminates the $\beta = \alpha$ term, and after the sum is taken ϵ can be set equal to zero. We will proceed as if we were doing this, though not explicitly, by simply ignoring the condition $\beta \neq \alpha$ on the sum. Then, according to Eq. (6.67c), the PB is

$$B = -2 \sum_{\beta,\gamma} k_\gamma x_\gamma \frac{J_{\alpha\beta}(x_\alpha \delta_{\gamma\beta} - x_\beta \delta_{\gamma\alpha})}{k_\alpha - k_\beta}$$

$$= -2 \sum_\gamma k_\gamma x_\gamma \frac{x_\alpha J_{\alpha\gamma}}{k_\alpha - k_\gamma} + 2 \sum_\beta k_\alpha x_\alpha \frac{x_\beta J_{\alpha\beta}}{k_\alpha - k_\beta}$$

$$= 2 \sum_\beta x_\alpha x_\beta \frac{(k_\alpha - k_\beta)J_{\alpha\beta}}{k_\alpha - k_\beta} = 2x_\alpha \sum_\beta x_\beta J_{\alpha\beta}.$$

This cancels (6.69) because of the antisymmetry of the $J_{\alpha\beta}$, so that $\{I_\alpha, H\} = 0$: the I_α are constants of the motion.

We now show that the I_α are in involution: $\{I_\alpha, I_\beta\} = 0$. The calculations are similar to those for $\{I_\alpha, H\}$, and we present them only in outline. What needs to be calculated is

$$\{I_\alpha, I_\beta\} = \{x_\alpha^2 + \Phi_\alpha, x_\beta^2 + \Phi_\beta\}. \tag{6.71}$$

There are three kinds of terms here:

$$\{x_\alpha^2, x_\beta^2\}, \quad \{x_\alpha^2, \Phi_\beta\}, \quad \{\Phi_\alpha, \Phi_\beta\}.$$

The first kind vanishes. The second kind can be calculated with the aid of Eq. (6.67c) and turns out to be symmetric in α and β, which means that the two terms of the second kind cancel each other. The third kind can be calculated with the aid of Eq. (6.70a). We have (again, we ignore the condition on the sums)

$$\{\Phi_\alpha, \Phi_\lambda\} = \sum_{\beta,\mu} \frac{1}{(k_\alpha - k_\beta)(k_\lambda - k_\mu)} \{J_{\alpha\beta}^2, J_{\lambda\mu}^2\}$$

$$= 4 \sum_{\beta,\mu} \frac{J_{\alpha\beta} J_{\lambda\mu}}{(k_\alpha - k_\beta)(k_\lambda - k_\mu)} \{J_{\alpha\beta}, J_{\lambda\mu}\}$$

$$= 4 \sum_{\beta,\mu} \frac{J_{\alpha\beta} J_{\lambda\mu}}{(k_u - k_\beta)(k_\lambda - k_\mu)} (J_{\lambda\beta}\delta_{\alpha\mu} + J_{\mu\alpha}\delta_{\beta\lambda} + J_{\alpha\lambda}\delta_{\beta\mu} + J_{\beta\mu}\delta_{\alpha\lambda}).$$

Consider the first two terms:

$$S_{12} = \sum_\beta \frac{J_{\alpha\beta} J_{\lambda\alpha} J_{\lambda\beta}}{(k_\alpha - k_\beta)(k_\lambda - k_\alpha)} + \sum_\mu \frac{J_{\alpha\lambda} J_{\lambda\mu} J_{\mu\alpha}}{(k_\alpha - k_\lambda)(k_\lambda - k_\mu)}$$

$$= \frac{J_{\lambda\alpha}}{k_\lambda - k_\alpha} \sum_\beta \frac{J_{\alpha\beta} J_{\lambda\beta}}{k_\alpha - k_\beta} + \frac{J_{\alpha\lambda}}{k_\alpha - k_\lambda} \sum_\beta \frac{J_{\lambda\beta} J_{\beta\alpha}}{k_\lambda - k_\beta}$$

$$= \frac{J_{\lambda\alpha}}{k_\lambda - k_\alpha} \sum_\beta J_{\lambda\beta} J_{\beta\alpha} \left[\frac{1}{k_\lambda - k_\beta} - \frac{1}{k_\alpha - k_\beta} \right] = J_{\alpha\lambda} \sum_\beta \frac{J_{\lambda\beta} J_{\beta\alpha}}{(k_\lambda - k_\beta)(k_\alpha - k_\beta)}.$$

Now consider the third term:

$$S_3 = \sum_{\beta,\mu} \frac{J_{\alpha\lambda} J_{\lambda\mu} J_{\alpha\lambda}\delta_{\beta\mu}}{(k_\alpha - k_\beta)(k_\lambda - k_\beta)} = J_{\alpha\lambda} \sum_\mu \frac{J_{\lambda\mu} J_{\alpha\lambda}}{(k_\alpha - k_\mu)(k_\lambda - k_\mu)} = -S_{12}.$$

Hence

$$\{\Phi_\alpha, \Phi_\lambda\} = 4\delta_{\alpha\lambda} \sum_{\beta,\mu} \frac{J_{\alpha\beta} J_{\lambda\mu} J_{\beta\mu}}{(k_\alpha - k_\beta)(k_\lambda - k_\mu)}.$$

But since the calculation is performed only for $\alpha \neq \lambda$, this vanishes, so $\{\Phi_\alpha, \Phi_\lambda\} = 0$ and the I_α are in involution.

Since the I_α are in involution, (x, p) can be replaced by a new set of canonical coordinates (z, I), where the z_α, yet to be found, are canonically conjugate to the I_α. Moreover, because the I_α are constants of the motion, the Hamiltonian will be independent of the z_α, or $H = H(I)$. In fact it turns out that

$$H = \frac{1}{2} \sum_\alpha k_\alpha I_\alpha. \tag{6.72}$$

Indeed, since the $\frac{1}{2} \sum k_\alpha x_\alpha^2$ term of H can be obtained from Eq. (6.72) and the definition of the I_α, proof of (6.72) requires only showing that $\sum k_\alpha \Phi_\alpha = \frac{1}{2} \sum J_{\alpha\beta}^2$. This can be shown by ignoring the condition on the sums in the definition of the Φ_α and using the symmetry of $J_{\alpha\beta}^2$:

$$\sum_\alpha k_\alpha \Phi_\alpha = \sum_{\alpha,\beta} \frac{k_\alpha J_{\alpha\beta}^2}{k_\alpha - k_\beta} = \sum_{\beta,\alpha} \frac{k_\beta J_{\alpha\beta}^2}{k_\beta - k_\alpha} = \frac{1}{2} \sum_{\alpha,\beta} \frac{(k_\alpha - k_\beta) J_{\alpha\beta}^2}{k_\alpha - k_\beta} = \frac{1}{2} \sum_{\alpha,\beta} J_{\alpha\beta}^2.$$

6.2.3 MOTION ON THE TORI

RATIONAL AND IRRATIONAL WINDING LINES

The LI theorem would appear to imply that completely integrable Hamiltonian dynamical systems are quite common; after all, they require only the existence of n independent constants of the motion in involution. The fact is, however, that such systems are far from common and hence it may seem strange that we spend so much time on them. Nevertheless integrable systems are important because they are simpler to deal with and because there are many other systems that can be approximated as small enough perturbations of them. Later we will discuss what is meant by "small enough" and how to perform the approximations. It is important, in the meantime, to understand details of motion of integrable systems.

Equation (6.40) gives the motion of an n-dimensional completely integrable system on each of its n-tori $\mathbb{T}^n$ as

$$\phi^\alpha(t) = \nu^\alpha(J)t + \Theta^\alpha, \tag{6.73}$$

where the ϕ^α, running from 0 to 2π, are coordinates in a basis of principal directions. For each α there is a time $t_\alpha = 2\pi/\nu^\alpha$ in which ϕ^α returns to its original value modulo 2π; that is, each α projection of the motion is periodic.

But the full motion on $\mathbb{T}^n$ is not always periodic. It is periodic iff there is a time T when all of the ϕ^α return simultaneously to their original values, that is, a time T such that

$$\phi^\alpha(t + T) = \phi^\alpha(t) \bmod (2\pi) \quad \forall \alpha. \tag{6.74}$$

According to Eq. (6.73) this means that

$$\nu^\alpha T = 2\pi k^\alpha \quad \forall \alpha,$$

where the k^α are integers, and this, in turn, implies that all the ratios

$$\frac{\nu^\alpha}{\nu^\beta} \equiv \frac{k^\alpha}{k^\beta} \tag{6.75}$$

are rational: there must exist nonzero integers l^α such that $\sum \nu^\alpha l^\alpha = 0$. In other words, the orbit of the dynamical system is periodic iff the frequencies are *commensurate* (Section 4.2.2). Otherwise it is not. The frequencies are functions of the J_α, so whether or not they are commensurate depends in general on the particular torus: on a given torus either all orbits are periodic or none are. A system can be periodic on some of its tori (i.e., for certain initial conditions), but not on others. Since the rational numbers have measure zero over the reals, completely integrable systems are *almost never* periodic.

As was described in Section 4.2.2, this is related to rationality and irrationality of winding lines on the torus. We will first discuss these questions on the 2-torus $\mathbb{T}^2$, which is much easier to visualize than $\mathbb{T}^n$ for $n > 2$.

Suppose that $\mathbb{T}^2$ is slit open along the x^1 and x^2 axes of Fig. 6.10 and laid flat on the plane $\mathbb{R}^2$. Assuming that x^α varies from 0 to 2π, the torus then occupies a square of side 2π, whose opposite edges represent the same line on the torus, as in Fig. 6.11. To reconstitute the torus, just glue the right edge of the square to the left, and the top to the bottom. Hence a square on $\mathbb{R}^2$ with opposite edges *identified* (potentially glued together) represents $\mathbb{T}^2$.

Now draw a line of positive slope through the origin of the square. This line hits one of the edges of the square that do not pass through the origin at the point A of Fig. 6.11. The point A' on the opposite edge is actually the same point as A on $\mathbb{T}^2$, so the line also arrives at A' and continues with the same slope until it hits some edge at another point B. The two lines in the square represent just one line on $\mathbb{T}^2$, starting at the origin and winding around to B. This line can be extended further by finding the other representation B' of B, etc. As the process is repeated, a set of lines appears on the square, representing a single line that winds around $\mathbb{T}^2$. If one line of the set passes through $(2\pi, 2\pi)$, the winding line on $\mathbb{T}^2$ comes back to its starting point, for $(0, 0)$ and $(2\pi, 2\pi)$ represent the same point on the torus. If the processes is repeated ad infinitum and no line ever passes through $(2\pi, 2\pi)$,

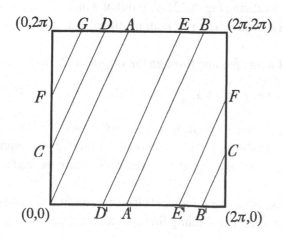

FIGURE 6.11
The torus $\mathbb{T}^2$ represented by a square in $\mathbb{R}^2$ with opposite sides identified. A winding line starts at the origin, goes to A, continues at A' (which represents the same point as A on $\mathbb{T}^2$), goes to B, continues at B', etc., to G. The line continues past G, but that is not shown. If the slope is rational, the line will eventually hit one of the corners of the square, all of which represent the same point on $\mathbb{T}^2$.

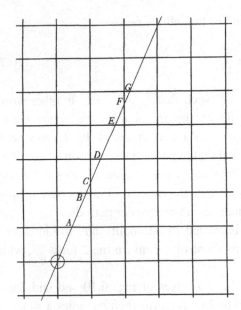

FIGURE 6.12
The torus $\mathbb{T}^2$ represented by covering $\mathbb{R}^2$ with squares of side 2π, so the coordinates of the lattice points (not labeled in the figure) are all $(2k\pi, 2l\pi)$ for k, l integers. The winding line of Fig. 6.11 is shown, together with the points A to G. If the slope is rational, the line will eventually hit one of the lattice points.

the winding line is *irrational*: it never returns to its starting point and covers the torus densely, as described in Section 4.2.2.

To see explicitly what this has to do with rational and irrational numbers, represent $\mathbb{T}^2$ on $\mathbb{R}^2$ by *periodic boundary conditions*: cover the plane by adjacent squares, each of which separately represents the torus (Fig. 6.12). Then all the *lattice points*, whose coordinates are $(2\pi k, 2\pi l)$ with integer k and l, represent the origin on $\mathbb{T}^2$. The winding line on $\mathbb{T}^2$ is then represented by a single line on $\mathbb{R}^2$, and it returns to the origin on $\mathbb{T}^2$ iff its representation on $\mathbb{R}^2$ passes through a lattice point. But that happens iff its slope is a rational number l/k. Since the rational numbers have *measure* zero (discussed in more detail in Chapter 7), *almost all* winding lines on $\mathbb{T}^2$ are irrational! It should now be clear that it is not frivolous to discuss irrational flow on a torus.

This procedure is easily generalized to the n-torus $\mathbb{T}^n$, $n > 2$: simply replace the squares on $\mathbb{R}^2$ by right parallelepipeds of dimension n on $\mathbb{R}^n$ and represent $\mathbb{T}^n$ either by one such parallelepiped with opposite $(n-1)$-dimensional faces identified (generalizing Fig. 6.11) or by periodic boundary conditions (generalizing Fig. 6.12). A rational winding line passes through a lattice point (coordinates $2\pi k^\alpha$ with integer k^α); an irrational one misses them all.

In parametric form the equation of a line passing through the origin of $\mathbb{R}^n$ is

$$x^\alpha(\tau) = \mu^\alpha \tau \quad \forall \alpha, \tag{6.76}$$

and the line will pass through a lattice point iff there is a value of τ such that $\mu^\alpha \tau = 2\pi k^\alpha \, \forall \alpha$. This, in turn, can happen iff $\mu^\alpha/\mu^\beta = k^\alpha/k^\beta$ is rational for all α, β (or, equivalently, if there exist nonzero integers l^α such that $\sum \mu^\alpha l^\alpha = 0$). This clearly reproduces the argument of Eq. (6.75).

The point of this discussion is to make clear that for *almost all* initial conditions the orbit of an integrable dynamical system is an irrational winding line on a torus: given a point on the orbit [e.g., the initial point $\xi(0)$] and a neighborhood of that point no matter how small, there is a time τ when the orbit will again enter that neighborhood. In the unlikely case $\xi(0) = \xi(\tau)$ for some τ, the orbit would be periodic with period τ. Otherwise the orbit is called *quasiperiodic*: the generic orbit of a completely integrable system is quasiperiodic.

FOURIER SERIES

We now discuss how the motion on the tori is translated to (q, p) coordinates. In general, that requires the CT from (ϕ, J) to (q, p), but even before it is known, the nature of AA variables can be used to establish some of the properties of the CT and thereby some general properties of the motion on $\mathbf{T}^*\mathbf{Q}$.

First consider the CT itself, without reference to the dynamics. Pick a point $\xi \in \mathbf{T}^*\mathbf{Q}$. If all of the J_α coordinates of ξ are held fixed (by remaining on the torus passing through ξ) and all of its ϕ^α coordinates are simultaneously changed by 2π (which does not in general happen in a motion), ξ is not changed. That means that none of the (q, p) coordinates are changed; hence the CT (we emphasize: the CT, not the dynamics) has the property that

$$q^\alpha(\phi, J) = q^\alpha(\phi + 2\pi k, J) \quad \text{and} \quad p^\alpha(\phi, J) = p^\alpha(\phi + 2\pi k, J),$$

where $\phi + 2\pi k$ is the set $\{\phi^\alpha + 2\pi k^\alpha\}$ with integer k^α and $\alpha = 1, 2, \ldots, n$. Thus for fixed J_α (as usual writing ξ for the q and p coordinates) the CT can be written as a multiple Fourier series of the form

$$\xi^j(\phi, J) = \sum A^j_{k^1 \ldots k^n} \exp[i(k^1\phi^1 + \cdots + k^n\phi^n)]. \tag{6.77}$$

The sum is over all positive and negative integer values of the k^α. For the full CT, the $A^j_{k^1 \ldots k^n}$ are (as yet unknown) functions of the J_α. If the CT is known, the $A^j_{k^1 \ldots k^n}$ can be obtained by the inverse Fourier transform:

$$A^j_{k^1 \ldots k^n} = \frac{1}{(2\pi)^n} \int_0^{2\pi} d\phi^1 \int_0^{2\pi} d\phi^2 \cdots \int_0^{2\pi} d\phi^n \xi^j(\phi, J) \exp[-i(k^1\phi^1 + \cdots + k^n\phi^n)].$$

Any dynamical variable $F \in \mathcal{F}(\mathbf{T}^*\mathbf{Q})$ can be written in a similar multiple Fourier series, for when the system returns to its initial point in $\mathbf{T}^*\mathbf{Q}$, so does the value of F:

$$F(\xi) = \sum F_{k^1 \ldots k^n} \exp[i(k^1\phi^1 + \cdots + k^n\phi^n)].$$

That has to do with the CT but not with the dynamics.

Now fold in the dynamics by using Eqs. (6.73) and (6.77). This gives a multiple Fourier series for the time development of the ξ^j:

$$\xi^j(J, \Theta; t) = \sum B^j_{k^1 \ldots k^n} \exp[it(k^1\nu^1 + \cdots + k^n\nu^n)], \tag{6.78}$$

where

$$B^j_{k^1 \ldots k^n} = \sum A^j_{k^1 \ldots k^n} \exp[i(k^1\Theta^1 + \cdots + k^n\Theta^n)].$$

The same is true of a dynamical variable F:

$$F(\xi(t)) = \sum G^j_{k^1 \ldots k^n} \exp[it(k^1 v^1 + \cdots + k^n v^n)].$$

Although these equations exhibit some general properties of the motion they are incomplete in that the Fourier coefficients are unknown functions that depend on the actual canonical transformation between (q, p) and (ϕ, J).

6.3 PERTURBATION THEORY

The equations of most dynamical systems cannot be solved exactly, so for the vast majority it is important to develop methods for finding approximate solutions. One of the most useful approaches, often the only analytic one, is through perturbation theory, which is widespread also in many other branches of physics. In this approach a solvable system is found that differs in small ways from the insoluble one, and then the solvable one is perturbed to approximate the other.

By their very nature, perturbative calculations always involve small parameters. Modern computers, on the other hand, can in principle be used to solve problems whose parameters are far from the perturbation range, which seems to eliminate the need for perturbation theory. Nevertheless, analytic perturbation theory remains necessary: it is used for checking the results of numerical calculations, at least for small values of the parameters, and a complete understanding of the solutions often requires a combination of analytical and numerical analyses.

6.3.1 EXAMPLE: THE QUARTIC OSCILLATOR; SECULAR PERTURBATION THEORY

In this section the *quartic oscillator* is used to exhibit some properties and difficulties of perturbative calculation. The Hamiltonian of the one-dimensional quartic oscillator is

$$H = \frac{p^2}{2m} + \frac{1}{2}m\omega_0^2 x^2 + \frac{1}{4}\epsilon m x^4 \equiv H_0 + \epsilon U, \tag{6.79}$$

where H_0 is the ordinary harmonic-oscillator Hamiltonian, and $U(x) = \frac{1}{4}mx^4$ contains the quartic term; ϵU is the *perturbation* of the harmonic oscillator. As long as $\epsilon > 0$ the motion is bounded, as is clear from the potential-energy diagram (Fig. 6.13). Solutions exist and are periodic, but their frequencies will not all be the same. As is well known, only for the harmonic oscillator is the frequency independent of amplitude.

Solution by perturbation theory is based on the hope that for small enough $\epsilon > 0$ the motion $x(t)$ on $\mathbb{Q}$ can be written as a convergent series in the perturbation parameter, of the form

$$x(t) = x_0(t) + \epsilon x_1(t) + \epsilon^2 x_2(t) + \cdots, \tag{6.80}$$

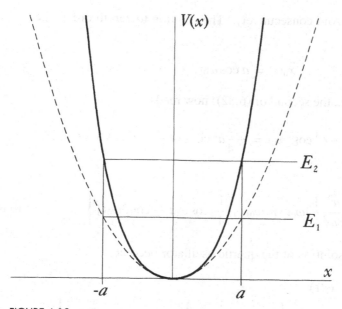

FIGURE 6.13

The potential $V(x) = \frac{1}{2}m\omega_0^2 x^2 + \epsilon m x^4$ of the quartic oscillator (heavy curve). The dashed curve represents the potential $\frac{1}{2}m\omega_0^2 x^2$ of the unperturbed harmonic oscillator. For motion with amplitude a the total energy for the harmonic oscillator is E_1, and for the quartic oscillator it is E_2.

where $x_0(t)$ is a solution on $\mathbb{Q}$ of the *unperturbed* dynamical system whose Hamiltonian is H_0. Equations for the $x_k(t)$ are obtained by inserting (6.80) into the second-order differential equation of motion

$$\ddot{x} + \omega_0^2 x + \epsilon x^3 = 0 \qquad (6.81)$$

obtained from Eq. (6.79) and assuming that this equation is satisfied for each power of ϵ separately. The resulting equations are

$$\left.\begin{array}{l} \ddot{x}_0 + \omega_0^2 x_0 = 0, \\[2pt] \ddot{x}_1 + \omega_0^2 x_1 = -x_0^3, \\[2pt] \ddot{x}_2 + \omega_0^2 x_2 = -3x_0^2 x_1, \\[2pt] \vdots \qquad\qquad \vdots \end{array}\right\} \qquad (6.82)$$

This is an infinite set of coupled second-order ordinary differential equations whose solution requires an infinite set of initial conditions. In order to set $x(0) = a$ and $\dot{x}(0) = 0$ for all values of ϵ, we take the conditions to be

$$x_0(0) = a, \; x_k(0) = 0, \quad k > 0, \quad \text{and} \quad \dot{x}_k(0) = 0 \; \forall k. \qquad (6.83)$$

Incidentally, note that (6.81) implies that ϵ is not dimensionless, so for $k \neq 0$ the x_k do not have the dimensions of length and the ω_k (defined below) do not have the dimensions of frequency.

Equations (6.82) are solved consecutively. The solution to zeroth order, (i.e., of the first of (6.82)) is

$$x_0(t) = a \cos \omega_0 t.$$

The first-order equation (i.e., the second of (6.82)) now reads

$$\ddot{x}_1 + \omega_0^2 x_1 = -a^3 \cos^3 \omega_0 t = -\frac{3}{4} a^3 \cos \omega_0 t - \frac{1}{4} a^3 \cos 3\omega_0 t,$$

and its solution is

$$x_1(t) = -\frac{a^3}{8\omega_0^2} \left\{ 3\omega_0 t \sin \omega_0 t + \frac{1}{4} [\cos \omega_0 t - \cos 3\omega_0 t] \right\}. \tag{6.84}$$

Then to first order in ϵ the solution of the quartic oscillator becomes

$$
\begin{aligned}
x(t) &\approx x_0(t) + \epsilon x_1(t) \\
&= a \cos \omega_0 t - \epsilon \frac{a^3}{8\omega_0^2} \left\{ 3\omega_0 t \sin \omega_0 t + \frac{1}{4} [\cos \omega_0 t - \cos 3\omega_0 t] \right\}.
\end{aligned} \tag{6.85}
$$

This satisfies the initial condition $x_1(t) = 0$, $\dot{x}_1(t) = 0$ (Problem 5).

A serious problem with this approximation is the linear dependence on t in the first term in the braces. It implies that the amplitude of oscillation grows without bound for any $\epsilon > 0$, no matter how small, diverging wildly from the unperturbed oscillator. Moreover, it was established right after (6.79) that the motion is bounded, so the solution of (6.85) makes no sense at all. That this would happen is evident even from the differential equation for x_1: it is the equation of an undamped oscillator driven at its resonant frequency, whose solution is known to diverge [see Eq. (4.73)]. Higher-order approximations get even worse; they involve terms of the form $(\epsilon t)^n$ (called *secular terms*).

The cause of the secular terms is that to all orders the right-hand sides of Eqs. (6.82) contain driving terms of the same frequency ω_0 as the unperturbed motion. Yet that the frequency will change if ϵ is nonzero is almost obvious. Figure 6.13 shows that for a given amplitude a the total energy of the harmonic oscillator is E_1 and that for the quartic oscillator is E_2. The vertical distance between the potential energy curve and the total energy represents the kinetic energy, and since this is greater for the quartic potential at every value of x between $-a$ and a, the particle in the quartic potential moves faster than the one in the harmonic potential. Therefore it covers the distance $2a$ more rapidly, so its period is lower and its frequency ω is greater than ω_0.

An attempt to get around this problem is to impose an ϵ-dependent frequency $\omega(\epsilon) \neq \omega_0$ on the solution. As ϵ approaches zero ω should approach ω_0, so one possibility might be to write $\omega(\epsilon) = \omega_0 + \epsilon \omega_1$, that is, to set

$$x(t) = a(\epsilon)\{\cos(\omega_0 + \epsilon \omega_1)t\} \approx a(\epsilon)\{\cos \omega_0 t - \epsilon \omega_1 t \sin \omega_0 t\}.$$

But this clearly again gives rise to secular terms.

A more successful approach (Poincaré, 1992; see also Tabor, 1989) generalizes the ϵ dependence of the frequency to a power series. There are then two expansions, namely Eq. (6.80) and

$$\omega = \omega_0 + \epsilon\omega_1 + \epsilon^2\omega_2 + \cdots. \tag{6.86}$$

In this approach the solution is explicitly forced to be periodic with frequency ω by assuming that it is of the form $x(\omega t)$ and by replacing the derivatives with respect to time by derivatives with respect to $\tau \equiv \omega t$, producing additional ϵ dependence in the second derivative:

$$\ddot{x} = \frac{d^2}{dt^2}x = \omega^2\frac{d^2}{d\tau^2}x \equiv \omega^2 x'' = (\omega_0 + \epsilon\omega_1 + \epsilon^2\omega_2 + \cdots)^2 x''. \tag{6.87}$$

The set of equations obtained by equating powers of ϵ on both sides of Eq. (6.87) becomes

$$\omega_0^2(x_0'' + x_0) = 0,$$
$$\omega_0^2(x_1'' + x_1) = -x_0^3 - 2\omega_0\omega_1 x_0'',$$
$$\vdots \qquad\qquad \vdots$$

The initial conditions are again given by Eq. (6.83) and the solution to the zeroth-order equation is the same as before, but the first-order equation is now

$$\omega_0^2(x_1'' + x_1) = \left(2\omega_1\omega_0 - \frac{3}{4}a^2\right)a\cos\tau - \frac{1}{4}a^3\cos 3\tau. \tag{6.88}$$

Here we have written $\tau = \omega_0 t$ rather than $\tau = \omega t$, to make the equation ϵ independent. Anyway, x_1 will eventually be multiplied by ϵ and added to x_0 to give the first-order solution, and the difference between ω and ω_0 in the arguments of these cosines is of higher order.

The independent variable is now τ and the resonant frequency is 1, so Eq. (6.88) is an undamped harmonic oscillator driven in part at its resonant frequency. This again gives rise to secular terms, but now the resonant term in the driver can be annihilated by choosing ω_1 to satisfy

$$\omega_1 = \frac{3a^2}{8\omega_0}.$$

That leaves only the $3\tau \equiv 3\omega_0 t$ term in the driver. The solution that satisfies the boundary conditions is

$$x_1(t) = -\frac{a^3}{32\omega_0^2}(\cos\omega_0 t - \cos 3\omega_0 t).$$

Then to first order in ϵ the solution for the quartic oscillator is

$$x(t) \approx a\cos\left\{\omega_0 + \epsilon\frac{3a^2}{8\omega_0}\right\}t - \epsilon\frac{a^2}{32\omega_0^2}(\cos\omega_0 t - \cos 3\omega_0 t). \tag{6.89}$$

In the first term ω_0 has been replaced by $\omega \equiv \omega_0 + \epsilon\omega_1$. Since the second term is of first order in ϵ, it does not matter whether ω or ω_0 is used.

We verify that (6.89) satisfies (6.81) to first order in ϵ. The second time derivative of $x(t)$ is

$$
\begin{aligned}
\ddot{x} &= -a\{\omega_0 + \epsilon\omega_1\}^2 \cos\{\omega_0 + \epsilon\omega_1\}t + \epsilon\left[\frac{a\omega\omega_1}{12}\cos\omega t - \frac{3a\omega\omega_1}{4}\cos 3\omega t\right] \\
&= -a\omega_0^2 \cos\{\omega_0 + \epsilon\omega_1\}t + O(\epsilon^2) \\
&\quad + \epsilon\omega_0^2\frac{a\omega_1}{12\omega_0}(\cos\omega_0 t - \cos 3\omega_0 t) - \epsilon 2a\omega_0\omega_1\left(\frac{1}{3}\cos 3\omega_0 t + \cos\omega_0 t\right) \\
&= -\omega_0^2 x - \epsilon\frac{3}{4}a^3\left(\frac{1}{3}\cos 3\omega_0 t + \cos\omega_0 t\right).
\end{aligned}
$$

If (6.89) is a first-order solution, the second term here should equal ϵx^3 to first order in ϵ. All that is needed to verify this is the zeroth-order term Z in x^3, namely

$$
Z = a^3[\cos^3\{\omega_0 + \epsilon\omega_1\}t]_{\epsilon=0} = a^3\cos^3\omega_0 t = a^3\left[\frac{1}{4}\cos 3\omega_0 t + \frac{3}{4}\cos\omega_0 t\right].
$$

This shows that $x(t)$ of Eq. (6.89) fulfills all the requirements: it satisfies the initial conditions, is bounded, and approaches the unperturbed solution in the limit $\epsilon \to 0$. It differs from the harmonic oscillator in two ways. First, it oscillates at a combination of the frequencies ω and (to first order in ϵ) 3ω. Second, ω depends through ω_1 on the amplitude a and on ϵ. Both the amplitude dependence of ω and the appearance of *harmonics* (the 3ω term in this case) are characteristic of nonlinear oscillators.

The procedure we used here is called *secular perturbation theory*. We do not go on to higher approximations.

The moral of this example is: *beware of perturbative calculations*. They require alertness and testing, for they often lead to nonphysical results. In our example the testing was relatively easy, for this is a one-freedom system. In higher freedoms the situation is often much cloudier, and keeping strict control of the perturbative analysis becomes more difficult and more important. As in the one-freedom case, perturbative calculations lead to divergent results brought about by resonances, but in general the resonances are between frequencies in different freedoms. They lead to breakdowns in perturbation theory and to chaotic solutions.

The next section is devoted to two perturbation methods adapted specifically to Hamiltonian dynamical systems, in which such problems are at least partially understood in a general way.

6.3.2 HAMILTONIAN PERTURBATION THEORY

The secular perturbation theory of Section 6.3.1 makes no use of the properties of the carrier manifold $\mathbf{TQ}$ or $\mathbf{T^*Q}$. In this section we discuss the perturbation of Hamiltonian dynamical systems on $\mathbf{T^*Q}$. Usually this will involve perturbations of completely integrable systems, but we begin from a more general point of view, without starting from completely integrable ones.

PERTURBATION VIA CANONICAL TRANSFORMATIONS

Suppose that the Hamiltonian H of a dynamical system Δ deviates only slightly from the Hamiltonian H_0 of a dynamical system Δ_0 whose solution is known. That is,

$$H(\xi, t, \epsilon) = H_0(\xi, t) + \epsilon H_1(\xi, t), \tag{6.90}$$

where ϵ is a small parameter. H_0 is called the *unperturbed Hamiltonian* and H the *perturbed* one. A perturbative calculation is based on the assumption that the difference between Δ and Δ_0 is slight: Δ_0 is taken as the zeroth approximation to Δ and then successive approximations improve the accuracy in stages.

This is accomplished by an ϵ-dependent canonical transformation from the initial set of variables ξ to a new set $\eta(\xi, \epsilon)$ in which the dynamics becomes in some sense more tractable. At $\epsilon = 0$ the CT is the identity transformation, and for $\epsilon \neq 0$ the CT is chosen in an attempt to make the new Hamiltonian, the one for the η variables, easier to handle than $H(\xi)$ (we will suppress the ϵ dependence of H).

We will first study how dynamical variables, in particular the Hamiltonian, are changed by such a CT. Only later will we apply the results to the motion. The changed, or perturbed, dynamical variables will be written as power series in ϵ whose leading terms are their unperturbed forms. As in all perturbative calculations, it is assumed that the first few terms of such a series give reasonably good approximations.

The CT is treated by means of either its global generating function (Section 5.3.3) or its infinitesimal one (Section 5.3.4). In the global treatment, ϵ is taken as fixed, though small. This is the method called *canonical perturbation theory*, and it is useful to about second order in ϵ. In the infinitesimal treatment, ϵ is taken to vary from 0 to some finite value, also small. This is called the *Lie transformation method*, and in principle it is easier to take to higher orders.

At this point we consider only the global method [the Lie method is discussed later in this section]. The CT is taken to be, like the identity transformation, of Type 2. Call the Type 2 generator $S(q, P, t; \epsilon)$ or simply $S(q, P, t)$. The ξ variables are represented by (q, p), and the η variables by (Q, P). According to Section 5.3.3 the new Hamiltonian for the η variables is then

$$\hat{H}(Q, P, t) = H(q, p, t) + \frac{\partial S}{\partial t}(q, P, t), \tag{6.91}$$

where q and p must be written as functions of (Q, P) on the right-hand side. Now expand in powers of ϵ:

$$\left.\begin{aligned}
\hat{H} &= \hat{H}_0 + \epsilon \hat{H}_1 + \epsilon^2 \hat{H}_2 + \cdots, \\
q &= Q + \epsilon q_1 + \epsilon^2 q_2 + \cdots, \\
p &= P + \epsilon p_1 + \epsilon^2 p_2 + \cdots;
\end{aligned}\right\} \tag{6.92}$$

everything here is a function of (Q, P). Now expand S in powers of ϵ, starting with qP, the Type 2 generator of the identity transformation:

$$S = qP + \epsilon S_1 + \epsilon^2 S_2 + \cdots, \tag{6.93}$$

where all the S_k are functions of (q, P). At present all we assume is that ϵ is small and that the power-series expansions are valid. No particular properties are assigned to S. Later, when this is used for canonical perturbation theory, S will be chosen to yield an $\hat{H}$ that makes the dynamics relatively easy to handle.

The expansions for q, p, and S are now inserted into the CT equations

$$Q = \frac{\partial S}{\partial P}, \quad p = \frac{\partial S}{\partial q},$$

to yield (up to first order in ϵ)

$$\left.\begin{array}{l} Q = Q + \epsilon q_1 + \cdots + \epsilon \dfrac{\partial S_1}{\partial P} + \cdots, \\[2mm] P + \epsilon p_1 + \cdots = P + \epsilon \dfrac{\partial S_1}{\partial q} + \cdots. \end{array}\right\} \tag{6.94}$$

Equating the coefficients of equal powers of ϵ leads to the first-order equations

$$p_1 = \frac{\partial S_1}{\partial q}, \quad q_1 = -\frac{\partial S_1}{\partial P}.$$

Equations of higher order are more complicated.

Now use this to obtain a first-order expression for $H(q, p, t)$ from (6.90) (remember that everything is a function now of (Q, P)):

$$H(q, p, t) \approx H_0(q, p, t) + \epsilon H_1(q, p, t)$$

$$\approx H_0 + \epsilon \left[\frac{\partial H_0}{\partial q} \frac{dq}{d\epsilon} + \frac{\partial H_0}{\partial p} \frac{dp}{d\epsilon} + H_1 \right]$$

$$\approx H_0 + \epsilon \left[-\frac{\partial H_0}{\partial q} \frac{\partial S_1}{\partial P} + \frac{\partial H_0}{\partial P} \frac{\partial S_1}{\partial q} + H_1 \right],$$

where H_0 in the second and third lines is taken at $\epsilon = 0$, when $(Q, P) = (q, p)$. This can be inserted into the right-hand side of (6.91), which then becomes, to first order in ϵ,

$$\hat{H}_0 + \epsilon \hat{H}_1 = H_0 + \epsilon \left[H_1 - \frac{\partial H_0}{\partial q} \frac{\partial S_1}{\partial P} + \frac{\partial H_0}{\partial P} \frac{\partial S_1}{\partial q} + \frac{\partial S_1}{\partial t} \right].$$

This is a first-order equation. In the partial derivatives of S_1, already multiplied by ϵ, to zeroth order (Q, P) can be replaced by (q, p). Hence

$$\left.\begin{array}{l} \hat{H}_0 = H_0, \\[2mm] \hat{H}_1 - H_1 = \{S_1, H_0\} + \dfrac{\partial S_1}{\partial t}. \end{array}\right\} \tag{6.95}$$

This shows how S_1 determines $\hat{H}_1$. We state without proof that the S_k with $k > 1$ affect only $\hat{H}_k$ with $k > 1$: although a complete theory must go on to higher k, the first-order results are not altered thereby.

AVERAGING

Equation (6.95) gives $\hat{H}_1$ if S_1 is known, but in the application of all this to dynamics, *both* H and S need to be found, so none of the S_k are known a priori. This means that even in first order there are more unknown functions than equations: the single Eq. (6.95) and the two unknown functions $\hat{H}_1$ and S_1. The way to get around this problem is to replace some of the dynamical variables by their averages over invariant tori, which averages out fluctuations about periodic motion of the unperturbed system. We will illustrate this procedure in first order on a system in one freedom (a $\mathbf{T}^*Q$ of two dimensions).

Suppose that (ϕ, J) are the AA variables of a one-freedom system Δ_0 whose Hamiltonian is $H_0(J)$. Suppose we want to integrate another system Δ, whose Hamiltonian is given, in analogy with Eq. (6.90), by

$$H(\phi, J) = H_0(J) + \epsilon H_1(\phi, J). \tag{6.96}$$

This equation is simpler than Eq. (6.90) because there is no t dependence. We assume explicitly that H_1 is a single-valued function and is therefore periodic in ϕ.

The equations of motion for Δ are

$$\dot{\phi} = \nu_0(J) + \epsilon \frac{\partial H_1}{\partial J}, \quad \dot{J} = -\epsilon \frac{\partial H_1}{\partial \phi} \equiv -\epsilon F(\phi, J), \tag{6.97}$$

where $\nu_0 \equiv \partial H_0/\partial J$ is the frequency of Δ_0. Since J is a generalized momentum, ϵF can be thought of as generalized force. Suppose for a moment that this force is turned off (ϵ is set equal to zero), so that the dynamics returns to Δ_0. Under the flow of Δ_0 the dynamical variable F is periodic because H_0 is periodic, and its average over one circuit on an invariant torus $\mathbb{T}$ (in this case $\mathbb{T}$ is just a closed C_α curve like those of Fig. 6.5) is

$$\langle F(J) \rangle \equiv \frac{1}{2\pi} \int_0^{2\pi} F(\phi, J) \, d\phi.$$

If ϵ is small, ϵF does not change much from its average in each period, and then as a first approximation (6.97) can be replaced by the new equations

$$\dot{\phi} = \nu_0(J) + \epsilon \frac{\partial H_1}{\partial J}, \quad \dot{J} = -\epsilon \langle F(J) \rangle.$$

These are the equations of motion of a new dynamical system $\langle \Delta \rangle$ which serves as an approximation to Δ (at this time we do not discuss how good an approximation it is). Not too surprisingly, the action variable J of Δ_0 is not a constant of the motion for $\langle \Delta \rangle$, which flows therefore from torus to torus (from one C_α to others) of Δ_0. Because J is not constant, ϕ is not linear in the time. Although $\langle \Delta \rangle$ does not flow on the C_α curves of Δ_0, it may turn out that $\langle \Delta \rangle$ has its own invariant tori (its own $\langle C \rangle_\alpha$ curves crossing those of Δ_0) and hence that $\langle \Delta \rangle$ is also completely integrable.

When this sort of averaging is applied to canonical perturbation theory, it is assumed that the perturbed motion is in fact completely integrable.

CANONICAL PERTURBATION THEORY IN ONE FREEDOM

Canonical perturbation theory is sometimes also called classical perturbation theory, both because it refers to classical mechanics and because it is a time-honored procedure that has led to many important results. It was developed by Poincaré and further extended by von Zeipel (1916). Among its triumphs has been the calculation of deviations from ellipticity of the orbits of planets around the Sun caused by nonsphericity of the Sun, the presence of other planets, and relativistic effects. In particular Delaunay (1867) used this method to study the detailed motion of the Moon. His extremely meticulous work, performed in the nineteenth century, took several decades. In this computer age it is impressive that all of his very accurate calculations were performed by hand! For a recent computer-based numerical calculation of the entire solar system, see Sussman and Wisdom (1992).

In general the technique is the following: As in the preceding subsections, the zeroth approximation to the dynamical system Δ is a completely integrable system Δ_0. Then AA theory and canonical transformations are used to improve the approximation in stages. Estimates are made for the accuracy at each stage, and the procedure is continued until the desired accuracy is reached. We will first explain this in detail for a Δ in one freedom. Later we will show how to extend the procedure to more freedoms.

Assume that $H_0(\xi)$, the Hamiltonian of Δ_0, is time independent, and write Eq. (6.90) in the form

$$H(\xi) = H_0(\xi) + \epsilon H_1(\xi).$$

Although H_0 is completely integrable, it may not be given initially in terms of its AA variables. Let (ϕ_0, J_0) be its AA variables and K_0 be its Hamiltonian in terms of these variables, a function only of J_0:

$$K_0(J_0) = H_0\left(\xi(\phi_0, J_0)\right),$$

where $\xi(\phi_0, J_0)$ is the CT from (ϕ_0, J_0) to $\xi \equiv (q, p)$. The subscript 0 is a reminder that all this refers to the unperturbed system. The perturbed Hamiltonian, like any dynamical variable, can also be written in terms of the unperturbed AA variables:

$$K(\phi_0, J_0) = H(\xi(\phi_0, J_0)) \equiv K_0(J_0) + \epsilon K_1(\phi_0, J_0), \tag{6.98}$$

where $K_1(\phi_0, J_0) = H_1(\xi(\phi_0, J_0))$. This is the analog of Eq. (6.90): (ϕ_0, J_0) replaces (q, p), and K replaces H. It is important to bear in mind that $K(\phi_0, J_0)$ is in principle a known function, and so are $K_0(J_0)$ and $K_1(\phi_0, J_0)$, for $H(\xi)$ is given in the statement of the problem and the CT from (q, p) to (ϕ_0, J_0) is obtained by applying to Δ_0 the general procedure for finding AA variables.

Assume now that Δ is also completely integrable. Then it has its own AA variables (ϕ, J), and since there must exist CTs from (q, p) to both (ϕ, J) and (ϕ_0, J_0), there must exist another CT, call it Φ, from (ϕ, J) to (ϕ_0, J_0). If Φ were known, the perturbed Hamiltonian (call it E) could be written as a function of J alone:

$$E(J) \equiv K(\phi_0, J_0) \equiv H(\xi). \tag{6.99}$$

This equation (or rather its first equality) is the analog of Eq. (6.91): E replaces $\hat{H}$ and (ϕ, J) replaces (Q, P).

The CT that must be found is Φ, and the condition on the CT is that the new Hamiltonian $E(J)$ be a function of J alone. We will soon see the extent to which this determines Φ. In any case, Φ determines $E(J)$, and once $E(J)$ is known the system can be integrated. Moreover, the CT from (ϕ, J) to (q, p) will then be obtained by composing Φ with $\xi(\phi_0, J_0)$, and that will yield the motion in terms of the original variables. The sequence of CTs may be illustrated as

$$(\phi, J) \overset{\Phi}{\to} (\phi_0, J_0) \overset{\xi}{\to} (q, p).$$

Many of the functions (as well as Φ) depend on ϵ, but this dependence is suppressed.

Successive approximations of Φ are obtained from successive approximations of its ϵ-dependent Type 2 generator $S(\phi_0, J)$. As in Eq. (6.93), S is expanded in a power series in ϵ,

$$S(\phi_0, J) = \phi_0 J + \epsilon S_1(\phi_0, J) + \epsilon^2 S_2(\phi_0, J) + \cdots,$$

and the problem is to find expressions for the S_k. Note, incidentally, that if a and b are constants, S and $\tilde{S} = S + a\phi_0 + bJ$ yield ϕ and J that differ only by additive constants, so terms in S that are linear in either ϕ_0 or J can be dropped.

We skip some of the steps analogous to those of the first subsection in obtaining the second-order approximations

$$J_0 \equiv \frac{\partial S}{\partial \phi_0} = J + \epsilon \frac{\partial S_1}{\partial \phi_0} + \epsilon^2 \frac{\partial S_2}{\partial \phi_0} + \cdots, \tag{6.100a}$$

$$\phi \equiv \frac{\partial S}{\partial J} = \phi_0 + \epsilon \frac{\partial S_1}{\partial J} + \epsilon^2 \frac{\partial S_2}{\partial J} + \cdots. \tag{6.100b}$$

Equation (6.99) can also be expanded in a power series to obtain successive approximations for $E(J)$:

$$E(J) \equiv E_0(J) + \epsilon E_1(J) + \epsilon^2 E_2(J) + \cdots$$
$$= K(\phi_0, J_0) \equiv K_0(J_0) + \epsilon K_1(\phi_0, J_0). \tag{6.101}$$

The reason there are higher orders of ϵ on the first line of (6.101) than on the second is that the ϵ dependence in the relation between J and J_0 has not been made explicit. This is overcome, as in the first subsection, by expanding K in a Taylor series (in J_0) about J and inserting the expansion of J_0 from Eq. (6.100a):

$$K(\phi_0, J_0) = K(\phi_0, J) + \frac{\partial K}{\partial J}(J_0 - J) + \frac{1}{2}\frac{\partial^2 K}{\partial J^2}(J_0 - J)^2 + \cdots$$

$$= K(\phi_0, J) + \frac{\partial K}{\partial J}\left(\epsilon\frac{\partial S_1}{\partial \phi_0} + \epsilon^2\frac{\partial S_2}{\partial \phi_0} + \cdots\right)$$

$$+ \frac{1}{2}\frac{\partial^2 K}{\partial J^2}\left(\epsilon\frac{\partial S_1}{\partial \phi_0} + \epsilon^2\frac{\partial S_2}{\partial \phi_0} + \cdots\right)^2 + \cdots$$

$$= K(\phi_0, J) + \epsilon\frac{\partial K}{\partial J}\frac{\partial S_1}{\partial \phi_0} + \epsilon^2\left\{\frac{\partial K}{\partial J}\frac{\partial S_2}{\partial \phi_0} + \frac{1}{2}\frac{\partial^2 K}{\partial J^2}\left(\frac{\partial S_1}{\partial \phi_0}\right)^2\right\} + \cdots.$$

In the K that appears on the right-hand side J_0 is replaced everywhere by J. The $\partial^j K/\partial J^j$ are jth derivatives of K with respect to its second variable, again with J_0 replaced by J. All of the derivatives of K, as well as K itself, are known functions linear in ϵ, so one more step is needed. That step yields

$$K(\phi_0, J_0) = K_0(J) + \epsilon K_1(\phi_0, J) + \epsilon\frac{\partial S_1}{\partial \phi_0}\left(\frac{\partial K_0}{\partial J} + \epsilon\frac{\partial K_1}{\partial J}\right)$$

$$+ \epsilon^2\left\{\frac{\partial K_0}{\partial J}\frac{\partial S_2}{\partial \phi_0} + \frac{1}{2}\frac{\partial^2 K_0}{\partial J^2}\left(\frac{\partial S_1}{\partial \phi_0}\right)^2 + \cdots\right\} + \cdots.$$

This is the full development up to second order in ϵ: all of the ϵ dependence has been made explicit in this expression.

The terms in the expansion of $E(J)$ can now be obtained by equating the coefficients of the powers of ϵ in this equation to those in the top line of Eq. (6.101):

$$E_0(J) = K_0(J), \tag{6.102a}$$

$$E_1(J) = K_1(\phi_0, J) + \frac{\partial K_0}{\partial J}\frac{\partial S_1}{\partial \phi_0} \equiv K_1(\phi_0, J) + \nu_0(J)\frac{\partial S_1}{\partial \phi_0}, \tag{6.102b}$$

$$E_2(J) = \frac{\partial K_1}{\partial J}\frac{\partial S_1}{\partial \phi_0} + \nu_0(J)\frac{\partial S_2}{\partial \phi_0} + \frac{1}{2}\frac{\partial \nu_0}{\partial J}\left(\frac{\partial S_1}{\partial \phi_0}\right)^2. \tag{6.102c}$$

Equation (6.102a) has been used to write $\partial K_0/\partial J = \partial E_0/\partial J \equiv \nu_0(J)$ in Eqs. (6.102b,c) [$\nu_0(J_0)$ is the unperturbed frequency, and $\nu_0(J)$ is obtained by replacing J_0 by J]. Equations (6.102a,b) are the analogs of Eqs. (6.95). The analogy is not obvious because now everything is independent of t and because K_0 is a function only of J_0, not also of ϕ_0. These equations give approximations to $E(J)$ up to second order in ϵ, and in principle the procedure can be extended to give approximations to higher orders. According to Eqs. (6.102a,b,c) what are needed are not the S_k, but only their derivatives.

As in the first subsection, some of these equations can be written in terms of PBs: $\partial K_0/\partial J$ is actually $\partial K_0/\partial J_0$ with J_0 replaced by J, so Eq. (6.102b) can be put in the form

$$E_1 = K_1 - \{S_1, K_0\}. \tag{6.103}$$

As before there is the problem of more unknown functions than equations, so this is where averaging comes in. In the previous subsection the average was taken over the unperturbed torus, but it will now be taken over the perturbed one. This is unavoidable, because the S_k are functions not of (ϕ_0, J_0), but of (ϕ_0, J). The averaging is simplified when the functional form of the $S_k(\phi_0, J)$ is understood. When ϕ_0 changes by 2π while the unperturbed action variable J_0 is fixed, a closed path C_0 is traversed, ending at the starting point. Since (ϕ_0, J) is a set of independent coordinates on $\mathbf{T}^*\mathbb{Q}$ (the CT is of Type 2), J is the same at the start as at the end of a trip around C_0. Hence holding J fixed and changing ϕ_0 by 2π also returns to the same starting point: it yields another closed path C. In one circuit around C the generating function S, whose leading term is $\phi_0 J$, changes by $2\pi J$, but Eqs. (6.100a,b) show that $\partial S/\partial \phi_0$ and $\partial S/\partial J$ do not change. Thus for fixed J both derivatives of S are periodic in ϕ_0. If this is true independent of ϵ, it must be true for each of the S_k: both derivatives of each S_k are periodic in ϕ_0. For fixed J, therefore, each S_k is the sum of a function periodic in ϕ_0 and another, which is at most linear in ϕ_0 and can therefore be discarded. That is what will simplify the averaging.

Because of their periodicity, the S_k can be expanded in Fourier series of the form

$$S_k(\phi_0, J) = \sum_{m=-\infty}^{\infty} \mathcal{S}_k(J; m)\exp(im\phi_0), \quad k \neq 0, \tag{6.104}$$

and the $\partial S_k/\partial \phi_0$ can be expanded similarly, except that their expansions will contain no constant (i.e., $m = 0$) terms. That means, in particular, that the averages of the $\partial S_k/\partial \phi_0$ vanish:

$$\left\langle \frac{\partial S_k}{\partial \phi_0} \right\rangle \equiv \frac{1}{2\pi}\int_0^{2\pi} \frac{\partial S_k}{\partial \phi_0} d\phi_0 = 0.$$

Now we are ready to take the average of Eq. (6.102b). Since E_1 and ν_0 are functions of J alone, and since $\langle f(J)\rangle = f(J)$, averaging yields

$$E_1(J) = \langle K_1 \rangle. \tag{6.105}$$

The average of K_1 can be calculated because $K_1(\phi_0, J)$ is a known function, so (6.105) gives E_1, a first approximation for $E(J)$. With this expression for E_1, Eq. (6.102b) becomes a simple differential equation for S_1:

$$\frac{\partial S_1}{\partial \phi_0} = \frac{\langle K_1 \rangle - K_1}{\nu_0}. \tag{6.106}$$

Its solution may be written in the form $S_1(\phi_0, J) + \sigma(J)$, where σ is an arbitrary function. But Eq. (6.100a) shows that σ does not affect the relation between J and J_0, and (6.100b) shows that it adds $\epsilon\sigma'(J)$ to $\phi - \phi_0$. Since J is a constant of the motion labeling the invariant torus, this σ merely resets the zero point for ϕ (the phase angle) on each torus. We will set $\sigma = 0$ to obtain a first approximation for $S(\phi_0, J)$.

A similar procedure applied to Eq. (6.102c) leads to

$$E_2(J) = \left\langle \frac{\partial K_1}{\partial J} \frac{\partial S_1}{\partial \phi_0} \right\rangle + \frac{1}{2} \frac{\partial v_0}{\partial J} \left\langle \left(\frac{\partial S_1}{\partial \phi_0} \right)^2 \right\rangle$$

$$= \frac{1}{v_0} \left\{ \left\langle \frac{\partial K_1}{\partial J} \right\rangle \langle K_1 \rangle - \left\langle \frac{\partial K_1}{\partial J} K_1 \right\rangle \right\} + \frac{1}{2v_0^2} \frac{\partial v_0}{\partial J} \left\{ \langle K_1^2 \rangle - \langle K_1 \rangle^2 \right\}$$

and (using this expression for E_2)

$$\frac{\partial S_2}{\partial \phi_0} = \frac{1}{v_0^2} \left\{ \left\langle \frac{\partial K_1}{\partial J} \right\rangle \langle K_1 \rangle - \left\langle \frac{\partial K_1}{\partial J} K_1 \right\rangle - \frac{\partial K_1}{\partial J} \langle K_1 \rangle + \frac{\partial K_1}{\partial J} K_1 \right\}$$

$$+ \frac{1}{2v_0^3} \frac{\partial v_0}{\partial J} \left\{ \langle K_1^2 \rangle - 2\langle K_1 \rangle^2 + 2\langle K_1 \rangle K_1 - K_1^2 \right\}.$$

The requirement that E be a function of J has led to differential equations for the first and second approximations to S, and in principle the procedure can be carried to higher orders. It is interesting, however, that E can be written as a function of J without first calculating S. Indeed, in the second approximation the Hamiltonian is

$$E(J) = K_0(J) + \epsilon \langle K_1 \rangle$$

$$+ \epsilon^2 \frac{1}{v_0} \left[\left\{ \left\langle \frac{\partial K_1}{\partial J} \right\rangle \langle K_1 \rangle - \left\langle \frac{\partial K_1}{\partial J} K_1 \right\rangle \right\} + \frac{1}{2v_0} \frac{\partial v_0}{\partial J} \left\{ \langle K_1^2 \rangle - \langle K_1 \rangle^2 \right\} \right] + \cdots.$$

Once $E(J)$ has been found in this way to kth order, the perturbed frequency is given to the same order by $v = \partial E / \partial J$. This is useful information, and sometimes it is all that is desired. Nevertheless, it is possible in principle to go further and to obtain the motion in terms of the initial ξ variables, also to order k. First the kth approximation for $S(\phi_0, J)$ is used to find Φ and then Φ is composed with $\xi(\phi_0, J_0)$ to yield the CT from (ϕ, J) to ξ, all to order k. Since $\phi(t) = vt + \phi(0)$ and J is a constant of the motion, this gives the ξ motion.

This describes canonical perturbation theory to second order for systems in one freedom. We do not go on to higher orders; in principle the procedure is the same, but it involves more terms, making it much more complicated. We go on to more than one freedom in the next subsection.

WORKED EXAMPLE 6.5

Solve the quartic oscillator of Eq. (6.79) by canonical perturbation theory to first order in ϵ. **(a)** Find the perturbed frequency v for the same initial conditions as in those used at Eq. (6.79). **(b)** Find the CT connecting (ϕ_0, J_0) and (ϕ, J).

Solution. The Hamiltonian is (H_1 and v_0 were previously called U and ω_0)

$$H = \frac{p^2}{2m} + \frac{1}{2} m v_0^2 q^2 + \frac{1}{4} \epsilon m q^4 \equiv H_0 + \epsilon H_1. \tag{6.107}$$

(a) The AA variables for the harmonic oscillator are (see Problem 7)

$$\phi_0 = \tan^{-1}\left\{mv_0\frac{q}{p}\right\}, \quad J_0 = \frac{1}{2}\left(\frac{p^2}{mv_0} + mv_0q^2\right),$$

and in terms of the AA variables $H_0 = v_0 J_0 \equiv K_0(J_0)$. The perturbed Hamiltonian is

$$K(\phi_0, J_0) \equiv K_0(J_0) + \epsilon K_1(\phi_0, J_0) = v_0 J_0 + \epsilon \mu J_0^2 \sin^4 \phi_0, \tag{6.108}$$

where $\mu = 1/mv_0^2$. In first order the new Hamiltonian E is (remember to replace J_0 by J)

$$E_1(J) = \langle K_1 \rangle = \frac{\mu J^2}{2\pi}\int_0^{2\pi} \sin^4 \phi_0 \, d\phi_0 = \frac{3\mu}{8}J^2.$$

Use $\int \sin^4 x \, dx = \frac{1}{8}[3x - \cos x\{2\sin^3 x + 3\sin x\}]$, so that to first order

$$E(J) \approx v_0 J + \epsilon\frac{3\mu}{8}J^2$$

and the frequency is

$$v = \frac{\partial E}{\partial J} \approx v_0 + \epsilon\frac{3\mu}{4}J. \tag{6.109}$$

To first order in ϵ the J in this equation can be replaced by J_0. As in Section 6.3.1, consider motion whose initial conditions are $q(0) = a$, $p(0) = 0$. Then $J_0 = \frac{1}{2}mv_0a^2$, and to first order the perturbed frequency is

$$v \approx v_0 + \epsilon\frac{3a^2}{8v_0}.$$

This is the same correction to the frequency as was found in Section 6.3.1. For more general initial conditions the perturbed frequency depends on the energy.

(b) The CT from (ϕ_0, J_0) to (ϕ, J) is obtained by first finding the generator $S(\phi_0, J)$. To first order this requires Eq. (6.106), according to which

$$S_1 = \frac{\mu J^2}{v_0}\int\left(\frac{3}{8} - \sin^4 \phi_0\right)d\phi_0,$$

and then

$$S \approx \phi_0 J + \epsilon\frac{\mu J^2}{8v_0}(2\sin^2 \phi_0 + 3)\sin \phi_0 \cos \phi_0,$$

from which

$$J_0 = \frac{\partial S}{\partial \phi_0} \approx J + \epsilon \frac{\mu J^2}{8 v_0} (4 \cos 2\phi_0 - \cos 4\phi_0),$$

$$\phi = \frac{\partial S}{\partial J} \approx \phi_0 + \epsilon \frac{\mu J}{4 v_0} (2 \sin^2 \phi_0 + 3) \sin \phi_0 \cos \phi_0.$$

The first of these is a quadratic equation for J, but to first order in ϵ its solution can also be obtained simply by replacing J^2 by J_0^2 (see also Problem 9). The same replacement can be made in the equation for ϕ, so to first order

$$J \approx J_0 - \epsilon \frac{\mu J_0^2}{8 v_0} (4 \cos 2\phi_0 - \cos 4\phi_0),$$

$$\phi \approx \phi_0 + \epsilon \frac{\mu J_0}{4 v_0} (2 \sin^2 \phi_0 + 3) \sin \phi_0 \cos \phi_0.$$

This gives Φ, the CT connecting (ϕ, J) and (ϕ_0, J_0). The expressions for (ϕ_0, J_0) in terms of (q, p) can be inserted into these, but the resulting transcendental equations cannot be inverted analytically to obtain (q, p) in terms of (ϕ, J). To go further requires numerical calculation.

CANONICAL PERTURBATION THEORY IN MANY FREEDOMS

We now move on to several freedoms. As before, $H(\xi) = H_0(\xi) + \epsilon H_1(\xi)$ where H_0 is a completely integrable Hamiltonian for the dynamical system Δ_0, but now $\xi \equiv \{\xi^1, \xi^2, \ldots, \xi^{2n}\}$. Also, $\phi_0 \equiv \{\phi_0^1, \ldots, \phi_0^n\}$ and $J_0 \equiv \{J_{01}, \ldots, J_{0n}\}$. As before, in terms of the AA variables of the unperturbed problem the Hamiltonian is given by Eq. (6.98), and it is assumed that the perturbed system is also completely integrable, with its own AA variables (ϕ, J). The Hamiltonian of the perturbed system is written in the form of Eq. (6.101), and calculations similar to those for one freedom lead to analogs of Eqs. (6.102a,b,c) (the sums go from 1 to n):

$$E_0(J) = K_0(J), \tag{6.110a}$$

$$E_1(J) = K_1(\phi_0, J) + \sum_\alpha v_0^\alpha(J) \frac{\partial S_1}{\partial \phi_0^\alpha}, \tag{6.110b}$$

$$E_2(J) = \sum_\alpha \left(\frac{\partial K_1}{\partial J_\alpha} \frac{\partial S_1}{\partial \phi_0^\alpha} + v_0^\alpha(J) \frac{\partial S_2}{\partial \phi_0^\alpha} \right) + \frac{1}{2} \sum_{\alpha\beta} \frac{\partial v_0^\alpha}{\partial J_\beta} \frac{\partial S_1}{\partial \phi_0^\alpha} \frac{\partial S_1}{\partial \phi_0^\beta}. \tag{6.110c}$$

Averages are now taken over the n variables ϕ_0^α: for a function $f(\phi_0, J)$,

$$\langle f \rangle \equiv \frac{1}{(2\pi)^n} \int_0^{2\pi} d\phi_0^1 \int_0^{2\pi} d\phi_0^2 \cdots \int_0^{2\pi} d\phi_0^n \, f(\phi_0, J).$$

As in the one-freedom case, $E_1(J) = \langle K_1 \rangle$, and the analog of Eq. (6.106) becomes

$$\sum_\alpha v_0^\alpha \frac{\partial S_1}{\partial \phi_0^\alpha} = \langle K_1 \rangle - K_1. \tag{6.111}$$

As before, K, K_1, and their averages are known functions, and the right-hand side can be written in a Fourier series of the form

$$\langle K_1 \rangle - K_1 = \sum_m \mathcal{K}_m(J) \exp\{im \cdot \phi_0\},$$

where the sum is over all sets of positive and negative integers $m = \{m^1, m^2, \ldots, m^n\}$, and $m \cdot \phi_0 \equiv \sum_\alpha m^\alpha \phi_0^\alpha$. The $\mathcal{K}_m$ are given by

$$\mathcal{K}_m = \frac{1}{(2\pi)^n} \int_0^{2\pi} d\phi_0^1 \int_0^{2\pi} d\phi_0^2 \cdots \int_0^{2\pi} d\phi_0^n \{\langle K_1 \rangle - K_1\} \exp\{-im \cdot \phi_0\},$$

which can therefore all be calculated in principle. That means that the Fourier coefficients of S_1 can also be calculated. The Fourier series for S_1 is

$$S_1 = \sum_m \mathcal{S}_1(J; m) \exp\{im \cdot \phi_0\},$$

so that

$$\sum_\alpha \nu_0^\alpha \frac{\partial S_1}{\partial \phi_0^\alpha} = i \sum_m (\nu_0 \cdot m) \mathcal{S}_1(J; m) \exp\{im \cdot \phi_0\}.$$

Now use Eq. (6.111) and equate these Fourier coefficients with those of $\langle K_1 \rangle - K_1$; then

$$\mathcal{S}_1(J; m) = \frac{\mathcal{K}_m(J)}{i(\nu_0 \cdot m)}. \tag{6.112}$$

Equation (6.112) is an explicit expression for the Fourier coefficients of S_1 and provides a solution to the problem in first order *as long as none of the denominators vanish!* If a denominator vanishes, the corresponding Fourier coefficient blows up and the entire perturbation procedure fails. Then the perturbed system cannot be described by AA variables that are related through a CT to the AA variables of the unperturbed system. This problem is very serious, for the Fourier series involve sums over all integer sets m, and if the unperturbed frequencies are commensurate there is *definitely* an m for which $\nu_0 \cdot m = 0$. Thus canonical perturbation theory does not work for commensurate frequencies unless the corresponding $\mathcal{K}_m$ happen to vanish. But even if the frequencies are not commensurate, there are bound to be integer sets m for which $\nu_0 \cdot m$ is very small, and then the corresponding $\mathcal{S}_1$ can be very large and the corrections become large, so that the perturbation calculation may work only for an extremely small, restricted range of ϵ.

Systems with commensurate frequencies, or *resonances*, are sometimes called *degenerate*. Canonical perturbation theory does not work in general for degenerate systems, and it must be applied with caution to nearly degenerate systems (those for which $\nu_0 \cdot m$ gets very small compared to $\mathcal{K}_m$). When the method works, the perturbed system is completely integrable and its invariant tori are slight distortions of the tori of the unperturbed system.

When it does not, the original invariant tori not only distort, they may actually break apart. This will be discussed in more detail in connection with the KAM theorem in Chapter 7.

There are other problems associated with canonical perturbation theory. The method assumes that the perturbed system, like the unperturbed one, is completely integrable. Without this assumption, the strategy of averaging could not work, for it depends on E being a function only of J. But averaging is a self-fulfilling strategy, for it forces the approximations to the perturbed system to be completely integrable in every order. The exact perturbed dynamics Δ, even if it can't be integrated, exists on its own: it is the dynamical system belonging to the Hamiltonian $H = H_0 + \epsilon H_1$, and it may simply not be completely integrable. The theory, however, constructs a sequence of completely integrable dynamical systems with which to approximate Δ. It is far from clear that such a sequence will converge even to a completely integrable Δ (or for that matter that it converges to anything!), and it is clear that it can not if Δ is not completely integrable. Thus canonical perturbation theory, through averaging, generally masks many details of Δ: it never shows how it deviates from complete integrability, nor in particular how it may become chaotic. Nevertheless, when used judiciously, it is often a useful technique for studying many properties of perturbed dynamical systems. The KAM theorem will show how useful. (For systems in one freedom the situation is not so stark, for all time-independent Hamiltonian one-freedom systems are completely integrable in the sense of the LI theorem: the Hamiltonian is a constant of the motion.)

WORKED EXAMPLE 6.6

Consider two one-freedom harmonic oscillators coupled by a nonlinear interaction, with Hamiltonian given by

$$H = \frac{1}{2}(p_1^2 + p_2^2) + \frac{1}{2}[(v_0^1)^2(q^1)^2 + (v_0^2)^2(q^2)^2] + \epsilon(v_0^1 v_0^2)^2(q^1 q^2)^2.$$

In first-order canonical perturbation theory: **(a)** Find the two frequencies v^1, v^2, the Hamiltonian $E(J)$, and the generating function $S(\phi, J_0)$. **(b)** Obtain the CT from (ϕ_0, J_0) to (ϕ, J). **(c)** Discuss the CT from (q, p) to (ϕ, J).

Solution. **(a)** We will not use the summation convention: all sums will be indicated by summation signs. The unperturbed Hamiltonian for the two uncoupled oscillators is

$$H_0 = \frac{1}{2}(p_1^2 + p_2^2) + \frac{1}{2}[(v_0^1)^2(q^1)^2 + (v_0^2)^2(q^2)^2],$$

and the coupling perturbation is

$$\epsilon H_1 = \epsilon(v_0^1 v_0^2)^2(q^1 q^2)^2,$$

which is of fourth order in the q^α, not quadratic as in small-vibration theory.

The CT from (q, p) to (ϕ_0, J_0) for the one-freedom harmonic oscillator was found in Worked Example 6.5 and in Problem 7 (here the mass is set equal to 1):

$$\phi_0^\alpha = \tan^{-1}\left\{\nu_0^\alpha \frac{q^\alpha}{p_\alpha}\right\}, \quad J_{\alpha 0} = \frac{1}{2}\left[\frac{p_\alpha^2}{\nu_0^\alpha} + \nu_0^\alpha (q^\alpha)^2\right], \quad \alpha = 1, 2. \quad (6.113)$$

The inverse CT is

$$q^\alpha = \sqrt{\frac{2J_{\alpha 0}}{\nu_0^\alpha}} \sin \phi_0^\alpha, \quad p_\alpha = \sqrt{2\nu_0^\alpha J_{\alpha 0}} \cos \phi_0^\alpha. \quad (6.114)$$

In terms of the unperturbed AA variables, the Hamiltonian is

$$K(\phi_0, J_0) \equiv K_0 + \epsilon K_1 = \sum_{\alpha=1}^{2} \nu_0^\alpha J_{\alpha 0} + 4\epsilon \nu_0^1 J_{10} \nu_0^2 J_{20} \left(\sin^2 \phi_0^1\right)\left(\sin^2 \phi_0^2\right).$$

The unperturbed frequencies are the ν_0^α. In first order the perturbed Hamiltonian E is (remember to replace each $J_{\alpha 0}$ by J_α)

$$E_1(J) = \langle K_1 \rangle = \frac{4\nu_0^1 J_1 \nu_0^2 J_2}{(2\pi)^2} \int_0^{2\pi} \sin^2 \phi_0^1 \, d\phi_0^1 \int_0^{2\pi} \sin^2 \phi_0^2 \, d\phi_0^2 = \nu_0^1 J_1 \nu_0^2 J_2,$$

so in first order the new Hamiltonian $E(J)$ is

$$E(J) \approx K_0(J) + \epsilon E_1(J) = \sum \nu_0^\alpha J_\alpha + \epsilon \nu_0^1 J_1 \nu_0^2 J_2.$$

The first-order perturbed frequencies obtained from this are

$$\nu^1 \approx \frac{\partial E}{\partial J_1} = \nu_0^1 + \epsilon \nu_0^1 \nu_0^2 J_2, \quad \nu^2 \approx \frac{\partial E}{\partial J_2} = \nu_0^2 + \epsilon \nu_0^1 \nu_0^2 J_1.$$

(b) The first-order CT between (ϕ_0, J_0) and (ϕ, J) requires finding

$$S \approx \sum_\alpha \phi_0^\alpha J_\alpha + S_1(\phi_0, J).$$

The function S_1 is obtained from its Fourier series, whose components $S_1(m)$ (we suppress the J dependence) are given in terms of the Fourier components $\mathcal{K}_m$ of $\langle K_1 \rangle - K_1$ by Eq. (6.112). Thus what must first be found are the $\mathcal{K}_m$, where $m \equiv (m_1, m_2)$ is a vector with two integer components, corresponding to the two freedoms. The $\mathcal{K}_m$ are given by

$$\mathcal{K}_m = \frac{1}{(2\pi)^2} \int_0^{2\pi} d\phi_0^1 \int_0^{2\pi} d\phi_0^2 \{\langle K_1 \rangle - K_1\} \exp\{-im_1\phi_0^1 - im_2\phi_0^2\}$$

$$= \frac{1}{(2\pi)^2} \int_0^{2\pi} d\phi_0^1 \int_0^{2\pi} d\phi_0^2 \nu_0^1 J_1 \nu_0^2 J_2 \left\{1 - 4\left(\sin^2 \phi_0^1\right)\left(\sin^2 \phi_0^2\right)\right\}$$

$$\times \exp\{-im_1\phi_0^1 - im_2\phi_0^2\}.$$

The integrations are straightforward; the only nonzero coefficients are those with $(m_1, m_2) = (\pm 2, 0), (0, \pm 2),$ or $(\pm 2, \pm 2)$. Write $A \equiv \frac{1}{4} v_0^1 J_1 v_0^2 J_2$, and then

$$\mathcal{K}_{\pm 2, \pm 2} = -A, \mathcal{K}_{0, \pm 2} = \mathcal{K}_{\pm 2, 0} = 2A.$$

According to Eq. (6.112), $\mathcal{S}_1(m) = -i\mathcal{K}_m/(v_0^1 m_1 + v_0^2 m_2)$, so that the only nonzero $\mathcal{S}_1(m)$ are

$$\mathcal{S}_1(2, 2) = -\mathcal{S}_1(-2, -2) = \frac{iA}{2(v_0^1 + v_0^2)},$$

$$\mathcal{S}_1(2, -2) = -\mathcal{S}_1(-2, 2) = \frac{iA}{2(v_0^1 - v_0^2)},$$

$$\mathcal{S}_1(2, 0) = -\mathcal{S}_1(-2, 0) = -\frac{iA}{v_0^1},$$

$$\mathcal{S}_1(0, 2) = -\mathcal{S}_1(0, -2) = -\frac{iA}{v_0^2}.$$

With these, the first-order expression for S becomes

$$S \approx \sum_\alpha \phi_0^\alpha J_\alpha + \epsilon \sum_m \mathcal{S}_1(J; m) \exp\{im_1\phi_0^1 + m_2\phi_0^2\}$$

$$= \sum_\alpha \phi_0^\alpha J_\alpha + \frac{1}{2}\epsilon J_1 J_2 \left\{ v_0^2 \sin 2\phi_0^1 + v_0^1 \sin 2\phi_0^2 \right.$$

$$\left. - \frac{v_0^1 v_0^2}{2(v_0^1 + v_0^2)} \sin 2(\phi_0^1 + \phi_0^2) - \frac{v_0^1 v_0^2}{2(v_0^1 - v_0^2)} \sin 2(\phi_0^1 - \phi_0^2) \right\}.$$

The first-order CT between (ϕ, J) and (ϕ_0, J_0) can now be found. Write $S \approx \sum_\alpha \phi_0^\alpha J_\alpha + \epsilon J_1 J_2 F(\phi_0)$. Then

$$\phi^1 = \frac{\partial S}{\partial J_1} \approx \phi_0^1 + \epsilon J_2 F(\phi_0), \quad \phi^2 = \frac{\partial S}{\partial J_2} \approx \phi_0^2 + \epsilon J_1 F(\phi_0),$$

$$J_{10} = \frac{\partial S}{\partial \phi_0^1} \approx J_1 + \epsilon J_1 J_2 \frac{\partial F}{\partial \phi_0^1}, \quad J_{20} = \frac{\partial S}{\partial \phi_0^2} \approx J_2 + \epsilon J_1 J_2 \frac{\partial F}{\partial \phi_0^2}.$$

These equations are good only to first order in ϵ, so wherever J_α appears multiplied by ϵ, it can be replaced by $J_{\alpha 0}$. That leads to the following solution of these equations for the first-order CT:

$$\phi^1 \approx \phi_0^1 + \epsilon J_{20} F(\phi_0), \quad \phi^2 \approx \phi_0^2 + \epsilon J_{10} F(\phi_0),$$

$$\tag{6.115}$$

$$J_1 \approx J_{10} - \epsilon J_{10} J_{20} \frac{\partial F}{\partial \phi_0^1}, \quad J_2 \approx J_{20} - \epsilon J_{10} J_{20} \frac{\partial F}{\partial \phi_0^2}.$$

(c) Finding the CT between (q, p) and (ϕ, J) requires using Eqs. (6.113) to replace (ϕ_0, J_0) in Eqs. (6.115) by their expressions in terms of (q, p). The $J_{\alpha 0}$ appear explicitly in Eqs. (6.115), whereas the ϕ_0^α are buried in F and its partial derivatives. The calculation requires writing sines and cosines of multiples and sums of the ϕ_0^α

as functions of (q, p). We show two examples of how to do this:

$$\sin 2\phi_0^\alpha = 2 \sin \phi_0^\alpha \cos 2\phi_0^\alpha = \frac{q^\alpha p_\alpha}{J_{\alpha 0}} = \frac{2v_0^\alpha q^\alpha p_\alpha}{p_\alpha^2 + v_0^{\alpha 2} q^{\alpha 2}},$$

$$\cos 2\phi_0^\alpha = \cos^2 \phi_0^\alpha - \sin^2 \phi_0^\alpha = \frac{1}{2v_0^\alpha J_{\alpha 0}} p_\alpha^2 - \frac{v_0^\alpha}{2 J_{\alpha 0}} q^{\alpha 2} = \frac{p_\alpha^2 - v_0^{\alpha 2} q^{\alpha 2}}{p_\alpha^2 + v_0^{\alpha 2} q^{\alpha 2}}.$$

These and similar expressions can be inserted into Eqs. (6.115) to yield the CT from (q, p) to (ϕ, J). Inverting the resulting transcendental equations analytically to find the CT from (ϕ, J) to (q, p) is all but impossible. This is as far as we go with this problem.

We add the following remarks. If $v_0^1 = v_0^2$, the resonance condition, the equation for S shows that canonical perturbation theory blows up for this dynamical system even in first order. But as long as $v_0^1 \neq v_0^2$, the calculation makes sense. Even if there is some integer n such that $v_0^1 = n v_0^2$, there is no problem, for there are no nonzero $\mathcal{K}_m$ for $m \equiv (m_1, m_2) = l(1, -n)$, where l is some integer. In higher orders, however, this problem might arise.

We have proceeded on the assumption that this system is integrable and that perturbation theory works. In fact the system has been shown to be chaotic for certain parameter values in a sense that will be described more fully in the next chapter. Specifically, if we set $v_0^1 = v_0^2 = 1$ and $\epsilon = 0.05$, and then vary the energy (in our case H) from say $H = 10$ to $H = 20$ and then to $H = 50$, the solutions become completely chaotic (see Pullen and Edmonds, 1981).

THE LIE TRANSFORMATION METHOD

Because canonical perturbation theory uses the global generating function S, it mixes the old and new variables and thereby makes higher order corrections extremely cumbersome. There is no simple and systematic way to disentangle the mixed dependence within S, and in each order this mixing gets more convoluted. The technique then becomes too complicated to be of practical use, and we have stopped at the second order.

If the CT is handled through its infinitesimal generating function, however, the new AA variables are given directly in terms of the old ones, and it should be easier in principle to carry the calculations to higher orders. This is what is done in the *Lie transformation method* (Hori, 1966; Deprit, 1969; Giacaglia, 1972; Cary, 1981). In this book we do not actually carry perturbative calculations past the second order, but we will show in this subsection that the Lie method moves systematically even to the fourth order. The Lie method will be used in Chapter 7 to discuss the KAM theorem.

In the Lie transformation method, as before, the unperturbed system Δ_0 and the perturbed one Δ are both assumed to be completely integrable. The Hamiltonian for Δ is $H(\xi, t, \epsilon)$, and at $\epsilon = 0$ it becomes the unperturbed Hamiltonian for Δ_0. We will now start with H already written in terms of its AA variables, not in terms of (q, p). (Since the ξ are AA variables, H is what was called K in the canonical perturbation method.) Then $\xi \equiv (\phi_0, J_0)$ denotes the AA variables of Δ_0 and $\eta \equiv (\phi, J)$ denotes the AA variables of

Δ. The new Hamiltonian will be called $E(J)$ because it is written in terms of the new AA variables for the η flow [this is the notation of the previous two subsections]. For each value of ϵ there is a canonical transformation $\eta(\xi, t, \epsilon)$, in general time-dependent, from the ξs to the ηs, and since both sets are canonical variables on $\mathbf{T}^*\mathbb{Q}$ these transformations form an ϵ family of CTs. They therefore satisfy PB equations of the form

$$\frac{d\eta^k}{d\epsilon} = \{\eta^k, G\}, \tag{6.116}$$

where $G(\xi, t, \epsilon)$ is the infinitesimal generator of the one-parameter group φ^G of CTs (Section 5.3.4). In canonical perturbation theory the problem was to find S, and in the Lie method the problem is to find G.

The dynamical system itself forms a one-parameter group φ^H of CTs, this one generated by the Hamiltonian H. If we write ξ (rather than ξ_0) for the initial point in $\mathbf{T}^*\mathbb{Q}$ and $\zeta(\xi, t)$ for the moving point, the action of the dynamical group φ^H (the motion) is given by the similar PB equations

$$\frac{d\zeta^k}{dt} = \{\zeta^k, H\}. \tag{6.117}$$

From the point of view of perturbation theory, the essential difference between (6.116) and (6.117) is that H is known, but G is to be found. However, we start the discussion by treating them both as known functions. Both G and H are assumed to depend explicitly on t and ϵ.

Figure 6.14 illustrates the relation between the flows of φ^G and φ^H. Start with some initial point $\xi \in \mathbf{T}^*\mathbb{Q}$. For fixed $t = 0$ the flow φ^G generated by $G(\eta, 0, \epsilon)$, starting at $\epsilon = 0$, carries ξ along the curve labeled ϵ on the left in the figure. At some $\epsilon > 0$ the flow brings ξ to $\eta_0(\xi, \epsilon)$ (the subscript 0 on η indicates that $t = 0$ on this trajectory). For fixed ϵ the flow of φ^H can then carry η_0 along the upper t curve to the point $\zeta_\epsilon(\eta_0, t)$ (the subscript ϵ on ζ indicates that $\epsilon > 0$ is fixed on this trajectory). Now consider the two flows in the opposite order. For fixed $\epsilon = 0$ the flow of $H(\zeta, t, 0)$ carries ξ along the lower t curve to $\zeta_0(\xi, t)$, and then the flow of φ^G carries ζ_0 along the right-hand ϵ curve to $\eta_t(\zeta_0, \epsilon)$. Notice that even though the two paths end up at the same values of t and ϵ, in general they will end up at different points. The exact shape of the trajectories in the figure depend on the starting time t, here taken to be $t = 0$.

Along a t curve the η^k vary according to (we omit the index k)

$$\frac{d\eta}{dt} = \frac{\partial \eta}{\partial t} + \{\eta, H\}.$$

The partial derivative here is taken holding ϵ fixed, and because the transformations are canonical the PB can be taken with respect to either the ξ, ζ, or η variables on $\mathbf{T}^*\mathbb{Q}$. Since the Hamiltonian for the η is E,

$$\frac{d\eta}{dt} = \{\eta, E\}.$$

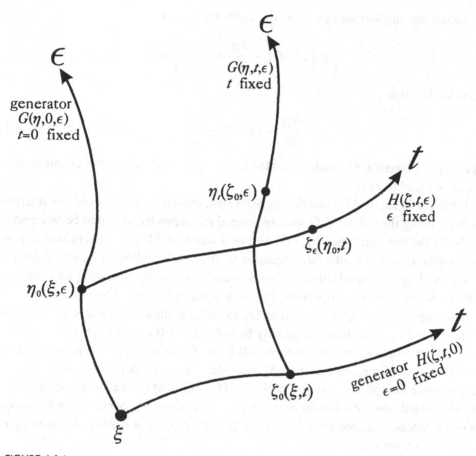

FIGURE 6.14

The relation between the flows generated by H and by G. Explanation in the text.

Hence

$$\frac{\partial \eta}{\partial t} = \{\eta, E - H\} \equiv \{\eta, R\}, \tag{6.118}$$

defining the function $R = E - H$. So far this definition ignores the fact that E is a function of the η^k and H of the ζ^k.

The next step is to obtain a differential equation for R, and from that another for E. For the R equation, take the derivative of Eq. (6.116) with respect to t and the derivative of Eq. (6.118) with respect to ϵ (holding t fixed). The $d/d\epsilon$ of Eq. (6.116) is taken holding t fixed, so for the purposes of this argument it should be written $\partial/\partial\epsilon$. This gives

$$\frac{\partial^2 \eta}{\partial t \, \partial \epsilon} = \frac{\partial}{\partial t}\{\eta, G\} = \frac{\partial}{\partial \epsilon}\{\eta, R\}.$$

Both partial derivatives can be taken inside the PBs [see Eq. (5.62)], and then with the aid of Eqs. (6.116) and (6.118) this can be put in the form

$$\{\{\eta, R\}, G\} + \left\{\eta, \frac{\partial G}{\partial t}\right\} = \{\{\eta, G\}, R\} + \left\{\eta, \frac{\partial R}{\partial \epsilon}\right\}.$$

The Jacobi identity and antisymmetry of the PB then yield

$$\left\{ \eta, \left(\{R, G\} - \frac{\partial R}{\partial \epsilon} + \frac{\partial G}{\partial t} \right) \right\} = 0,$$

which implies that

$$\frac{\partial R}{\partial \epsilon} - \{R, G\} = \frac{\partial G}{\partial t}. \tag{6.119}$$

In principle a constant C should be added to the right-hand side of this equation, but we have set it equal to zero.

If the PB in Eq. (6.119) had the opposite sign, the left-hand side could be interpreted as $dR/d\epsilon$ along the ϵ flow: if G were in general ϵ independent, it could be interpreted as $dR/d\epsilon$ for the flow generated by $-G$. But as it stands (6.119) is not very transparent. It does, however, lead to a differential equation for the new Hamiltonian E, whose derivation is simplified by the introduction of two operators, $\mathcal{U}$ and $\mathcal{P}_G$ operating on functions in $\mathcal{F}(\mathbf{T}^*\mathbb{Q})$ (in this subsection operators will be designated by cursive letters).

The operator $\mathcal{U}$ is φ_{fn}^G as defined at Eq. (5.105): it shows the change φ^G induces on functions on $\mathbf{T}^*\mathbb{Q}$. It is defined explicitly by $\mathcal{U}F(\xi) = F((\varphi^G)^{-1}\xi)$, where $F \in \mathcal{F}(\mathbf{T}^*\mathbb{Q})$. In our present application, the functions will be on $\mathbf{T}^*\mathbb{Q} \times \mathbb{R}(\epsilon)$ (they depend on ϵ as well as on the point of the phase manifold), and because G is ϵ dependent, so is φ^G. This ϵ dependence complicates the description of $\mathcal{U}$ considerably, even more so because φ^G depends not only on ϵ, but also on its starting value ϵ_0. However, our notation will suppress this extra dependence, and we will write simply φ^G: instead of writing $\mathcal{U}(\epsilon, \epsilon_0)F(\eta, \epsilon) = F((\varphi^G)^{-1}\eta, \epsilon)$, we write simply

$$\mathcal{U}(\epsilon)F(\eta) = F((\varphi^G)^{-1}\eta) \equiv F(\xi), \tag{6.120}$$

where $\eta = \xi$ at $\epsilon = \epsilon_0$. From this equation it follows that

$$F(\eta) = \mathcal{U}^{-1}F(\xi), \tag{6.121}$$

which describes how such a function changes under the Hamiltonian ϵ-flow generated by G. In these equations ξ is independent of ϵ, but $\eta = \varphi^G\xi$ depends on ϵ.

Here is a useful way to understand $\mathcal{U}^{-1}$ and $\mathcal{U}$. View φ^G as a passive transformation, and then $\mathcal{U}^{-1}$ is the operator that rewrites functions of the ξ variables as functions of the η variables and $\mathcal{U}$ rewrites functions of η as functions of ξ.

Under the flow generated by G the total ϵ derivative of any function $F(\eta, \epsilon)$ satisfies

$$\frac{dF(\eta)}{d\epsilon} = \{F(\eta), G\} + \frac{\partial F(\eta)}{\partial \epsilon}, \tag{6.122}$$

where $\partial F/\partial \epsilon$ is the derivative of the explicit ϵ dependence. According to Eq. (6.121) this derivative can also be written as

$$\frac{dF(\eta)}{d\epsilon} = \frac{d\mathcal{U}^{-1}}{d\epsilon}F(\xi) + \mathcal{U}^{-1}\frac{\partial F(\xi)}{\partial \epsilon}.$$

But $\mathcal{U}^{-1}\partial F(\xi)/\partial\epsilon = \partial F(\eta)/\partial\epsilon$ and $F(\xi) = \mathcal{U}F(\eta)$, so

$$\frac{dF(\eta)}{d\epsilon} = \frac{d\mathcal{U}^{-1}}{d\epsilon}\mathcal{U}F(\eta) + \frac{\partial F(\eta)}{\partial\epsilon},$$

and hence it follows from (6.122) that

$$\{F(\eta), G\} = \frac{d\mathcal{U}^{-1}}{d\epsilon}\mathcal{U}F(\eta). \tag{6.123}$$

Since this equation is true for any function $F \in \mathcal{F}(\mathbf{T}^*\mathbb{Q})$, it gives a relation between $\mathcal{U}$ and the PB with respect to G. That leads to the definition of what we will call the *Poisson operator* $\mathcal{P}_G$:

$$\mathcal{P}_G \equiv \{\bullet, G\}; \tag{6.124}$$

that is, $\mathcal{P}_G F = \{F, G\}$.

In terms of $\mathcal{P}_G$, the relation (6.123) between $\mathcal{U}$ and the PB is

$$\mathcal{P}_G = \frac{\partial\mathcal{U}^{-1}}{\partial\epsilon}\mathcal{U} \quad \text{or} \quad \frac{\partial\mathcal{U}^{-1}}{\partial\epsilon} = \mathcal{P}_G\mathcal{U}^{-1}. \tag{6.125}$$

These are operator equations; for instance, the first one says that applying $\mathcal{P}_G$ gives the same result as applying first $\mathcal{U}$ and then $\partial\mathcal{U}^{-1}/\partial\epsilon$. We have seen that $\mathcal{U}$ can be understood as changing variables from η to ξ, and $\mathcal{P}_G$ is essentially the Poisson bracket, so in a way this relation gives meaning to $\partial\mathcal{U}/\partial\epsilon$. That meaning can be made clearer by using the fact that $\partial(\mathcal{U}^{-1}\mathcal{U})/\partial\epsilon = \partial\mathbb{I}/\partial\epsilon \equiv 0$, or

$$\frac{\partial\mathcal{U}^{-1}}{\partial\epsilon}\mathcal{U} = -\mathcal{U}^{-1}\frac{\partial\mathcal{U}}{\partial\epsilon}.$$

Then Eq. (6.125) can be put in the form

$$\mathcal{P}_G = -\mathcal{U}^{-1}\frac{\partial\mathcal{U}}{\partial\epsilon} \quad \text{or} \quad \frac{\partial\mathcal{U}}{\partial\epsilon} = -\mathcal{U}\mathcal{P}_G. \tag{6.126}$$

The second of Eqs. (6.126) clarifies the meaning of $\partial\mathcal{U}/\partial\epsilon$: applying $\partial\mathcal{U}/\partial\epsilon$ to a function $F(\eta)$ is the same as taking the PB of F with G, rewriting the result as a function of ξ, and changing its sign.

Recall that we intend to find a differential equation for E. The first step is to rewrite Eq. (6.119) in terms of $\mathcal{U}$ and $\mathcal{P}_G$. The PB in Eq. (6.119) is $\{R, G\} \equiv \mathcal{P}_G R$, so (6.119) becomes

$$\frac{\partial R}{\partial\epsilon} - \mathcal{P}_G R = \frac{\partial G}{\partial t}. \tag{6.127}$$

According to its definition, R is $E - H$. But E is usually written as a function of the new η variables, and H as a function of the old ξ variables. To rewrite H as a function of η apply $\mathcal{U}^{-1}(\epsilon)$, so R as a function of η is given by $R = E - \mathcal{U}^{-1}H$. Then (6.127) becomes

$$\frac{\partial E}{\partial\epsilon} - \frac{\partial\mathcal{U}^{-1}}{\partial\epsilon}H - \mathcal{U}^{-1}\frac{\partial H}{\partial\epsilon} - \left[\mathcal{P}_G E - \frac{\partial\mathcal{U}^{-1}}{\partial\epsilon}H\right] = \frac{\partial G}{\partial t},$$

where the expression in brackets is $\mathcal{P}_G R$ with (6.125) used for the second term. Two of the terms cancel out and the result may be put in the form

$$\frac{\partial G}{\partial t} = \frac{\partial E}{\partial \epsilon} - \mathcal{P}_G E - \mathcal{U}^{-1} \frac{\partial H}{\partial \epsilon}. \tag{6.128}$$

This is the desired differential equation for E (recall that G is still being treated as a known function).

A formal solution for E can be obtained (and is stated at Problem 10) if G and H are known and if $\mathcal{U}^{-1}$ can be found. However, in perturbation theory G is not a known function, and therefore $\mathcal{U}$, defined in terms of φ^G, is not a known operator. In fact the essence of the problem is to find φ^G and its generator G. One proceeds as usual in perturbation theory, by expanding all functions and operators in power series in ϵ. Assume as before that H is of the form $H(\zeta, t, \epsilon) = H_0(\zeta, t) + \epsilon H_1(\zeta, t)$. The time t is included in H_0 only for generality, for if, as assumed, the unperturbed system is completely integrable and written in AA variables, H_0 is a function of the Js alone, independent of t. As for the rest, write (note the indices on the G_k and $\mathcal{P}_k$)

$$\left. \begin{array}{ll} E(\zeta, t, \epsilon) = \displaystyle\sum_0^\infty \epsilon^n E_n(\zeta, t), & G = \displaystyle\sum_0^\infty \epsilon^n G_{n+1}(\zeta, t), \\[2mm] \mathcal{U}^{-1}(\epsilon) = \displaystyle\sum_0^\infty \epsilon^n V_n, & \mathcal{P}_G = \displaystyle\sum_0^\infty \epsilon^n \mathcal{P}_{n+1}. \end{array} \right\} \tag{6.129}$$

The reason $\mathcal{U}^{-1}$ and not $\mathcal{U}$ is expanded is that $\mathcal{U}^{-1}$ is what appears in Eq. (6.128). At $\epsilon = 0$ the CT is assumed to be the identity, so $V_0 = \mathbb{I}$ and $E_0 = H_0$.

The four power series of Eq. (6.129) are now inserted into Eq. (6.128), which becomes

$$\sum_n \epsilon^n \frac{\partial G_{n+1}}{\partial t} = \sum_n n \epsilon^{n-1} E_n - \sum_n \epsilon^n \mathcal{P}_{n+1} \sum_m \epsilon^m E_m - \left(\sum_n \epsilon^n V_n \right) H_1.$$

Now coefficients of equal powers of ϵ are equated, and some manipulation of indices yields the general result

$$\frac{\partial G_n}{\partial t} = n E_n - \sum_{m=0}^{n-1} \mathcal{P}_{n-m} E_m - V_{n-1} H_1. \tag{6.130}$$

The V_k can be eliminated from these equations in favor of the $\mathcal{P}_m$ by using the second of Eqs. (6.125) to obtain the relation

$$V_n = \frac{1}{n} \sum_{m=0}^{n-1} \mathcal{P}_{n-m} V_m.$$

This is a set of recursion relations, whose first four solutions are ($V_0 = \mathbb{I}$ is added; see Problem 11)

$$V_0 = \mathbb{I}, \quad V_1 = \mathcal{P}_1, \quad V_2 = \frac{1}{2}(\mathcal{P}_1^2 + \mathcal{P}_2),$$

$$V_3 = \frac{1}{6}\mathcal{P}_1^3 + \frac{1}{6}\mathcal{P}_1\mathcal{P}_2 + \frac{1}{3}\mathcal{P}_2\mathcal{P}_1 + \frac{1}{3}\mathcal{P}_3. \tag{6.131}$$

The reason the $\mathcal{P}_k$ were indexed with $n + 1$ in Eq. (6.129) is so that the indices on the $\mathcal{P}_k$ operators in each term sum up to the order (index) of the $\mathcal{V}$ operator in these equations. This procedure can be extended to give a general expression for $\mathcal{V}_n$ in terms of $\mathcal{P}_m$ operators with $m \leq n$, but we omit that.

Equation (6.124) can be used to show that

$$\mathcal{P}_n = \{\bullet, G_n\},$$

and this, together with the recursion relations for the $\mathcal{V}_n$ and Eq. (6.130), leads to an infinite set of coupled equations. Up to $n = 4$ these are ($E_0 = H_0$ is included as the equation of zeroth order and is used in all the others)

$$E_0 = H_0, \tag{6.132a}$$

$$\frac{\partial G_1}{\partial t} - \{G_1, H_0\} = E_1 - H_1, \tag{6.132b}$$

$$\frac{\partial G_2}{\partial t} - \{G_2, H_0\} = 2E_2 - \mathcal{P}_1(E_1 + H_1), \tag{6.132c}$$

$$\frac{\partial G_3}{\partial t} - \{G_3, H_0\} = 3E_3 - \mathcal{P}_1 E_2 - \mathcal{P}_2\left(E_1 + \frac{1}{2}H_1\right) - \frac{1}{2}\mathcal{P}_1^2 H_1, \tag{6.132d}$$

$$\frac{\partial G_4}{\partial t} - \{G_4, H_0\} = 4E_4 - \mathcal{P}_1 E_3 - \mathcal{P}_3\left(E_1 + \frac{1}{3}H_1\right)$$
$$- \frac{1}{3}\left(\frac{1}{2}\mathcal{P}_1\mathcal{P}_2 + \mathcal{P}_2\mathcal{P}_1\right) - \frac{1}{6}\mathcal{P}_1^3 H_1. \tag{6.132e}$$

The indices are grouped as in Eq. (6.131). Because the unperturbed system is completely integrable, H_0 is time independent. That makes it possible to interpret the left-hand sides of all of Eqs. (6.132) as the rate of change of the G_k backward in time, as Eq. (6.119) could have been interpreted in terms of a flow in the $-\epsilon$ direction.

As in canonical perturbation theory, there are too many unknown functions and not enough equations. In Eqs. (6.132) only H_0 and H_1 (and therefore E_0) are known functions. For each order $k > 0$ there is one equation for two functions, G_k and E_k (the $\mathcal{P}_k$ are simply PBs with respect to the G_k). Thus some additional strategy is needed. We do not describe it in general, but illustrate it in the example of the next subsection.

EXAMPLE: THE QUARTIC OSCILLATOR

We return to the quartic oscillator of Worked Example 6.5 and show how to treat it by the Lie method. Equation (6.108) gives the perturbed Hamiltonian in terms of the initial (unperturbed) AA variables (remember that K is now called H and recall that $\mu = 1/mv_0^2$):

$$H(\phi_0, J_0) \equiv H_0(J_0) + \epsilon H_1(\phi_0, J_0) = v_0 J_0 + \epsilon \mu J_0^2 \sin^4 \phi_0. \tag{6.133}$$

This H_0 is the starting point for Eqs. (6.132).

It is shown in Problem 10 that if $\mathcal{V}(\epsilon)$ is an operator such that $\partial \mathcal{V}/\partial \epsilon = \mathcal{P}_G \mathcal{V}$, then the function F defined by

$$F = \mathcal{V}(\epsilon) \int_{\epsilon_0}^{\epsilon} \mathcal{V}^{-1}(\epsilon') g(\epsilon')\, d\epsilon' \tag{6.134}$$

satisfies the equation $\partial F/\partial \epsilon - \mathcal{P}_G F = g$. This fact is used to deal with Eqs. (6.132) (observe that the PB with respect to H_0 is just $\mathcal{P}_{H_0}$) by finding an operator $\mathcal{V}_0$ that satisfies $\partial \mathcal{V}_0/\partial t = \mathcal{P}_{H_0} \mathcal{V}_0$ so that the relation between $\mathcal{V}_0$ and H_0 will be analogous to the relation between $\mathcal{U}^{-1}(\epsilon)$ and G in Eq. (6.125). What will actually be found is not $\mathcal{V}_0$, but $\mathcal{V}_0^{-1}$, the analog of $\mathcal{U}(\epsilon)$, and then that can be inverted to yield $\mathcal{V}_0$.

Just as $\mathcal{U}(\epsilon)$ is the ϵ-translation operator belonging to G, so $\mathcal{V}_0^{-1}(t)$ is the time-translation operator belonging to H_0. The initial time t_0 can be omitted from the argument of $\mathcal{V}_0^{-1}(t)$, for H_0 is t independent. Since Δ_0 is completely integrable, the time translation of its AA variables (ϕ_0, J_0) is known and from that $\mathcal{V}_0$ can be found. The motion of the AA variables is $\phi_0(t) = \phi_0(0) + v_0 t$ and $J_0(t) = J_0 = \text{const.}$, so according to Eq. (6.120),

$$\mathcal{V}_0(t) F(\phi_0, J_0, t) = F(\phi_0(t), J_0, t) \tag{6.135}$$

for any function $F \in \mathcal{F}(\mathbf{T}^*\mathbb{Q}) \times \mathbb{R}(t)$.

Each of Eqs. (6.132) is of the form

$$\frac{\partial G_k}{\partial t} - \{G_k, H_0\} = g_k, \tag{6.136}$$

and in accordance with the preceding discussion this is equivalent to

$$G_k = \mathcal{V}_0(t) \int_0^t \mathcal{V}_0^{-1}(t') g_k(t')\, dt',$$

where $\mathcal{V}_0(t)$, and hence also $\mathcal{V}_0^{-1}(t)$, is defined in Eq. (6.135). To proceed, use the fact that since $\mathcal{V}_0^{-1}(t)$ translates forward in time and $\mathcal{V}_0(t)$ translates backward in time by t, it follows that $\mathcal{V}_0(t)\mathcal{V}_0^{-1}(t') = \mathcal{V}_0^{-1}(t' - t) = \mathcal{V}_0(t - t')$.

For $k = 1$ the result is (we put only the upper limit on the integral)

$$G_1 = \int^t \mathcal{V}_0(t - t')[E_1 - H_1]\, dt'$$

$$= \int^t \left[E_1(\phi_0(t - t'), J_0, t') - \mu J_0^2 \sin^4\{\phi_0(t - t')\} \right] dt'. \tag{6.137}$$

This equation, like (6.132b), still contains the two unknown functions G_1 and E_1. Some simplification results when it is assumed that the perturbed system is also completely integrable so that E_1 is independent of both ϕ and t, but even that is not enough. Again, the strategy is to use averaging.

Because H_1 is a positive function, its integral in (6.137) grows as t increases, which leads to a secular term in G_1, one that grows without bound. If E_1 could be made to cancel out this growth in each period, the secular term would be eliminated. Therefore E_1 is chosen equal to the average of H_1 over each period. Now, t varies from some initial value t_0 to $t_0 + 2\pi/\nu_0$ in one period, so (as before, use $\int \sin^4 x \, dx = \frac{1}{8}[3x - \cos x(2\sin^3 x + 3\sin x)]$)

$$E_1(J_0) = \frac{\nu_0}{2\pi} \int^{2\pi/\nu_0} dt' H_1(J_0, \phi_0(0) + \nu_0 t') = \frac{3\mu}{8} J_0^2.$$

To obtain $E_1(J)$, replace J_0 by J. This result (and hence also the first-order perturbed frequency) is the same as in canonical perturbation theory: the two procedures yield the same result in first order. Integration of $\sin^4 x$ again leads to an expression for G_1:

$$G_1 = -\frac{1}{8\nu_0} J_0^2 \mu \cos \phi_0 (2\sin^3 \phi_0 + 3\sin \phi_0).$$

This can be used to find the first-order transformation between (ϕ, J) and (ϕ_0, J_0). However, we do not go into that.

The next step is to proceed to second order, but except for the following comment, we leave that to Problem 13. The second term on the right-hand side of Eq. (6.132c) is

$$\mathcal{P}_1(E_1 + H_1) \equiv \frac{\partial G_1}{\partial J_0} \frac{\partial(E_1 + H_1)}{\partial \phi_0} - \frac{\partial G_1}{\partial \phi_0} \frac{\partial(E_1 + H_1)}{\partial J_0}.$$

The expressions already obtained for E_1 and H_1 can be inserted into this equation and then into Eq. (6.132c) to obtain g_2, the right-hand side of Eq. (6.136) with $k = 2$. The g_2 so obtained depends on (ϕ_0, J_0) and on the unknown function E_2, which is similar to the dependence of g_1 in first order. As E_1 was chosen in that case, E_2 is chosen to cancel the one-period average of the potentially secular term.

The example of the quartic oscillator shows how averaging can be used to eliminate the secular terms in the Lie method. We will not discuss averaging more generally.

This completes our general treatment of Hamiltonian perturbation theory, but in the next chapter we will return to some of these topics. It will be seen then that the same problems arise in the Lie method as in canonical perturbation theory: secular terms, small denominators, and all that entails. But the Lie method has some advantages. One is that the equations to any order can be written out explicitly. We wrote them out to fourth order without too much difficulty, which would have been much more complicated in the canonical method. Another advantage is that the general results we obtained are valid for any number of freedoms. In the canonical method, freedoms higher than one become very cumbersome. The Lie method is also more adapted to proving general results and theorems, like the KAM theorem of Chapter 7.

6.4 ADIABATIC INVARIANCE

When the Hamiltonian depends on the time, even in one freedom, it can be quite difficult to solve for the motion. This results in part because in time-dependent systems constants of the motion are hard to find. But when the time dependence is sufficiently slow (what this means will be explained later) there often exist dynamical variables that are almost constant, and these can be used almost like exact constants to analyze properties of the motion. In particular, such approximate constants are the familiar action variables of completely integrable systems.

6.4.1 THE ADIABATIC THEOREM

We start with one-freedom systems. Suppose that a completely integrable one-freedom Hamiltonian

$$H(q, p, \lambda) \tag{6.138}$$

contains a constant parameter λ. The AA variables will depend on the value of λ: they become $J(q, p, \lambda)$, $\phi(q, p, \lambda)$. If λ is replaced by a time-dependent function $\lambda(t)$, the system will in general no longer be completely integrable, but if the usual prescription for finding AA variables is followed (treating λ as if it were time independent), $J(q, p, \lambda)$, $\phi(q, p, \lambda)$ become functions of the time through $\lambda(t)$. Then J will no longer be a constant of the motion. It will be shown in what follows, however, that if λ varies very slowly J is very nearly constant. More specifically, if λ changes by an amount $\Delta\lambda$ in a time ΔT, the change ΔJ in J is (roughly) proportional to $\Delta\lambda/\Delta T$, so that for any $\Delta\lambda$ no matter how large, ΔJ can be made as small as desired by forcing the change in λ to take place over a long enough time ΔT. Such a J is called an *adiabatic invariant*.

OSCILLATOR WITH TIME-DEPENDENT FREQUENCY

For an example of this kind of adiabatic invariant consider a particle on a spring (a harmonic oscillator) whose force constant k (or natural frequency $\omega = \sqrt{k/m}$) is made to change with time. For instance, the spring could be heated, which changes k. The Hamiltonian is $H = \frac{1}{2}p^2 + \frac{1}{2}\omega(t)^2 q^2$, so for a given energy and constant ω the trajectory in $\mathbf{T}^*\mathbb{Q}$ is an ellipse whose equation is

$$p^2 + \omega^2 q^2 = 2E.$$

The action,

$$J = \frac{1}{2\pi} \oint p \, dq = \frac{1}{2\pi} \oint \sqrt{\frac{2E}{\omega} - q^2} \, dq = \frac{E}{\omega},$$

is the area of that ellipse. As t increases, changing ω, both J and E vary, but it is not clear a priori how they vary or even how their variations are related.

As we will show in the next subsection, however, J is an adiabatic invariant, so if $\omega(t)$ changes slowly enough, the area J of the ellipse in $\mathbf{T}^*\mathbb{Q}$ remains approximately constant. Suppose that at $t = 0$ the frequency $\omega(0)$ is high (the force constant is large). Then the major axis of the ellipse starts out in the p direction. If the spring is heated, then at $t = \Delta T$ the frequency $\omega(\Delta T)$ is lower; suppose that it is so low at $t = \Delta T$ that the major axis is in the q direction. Then although the area of the ellipse remains approximately constant in the time interval ΔT, the ellipse itself gets distorted. The trajectory in $\mathbf{T}^*\mathbb{Q}$ is strictly an ellipse only if ω is constant, but if $\omega(t)$ varies slowly enough, the trajectory looks like a slowly distorting ellipse throughout the time interval ΔT. More to the point, however, is that the energy change can be calculated from the constancy of $J = E(t)/\omega(t)$: the energy must increase to compensate for the change in ω, or $E(\Delta T) = E(0)\omega(\Delta T)/\omega(0)$.

THE THEOREM

We now show that J is indeed an adiabatic invariant in any one-freedom completely integrable system (Lichtenberg, 1969). Insert the time dependence into the λ of (6.138), and calculate

$$\dot{H} = \frac{\partial H}{\partial q}\dot{q} + \frac{\partial H}{\partial p}\dot{p} + \frac{\partial H}{\partial \lambda}\dot{\lambda}. \tag{6.139}$$

For a physical trajectory, the first two terms cancel by Hamilton's canonical equations; therefore along the motion the change in H in a time interval ΔT is

$$\Delta H = \int_0^{\Delta T} \frac{\partial H}{\partial \lambda}\dot{\lambda}\, dt,$$

where we have set the initial time to zero. We now make the crucial *adiabatic assumption*, namely that $\lambda(t)$ varies so slowly that $\dot{\lambda} \approx \Delta\lambda/\Delta T$ is constant and small during the time of integration (how small has yet to be established). Then the change in H is

$$\Delta H = \frac{\Delta\lambda}{\Delta T} \int_0^{\Delta T} \frac{\partial H}{\partial \lambda}\, dt. \tag{6.140}$$

The argument now proceeds in two steps. Step 1 shows that for a given $\Delta\lambda$ and for large enough ΔT the right-hand side of (6.140) is independent of the initial conditions, as long as they all belong to the same energy. Step 2 shows that this independence implies the adiabatic invariance of J. Both steps make use of some of the reasoning of the Liouville volume theorem [Section 5.4.1], in particular some aspects of Fig. 5.9. That figure is redrawn and somewhat altered in Fig. 6.15.

Step 1. In Fig. 5.9, $R(0)$ at time $t = 0$ is bordered by an arbitrary closed curve in $\mathbf{T}^*\mathbb{Q}$, but in Fig. 6.15 we take it to be a curve $\mathcal{C}(0)$ of constant energy

$$E(0) = H(q, p, \lambda(0)). \tag{6.141}$$

This is an implicit equation for $\mathcal{C}(0)$; it can be made explicit only if it can be solved for $p(q, E(0), \lambda(0))$. The "tube" in the figure, rising from the $\mathbf{T}^*\mathbb{Q}$ plane at $t = 0$ to the $\mathbf{T}^*\mathbb{Q}$ plane at $t = \Delta t$, is mapped out by $\mathcal{C}(t)$ as time progresses [$\mathcal{C}(0)$ is partially seen at $t = 0$, but $R(0)$ is obscured by the tube]: as time develops, each initial point on $\mathcal{C}(0)$ moves, and

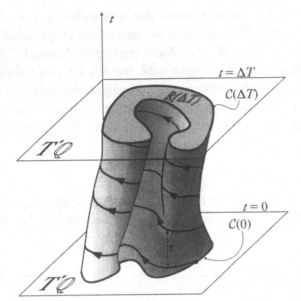

FIGURE 6.15
The tube formed by the development in time of a constant-energy curve $C(0)$ bounding a region $R(0)$ in $\mathbf{T}^*\mathbb{Q}$ at $t = 0$ [$R(0)$ is obscured by the tube.] At $t = \Delta T$ the curve and the region it bounds become $C(\Delta T)$ and $R(\Delta T)$. The trajectory whose initial conditions belong to the dot on $C(0)$ is the spiral winding up the tube. The distortion of the tube in each circuit of C is very slight if the adiabatic condition is satisfied; it is exaggerated in the figure. τ is the time it takes the system to go once around C. Because C changes in time, τ can be defined only approximately, but the better the adiabatic condition is satisfied, the more accurately τ can be defined.

the collection of such initial points map out the tube. Although all the points start out at the same energy, there is no guarantee that at a later time they will also all have the same energy, so $C(\Delta T)$ is not necessarily a curve of constant energy. What Step 1 will show is that $C(\Delta T)$ is *approximately* a curve of constant energy if the t dependence of λ is slow enough. In fact if λ were entirely time independent, there would be no problem, for the energy would not change and all points that start on $C(0)$ would continue to have the same energy. As time went on $C(t)$ would remain the same and the tube would rise vertically undistorted: $R(\Delta T)$ and $C(\Delta T)$ would look exactly the same as $R(0)$ and $C(0)$. The time dependence of λ is what causes the tube of Fig. 6.15 to distort and raises the question of the energy on $C(\Delta T)$.

Consider an initial condition on $C(0)$, indicated by the dot in the $t = 0$ plane. If λ were constant, the system would run around $C(0)$, and as time proceeded from $t = 0$ to $t = \Delta T$, it would spiral simply up the surface of the tube to some point on $C(\Delta T)$. Different initial conditions would start at different points on $C(0)$ and yield different nonintersecting spirals. If the ΔT in the figure represented a long time in which the system ran many times around $C(0)$, the spirals would pack tightly on the undistorted tube. Almost the same is true on the distorted tube when $\lambda = \lambda(t)$: the system still spirals up the surface of the tube to $C(\Delta T)$.

To study the energy on $C(\Delta T)$ we return to the integral of Eq. (6.140), which is taken along the motion, or along one of the spirals. Since different spirals lie on different parts of the distorted tube, the integral is different along different spirals (i.e., for different initial condition). The point of Step 1 is that if the tube distorts slowly enough ΔH is almost the same for all initial conditions. Indeed, if the tube hardly distorts in the time τ that it takes the system to cycle once around, the spirals pack together so closely that the integral is almost the same on all of them. Now, the distortion of the tube depends on how the expression for H changes with time, and that depends on how λ changes with time. The distortion can be made small in one period τ by forcing the one-period change in λ, namely

$\dot{\lambda}\tau \equiv \tau \Delta\lambda/\Delta T$, to be much smaller than λ:

$$\tau\frac{\Delta\lambda}{\Delta T} \ll \lambda, \quad \text{or} \quad \frac{\Delta T}{\tau} \gg \frac{\Delta\lambda}{\lambda}. \tag{6.142}$$

This is the *adiabatic condition*: the rate of change of λ should be slow, or the time ΔT it takes to achieve a given change $\Delta\lambda$ should be long, in accordance with this equation.

If the adiabatic condition (6.142) is satisfied, the integral ΔH is almost the same for all initial conditions on $\mathcal{C}(0)$, making the energy almost the same for all points on $\mathcal{C}(\Delta T)$. Call the new energy $E(\Delta T) \equiv H(q, p, \lambda(\Delta T))$. In general, of course, $E(\Delta T) \neq E(0)$. In particular, the integral and the denominator of $\Delta\lambda/\Delta T$ both grow with ΔT. A useful mnemonic for remembering the energy change according to (6.140) is that ΔH is $\Delta\lambda$ times the average of $\partial H/\partial\lambda$ in the time ΔT.

Step 2. The remainder of the theorem is quite short. Adiabatic invariance of J is proven almost immediately by invoking the Liouville volume theorem, according to which $R(0)$ and $R(\Delta T)$ have equal areas. Because both $\mathcal{C}(\Delta T)$ and $\mathcal{C}(0)$ are equal-energy curves, the areas at both times can be calculated by solving $E = H(q, p, \lambda)$ for $p(q, E, \lambda)$ and integrating to obtain $J \equiv (2\pi)^{-1} \oint p\, dq$. Hence the LV theorem implies that

$$J(\Delta T) = J(0). \tag{6.143}$$

To the extent that the adiabatic assumption is valid, J is therefore constant under the motion: J is an adiabatic invariant, as was asserted. This result is the *adiabatic theorem*.

REMARKS ON $N > 1$

There is a way to apply our one-freedom results to certain systems with many freedoms. Suppose some of the freedoms of a multiple-freedom system are known a priori to change significantly more slowly than the others. Then it is often possible to treat the slow ones as adiabatically varying parameters on a reduced system consisting only of the rapid freedoms. Two considerations are important in this connection. The first is that the slow freedoms must vary adiabatically, that is, that significant changes in them take place over times that are long with respect to any period ΔT of the reduced system.

The second consideration is connected to complete integrability. One of the explicit assumptions of the adiabatic theorem is that the system is completely integrable if λ is time independent (this is necessary for the action variable J to be defined at each value of λ). But complete integrability is the exception rather than the rule for systems with more than one freedom, and therefore adiabatic considerations cannot usually be applied to systems with many freedoms. Nevertheless, systems exist such that if some variables are artificially held fixed, the reduced system formed by the remaining variables is completely integrable. If it is known a priori that the reduced system oscillates many times in the time it takes the slowly varying variables to undergo significant changes, then the adiabatic theorem can be applied to the reduced system. This is a particularly fruitful technique if the reduced system has just one freedom, for essentially all one-freedom (oscillating) systems are completely integrable. This will be made clearer in some of the examples to be discussed later.

6.4.2 HIGHER APPROXIMATIONS

Even for one-freedom systems such as harmonic oscillators with slowly varying ω, adiabatic invariance is only approximate.

In order to show how approximate, we use general canonical perturbation theory to start to calculate the perturbation series for J in the case of an oscillator with a time-dependent frequency. The Hamiltonian may be written in the form

$$H = \frac{1}{2}p^2 + \rho(\epsilon t)q^2,$$

where $\rho = \omega^2$ and the small parameter ϵ is inserted to ensure the slow time dependence of ρ and to have a perturbation parameter. In the $\epsilon \to 0$ limit the Hamiltonian is time independent.

Transform to unperturbed AA variables by using a generating function similar to the one of Problem 7:

$$S = \frac{1}{2}\rho q^2 \cot \phi_0.$$

Straightforward calculation yields the results of Problem 7, slightly modified because of the different parameters and the different way the Hamiltonian is written:

$$\phi_0 = \tan^{-1}\left\{\rho\frac{q}{p}\right\}, \quad J_0 = \frac{1}{2}\left(\rho q^2 + \frac{p^2}{\rho}\right).$$

Because ρ is time dependent, however, the new Hamiltonian is not simply H written in the new coordinates, but

$$K = \nu_0 J_0 + \frac{\partial S}{\partial t} = \nu_0 J_0 + \frac{1}{2}\epsilon J_0 \frac{\rho'}{\rho} \sin 2\phi_0, \qquad (6.144)$$

where ρ' is the derivative of ρ with respect to ϵt (this makes the ϵ dependence explicit).

The first term $\nu_0 J_0$ of (6.144) is the unperturbed, completely integrable Hamiltonian in AA variables. The ϵ-dependent term is the perturbation, called K_1 in canonical perturbation theory. The next step requires using (6.106) to find the first-order correction to the generating function and then (6.100a and b) to calculate the new AA variables to first order. We write $S(\phi_0, J)$ for the generating function of the transformation from (ϕ_0, J_0) to the new variables (ϕ, J). The calculation is simplified by the fact that the average of K_1 over ϕ vanishes, and then to first order

$$S \approx \phi_0 J + \epsilon \frac{\rho'}{4\nu\rho} J_0 \cos 2\phi_0,$$

where, as in the general discussion, J is replaced by J_0 in the second term. Then

$$J_0 = \frac{\partial S}{\partial \phi_0} = J - \epsilon \frac{\rho'}{2\nu\rho} J \sin 2\phi_0,$$

or

$$J = J_0 \left(1 + \epsilon \frac{\rho'}{2\nu\rho} \sin 2\phi_0 \right). \tag{6.145}$$

This equation gives the first-order correction to the action. It shows the first-order departure of the adiabatically invariant J from strict invariance. The time dependence of J can be calculated if $\rho(\epsilon t)$ is a known function. Because the time derivative of ρ adds a factor of ϵ, the lowest order in dJ/dt is the second, so that to first order J is actually invariant. In other words, adiabatic invariance, at least for this problem, gives a pretty good approximation.

6.4.3 THE HANNAY ANGLE

For many years only the adiabatic behavior of the action J had been studied, but recently (Hannay, 1985), motivated by the quantum analog of the adiabatic theorem (Aharonov & Bohm, 1959; Berry, 1984 and 1985), interest has also turned to the angle ϕ, and we now discuss that. For the purposes of this discussion it is assumed that there are $N \geq 1$ time-dependent parameters, so the Hamiltonian is of the form $H(q, p, \lambda)$, where $\lambda(\epsilon t) = \{\lambda_1(\epsilon t), \lambda_2(\epsilon t), \ldots, \lambda_N(\epsilon t)\}$. As in Section 6.4.1, ϵ is a small parameter and the system is completely integrable for constant λ. Suppose further that the λ_k are all periodic in t (and hence in ϵt) with commensurate periods. Then there is a *Hamiltonian period T* such that $\lambda_k(\epsilon T) = \lambda_k(0)$ for all k and the Hamiltonian returns to its original form. If $N = 1$, we will assume the λ manifold $\mathbb{L}$ to be a circle, not a line, so that the path from $\lambda(0)$ to $\lambda(\epsilon T)$ is a noncontractible closed loop.

The transformation to AA variables for constant λ yields a time-dependent Hamiltonian

$$K(J, \phi, \lambda) = \nu(\lambda)J + \frac{\partial S(q, J, \lambda)}{\partial t} \equiv \nu(\lambda)J + K_1(J, \phi, \lambda). \tag{6.146}$$

Although we do not prove it, the argument that led to Eq. (6.143) can be generalized to λ replaced by $\boldsymbol{\lambda}$, so that for the present case J is also an adiabatic invariant. We now show how to calculate the change $\Delta\phi$ of ϕ in the Hamiltonian period T.

The total change $\Delta\phi$ in one period can be found by calculating $\dot{\phi}$ in two ways. First, use

$$\dot{\phi} = \{\phi, H\}^{q,p} + \frac{\partial \phi(q, p, \lambda)}{\partial t}.$$

The PB here is the rate of change of ϕ when it is not an explicit function of t (i.e., for a fixed value of λ). But that is known: it is the $\dot{\phi}$ obtained from the transformation to AA variables of the system in which H is time independent; therefore

$$\dot{\phi} = \nu + \frac{\partial \phi(q, p, \lambda)}{\partial t}.$$

Second, use Eq. (6.146) to get

$$\dot{\phi} = \frac{\partial K}{\partial J} = \nu + \frac{\partial K_1}{\partial J} \equiv \nu + \delta\dot{\phi}. \tag{6.147}$$

From these two expressions, we obtain

$$\delta\dot{\phi} = \frac{\partial\phi(q,p,\lambda)}{\partial t} = \frac{\partial\phi(q,p,\lambda)}{\partial\lambda_k}\dot{\lambda}_k \equiv \alpha^k\dot{\lambda}_k. \tag{6.148}$$

In a motion, λ varies, so ν changes in time, and hence the total change of ϕ in one period contains two parts:

$$\Delta\phi = \int_0^T \nu(\lambda)\,dt + \int_0^T \delta\dot{\phi}\,dt \equiv \delta\nu + \delta\phi. \tag{6.149}$$

The part due to ν is called the *dynamical* change, and the part due to $\delta\dot{\phi}$ the *geometric* change. The dynamical change can be calculated when $\lambda(\epsilon t)$ is known. The geometric change $\delta\phi$, called the *Hannay angle*, is more interesting, as we will explain shortly. The geometric change can also be calculated by simply integrating over T: find $\delta\dot{\phi}$ from (6.147) and calculate

$$\delta\phi = \int_0^T \delta\dot{\phi}\,dt.$$

This is how it will be calculated in treating the Foucault pendulum in the next worked example.

Equation (6.147) cannot be used to calculate $\delta\dot{\phi}$ if K_1 is not known. One can, however, calculate $\delta\dot{\phi}$ from (6.148) and write out the α_k as functions of (ϕ, J). But to use that to calculate $\delta\phi$ requires knowing how ϕ varies in time, or knowing the answer to the problem before solving it. As is often the case in perturbative calculations, the difficulty is avoided by averaging. Let

$$A^k(J,\lambda) \equiv \langle\alpha^k\rangle = \frac{1}{2\pi}\int_0^{2\pi}\alpha^k(\phi,J,\lambda)\,d\phi.$$

Then the average of $\delta\phi$ is

$$\langle\delta\phi\rangle = \int_0^T A^k\dot{\lambda}_k\,dt = \int_{\lambda(0)}^{\lambda(\epsilon T)} A^k\,d\lambda_k.$$

By assumption $\lambda(\epsilon T) = \lambda(0)$, so this is an integral over a closed path C in the λ manifold $\mathbb{L}$:

$$\langle\delta\phi\rangle = \oint_C A^k\,d\lambda_k. \tag{6.150}$$

The result is that $\langle\delta\phi\rangle$ does not depend on T but only on the path C in $\mathbb{L}$, as long as λ changes adiabatically (slowly). That is why it is called *geometric*. An expression such as $A^k\,d\lambda_k$ is a one-form, this one on $\mathbb{L}$, and the integration in Eq. (6.150) is an integral over a one-form, like the integral Poincaré invariants of the Liouville volume theorem. Although λ returns to its original value in time T, in general $\langle \delta\phi \rangle$ does not (it is then called *nonholonomic*). An n-form α is *exact* if there is an $(n-1)$-form β such that $\alpha = d\beta$. In particular, a one-form α is exact if $\alpha = df$ for some function f. If α is exact, $\oint \alpha = 0$; therefore $\langle \delta\phi \rangle$ is nonholonomic iff $A^k \, d\lambda_k$ is not exact. Berry (1988) points out that the Hannay angle is related to the concept of parallel transport on manifolds.

If $N = 3$, in particular if $\mathbb{L} = \mathbb{R}^3$, Eq. (6.150) can be written

$$\langle \delta\phi \rangle = \oint_C \mathbf{A} \cdot d\lambda,$$

and Stokes's theorem can be used to convert this to

$$\langle \delta\phi \rangle = \int_\Sigma \nabla \wedge \mathbf{A} \cdot d\sigma,$$

where Σ is a surface bounded by C and $d\sigma$ is the vector surface element on Σ. Writing $\nabla \wedge \mathbf{A} = \mathbf{B}$ makes this reminiscent of electrodynamics: $\langle \delta\phi \rangle$ looks like the magnetic flux through Σ. The integrand $\mathbf{B} \cdot d\sigma \equiv \nabla \wedge \mathbf{A} \cdot d\sigma$ is a two-form.

WORKED EXAMPLE 6.7

The point of suspension of a *Foucault pendulum* is attached to the rotating Earth, so its motion depends on, and has been used to measure, the Earth's rate of rotation. Although the Hamiltonian will be seen not to depend explicitly on t, the motion in the rotating noninertial frame yields a Hannay angle, and the Foucault pendulum is often used to illustrate it (Berry, 1985); the Earth's rotation provides a natural period. Treat the pendulum in the small-angle limit, i.e., as a harmonic oscillator in two freedoms (in the horizontal plane). See Fig. 6.16. If the Earth's angular speed is constant and equal to Ω_E, the horizontal plane is rotating at $\Omega = \Omega_E \sin\alpha$, where α is the latitude. It is this Ω, zero at the equator and equal to $\pm 2\pi$ day^{-1} at the poles, that enters the equations. (This is actually the z component, perpendicular to the plane, of the angular velocity of the plane. There is also a component parallel to the plane, but this plays no role in the motion we are describing.) **(a)** Find the Lagrangian and Hamiltonian of the system in polar coordinates in the rotating frame. **(b)** Transform to action–angle variables and write down the Hamiltonian in terms of AA variables. **(c)** Use Eqs. (6.147) and (6.149) to find the Hannay angles for the radial and angular motion of the small-angle Foucault pendulum. (Take the Earth's gravitational field to be uniform.)

Solution. (a) In polar coordinates (r, θ) in an inertial frame the Lagrangian is

$$L = \frac{m}{2}(\dot{r}^2 + r^2\dot{\theta}^2) - \frac{\omega_0^2 m}{2} r^2,$$

where ω_0 is the natural frequency of the oscillator. The transformation between (r, θ)

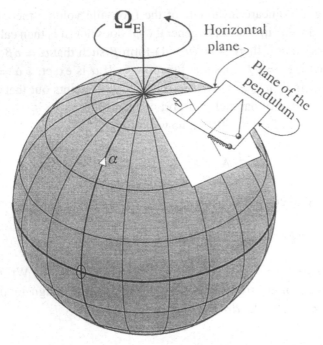

FIGURE 6.16
The Foucault pendulum and the equivalent harmonic oscillator, both in the same plane, which makes
an angle ϑ with respect to the North–South line.

and the rotating coordinates (r, ϑ) is

$$r = r,$$
$$\vartheta = \theta + \Omega t.$$

The Lagrangian is then

$$\left. \begin{aligned} L &= \frac{m}{2}[\dot{r}^2 + r^2(\dot{\vartheta} - \Omega)^2] - \frac{\omega_0^2 m}{2} r^2 \\ &= \frac{m}{2}[\dot{r}^2 + r^2\dot{\vartheta}^2] - m\Omega r^2\dot{\vartheta} - \frac{\omega^2 m}{2} r^2, \end{aligned} \right\} \tag{6.151}$$

where $\omega^2 = \omega_0^2 - \Omega^2$.

The period of the Earth's rotation is much longer than that of a typical pendulum,
so $\Omega \ll \omega_0$, and hence ω and ω_0 are equal for all practical purposes. Recall from
the discussion at Eq. (2.54) that in a rotating system the free-particle Lagrangian
picks up two terms: one, $2\Omega(x\dot{y} - \dot{x}y)$, mimics an equivalent uniform magnetic
field perpendicular to the plane of rotation, and another, $\Omega^2(x^2 + y^2)$, mimics an
inverted harmonic oscillator potential and provides the so-called centrifugal force.
In the second line of Eq. (6.151) the $m\Omega r^2\dot{\vartheta}$ term is the equivalent magnetic field,
and the weak centrifugal force is buried in the third term (that $\Omega \ll \omega_0$ means that
the inverted harmonic oscillator weakens the harmonic oscillator only slightly). In
the rotating coordinates the system looks like a slightly weaker harmonic oscillator
for a charged particle in a uniform magnetic field.

To obtain the Hamiltonian, first calculate the momenta:

$$p_r \equiv \frac{\partial L}{\partial \dot{r}} = m\dot{r},$$

$$p_\vartheta \equiv \frac{\partial L}{\partial \dot{\vartheta}} = mr^2(\dot{\vartheta} - \Omega).$$

In the rotating frame the angular momentum is not simply $mr^2\dot{\vartheta}$, as it usually is in inertial polar coordinates. The canonical procedure then gives the Hamiltonian

$$H = \frac{p_r^2}{2m} + \frac{p_\vartheta^2}{2mr^2} + \Omega p_\vartheta + \frac{m\omega_0^2}{2} r^2. \tag{6.152}$$

Note that ω_0^2 reappears in this equation: the rotation of the frame is entirely contained in the Ωp_ϑ term. Since ϑ is an ignorable coordinate, $p_\vartheta \equiv j$ is a constant of the motion.

For completeness we include the expression for the Hamiltonian in rotating Cartesian coordinates (we leave out the definitions of the momenta p_x and p_y):

$$H = \frac{1}{2m}\{(p_x - m\Omega y)^2 + (p_y + m\Omega x)^2\} + \frac{\omega^2 m}{2}(x^2 + y^2).$$

Compare this with Eq. (6.24).

(b) To find the AA variables, first calculate J_ϑ and J_r:

$$J_\vartheta \equiv \frac{1}{2\pi} \oint p_\vartheta \, d\vartheta = p_\vartheta \equiv j.$$

To find J_r, first solve (6.152) for p_r:

$$p_r = \frac{1}{r}\sqrt{-m^2\omega_0^2 r^4 + 2mr^2(E - \Omega j) - j^2},$$

where we have written $E = H$. Keep in mind that there are two solutions here, the positive and negative square roots. Then

$$J_r \equiv \frac{1}{2\pi} \oint p_r \, dr = \frac{1}{2\pi} \oint \sqrt{-m^2\omega_0^2 r^4 + 2mr^2(E - \Omega j) - j^2} \, \frac{dr}{r},$$

where the integral is taken around the roots of the radical. It is convenient to perform this integration in the complex plane. We outline the calculation. Write $a = m^2\omega_0^2$, $b = 2m(E - \Omega j)$, $c = j^2$ (all positive). Then the integral becomes

$$\oint \sqrt{-ar^2 + b - \frac{c}{r^2}} \, dr$$

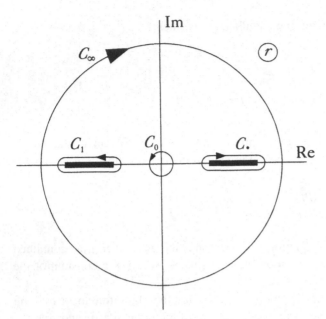

FIGURE 6.17
The complex r plane for calculating J_r for the Foucault pendulum.

between the roots of the radical. The squares of the roots are

$$r_\pm^2 = \frac{1}{2a}\{b \pm \sqrt{b^2 - 4ac}\}, \tag{6.153}$$

which are both real and positive (prove this!). Thus there are four real roots, two positive and two negative. Figure 6.17 shows the complex r plane with two cuts (dark lines) on the real axis, one between the positive roots and one between the negative ones. The integration takes place around the positive roots, along the contour labeled $C_\bullet$. For the integration, $C_\bullet$ is distorted to the combination of C_0 around the origin, C_∞ at infinity, and C_1 around the other cut. The integral around C_1 is the negative of that around $C_\bullet$, so

$$\oint_{C_\bullet} = \oint_{C_1} + \oint_{C_0} + \oint_{C_\infty} = -\oint_{C_\bullet} + \oint_{C_0} + \oint_{C_\infty} = \frac{1}{2}\left[\oint_{C_0} + \oint_{C_\infty}\right].$$

To perform the integrations around C_0 and C_∞, use the Cauchy integral theorem. The residue around C_0 is $\sqrt{-c}$. For the integral around C_∞ write $u = 1/r$. The integral then takes place around the origin in the u plane and is

$$\sqrt{-a} \oint \sqrt{1 - \frac{b}{a}u^2 + \frac{c}{a}u^4}\frac{du}{u^3} \approx \sqrt{-a} \oint \left[1 - \frac{b}{2a}u^2 + \cdots\right]\frac{du}{u^3},$$

which leaves a residue of $-\frac{1}{2}b(-a)^{1/2}$. From the definitions of a, b, and c, the result is then

$$J_r = \frac{E - \Omega j}{2\omega_0} - \left|\frac{j}{2}\right|.$$

To write down the Hamiltonian in terms of the AA variables, one need only solve this equation for $E \equiv H$:

$$H = \omega_0(2J_r + |j|) - \Omega j.$$

(c) To find the Hannay angles using Eqs. (6.147) and (6.149), first find the frequency for the r motion:

$$\dot{\phi}_r \equiv \frac{\partial H}{\partial J_r} = 2\omega_0,$$

which is the same as in the nonrotating frame (remember that $r = \sqrt{x^2 + y^2}$ is always positive, so its frequency is twice that of x or y; see Problem 29). Therefore, according to (6.147) $\delta\dot{\phi}_r = 0$, and then (6.149) implies that the Hannay angle for the r motion is zero. Next, find the frequency for the ϑ motion:

$$\dot{\phi}_\vartheta \equiv \frac{\partial H}{\partial j} = \pm\omega_0 + \Omega$$

(the sign of the first term depends on whether j is positive or negative). Therefore, according to (6.147) $\delta\dot{\phi}_\vartheta = +\Omega$. To use (6.149), note that T is one day. Thus

$$\delta\phi_\vartheta = \int_0^T \Omega \, dt = \Omega T = \Omega_E T \sin\alpha.$$

Since $\Omega_E = 2\pi$ day^{-1}, this means that $\delta\phi_\vartheta = 2\pi \sin\alpha$. Hence after one day the angle at which the pendulum is swinging will have changed by an amount that depends on the latitude. This is the Hannay angle for the ϑ motion.

6.4.4 MOTION OF A CHARGED PARTICLE IN A MAGNETIC FIELD

THE ACTION INTEGRAL

The motion of a charged particle in a variable magnetic field can be very complicated, even chaotic, and more often than not can be studied only by numerical methods. This is true, for instance, of the general Störmer problem (Section 4.1.4) (De Almeida et al., 1992). However, if the magnetic field changes slowly in space or time, useful results can be obtained by adiabatic approximation. It has been fruitful in explaining the long-term trapping of charged particles in the Earth's magnetosphere (Alfvén, 1950) and has been applied to magnetic confinement of hot plasmas to bounded space regions, especially in attempts to control thermonuclear reactions.

Before getting to varying fields, a reminder about the motion of a charged particle in a constant and uniform magnetic field $\mathbf{B}$ (Section 2.2.4). The particle's velocity can be decomposed into components parallel and perpendicular to the field lines: $\mathbf{v} = \mathbf{v}_{\parallel} + \mathbf{v}_{\perp}$. Then $\mathbf{v}_{\parallel}$ is constant, and the orbit projects as a circle onto the plane perpendicular to $\mathbf{B}$ [taken along the 3 direction in Chapter 2]. The *radius of gyration* ρ of the circle is given by

$$\rho = \frac{m v_{\perp}}{eB}, \tag{6.154}$$

where m is the mass of the particle and e its charge. As was shown in Section 6.1.2, the particle traverses its circular orbit at the cyclotron frequency $\omega_C = eB/m$. The circle is called a *cyclotron orbit*.

Now consider a varying field, but with the adiabatic assumption: the magnetic field $\mathbf{B}$ felt by the particle changes slowly compared to one cyclotron period $\tau_C = 2\pi/\omega_C$. This is in keeping with the criterion established around Fig. 6.15: the $\mathcal{C}$ curve in $\mathbf{T}^*\mathbb{Q}$ distorts slowly compared with the rate at which the system goes around it. If the change of $\mathbf{B}$ is small in τ_C, it is also small in one radius of gyration ρ. What this means is that the particle sees an approximately constant magnetic field in each gyration and moves in an approximate cyclotron orbit.

The nonuniformity of $\mathbf{B}$ causes the orbit to accelerate, however, along the magnetic field lines. This can be explained by considering the Lorentz force

$$\mathbf{F} = e\mathbf{v} \wedge \mathbf{B}.$$

If $\mathbf{B}$ is uniform and in the 3 direction, the Lorentz force is in the $(1, 2)$ plane and hence causes no acceleration along the field. But suppose that $\mathbf{B}$ is not uniform. Maxwell's equations imply that a nonuniform static B cannot be unidirectional. Figure 6.18 illustrates the static situation when $\mathbf{B}$ differs slightly from uniformity and B is approximately constant on surfaces perpendicular to $\mathbf{B}$. The Lorentz force then has a component along the average field, and the particle accelerates in that direction. More generally, when B is not constant

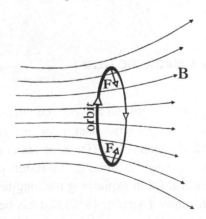

FIGURE 6.18
The approximately circular orbit of a particle in a nonuniform magnetic field $\mathbf{B}$. The Lorentz force $\mathbf{F}$, shown at two positions on the orbit, has a component along the average field. This orbit is associated with J_1.

on surfaces perpendicular to **B** or when there are other external forces acting on the particle, the orbit will also drift *across* field lines. We do not discuss external forces at this point.

The way nonuniform *B* causes drift across field lines (this has already been seen in the Störmer problem) is as follows: Even though the speed of the particle remains constant (the magnetic force is always perpendicular to the velocity), at points of stronger *B* the Lorentz force is stronger than at others. Then the net force on the approximate cyclotron orbit becomes nonzero in the plane of the orbit, causing the entire orbit to drift across lines of force. Another way to say this is that the radius of curvature is different at different points of the orbit.

Now, $\mathbf{v}_\perp$ has contributions from both the gyration and the drift. We make the assumption (this will be justified later) that the contribution from the drift is much smaller than that from gyration, and then Eq. (6.154) still applies, as do the definitions of ω_C and τ_C. Then the particle moves on an approximate cyclotron orbit, and its position is described by the location of the center of the orbit plus the angular location of the charge on it. The condition for applicability of the adiabatic theorem is that changes in the relevant physical quantities be small in the time τ_C or within one radius of gyration. When this condition is satisfied, the motion on the cyclotron orbit can be thought of as the reduced system of Section 6.4.2 (such a treatment is sometimes called the *guiding center approximation*). The rest of the motion is of two types: motion of the orbit along the magnetic field lines and drift across them. Each of the resulting three types of motion (on the orbit, along the field lines, and across them) is associated with its own action integral, and under certain conditions these can become adiabatic invariants. Thus we will discuss three adiabatic invariants.

For cyclic motion in general the action integral J is

$$J = \frac{1}{2\pi} \oint_C p \, dq,$$

where C is a closed curve in $\mathbf{T}^*\mathbb{Q}$. For the purposes of this section we write this as

$$J = \frac{1}{2\pi} \oint_C \mathbf{p} \cdot d\mathbf{l}, \tag{6.155}$$

where C is the orbit in $\mathbb{R}^3$ and $d\mathbf{l}$ is the element of length around the orbit. For example, if the particle moves in a circle about a fixed center, as in cyclotron motion, dq is $d\phi$ and p is the angular momentum $mv\rho$, so

$$J = \frac{1}{2\pi} \oint_C p \, dq = \frac{1}{2\pi} \oint_C mv\rho \, d\phi = \frac{1}{2\pi} \oint_C mv \, dl = \frac{1}{2\pi} \oint_C \mathbf{p} \cdot d\mathbf{l},$$

for $\rho \, d\phi$ is the element of arc length along the circle. The momentum $\mathbf{p}$ must be written as a function of position on the orbit.

In application to a particle of charge e in a magnetic field (6.155) becomes

$$J = \frac{1}{2\pi} \oint_C [m\mathbf{v} + e\mathbf{A}] \cdot d\mathbf{l} \equiv \frac{1}{2\pi} \oint_C m\mathbf{v} \cdot d\mathbf{l} + \frac{1}{2\pi} \oint_C e\mathbf{A} \cdot d\mathbf{l},$$

where $\mathbf{A}$ is the vector potential: $\mathbf{B} = \nabla \wedge \mathbf{A}$ [use the canonical momentum of Eq. (5.6)]. According to Stokes's theorem, this can be written

$$J = \frac{1}{2\pi} \oint_C m\mathbf{v} \cdot d\mathbf{l} + \frac{1}{2\pi} \int_\Sigma e\mathbf{B} \cdot d\boldsymbol{\sigma}, \qquad (6.156)$$

where Σ is a surface of which C is the boundary and $d\boldsymbol{\sigma}$ is the vector element of area on that surface. We will apply Eq. (6.156) to all three types of motion mentioned above.

THREE MAGNETIC ADIABATIC INVARIANTS

First consider the action integral J_1 for gyration about the magnetic field lines, illustrated in Fig. 6.18. The first integral on the right-hand side of Eq. (6.156) is just $mv_\perp$ times the circumference $2\pi\rho$ of the orbit, or $(mv_\perp)^2/eB$. The second integral is eB times the area, $\frac{1}{2}\rho^2 eB \equiv (mv_\perp)^2/2eB$. According to the right-hand rule these contributions have opposite signs: if $\mathbf{B}$ is out of the paper, $\mathbf{v}_\perp$ must rotate counterclockwise for a centripetal force, while $\mathbf{A}$ goes clockwise. Thus

$$J_1 = \frac{(mv_\perp)^2}{2eB} = \frac{p_{d\perp}^2}{2eB},$$

where we write p_d for the dynamical momentum mv. If B changes slowly enough, J_1 is an adiabatic invariant and $p_{d\perp}^2$ is proportional to B. If in addition $p_{d\parallel} \equiv mv_\parallel$ is much smaller than $p_{d\perp}$, essentially the entire kinetic energy is $p_{d\perp}^2/2m$ and hence the kinetic energy is proportional to B. The increase in energy (actually the increase in speed) with increasing B is known as *betatron acceleration*. The quantity

$$M = \frac{eJ_1}{m} \equiv \frac{p_{d\perp}^2}{2mB}, \qquad (6.157)$$

called the *first adiabatic invariant* (Alfvén, 1950), is the magnetic moment of the cyclotron orbit. Because J_1 is an adiabatic invariant, so is M. This is important in plasma physics, in attempting to confine charged particles to bounded regions.

> **REMARK:** Strictly speaking, this M is an adiabatic invariant only in the nonrelativistic limit, when m can be treated as a constant. ☐

Next, consider a nonuniform field in which the cyclotron orbit can accelerate along the magnetic lines of force. This type of motion occurs, for example, in the Störmer problem, and more generally, as pointed out above, whenever the magnetic field is nonuniform. If the motion along lines of force also oscillates (and we show below that it does), there is a second action integral, called J_2. Since the contour of integration in this case is along the magnetic field lines, $\oint_\Sigma \mathbf{B} \cdot d\boldsymbol{\sigma} = 0$ and the second term of (6.156) vanishes. The component of the velocity $\mathbf{v}$ along $d\mathbf{l}$ is parallel to $\mathbf{B}$ ($d\mathbf{l}$ is along the field lines), so that

$$J_2 = \frac{1}{2\pi} \oint mv_\parallel \, dl \equiv \frac{1}{2\pi} \oint p_{d\parallel} \, dl. \qquad (6.158)$$

The $p_{d\parallel}$ in the integrand can be written in terms of the energy E of the particle and the field strength B. The energy is

$$E = \frac{1}{2m}\left(p_{d\perp}^2 + p_{d\parallel}^2\right),$$

and then from (6.157),

$$p_{d\parallel} = \sqrt{2mE - p_{d\perp}^2} = \sqrt{2m\,[E - MB]}. \tag{6.159}$$

Assume that the magnetic field changes adiabatically as the orbit moves, so that M is an adiabatic invariant. Then $p_{d\parallel}$ goes to zero when $B \equiv B_{max} = E/M$, and that gives rise to a turning point: the cyclotron orbit will start moving in the opposite direction. At first B will decrease, but eventually the field lines must converge again and B will increase. The motion will continue until $p_{d\parallel} = 0$ again, when B reaches B_{max}. (Such turning points are called *magnetic field mirrors* in plasma physics.) Thus, as asserted, the orbit oscillates back along the lines of force, and the particle completes closed orbits in $\mathbf{T}^*\mathbb{Q}$, analogous to motion in an oscillator potential. If B changes slowly enough (in the analogous oscillator, if the potential changes slowly enough), J_2 is also an adiabatic invariant. The particle experiences changes in B because the field is time dependent and/or because it is nonuniform. Nonuniformity will also cause drift across field lines.

To see this in more detail, consider for example a dipole field (i.e., the Störmer problem). A complete treatment requires numerical calculations (Dragt, 1965; see also the account in Lichtenberg, 1969), but a qualitative description of the physics is not too involved. A magnetic dipole field is illustrated in Fig. 6.19. If B did not vary transversely (i.e., in the plane perpendicular to the field lines), the guiding center about which the particle performs cyclotron motion would move down one field line, say S_0, to the point where $B = B_{max}$ on that line. Transversal nonuniformity causes the guiding center to drift to another line

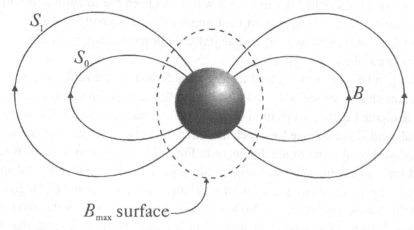

FIGURE 6.19

Some field lines of a magnetic dipole field. Only field lines in the plane of the paper are shown. The dotted curve indicates the intersection of the $B = B_{max}$ surface with the plane of the paper. The sphere represents the Earth, which is the source of the dipole field in the Störmer problem.

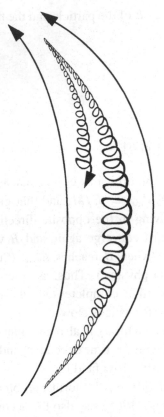

FIGURE 6.20
A particle spiraling along a magnetic field line to $B = B_{max}$ and starting back along a neighboring line (magnetic field mirror). This orbit is associated with J_2.

such as S_1. But B_{max} depends only on E, which remains constant, and on M, which is an adiabatic invariant, so for adiabatic motion B_{max} is a constant. While drifting across the field lines, the cyclotron orbit also oscillates along them, bouncing off the field mirrors, more or less as illustrated in Fig. 6.20. If B changes slowly compared to the rate of cycling around lines of force, J_2 is an adiabatic invariant for this motion. For this to be true, the orbit must move across field lines in such a way as to keep the integral constant in Eq. (6.158). Because $B(S_1) > B(S_0)$ Eq. (6.159) implies that $v(S_1) > v(S_0)$. This is what compensates for the extra length of S_1 in the integral and allows the integrals along S_1 and S_0 to be equal between the two turning points.

While drifting in latitude between field lines such as S_0 and S_1, the orbit can also drift in longitude (azimuthally) around the dipole, as in the Störmer problem. For one such cycle around the dipole Eq. (6.156) yields a third action integral J_3, in which C is a circuit around the dipole and Σ is a surface for which that circuit is the border. If, for example, C is a circle in the equatorial plane of the dipole, as in Fig. 6.21, the second term of (6.156) is the flux of **B** through a hemisphere capping C. Suppose the radius of C is R, and write $\omega_D = 2\pi v_\perp / R$ for the (circular) frequency with which the orbit goes around C. In general $\omega_D \ll \omega_C$, and this can be used (Problem 30) to estimate the relative sizes of the two terms of (6.156). It turns out that the second integral is much greater than the first, so that to a good approximation J_3 is the total flux enclosed by C. When the field changes so slowly that J_3 is an adiabatic invariant, the total flux of **B** through the drift orbit therefore remains constant.

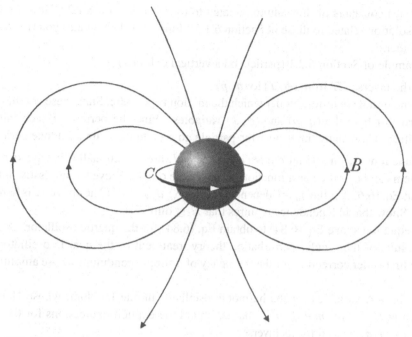

FIGURE 6.21
A circular orbit C in the equatorial plane of the dipole. This orbit is associated with J_3.

We conclude by outlining the picture that emerges from this section. In a magnetic field (with the proper symmetry) that varies slowly enough, there are three adiabatic invariants associated with three distinct periodic motions of the charge. The fastest of these is cyclotron motion about the magnetic field lines. Next is periodic longitudinal motion along the field lines. The slowest is drift around a trajectory that encloses part of the magnetic field. The actual calculation of the integrals can be quite difficult, usually requiring numerical methods. We do not go into that, but see Dragt (1965). In addition, we do not discuss the motion that arises when the magnetic field varies in time so rapidly that the induced electric field is large enough to have to be taken into account.

PROBLEMS

1. **(a)** Show that both S_a and S_b of Eqs. (6.3a) and (6.3b) satisfy the HJ equation for the Hamiltonian $H = \frac{1}{2}p^2$. For each of the S_k functions, solve the problem completely. That is, find $q(t)$ and $p(t)$ and identify the constants P_k, $k = a, b$. (You should convince yourself that the two solutions are really different, that there is no way to choose Q_a in S_a so as to turn S_a into S_b.)

   (b) Obtain S_a by reversing this procedure. That is, take Q_a to be the initial position and P_a to be the constant of the motion found in Part (a), write the CT from (q, p) to (Q_a, P_a), and solve Eqs. (6.2) for S_a. (S_b is the standard solution obtained by separation of variables.)

2. The HJ equation for the Hamiltonian of Eq. (6.24) does not separate in the usual way. Try writing $W = \frac{1}{2}bxy + X(x) + Y(y)$ to separate it and then solve the HJ equation. How

are the (P, Q) constants of this solution related to those of Section 6.1.2? How are the p_α of this solution related to those of Section 6.1.2? Find the CT that takes you from one set to the other.

3. For the example of Section 6.2.1 (particle on a vertical cylinder):
   (a) Find the inverse CT from (ϕ, J) to (q, p).
   (b) Find the initial conditions under which the motion is periodic. State these conditions in terms of both the (q, p) and (ϕ, J) variables. Find the period for each initial condition. How many times does the particle go around the cylinder in one period?

4. A set of functions f_α on a manifold $\mathbb{M}$, $\alpha = 1, \ldots, N$, are (functionally) independent if none of them can be written as a function of any of the others. Prove the necessity of the assertion at Eq. (6.62): if the f_α are dependent, then $\bigwedge_\alpha df_\alpha = 0$. (The converse is harder to prove.) Show that independence requires that $N \le \dim \mathbb{M}$.

5. Solve the equation above Eq. (6.84) to obtain Eq. (6.84) for the quartic oscillator.

6. Use the results of the secular perturbation theory treatment of the quartic oscillator to obtain the first-order correction for the frequency of a simple pendulum whose amplitude is Θ.

7. (a) Find the AA variables for the harmonic oscillator in one freedom, whose Hamiltonian is $H = \frac{1}{2}p^2/m + \frac{1}{2}kq^2$. That is, obtain the explicit expressions for the CT $\phi(q, p)$, $J(q, p)$, and for its inverse.
   (b) Show that the Type 1 generator for the CT is $F^1 = \frac{1}{2}m\omega q^2 \cot\phi$.

8. Obtain expressions in terms of elliptic integrals for the action variable of a simple pendulum. [For the angle variable see Lichtenberg and Lieberman (1992, Section 1.3).]

9. In Worked Example 6.5 solve the quadratic equation for J in terms of (ϕ_0, J_0) and show that to first order in ϵ it is equivalent simply to replacing J^2 in the quadratic equation by J_0^2.

10. Prove the assertion at Eq. (6.134): suppose that $\mathcal{V}(\epsilon)$ is an operator that, like $\mathcal{U}^{-1}(\epsilon)$, satisfies the equation $\partial\mathcal{V}/\partial\epsilon = \mathcal{P}_G\mathcal{V}$, where $\mathcal{P}_G$ is defined in Eq. (6.124). Show that the function

$$F(\epsilon) = \mathcal{V}(\epsilon) \int_{\epsilon_0}^{\epsilon} \mathcal{V}^{-1}(\epsilon')g(\epsilon')\,d\epsilon',$$

satisfies the equation $\partial F/\partial\epsilon - \mathcal{P}_G F = g$. [Hint: Use the Leibnitz rule for derivatives of integrals and remember that $\mathcal{V}$ is an operator; thus $\mathcal{V}(\epsilon)g(\epsilon)$ is a function of ϵ.]

11. Obtain the results of Eq. (6.131) as well as an expression for $\mathcal{V}_4$ in terms of the $\mathcal{P}_k$.

12. For the quartic oscillator in the Lie transform method, use the expression for G_1 to find the first-order transformation between (ϕ, J) and (ϕ_0, J_0). [Hint: recall that G is the infinitesimal generator of the ϵ transformation.]

13. For the quartic oscillator in the Lie transform method, find E_2 and the second-order approximation to the perturbed frequency ν. Obtain an explicit expression for ν for motion starting at $q = a$ and $p = 0$.

14. In Worked Example 6.3 go to the limit $c \to 0$ and $\alpha_2 \to -\alpha_1 \equiv \alpha$ so that $\alpha c = \mu/2 < \infty$. In this limit the system becomes that of a charged particle in the field of a point dipole in two freedoms. Obtain an equation for S.

15. Consider the *three*-freedom problem of a charged particle in the field of a dipole (see Problem 14) by using polar coordinates to separate the HJ equation. Obtain an expression for S in terms of integrals. (Have the dipole point along one of the axes.)

16. In Worked Example 6.3 go to the limit $c \to \infty$ and $\alpha_2 \to \infty$ so that $\alpha_2/4c^2 = l < \infty$. In this limit the system becomes that of a particle in a Coulomb field subjected to an additional constant electric field (the two-freedom Stark problem). Obtain an equation for S.

17. Solve the *three*-freedom Stark problem (see Problem 16) directly by using parabolic polar coordinates to separate the HJ equation. Obtain an integral expression for S. [Parabolic polar coordinates (ξ, η, ϕ) are related to Cartesian coordinates (x, y, z) by $x = \xi \eta \cos \phi$, $y = \xi \eta \sin \phi$, and $z = \frac{1}{2}(\xi^2 - \eta^2)$. Take the constant electric field pointing in the z direction.]

18. Show that H, j_z, and $|\mathbf{j}|^2$ are independent constants of the motion for the three-freedom Kepler problem.

19. In the geometric explanation (Section 6.1.3) of the origin of the Q^α no account was taken of the necessity for the PB relation $\{Q^\mu, Q^\nu\} = 0$. Prove that if the Hessian condition $|\partial^2 W_Q/\partial q^\alpha \partial Q^\beta| \neq 0$ is satisfied, this PB relation follows from Eq. (6.31). [Answer: Write $\partial W_Q/\partial q^\alpha = F_\alpha(q, Q)$. Equation (6.31) states that $p_\alpha = F_\alpha(q, Q)$. The Hessian condition $|\partial F_\alpha/\partial Q^\beta| \neq 0$ implies invertibility: $Q^\mu = G^\mu(q, p)$. Note that $G^\mu(q, F(q, Q)) \equiv Q^\mu$ and $F_\alpha(q, G(q, p)) \equiv p_\alpha$. Then

$$\frac{\partial F_\alpha}{\partial Q^\nu} \frac{\partial G^\nu}{\partial p_\beta} = \delta_\alpha^\beta, \quad \frac{\partial F_\alpha}{\partial Q^\nu} \frac{\partial G^\nu}{\partial q^\mu} = -\frac{\partial F_\alpha}{\partial q^\mu}, \quad \frac{\partial G^\mu}{\partial p_\alpha} \frac{\partial F_\alpha}{\partial q^\nu} = -\frac{\partial G^\mu}{\partial q^\nu}.$$

Calculating, we get

$$\frac{\partial F_\rho}{\partial Q^\nu} \{Q^\nu, Q^\mu\} = \frac{\partial F_\rho}{\partial Q^\nu} \left\{ \frac{\partial G^\nu}{\partial q^\alpha} \frac{\partial G^\mu}{\partial p_\alpha} - \frac{\partial G^\nu}{\partial p_\alpha} \frac{\partial G^\mu}{\partial q^\alpha} \right\}$$

$$= -\left\{ \frac{\partial G^\mu}{\partial q^\rho} + \frac{\partial G^\mu}{\partial p_\alpha} \frac{\partial F_\rho}{\partial q^\alpha} \right\} = 0.$$

We used $\partial F_\rho/\partial q^\alpha = \partial F_\alpha/\partial q^\rho$. The required result follows from the Hessian condition.]

20. A particle of mass m is constrained to move on the x axis subject to the potential $V = a \sec^2(x/l)$.

    (a) Solve the HJ equation, obtaining an integral expression for S. Find $x(t)$ from the solution of the HJ equation.

    (b) Find the AA variables (ϕ, J) and the frequency ν. Obtain the amplitude dependence of ν. Check that the small-amplitude limit of ν is the same as that given by the theory of small vibrations (Section 4.2.1).

21. Find the AA variables for a free particle in two freedoms constrained to move inside a circular region of radius a (i.e., subjected to the potential $V(r) = 0$, $r \leq a$, and $V(r) = \infty$, $r > a$). The particle makes elastic collisions (the angle of incidence equals the angle of reflection) with the circular boundary.

22. An otherwise free particle is constrained inside an ellipsoid of revolution.

    (a) Show that the (nonholonomic) constraint conditions can be stated very simply in terms of suitably chosen prolate spheroidal coordinates. [Prolate spheroidal coordinates (ξ, η, ϕ) are an extension to three dimensions of confocal elliptical coordinates of Worked Example 6.3. They are related to Cartesian (x, y, z) coordinates by $x = c \sinh \xi \sin \eta \cos \phi$, $y = c \sinh \xi \sin \eta \sin \phi$, and $z = c \cosh \xi \cos \eta$.]

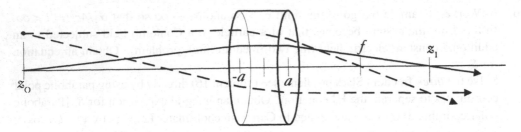

FIGURE 6.22
The solid line is an orbit starting on the axis. The dashed lines are orbits starting from a point off the axis; they show how an image is formed.

(b) Show that the HJ equation for the free particle is separable in prolate spheroidal coordinates. Obtain integral expressions for Hamilton's characteristic function.

(c) The particle makes elastic collisions (energy is conserved and angle of incidence equals angle of reflection) with the walls of the ellipsoid. Write down the boundary conditions on the momenta canonical to the ellipsoidal coordinates.

23. Solve the HJ equation for the Hamiltonian

$$H(q, p, t) = f(t)\left(\frac{p^2}{2m} + \frac{kq^2}{2}\right),$$

where m and k are constants and $f(t)$ is an integrable function. Find $q(t)$ and $p(t)$ and the phase-space trajectory. Find the kinetic energy as a function of the time for the three special cases (a) $f(t) = e^{\alpha t}$, (b) $f(t) = e^{-\alpha t}$, and (c) $f(t) = \cos \Omega t$, where $\alpha > 0$ and Ω are constants. Describe the motion for these three cases.

24. A magnetic lens for a beam of charged particles is formed by an appropriately designed magnetic field $\mathbf{B} = \nabla \times \mathbf{A}$ restricted to a finite region. The vector potential $\mathbf{A}$ is given in cylindrical polar coordinates by $\mathbf{A} \equiv (A_r, A_\theta, A_z) = (0, \frac{1}{2}rh(z), 0)$, where $h(z)$ is nonzero in a region $r < R$ and $-a < z < a$ (Fig. 6.22). Consider an electron that starts out from an *object point* near the axis with z coordinate $z_0 \ll -a$ (see Fig. 6.22), passes through the magnetic lens, and issues from the other side, as shown in the figure.

(a) Show that t and θ can be separated in this equation but that r and z cannot, that is, that Hamilton's principal function can be written in the form $W(r, \theta, z, t) = -Et + p_\theta\theta + s(r, z)$.

(b) If an electron is to pass through the lens, it must start out moving close to the z direction and cross the z axis. Obtain an approximate expression for $s(r, z)$ for such an electron by expanding s in powers of r up to second order. Show that in this approximation the setup behaves like a lens for monoenergetic electrons, that is, that there exists an *image point* whose z coordinate z_1 satisfies the *lens equation* $1/z_0 + 1/z_1 = 1/f$, where f is proportional to $\int_{-a}^{a} h(z)^2 \, dz$. Find the constant of proportionality. [Hints: Do this for the solid-line orbit in Fig. 6.22. The function s must be continuous at $r = 0$ (so it has no linear term in r) and at $z = \pm a$. Assume that s does not depend strongly on z in the region $|z| \leq a$.]

25. An object is bouncing vertically and perfectly elastically in an accelerating elevator. If the time dependence of the acceleration $a(t)$ is slow enough to satisfy the adiabatic assumption, find the maximum heights $h_{max}(t)$ that the object reaches on its bounces [$h_{max}(t)$ is measured relative to the floor of the elevator].

26. **(a)** In kinetic theory the model of a gas is sometimes taken as a collection of hard spheres confined in a box. Consider a single free particle of mass m confined to a finite interval of the x axis (a one-dimensional box), moving rapidly and bouncing back and forth between the ends of the interval (the walls). Suppose the walls are now slowly brought together. Use the adiabatic theorem to find out how the *pressure* varies as a function of the length of the box. How does the *temperature* (energy) vary? In an adiabatic process PV^γ is constant. From your results, calculate γ for such a *one-dimensional* gas. Compare the result from the usual kinetic theory.

   (b) Now suppose the length changes suddenly. Which thermodynamic variables (P, V, T) remain constant? Explain.

27. Return to Problem 21.

   (a) Show that if the radius a of the circle changes adiabatically, the energy goes as $1/a^2$.

   (b) Find the angle Φ that the particle's velocity makes with the wall (as a function of E and the Js). Show that Φ is an adiabatic invariant.

   (c) Find the average force the particle applies to the wall.

   (d) Assume that the circle is filled with a gas of particles all at the same energy moving with equal probability at all angles with respect to the wall. Find how the pressure the gas exerts on the wall depends on a when a changes adiabatically. Find γ for this system.

28. Consider the one-freedom Hamiltonian

$$H = \frac{p^2}{2m} + G \tan^2 \mu x,$$

where μ is a function of the time t. Assuming that $\mu(t)$ is given and varies slowly enough to satisfy the adiabatic hypothesis, find how the energy and frequency of oscillation vary with time. [Hint: The integral $I = \int_{-\rho}^{\rho} \sqrt{a^2 - \tan^2 \xi} \, d\xi$, where ρ is the root of the radical, can be performed in the complex plane. Show that $\oint_C \sqrt{a^2 - z^2}[1 + z^2]^{-1} dz = -2\pi(1 + \sqrt{a^2 + 1})$, where C is a contour that goes once around a cut on the real axis from $z = -a$ to $z = a$.]

29. Write the general solution of the two-dimensional isotropic oscillator in polar coordinates (to observe that r oscillates at twice the frequency of x or y).

30. Estimate the sizes of the two integrals of Eq. (6.156) and show that $\omega_D \ll \omega_C$ implies that the second integral dominates.

CHAPTER 7

NONLINEAR DYNAMICS

CHAPTER OVERVIEW

This chapter is devoted almost entirely to nonlinear dynamical systems, whose equations of motion involve nonlinear functions of positions and velocities. We concentrate on some general topics, like stability of solutions, behavior near fixed points, and the extent to which perturbative methods converge. Because nonlinear systems involve complicated calculations, the chapter stresses mathematical detail. It also presents results of some numerical calculations. The importance of numerical methods can not be overemphasized: because nonlinear systems are inherently more complicated than linear ones, it is often impossible to handle them by purely analytic methods. Some results depend also on topics from number theory, and these are discussed in an appendix at the end of the chapter.

We have already dealt with nonlinear systems perturbatively (e.g., the quartic oscillator of Section 6.3.1), but now we will go into more detail. It will be shown, among other things, that in the nonperturbative regime nonlinear systems often exhibit the kind of complicated behavior that is called chaos.

The first four sections of the chapter do not use the Lagrangian or Hamiltonian description to deal with dynamical systems. In them we discuss various kinds of systems, largely oscillators, which are often dissipative and/or driven. We go into detail concerning some matters that have been touched on earlier (e.g., stability, the Poincaré map), introducing ideas and terminology that are important in understanding the behavior of nonlinear systems of all kinds. In particular, considerable space is devoted to discrete maps. (Section 7.3 differs from the rest of the chapter in that it concerns *parametric oscillators*, which are not, strictly speaking, nonlinear.)

Section 7.5 deals with Hamiltonian dynamics. In it we use some of the material from the first four sections and, of course, from previous chapters, to discuss Hamiltonian chaos. In Section 7.5.4, one of the most important parts of the book, we take a second look at perturbation theory through the KAM theorem. A reader interested primarily in the behavior of Hamiltonian systems, or one familiar with nonlinear oscillators, may want to start with Section 7.5, referring to the earlier sections only when necessary.

7.1 NONLINEAR OSCILLATORS

This section is devoted to nonlinear oscillating systems. Nonlinear oscillators are often treated by approximations that start with linear ones, but when the nonlinearities are large the motion often differs drastically from any conceivable linear approximation. Analytic approximation techniques then fail entirely and different methods are needed. The computer then becomes essential, and computer results often provide important insights.

In any case, the physics of nonlinear oscillators is still far from being completely understood. Nevertheless, it has been studied analytically with fruitful results, and we start our discussion with some analytic approximation methods for nonlinear oscillators. We will be interested largely in the *periodic* motion of such oscillators, that is, in periodic solutions of their equations, assuming that they exist.

The typical equation of the kind we concentrate on is

$$\ddot{x} = g(x, \dot{x}, t),$$

where g is a function with some period T: that is, $g(x, \dot{x}, t + T) = g(x, \dot{x}, t)$. There are no general theorems that guarantee that such equations have periodic solutions, and the methods for finding them tend to be somewhat *ad hoc*. One important difference from the linear case is that the response and driver may have different frequencies, even incommensurate ones. A simple example (Forbat, 1966) is

$$\ddot{x} = -x + (x^2 + \dot{x}^2 - 1) \sin(\sqrt{2}t). \tag{7.1}$$

The driving frequency here is $\Omega = \sqrt{2}$. A particular periodic solution is $x(t) = \sin t$, with response frequency $\omega = 1$, as is easily verified by inserting $x(t)$ into Eq. (7.1). Another is $\cos t$. Because the equation is not linear, solutions cannot be constructed from linear combinations, *even simple multiples*, of $\sin t$ and $\cos t$. These two solutions belong to particular initial conditions ($x(0) = 0$, $\dot{x}(0) = 1$ and $x(0) = 1$, $\dot{x}(0) = 0$, respectively), and according to the theory of ordinary differential equations they are the unique solutions for those initial conditions. If the equation had been linear, these solutions could have been used easily to construct solutions for other initial conditions, but in this nonlinear example they cannot. The response and driving frequencies for these solutions are incommensurate: $\omega = \Omega/\sqrt{2}$.

7.1.1 A MODEL SYSTEM

The one-freedom system of Fig. 7.1 will be used to illustrate some properties of nonlinear oscillations. A bead of mass m slides on a horizontal bar and is acted on by a spring attached to a fixed point a distance a from the bar. The force constant of the spring is k.

The Lagrangian of the system is

$$L \equiv T - V = \frac{1}{2}m\dot{x}^2 - \frac{1}{2}k(\sqrt{a^2 + x^2} - l)^2, \tag{7.2}$$

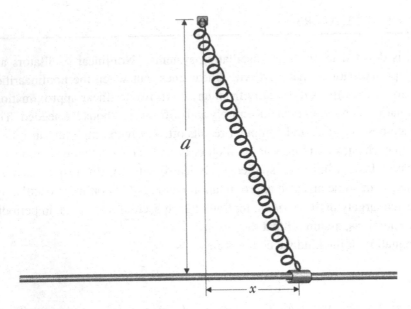

FIGURE 7.1

A spring-loaded bead free to slide on a horizontal bar. The distance a from the bar to the fixed end of the spring may be less than, more than, or equal to the unstretched length l of the spring.

where l is the natural length of the spring: $\sqrt{a^2 + x^2} - l$ is its *stretch*. The equation of motion is

$$m\ddot{x} + kx[1 - l(a^2 + x^2)^{-1/2}] = 0. \tag{7.3}$$

Later, damping and a driving force will be added, but first we make some remarks about the system in the simple form of Eq. (7.3). It is clear from the physics that at $x = 0$ there is no force on the bead: $x = 0$ is an equilibrium position. If $a \geq l$, the spring is stretched or at its natural length at $x = 0$ and the equilibrium is stable, for if the bead is displaced the spring pulls it back to $x = 0$. If $a < l$, the spring is compressed at $x = 0$ and the equilibrium is unstable, for if the bead is displaced the spring pushes it farther. In addition, if $a < l$ there are two values of x, at $x = \pm\sqrt{l^2 - a^2}$, at which the spring reaches its natural length. These are both stable equilibrium points. All the equilibrium points and their stability can also be understood by studying the potential (see Fig. 7.2).

Exact analytic solutions for the motion of this system are essentially impossible to obtain. When $x = 0$ is a stable point, the motion for small displacements from equilibrium can be approximated by expanding the force term of Eq. (7.3) in a Taylor series. Then the equation of motion becomes

$$m\ddot{x} + k\left(1 - \frac{l}{a}\right)x + \frac{kl}{2a^3}x^3 - \frac{3kl}{8a^5}x^5 \approx 0. \tag{7.4}$$

We will keep only terms up to the cubic one, with several relations between a and l. If only the first term is kept, the system becomes the harmonic oscillator, the linear approximation to the problem.

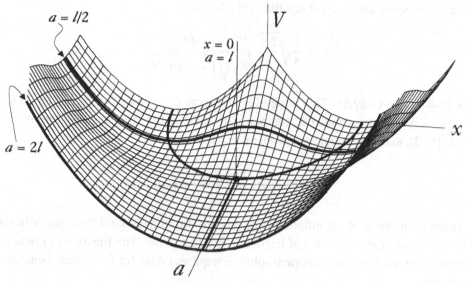

FIGURE 7.2

The potential $V(x) = \frac{1}{2}k(\sqrt{a^2 + x^2} - l)^2 - Z(a)$ for $0 \le a \le 2l$ in the system of Fig. 7.1. $V(x)$ graphs for $a = l/2$ and $a = 2l$ are emphasized in the diagram. The minima of V are the stable equilibrium positions x_0 in the one-dimensional configuration space, and the function $Z(a)$ is chosen to make $V(x_0) = 0$ for every value of a. As a result, the stable equilibrium positions lie on the heavy curve (in the valley of the graph). The curve bifurcates at $x = 0$, $a = l$, showing that for each $a > l$ there is one stable equilibrium position and for each $a < l$ there are two (compare with the ω-dependent bifurcation of the stable equilibrium position in Worked Example 2.2.1). There is also an unstable equilibrium position for each $a < l$ on the ridge that runs from the bifurcation point to the V axis. (The origin, where the x, a, and V axes meet, is hidden by the surface.)

First consider the case in which $a = l$. Then the linear term vanishes, and the approximate equation of motion becomes

$$m\ddot{x} + \frac{k}{2l^2}x^3 = 0.$$

This equation describes an oscillator so radically anharmonic that even its zeroth approximation is not harmonic: there is no linear approximation of which this is a perturbation. Choose the initial conditions to be $\dot{x}(0) = 0$, $x(0) = -A$, so the amplitude of the motion is A and the total energy is $E = kA^4/8l^2$. According to Eq. (1.46) (we set $t_0 = 0$)

$$t = 2l\sqrt{\frac{m}{k}} \int_{-A}^{x} \frac{d\xi}{\sqrt{A^4 - \xi^4}}.$$

The variable of integration here is called ξ rather than x or x'. The integral can be converted to a standard form by the substitution

$$\xi = -A\cos\theta,$$

and then after some algebra, the equation for t becomes

$$t = \frac{l}{A}\sqrt{\frac{2m}{k}} \int_0^\Theta \frac{d\theta}{\sqrt{1 - \frac{1}{2}\sin^2\theta}},$$

where $\Theta = \arccos(-\xi/A)$. This integral is known as the *elliptic integral of the first kind* $F(\Theta, \frac{1}{2})$. Like the trigonometric functions, it is tabulated (see, e.g., Abramowitz and Stegun, 1972), and in principle the problem is solved:

$$t = \frac{l}{A}\sqrt{\frac{2m}{k}} F\left(\Theta, \frac{1}{2}\right).$$

The solution in terms of the elliptic integral is not different in kind from the solution of the harmonic oscillator in terms of trigonometric functions. The functions in both cases are transcendental, and any numerical solution requires tables (or their equivalent through computers).

Setting the upper limit Θ to $\pi/2$ yields $t = P/4$, where P is the period. It is found in this way that

$$P = 10.488\frac{l}{A}\sqrt{\frac{m}{k}}.$$

The period is inversely proportional to the amplitude; compare with Problem 1.7. One of the hallmarks of nonlinear oscillators as opposed to linear ones is that their periods vary with amplitude.

7.1.2 DRIVEN QUARTIC OSCILLATOR

In this section we study what happens when damping and a driving force are added to the system described by Eq. (7.4), again restricted to third order in x. The dynamics of the systems for $a < l$ and $a > l$ are quite different, so they will be treated separately.

For $a < l$ the linear term is negative and the force in the vicinity of the origin is repulsive (this is clear in Fig. 7.2). At large values of x, however, the positive cubic term in the force (quartic in the potential) begins to dominate and the motion changes character. In the same approximation as was made above for $a = l$ (i.e., up to terms of order x^3) this system can also be solved exactly in terms of elliptic integrals, and we discuss some of its aspects in Section 7.2.

For $a > l$, which is all we will deal with here, the linear term is positive, so the force in the vicinity of the origin is attractive (see Crutchfield and Huberman, 1979). The smaller the displacements, the more nearly the system approaches the harmonic oscillator, but as in the case of $a < l$ the cubic term dominates when x becomes large. In the absence of damping and a driving force this is the problem of the quartic oscillator, treated several times in Chapter 6. This is the case, with damping and without, that will be discussed in this section.

There exist other systems related to these, but with a negative cubic term. At large displacements the force in those systems becomes repulsive, so oscillations take place only for relatively small displacements and only if the linear term is positive (otherwise the force is repulsive at all values of x). We do not discuss those systems.

DAMPED DRIVEN QUARTIC OSCILLATOR; HARMONIC ANALYSIS

We now add damping (a term linear in $\dot{x}$) and a sinusoidal driving term to Eq. (7.4). Then the third-degree equation, divided by m, becomes

$$\ddot{x} + 2\beta\dot{x} + \omega_0^2 x + \epsilon x^3 = f_1 \sin \Omega t + f_2 \cos \Omega t, \qquad (7.5)$$

where Ω is the driving frequency. The driver is written as the sum of two terms to allow the response to contain a term of the simple form $\Gamma \sin \Omega t$, which may be out of phase with the driver. The driver can also be written as $f_0 \sin(\Omega t + \gamma)$, where γ is the phase difference between the driver and the response:

$$\tan \gamma = f_2/f_1. \qquad (7.6)$$

The amplitude $f_0 = \sqrt{f_1^2 + f_2^2}$ and frequency Ω of the driving term are given in the statement of the problem, but γ (or equivalently f_1 and f_2 separately) have to be found.

No closed-form solution of Eq. (7.5) is known. The method of successive approximations that we now describe, sometimes called the *method of harmonic analysis*, is due to Duffing (1918). In fact the problem is known also as the Duffing oscillator. We will go only to the second approximation in ϵ.

The equation is written in the form

$$\ddot{x} = -2\beta\dot{x} - \omega_0^2 x - \epsilon x^3 + f_1 \sin \Omega t + f_2 \cos \Omega t, \qquad (7.7)$$

and the first approximation $x_1(t)$ is taken to be a response at the driving frequency, almost as though this were simply a forced harmonic oscillator:

$$x_1(t) = \Gamma \sin \Omega t.$$

The second approximation $x_2(t)$ is obtained by inserting $x_1(t)$ into the right-hand side of (7.7). The equation then becomes

$$\ddot{x}_2(t) = (f_2 - 2\beta\Gamma\Omega)\cos \Omega t + \left(f_1 - \omega_0^2\Gamma - \frac{3}{4}\epsilon\Gamma^3\right)\sin \Omega t + \frac{1}{4}\epsilon\Gamma^3 \sin 3\Omega t,$$

which is an explicit equation for the second derivative of $x_2(t)$ and can therefore be integrated (twice) directly. The constants of integration lead to an additive expression of the form $x_0 + v_0 t$, which does not lend itself to successive approximation unless $x_0 = v_0 = 0$ (moreover, the solution is not even periodic unless $v_0 = 0$). This corresponds to the zeroth mode, and we leave it out, so the second approximation becomes

$$x_2(t) = \frac{(2\beta\Gamma - f_2/\Omega)}{\Omega}\cos \Omega t + \frac{\left(\omega_0^2\Gamma + \frac{3}{4}\epsilon\Gamma^3 - f_1\right)}{\Omega^2}\sin \Omega t - \frac{\epsilon\Gamma^3}{36\Omega^2}\sin 3\Omega t. \quad (7.8)$$

This second approximation has two parts. The first, an oscillation at the driving frequency Ω, itself has two terms; this is called the *isochronous* part. The other part is a *harmonic* at a frequency of 3Ω. Although we go only to the second approximation, the third can be obtained similarly by inserting the second into Eq. (7.7), and so on. The convergence properties of this procedure are not known, but the second approximation already gives reasonably good results.

To proceed, assume that the second approximation differs from the first only in the harmonic term. That is, set the isochronous part of the second approximation equal to $x_1(t)$. This leads to two equations:

$$f_1 = \left(\omega_0^2 - \Omega^2\right)\Gamma + \frac{3}{4}\epsilon\Gamma^3,$$
$$f_2 = 2\beta\Gamma\Omega.$$

Squaring and adding yields

$$f_0^2 = \Gamma^2\left\{\left[\omega_0^2 - \Omega^2 + \frac{3}{4}\epsilon\Gamma^2\right]^2 + 4\beta^2\Omega^2\right\},$$

which establishes a relation, to first order in ϵ, between the amplitude Γ of the steady state and the parameters f_0 and Ω of the driver.

To understand this result it helps to graph this equation. This can be done by first solving for Ω^2:

$$\Omega^2 = \omega_0^2 - 2\beta^2 + \frac{3}{4}\epsilon\Gamma^2 \pm \left[4\beta^2\left(\beta^2 - \omega_0^2\right) - 3\epsilon\beta^2\Gamma^2 + \left(f_0/\Gamma\right)^2\right]^{1/2}.$$

This can be written in the form

$$\left(\frac{\Omega}{\omega_0}\right)^2 = 1 - 2\left(\frac{\beta}{\omega_0}\right)^2 + \frac{3\epsilon f_0^2}{4\omega_0^6}\left(\frac{\Gamma}{\Gamma_0}\right)^2$$

$$\pm\left\{4\left(\frac{\beta}{\omega_0}\right)^2\left[\left(\frac{\beta}{\omega_0}\right)^2 - 1\right] - \frac{3\epsilon f_0^2}{\omega_0^6}\left(\frac{\beta}{\omega_0}\right)^2\left(\frac{\Gamma}{\Gamma_0}\right)^2 + \left(\frac{\Gamma_0}{\Gamma}\right)^2\right\}^{1/2}, \quad (7.9)$$

where $\Gamma_0 = f_0/\omega_0^2$. According to Eq. (7.6), the phase difference γ satisfies

$$\tan\gamma = \frac{2\beta\Omega}{\omega_0^2 - \Omega^2 + \frac{3}{4}\epsilon\Gamma^2}. \quad (7.10)$$

Before studying the graph, we make two digressions. First we check the result by going to the $\epsilon = 0$ limit, in which the amplitude and phase difference should agree with the results of Section 4.2.4 for the damped harmonic oscillator. This is easily verified: the phase difference reduces immediately to the result of Section 4.2.4, whereas the amplitude takes a bit of algebra. (Later we will go to the $\beta = 0$ limit to study the undamped case.)

The second digression involves initial conditions. Equation (7.9) can be treated as a cubic equation, solvable in principle, for $(\Gamma/\Gamma_0)^2$ as a function of the parameters of the driver. Suppose it has been solved. Then Eq. (7.8) can be written in the form

$$x_2(t) = \Gamma\sin\Omega t - \frac{\epsilon\Gamma^3}{36\Omega^2}\sin 3\Omega t, \quad (7.11)$$

which gives the steady-state behavior of the Duffing oscillator to second order in ϵ. This particular solution belongs to particular initial conditions, found by setting $t = 0$:

$$x_2(0) = 0, \quad \dot{x}_2(0) = \Omega\Gamma\left[1 - \frac{\epsilon\Gamma^2}{12\Omega^2}\right].$$

Both Γ and γ are determined by f_0 and Ω, so these initial conditions are determined entirely by the driver. Since the system is nonlinear there is no way to use this solution to obtain other solutions with the same driver but with different initial conditions. In this book we do not discuss other initial conditions for the Duffing oscillator (but see Nayfeh and Mook, 1979).

We now turn to the graph. Rather than solve for Γ as a function of Ω, we use Eq. (7.9) to plot $(\Omega/\omega_0)^2$ as a function of $|\Gamma/\Gamma_0|$ by adding together the terms on the right-hand side. Then the graph can be turned on its side and read as a graph for $|\Gamma/\Gamma_0|$ as a function of $(\Omega/\omega_0)^2$. Figure 7.3 is the result for fixed f_0 and three values of β. This graph looks like a distortion of Fig. 4.22a for the driven damped *linear* oscillator, with the distortion giving rise to regions of Ω in which Γ takes on multiple values, at least for small values of β.

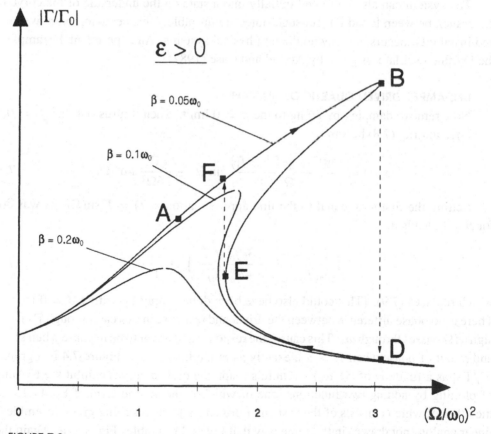

FIGURE 7.3
The response $|\Gamma/\Gamma_0|$ of the driven, damped, quartic oscillator as a function of $(\Omega/\omega_0)^2$ for $\epsilon > 0$ and for fixed f_0 and three values of β. It is as though the nonlinearity causes the curves of Fig. 4.22a to bend over. Hysteresis can occur for sufficiently small β, as described in the text.

Multiple-valued regions lead to discontinuities in the steady-state behavior of the system. We demonstrate this on the $\beta = 0.05 \, \omega_0$ curve of Fig. 7.3. Suppose the oscillator is first driven at a value of Ω that lies below the triple-valued region, say at the frequency of the point A on Fig. 7.3. If Ω is then raised very slowly, so slowly that at each driving frequency the system remains in its steady state or very close to it, Γ will climb along the curve as far as point B. If Ω continues to increase, Γ must jump discontinuously to the low value at D. Then as Ω continues to rise slowly, Γ will decrease slowly along the rest of the curve. Suppose now that the system starts at an Ω that lies above the triple-valued region (an Ω higher than at D). If Ω is then lowered very slowly, Γ will follow the curve to E, where it must jump discontinuously to the high value at F. This type of behavior is called *hysteresis*. By changing the driving frequency appropriately, the system can be forced to move from D to E to F to B and back to D again, sweeping out a finite area in the plane of the graph. The region EFBD swept out in this way is called a *hysteresis loop*. At larger values of β the loop gets quite small, as is evident from Fig. 7.3, and for sufficiently large β there is no hysteresis. The smaller the hysteresis loop, the easier it is to make the system jump from one branch of the curve to another, and its area is a measure of the stability of a state that lies on the loop.

The system can also be forced initially into a state on the underside of the curve (on the branch between E and B), but such states are unstable. This behavior, both hysteresis and instability, occurs in many nonlinear physical systems. An experimental example for the Duffing oscillator is given by Arnold and Case (1982).

UNDAMPED DRIVEN QUARTIC OSCILLATOR

Now remove damping by going to the $\beta = 0$ limit. Then it turns out that $f_2 = 0$, so $f_0 = f_1$, and Eq. (7.8) becomes

$$x_2(t) = \frac{\left(\omega_0^2 \Gamma + \frac{3}{4}\epsilon\Gamma^3 - f_1\right)}{\Omega^2} \sin \Omega t - \frac{\epsilon \Gamma^3}{36\Omega^2} \sin 3\Omega t. \tag{7.12}$$

Setting the first term equal to the first approximation $x_1(t) = \Gamma \sin \Omega t$, as was done for $\beta \neq 0$, leads to

$$\left(\frac{\Omega}{\omega_0}\right)^2 = 1 + \frac{3\epsilon f_0^2}{4\omega_0^6}\left(\frac{\Gamma}{\Gamma_0}\right)^2 + \frac{\Gamma_0}{\Gamma}, \tag{7.13}$$

which replaces (7.9). [This could also have been done simply by setting $\beta = 0$ in (7.9).] There is no phase difference between the driver and response in this case, though Γ changes sign as Ω passes through ω_0. This causes the response and driver to be in phase when $\Omega < \omega_0$ and π out of phase when $\Omega > \omega_0$, exactly as in the linear case. Figure 7.4 is a graph of $|\Gamma/\Gamma_0|$ as a function of $(\Omega/\omega_0)^2$. [On this graph the dashed curves exhibit the technique of plotting by adding two functions, one of which is the second term of Eq. (7.13) and the other of which consists of the first and third terms.] The resulting graph resembles its linear analog (not drawn) in the same way that Fig. 7.3 resembles Fig. 4.22a. Again there is a range of Ω for which Γ is multiple valued, but because it extends to unboundedly large Ω, hysteresis does not occur. If Γ starts at large Ω on the out-of-phase branch of the graph

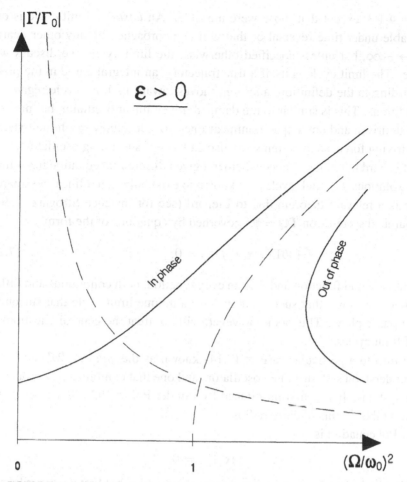

FIGURE 7.4
The response $|\Gamma/\Gamma_0|$ of a driven undamped quartic oscillator as a function of $(\Omega/\omega_0)^2$. The dotted curves correspond to the functions $(\Omega/\omega_0)^2 = |\Gamma_0/\Gamma|$ and $(\Omega/\omega_0)^2 = A(\Gamma/\Gamma_0)^2$, where A is found in Eq. (7.13). The response is in phase with the driving frequency for $\Omega < \omega_0$ and out of phase for $\Omega > \omega_0$.

and slowly decreases, it will eventually jump discontinuously to the in-phase branch. But then if Γ increases, it will not jump back to the out-of-phase branch.

We have dealt only with the case of positive ϵ. In our model system of Section 7.1.1, that corresponds to $a > l$ with damping and a driving force added.

As is seen in the hysteresis of the driven damped Duffing oscillator, nonlinear systems can jump from one state of motion to another. Before discussing stability more generally in Section 7.2, we present an example of a nonlinear oscillator.

7.1.3 EXAMPLE: THE VAN DER POL OSCILLATOR

An important class of nonlinear oscillators comprises those that exhibit *limit cycles* (the term is Poincaré's). A *stable* limit cycle is an isolated periodic trajectory that will be approached by any other nearby trajectory as $t \to \infty$, something like the periodic trapped orbits of Section 4.1.3, except that those were unstable. An *unstable* limit cycle is one that becomes stable under time reversal or that will be approached by any other nearby trajectory as $t \rightarrow -\infty$, but unless specified otherwise, the limit cycles we discuss will always be stable. The limit cycle is itself a true trajectory, an integral curve in the phase manifold. According to the definition, a system moves toward its limit cycle regardless of its initial conditions. This is similar to the damped driven linear oscillator, but now the discussion is of undriven and undamped nonlinear ones. Because they reach their steady state without a driving force, such systems are also called self-sustaining oscillators.

In general it is hard to tell in advance whether a given differential equation has a limit cycle among its solutions, but such cycles are known to exist only in nonlinear dissipative systems. There is a relevant theorem due to Liénard (see for instance Strogatz, 1994), applying to dynamical systems on $\mathbf{TQ} = \mathbb{R}^2$ governed by equations of the form

$$\ddot{x} + \beta \Gamma(x) \dot{x} + f(x) = 0, \tag{7.14}$$

where Γ is a nonlinear odd function and f is an even function, both continuous and differentiable. The theorem shows that such systems have a unique limit cycle that surrounds the origin in the phase plane. This book, however, will not treat the general questions of the existence of limit cycles.

Instead, we turn to a particular case of (7.14) known as the *van der Pol oscillator*, one of the best understood self-sustained oscillators and one that exhibits many properties generic to such systems. It was first introduced by van der Pol in 1926 in a study of the nonlinear vacuum tube circuits of early radios.

The van der Pol equation is

$$\ddot{x} + \beta(x^2 - 1)\dot{x} + x = 0, \tag{7.15}$$

where $\beta > 0$. We will deal with this equation mostly numerically, but first we argue heuristically that a limit cycle exists. The term involving $\dot{x}$ looks like an x-dependent frictional term, positive when $|x| > 1$ and negative when $|x| < 1$. That implies that the motion is damped for large values of x and gains energy when x is small. At $|x| = 1$ these effects cancel out – the system looks like a simple harmonic oscillator. It is therefore reasonable to suppose that a steady state can set in around $|x| = 1$, leading to a limit cycle. Another way to see this is to study the energy $E \equiv \frac{1}{2}(x^2 + v^2)$, which is the distance from the origin in the phase plane (we write $v \equiv \dot{x}$ and $\dot{v} \equiv \ddot{x}$). The time rate of change of E is given by

$$\frac{dE}{dt} = x\dot{x} + v\dot{v} = xv + \beta(1 - x^2)v^2 - vx = \beta(1 - x^2)v^2.$$

When $|x| > 1$ this is negative, meaning that E decreases, and when $|x| > 1$ it is positive, meaning that E increases. When $|x| = 1$ the energy is constant, so a limit cycle seems plausible.

In the $\beta = 0$ limit Eq. (7.15) becomes the equation of a harmonic oscillator with frequency $\omega = 1$. Apparently, then, the phase portrait for small β should look something like the harmonic oscillator, consisting of circles about the origin in the phase plane. The extent to which this is true will be seen as we proceed.

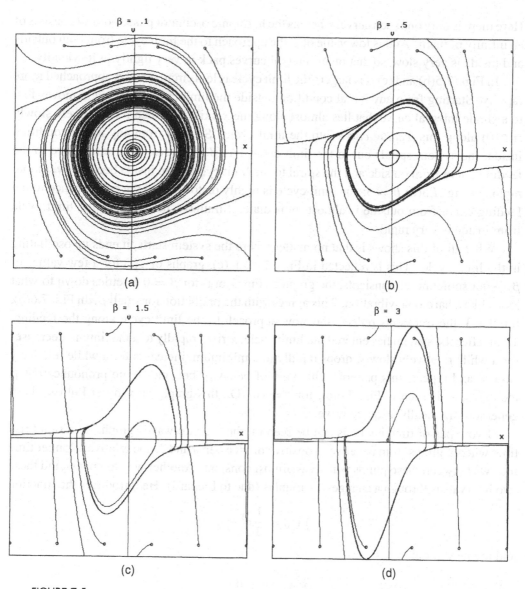

FIGURE 7.5

Phase portraits of the van der Pol oscillator. Initial points, indicated by small circles, are at $v = \pm4.5$, $x = -5, -3, -1, 1, 3, 5$. (a) $\beta = 0.1$. Integral curves pack tightly about the almost circular limit cycle. (b) $\beta = 0.5$. The limit cycle is no longer circular, and all external initial points converge to the feeding curve before entering the limit cycle. (c) $\beta = 1.5$. The limit cycle is more distorted. (d) $\beta = 3$. The limit cycle is highly distorted and the approach to it is more rapid from inside. From outside, the approach to the feeding curve is more rapid.

We now turn to numerical computations. Figures 7.5(a)–(d) are phase portraits of (7.15) with several initial conditions and for various values of β, all obtained on a computer. At each value of β all the solutions converge to a final trajectory, the limit cycle. In Fig. 7.5(a) $\beta = 0.1$ is small, and the limit cycle is almost a circle around the origin. As was predicted above, this resembles an integral curve, but not the phase portrait, of the harmonic oscillator.

Here there is only one such curve, whereas the harmonic oscillator phase portrait consists of an infinity of them. At this low value of β the approach to the limit cycle from both outside and inside is very slow, so that other integral curves pack in very tightly in its vicinity.

In Figs. 7.5(b) and (c) β is larger: the limit cycle is less circular and is approached more rapidly. Starting from any initial condition outside the limit cycle, the system moves first to a single integral curve that lies almost along the x axis. It is then fed relatively slowly ($v \approx 0$) along this *feeding curve* into the limit cycle. Starting from any initial condition inside, the system spirals out to the limit cycle. As β increases, the convergence to the feeding curve from outside and the spiral to the limit cycle from inside become ever more rapid. In Fig. 7.5(d) $\beta = 3$: the limit cycle is highly distorted and the convergence to the feeding curve from outside is almost immediate. Similarly, the spiral to the limit cycle from inside is very rapid.

What all of this shows is that no matter where the system starts, it ends up oscillating in the limit cycle. This is reflected in Figs. 7.6(a)–(c), graphs of $x(t)$ for a few values of β. After some initial transients, the graph of Fig. 7.6(a) for $\beta = 0.1$ settles down to what looks like a harmonic vibration. This agrees with the prediction for small β. In Fig. 7.6(b), for $\beta = 3$, the transients reflects the slow approach to the limit cycle along the feeding. Then, after the system has entered the limit cycle, x rises rapidly to a maximum, decreases for a while relatively slowly, drops rapidly to a minimum, increases for a while relatively slowly, and repeats this pattern. This kind of behavior becomes more pronounced as β increases, as is seen in Fig. 7.6(c), for $\beta = 10$. On this graph the van der Pol oscillator generates practically a square wave.

Two types of time intervals can be seen on the "square-wave" graph: 1. a rapid rise time when x jumps from negative to positive or vice versa and 2. the relatively longer time interval between these jumps when x is almost constant. Another way to understand these two intervals is through a change of variables (due to Liénard). He introduces the function

$$Y(x) = \frac{1}{3}x^3 - x$$

and the new variable

$$s = \beta(v + \beta Y).$$

Note that (x, s) form coordinates on the phase plane, and in their terms Eq. (7.15) is

$$\dot{x} = \beta[s - Y(x)],$$
$$\dot{s} = -\beta^{-1}x.$$

Now consider the curve $s = Y(x)$ in the (x, s) plane (Fig. 7.7). If the phase point is above the curve, $s - Y(x)$ is positive and hence so is $\dot{x}$, and conversely if the phase point is below the curve, $\dot{x}$ is negative. Suppose that initially the phase point is at $s = 0$, $x < -\sqrt{3}$ (i.e., on the x axis to the left of the curve). To understand the square wave, consider only relatively large values of β. Then $\dot{s}$ is small (because β is large) and positive, and $\dot{x}$ is positive: the phase point approaches the curve. As it comes closer to the curve $s - Y(x)$ decreases, and so does $\dot{x}$. The phase point doesn't quite get to the curve but climbs slowly (for $\dot{s}$ and $\dot{x}$

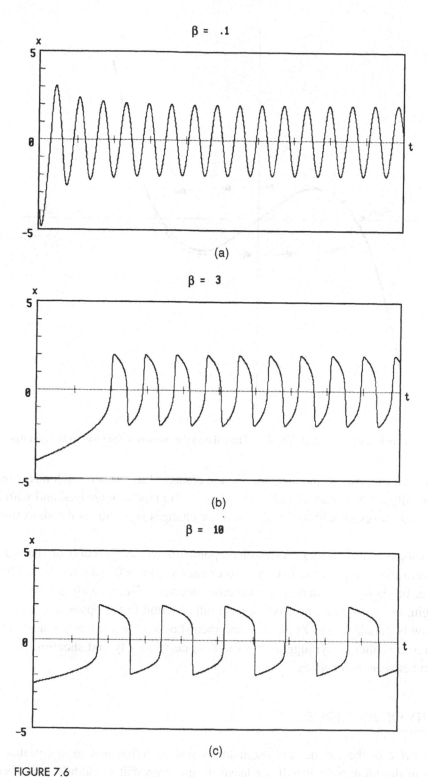

FIGURE 7.6

$x(t)$ curves for the van der Pol oscillator. $x(0) = -4$, $v(0) = -4$. (a) $\beta = 0.1$. (b) $\beta = 3$. The first part of the graph shows the motion along the feeding curve in the phase plane. (c) $\beta = 10$ (extremely high β). In the limit cycle (after the feeding curve) the graph is almost a square wave.

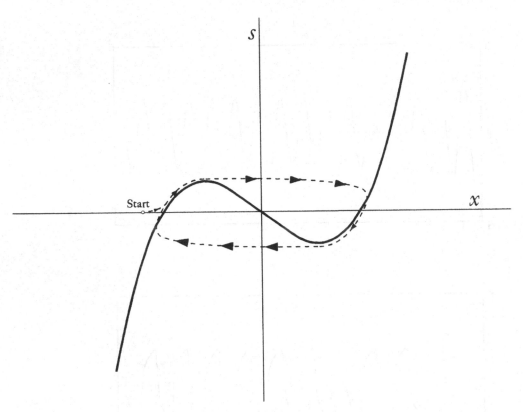

FIGURE 7.7

The $s = Y(x)$ curve in the (x, s) plane. The dotted line shows the motion of the phase point for large β.

are small) next to it as shown in the diagram while x changes little; this is the longer time interval. Eventually the $Y(x)$ curve bends away, the $s - Y(x)$ factor increases, and with it $\dot{x}$: the phase point moves rapidly to the right while x changes sign; this is the short time interval.

When x changes sign, so does $\dot{s}$. As the phase point approaches the curve on the right, $\dot{x}$ approaches zero and eventually the $Y(x)$ curve is crossed with $\dot{x} = 0$ and $\dot{s}$ negative. Then the phase point lies below the curve and $\dot{x}$ becomes negative. Because all is symmetric about the origin, the phase point now performs similar motion in the opposite direction. It descends slowly parallel to the $Y(x)$ curve and then shoots across to the left side. The motion then repeats, alternately hugging the $s = Y(x)$ curve slowly and shooting across back and forth between its branches.

7.2   STABILITY OF SOLUTIONS

The limit cycle of the van der Pol oscillator is *stable*. What that means is that if the system is in the limit cycle it will not leave it, and even if it is slightly perturbed, pushed out of the limit cycle, it will return to it. In general, however, the motion of any dynamical system, not necessarily an oscillation, may be unstable: it might change in some fundamental way, especially if perturbed. In this section we discuss stability and instability and define them more completely.

The model system of Fig. 7.1 with $a < l$ provides some simple examples of stability. The equilibrium positions at $x = \pm b$, where $b = \sqrt{l^2 - a^2}$, are stable, and in $\mathbf{TQ}$ the points with coordinates $(x, \dot{x}) = (\pm b, 0)$ are stable equilibrium points. If the system is in the neighborhood of $(b, 0)$, for instance, it will remain in that neighborhood; in configuration space it will oscillate about b, never moving very far away. In contrast, $x = 0$ [or $(x, \dot{x}) = (0, 0)$] is unstable. If the system is at rest at $x = 0$, the slightest perturbation will cause it to move away. Such examples have arisen earlier in the book, but we now go into more detail. Another kind of instability occurs at the bifurcation in Fig. 7.2. If the parameter a is allowed to decrease, when it passes through the value $a = l$ the motion changes character. The stable equilibrium point becomes unstable and two new stable ones appear.

7.2.1 STABILITY OF AUTONOMOUS SYSTEMS

Every equation of motion is of the form

$$\dot{\xi} = f(\xi, t), \tag{7.16}$$

where $f \equiv \{f^1, f^2, \ldots, f^{2n}\}$ is made up of the components of the dynamical vector field Δ. We start the discussion with *autonomous* systems, those for which f does not depend on the time.

DEFINITIONS

A *fixed point* (a *stationary point*, *equilibrium point*, or *critical point*) of the dynamical system of the time-independent version of Eq. (7.16) is ξ_f such that $\Delta(\xi_f) = 0$, in other words such that $f^k(\xi_f) = 0 \, \forall k$. Then the solution with the initial condition ξ_f satisfies the equation $\dot{\xi}(t; \xi_f) = 0$, or $\xi(t; \xi_f) = \text{const.}$ (we take $t_0 = 0$ for autonomous systems). Such a fixed point can be stable or unstable, as was described in Chapters 1 and 2 (elliptic and hyperbolic points). We will define three different kinds of stability.

1. **Lyapunov Stability.** Roughly speaking, a fixed point is stable in the sense of Lyapunov if solutions that start out near the fixed point stay near it for long times. More formally, ξ_f is a stable fixed point if given any $\epsilon > 0$, no matter how small, there exists a $\delta(\epsilon) > 0$ such that if $|\xi_f - \xi_0| \equiv |\xi(0; \xi_f) - \xi(0; \xi_0)| < \delta$ then $|\xi(t; \xi_f) - \xi(t; \xi_0)| < \epsilon$ for all $t \geq 0$. Real rigor requires a definition of distance, denoted by the magnitude symbol as in $|\xi_f - \xi_0|$, but we will simply assume that this is Euclidean distance unless specified otherwise. This definition of stability resembles the usual definition of continuity for functions. It means that if you want a solution that will get no further than ϵ from the fixed point, all you need do is find one that starts at a ξ_0 no further from ξ_f than δ (clearly $\delta \leq \epsilon$, or the condition would be violated at $t = 0$). See Fig. 7.8.

Figure 7.9 shows the phase diagram of the model system for $a < l$ (the diagram in $\mathbf{TQ}$ is almost the same as in $\mathbf{T^*Q}$ because $p = m\dot{x}$ for this system; we will write $\mathbb{M}$ in general). The fixed points at $\xi_f = (\pm b, 0)$ are stable (or elliptic): close to the equilibrium point the

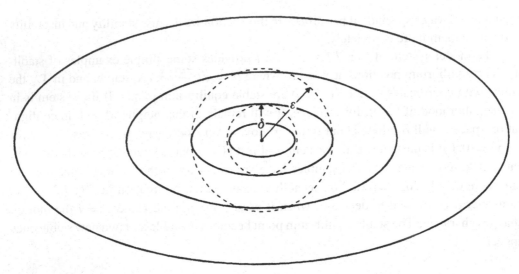

FIGURE 7.8
Illustration of Lyapunov stability. All motions that start within the δ circle remain within the ϵ circle.

orbits are almost ellipses (in the small-vibrations approximation they *are* ellipses) and the situation is similar to that of Fig. 7.8. In contrast, $\xi_f = (0, 0)$ is unstable (hyperbolic): given a small ϵ, almost every orbit that starts out within a circle of radius ϵ about this ξ_f moves out of the circle.

2. Orbital Stability. Although any motion that starts out near $\xi_f = (b, 0)$ remains near ξ_f, it does not follow that two motions that start out near ξ_f remain near each other, even if the distance between their integral curves never gets very large. Two motions that start close together, but on different ellipses, will stay close together only if their periods are the same. If they cover their ellipses at different rates, one motion may be on one

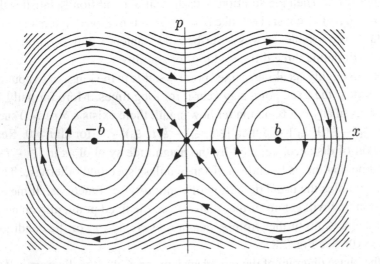

FIGURE 7.9
The phase diagram of the model system of Fig. 7.1 for $a < l$. The three stationary points are indicated by dots. $(0, \pm b)$ are both Lyapunov stable. $(0, 0)$ is unstable.

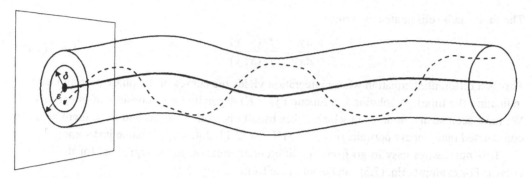

FIGURE 7.10
Lyapunov stability for a time-dependent system. The stable solution ξ_s is represented by the solid curve with the tube of radius ϵ surrounding it. The dotted curve is a solution that starts out within a circle of radius δ and never leaves the radius-ϵ tube.

side on its ellipse when the other is on the other side of its. (An example is the $a = l$ case of the model system, in which the period depends on the amplitude.) Nevertheless, the motion on either one of the ellipses will at some time pass close to any given point on the other. Orbits passing close to *unstable* fixed points, however, characteristically move far apart. This leads to the concept of orbital stability: the solution $\xi(t; \xi_1)$ starting at ξ_1 is orbitally stable if given an $\epsilon > 0$ there exists a $\delta > 0$ and another time $\tau(t)$ such that if $|\xi_1 - \xi_0| < \delta$ then $|\xi(\tau(t); \xi_0) - \xi(t, \xi_1)| < \epsilon$. In short, if the *orbits*, not necessarily the *motions* themselves, remain close, the solution is orbitally stable. The solutions in the vicinity of $(0, 0)$ for $a = l$ and of the stable fixed points at $(b, 0)$ of Fig. 7.9 are orbitally stable.

3. Asymptotic Stability. A point $\xi_f \in \mathbb{M}$ is asymptotically stable if all orbits that start out close enough to ξ_f converge to it or if it attracts all orbits near it. More formally, ξ_f is asymptotically stable if it is Lyapunov stable and if there exists a $\gamma > 0$ such that $|\xi_f - \xi_0| < \gamma$ implies that $\lim_{t \to \infty} |\xi_f - \xi(t; \xi_0)| = 0$.

For later use we indicate how these ideas must be altered for *nonautonomous* systems, those for which f depends on t. For them it is not *points*, but *solutions* that are stable or unstable, and the initial time t_0 enters into the definitions. For example, a solution $\xi_s(t; \xi_1, t_0)$ starting at ξ_1 is Lyapunov stable if given an $\epsilon > 0$ there exists a $\delta(\epsilon) > 0$ such that $|\xi_1 - \xi_0| \equiv |\xi_s(t_0; \xi_1, t_0) - \xi(t_0; \xi_0, t_0)| < \delta$ implies that $|\xi_s(t; \xi_1, t_0) - \xi(t; \xi_0, t_0)| < \epsilon$ for all $t > t_0$. For nonautonomous systems, Fig. 7.10 replaces Fig. 7.8 as the diagram that goes with this definition.

THE POINCARÉ–BENDIXON THEOREM

Fixed points of autonomous systems with dim $\mathbb{M} = 2$ can be read off the phase portrait even when t is eliminated from consideration. For such a system Eq. (7.16) can be put in the form [write (x, y) in place of $(q, \dot{q})$ or (q, p) and use x, y to index the f functions]

$$\frac{dx}{dt} = f^x(x, y), \quad \frac{dy}{dt} = f^y(x, y). \tag{7.17}$$

The time can be eliminated by writing

$$\frac{dy}{dx} = \frac{f^y(x, y)}{f^x(x, y)}.$$ (7.18)

This is a differential equation whose integration yields the curves of the phase portrait (without indicating the time). A solution is a function $y(x; E)$ depending on a constant of integration E. We have seen such solutions, in which E has usually been the energy, and from them we have constructed many phase portraits (Figs. 1.5, 1.7, 2.7, 2.11, 3.4, 4.14, to name just some).

It is not always easy to go from the differential equation to the equations for the integral curves. For example, Eq. (7.3) can be put in the form

$$p \equiv m\dot{x} = y \equiv mf^x(x, y), \dot{y} = -kx[1 - l(a^2 + x^2)^{-1/2}] \equiv f^y(x, y).$$ (7.19)

The resulting differential equation for $y(x)$, namely

$$\frac{dy}{dx} = -\frac{mkx[1 - l(a^2 + x^2)^{-1/2}]}{y},$$

is not easily integrated [anyway, we already have its phase portrait for $a < l$ (Fig. 7.9)]. Points where $f^x = 0$ or $f^y = 0$ are not in general fixed points of the phase portrait; they are points where the integral curves are vertical (parallel to the y axis) or horizontal (parallel to the x axis). Fixed points occur where both $f^x = 0$ and $f^y = 0$. Stable fixed points occur within closed orbits, and the functions f^x and f^y of Eq. (7.17) can be used to tell whether there are closed orbits in a finite region D of the (x, y) plane. Write ∂D for the closed curve that bounds D. Then it follows from (7.18) that

$$\oint_{\partial D} (f^x dy - f^y dx) = 0.$$

But then Stokes's theorem implies that

$$\int_D \left(\frac{\partial f^x}{\partial x} + \frac{\partial f^y}{\partial y} \right) d\sigma = 0,$$

where σ is the element of area on the (x, y) plane. This means that a necessary, but not sufficient, condition for there to be closed orbits in D is that $(\partial f^x/\partial x + \partial f^y/\partial y)$ either change sign or be identically zero in D. This is known as the Poincaré–Bendixson theorem and can prove useful in looking for closed orbits in regions of the phase manifold for one-freedom systems. In particular, periodic orbits are closed, so this is also a necessary condition for the existence of periodic orbits and could have been used in the analysis of the van der Pol oscillator.

LINEARIZATION

Detailed information about the motion of an autonomous system close to a fixed point ξ_f can be obtained by *linearizing* the equations of motion. This is done as follows: First the origin is moved to the fixed point by writing

$$\zeta(t) = \xi(t; \xi_0) - \xi_f.$$ (7.20)

Second, Eq. (7.16) is written for ζ rather than for ξ:

$$\dot{\zeta} = f(\zeta + \xi_f) \equiv g(\zeta) \tag{7.21}$$

(g is independent of t in the autonomous case). Third, g is expanded in a Taylor series about $\zeta = 0$. Written out in local coordinates, the result is

$$\dot{\zeta}^j = \left.\frac{dg^j}{d\zeta^k}\right|_{\xi_f} \zeta^k + o(\zeta^2) \equiv A_k^j \zeta^k + o(\zeta^2),$$

where $o(\zeta^2)$ contains higher order terms in ζ. We assume that the A_k^j do not all vanish. Finally, the linearization of Eq. (7.21) is obtained by dropping $o(\zeta^2)$. The result is written in the form

$$\dot{\mathbf{z}} = \mathbf{A}\mathbf{z}, \tag{7.22}$$

where $\mathbf{z}$ is the vector whose components are the ζ^k and $\mathbf{A}$ is the constant matrix whose elements are the A_k^j.

Linearization is equivalent to going to infinitesimal values of the ζ^k or to moving to the tangent space $\mathbf{T}_0\mathrm{M}$ at $\zeta = 0$. Equation (7.22) is similar in structure to (5.56) and can be handled the same way. Its general solution is

$$\mathbf{z}(t) = e^{\mathbf{A}t}\mathbf{z}_0, \tag{7.23}$$

where $\mathbf{z}_0$ represents the $2n$ initial conditions and

$$e^{\mathbf{A}t} \equiv \exp\{\mathbf{A}t\} \equiv \sum_{n=0}^{\infty} \mathbf{A}^n t^n/n!.$$

Clearly, if $\mathbf{z}_0 = 0$ then $\mathbf{z}(t) = 0$ – that is the nature of the fixed point. What is of interest is the behavior of the system for $\mathbf{z}_0 \neq 0$, although $\mathbf{z}_0$ must be small and $\mathbf{z}(t)$ remain small for (7.22) to approximate (7.21) and for (7.23) to approximate the actual motion.

The linearized dynamics in $\mathbf{T}_0\mathrm{M}$ depends on the properties of $\mathbf{A}$, in particular on its eigenvalue structure. For instance, suppose that $\mathbf{A}$ is diagonalizable over the real numbers; let $\lambda_1, \ldots, \lambda_{2n}$ be its eigenvalues and $\mathbf{z}_1 \ldots \mathbf{z}_{2n}$ be the corresponding eigenvectors. If $\mathbf{z}_0$ happens to be one of the $\mathbf{z}_k$, Eq. (7.23) reads $\mathbf{z}(t) = e^{\lambda_k t}\mathbf{z}_k$ (no sum on k): $\mathbf{z}(t)$ is just a time-dependent multiple of $\mathbf{z}_k$. If $\lambda_k > 0$, then $\mathbf{z}(t)$ grows exponentially and this points to instability. If $\lambda_k < 0$, then $\mathbf{z}(t)$ approaches the origin exponentially and this points to stability. If $\lambda_k = 0$, then $\mathbf{z}(t)$ is a constant vector (does not vary in time). This is a special *marginal* case, and stability can be determined only by moving on to higher order terms in the Taylor series. More generally, $\mathbf{A}$ may not be diagonalizable over the reals, so Eq. (7.23) requires a more thorough examination.

If $\mathbf{A}$ is not diagonalizable over the reals, any complex eigenvalues it may have must occur in complex conjugate pairs. Then it is clearly reasonable to study the two-dimensional eigenspace of $T_0 M$ that is spanned by the two eigenvectors belonging to such a pair of eigenvalues. It turns out that also for real eigenvalues it is helpful to study two-dimensional subspaces spanned by eigenvectors [remember that $\dim(T_0 M) = 2n$ is even, so $T_0 M$ can be decomposed into n two-dimensional subspaces]. It is much easier to visualize the motion projected onto a plane than to try to absorb it in all the complications of its $2n$ dimensions. In fact it is partly for this reason that we discussed the $n = 2$ case even before linearization.

Let us therefore concentrate on two eigenvalues, which we denote by λ_1, λ_2, and for the time being we take $T_0 M$ to be of dimension 2. When the two λ_k are complex, they have no eigenvectors in the real vector space $T_0 M$. When they are real, they almost always have two linearly independent eigenvectors z_1, z_2, although if the λ_k are equal there may be only one eigenvector (that is, A may not be diagonalizable; see the book's appendix).

We start with three cases of real eigenvalues.

Case 1. $\lambda_2 \geq \lambda_1 > 0$, linearly independent eigenvectors. Figure 7.11(a) shows the resulting phase portrait of the dynamics on $T_0 M$. The fixed point at the origin is an *unstable node*. (The eigenvectors z_1 and z_2 are not in general along the initial ζ coordinate directions, but we have drawn them as though they were, at right angles. Equivalently, we have performed a ζ coordinate change to place the coordinate directions along the eigenvectors.) All of the integral curves of Fig. 7.11(a) pass through the origin, and the only curve that is not tangent to z_1 at the origin is the one along z_2. This is because $\lambda_1 > \lambda_2$, so the growth in the z_1 direction is faster than in the z_2 direction. Figure 7.11(b) shows the phase portrait when $\lambda_1 = \lambda_2$.

Case 2. $\lambda_1 = \lambda_2 \equiv \lambda > 0$, but A is not diagonalizable. In that case A is of the form

$$A = \begin{vmatrix} \lambda & 0 \\ \mu & \lambda \end{vmatrix},$$

and calculation yields

$$\exp At = e^{\lambda t} \begin{vmatrix} 1 & 0 \\ \mu t & 1 \end{vmatrix}.$$

If the initial vector z_0 has coordinates (ζ_1, ζ_2) in the (z_1, z_2) basis, that is, if $z_0 = \zeta_1 z_1 + \zeta_2 z_2$, then $z(t) = e^{\lambda t} \{\zeta_1 z_1 + (\mu t \zeta_1 + \zeta_2) z_2\}$. Figure 7.12 is a phase portrait of such a system on $T_0 M$ for positive μ. The fixed point at the origin is again an unstable node.

Case 3. $\lambda_1 < 0, \lambda_2 > 0$. In this case the fixed point at the origin is stable along the z_1 direction and unstable along the z_2 direction. Figure 7.13 is the phase portrait on $T_0 M$ for this kind of system. The origin is a hyperbolic point (Section 1.5.1), also called a *saddle point*. The z_1 axis is its stable manifold (Section 4.1.3): all points on the z_1 axis converge to the origin as $t \to \infty$. The z_2 axis is its unstable manifold: all points on the z_2 axis converge to the origin as $t \to -\infty$.

If the signs of the eigenvalues are changed, the arrows on the phase portraits reverse. The unstable nodes of Cases 1 and 2 then become stable nodes. We do not draw the corresponding phase portraits.

We now move on to complex eigenvalues. They always occur in complex conjugates pairs, and we will write $\lambda_1 = \lambda_2^* \equiv \lambda$. In this case there are no eigenvectors in the real

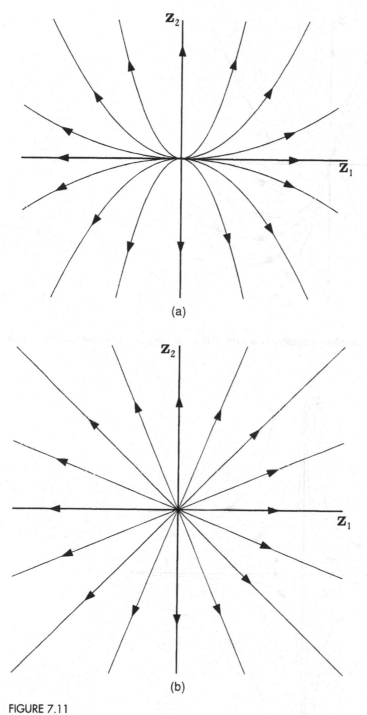

FIGURE 7.11

(a) Unstable fixed point for real $\lambda_2 > \lambda_1 > 0$. In this and the subsequent four figures, the z_1 and z_2 directions are indicated, but not the vectors themselves. (b) Unstable fixed point for real $\lambda_2 = \lambda_1 > 0$.

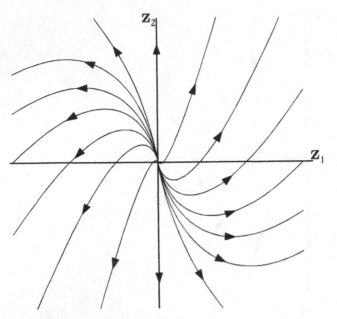

FIGURE 7.12
Unstable fixed point for nondiagonalizable $\mathbf{A}$ matrix. All of the integral curves are tangent to $\mathbf{z}_2$ at the fixed point.

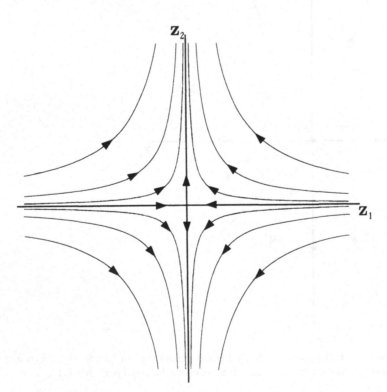

FIGURE 7.13
Hyperbolic fixed point for real $\lambda_1 < 0, \lambda_2 > 0$.

vector space T_0M, but there are eigenvectors in its *complexification*. Let $z = u + iv$ be an eigenvector of A belonging to eigenvalue $\lambda \equiv \alpha + i\beta$, where u and v are real vectors in T_0M and α and β are real numbers. Then

$$Az = \lambda z \equiv A(u + iv) = \alpha u - \beta v + i(\beta u + \alpha v).$$

The real and imaginary parts of all expressions here are separated, so

$$Au = \alpha u - \beta v, \quad Av = \beta u + \alpha v. \tag{7.24}$$

From this it follows that $Az^* = \lambda^* z^*$, so that z^* is an eigenvector belonging to λ^* and is therefore linearly independent of z. In terms of z and z^* the vectors u and v are given by

$$u = \frac{z + z^*}{2}, \quad v = \frac{z - z^*}{2i},$$

which shows that u and v are linearly independent and therefore that (7.24) completely defines the action of A on T_0M. That λ is complex implies that $\beta \neq 0$.

Now apply Eq. (7.23) to the initial condition $z_0 = z$, where z is an eigenvector belonging to λ:

$$
\begin{aligned}
z(t) = \exp(At)z = e^{\lambda t}z &= e^{\alpha t}(\cos \beta t + i \sin \beta t)(u + iv) \\
&= e^{\alpha t}[u \cos \beta t - v \sin \beta t + i(u \sin \beta t + v \cos \beta t)] \\
&= e^{At}u + ie^{At}v.
\end{aligned}
$$

Equating real and imaginary parts we obtain

$$
\begin{aligned}
e^{At}u &= e^{\alpha t}(u \cos \beta t - v \sin \beta t), \\
e^{At}v &= e^{\alpha t}(u \sin \beta t + v \cos \beta t).
\end{aligned}
\tag{7.25}
$$

Since u and v are linearly independent, this equation describes the action of $\exp(At)$ on all of T_0M. We discuss two cases.

Case 4. $\alpha > 0$. For the purposes of visualizing the results, let u and v be at right angles (see the parenthetical remarks under Case 1). The trigonometric parts of Eq. (7.25) describe a rotation of T_0M through the angle βt, which increases with time. The $e^{\alpha t}$ factor turns the rotation into a growing spiral. Figure 7.14 is the phase portrait of the resulting dynamics on T_0M. The fixed point at the origin is an *unstable* focus. It is assumed in Fig. 7.14 that u and v have equal magnitudes. If the magnitudes were not equal, the spirals would be built on ellipses rather than on circles.

Case 5. Here $\alpha = 0$. In this case only the trigonometric terms of Eq. (7.25) survive, and the phase portrait, Fig. 7.15, consists of ellipses (circles if the magnitudes of u and v are equal). The origin is an elliptic point, also called a *center*.

Changing the signs of the eigenvalues leads to altered phase portraits. If $\alpha < 0$ the origin in Case 4 becomes a *stable focus*. Changing the sign of β changes the rotations in both Cases 4 and 5 from counterclockwise to clockwise. This demonstrates that stability

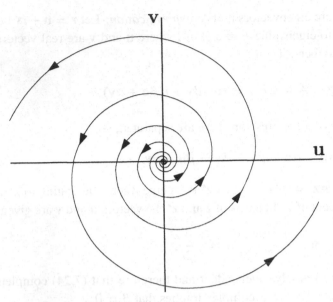

FIGURE 7.14
Unstable fixed point for complex λ with $\Re(\lambda) \equiv \alpha > 0$. The magnitudes of **u** and **v** are assumed equal. The **u** and **v** directions are indicated, but not the vectors themselves. The sense of rotation is determined by the sign of $\Im(\lambda) \equiv \beta$. (Is β positive or negative here and in the subsequent two figures?)

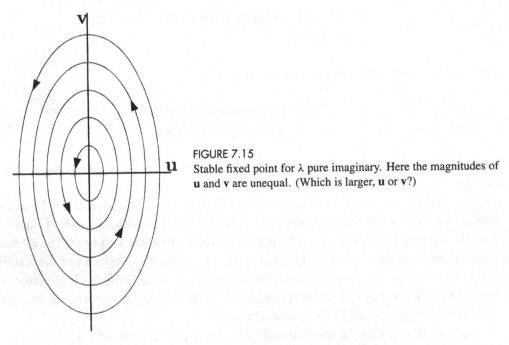

FIGURE 7.15
Stable fixed point for λ pure imaginary. Here the magnitudes of **u** and **v** are unequal. (Which is larger, **u** or **v**?)

and instability at a fixed point can be classified by the real part of the eigenvalues of **A**. The subspace spanned by eigenvectors belonging to eigenvalues for which $\Re(\lambda_k) < 0$ is the stable manifold $\mathbb{E}^s$ (actually stable *space* in the linearization) of $T_0 M$, and the one spanned by those for which $\Re(\lambda_k) > 0$ is the unstable manifold $\mathbb{E}^u$. The subspace spanned by those for which $\Re(\lambda_k) = 0$ is called the *center manifold* $\mathbb{E}^c$.

WORKED EXAMPLE 7.1

Find the $\mathbf{A}$ matrix for the model of Fig. 7.1 with $a < l$ and find its linearized forms at the three fixed points. Find the eigenvalues and eigenvectors of $\mathbf{A}$ and the stable, unstable, and center manifolds in all three cases. Draw phase portraits in $\mathbf{T}_0 M$ for all three cases. Discuss the direction of motion in the case of the two elliptic points.

Solution. The only matrix elements of $\mathbf{A}$ that are nonzero are (we continue to write $y \equiv \dot{x}$)

$$A_y^x \equiv \frac{df^x}{dy} = 1 \tag{7.26}$$

and

$$A_x^y \equiv \frac{df^y}{dx} = k\left[\frac{la^2}{(a^2 + x^2)^{3/2}} - 1\right]. \tag{7.27}$$

There is a fixed point at the origin of Fig. 7.9, and at $x = y = 0$ the $\mathbf{A}$ matrix is

$$\begin{bmatrix} A_x^x & A_y^x \\ A_x^y & A_y^y \end{bmatrix} = \begin{bmatrix} 0 & 1 \\ \lambda^2 & 0 \end{bmatrix}, \tag{7.28}$$

where $\lambda^2 = k(l - a)/a > 0$. The eigenvalues are $\pm\lambda$, so this is Case 3. The (unnormalized) eigenvectors are

$$\mathbf{z}_+ = \begin{bmatrix} 1 \\ \lambda \end{bmatrix}, \quad \mathbf{z}_- = \begin{bmatrix} 1 \\ -\lambda \end{bmatrix},$$

belonging to $+\lambda$ and $-\lambda$, respectively; $\mathbf{z}_+$ spans $\mathbb{E}^u$, $\mathbf{z}_-$ spans $\mathbb{E}^s$, and $\mathbb{E}^c$ is empty. In the initial x, y coordinates the eigenvectors are not orthogonal: the directions in which they point depend on λ, for the y component is λ times as large as the x component. Figure 7.16 is the phase portrait in $\mathbf{T}_0 M$ for $\lambda \approx 0.75$.

At the other two fixed points (i.e., at $x = \pm b \equiv \pm\sqrt{l^2 - a^2}$) the $\mathbf{A}$ matrix is

$$\begin{bmatrix} A_x^x & A_y^x \\ A_x^y & A_y^y \end{bmatrix} = \begin{bmatrix} 0 & 1 \\ -\beta^2 & 0 \end{bmatrix}, \tag{7.29}$$

where $\beta^2 = kb^2/l^2 > 0$. The eigenvalues are $\pm i\beta$, so this is Case 5: $\mathbb{E}^c$ is two dimensional and is hence the entire space $\mathbf{T}_0 M$. The (unnormalized) eigenvectors are

$$\mathbf{z} = \begin{bmatrix} 1 \\ i\beta \end{bmatrix}, \quad \mathbf{z}^* = \begin{bmatrix} 1 \\ -i\beta \end{bmatrix},$$

belonging to $i\beta$ and $-i\beta$, respectively. If $\mathbf{z} = \mathbf{u} + i\mathbf{v}$, as above, then

$$\mathbf{u} = \begin{bmatrix} 1 \\ 0 \end{bmatrix}, \quad \mathbf{v} = \begin{bmatrix} 0 \\ \beta \end{bmatrix}.$$

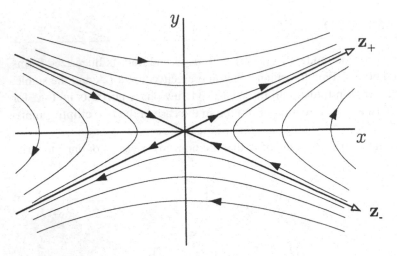

FIGURE 7.16

The linearization of the unstable, hyperbolic fixed point at the origin of the model system with $a < l$. Here $\lambda \approx 0.75$. The eigenvectors $\mathbf{z}_+$ and $\mathbf{z}_-$ are shown. The arrows on the eigenvectors do not necessarily point in the direction of the motion, for any multiples of them are equally valid eigenvectors. For instance, $\mathbf{z}_-$ could have been drawn pointing in the opposite direction.

Figure 7.15 can be used for the phase portrait in both cases. Which of the elliptical axes is minor and which major depends on $|\beta|$. In Fig. 7.15, $|\beta| > 1$. If β is positive, $\mathbf{v}$ points in the $y \equiv \dot{x}$ direction and the motion is clockwise, opposite to that in Fig. 7.15. Therefore Fig. 7.15 is for negative β. If $\beta = 1$, the ellipse is a circle.

To what extent does the linearized dynamics approximate the true nonlinear dynamics in the vicinity of a fixed point? The answer is given by the Hartman–Grobman theorem (see Hartman, 1982; also Wiggins, 1990, around p. 234). This theorem shows that there is a continuous and invertible (but not necessarily differentiable) map that carries the linearized dynamics into the actual dynamics in a finite neighborhood of the fixed point. This means that close to the fixed point they look similar. We do not go further into this question.

Worked Example 7.1 involved only elliptic and hyperbolic points, but when damping is added, the two stable fixed points become foci. With damping the second of Eqs. (7.19) becomes

$$\dot{y} = -kx[1 - l(a^2 + x^2)^{-1/2}] - 2\gamma y,$$

where $\gamma > 0$ is the damping factor. The $\mathbf{A}$ matrix now has a third nonzero term: $A_y^y = -2\gamma$, so that for the fixed point at the origin

$$\begin{bmatrix} A_x^x & A_y^x \\ A_x^y & A_y^y \end{bmatrix} = \begin{bmatrix} 0 & 1 \\ \lambda^2 & -2\gamma \end{bmatrix},$$

where λ is the same as without damping. The two eigenvalues are now

$$\lambda_{d\pm} = -\gamma \pm \sqrt{\gamma^2 + \lambda^2}$$

(subscript d for "damping"), which are both real. The radical is larger than γ, so $\lambda_{d+} > 0$ and $\lambda_{d-} < 0$. The eigenvalues lie on either side of zero, so this fixed point remains hyperbolic. The eigenvectors are

$$\mathbf{z}_+ = \begin{bmatrix} 1 \\ \lambda_{d+} \end{bmatrix}, \quad \mathbf{z}_- = \begin{bmatrix} 1 \\ \lambda_{d-} \end{bmatrix},$$

and since $|\lambda_{d+}| \neq |\lambda_{d-}|$ the eigenvectors are not symmetric about the x axis.

At the other two fixed points the matrix becomes

$$\begin{bmatrix} A_x^x & A_y^x \\ A_x^y & A_y^y \end{bmatrix} = \begin{bmatrix} 0 & 1 \\ -\beta^2 & -2\gamma \end{bmatrix},$$

where β is the same as without damping. The eigenvalues are now

$$\tau_d = -\gamma + i\sqrt{\beta^2 - \gamma^2}, \quad \tau_d^* = -\gamma - i\sqrt{\beta^2 - \gamma^2},$$

a complex conjugate pair (we consider only the case of weak damping: $\gamma < |\beta|$). This is Case 4 with negative real part, so the two points are stable foci. The eigenvectors are

$$\mathbf{z} = \begin{bmatrix} 1 \\ -\gamma + i\mu \end{bmatrix}, \quad \mathbf{z}^* = \begin{bmatrix} 1 \\ -\gamma - i\mu \end{bmatrix},$$

where $\mu = \sqrt{\beta^2 - \gamma^2} > 0$. The real and imaginary parts of $\mathbf{z} = \mathbf{u} + i\mathbf{v}$ are

$$\mathbf{u} = \begin{bmatrix} 1 \\ -\gamma \end{bmatrix}, \quad \mathbf{v} = \begin{bmatrix} 0 \\ \mu \end{bmatrix}.$$

Now $\mathbf{u}$ does not point along the x axis, as it did in the absence of damping.

The linearizations with damping and without differ, and their difference reflects changes in the global phase portrait. Figure 7.17 is the phase portrait (without arrows) obtained

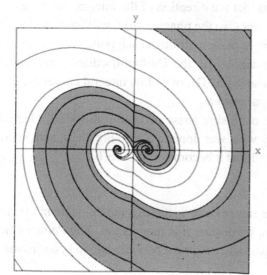

FIGURE 7.17

The phase portrait of the model system with damping. All the initial points whose motions converge to $(b, 0)$ are shaded dark [this is the *basin* of $(b, 0)$]. The basin of $(-b, 0)$ is not shaded. The stable and unstable invariant manifolds of the hyperbolic point at the origin form the separatrix between the two basins.

from a numerical solution of the full nonlinear differential equation. The origin is a saddle point and almost every integral curve converges to one or the other two fixed points at $(x, y) = (\pm b, 0)$. There are three exceptional solutions: the saddle point itself and the two integral curves that form its stable manifolds. These two integral curves also form the separatrix between the motions that converge to $(b, 0)$ and to $(-b, 0)$. Although the phase portrait of Fig. 7.17 has been obtained by computer calculation, it could have been sketched roughly without it. The linearizations at the fixed points, together with the phase portrait of the undamped system, would have been enough for a rough sketch. In other words, even for the damped system, linearization at the fixed points gives enough information to yield the general properties of the global phase portrait. This is essentially the content of the Hartman–Grobman theorem: the linearization really looks like the real thing in the vicinity of a fixed point.

On the other hand, numerical calculation provides about the only way to obtain an accurate phase portrait of this system, as analytic results shed little light on it. For truly nonlinear systems, except in some special cases, analytic closed-form solutions can almost never be obtained. The situation is even more complicated if a driving force is added.

7.2.2 STABILITY OF NONAUTONOMOUS SYSTEMS

When a time-dependent driving force is added, the system is no longer autonomous. Stationary states of nonautonomous systems are not necessarily fixed points of $\mathbb{M}$ but are orbits. In the case of oscillators, when such orbits are periodic and isolated from each other they are limit cycles, as in the van der Pol oscillator of Section 7.1.3. Like fixed points, limit cycles can be stable or unstable. We do not present phase portraits of systems with limit cycles, as Figs. 7.5 show their general characteristics.

In general, moreover, phase portraits are not as useful for nonautonomous systems as for autonomous ones. On $\mathbb{TQ}$, for instance, a driven oscillator may have the same values of x and $\dot{x}$ at two different times, but since the driving force may be different at the two times, so may the values of $\ddot{x}$. That means that the direction of the integral curve passing through $(x, \dot{x})$ will be different at the two times, so the phase portrait will fail to satisfy the condition that there is only one integral curve passing through each point of the manifold. That can become quite confusing. For example, for the Duffing oscillator, because the $x_2(t)$ of Eq. (7.11) depends on both $\sin \Omega t$ and $\sin 3\Omega t$, one of its integral curves can look like Fig. 7.18(a) for certain values of the parameters.

There are several ways to deal with a confusing phase portrait of this kind. One way is to develop it in time. This is essentially what was done to obtain Fig. 7.10, and when it is done for the Duffing oscillator of Fig. 7.18(a) it becomes Fig. 7.18(b).

THE POINCARÉ MAP

Another way to deal with nonautonomous systems is through the Poincaré map (mentioned in Section 4.2.2). In this subsection we discuss it in more detail. The Poincaré map is particularly useful in systems that are periodic or almost periodic and in systems with a periodic driving force.

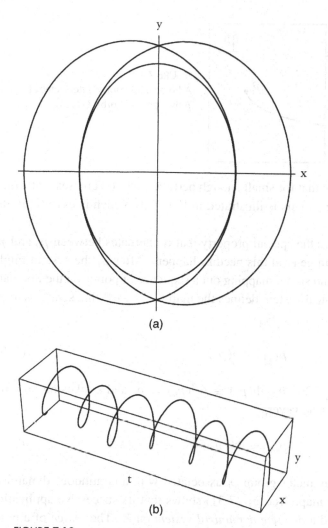

FIGURE 7.18

A single integral curve of the Duffing oscillator, obtained from the $x_2(t)$ of Eq. (7.11). (a) The integral curve intersects itself at several points. (b) If the integral curve is developed in time, there are no intersections.

Consider a periodic orbit of a dynamical system Δ on a carrier manifold $\mathbb{M}$ of dimension n. If $\xi(t, \xi_0)$ is a solution of period τ, then

$$\xi(t + \tau, \xi_0) = \xi(t, \xi_0).$$

Let $\mathbb{P}$ be an $(n - 1)$-dimensional submanifold of $\mathbb{M}$, called a *Poincaré section*, that is transverse to this orbit at time t_0, that is, that intersects the orbit at the point $\xi(t_0, \xi_0) \equiv p \in \mathbb{P}$, as shown in Fig. 7.19.

Now suppose that Δ is subjected to some perturbation, which forces $\xi(t, \xi_0)$ to change to some nearby orbit $\xi(t)$. Or suppose that Δ is not perturbed, but that the initial conditions are slightly changed, again causing $\xi(t, \xi_0)$ to change to a different orbit that we will also call $\xi(t)$. Then $\xi(t + \tau)$ will not in general be the same point in $\mathbb{M}$ as $\xi(t)$. Pick t_0 so that

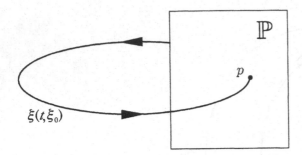

FIGURE 7.19
A Poincaré section $\mathbb{P}$ intersected at a point p by a periodic orbit $\xi(t, \xi_0)$.

$\xi(t_0) \equiv p_0$ lies on $\mathbb{P}$ and assume that for small enough perturbation $\xi(t)$ crosses $\mathbb{P}$ at a time close to $t_0 + \tau$ at some point p_1. This is illustrated in Fig. 7.20, which is essentially the same as Fig. 4.19.

The system of Fig. 7.20 has the special property that it alternates between p_0 and p_1 every time it arrives at $\mathbb{P}$, but in general this needn't happen. After p_1 the system might arrive at a third point $p_2 \in \mathbb{P}$, and so on, mapping out a sequence of points on the Poincaré section. In this way the dynamical system defines the map $\Phi : \mathbb{P} \to \mathbb{P}$ that sends p_0 to p_1 and p_1 to p_2, and more generally p_k to p_{k+1}:

$$p_{k+1} = \Phi(p_k). \tag{7.30}$$

In the special example of Fig. 7.20, $\Phi \circ \Phi(p_k) \equiv \Phi^2(p_k) = p_k$ and k takes on only the values 0 and 1. More generally, however,

$$p_{k+l} = \Phi^k(p_l). \tag{7.31}$$

Unlike the flow or one-parameter group φ_t associated with a continuous dynamical system, Φ is a single discrete map. Yet Eq. (7.31) shows that its successive application maps out what might be called a *discrete dynamical system* on $\mathbb{P}$. The *image* of a point $p \in \mathbb{P}$ under successive applications of Φ (i.e., the set of points obtained in this way), called Σ in Section 4.2.2, is analogous to an integral curve of a continuous dynamical system. The set of such images on $\mathbb{P}$ is called a *Poincaré map*. Although it *restricts* the continuous

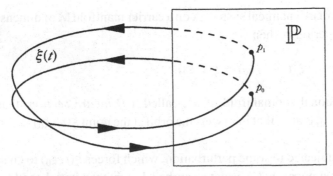

FIGURE 7.20
A Poincaré section $\mathbb{P}$ intersected at two points by an orbit $\xi(t)$. The Poincaré map Φ satisfies $\Phi(p_0) = p_1$ and $\Phi(p_1) = p_0$.

dynamical system Δ from which it is formed to the Poincaré section $\mathbb{P} \subset \mathbb{M}$, it can give useful and often important information about Δ. For example, the simplest situation is one in which $p \equiv p_f$ is a fixed point of the Poincaré map: $\Phi(p_f) = p_f$. Then the orbit of Δ that passes through p_f is periodic, as in Fig. 7.19. A limit cycle of Δ is a periodic orbit, and whether it is stable or unstable can also be answered entirely within the context of the Poincaré map. Indeed, suppose that $p_f \in \mathbb{P}$ corresponds to a limit cycle of Δ, and suppose that $p_0 \in \mathbb{P}$ is close to p_f. Then if the Poincaré map of p_0 approaches p_f, the limit cycle is stable, and if it diverges from p_f, it is unstable. In other words, the stability or instability of the limit cycle of Δ is reflected in the stability or instability of p_0 under the Poincaré map.

> **REMARK:** We came across a discrete dynamical system in connection with scattering by two hard disks in Section 4.1.3. There the role of Φ was played by M. □

In Fig. 7.20 neither p_0 nor p_1 is a fixed point of Φ: the map alternates between the two of them. As has been pointed out, however, both p_0 and p_1 separately are fixed points of Φ^2. Similar situations often arise. Discrete maps, like autonomous systems, can have stable orbits, and in the example of Fig. 7.20 the two points $\{p_0, p_1\}$ form such a stable orbit. In general, a stable orbit of a discrete map can consist of more than two points, even an infinite number, and they even can form a dense set or a fractal set (Section 4.1.3). Often a stable isolated orbit is a limit cycle of the discrete map; an example of this is the logistic map, to be discussed in Section 7.4.

LINEARIZATION OF DISCRETE MAPS

The preceding discussion shows the importance of the stability of the Poincaré map (and of other discrete maps that are useful in describing dynamical systems). Analyzing stability of such maps is analogous to analyzing stability of continuous dynamical systems, and we will proceed in a similar way.

The general discrete map that we treat is of the form

$$p_{k+1} = \Phi(p_k, a), \tag{7.32}$$

where a represents a set of what are called *tuning parameters* (e.g., the frequency and strength of the driving force, the damping factor, etc.). Let p_f be a fixed point of Φ. The first step is to move the origin to the fixed point, much as was done at Eq. (7.20): write $p_k' = p_k - p_f$. Then

$$p_{k+1}' \equiv p_{k+1} - p_f = \Phi(p_k, a) - p_f = \Phi(p_k' + p_f, a) \equiv \Phi'(p_k', a).$$

Drop the primes and simply use Eq. (7.32), remembering that the fixed point is now at the origin.

The next step is to linearize about the fixed point by expanding the right-hand side of (7.32) in a Taylor series. Let $m \equiv n - 1$ be the dimension of $\mathbb{P}$. Then p_k and Φ have m

components each, and the ith component of Φ expands to

$$\Phi^i(p, a) = \frac{\partial \Phi^i}{\partial p^j} p^j + o(p^2) \approx A^i_j(a) p^j,$$

defining the matrix $\mathbf{A}(a)$, called the *tangent map*, whose components are the $A^i_j(a)$. In fact, as in the continuous case, linearization shifts from $\mathbb{P}$ to its tangent space $\mathbf{T}_{p_f}\mathbb{P}$ at the fixed point. But since $\mathbb{P}$ is itself often a vector space, the distinction between $\mathbb{P}$ and $\mathbf{T}_{p_f}\mathbb{P}$ can usually be neglected. We suppress the dependence on a (although it will become important when we come to bifurcations) and write p as an m-component vector $\mathbf{p}$ to write the linearized form of Eq. (7.32) as the vector equation

$$\mathbf{p}_{k+1} = \mathbf{A}\mathbf{p}_k.$$

Note the differences between this equation and (7.22). What appears on the left-hand side here is not the time derivative, but the new point itself, which means that the solution depends not on the exponential of $\mathbf{A}$, as in (7.23), but on a power of it:

$$\mathbf{p}_k = \mathbf{A}^k \mathbf{p}_0, \tag{7.33}$$

where $\mathbf{p}_0$ is the initial point (necessarily close to the fixed point $\mathbf{p}_f = 0$). Another difference, even more evident, is that (7.33) yields not a continuous curve, but a discrete set of points.

Assume that $\mathbf{A}$ can be diagonalized. Then if $\mathbf{p}_0$ is an eigenvector of $\mathbf{A}$ belonging to eigenvalue λ,

$$\mathbf{p}_k = \lambda^k \mathbf{p}_0 \tag{7.34}$$

and the modulus of $\mathbf{p}_k$, its magnitude, is given by $|\mathbf{p}_k| = |\lambda|^k |\mathbf{p}_0|$. The modulus of $\mathbf{p}_f$ is zero (the origin), so whether or not $\mathbf{p}_k$ approaches $\mathbf{p}_f$ (i.e., whether or not $\mathbf{p}_f$ is stable) depends on $|\lambda|$. The stable submanifold $\mathbb{E}^s$ is spanned by the eigenvectors belonging to eigenvalues for which $|\lambda| < 1$, the center manifold $\mathbb{E}^c$ is spanned by those for which $|\lambda| = 1$, and the unstable manifold $\mathbb{E}^u$ is spanned by those for which $|\lambda| > 1$. For continuous systems stability depends on the real part of the eigenvalues, but for discrete maps it depends on the modulus of the eigenvalues. This is an important difference. Nevertheless, linearization of discrete maps gives some results that are similar to the ones for linearization of autonomous systems. To see how similar, write $\lambda = e^\alpha$, so that Eq. (7.34) becomes

$$\mathbf{p}_k = e^{\alpha k} \mathbf{p}_0.$$

This looks like the diagonalized version of Eq. (7.23), except that here k is a discrete variable taking on integer values, whereas in (7.23) t is a continuous variable. As a result, all of the different cases of the autonomous system occur also here, with similar diagrams, but only for discrete sets of points and in the Poincaré section $\mathbb{P}$ (for the Poincaré map, α plays the role that λ plays for autonomous systems). In addition, as we shall see, there is a new type of diagram.

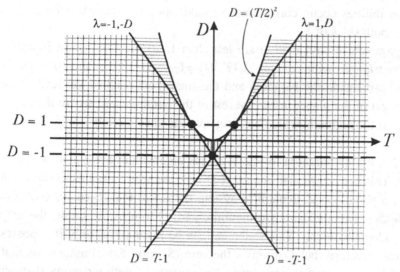

FIGURE 7.21

The (T, D) plane for a two-dimensional discrete map. The horizontal hatching indicates the region in which both eigenvalues are real and at least one of the $|\lambda| > 1$. The vertical hatching indicates the region in which both eigenvalues are real and at least one of the $|\lambda| < 1$. In the cross-hatched region, $|\lambda_1| < 1 < |\lambda_2|$. Inside the parabola $D = (T/2)^2$, the two eigenvalues are a complex conjugate pair. The only region in which $|\lambda| < 1$ for both eigenvalues is inside the stability triangle with vertices $(-2, 1), (2, 1), (0, -1)$ indicated by black dots. The only place where $|\lambda|$ can be 1 is on the two lines $D = \pm T - 1$.

In two dimensions the eigenvalues of $\mathbf{A}$ can be written entirely in terms of the determinant $D \equiv \det \mathbf{A}$ and the trace $T \equiv \text{tr } \mathbf{A}$, which are both real. The eigenvalues satisfy the equations $T = \lambda_1 + \lambda_2$ and $D = \lambda_1 \lambda_2$, and thus they are the roots of the quadratic equation

$$\lambda = \frac{T}{2} \pm \sqrt{\left(\frac{T}{2}\right)^2 - D}. \tag{7.35}$$

The eigenvalues determine the nature of the fixed point; D and T determine the eigenvalues; hence D and T determine the nature of the fixed point. Figure 7.21 maps out the (T, D) plane and shows the various possibilities.

If $D > (T/2)^2$, in the (blank) interior of the parabola in Fig. 7.21, the eigenvalues are a complex conjugate pair and, as in autonomous continuous systems, the Poincaré map involves rotation about the fixed point. The product of the eigenvalues is $\lambda\lambda^* = D$, so there is a real number θ such that $\lambda = \sqrt{|D|}e^{i\theta}$. If $|D| < 1$, the fixed point is stable (dim $\mathbb{E}^s = 2$), and if $|D| > 1$, it is unstable (dim $\mathbb{E}^u = 2$).

If $D < (T/2)^2$, the eigenvalues are both real. The stability of the fixed point (i.e., whether $|\lambda|$ is greater or less than 1) depends on the particular values of T and D. Because $\lambda_2 = D/\lambda_1$, the moduli of the eigenvalues can both be greater than 1 (unstable, dim $\mathbb{E}^u = 2$, $|D| > |\lambda_1|, |\lambda_2|$), both less than 1 (stable, dim $\mathbb{E}^s = 2, |D| < |\lambda_1|, |\lambda_2|$), and on both sides of 1 (unstable, dim $\mathbb{E}^s = \dim \mathbb{E}^u = 1$, $|\lambda_1| < |D| < |\lambda_2|$). In Fig. 7.21 the horizontal hatching indicates the region in which $|\lambda| > 1$ for at least one of the eigenvalues, and the vertical hatching indicates the region in which $|\lambda| < 1$ for one of them. The cross-hatched region, in which both possibilities occur, corresponds to saddle points, where the moduli of the eigenvalues lie on both sides of 1.

The only region where both moduli are less than 1, so the fixed point is stable, is the small *stability triangle* with vertices at $(T, D) = (-2, 1)$, $(2, 1)$, and $(0, -1)$. This includes both the part inside the parabola and the small part hatched vertically. There $\dim \mathbb{E}^s = 2$. The rest of the plane, inside the rest of the parabola and where the hatching is only horizontal, is the region in which both eigenvalues have moduli 1 or greater. There $\dim \mathbb{E}^u = 2$.

If $D = (T/2)^2$, on the parabola in Fig. 7.21, the eigenvalues are equal: $\lambda_1 = \lambda_2 \equiv \lambda = T/2 \equiv \sqrt{D}$. This is of particular interest if $D = 1$, because it applies to Hamiltonian systems. If Δ is a Hamiltonian dynamical system on $\mathbb{M} \equiv \mathbf{T}^*\mathbb{Q}$, then for each time t the dynamics yields a canonical transformation on $\mathbb{M}$. More to the point, the map Φ of $\mathbb{P}$ into itself is also canonical. This statement can be meaningful only if $\mathbb{P}$ possesses a (local) canonical structure, but Darboux's theorem (Section 5.4.2) assures us that an even-dimensional Poincaré section can always be constructed with a canonical structure inherited from $\mathbf{T}^*\mathbb{Q}$. Then Φ is canonical because it represents the development of Δ over a finite time. The Liouville volume theorem (Section 5.4.1) then tells us, moreover, that Φ preserves areas, and so does the tangent map $\mathbf{A}$, for it is the infinitesimal form of Φ. If $\mathbf{p}_1, \mathbf{p}_2 \in \mathbb{P}$ are two vectors, the area they subtend is, *up to sign*,

$$\det \mathbf{P} \equiv \det \begin{bmatrix} p_1^1 & p_2^1 \\ p_1^2 & p_2^2 \end{bmatrix},$$

where p_i^j is the jth component of $\mathbf{p}_i$. The transformed area is $\det \mathbf{AP} = \det \mathbf{A} \det \mathbf{P}$, and since $\mathbf{A}$ is area preserving, it follows that $D \equiv \det \mathbf{A} = \pm 1$ (the sign ambiguity enters here). This is why $|D| = 1$ is of particular interest for Hamiltonian systems. Area-preserving maps, because they correspond to Hamiltonian systems, are called *conservative*. Maps for which $|D| < 1$ are called *dissipative*.

Hence for a two-dimensional area-preserving tangent map, $D = 1$ (for the purposes of this discussion we choose the positive sign) and Eq. (7.35) becomes

$$\lambda = \frac{T}{2} \pm \sqrt{\left(\frac{T}{2}\right)^2 - 1}.$$

Then $\lambda_2 = 1/\lambda_1$, and the eigenvalues can be calculated entirely in terms of T.

The various possibilities can now be classified as follows:

Case 1: $|T| > 2$. Then λ_1, λ_2 are real, and $|\lambda_1| > 1 \Rightarrow |\lambda_2| < 1$ (this case lies in the cross-hatched area of Fig. 7.21). The fixed point is hyperbolic: $\mathbb{E}^u$ and $\mathbb{E}^s$ are each one dimensional. There are two possibilities. If λ_1 is positive (and hence so is λ_2), the Poincaré map develops very much like the phase portrait in Case 3 of the continuous autonomous system (Fig. 7.13). If λ_1 is negative (and hence so is λ_2), the Poincaré map jumps from one branch of its hyperbola to the other each time k increases by 1 in (7.33).

Case 2: $|T| < 2$. The map lies inside the parabola of Fig. 7.21: the eigenvalues occur in complex conjugate pairs and their product is 1, so $\lambda_1 = e^{i\theta}$, $\lambda_2 = e^{-i\theta}$, where θ is real.

This is a pure rotation by the angle θ each time k increases by 1 in (7.33). The fixed point is elliptic: $\mathbb{E}^c$ is two dimensional. If θ/π is irrational, the points of the Poincaré map are dense on circles.

Case 3: $|T| = 2$. The map lies on one of the two upper vertices of the stability triangle in Fig. 7.21. Then $\lambda_1 = \lambda_2 \equiv \lambda = \pm 1$. (This is the new type of diagram that does not occur in autonomous systems.) There are two possibilities: $\mathbf{A}$ is diagonalizable or it isn't.

If $\mathbf{A}$ is diagonalizable, the Poincaré map depends on the sign of λ. If $\lambda = 1$ all points of the Poincaré map are fixed. If $\lambda = -1$, each point of the Poincaré map jumps to its reflection through the origin each time k increases by 1 in (7.33). $\mathbb{E}^c$ is two dimensional.

If $\mathbf{A}$ is not diagonalizable, then it is similar to the $\mathbf{A}$ of Case 2 for autonomous systems, namely

$$\mathbf{A} = \begin{bmatrix} \pm 1 & 0 \\ \mu & \pm 1 \end{bmatrix}.$$

The Poincaré map traces out points on lines parallel to the 1 axis (or the 2 axis if μ is in the upper right-hand corner of $\mathbf{A}$). The fixed point is said to be *parabolic*. We leave the details of the motion to Problem 3. See also Problem 4.

EXAMPLE: THE LINEARIZED HÉNON MAP

We end this section with a brief introduction to a particular discrete map, the *Hénon map* (Hénon, 1976, Devaney, 1986), which exhibits many properties common to two-dimensional discrete maps. Later, in Section 7.5.2, it will be related to a physical system, but at this point we neglect that and study it as a typical example. Although this section will deal only with its linearization, we start with the general expression.

The Hénon map $H : \mathbb{R}^2 \to \mathbb{R}^2$ is defined by the equations

$$\left. \begin{array}{l} x_{n+1} = 1 - ax_n^2 + y_n, \\ y_{n+1} = bx_n, \end{array} \right\} \tag{7.36}$$

where a and b make up the tuning parameters [defined at Eq. (7.32)]. Here we have written $\mathbf{p} = (x, y)$ rather than $\mathbf{p} = (p^1, p^2)$ in order to avoid a profusion of indices. This nonlinear map will be discussed in Section 7.5.2; at this point we linearize about its fixed points.

The fixed points are obtained by setting $x_{n+1} = x_n \equiv X$ and $y_{n+1} = y_n \equiv Y$. This results in a quadratic equation for X, namely

$$aX^2 + (1 - b)X - 1 = 0,$$

whose solutions are

$$X_\pm = \frac{b - 1 \pm \sqrt{(b - 1)^2 + 4a}}{2a}.$$

The corresponding values for Y are

$$Y_\pm = bX_\pm.$$

There are two fixed points, $\mathbf{p}_+ = (X_+, Y_+)$ and $\mathbf{p}_- = (X_-, Y_-)$.

The prescription for analyzing the stability about any fixed point $\mathbf{p}_f = (X, Y)$ starts by moving the origin to $\mathbf{p}_f$. For (X, Y) either fixed point, (X_+, Y_+) or (X_-, Y_-), write $x = X + x'$ and $y = Y + y'$, which yields the following equations for (x', y'), the deviation from (X, Y):

$$x'_{n+1} = -2aXx'_n - a(x'_n)^2 + y'_n,$$
$$y'_{n+1} = bx'_n.$$

To linearize, drop the quadratic term in the first of these equations:

$$x'_{n+1} = -2aXx'_n + y'_n, \tag{7.37}$$
$$y'_{n+1} = bx'_n.$$

The tangent map is thus

$$\mathbf{A} = \begin{pmatrix} -2aX & 1 \\ b & 0 \end{pmatrix}. \tag{7.38}$$

The trace and determinant of the tangent map are $\operatorname{tr}\mathbf{A} \equiv T = -2aX$ and $\det\mathbf{A} \equiv D = -b$. The nature of the fixed points depends on a and b (recall that $X_\pm$ depend on a and b). Whether the fixed point is stable or not is answered by referring to Fig. 7.21: stability occurs only inside a region that corresponds to the stability triangle of that figure (see Problem 5). All other points on the (a, b) plane lead to fixed points that are unstable or have center manifolds.

7.3 PARAMETRIC OSCILLATORS

This chapter is devoted to nonlinear dynamics, but most of it concerns oscillating systems. In this section we discuss another type of oscillating system even though we will consider only its linear approximation.

Harmonic oscillators can be generalized not only by explicitly nonlinear terms in their equations of motion, but also by frequencies that depend on time. We have already treated time-dependent frequencies in Section 6.4.1 on adiabatic invariance, but the systems we will now discuss have frequencies that depend periodically on the time. Because such systems occur in several physical contexts (e.g., astronomy, solid state physics, plasma physics, stability theory) the general properties of their solutions are of some interest.

The general equation of this type, called *Hill's equation* (Hill, 1886), is

$$\ddot{x} + G(t)x = 0, \quad G(t + \tau) = G(t). \tag{7.39}$$

Systems satisfying (7.39) are called *parametric oscillators*. Although these are linear equations, what they have in common with nonlinear ones is that they have no known general solutions in closed form, and actual solutions are found by approximation and perturbative techniques. Nevertheless, some general results can be obtained analytically. Our approach will be through *Floquet theory* (Floquet, 1883; see also Magnus and Winkler, 1966, or McLachlan, 1947).

7.3.1 FLOQUET THEORY

THE FLOQUET OPERATOR R

Equation (7.39) is actually more general than it may seem at first. Indeed, a *damped* parametric oscillator equation

$$\ddot{y} + \eta_1(t)\dot{y} + \eta_2(t)y = 0,$$

where $\eta_i(t + \tau) = \eta_i(t)$, $i = 1, 2$, can be brought into the form of (7.39). To do this, make the substitution $y = x \exp\{-\frac{1}{2}\int \eta_1(t)\, dt\}$. It is then found that x satisfies (7.39) with $G(t) = [\eta_2(t) - \frac{1}{2}\dot{\eta}_1 - \frac{1}{4}\eta_1^2]$, which is also periodic with period τ. Thus the undamped-oscillator Eq. (7.39) can be used also to analyze the damped parametric oscillator.

Equation (7.39) is general also because it represents many possible systems, one for each function G, and each G will lend the system different properties. For example, the stability of oscillatory solutions of these equations depends on G. We will describe oscillatory solutions and will then describe a way of parametrizing G in order to find the conditions for stability and instability. We consider only real functions G.

For any G Eq. (7.39) is a linear ordinary differential equation of second order with real periodic coefficients, whose solutions $\mathbf{x}(t)$ form a two-dimensional vector space $\mathbb{V}_G$ (boldface to emphasize the vector-space nature of the solutions). Although $G(t)$, and hence also Eq. (7.39), are periodic with period τ, the general solution $\mathbf{x}(t)$ need not be periodic. Nevertheless, the periodicity of (7.39) implies that if $\mathbf{x}(t)$ is a solution, so is $\mathbf{x}(t + \tau)$. Thus $\mathbf{x}(t + \tau)$ is also in $\mathbb{V}_G$, and hence there is an operator $\mathbf{R} : \mathbb{V}_G \to \mathbb{V}_G$, called the Floquet operator, that maps $\mathbf{x}(t)$ onto $\mathbf{x}(t + \tau)$. Moreover, $\mathbf{R}$ is linear. Indeed, suppose $\mathbf{x}_1(t)$ and $\mathbf{x}_2(t)$ are solutions; then because (7.39) is linear, $\alpha\mathbf{x}_1(t) + \beta\mathbf{x}_2(t) \equiv \mathbf{y}(t)$ is also a solution. Hence

$$\mathbf{y}(t + \tau) \equiv \mathbf{R}\mathbf{y}(t) = \mathbf{R}[\alpha\mathbf{x}_1(t) + \beta\mathbf{x}_2(t)].$$

But

$$\mathbf{y}(t + \tau) \equiv \alpha\mathbf{x}_1(t + \tau) + \beta\mathbf{x}_2(t + \tau) \equiv \alpha\mathbf{R}\mathbf{x}_1(t) + \beta\mathbf{R}\mathbf{x}(t).$$

Hence

$$\mathbf{R}[\alpha\mathbf{x}_1(t) + \beta\mathbf{x}_2(t)] = \alpha\mathbf{R}\mathbf{x}_1(t) + \beta\mathbf{R}\mathbf{x}(t),$$

or $\mathbf{R}$ is linear.

If the Floquet operator $\mathbf{R}$ is known, the solution at any time t is determined by the solution during one τ period, for instance $0 \le t < \tau$. Hence one-period solutions can be extended to longer times by applying $\mathbf{R}$, so the stability of a solution can be understood in terms of $\mathbf{R}$. As in Section 7.2.2, this involves the eigenvalues of $\mathbf{R}$ and its stable and unstable manifolds. Different instances of Eq. (7.39) (i.e., different G functions) have different solution spaces $\mathbb{V}_G$ in which the solutions have different long-term behavior, so $\mathbf{R}$ depends on G (we write $\mathbf{R}$ rather then $\mathbf{R}_G$ only to avoid too many indices in the equations). Since stability depends on $\mathbf{R}$, it depends through $\mathbf{R}$ on G. We will turn to this dependence later.

STANDARD BASIS

First we present a standard representation of $\mathbf{R}$ in a *standard basis* in $\mathbb{V}_G$. In general, solutions are determined by their initial conditions, and we choose the standard basis $\{\mathbf{e}_1(t), \mathbf{e}_2(t)\}$ to satisfy the conditions (we use boldface for the functions, but not for the values they take on at specific values of t)

$$\begin{aligned} e_1(0) &= 1, & \dot{e}_1(0) &= 0, \\ e_2(0) &= 0, & \dot{e}_2(0) &= 1 \end{aligned} \tag{7.40}$$

The standard basis is the analog of the sine and cosine basis for the harmonic oscillator. Incidentally, there is no implication that the $\mathbf{e}_k(t)$ are in any sense normalized or orthogonal. The functions $\{\mathbf{e}_1(t), \mathbf{e}_2(t)\}$ form a basis iff they are linearly independent. Their linear independence follows from Eqs. (7.40). Indeed, consider the equations

$$\begin{aligned} \alpha_1 \mathbf{e}_1(t) + \alpha_2 \mathbf{e}_2(t) &= 0, \\ \alpha_1 \dot{\mathbf{e}}_1(t) + \alpha_2 \dot{\mathbf{e}}_2(t) &= 0; \end{aligned}$$

$\mathbf{e}_1$ and $\mathbf{e}_2$ are linearly dependent iff these equations yield nonzero solutions for α_1 and α_2. But that can happen iff the *Wronskian determinant*

$$W(\mathbf{e}_1, \mathbf{e}_2) \equiv \begin{bmatrix} \mathbf{e}_1 & \mathbf{e}_2 \\ \dot{\mathbf{e}}_1 & \dot{\mathbf{e}}_2 \end{bmatrix} \equiv \mathbf{e}_1\dot{\mathbf{e}}_2 - \mathbf{e}_2\dot{\mathbf{e}}_1$$

of the two functions, vanishes. Equations (7.40) show that $W(\mathbf{e}_1, \mathbf{e}_2) = 1$ at time $t = 0$, and Eq. (7.39) can be used to show that $dW(\mathbf{e}_1, \mathbf{e}_2)/dt = 0$. Hence $W(\mathbf{e}_1, \mathbf{e}_2) = 1$ for all time t. Thus $W(\mathbf{e}_1, \mathbf{e}_2) \neq 0$, so $\mathbf{e}_1$ and $\mathbf{e}_2$ are linearly independent as asserted, and they therefore form a basis for $\mathbb{V}_G$. It is interesting, moreover, that $W(\mathbf{e}_1, \mathbf{e}_2) = 1$: since $\mathbf{e}_1$ and $\mathbf{e}_2$ are solutions of (7.39) with a particular G function, they depend on G, and yet $W(\mathbf{e}_1, \mathbf{e}_2) = 1$ is independent of G.

Because $\{\mathbf{e}_1(t), \mathbf{e}_2(t)\}$ form a basis, every solution $\mathbf{x}(t) \in \mathbb{V}_G$ can be written in the form

$$\mathbf{x}(t) = a_1 \mathbf{e}_1(t) + a_2 \mathbf{e}_2(t). \tag{7.41}$$

In particular, the $\mathbf{e}_k(t + \tau) \equiv \mathbf{R}\mathbf{e}_k(t)$ are solutions. The right-hand side here is merely the result of $\mathbf{R}$ acting on the basis vectors, so

$$\mathbf{e}_k(t + \tau) = r_{jk}\mathbf{e}_j(t), \tag{7.42}$$

where the r_{jk} are the matrix elements of R, the representation of $\mathbf{R}$ in the standard basis (see the book's appendix). Equation (7.42) is the expansion of $\mathbf{e}_k(t + \tau)$ in the standard basis according to (7.41). The r_{jk} can be calculated from the initial values (7.40) of $\mathbf{e}_1$ and $\mathbf{e}_2$. For instance (as above, we do not use boldface for values that the functions take on at specific values of t)

$$e_1(\tau) = r_{11}e_1(0) + r_{21}e_2(0) = r_{11}.$$

In this way it is found that

$$\begin{aligned}\mathbf{e}_1(t+\tau) &= e_1(\tau)\mathbf{e}_1(t) + \dot{e}_1(\tau)\mathbf{e}_2(t), \\ \mathbf{e}_2(t+\tau) &= e_2(\tau)\mathbf{e}_1(t) + \dot{e}_2(\tau)\mathbf{e}_2(t),\end{aligned}\Big\}$$

or that

$$R = \begin{bmatrix} e_1(\tau) & e_2(\tau) \\ \dot{e}_1(\tau) & \dot{e}_2(\tau) \end{bmatrix}. \tag{7.43}$$

(See the book's appendix to understand the order in which the matrix elements are written.)

EIGENVALUES OF R AND STABILITY

Stability of the solutions depends on the eigenvalues of $\mathbf{R}$. Because dim $\mathbb{V}_G = 2$, the eigenvalue condition $\mathbf{Rx} = \rho\mathbf{x}$ yields a quadratic equation for ρ. This equation can be obtained in the standard basis from Eq. (7.43); it is (we will often write simply e_k and $\dot{e}_k$ for $e_k(\tau)$ and $\dot{e}_k(\tau)$)

$$\rho^2 - (e_1 + \dot{e}_2)\rho + 1 = 0, \tag{7.44}$$

whose solutions are [using the fact that $W(\mathbf{e}_1, \mathbf{e}_2) = 1$]

$$\rho_{\pm} = \frac{(e_1 + \dot{e}_2) \pm \sqrt{(e_1 + \dot{e}_2)^2 - 4}}{2}. \tag{7.45}$$

Notice that $\rho_{\pm}$ can be complex and that

$$\operatorname{tr}\mathbf{R} \equiv \rho_+ + \rho_- = e_1(\tau) + \dot{e}_2(\tau) \equiv T \tag{7.46}$$

[T is sometimes also called the *discriminant* Δ of (7.39)]; also,

$$\det\mathbf{R} \equiv \rho_+\rho_- = 1. \tag{7.47}$$

The determinant condition is essentially a consequence of the Wronskian condition. The trace condition will allow the calculation of certain special solutions, as will be seen in what follows.

Since $\mathbf{R}$ depends on G, so do $\rho_{\pm}$. Some G functions yield eigenvalues that are distinct, and some yield eigenvalues that are equal. We take up these two cases separately.

Case 1. R *has distinct eigenvalues.* Then **R** is diagonalizable, or its eigenfunctions, which we will call $e_{\pm}(t)$, form a basis in $\mathbb{V}_G$. (This is the analog of the exponential form of the solutions of the harmonic oscillator.) If G is such that the eigenvalues do not lie on the unit circle, one of them has magnitude greater than one and the other less than one. As in Section 7.2.2, **R** then has a stable and unstable manifold $\mathbb{E}^s$ and $\mathbb{E}^u$, each of dimension one. If G is such that the eigenvalues lie on the unit circle, they are complex conjugates and **R** has only a center manifold $\mathbb{E}^c$, of dimension two. It is usual to write this in terms of the *Floquet characteristic exponent* λ (also known as the quasi-energy eigenvalue) given by

$$\rho_{\pm} = \exp\{\pm i\lambda\tau\}; \tag{7.48}$$

λ is defined up to addition of $2n\pi/\tau$, where n is a nonzero integer. The eigenfunctions $e_{\pm}$ can be written at any time in terms of λ (or $\rho_{\pm}$) and of $e_{\pm}(t)$ for $0 \leq t < \tau$:

$$e_{\pm}(t + n\tau) = \rho_{\pm}^n e_{\pm}(t) \equiv \exp\{\pm in\lambda\tau\}\, e_{\pm}(t). \tag{7.49}$$

In other words, if $e_{\pm}(t)$ is known for $0 \leq t < \tau$ it is known for any other time. If $e_{\pm}(t)$ for all t is written in the form

$$e_{\pm}(t) = \exp\{\pm i\lambda t\}\, u_{\pm}(t), \tag{7.50}$$

then the $u_{\pm}(t)$, not themselves solutions of (7.39), are both periodic with period τ. Indeed, from (7.49) and (7.50) it follows that

$$e_{+}(t + \tau) \equiv \exp\{i\lambda(t + \tau)\}\, u_{+}(t + \tau) = \exp\{i\lambda\tau\}[\exp\{i\lambda t\}\, u_{+}(t)],$$

so that $u_{+}(t + \tau) = u_{+}(t)$. Similarly, $u_{-}(t + \tau) = u_{-}(t)$.

Stability can be stated in terms of λ. The Floquet exponent λ is real iff $\rho_{\pm}$ are a complex-conjugate pair on the unit circle. A necessary and sufficient condition for this (Problem 16) is that $|T| \equiv |e_1(\tau) + \dot{e}_2(\tau)| < 2$. This is the situation in which dim $\mathbb{E}^c = 2$, when the solutions oscillate and are bounded and hence stable. The operator **R** cannot be diagonalized over the reals: it is a rotation in the real space $\mathbb{V}_G$. Although the solutions oscillate, Eq. (7.50) shows that they are not strictly periodic unless $\lambda\tau/2\pi$ is rational. We call this Case 1S (for "Stable"). A limiting subcase is $\lambda\tau = n\pi$ (or $\rho_{\pm} = \pm 1$), in which **R** $= \pm \mathbb{I}$.

If G is such that λ is complex or imaginary, (7.49) shows that one of $e_{\pm}$ diverges exponentially (is unstable), while the other converges to zero (is stable) in the solution space: $\mathbb{E}^s$ and $\mathbb{E}^u$ are each one dimensional. In terms of the standard basis, this occurs iff $|T| > 2$. The divergent solution is called *parametrically resonant.* We call this Case 1U (for "Unstable"). It also has the limiting subcase in which **R** $= \pm \mathbb{I}$. In fact Cases 1S and 1U are separated by this limiting subcase, for which $|T| = 2$.

Case 2. R *has equal eigenvalues* Then the determinant condition implies that $\rho_{+} = \rho_{-} \equiv \rho = \pm 1$ and the trace condition implies that $2\rho = e_1 + \dot{e}_2$, or $|T| = 2$. We exclude

$\mathbf{R} = \pm\mathbb{I}$, which is the subcase separating Case 1S from Case 1U. The only other possibility is that G is such that $\mathbf{R}$ is not diagonalizable and therefore can be represented by

$$R = \begin{bmatrix} \rho & \mu \\ 0 & \rho \end{bmatrix} \tag{7.51}$$

with $\mu \neq 0$. For this situation consider the pair of vectors whose representation in the basis of (7.51) is

$$\tilde{e}_1 = \begin{bmatrix} \mu \\ 0 \end{bmatrix} \quad \text{and} \quad \tilde{e}_2 = \begin{bmatrix} \mu \\ 1 \end{bmatrix}. \tag{7.52}$$

Because these are linearly independent and hence also form a basis for $\mathbb{V}_G$, they can be used to study the stability of solutions. From their definition, $\mathbf{R}\tilde{e}_1(t) \equiv \tilde{e}_1(t + \tau) = \rho\tilde{e}_1(t)$ and $\mathbf{R}\tilde{e}_2(t) \equiv \tilde{e}_2(t + \tau) = \tilde{e}_1(t) + \rho\tilde{e}_2(t)$. Iterating this operation leads to

$$\left. \begin{array}{l} \tilde{e}_1(t + n\tau) \equiv \mathbf{R}^n\tilde{e}_1(t) = \rho^n\tilde{e}_1(t), \\ \tilde{e}_2(t + n\tau) \equiv \mathbf{R}^n\tilde{e}_2(t) = n\rho^{n-1}\tilde{e}_1(t) + \rho^n\tilde{e}_2(t). \end{array} \right\} \tag{7.53}$$

This shows that $\tilde{e}_1(t)$ oscillates with period τ if $\rho = 1$ and with period 2τ if $\rho = -1$ and that $\tilde{e}_2(t)$ diverges for both values of ρ. The divergence of $\tilde{e}_2(t)$ is not exponential as in Case 1U, but linear (or algebraic). As in Case 1U, nevertheless, dim $\mathbb{E}^u = 1$, but now dim $\mathbb{E}^c = 1$.

We show briefly how the $\{\tilde{e}_1, \tilde{e}_2\}$ basis can be expressed in terms of the $\{e_1, e_2\}$ basis. Write $\tilde{e}_1 = \alpha_1 e_1 + \alpha_2 e_2$. Remember that $\tilde{e}_1$ is an eigenvector of $\mathbf{R}$; when that is written in the $\{e_1, e_2\}$ basis it leads to the equation

$$(r_{11} - \rho)\alpha_1 + r_{12}\alpha_2 = (e_1 - \rho)\alpha_1 + e_2\alpha_2 = 0,$$

where Eq. (7.43) has been used [recall that $e_k \equiv e_k(\tau)$]. It follows that (up to a constant multiple)

$$\tilde{e}_1(t) = e_2 e_1(t) - [e_1 - \rho] e_2(t).$$

If $e_2 \neq 0$, this has nonzero component along e_1, so $\tilde{e}_2(t)$ can be chosen as $\tilde{e}_2(t) = e_2(t)$. Then after some algebra it is found that

$$\tilde{e}_2(t + \tau) \equiv \mathbf{R}\tilde{e}_2(t) = \tilde{e}_1(t) + \rho\tilde{e}_2(t). \tag{7.54}$$

If $e_2 = 0$, the situation is slightly different. We leave that to Problem 17.

In sum, stability is determined largely by $|T|$. If $|T| < 2$, all solutions oscillate and are stable (Case 1S). If $|T| > 2$, there is one stable solution converging to zero, but the general solution is exponentially unstable (Case 1U). These two possibilities are separated by $|T| = 2$, for which there is one stable oscillating solution and the general solution is either also stable and oscillating (limit of Cases 1S and 1U) or is algebraically unstable (Case 2). Further, when $|T| = 2$, the oscillating solutions have period τ if $T = 2$ and period 2τ if $T = -2$.

DEPENDENCE ON G

So far it has been shown that stability of the solutions depends on the G function of Eq. (7.39). We now describe how this stability changes within a particular kind of one-parameter set of G functions: $G \equiv \epsilon + G_0$, where G_0 is a fixed function and ϵ is a real parameter. Equation (7.39) now becomes

$$\ddot{x} + [\epsilon + G_0(t)] x = 0, \quad G_0(t + \tau) = G_0(t), \tag{7.55}$$

which depends explicitly on the parameter ϵ. Everything will now depend on ϵ: for example, the standard basis will be written in the form $\{\mathbf{e}_1(t, \epsilon), \mathbf{e}_2(t, \epsilon)\}$ and tr $\mathbf{R}$ in the form $T(\epsilon) \equiv e_1(\epsilon) + \dot{e}_2(\epsilon)$.

We state a theorem without proof (see Magnus and Winkler, 1966; McLachlan, 1974) that tells how the stability of solutions depends on ϵ. Specifically, the theorem tells how Cases 1S, 1U, and 2 are distributed throughout the range of ϵ for fixed τ. It states that there are two infinite sequences of ϵ values, ϵ_n and ϵ'_{n+1}, $n = 0, 1, 2, 3, \ldots$, ordered according to

$$\epsilon_0 < \epsilon'_1 \le \epsilon'_2 < \epsilon_1 \le \epsilon_2 < \epsilon'_3 \le \epsilon'_4 \ldots, \tag{7.56}$$

that divide the entire range of ϵ in the following way: both the ϵ_k and the ϵ'_k increase without bound as $k \to \infty$. The intervals

$$(\epsilon_0, \epsilon'_1), \ (\epsilon'_2, \epsilon_1), \ (\epsilon_2, \epsilon'_3), \ (\epsilon'_4, \epsilon_3), \ \ldots$$

(those with an unprimed ϵ_k at one end and an ϵ'_k at the other) correspond to Case 1S. The intervals

$$(-\infty, \epsilon_0), (\epsilon'_1, \epsilon'_2), (\epsilon_1, \epsilon_2), (\epsilon'_3, \epsilon'_4), \ \ldots$$

(those with either unprimed ϵ_ks or ϵ'_ks at both ends) correspond to Case 1U. The points $\epsilon = \epsilon_k$ and $\epsilon = \epsilon'_k$ that separate the intervals correspond either to the limiting subcase of Case 1, in which the solutions are stable and periodic with period τ or 2τ, or to Case 2, in which the solutions are unstable.

In general the trace $T(\epsilon)$ depends not only on ϵ, but also on the periodicity τ of G_0, and so the transition values ϵ_n and ϵ'_n depend on τ. There are regions in the (τ, ϵ) plane in which $|T|$ is less than 2 and others in which it is greater. Those in which $|T| > 2$ are regions of parametric resonance, and in principle they can be mapped out in the (τ, ϵ) plane.

In the next section we make all this concrete by moving on to a physical example.

7.3.2 THE VERTICALLY DRIVEN PENDULUM

THE MATHIEU EQUATION

The physical example to which we turn is nonlinear, but it will be linearized and transformed by changes of variables to the form of (7.55) in which G is a cosine function. The result is known as the *Mathieu equation* (Whittaker & Watson, 1943).

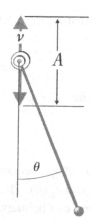

FIGURE 7.22
A pendulum whose support vibrates with amplitude A and frequency ν.

Consider a frictionless plane pendulum consisting of a light stiff rod of a fixed length l with a mass m attached to its end and whose pivot oscillates vertically with frequency ν (see Fig. 7.22). The vertical position of the pivot is given by $y = A \cos(\nu t)$. Although the equation of the pendulum can be obtained from the appropriate Lagrangian (Problem 18), we use the equivalence principle (Section 1.5.3). Since the acceleration of the pivot is $-A\nu^2 \cos \nu t$, the acceleration of gravity g in the usual equation for a pendulum should be replaced by $g \pm A\nu^2 \cos \nu t$ (the sign, which can be chosen for convenience, depends on when t is taken to be zero and which direction is counted as positive). The equation of motion of this driven pendulum is therefore

$$\ddot{\theta} + [\omega^2 - \alpha\nu^2 \cos \nu t] \sin \theta = 0, \tag{7.57}$$

where $\alpha = A/l$ and $\omega = \sqrt{g/l}$ is the natural (circular) frequency of the pendulum.

We want to study how the behavior of the pendulum depends on the amplitude A and (circular) frequency $\nu \equiv 2\pi/\tau$ of the pivot's vibration and will treat the system in two regimes: one in which the pendulum hangs downward the way a pendulum normally does and the other in which it is inverted (i.e., for angles θ close to π). It will be seen that contrary to intuition there are certain conditions under which the inverted vibrating pendulum is stable. When θ is replaced by $\phi = \pi - \theta$, so that θ close to π means that ϕ is small, Eq. (7.57) becomes

$$\ddot{\phi} - [\omega^2 - \alpha\nu^2 \cos \nu t] \sin \phi = 0, \tag{7.58}$$

which differs from (7.57) only in sign. Both equations are nonlinear. They can be fully solved only numerically, and the solutions can exhibit chaotic behavior. We will treat them, however, in the small-angle approximation by linearizing them.

The two equations are first written in a unified way. Start by changing the independent variable to $t' = \nu t/2$ and dropping the prime. The driving frequency ν is then partially hidden in the new time variable and both equations become

$$\ddot{\eta} + [\epsilon - 2h \cos 2t] \sin \eta = 0, \tag{7.59}$$

where η can be either θ or ϕ, and $h = \pm 2A/l$ (the sign of the cosine term is irrelevant). In (7.57) $\epsilon = (2\omega/\nu)^2$, and in (7.58) $\epsilon = -(2\omega/\nu)^2$, so positive ϵ describes the normal pendulum and negative ϵ the inverted one. In the small-angle approximation Eq. (7.59) takes on the linearized form

$$\ddot{\eta} + [\epsilon - 2h \cos 2t]\, \eta = 0. \tag{7.60}$$

This is the *Mathieu equation* (Abramowitz and Stegun, 1972), an instance of Eq. (7.55), with $G_0(t) = 2h \cos 2t$, a function of period π.

STABILITY OF THE PENDULUM

According to (7.56), the stability of solutions of Eq. (7.60) is determined by a sequence of ϵ_k and ϵ_k', which are now functions depending on h; we write $\epsilon_k(h)$ and $\epsilon_k'(h)$. Hence the $(\epsilon_k, \epsilon_{k+1})$ etc. intervals depend on h, so the stability of the pendulum can be mapped out in the (ϵ, h) plane. Since l and ω are fixed, $h \propto A$ and $\epsilon \propto \nu^{-2}$, so a stability map in the (ϵ, h) plane is equivalent to one in the (ν, A) plane.

Mapping out the regions of stability and instability requires only finding their boundaries, where $|T| = 2$. Case 2 does not occur in the Mathieu equation (7.60), so its solutions at the boundaries are periodic with period $\tau = \pi$ or $2\tau = 2\pi$. This leads to an eigenvalue problem: to find the values of $\epsilon(h)$ that correspond to solutions of (7.60) with period π or 2π.

Solutions of the Mathieu equation have been found by many authors. A convenient place to find the needed results is in Section 20 of the Abramowitz and Stegun book cited above (Fig. 7.23). The $|T| > 2$ regions of parametric resonance or instability are shaded, and the $|T| < 2$ regions of stability are unshaded. The curves separating them are the $\epsilon_n(h)$ ($T = 2$; solid curves) and $\epsilon_n'(h)$ ($T = -2$; dashed curves). Only positive h is shown because the graph is symmetric about the ϵ axis.

First consider positive ϵ. The figure shows that the regions of parametric resonance (we will call them *tongues*) touch the ϵ axis wherever ϵ is the square of an integer n. At

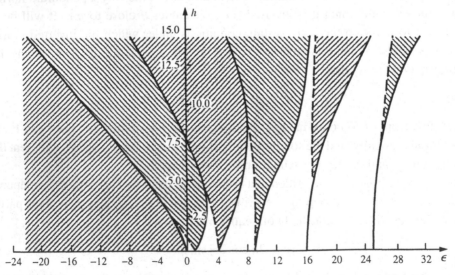

FIGURE 7.23
The stability and instability regions for the inverted pendulum (from Magnus and Winkler, 1966).

these points $v = 2\omega/n$ (compare Problem 19). The tongues get broader in the direction of increasing A (increasing h), but the lower the driving frequency v (the larger ϵ) the narrower the tongue. At high v and high A (low ϵ and high h) the tongues coalesce into a large region of instability. In between the tongues at relatively low A are stable (unshaded) regions. The unstable regions become narrower as ϵ increases (as v decreases), eventually becoming indistinguishable from curves at the scale of this graph. Physical considerations alone predict that the positive ϵ axis itself is a region of stability, since it represents the pendulum hanging down undisturbed.

> **REMARK:** The linearized theory predicts that the angle η will grow without bound in unstable motion. Of course this ignores dissipation and nonlinear effects in a real pendulum. ☐

THE INVERTED PENDULUM

Now consider negative ϵ. Physical considerations predict that the entire negative-ϵ axis is a region of instability, since it represents the inverted pendulum at rest: the slightest disturbance causes the pendulum to swing. In fact it might be expected on such considerations that the entire negative-ϵ half plane is unstable. Yet it turns out that a small part of the stable region that intersects the ϵ axis between $\epsilon = 0$ and $\epsilon = +1$ extends across the h axis into the negative-ϵ half plane. That may be surprising, for it represents a small region of stability for the inverted pendulum at low values of $|\epsilon|$ (high v) and amplitudes that are not too high. The boundaries of this region, namely $\epsilon_0(h)$ and $\epsilon_1'(h)$, have been calculated numerically and are reported in Abramowitz and Stegun to eighth order in h:

$$\epsilon_0(h) = -\frac{h^2}{2} + \frac{7h^4}{128} - \frac{29h^6}{2304} + \frac{68687h^8}{18874368} + \cdots,$$

$$\epsilon_1(h) = 1 - h - \frac{h^2}{8} + \frac{h^3}{64} - \frac{h^4}{1536} - \frac{11h^5}{36864} + \frac{49h^6}{589824} - \frac{55h^7}{9437184} - \frac{83h^8}{35389440} + \cdots.$$

These formulas are good only for relatively small values of h; we give some numerical results where the approximations are valid.

The $\epsilon_1'(h)$ curve crosses the h axis at $h \approx 0.90805$, or at $A \approx 0.45403\,l$, at which value $\epsilon_0(0.90805) \approx -0.38047$ or $v \approx 5.2567\,\omega$. That means that if the pivot amplitude is $0.45403\,l$, the inverted pendulum is stable at all frequencies higher than $v_{min} = 5.2567\,\omega$. At lower amplitudes the minimum frequency v_{min} for stability increases [$\epsilon_0(h)$ decreases] as h approaches zero. It is shown in Problem 21 that v_{min} grows as $1/A$ in the limit as $A \to 0$.

At higher amplitudes there is also a maximum v for stability. For instance, at $h = 1$ (or $A = l/2$) the two curves take on the values $\epsilon_0(1) \approx -0.45426$ (where $v \approx 4.4028\omega$) and $\epsilon_1'(1) \approx -0.11025$ (where $v \approx 18.141\omega$). That means that if the pivot amplitude is $l/2$, the inverted pendulum is stable at all frequencies between $v_{min} = 4.4028\omega$ and $v_{max} = 18.141\omega$. The graph shows that the stability region gets narrower as h increases, essentially vanishing by the time $h = 3$.

A reasonable estimate of the small-h shape of the first tongue can be obtained in the following way. It is shown in Problem 19 that the first tongue reaches the ϵ axis at $(h, \epsilon) = (0, 1)$. With those values of the parameters the solution of Eq. (7.60) is $\eta(t) = \eta_0 \cos(t + \delta)$. Assume from considerations of continuity that for values of (h, ϵ) close to $(0, 1)$ the solution is similar:

$$\eta(t) = \eta_0 \exp(i \mu t) \cos(t + \delta), \tag{7.61}$$

where μ is a constant yet to be determined. There will be two values of μ, for it is given by a quadratic equation. If they are real, the solutions are stable, whereas if they are complex, the general solution is unstable. This is similar to the Floquet-theory expression $\exp(\pm i \lambda t) u_{\pm}(t)$ in Eq. (7.50), but there is a difference: μ is not λ and $\cos(t + \delta)$ does not have the same period as $u_{\pm}(t)$.

The $\eta(t)$ of (7.61) is now inserted into (7.60), which leads to

$$(\mu^2 + 1 - \epsilon) \cos(t + \delta) + 2i \mu \sin(t + \delta) + 2h \cos 2t \cos(t + \delta) = 0.$$

The three terms of this equation can be transformed by using trigonometric identities. In the last term, $\cos 2t \cos(t + \delta) = \frac{1}{2}[\cos(3t + \delta) + \cos(t - \delta)]$, the third harmonic is neglected to this order of approximation (recall the quartic oscillator of Section 6.3.1). The result is

$$\cos t [(\mu^2 + 1 - \epsilon + h) \cos \delta + 2i \mu \sin \delta]$$
$$+ \sin t [2i \mu \cos \delta - (\mu^2 + 1 - \epsilon - h) \sin \delta] = 0.$$

If this equation is to hold at all times t, the coefficients of $\cos t$ and $\sin t$ must vanish separately. Therefore

$$(\mu^2 + 1 - \epsilon + h) \cos \delta + 2i \mu \sin \delta = 0,$$
$$2i \mu \cos \delta - (\mu^2 + 1 - \epsilon - h) \sin \delta = 0.$$

There can be no trivial solution of these equations, for it is impossible that $\cos \delta = \sin \delta = 0$. A nontrivial solution requires that the determinant of the coefficients vanish, which yields the equation

$$-(\mu^2 + 1 - \epsilon)^2 + h^2 + 4\mu^2 = 0. \tag{7.62}$$

If μ^2 is positive, μ is real and the solutions of (7.61) are stable. If μ^2 is negative, μ is complex and the solutions are in general unstable. Therefore the curve in the (h, ϵ) plane that separates the stable from the unstable solutions is determined by the condition that $\mu^2 = 0$. This is the boundary of the tongue and is given by

$$h = \pm(1 - \epsilon), \tag{7.63}$$

which represents two straight lines in the plane. This calculation holds only for small values of h and thus gives essentially only the tangents to the tongue where it reaches the ϵ axis.

DAMPING

If the pendulum is damped, that adds a term of the form $\beta\dot{\eta}$ to Eq. (7.60). We continue to assume a solution of the form of (7.61). Then after some algebra (Problem 20) Eq. (7.62) is replaced by

$$-(\mu^2 + 1 - \epsilon - i\mu\beta)^2 + h^2 - (2i\mu + \beta)^2 = 0,$$

and the condition for $\mu^2 = 0$ becomes

$$h^2 - (1 - \epsilon)^2 = \beta^2. \tag{7.64}$$

This can be recognized as the equation of a hyperbola opening in the h direction, whose asymptotes are the two lines of Eq. (7.63): the region of instability does not reach the ϵ axis. This result is reminiscent of the driven oscillator. In the driven oscillator, damping completely removes the instability of the undamped system. In the present case even the damped system has unstable solutions, but they require larger values of h (i.e., larger amplitudes). Indeed, the minimum value of h for instability is $h = \beta$ (at $\epsilon = 1$).

We conclude with some indications to the literature. Landau and Lifshitz (1976) discuss the inverted pendulum twice, both in the linear limit and in a weakly nonlinear form (their Section 6.30). They calculate some of its parameters by using a different approach from the one we present here. When the general case in which the oscillations are not small is studied by numerical methods, it is found that the forced pendulum can become completely chaotic. If dissipation is included, a strange attractor appears (Blackburn, Smith, and Jensen, 1992). Acheson (1993) and Acheson and Mollin (1993) have shown both experimentally and theoretically that a multiple pendulum also has regions of stability. The latter paper has a graph that successfully compares theoretical and experimental results for the single inverted pendulum.

WORKED EXAMPLE 7.2

A uniform, but time-dependent magnetic field $\mathbf{B}(t) = \hat{\mathbf{k}}B(t)$ in the z direction acts on a particle of mass m and charge e. **(a)** Find the induced electric field and show that the total Lorentz force on the particle lies in the (x, y) plane. Write the Lorentz force in terms of the cyclotron frequency (Section 6.4.4) ω_C. **(b)** By transforming to an appropriately rotating coordinate system (see Section 2.2.4 and Problem 2.8) show that the equation obtained is that of a parametric oscillator. [Hint: In Chapter 2 the magnetic field was constant in time. Here it isn't. That means that ωt of the old treatment must be replaced by an integral here.] **(c)** Show that the magnetic moment of Eq. (6.157) is an adiabatic invariant.

Solution. **(a)** Start by writing a vector potential $\mathbf{A}$ that yields a uniform magnetic field $\mathbf{B}$ (see Section 2.2.4): $\mathbf{A} = -\frac{1}{2}\mathbf{r} \wedge \mathbf{B}$. Since $\mathbf{B}$, although uniform (the same at all points in space), depends on t, so does $\mathbf{A}$, and in accordance with Eq. (2.45), $\mathbf{E}$ is given by

$$\mathbf{E} \equiv -\frac{\partial \mathbf{A}}{\partial t} = \frac{1}{2}\mathbf{r} \wedge \dot{\mathbf{B}} = -\frac{1}{2}\dot{B}\hat{\mathbf{k}} \wedge \mathbf{r},$$

where $\mathbf{r}$ is the position vector relative to an arbitrarily chosen origin. We write $\mathbf{r}$, rather than $\mathbf{x}$ as in Section 2.2 in order to call the components x, y without confusing x with the magnitude of the position vector.

Now write the Lorentz force equation, using $\omega_C \equiv \omega = eB/m$ (we drop the subscript C on ω, for this is the only frequency that enters here) and $\mathbf{v} = \dot{\mathbf{r}}$:

$$\ddot{\mathbf{r}} + \omega \hat{\mathbf{k}} \wedge \dot{\mathbf{r}} + \frac{1}{2}\dot{\omega}\hat{\mathbf{k}} \wedge \mathbf{r} = 0.$$

Because B depends on time, ω also depends on time. The total force is perpendicular to $\hat{\mathbf{k}}$; hence it is in the (x, y) plane. Ignore the motion in the z direction. In Cartesian coordinates this equation becomes

$$\frac{d^2x}{dt^2} - \frac{dy}{dt}\omega(t) - \frac{1}{2}\dot{\omega}(t)y = 0,$$

$$\frac{d^2y}{dt^2} + \frac{dx}{dt}\omega(t) + \frac{1}{2}\dot{\omega}(t)x = 0.$$

(b) It was seen in Section 2.2.4 that one way to deal with charged particles in *constant* magnetic fields is to transform to systems rotating at the Larmor frequency $\omega/2$. As suggested in the statement of the problem, $\frac{1}{2}\omega t$ should be replaced in the trigonometric functions by an integral over time. Write

$$\Omega(t) \equiv \frac{1}{2}\int^t \omega(t')dt'$$

[note that $\dot{\Omega}(t) = \frac{1}{2}\omega(t)$] and transform to a new coordinate system rotating at the rate $\Omega(t)$ with respect to (x, y).

The easiest way to proceed is probably to transform to a complex variable. Write $Z = x + iy$, and then the equation of motion becomes

$$\ddot{Z} + i\omega\dot{Z} + \frac{i\dot{\omega}}{2}Z = 0.$$

The rotated system is obtained by defining ζ in accordance with

$$Z = \zeta \exp\left\{-\frac{i}{2}\int^t \omega(t')\,dt'\right\} \equiv \zeta e^{-i\Omega(t)}.$$

(The sign of the exponential here is important: it determines the direction of rotation. See Section 2.2.4 and Problem 2.8). The equation for ζ is then

$$\ddot{\zeta} + \frac{1}{4}\omega^2(t)\zeta = 0.$$

This is the harmonic oscillator with a time-dependent frequency. If ω^2 is periodic it is the equation of a parametric oscillator. Now, however, the variable is complex. Nevertheless, there will be regions of stability and instability, as described for the inverted pendulum.

(c) Since ω^2 is a real function, the real and imaginary parts of $\zeta \equiv \xi + i\eta$ separately satisfy the same equation:

$$\ddot{\xi} + \frac{1}{4}\omega^2(t)\xi = 0,$$

$$\ddot{\eta} + \frac{1}{4}\omega^2(t)\eta = 0.$$

If B and hence also ω vary slowly enough, each of the action variables of these two equations are adiabatic invariants. It has already been seen (e.g., in Section 6.4.1) that the action is

$$J_\alpha = \frac{E_\alpha}{\omega} \equiv \frac{p_\alpha^2 + \omega^2\alpha^2}{2\omega} \equiv \frac{\dot{\alpha}^2 + \omega^2\alpha^2}{2\omega},$$

where α is either ξ or η [the Hamiltonian is of the form $H = \frac{1}{2}(p^2 + \omega^2\alpha^2)$, so that $\dot{\alpha} = p$]. Write $J \equiv J_\xi + J_\eta$. Then J is also an adiabatic invariant, and it can be written as

$$J = \frac{\dot{\xi}^2 + \dot{\eta}^2 + \omega^2(\xi^2 + \eta^2)}{2\omega} = \frac{|\dot{\zeta}|^2 + \omega^2|\zeta|^2}{2\omega} = \frac{|\dot{Z}|^2 + \omega^2|Z|^2}{2\omega}.$$

The last equality follows from the definition of ζ in terms of Z and from the fact that $\Omega(t)$ is a real function.

Now, $|\dot{Z}|^2 = \dot{x}^2 + \dot{y}^2 \equiv v^2$ and $|Z|^2 = r^2$. As a result, J looks just like the action for a harmonic oscillator (or rather the sum of two harmonic oscillators), namely E/ω: the first term in the numerator is twice the kinetic energy and the second is twice the potential energy. Since B varies slowly in time, the motion is roughly circular motion in the plane perpendicular to B, and $\dot{x}_{\max} = \dot{y}_{\max} = v_\perp$. The energy E is the sum of the maximum kinetic energies in the two coordinates, or $E = v_\perp^2$. Hence (putting back the time dependence)

$$J = \frac{v_\perp^2(t)}{\omega(t)} = \frac{mv_\perp^2(t)}{eB(t)} = 2M/e,$$

where M is the magnetic moment, as defined in Eq. (6.157). Since J is an adiabatic invariant, so is M.

7.4 DISCRETE MAPS; CHAOS

In this section we return to a discussion of discrete maps.

7.4.1 THE LOGISTIC MAP

Even seemingly simple discrete maps can give rise to really complicated and interesting behavior. In this section we present an example of this in one dimension, the *logistic map*. This example has particular interest because it will lead to our second encounter with chaos in dynamical systems (the first was chaotic scattering in the Chapter 4). It can also lead to some intuitive feel for behavior that can be expected in other dynamical systems of higher dimensions. In fact some one-dimensional results can even be quantitatively correct in higher dimensions.

DEFINITION

We start by reducing the Hénon map to one dimension by setting $y_n = ax_n - 1$ and $b = 0$ in Eq. (7.36), which then becomes

$$x_{n+1} = ax_n(1 - x_n) \equiv F_a(x_n). \tag{7.65}$$

This is the *logistic equation* (May, 1976; Feigenbaum, 1978; Devaney, 1986; Ott, 1993; also the book by Hao, 1984). It arises in many disciplines as well as in physics, for example, biology, engineering, and economics. The discrete dynamical system it describes is in a sense universal: its behavior, which changes dramatically depending on the tuning parameter a, is characteristic of many such one-dimensional systems. Not only the qualitative aspects of the behavior, but also many of the quantitative results depend very little on the details. That is, if F_a were replaced by some other function with a quadratic maximum, the behavior we are about to describe would be very similar. In his 1976 paper May names several functions other than F_a that give similar qualitative and quantitative results. The functions he names are

$$x \exp\{a(1 - x)\};$$
$$ax \quad \text{if } x < \frac{1}{2}, \quad a(1 - x) \quad \text{if } x > \frac{1}{2};$$
$$\lambda x \quad \text{if } x < 1, \quad \lambda x^{1-a} \quad \text{if } x > 1.$$

We will not discuss these functions further, except to mention that the range of the tuning parameter a is not the same in all of them. The equality of quantitative results for such different functions is called *universality*.

FIXED POINTS

The logistic map is generally applied only to x in the interval $I = [0, 1]$. Figure 7.24 shows graphs of $F_a(x)$ on this interval for four values of a. The graphs show that $F_a(0) = F_a(1) = 0$ for all a and that the maximum is always at $x = \frac{1}{2}$, where $F_a(\frac{1}{2}) = a/4$. Thus F_a maps I into itself for $a \leq 4$, so that at these values of a, iterations of F_a according to Eq. (7.65) remain within I. For $a > 4$, F maps I out of the interval and its iterations diverge; we will not deal with $a > 4$.

To study F_a, we first find its fixed points for various values of a by setting $x_{n+1} = x_n \equiv x_f$ or simply $x_f = F_a(x_f)$. The solutions are

$$x_f = 0, \quad x_f = 1 - \frac{1}{a}.$$

We will generally ignore the trivial $x_f = 0$ solution. The other x_f is in I only if $a > 1$, so we will consider only values of a in the interval $1 \leq a \leq 4$. On Fig. 7.24 x_f for each value

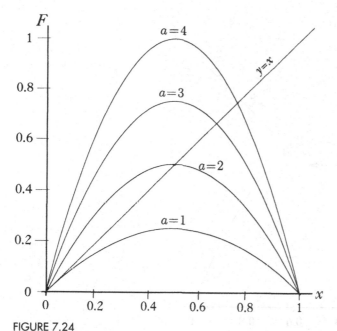

FIGURE 7.24

Graphs of $F_a(x) = ax(1-x)$ in the interval $0 \le x \le 1$ for four values of a in the interval $1 \le a \le 4$. For each a the fixed point is at the intersection of $F_a(x)$ and the line $y = x$.

of a is at the intersection of $F_a(x)$ and the line $y = x$. The only intersection with $F_1(x)$ is at $x = 0$.

Next, we investigate the stability of fixed points by linearizing about x_f. The map is one dimensional, so **A** is a 1×1 matrix and its only element is its eigenvalue λ. The derivative of F_a at the fixed point is **A**, so the stability condition $|\lambda| < 1$ can be restated in the form $|dF_a/dx|_{x_f} < 1$. At the origin $dF_a/dx = a > 1$, so the fixed point at the origin is unstable. At the other fixed point

$$\left.\frac{dF_a}{dx}\right|_{x_f} = a(1-2x)|_{x_f} = 2 - a. \qquad (7.66)$$

This implies that the nontrivial fixed point is stable for $a < 3$ and unstable for $a > 3$. In Fig. 7.24 the slope at the fixed point is steepest ($|dF_a/dx|$ is greatest) for $a = 4$, where $dF_4/dx = -2$.

The fixed point is not only stable but is an *attractor* for $1 < a < 3$. This can be seen very neatly graphically. Figure 7.25 is a graph of $F_{2.6}(x)$ (the slope at the fixed point $x_f \approx 0.6154$ is $dF_{2.6}/dx = -0.6$, so x_f is stable). Pick an arbitrary value of $x_0 \in I$ ($x_0 = 0.15$ in the figure) and draw a vertical line from $(x_0, 0)$ to the $F_{2.6}(x)$ curve (the arrow in the diagram). The ordinate at the intersection (at the arrow point) is $x_1 = F_{2.6}(x_0)$. To find x_1 on the x axis, draw a horizontal line from that intersection. Where it reaches the $y = x$ line (the next arrow in the diagram), both the ordinate and abscissa equal x_1. To find x_2, draw a vertical line at $x = x_1$ from the $y = x$ line to the curve (the third arrow). Repeat this procedure by following the arrows in Fig. 7.25: the sequence of intersections with the $F_{2.6}(x)$ curve is the sequence of x_n values; and it is seen that it converges rapidly to x_f at the intersection

4

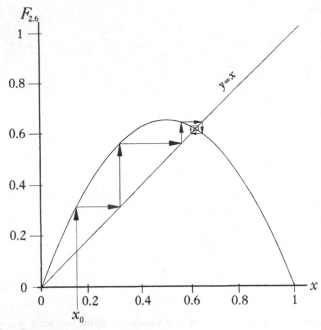

FIGURE 7.25

$F_a(x)$ for $a = 2.6$. Starting with an arbitrarily chosen $x_0 = 0.15$, the logistic map leads to the x_n sequence indicated in the curve. The x_n values for $n = 1, 2, 3, \ldots$ are the ordinates of the intersections on the $y = x$ line. The sequence closes in on the fixed point $x_f = 0.6154$ at the intersection of $F_{2.6}(x)$ and $y = x$.

of the $y = x$ line and the $F_{2.6}(x)$ curve. This is typical of the behavior of the logistic map for $a < 3$.

The same kind of construction will show that for all a the trivial fixed point at $x = 0$ is a *repeller* (Problem 6).

PERIOD DOUBLING

To see what happens if $a > 3$, for which the fixed point at $x_f = 1 - 1/a$ is unstable according to Eq. (7.66), consider $a = 3.5$, for which the fixed point is at $x_f \approx 0.7143$ (the slope at x_f is $dF_{3.5}/dx = -1.5$). This case is shown in Fig. 7.26. Starting at the same value $x_0 = 0.15$ and proceeding in the same way as for Fig. 7.25, the x_n sequence fails to converge to x_f. It settles down to a sort of limit cycle: from point A, it goes to B, then to C, D, E, F, G, H, and finally back to A again (or very close to it). In the limit it keeps repeating this cycle from A to H and back to A again, corresponding to four values of the x_n: one at the AB level, one at the CD level, one at the EF level, and one at the GH level.

Thus when n is large, x_n is one of the four values in the limit cycle, so if $F_{3.5}$ is applied four times to x_n, the sequence returns to x_n. For large n each x_n is a fixed point of $F_{3.5}^4$, defined by

$$F_{3.5}^4(x_n) \equiv F_{3.5}(F_{3.5}(F_{3.5}(F_{3.5}(x_n))))$$

$$\equiv F_{3.5} \circ F_{3.5} \circ F_{3.5} \circ F_{3.5}(x_n) = x_n.$$

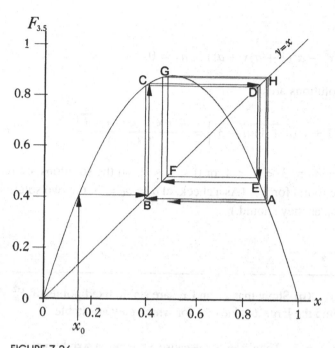

FIGURE 7.26

$F_a(x)$ for $a = 3.5$. Starting from the same $x_0 = 0.15$ as in Fig. 7.25, $\lim_{n \to \infty} x_n$ fails to converge to a single value. Instead it converges to a cycle of four values obtained from the limit cycle ABCDEFGH. Only some of the arrows are drawn.

Moreover, each x_n is a *stable* fixed point of $F_{3.5}^4$, as can be seen from the way the limit cycle is approached in Fig. 7.26 (see also Problem 6). Because $x_f = 0.7143$ is an unstable fixed point of $F_{3.5}$, it is also an unstable fixed point of $F_{3.5}^4$ (we continue to ignore the trivial unstable fixed point at $x = 0$). We have not written out the sixteenth degree polynomial $F_{3.5}^4$. To deal with it analytically is extremely difficult.

The fixed points of $F_a^n \equiv F_a \circ F_a \circ \cdots \circ F_a$ for general values of n and a are important in analyzing the logistic map. We start with $n = 2$. According to Eq. (7.66), up to $a = 3 \equiv a_0$ the fixed point at $x_f = 1 - 1/a$ is stable, but if $a > 3$, even if $a - 3 = \epsilon$ is very small, it is unstable. For small enough ϵ, however, there is a stable limit cycle consisting of two values of x. This is proven by showing that F_a^2 has three fixed points, two of which are new and stable for small ϵ. The fixed point at $x_f = 1 - 1/a$ is unstable for F_a^2 because it is unstable for F_a. If any others exist, they are solutions of the equation

$$x = F_a \circ F_a(x) = aF_a(x)[1 - F_a(x)] = a^2 x(1 - x)[1 - ax(1 - x)].$$

After some algebra (still ignoring the trivial $x = 0$ solution), this equation can be expressed as

$$a^3 x^3 - 2a^3 x^2 + a^2(1 + a)x - a^2 + 1 = 0.$$

Now, $x = x_f = 1 - 1/a$ is known to be a solution, so the factor $x - 1 + 1/a$ can be divided out. The resulting equation is

$$a^3 x^2 - a^2 (1 + a)x + a(1 + a) = 0,$$

a quadratic equation whose solutions are

$$x_\pm = \frac{1}{2a}\left[a + 1 \pm \sqrt{(a+1)(a-3)}\right] = \frac{4 + \epsilon \pm \sqrt{\epsilon(4+\epsilon)}}{2(3+\epsilon)}. \tag{7.67}$$

For the purposes of this calculation, $3 < a < 4$, or $0 < \epsilon < 1$, so the solutions are real. Thus two new fixed points are found for F_a^2. (As a check, at $a = a_0 = 3$, the two solutions coalesce to $x = \frac{2}{3} = 1 - 1/a_0$, as they should.)

WORKED EXAMPLE 7.3

(a) Show that $F_a(x_\pm) = x_\mp$. (b) Show that x_+ and x_- are stable fixed points of F_a^2 for small enough ϵ. (c) Find the largest value of ϵ for which they are stable.

Solution. (a) Write $F_a(x_+) \equiv \bar{x}$. Then $\bar{x} \neq x_+$ because x_+ is not a fixed point for F_a. But $F_a(\bar{x}) = x_+$ because x_+ is a fixed point for F_a^2, as has been shown above. Then $\bar{x} \neq x_f \equiv 1 - 1/a$ because x_f is a fixed point for F_a. But $F_a^2(\bar{x}) \equiv F_a(x_+) \equiv \bar{x}$, so $\bar{x}$ is a fixed point for F_a^2. Since $\bar{x}$ is not x_+ or x_f, it must be x_-. Moreover, $F_a(x_-) \equiv F_a(\bar{x}) = x_+$.

(b) Stability is determined by the derivatives of F_a^2 at $x_\pm$. The chain rule gives

$$\frac{dF_a^2}{dx} \equiv \frac{d}{dx} F_a(F_a(x)) = F_a'(F_a(x)) \cdot F_a'(x). \tag{7.68}$$

Now use Eqs. (7.66) and (7.68) to obtain

$$\begin{aligned}
\lambda = \left.\frac{dF_a^2}{dx}\right|_{x_\pm} &\equiv a\{1 - 2F_a(x_\pm)\} \cdot a\{1 - 2x_\pm\} \\
&= a^2 (1 - 2x_\mp)(1 - 2x_\pm) \\
&= a^2 [1 - 2(x_+ + x_-) + 4x_+ x_-].
\end{aligned}$$

This can be expressed entirely in terms of a or ϵ by using Eq. (7.67):

$$\lambda = 4 + 2a - a^2 = 1 - 4\epsilon - \epsilon^2. \tag{7.69}$$

The derivative is the same at both fixed points. Since ϵ is positive, as long as ϵ is small $|\lambda| < 1$, so $x_\pm$ are stable, which is what we set out to prove.

(c) The condition $-1 < \lambda < 1$ is found by solving the equations

$$1 - 4\epsilon - \epsilon^2 = 1$$

and

$$1 - 4\epsilon - \epsilon^2 = -1.$$

The allowed range of ϵ is $0 < \epsilon < 1$. The solutions of the quadratic equations are -4 and 0 for the first one and $-2 - \sqrt{6}$ and $-2 + \sqrt{6}$ for the second. There are two resulting regions where $-1 < \lambda < 1$. The first, namely $-2 - \sqrt{6} < \epsilon < -4$, is outside the allowed range of ϵ. The second, namely $0 < \epsilon < -2 + \sqrt{6} \approx 0.4495$, is allowed. Thus the maximum value of ϵ is $-2 + \sqrt{6}$.

As long as $a < a_0 \equiv 3$, the fixed point at $x_f = 1 - 1/a$ remains stable, but at a_0 it becomes unstable and the two-point limit cycle appears. This is called a *period-doubling bifurcation* (recall the similar phenomenon in Worked Example 2.2.1), and the map is said to undergo a *pitchfork bifurcation* from *period one* to *period two*. As a increases further, each of the two points of the bifurcation, fixed points of F_a^2, bifurcates to two fixed points of F_a^4, and each of these eventually also bifurcates. We now describe how, as a continues to increase, more and more such bifurcations take place at fixed points of F_a^n for higher n.

First we show how the next bifurcation arises. As ϵ (or a) increases in Eq. (7.69), λ decreases from $\lambda = 1$ at $\epsilon = 0$, passes through zero, and eventually reaches -1, after which x_+ and x_- also become unstable. Each of the two points then undergoes its own period-doubling bifurcation, and F_a undergoes a transition from period two to period four. The a value at this bifurcation, called a_1, was found analytically in the Worked Example 7.3 (see also Problem 7), but other bifurcations require numerical calculation (even $F_a^4(x)$ is already a polynomial of degree 16). Graphical analysis helps, and we now turn to some graphs.

Figure 7.27 shows graphs of F_a^2 for three different values of a. In Fig. 7.27(a) $a = 2.6$, corresponding to Fig. 7.25. The intersection of the curve with the $y = x$ line shows the fixed point, and the slope dF_a^2/dx at the intersection is clearly less than 1 (it is less than the slope of the line). That means that the fixed point is stable. As a increases, the curve changes shape, and Fig. 7.27(b) shows what happens at $a = 3.3$. There are now three fixed points, that is, three intersections. At the middle one dF_a^2/dx is greater than the slope of the line, hence greater than 1, and this point is unstable. At the other two fixed points the slope (negative in both cases) is of modulus less than one, so these are stable; they are the x_+ and x_- of Eq. (7.67). There must be a value of a between 2.6 and 3.3 where dF_a^2/dx at the period-one fixed point is equal to 1 and therefore the curve and the line are tangent. That value is $a = 3$, when the curve crosses the $y = x$ line in the opposite sense and the fixed point changes from stable to unstable. We have not drawn that graph.

As a increases, dF_a^2/dx at $x_\pm$ changes. Figure 7.27(c) shows the situation at $a = 3.5$, corresponding to Fig. 7.26. The slope at $x_\pm$ is found from (7.69) to be -1.25, so that x_+ and x_- are now unstable. In fact at $a \approx 3.4495 \equiv a_1$ they both bifurcated to yield a cycle of period 4. This is illustrated in Fig. 7.28, which is a graph of $F_{3.5}^4$, corresponding to Figs. 7.26 and 7.27(c). It is seen that $F_{3.5}^4$ has seven fixed points, of which three are

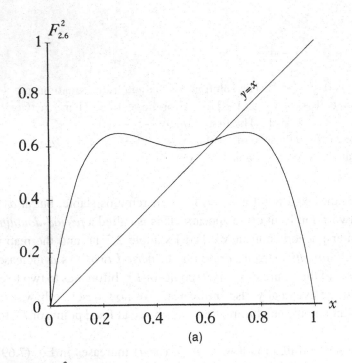

(a)

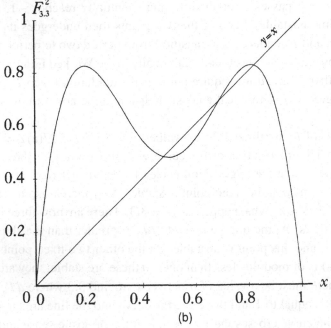

(b)

FIGURE 7.27
$F_a^2(x)$ for (a) $a = 2.6$, (b) $a = 3.3$, (c) $a = 3.5$.

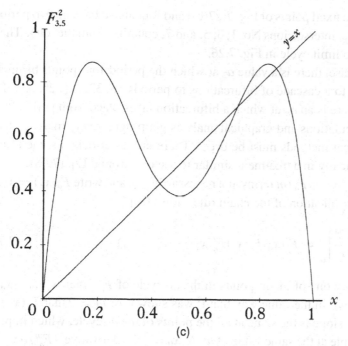

(c)

FIGURE 7.27
(Continued)

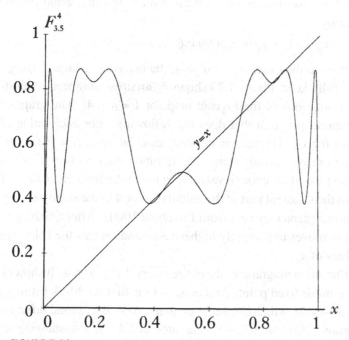

FIGURE 7.28
$F_a^4(x)$ for $a = 3.5$.

unstable (the three unstable fixed points of Fig. 7.27(c)) and four are stable (the two period-doubling bifurcations of $x_\pm$, intersections No. 1, 3, 5, and 7, counting from the left). These are the four x values of the limit cycle in Fig. 7.26.

If a continues to increase, there is a value a_2 at which the period-four points bifurcate to period eight, and so on to a cascade of bifurcations to periods 16, 32, $\ldots$, 2^n, $\ldots$. For every positive integer n there is an a_n at which a bifurcation takes place from period 2^n to period 2^{n+1}. Analytic calculations and graphical analysis get progressively more difficult as n increases, so numerical methods must be used. There are nevertheless some things that can be learned analytically in a treatment similar to the one around Eq. (7.68).

Consider $F_a^m \equiv F_a \circ F_a \circ \cdots \circ F_a$ (m terms, not necessarily 2^n), and write $F_a(x_n) = x_{n+1}$, as before. Then multiple application of the chain rule shows that

$$\frac{dF_a^m}{dx}\bigg|_{x_0} = F_a'(x_0)F_a'(x_1)F_a'(x_2)\cdots F_a'(x_{m-1}) \tag{7.70}$$

for arbitrary x_0. If x_0 is any one of the m points in the m-cycle of F_a^m, then $x_1, \ldots, x_{m-1}$ are the rest, as can be shown in the same way that it was shown for $m = 2$ that $F_a(x_\pm) = F_a(x_\mp)$. It follows that the slope is the same at all the points of the m-cycle, which implies that they all become unstable at the same value of a. At $x_0 = \frac{1}{2}$ the derivative $dF_a^m/dx = 0$ because $F_a'(x_0) = 0$. This means that $x = \frac{1}{2}$ is an extremum for all F_a^m. This is true even if $m \neq 2^n$.

Numerical calculation shows (see Feigenbaum) that as n increases, the period-doubling a_n converge to a limit given by

$$a_\infty = \lim_{n\to\infty} a_n \approx 3.5699456\ldots.$$

As a increases still further in the range $a_\infty < a < 4$, the behavior is largely irregular and will be described more fully later. Figure 7.29 shows *bifurcation diagrams* exhibiting the structure of the large-n solutions of the logistic map for $1 < a < 4$. The graphs are obtained by writing a computer program that does the following: For each value of a an arbitrary value is chosen for x_0. The map is iterated, calculating $x_{n+1} = F_a(x_n)$ from $n = 0$ as long as necessary to reach a steady state. The iteration then continues, but now successive values of x_{n+1} are plotted for enough values of n to exhibit the steady state. The number of iterations in both the transient part of the calculation and in the steady-state part depend on the value of a (in our graph they vary from 1 to about 1000). After finishing with one value of a, the computer moves to a slightly higher value and repeats the calculation, running through 2,000 values of a.

Figure 7.29(a) is the bifurcation diagram for the entire interval $1 \leq a \leq 4$. It shows that up to $a = 3$ there is just one stable fixed point. At $a = a_0 = 3$ the first pitchfork bifurcation takes place, resulting in a two-cycle, which continues up to a_1. At a_1 the second bifurcation takes place, and so on. Figure 7.29(b) stretches out the interval $3.4 < a < 4$, showing more bifurcations and some details of the structure between a_∞ and $a = 4$. A large period-three window is clearly visible in this figure, as well as smaller period-five and period-six windows. The small box in the period-three window is blown up in Fig. 7.29(c). The self-similarity of the graph is evident.

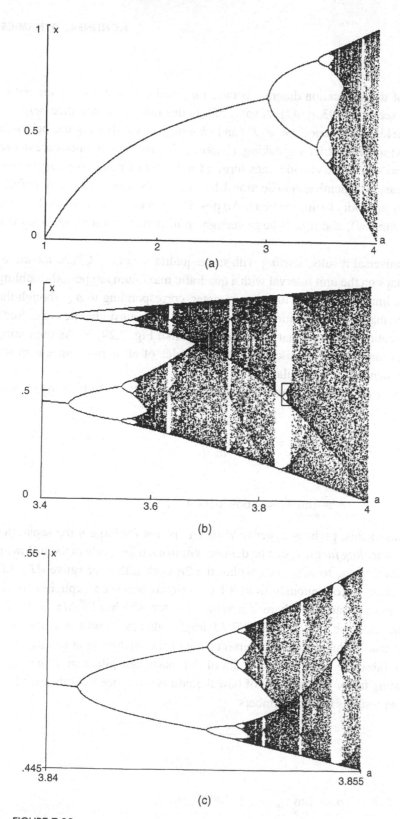

FIGURE 7.29

Bifurcation diagrams for the logistic map. (a) From $a = 1$ to $a = 4$. (b) From $a = 3.4$ to $a = 4$. (c) An enlargement of the small box in the period-3 window of (b): $3.84 < a < 3.855$ and $0.445 < x < 0.55$.

UNIVERSALITY

The structure of the bifurcation diagram between a_∞ and $a = 4$ has been studied in detail (for a review, see Ott, 1993, p. 42). A value of a in this range will be called *periodic* if iteration of F_a yields an m-period ($m \neq 2^n$) and *chaotic* if it yields irregular behavior (although the behavior is not, strictly speaking, chaotic). The periodic a values are dense between a_∞ and $a = 4$, and the chaotic ones form a Cantor set of nonzero measure (see Jacobson, 1981; measure is mentioned in Section 4.1.3 and discussed somewhat more fully in the number theory appendix to this chapter). All positive integer values of $m \neq 2^n$ occur (e.g., the period-3 window), and if m is large enough an m-period can look chaotic even though it is not.

We list some universal results, starting with some qualitative ones. Characteristic of one-dimensional maps on the unit interval with a quadratic maximum are period-doubling bifurcations up to a limit value of the tuning parameter, corresponding to a_∞, though the limit depends on the map. The bifurcation zone is followed by a quasi-chaotic zone. Such maps all have bifurcation diagrams that look very much like Fig. 7.29(a). As the tuning parameter increases through the quasi-chaotic range, cycles of all periods appear in the same order, independent of the particular map.

Next, we present two quantitative results (Feigenbaum, 1978): 1. In the range below a_∞, the numbers

$$\delta_n = \frac{a_{n+1} - a_n}{a_{n+2} - a_{n+1}}$$

converge to

$$\delta \equiv \lim_{n \to \infty} \delta_n \approx 4.6692016091 \ldots .$$

Another way of putting this, perhaps easier to visualize, is that for large n the separation $a_\infty - a_n \sim \delta^{-n}$. 2. A *scaling factor* s_n can be defined within each $2n$-cycle in the following way: as a increases from a_n to a_{n+1} (i.e., within the $2n$-cycle), the derivative $dF_a^{2^n}/dx$ on the limit cycle changes continuously from $+1$ to -1 (this was seen explicitly for the case $n = 1$). Continuity means that there is a value of a for which $dF_a^{2^n}/dx = 0$. Let $a = \bar{a}_n$ be that value. Because $F_a'(\frac{1}{2}) = 0$, Eq. (7.70) implies that the point $x = \frac{1}{2}$ lies on the $2n$-cycle at $a = \bar{a}_n$; that is, the line $x = \frac{1}{2}$ intersects one of the pitchforks at $a = \bar{a}_n$. Then s_n is defined as the (absolute value of the) width of that particular pitchfork at $a = \bar{a}_n$ or at $x = \frac{1}{2}$. This scaling factor is a measure of how the points are spaced within each limit cycle. The universal result is that the numbers

$$\alpha_n = \frac{s_n}{s_{n+1}}$$

converge to

$$\alpha \equiv \lim_{n \to \infty} \alpha_n \approx 2.5029078750 \ldots .$$

We emphasize that the universality of both δ and α means that they are independent of the exact form of F_a, as long as it has a quadratic maximum and is linear in a.

Lyapunov exponents were introduced in Section 4.1.3. They measure the rate at which two points, initially close together, are separated by the action of a given dynamical system and in general depend on the location of the two initial points. For the logistic map, the Lyapunov exponent $\lambda(x_0)$ is defined by

$$\lim_{\substack{N\to\infty \\ \epsilon\to 0}} \left| \frac{F_a^N(x_0 + \epsilon) - F_a^N(x_0)}{\epsilon} \right| \sim e^{N\lambda(x_0)}.$$

Therefore

$$\lambda(x_0) = \lim_{N\to\infty} \frac{1}{N} \log \left| \frac{dF_a^N(x_0)}{dx_0} \right|.$$

According to Eq. (7.70), then,

$$\lambda(x_0) = \lim_{N\to\infty} \frac{1}{N} \log \left| \prod_{k=0}^{N-1} F_a'(x_k) \right| = \lim_{N\to\infty} \frac{1}{N} \sum_{k=0}^{N-1} \log |F_a'(x_k)|, \qquad (7.71)$$

where $x_k = F_a^k(x_0)$. If λ is negative, two adjacent points come closer together and the system is stable. If λ is positive, the system is unstable.

Suppose that x_0 approaches a limit cycle. Then the $|F_a'(x_k)|$ with large k all lie between 0 and 1, and their logarithms are negative. In the limit of large N these negative values dominate in (7.71), and thus $\lambda(x_0)$ is negative for all x_0 for a values of limit cycles, in particular between the pitchfork bifurcation points. This reflects the stability of the map. Now suppose that x_0 approaches a pitchfork bifurcation point. Then $|F_a'(x_k)| \to 1$ for large k, and the logarithms go to zero. In the limit of large N the sum in (7.71) approaches zero and thus $\lambda(x_0) = 0$ at pitchfork bifurcation points. This is the border between stability and instability. Figure 7.30 shows $\lambda(0)$ as a function of a between $a = 3.4$ and $a = 4$. It is

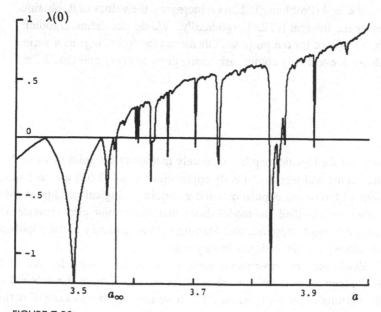

FIGURE 7.30

The Lyapunov exponent for the logistic map F_a as a function of a (from Shaw, 1981).

seen that although $\lambda \le 0$ for $a \le a_\infty$, it has positive values for $a \ge a_\infty$, where the map becomes chaotic. For values of a in the chaotic region the map is unstable: two initial x values that are close together are sent far apart by iterations of the map.

WORKED EXAMPLE 7.4

Consider F_4, and write $x_{n+1} = F_4(x_n)$. Make the transformation

$$\cos\theta_n = 1 - 2x_n \quad \forall n. \tag{7.72}$$

Since $0 < x < 1$, θ can be restricted to the range $-\pi < \theta < \pi$. Show that

$$\theta_n = 2^n\theta_0 (\text{mod } 2\pi). \tag{7.73}$$

Solution. Use F_4 to calculate x_{n+1} and insert that into (7.72):

$$\begin{aligned}
\cos\theta_{n+1} = 1 - 2x_{n+1} &= 1 - 8x_n(1 - x_n) \\
&= 2\left[1 - 4x_n + 4x_n^2\right] - 1 = 2(1 - 2x_n)^2 - 1 \\
&= 2\cos^2\theta_n - 1 = \cos 2\theta_n.
\end{aligned}$$

The restriction on the range of θ leads to

$$\theta_{n+1} = \cos^{-1}\cos 2\theta_n = 2\theta_n(\text{mod } 2\pi),$$

from which (7.73) follows immediately.

Comments: 1. Equation (7.73) can be used to obtain a closed-form expression for x_n in terms of x_0 at $a = 4$ (Problem 8). 2. As n increases, the values of θ_n obtained from Eq. (7.73) cover the interval $[0, 2\pi]$ *ergodically*. We do not define ergodicity exactly (see Rivlin, 1974), but for our purposes this means the following: as n varies, θ_n covers $[0, 2\pi]$ densely, eventually coming arbitrarily close to every point in $[0, 2\pi]$, independent of θ_0.

FURTHER REMARKS

So far we have discussed the logistic map from a purely mathematical point of view. Nevertheless, it has relevance in the real world. An early application of Eq. (7.65) was as a model for successive generations of biological populations under certain ecological conditions (May, 1976). Although probably oversimplified, the model shows that under some circumstances (for $a_\infty < a \le 4$) populations are almost unpredictable. Starting with an arbitrary initial population x_0, the populations x_n of subsequent generations can vary wildly.

Several numerical calculations and experiments have also established the relevance of the logistic map to physics. We mention only three. Careful numerical calculations (Crutchfield and Huberman, 1979) for the Duffing oscillator (Section 7.1.2) have uncovered a cascade of period-doubling bifurcations with increasing driving frequency. Our analytic treatment missed this. Two experiments on nonlinear driven LRC circuits (Testa et al., 1982, Linsay, 1981) have also found cascades of period-doubling bifurcations and even rough quantitative agreement between the measured and theoretical values of α and δ. Although explicit connections between physical systems and Eq. (7.65) may not be obvious, many real dynamical systems do in fact follow a similar period-doubling route to chaos.

7.4.2 THE CIRCLE MAP

Another important class of one-dimensional maps are the *circle maps*. These maps are widely used to model real physical systems that are driven, nonlinear, and dissipative, such as the Duffing oscillator discussed in Section 7.1. The behavior of such systems can be quite complicated, as we have seen, and it is remarkable that these one-dimensional discrete maps exhibit similar periodic, quasiperiodic, and even chaotic motion.

THE DAMPED DRIVEN PENDULUM

As an illustration, consider the example of the damped and periodically driven plane pendulum, say of a child on a swing (see Fig. 7.31). The equation of motion is

$$I\ddot{\theta} + \beta\dot{\theta} + mlg\sin\theta = \tau(t), \tag{7.74}$$

where I is the moment of inertia, l is the distance from the support to the center of mass, and β and $\tau(t)$ are the damping and periodic driving torques, respectively, provided here by friction and the "pumping" action of the child. What makes the system nonlinear is the $\sin\theta$ term.

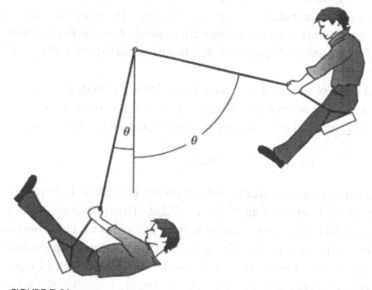

FIGURE 7.31

A child on a swing at two positions. By pumping, the child applies a driving torque τ at some pumping frequency Ω. Friction and air resistance provide the damping term β. The gravitational force provides the $mgl\sin\theta$ term.

The explicit form of the driving term in Eq. (7.74) is not very important as long as it is periodic, for most of the results we will obtain are universal in the same sense that the results of Section 7.4.1 are universal, independent of the explicit form of F_a. Therefore the driving torque may be taken to be of the simple form

$$\tau(t) = \tau_0 + \tau_1 \sin \Omega t.$$

When there is no friction and the child is not pumping, the swing oscillates like a simple undamped pendulum. With friction, but without pumping, the oscillations eventually damp out and the swing stops, independent of the initial conditions $\theta(t_0)$ and $\dot{\theta}(t_0)$. If the child pumps periodically, however, several different kinds of stationary states can arise, depending on the parameters. In this section we will show that the swing oscillates in a periodic stationary state if ω_0/Ω is rational, that is, if there exist integers j and k such that $\omega_0/\Omega = j/k$, where ω_0 is the natural frequency of the undamped pendulum. Moreover, periodicity prevails even if the frequency or strength of the driving torque are changed slightly. This is called *frequency locking* (sometimes also phase locking) and often also occurs when several nonlinear oscillators are coupled, even weakly. We will show also that there are *quasi-periodic* frequency-locking regimes, in which ω_0/Ω is irrational and that both the periodic and quasiperiodic regions broaden as the nonlinearity becomes stronger. When such regions overlap, the system can become chaotic.

THE STANDARD SINE CIRCLE MAP

Two possible approaches to trying to solve Eq. (7.74) are analytic approximation and numerical methods. Both these approaches have practical problems. The analytic approach, as was seen in Section 7.1, is very complicated for studying even relatively simple questions about a system as behaviorally rich as this one. The numerical approach suffers from the fact that all numbers on a computer are rational, whereas the behavior of this system is different for rationals and irrationals. It is more fruitful to deal with Eq. (7.74) by means of a Poincaré map.

Suppose that θ and $\dot{\theta}$ are observed periodically at the driving period, and let $(\theta_n, \dot{\theta}_n)$ be the nth observation at time $T_n = 2\pi n/\Omega$. Clearly $(\theta_{n+1}, \dot{\theta}_{n+1})$ is uniquely determined by the initial values $(\theta_n, \dot{\theta}_n)$ at the start of the period: there is some function f such that

$$(\theta_{n+1}, \dot{\theta}_{n+1}) = f(\theta_n, \dot{\theta}_n). \tag{7.75}$$

The function f can be calculated numerically, but we do not go into that. Periodic observation is a variation on the Poincaré map of Section 7.2.2. The difference is that now $\mathbb{P} \equiv \mathbf{T}\mathbb{Q}$ is the entire (cylindrical) velocity phase manifold, and the period of observation is determined not by the motion, but by the driving term. This Poincaré map establishes the basic properties of the motion. For instance, if the motion is periodic with a frequency that is a multiple of Ω (i.e., for periodic frequency locking), the map consists of a series of discrete points in $\mathbf{T}\mathbb{Q}$.

To the extent that the Poincaré map describes the dynamical system of Eq. (7.74), the entire motion is determined by the two-dimensional discrete map of Eq. (7.75). Assume now (and it is found in many cases to be true) that after some initial transient time the system reaches a steady state on an invariant curve in the phase manifold. Then the state of the system is completely determined by $\theta(t)$ alone, and the two-dimensional map of Eq. (7.75) collapses to a one-dimensional map of the form (Arnol'd & Avez, 1968, Herman, 1977, Jensen et al., 1984)

$$\theta_{n+1} = (\theta_n + \Theta + g(\theta_n)) \bmod 2\pi.$$

The mod 2π on the right-hand side emphasizes that θ lies on a circle. For the same reason, $g(\theta) = g(\theta + 2\pi)$ is a periodic function with period 2π, and Θ can be taken to lie between 0 and 2π. The general features of this map do not depend strongly on the form of $g(\theta)$ (recall the similar fact for the logistic map), so we make the further simplifying assumption that

$$g(\theta) = K \sin \theta. \tag{7.76}$$

Thus we have a somewhat simplified version of the damped driven pendulum.

The resulting map, known as the one-dimensional *standard sine-circle map* (we will call it simply the *circle map*) is

$$\theta_{n+1} = (\theta_n + \Theta + K \sin \theta_n) \bmod 2\pi \equiv C_{\Theta K}(\theta_n). \tag{7.77}$$

In words, in each iteration $C_{\Theta K}(\theta_n)$ changes θ_n by adding the constant Θ plus a sinusoidal term that depends on θ_n itself. The circle map is a model for the damped driven pendulum and for many similar systems. For the child on the swing, for instance, it gives the position at integer multiples of the pumping period $2\pi/\Omega$. Equation (7.77) has been obtained heuristically: it is important to emphasize that there is no rigorous result that establishes a connection between this equation, Eq. (7.74), and Eq. (7.75). Nevertheless, no cases have been found as of this writing without at least a qualitative connection.

ROTATION NUMBER AND THE DEVIL'S STAIRCASE

We now turn to the properties of $C_{\Theta K}$. Although Eq. (7.77), like the equation for the logistic map, looks quite simple, the discrete dynamics it describes has many complex features.

Consider first the case $K = 0$, for which $g(\theta) = 0$ and $C_{\Theta 0}$ is linear. If $\Theta/2\pi = j/k$, where j and k are incommensurate integers (i.e., if Θ is a rational multiple of 2π), iterating the map k times adds $2\pi j$ to θ, rotating the circle j times and returning every point to its initial position. A given θ_0 is carried by these k iterations to k equally spaced points on the circle. All that j determines is the order in which these k points are visited: in each application of $C_{\Theta 0}$ the circle undergoes the fraction $\rho = j/k$ of a complete rotation. The fraction ρ is called the *winding* or *rotation number* of the map. The definition of ρ is then extended to irrational values of $\Theta/2\pi$ (i.e., $\rho \equiv \Theta/2\pi$). In the irrational case the iterations carry an initial θ_0 to an infinite set of points, bringing it arbitrarily close to every point on the circle.

For $K \neq 0$ the definition of ρ is extended further, to a function $\rho(\Theta, K)$: it is taken to be the *average* amount by which the circle is rotated in each application of the map $C_{\Theta K}$. Suppose θ changes by $\Delta\theta_n$ in the nth iteration. Then the average change of θ during the first m iterations is

$$\Delta\theta_{\text{av}} = \frac{1}{m} \sum_{n=0}^{m} \Delta\theta_n.$$

The rotation number is defined as the average rotation $\Delta\theta_{\text{av}}/2\pi$ over all iterations:

$$\rho(\Theta, K) = \frac{1}{2\pi} \lim_{m\to\infty} \frac{1}{m} \sum_{n=0}^{m} \Delta\theta_n. \tag{7.78}$$

From the fact that $\Delta\theta = C_{\Theta K}(\theta) - \theta$ it follows (Problem 11) that

$$\rho(\Theta, K) = \frac{1}{2\pi} \lim_{m\to\infty} \frac{\hat{C}_{\Theta K}^m(\theta) - \theta}{m},$$

where $\hat{C}_{\Theta K}$ is $C_{\Theta K}$ without the mod 2π restriction, and $\hat{C}_{\Theta K}^m \equiv \hat{C}_{\Theta K} \circ \hat{C}_{\Theta K} \circ \cdots \circ \hat{C}_{\Theta K}$ (m terms). Note that $\rho \in [0, 1]$.

The rotation number $\rho(\Theta, K)$ has exotic properties. For the time being consider only nonnegative $K \leq 1$. As we have seen, $\rho(\Theta, 0) = \Theta/2\pi$ is a smooth continuous function. As soon as K is greater than 0, however, $\rho(\Theta, K)$ changes character. It remains everywhere continuous but is constant on finite Θ intervals about all points where $\rho(\Theta, K) = r$ is rational. Let $\Delta\Theta_{rK}$ be the width of the Θ interval for $\rho = r$ at a particular value of K. The $\Delta\Theta_{r0}$ all equal zero, but as K increases from 0 to 1, the $\Delta\Theta_{rK}$ intervals become finite and begin to grow, as shown in Fig. 7.32.

For each value of K there is some overlapping of the $\Delta\Theta_{rK}$, and their union $\bigcup_r \Delta\Theta_{rK}$ (taken over all rational values of $\rho(\Theta, K)$) occupies some part of the entire Θ range from 0 to 2π. The rest of the Θ range consists entirely of Θ values at irrational values of $\rho(\Theta, K)$. As the $\Delta\Theta_{rK}$ increase with increasing K their union leaves less and less room for irrational values, until at $K = 1$ there is no room left at all. Let $I_K \subset [0, 1]$ be the set of values of $\Theta/2\pi$ for which $\rho(\Theta, K)$ takes on irrational values: $\Theta/2\pi \in I_K$ implies that $\rho(\Theta, K)$ is irrational (I for "irrational"). Then the measure $\mu(I_K)$ of I_K on the interval $0 \leq \Theta/2\pi < 1$ is

$$\mu(I_K) = 1 - \frac{1}{2\pi} \bigcup_r \Delta\Theta_{rK}. \tag{7.79}$$

Then $\mu(I_0) = 1$ because I_0 consists of all the irrational numbers. For $K > 0$, however, $\mu(I_K)$ is less than 1 and decreases as K grows, reaching zero at $K = 1$; as each $\Delta\Theta_{rK}$ increases with K, the union on the right-hand side of Eq. (7.79) approaches 1, eventually reducing $\mu(I_1)$ to zero.

The graph of $\rho(\Theta, 1)$ is the near edge of the surface in Fig. 7.32, shown separately in Fig. 7.33. The function $\rho(\Theta, 1)$, the *Devil's Staircase*, is scale-invariant (self-similar) with

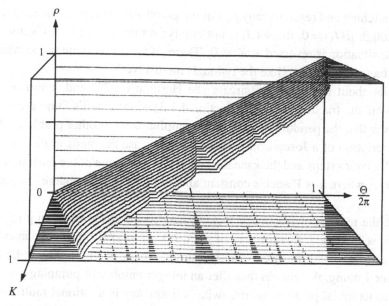

FIGURE 7.32

The rotation number $\rho(\Theta, K)$ for the circle map $C_{\Theta K}$ as a function of both K and Θ. The plateaus occur where $\rho(\Theta, K)$ takes on rational values, and they project down to the Arnol'd tongues, indicated by dark hatching on the $(K, \Theta/2\pi)$ plane. This graph was obtained numerically by iterating $C_{\Theta K}$ at each point between 100 and 600 times, depending on K. Numerical calculation obtains Arnol'd tongues even for some plateaus that are not visible to the eye. The K, $\Theta/2\pi$, ρ origin is at the far left lower corner.

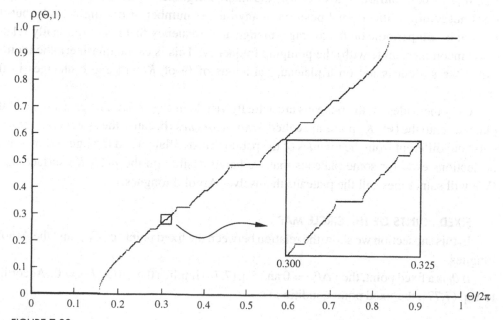

FIGURE 7.33

The Devil's Staircase: the rotation number $\rho(\Theta, 1)$ for the circle map $C_{\Theta 1}$ at $K = 1$. This is the near edge of Fig. 7.32. (The x axis is $\Theta/2\pi$ running from 0 to 1; the y axis is ρ, also running from 0 to 1.) The inset illustrates the self-similarity by enlarging the small window for $\Theta/2\pi$ between 0.300 and 0.325.

fractal structure: stretching and rescaling any part of the graph will reproduce the structure of the whole. Although $\mu(I_1) = 0$, the set I_1 is not empty; it turns out to be a Cantor set (Section 4.1.3). The situation is reversed at $K = 0$. There $\mu(I_0) = 1$, and the set on which $\rho(\Theta, 0)$ is rational has measure zero (like the rationals themselves).

For more details about the Devil's Staircase see Herman (1977) and Jensen et al. (1984). They calculate the fractal dimension (Section 4.1.3) of the Devil's Staircase to be 0.87. They also show that the pendulum equation has applications to other problems, for example, to the description of a Josephson junction. In that case the steps in the Devil's Staircase are called Shapiro steps and their measurement has led to one of the most precise measurements of h/e, where h is Planck's constant and e is the electron charge. See also Bak (1986).

The $\Delta\Theta_{rK}$ are the regions of frequency locking mentioned at the beginning of this section. This can be seen in the following way: if $\rho(\Theta, K) = r \equiv j/k$, then on the average after k iterations $C_{\Theta K}(\theta_n)$ rotates the circle exactly j times. To the extent that $C_{\Theta K}(\theta_n)$ is a model for the pumped swing, this means that after an integer number of pumping periods the swing is back in its initial position, so the swing's frequency is a rational multiple of the pumping frequency Ω. The plateaus in Fig. 7.32 are the $\Delta\Theta_{rK}$, and it is clear from the figure that if the system is on one of the plateaus, small variations of Θ or K leave it on the plateau. That is, small variations of Θ or K will not change ρ and hence will not change the average frequency: the frequency is "locked."

If $\rho(\Theta, K)$ is irrational, then on the average $C_{\Theta K}$ rotates the circle through an irrational multiple of 2π. Given any two points θ_1, θ_2 on the circle, there is an m high enough for $C_{\Theta K}(\theta_1)$ to be arbitrarily close to θ_2: the map is ergodic. This means that the swing does not return to its original position in any integer number of driving periods, but the motion is still periodic on the average, though at a frequency that is not rationally related to (is incommensurate with) the pumping frequency. This is called quasiperiodic motion. Since the system is not on a plateau, variations of Θ or K will cause changes in the frequency.

Only the widest of the plateaus are actually visible in Fig. 7.32. The projections of the plateaus onto the (Θ, K) plane are called *Arnol'd tongues* (because the system is nonlinear, these are different from the tongues of the parametric oscillator) and the figure shows these projections even for some plateaus that are barely visible on the $\rho(\Theta, K)$ surface itself. (We will sometimes call the plateaus themselves Arnol'd tongues.)

FIXED POINTS OF THE CIRCLE MAP

In this subsection we show the relation between the fixed points of $C_{\Theta K}$ and the Arnol'd tongues.

If θ_f is a fixed point, then $\Delta\theta_n = 0$ and Eq. (7.78) implies that $\rho(\Theta, K) = 0$. According to Eq. (7.77), the fixed point condition is

$$\Theta + K \sin \theta_f = 2m\pi, \tag{7.80}$$

where m is an integer (just another way to restate the mod 2π condition) and θ_f is the fixed point. This is a transcendental equation that can have more than one solution for given Θ, K, and m.

Typically m is either 0 or 1, for $\Theta \le 2\pi$ and $K \le 1$. Consider the case in which $m = 0$ (the $m = 1$ case can be treated similarly). Its solution is

$$\theta_f = -\sin^{-1}\left(\frac{\Theta}{K}\right). \tag{7.81}$$

Suppose that Θ and K are both positive. Then since $K \le 1$ Eq. (7.81) has solutions only if $\Theta \le K$. Consider a fixed Θ and K increasing from $K = 0$. There is no fixed point while $K < \Theta$, but when K reaches Θ a fixed point appears at $\theta_f = 3\pi/2$. As K continues to grow, this point bifurcates [there are two solutions of (7.81) close to $\theta_f = 3\pi/2$] and moves. Now consider fixed K and Θ decreasing from π. There is no fixed point while $\Theta > K$, but when Θ reaches K a fixed point appears. The border between the existence and nonexistence of fixed points is the line $K = \Theta$, the edge of the Arnol'd tongue whose apex is at the origin in Fig. 7.32, the lowest plateau in the figure. Thus the region in which $m = 0$ fixed points exist is the triangle on the (Θ, K) plane for which $\Theta \le K$ (i.e., the Arnol'd tongue itself).

This relation between fixed points and Arnol'd tongues applies also to periodic points with periods greater than 1. For instance, if there is a fixed point for $C_{\Theta K}^2 \equiv C_{\Theta K} \circ C_{\Theta K}$ (i.e., a point of period 2), on the average the circle undergoes a half rotation in each application of $C_{\Theta K}$. In that case $\rho(\Theta, K) = \frac{1}{2}$, and this corresponds to the Arnol'd tongue formed by the large plateau in the middle of Fig. 7.32. The other plateaus of Fig. 7.32 correspond to fixed points of other powers of $C_{\Theta K}$, and hence the figure may be viewed as an analog of a bifurcation diagram, now depending on the two variables K and Θ.

The stability of the fixed points is found by taking the derivative of $C_{\Theta K}(\theta)$ with respect to θ at $\theta = \theta_f$. The derivative is

$$C_{\Theta K}'(\theta) = 1 + K\cos\theta. \tag{7.82}$$

If θ_f is an $m = 0$ fixed point, this becomes

$$C_{\Theta K}'(\theta_f) = 1 + K\cos\theta_f = 1 + K\sqrt{1 - \frac{\Theta^2}{K^2}}, \tag{7.83}$$

and the condition for stability is that this have absolute value less than one. Consider Eq. (7.83) for fixed Θ and for K increasing from $K = \Theta$ (the value at which a fixed point appears) to $K = 1$. At $K = \Theta$ the derivative is 1, and $\cos\theta_f = 0$. As K grows, the fixed point bifurcates and the cosine takes on two values, one positive (for $\theta_f > 3\pi/2$) and the other negative (for $\theta_f < 3\pi/2$), corresponding to the two signs of the radical on the right-hand side of (7.83). Suppose that K is just slightly larger than Θ; for convenience we write $K^2 = \Theta^2/(1 - \alpha^2)$, where $0 < \alpha \ll 1$. Then the derivatives at the two new fixed points are

$$C_{\Theta K}'(\theta_f) = 1 \mp \frac{\alpha}{\sqrt{1 - \alpha^2}}\Theta;$$

one fixed point is stable (if $\theta_f > 3\pi/2$), and the other is unstable (if $\theta_f < 3\pi/2$). The stable one (the only one really of interest, corresponding to the positive square root) moves to higher values of θ as $K \to 1$. Similar results are obtained for fixed K and varying Θ.

We add some remarks, mostly without proof, for $K > 1$. First we cite a theorem (see Arnol'd, 1965) that guarantees that a smooth and invertible one-dimensional map cannot lead to chaos. According to Eq. (7.82) the derivative of the circle map is nonnegative for $K < 1$, so $C_{\Theta K}(\theta)$ is a monotonically increasing function of θ and hence is invertible. When $K = 1$ the derivative $C'_{\Theta 1}(\pi) = 0$, but $C''_{\Theta 1}(\pi) = 0$ as well, so this is a point of inflection (a cubic singular point), not a maximum: the function is still monotonically increasing, and hence invertible, and the theorem continues to apply. When $K > 1$, however, $C'_{\Theta K}(\theta)$ becomes negative for some values of θ, so the map has a maximum; hence it is no longer invertible, and the theorem no longer applies.

Lichtenberg & Lieberman (1992, p. 526) show that for $K > 1$ the circle map satisfies the conditions for chaos that are obtained for the logistic map at $a = 4$ (Worked Example 7.4). Chaos occurs, however, only for certain initial conditions. For instance, suppose that $K = 1 + \epsilon$ at a fixed point θ_f. From Eq. (7.83) it follows that

$$C'_{\Theta K}(\theta_f) \approx \frac{1}{2}[\Theta^2 - \epsilon(2 + \Theta^2)],$$

whose magnitude is less than 1 if Θ and ϵ are sufficiently small. This implies that θ_f can be stable; thus stable fixed points exist even for $K > 1$. In other words, $K > 1$ is a necessary, but not a sufficient condition for chaos. It turns out that when $K \leq 1$ the Lyapunov exponent is negative for rational values of $\rho(\Theta, K)$ (reflecting periodicity) and zero for irrational values (i.e., for quasiperiodicity). When $K > 1$, however, the Lyapunov exponent becomes positive for irrational rotation numbers. This is a signature of chaos.

7.5 CHAOS IN HAMILTONIAN SYSTEMS AND THE KAM THEOREM

It was pointed out in Section 6.3.2 that there are serious questions about applying perturbation techniques to completely integrable Hamiltonian systems: in both canonical perturbation theory and the Lie method it is assumed that the perturbed system is also completely integrable. In this section we want to scrutinize this assumption and study the extent to which it holds true. That will carry us back to the question of small denominators and the convergence of the Fourier expansions for the perturbed system. We will deal mostly with time-independent conservative Hamiltonian systems. However, time-independent Hamiltonian systems in n freedoms can be reduced to periodically time-dependent ones in $n - 1$ freedoms, so we will actually deal also with the latter kind (in fact the first system we consider is of this type). The first parts of this section discuss systems in two dimensions (one freedom), partly because it is easier to visualize what is going on in them, but also because more is known about them and because they are interesting in their own right. The last part, on the KAM theorem, is more general, but our presentation will be helped by the insights gained in the two-dimensional discussion. The appendix to this chapter discusses the number theory that is important to understanding much of the material, in particular to the KAM theorem.

7.5.1 THE KICKED ROTATOR

THE DYNAMICAL SYSTEM

The discussion of this section will focus on the important example of the *kicked rotator*. The unperturbed system is a plane rigid body of moment of inertia I free to rotate about one point; "free" means that the only force is applied at the pivot: no friction or gravity. The Hamiltonian of this system is

$$H_0 = \frac{J^2}{2I}, \tag{7.84}$$

where J is the angular momentum. This is clearly an integrable dynamical system. An obvious constant of the motion is J, and the tori on which the motion takes place are particularly simple: they are of one dimension only, circles of constant J running around the cylindrical $\mathbf{T}^*\mathbb{Q}$.

Now suppose the system is perturbed by an external, periodically applied *impulsive* force F, a "kick" of fixed magnitude and direction. The force is always applied to the same point on the body at a distance l from the pivot, as shown in Fig. 7.34, and its period is T. This means that between kicks the angular momentum is constant, and at each kick it increases by the amount of the impulsive torque applied. The impulsive torque τ provided by the force is given by $\tau = Fl \sin\phi \equiv \epsilon \sin\phi$. Assume that it is applied at times $t = 0, T, 2T, \ldots$.

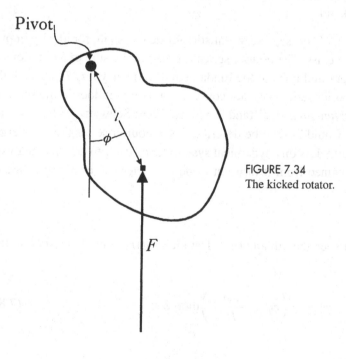

FIGURE 7.34
The kicked rotator.

The perturbed, explicitly time-dependent Hamiltonian that yields the correct equations of motion is

$$H = \frac{J^2}{2I} + \epsilon \cos \phi \sum \delta(t - nT). \tag{7.85}$$

We check that the equations obtained from this Hamiltonian are the correct ones. Hamilton's canonical equations are

$$\left. \begin{aligned} \dot{\phi} &= \frac{J}{I}, \\ \dot{J} &= \epsilon \sin \phi \sum \delta(t - nT). \end{aligned} \right\} \tag{7.86}$$

They imply, correctly, that J (and hence $\dot{\phi}$) is constant between kicks. They also give the correct change of J in each kick. Indeed, suppose that at time $t = nT$ the rotator arrives at $\phi \equiv \phi_n$ with momentum J_n. Then the nth kick, at time $t = nT$, changes the momentum to J_{n+1} given by

$$J_{n+1} - J_n = \lim_{\sigma \to 0} \int_{nT-\sigma}^{nT+\sigma} \dot{J} \, dt = \epsilon \sin \phi_n. \tag{7.87}$$

Thus H yields the motion correctly. Note that H is not the energy $J^2/2I$ of the rotator. Neither the energy nor the Hamiltonian are conserved: the energy is changed by the kicks, and $dH/dt = \partial H/\partial t \neq 0$. This dynamical system is called the *kicked rotator*.

> REMARK: Some authors call dynamical systems Hamiltonian only if they are time independent. According to those, this system is not Hamiltonian, but according to the terminology of this book, it is. □

The kicked rotator is not, of course, a very realistic physical system, for ideally impulsive δ-function forces do not exist. The Fourier series for the infinite sum of δ functions in Eq. (7.85) contains all integer multiples of the fundamental frequency $2\pi/T$, all with the same amplitude. Such a Fourier series does not converge, although it does in the sense of *distributions* or *generalized functions* (Gel'fand & Shilov, 1964; Schwartz, 1966). A realistic periodic driving force would have to be described by a Fourier series that converges in the ordinary sense. Nevertheless this dynamical system lends itself to detailed analysis, and many of the results obtained from it turn out to be characteristic of a large class of two-dimensional systems.

THE STANDARD MAP

During the interval between the nth and $(n + 1)$st kicks, $\phi(t) = \phi_n + J_{n+1}t/I$ so that at $t = (n + 1)T$,

$$\phi_{n+1} = \left(\phi_n + \frac{J_{n+1}}{I} T \right) \bmod 2\pi. \tag{7.88}$$

The mod 2π is needed for the same reason as in the circle map: because it is an angle variable, ϕ ranges only through 2π, and the mod 2π guarantees that. Equations (7.87) and (7.88) can be simplified if the units are chosen so that $T/I = 1$. They then become

$$
\left.
\begin{aligned}
\phi_{n+1} &= (\phi_n + J_{n+1}) \bmod 2\pi, \\
J_{n+1} &= \epsilon \sin \phi_n + J_n.
\end{aligned}
\right\}
\tag{7.89}
$$

These equations define what is known as the *standard map* (also the Chirikov–Taylor map; Chirikov, 1979). It is an important tool for understanding chaos in Hamiltonian systems, particularly in two freedoms. In particular, we will use it as a bridge to the KAM theorem, which concerns the onset of chaos in Hamiltonian systems. (For other examples of physical systems that lead to the standard map, see Jackson, 1990, vol. 2, pp. 33 ff.)

We denote the standard map of Eqs. (7.89) by Z_ϵ:

$$
Z_\epsilon : (\phi_n, J_n) \longmapsto Z_\epsilon(\phi_n, J_n) \equiv (\phi_{n+1}, J_{n+1}).
$$

Not only ϕ, but also J is periodic with period 2π under Z_ϵ, a fact that will prove significant later. The map Z_ϵ provides the same kind of Poincaré map for the kicked rotator as $C_{\Theta K}$ provided for the pendulum in Section 7.4.2: the system is observed once in each period of the applied force, and the Poincaré section $\mathbb{P}$ is the full phase manifold $\mathbf{T}^*\mathbb{Q}$. We list some properties of Z_ϵ (Problem 13). In spite of the fact that the Hamiltonian of Eq. (7.85) is not conserved, Z_ϵ yields a conservative discrete dynamical system. That is, it is area-preserving (has Jacobian determinant equal to 1; see the discussion in Section 7.2.2). Both $(0, 0)$ and $(\pi, 0)$ are fixed points of Z_ϵ for all values of ϵ. The origin is hyperbolic for $\epsilon > 0$ and the other point is elliptic, at least for small ϵ.

POINCARÉ MAP OF THE PERTURBED SYSTEM

If the rotator is not kicked (i.e., in the unperturbed system with $\epsilon = 0$) each constant-J circular orbit of the rotator, on which only ϕ varies, is replaced in the Poincaré map by a sequence of discrete values of ϕ on that orbit (remember that $\mathbb{P}$ is all of $\mathbf{T}^*\mathbb{Q}$). If $J = 2\pi r$, where r is a rational number (we will say J "is rational"), a finite number of ϕ values exist; if J is not rational, the Poincaré map is ergodic on the circle. Figure 7.35a shows the results of iterating Z_0 300 times numerically at each of 62 initial points of $\mathbb{P}$. The orbits, circles that go around $\mathbb{P}$ at constant J, are horizontal lines in the figure. The figure is drawn for $-\pi < \phi \le \pi$. According to Eq. (7.89) the map will be the same for initial values J_0 and $J_0 + 2\pi$ even for nonzero ϵ, so the Poincaré map will be identical at values of J that differ by 2π. Therefore the map for a J interval of 2π shows the behavior of the system for all J and it need be drawn only in the range $-\pi < J \le \pi$. (This periodicity of the map means that the cylindrical $\mathbb{P}$ can be thought of as the torus $\mathbb{T}^2$, but this $\mathbb{T}^2$ should not be confused with the one-dimensional invariant tori (Section 6.2.1) that are the orbits of the unperturbed system.) In this numerical calculation no more than 300 points can be drawn on an orbit of the Poincaré map, so they all look as though J is rational. Moreover, in a computer all numbers are rational.

Under the perturbation, as ϵ increases, the picture changes (Fig. 7.35b, $\epsilon = 0.05$). Most of the orbits of the Poincaré map remain slightly distorted versions of the $\epsilon = 0$ orbits and go completely around the cylinder, or around $\mathbb{T}^2$ in the ϕ direction. But now a new kind of orbit appears: closed curves centered on the fixed point $(\phi, J) = (\pi, 0)$ (they are clearly elliptic in the figure, verifying one of the results of Problem 13). Although they are closed orbits, they fail to go around the cylinder. They are *contractible*: each can be shrunk to a point (see the third subsection of Section 6.2.2). Contractible orbits represent motion in which the rotator oscillates about the angle $\phi = \pi$, whereas noncontractible

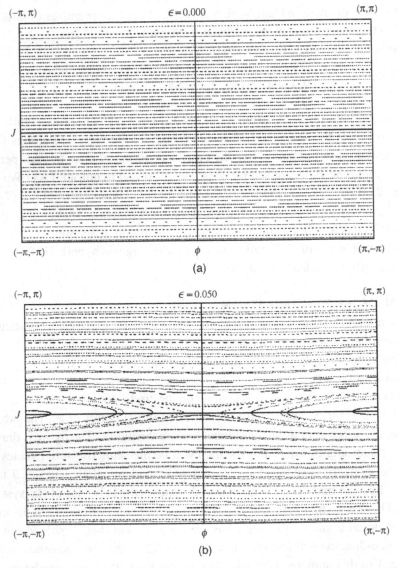

FIGURE 7.35

The standard map for several values of ϵ. In (a)–(c) the map is iterated 300 times; in the rest of the figures, 500 times. The initial points for the iterations are obtained by choosing ϕ randomly for each of 62 equally spaced values of J. In (e) ϵ is about 3% above the critical value $\epsilon_c \approx 0.9716$.

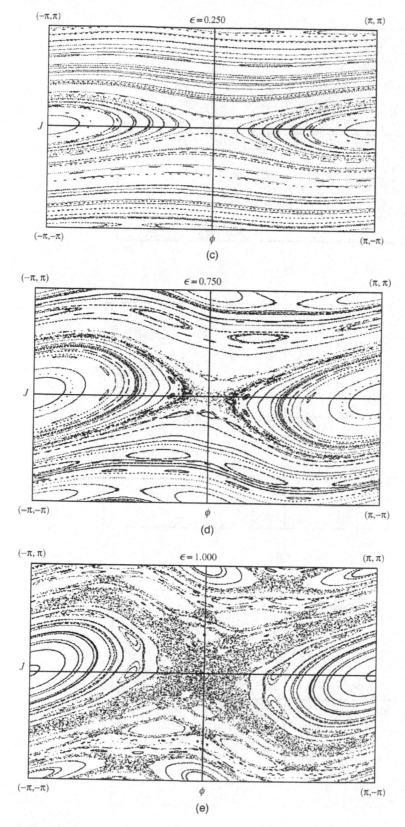

FIGURE 7.35
(*Continued*)

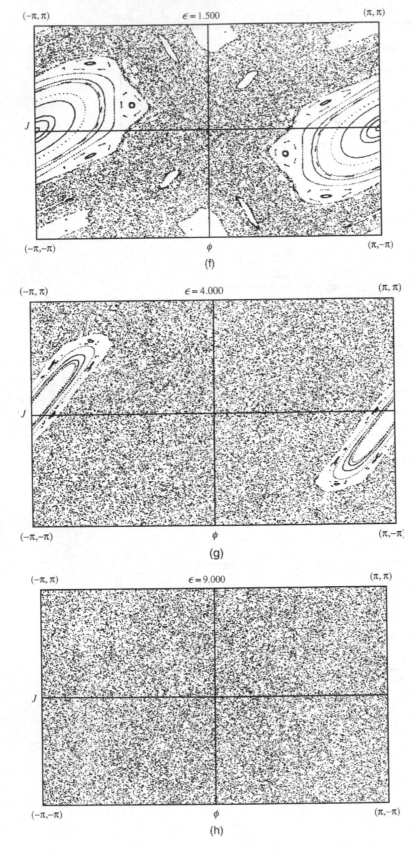

FIGURE 7.35
(*Continued*)

ones represent complete rotations about the pivot. Between these two kinds of orbits lies a separatrix passing through the origin, which is seen to be hyperbolic, in agreement with another result of Problem 13. Separatrices are important, as will be seen as we proceed.

All of the orbits in Fig. 7.35b are still (one-dimensional) tori. Thus to the extent the picture describes the perturbed dynamics with $\epsilon = 0.05$, the dynamical system is integrable, for the motion lies entirely on tori. According to the remarks at the end of the canonical perturbation theory section, that means that perturbation theory will work.

At higher values of ϵ (Fig. 7.35c, $\epsilon = 0.25$) other smaller contractible periodic orbits of the Poincaré map become visible, centered on the period-two points at $(0, \pi)$ and (π, π). They too are separated from adjacent noncontractible orbits by separatrices, these passing through a pair of period-two points that are located near $(-\pi/2, \pi)$ and $(\pi/2, \pi)$ (a conjugate pair of period-two points of Z_0.) These are still one-dimensional tori, and the system still looks integrable. As ϵ increases, more and more such contractible orbits appear, associated in the same way with other periodic points of higher orders.

Eventually (Fig. 7.35d, $\epsilon = 0.75$) disordered scattered points appear, not on a curve at all, but in two-dimensional regions of $\mathbb{P}$. The region near the origin of the figure is an example. It is located, characteristically, in the vicinity of a separatrix, this one between the largest contractible orbits and the noncontractible ones. This is the visible onset of chaos. At lower ϵ the Poincaré map has periodic orbits in this chaotic region, but as ϵ increases, the chaos destroys those periodic orbits. With higher ϵ (Fig. 7.35e, $\epsilon = 1.0$) more and more chaotic regions appear in the vicinity of separatrices, destroying more and more periodic orbits, both the contractible and noncontractible varieties.

At first (that is, for relatively low values of ϵ), the chaotic regions are bounded by noncontractible periodic Poincaré orbits. That means that J, although it jumps around chaotically within a two-dimensional region, cannot wander very far and thus cannot cross one of the periodic orbits that bound the region. In that case J will never change by more than 2π. But there is a critical value ϵ_c after which all of the noncontractible orbits are destroyed. Then J can change by more than 2π. If $\epsilon > \epsilon_c$, the momentum $|J|$, and hence also the energy $J^2/2I$, can increase without bound. It has been found (Lichtenberg and Lieberman, 1992, Section 4.1b) that $\epsilon_c \approx 0.9716$, so that the ϵ of Fig. 7.35e is just above the critical value.

As ϵ increases further (Figs. 7.35f,g) the standard map becomes progressively more chaotic, reflecting the chaotic nature of the kicked rotator. When the kicks get strong enough, the dynamics becomes completely chaotic, as is evident in Fig. 7.35h, where the last of the periodic orbits has been destroyed.

Because Figs. 7.35(a)–(h) are generated by a computer, which can deal only with rational numbers and only with finite precision, they do not show all the details. In reality even very small values of ϵ destroy some of the tori passing through the origin in the standard map. Hence it can be misleading to visualize the transition to chaos entirely in terms of pictures like those of Figs. 7.35. The next few sections discuss in some detail the mechanisms that lead to the destruction of periodic orbits, and then it will be seen that the difference between rational and irrational numbers is crucial and that Figs. 7.35 tell only part of the story.

7.5.2 THE HÉNON MAP

We return to the Hénon map, whose linearized version was discussed in Section 7.2.2. We will not go into great detail, relying on references to the literature, both original papers and books.

The Hénon map is generated by another kicked Hamiltonian system, the harmonic oscillator kicked with period τ by a force proportional to the cube of the displacement (Heagy, 1992). The Hamiltonian is

$$h = \frac{1}{2}(P^2 + X^2) + \frac{1}{3}X^3 \sum_{n=-\infty}^{\infty} \delta(t - n\tau).$$

We do not describe the technique for obtaining the map from this Hamiltonian, but it is similar to the technique used for the kicked rotator. The Hénon map is of interest not only because it is related to a Hamiltonian system, but also because it has a *strange attractor*. More about this is presented at the end of this section.

Hénon's (1976) original definition of his area-preserving map is

$$\begin{aligned}
X_{n+1} &= X_n \cos \tau + \left(P_n - X_n^2\right) \sin \tau, \\
P_{n+1} &= -X_n \sin \tau + \left(P_n - X_n^2\right) \cos \tau,
\end{aligned} \tag{7.90}$$

where τ is a constant. To show how this map is related to the Hénon map of Section 7.2.2, eliminate the P_k from this map. Then it is found that

$$X_{n+1} + X_{n-1} = 2X_n \cos \tau - X_n^2 \sin \tau. \tag{7.91}$$

Now write

$$\left.\begin{aligned}
X &= \lambda x + \cot \tau, \\
a &= \cos \tau(\cos \tau - 2), \\
\lambda &= a/\sin \tau.
\end{aligned}\right\} \tag{7.92}$$

Then some algebra leads to

$$x_{n+1} + x_{n-1} = 1 - ax_n^2,$$

which is the first line of Eq. (7.36) with $b = -1$. It follows from the discussion in Section 7.2 that the map is area-preserving (conservative) when $|b| = 1$. The second line of Eq. (7.36) can be obtained by setting

$$P = \frac{\lambda}{\sin \tau}\left(1 + \frac{\cot \tau}{\lambda} + x \cos \tau + y\right).$$

The more general case, with $|b| \neq 1$, can be identified with essentially the same physical system, but with damping. See Heagy (1992) for details.

Before proceeding we point out an interesting interpretation of the map in terms of folding and reflections in the plane (Hénon, 1976). Consider the following three maps:

$$M_1 : (x, y) \mapsto (x, y + 1 - ax^2),$$
$$M_2 : (x, y) \mapsto (bx, y),$$
$$M_3 : (x, y) \mapsto (y, x),$$

with $a > 0$, $0 < b < 1$. The first map M_1 is quadratic and bends the plane by raising all points on the y axis by one unit, raising them by less at values of x where $ax^2 < 1$, not moving them at $ax^2 = 1$, and lowering them by progressively greater amounts at each value of x for which $ax^2 > 1$. The second map M_2 squeezes the plane in toward the y axis. The third map M_3 reflects the plane in the line $y = x$. The Hénon map is obtained by combining these [compare Eq. (7.36)]:

$$H = M_3 M_2 M_1 : (x, y) \mapsto (1 - ax^2 + y, bx). \tag{7.93}$$

The reason we mention this here is that this kind of squeezing, folding, and reflection is a common feature of maps that lead to chaos.

For numerical calculation it is convenient to change coordinates by making the replacements $x \to ax$ and $y \to ay/b$. Then (7.36) becomes

$$\begin{aligned} x_{n+1} &= a - x_n^2 + by_n, \\ y_{n+1} &= x_n. \end{aligned} \tag{7.94}$$

In what follows we use this form for the Hénon map H.

As we know, the Hénon map is not area-preserving when $b \neq 1$. Indeed, its Jacobian matrix

$$J = \begin{bmatrix} -2x & b \\ 1 & 0 \end{bmatrix}$$

has absolute determinant $b \neq 1$ (for $0 < b < 1$). In Section 7.2.2 such a map was called dissipative. Repeated application of a dissipative map to a volume of initial conditions will shrink the area asymptotically to an attractor of lower dimension. Different initial conditions may converge to different attractors: those points that converge to a given attractor form its *basin of attraction*. For example, Fig. 7.17 illustrates two one-point attractors and their basins. Although many dissipative maps are not invertible, the Hénon map is. Its inverse is simply

$$x_n = y_{n+1},$$
$$y_n = \frac{1}{b} \left[x_{n+1} - a + y_{n+1}^2 \right].$$

The details of H, in particular the properties of its attractor, depend sensitively on the tuning parameters a and b. For example, if $1.42 < a < 2.65$ and $b = 0.3$, orbits for most initial conditions go off to infinity (Feit, 1978; see also Jackson, 1990). However, for $a = 1.4$ and $b = 0.3$, there is a finite attractor, namely the curve shown in Fig. 7.36, which has been obtained by numerical calculation. The point at infinity is another attractor.

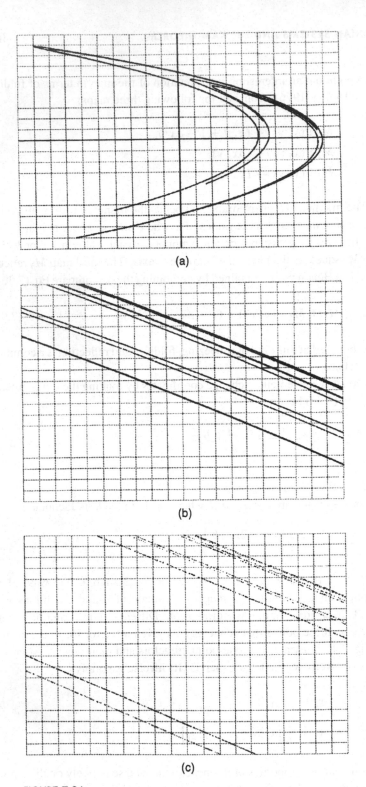

FIGURE 7.36

(a) The attractor for the Hénon map with $a = 1.4$ and $b = 0.3$. In this view x and y both run from -2 to 2. The small outlined box is enlarged in the next view. (b) Part of the same attractor. In this view x runs from 1.0 to 1.2, and y runs from 0.6 to 0.8. The small outlined box is enlarged in the next view. (c) Part of the same attractor. In this view x runs from 1.15 to 1.16, and y runs from 0.72 to 0.73. There are fewer points in this view because the computer plots only one in every 16,000 points it calculates.

The fixed points of H were discussed in Section 7.2. In the present coordinates the two fixed points, both hyperbolic, are at $x = y \approx 0.883896$ and -1.583896. The first of these lies on the finite attractor, and hence its neighborhood is in the basin of the attractor. The neighborhood of the second fixed point is in the basin of the attractor at infinity.

The Hénon map is of particular interest because of the nature of its finite attractor. In Fig. 7.36(a) it looks as though the attractor consists of three or four well-defined curves. In fact it is more complicated than that. On the figure there is a 20×20 grid, therefore consisting of 400 small rectangles. Figure 7.36(b) shows the result of enlarging one of the rectangles through which the attractor passes, the one defined by $1 < x < 1.2$, $0.6 < y < 0.8$. It is seen that what looked like three or four lines actually consists of sets of points distributed along about seven. This figure is also divided into 400 boxes, and Fig. 7.36(c) shows the result of enlarging one of them. Again, several lines of points can be seen. As smaller and smaller portions of the attractor are enlarged, showing more and more detail, the structure repeats itself, exhibiting self-similarity. The local structure of the attractor has been recognized (see Hénon and other references in Jackson) as a Cantor set of lines: it is as though transverse lines have been drawn through the points of a Cantor set, and the points of the attractor are distributed along those lines.

The finite attractor of the Hénon map is fractal (Section 4.1.3) and chaotic. Its fractal dimension has been found to be $d \approx 1.26$ (Russell et al., 1980). It is chaotic in the sense that its orbits are sensitive to initial conditions, that is, initial conditions that are very close to each other move very far apart under iteration of the map (see Ott, Fig. 1.15 for a clear illustration of this for H). This can be stated more precisely in terms of nonzero Lyapunov exponents, but we do not go into that.

Attractors that are fractal, localized, and chaotic are sometimes called *strange* (Feit uses that terminology), although there is some disagreement about the definition of strangeness. At any rate, many maps have such attractors, and we have presented the Hénon map as an important example.

7.5.3 CHAOS IN HAMILTONIAN SYSTEMS

The example of the kicked rotator illustrates the way chaos can arise in Hamiltonian dynamics. The assumption of canonical perturbation theory (i.e., when a completely integrable system is perturbed it remains completely integrable) implies that the perturbed system must have $n - 1$ constants of the motion in addition to the Hamiltonian. If this were true, the system would move on the tori that are the level sets of these other constants. But we have just seen on the example of the standard map that the perturbation destroys some of those tori, and the stronger the perturbation, the more tori it destroys.

According to Eq. (6.112), namely

$$S_1(J, m) = \frac{\mathcal{K}_m(J)}{i(\nu_0 \cdot m)}, \tag{7.95}$$

and the discussion following it, perturbation theory breaks down for small denominators (i.e., when the frequencies are commensurate or nearly so). Recall that the ν_0 are functions of J, and that J determines the invariant torus of the unperturbed dynamics. That means that perturbation theory breaks down for some tori before it does for others. This is just what we saw happening in the standard map: some tori were destroyed before others. These two phenomena are in fact the same, described first in terms of small denominators and then in terms of breakup of tori. Perturbation theory breaks down even in lowest order for commensurate frequencies, for which some of the denominators surely vanish. Commensurate frequencies correspond to what we have called rational J in the standard map, and we start our considerations from them.

POINCARÉ–BIRKHOFF THEOREM

Consider the unperturbed standard map Z_0 of Eq. (7.89) with $\epsilon = 0$. Choose $J \equiv J_r = 2\pi j/k$, a rational multiple of 2π (the subscript r stands for "rational"). Then

$$Z_0^k(\phi, J_r) = (\phi, J_r),$$

or the kth iteration of the unperturbed standard map sends every point of the J_r circle back to its original position: Z_0^k is the identity on the J_r circle. Since Z_0 is a continuous function, Z_0^k maps the J circle in one sense for $J \equiv J_+ > J_r$ and in the other for $J \equiv J_- < J_r$. To visualize this, distort the cylinder to a truncated cone and flatten it out, as shown in Fig. 7.37. The noncontractible circles become concentric, the J direction becomes radial, and then the J_-, J_r, and J_+ circles are arranged as illustrated in Fig. 7.38a.

Now move on to ϵ slightly larger than 0 (Fig. 7.38b). By continuity in ϵ, for small enough ϵ there must exist in the ring between J_+ and J_- a closed curve C consisting of points whose ϕ values, like those of the points on J_r, are not changed by Z_ϵ^k. In other words, this curve C is mapped purely along the J direction by Z_ϵ^k, or purely radially as seen in Fig. 7.39. Inside and outside C the value of ϕ is changed: inside points are mapped in the sense of J_- (counterclockwise in our example); outside, ones in the J_+ direction. We write D for the curve into which C is mapped: $Z_\epsilon^k(\phi, J_C) = (\phi, J_D)$, where $(\phi, J_C) \in C$ and $(\phi, J_D) \in D$.

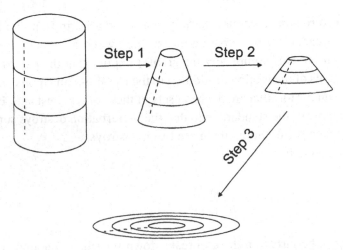

FIGURE 7.37
Distorting the cylindrical phase manifold of the standard map into a circular disk with a hole in it. In Step 1 the cylinder is distorted to a cone. In Step 2 the cone starts to be flattened. Step 3 finishes the flattening. The dotted line represents the J axis, which becomes the radial direction on the final disk.

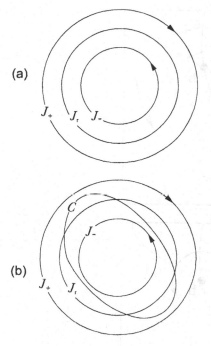

FIGURE 7.38

(a) Three orbits of the unperturbed standard map Z_0^k. The J_r orbit is left stationary by Z_0^k, and each other orbit is mapped by Z_0^k in the sense of its arrow. (b) The ϕ coordinate is left invariant on C by the perturbed map Z_ϵ^k. The ϕ coordinates of points inside and outside C are changed in the sense of J_- and J_+, respectively. In this and the next two figures the distances between curves are exaggerated in the interest of clarity.

In general, C and D will intersect, for the map is area-preserving, so D cannot lie entirely inside or outside C. Because they are both closed curves, they must intersect an even number of times, as in Fig. 7.39. At each intersection (and only at the intersections) $J_C = J_D$ and since ϕ is left unaltered, every intersection is a fixed point for Z_ϵ^k. Thus Z_ϵ^k has an even number of fixed points on C.

Whether these fixed points are hyperbolic or elliptic can be discovered by studying the motion in their neighborhoods. Consider the intersections marked **a** and **b** in Fig. 7.39 and enlarged in Fig. 7.40. The directions of the mappings in their neighborhoods are indicated in the upper half of the figure, and they imply that **a** is hyperbolic and **b** is elliptic, as indicated in the lower half. This is the general situation. There is an even number of fixed points, alternately elliptic and hyperbolic. In short, when $J \equiv J_r = 2\pi j/k$ labels an

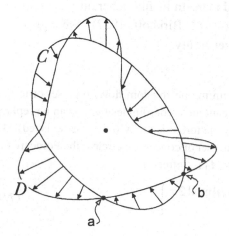

FIGURE 7.39

Z_ϵ^k maps C onto D in the J direction, along the arrows. The areas enclosed by C and D are equal. **a** and **b** are fixed points for Z_ϵ^k.

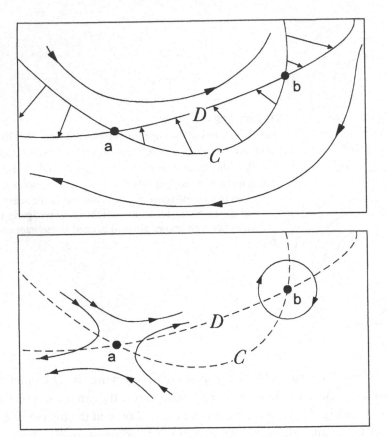

FIGURE 7.40
A closer look at the fixed points **a** and **b** in order to exhibit their nature.

invariant torus of the unperturbed standard map Z_0, the perturbation breaks up the torus
into an even number of alternating elliptic and hyperbolic fixed points of Z_ϵ^k. Their number
will be counted later in this section.

This is a general result that applies to many one-freedom (i.e., two-dimensional) Hamiltonian dynamical system other than the standard map. In its full generality it is known as
the Poincaré–Birkhoff theorem (Poincaré, 1909–1912; Birkhoff, 1927). We digress now
with another important example to outline this generality.

THE TWIST MAP

Every completely integrable Hamiltonian system in one freedom flows on a set of tori. In
the Poincaré-map description of such a system the canonical coordinates (ϕ, J) can be replaced
by other coordinates (ϕ, r), where r is the radius of the torus (actually only J is replaced). The
Poincaré manifold $\mathbb{P}$ is chosen to intersect the tori in a set of concentric circles (the Poincaré map
will be denoted by M). For the unperturbed system M_0 is defined by

$$\left.\begin{aligned}
\phi_{n+1} &= \{\phi_n + \Phi(r_n)\} \bmod 2\pi, \\
r_{n+1} &= r_n.
\end{aligned}\right\} \tag{7.96}$$

This is the Poincaré map for the most general completely integrable two-dimensional (one-freedom) system, in which each circle (torus) is rotated rigidly, even though the angle Φ through which it is rotated depends on the torus. This is called the (unperturbed) *twist map*. The unperturbed standard map Z_0 is seen to be a special case of M_0.

The perturbation adds ϵ-dependent terms to both lines of M_0: the general form of the perturbed map M_ϵ is

$$\left.\begin{aligned}
\phi_{n+1} &= \{\phi_n + \Phi(r_n) + \epsilon\theta(r_n, \phi_n)\}\bmod 2\pi, \\
r_{n+1} &= r_n + \epsilon\rho(r_n, \phi_n).
\end{aligned}\right\} \tag{7.97}$$

This is the *complete twist map*. The standard map Z_ϵ is also seen to be a special case of M_ϵ. Now choose a value of r for which Φ is rational; let $\Phi(r_r) = 2\pi j/k$. Then essentially the same argument that led to the Poincaré–Birkhoff theorem for the standard map can be modified to obtain the same result for the twist map. We do not go into that here, however, but see Moser (1973) and also Ott (1993, pp. 230 ff).

Moreover, the relevant properties of the standard map are possessed not only by the twist maps, but by more general area-preserving maps in two dimensions. They are Poincaré maps for Hamiltonian systems, including time-independent ones (even though we have used the example of the time-periodic driven pendulum). The Poincaré section is generally a submanifold whose dimension is less than that of the entire phase manifold, so even though the maps are of dimension two, the dimension of the phase manifold may be greater than two. In what follows we continue to denote the map by Z, even though it need not be the standard map, and we will continue to use the standard map as an example.

If the Poincaré section has more than two dimensions the situation is more complicated, for the elliptic–hyperbolic classification of fixed points breaks down. For example, in four freedoms a fixed point could be stable along two of its principal directions but hyperbolic along two others: $\mathbb{E}^s$ could be of dimension three and $\mathbb{E}^u$ of dimension one, or three of the eigenvalues of the linearized matrix could be smaller than 1 and the other larger. The extension of the Poincaré–Birkhoff theorem to higher dimensions is at present an active area of research.

NUMBERS AND PROPERTIES OF THE FIXED POINTS

We now proceed to count the number of elliptic and hyperbolic fixed points on the curve C of Figs. 7.39 and 7.40. If (ϕ_0, J_0) is a fixed point of Z_ϵ^k, so are $(\phi_s, J_s) \equiv Z_\epsilon^s(\phi_0, J_0)$, $s = 0, 1, 2, \ldots, k-1$, all of which are mapped into each other by Z_ϵ. This means that each of the (ϕ_s, J_s) is a periodic point of Z_ϵ with period k. Some could conceivably have period less than k, but in fact none do (except for the trivial period zero: Z_ϵ^0 is the identity). Indeed, suppose that $k > 0$ is the smallest positive integer for which $Z_\epsilon^k(\phi_0, J_0) = (\phi_0, J_0)$ and that $Z_\epsilon^m(\phi_s, J_s) = (\phi_s, J_s)$, $0 \le m < k$. Apply Z_ϵ^{-s} to both sides of the last equation, we obtain $Z_\epsilon^m(\phi_0, J_0) = (\phi_0, J_0)$. But by assumption there is no positive integer smaller than k for which this is true, so $m = 0$, as asserted. This does not mean that the $Z_\epsilon^s(\phi_0, J_0)$ exhaust the fixed points of Z_ϵ^k, but the same argument shows that if there are any others, they must also lie in sets of k distinct points mapped into each other by Z_ϵ. If there are l such sets, Z_ϵ^k has lk fixed points. Moreover, continuity in ϕ and J implies that the fixed point (ϕ_0, J_0) of Z_ϵ^k is of the same type (elliptic or hyperbolic) as $Z_\epsilon(\phi_0, J_0)$, so each of the l sets contains only points of a given type. Since the type alternates around C, there must be an even number of sets (replace l by $2l$) or an even number $2lk$, of fixed points, lk, of each type.

In conclusion, as soon as the perturbation is turned on, each one-dimensional *rational torus* (also called *resonant* or *periodic* tori) of the unperturbed system (i.e., with $J = 2\pi j/k$) breaks up into a set of k elliptic and k hyperbolic fixed points ($l = 1$ for ϵ very small). Figure 7.35b illustrates this for the standard map: at $\epsilon = 0.05$ the $J = 0$ torus (take $k = 1$ for $J = 0$) has been destroyed and replaced by the hyperbolic fixed point at $(\phi, J) = (0, 0)$ and the elliptic one at $(\phi, J) = (\pi, 0)$ surrounded by contractible orbits. In Fig. 7.35c, for $\epsilon = 0.25$, two hyperbolic and two elliptic fixed points appear on the $J = \pi$ circle (i.e., for $k = 2$) at the upper and lower edges of the diagram. Figure 7.35d shows many more such fixed points at other values of J (as well as the chaotic region around the origin).

The computer-generated plots of Fig. 7.35 are not detailed enough to show all of the elliptic and hyperbolic fixed points predicted by the Poincaré–Birkhoff theorem. According to the theorem fixed points must appear on *every* rational torus as soon as the perturbation is turned on for arbitrarily small $\epsilon > 0$. Clearly there must be some destruction also of tori near the rational ones. For instance, in Fig. 7.35b the contractible orbits at $(0, 0)$ and $(\pi, 0)$ overlap neighboring noncontractible tori belonging to many rational and irrational values of J (this will be discussed more fully in the next subsection). A similar phenomenon was observed with the circle map: at $K = 0$ the frequency-locked intervals have measure zero, but as K increases, their measure increases and they cover more and more of the Θ interval. At $\epsilon = 0$ (i.e., for the map Z_0) the orbits around elliptic and hyperbolic fixed points occupy regions of measure zero on the $\mathbb{T}^2$ torus of the phase manifold (the dynamical system is completely integrable), but as ϵ increases, their measure increases and they occupy more and more of $\mathbb{T}^2$.

The argument that showed how the initial rational tori break up into fixed points is valid for any other set of tori, in particular for the contractible ones. That is, if one of them has a rational period, it will break up, as ϵ changes, into a set of alternating elliptic and hyperbolic fixed points. For example, in Fig. 7.35d two of the orbits around $(\phi, J) = (\pi, 0)$ are broken up in this way. There are also closed orbits about the new elliptic points, and they too will break up into sets of alternating elliptic and hyperbolic orbits, and so on ad infinitum. The number $2lk$ of fixed points we found for Z_ϵ^k includes only the points in the first tier, lying in a narrow ring around J_r. Other elliptic fixed points show up in Figs. 7.35 as centers of elliptic orbits, and in Fig. 7.35d "islands" of them can be seen circling $(\pi, 0)$.

As for the new hyperbolic fixed points formed in this infinite process, each has its own stable and unstable manifolds, and the situation gets quite complicated. To study it further requires moving on to the irrational tori.

THE HOMOCLINIC TANGLE

As soon as the perturbation is turned on the rational invariant tori break apart, at first into chains of islandlike elliptic points separated by hyperbolic ones, and eventually, as illustrated in Figs. 7.35, into chaotic regions. This breakup swamps and destroys some of the irrational tori. To understand how, we turn to the hyperbolic fixed points created when the rational ones break apart.

Each hyperbolic fixed point created by the destruction of a periodic torus has its *stable* and *unstable manifolds* $\mathbb{W}^s$ and $\mathbb{W}^u$ in the Poincaré section $\mathbb{P}$. We write $\mathbb{W}^s$ and $\mathbb{W}^u$ to emphasize that unlike $\mathbb{E}^s$ and $\mathbb{E}^u$ of Section 7.2.1 they are nonlinear sets. They need not in general have manifold structure, but we assume that they do (recall the nonlinear stable and unstable manifolds of chaotic scattering in Section 4.1, as well as in Fig. 7.17). By definition the stable manifold $\mathbb{W}^s$ of a hyperbolic fixed point p_f consists of the set of points $x \in \mathbb{P}$ for which $Z_\epsilon^k x \rightarrow p_f$ in the limit as $k \rightarrow +\infty$. Similarly, the unstable manifold $\mathbb{W}^u$ consists of the set of points for which $Z_\epsilon^{-k} x \rightarrow p_f$ in the same limit. In the two-dimensional case, which is all we are considering now, these are one-dimensional curves, but in general they can be manifolds of higher dimension. If x is a point on $\mathbb{W}^s$, the $Z_\epsilon^k x$ are a *discrete* set of points, all on $\mathbb{W}^s$. At the fixed point, $\mathbb{W}^s$ and $\mathbb{W}^u$ are tangent to the stable and unstable spaces (straight lines) $\mathbb{E}^s$ and $\mathbb{E}^u$ of the linear approximation. We now show that $\mathbb{W}^s$ and $\mathbb{W}^u$ can not intersect themselves, but that they can and do intersect each other, and that this leads to what is called the *homoclinic tangle* and to chaos.

First we show that they can not intersect themselves. This follows from the invertibility of Z_ϵ (recall that Z_ϵ now stands for the general map in two dimensions obtained from the Poincaré section of a perturbed Hamiltonian system). Indeed, suppose that the unstable manifold $\mathbb{W}^u$ of a fixed point p_f intersected itself at a point $\tilde{p}$; then $\mathbb{W}^u$ would contain a loop $\mathbb{L}$ from $\tilde{p}$ back around to $\tilde{p}$ again. Applying Z^{-1} (we suppress the ϵ dependence) to all the points of $\mathbb{L}$ yields a set of points $Z^{-1}\mathbb{L}$, and the continuity of Z^{-1} implies that $Z^{-1}\mathbb{L}$ is some interval $\Delta\mathbb{W}^u$ of $\mathbb{W}^u$. Then $\Delta\mathbb{W}^u$ is mapped by Z onto $\mathbb{L}$ in such a way that both of its end points are mapped into $\tilde{p}$. But if Z is invertible it cannot map two distinct points into one, so there can be no such loop and hence no intersection. The same proof works for stable manifolds if Z is replaced by Z^{-1}.

Moreover, stable manifolds of different fixed points cannot intersect each other. Indeed, suppose the stable manifolds of p_{f1} and p_{f2} intersected at $\tilde{p}$. Then $Z^k(\tilde{p})$ would approach the distinct points p_{f1} and p_{f2} as $k \rightarrow \infty$, which is impossible if Z is invertible. Thus they can't intersect. This same proof, but with Z replaced by Z^{-1}, can be used to show that unstable manifolds of different fixed points also cannot intersect.

There is nothing, however, to prevent stable manifolds from intersecting with unstable ones. The stable and unstable manifolds of a single fixed point intersect in what are called *homoclinic* points, and those of two different fixed points, in *heteroclinic* points.

In Fig. 7.41, x_0 is a heteroclinic point that lies on the $\mathbb{W}^u$ of p_{f1} and the $\mathbb{W}^s$ of p_{f2} [Fig. 7.41(a)]. Since both $\mathbb{W}^u$ and $\mathbb{W}^s$ are invariant under Z, the $Z^k x_0$ are a set of discrete points that lie on both manifolds, so $\mathbb{W}^u$ and $\mathbb{W}^s$ must therefore intersect again. For instance, because $x_1 \equiv Z x_0$ is on both manifolds, $\mathbb{W}^u$ must loop around to meet $\mathbb{W}^s$. Similarly, the $x_k \equiv Z^k x_0$ for all positive integers k lie on both manifolds, so $\mathbb{W}^u$ must loop around over and over again, as illustrated in Fig. 7.41(b). The inverse map also leaves $\mathbb{W}^u$ and $\mathbb{W}^s$ invariant, and hence the $x_{-k} \equiv Z^{-k} x_0$ are intersections that force $\mathbb{W}^s$ to loop around similarly to meet $\mathbb{W}^u$; some of these loops of $\mathbb{W}^s$ are illustrated in Fig. 7.41(c). As $|k|$ increases and x_k approaches one of the fixed points, the spacing between the intersections gets smaller, so the loops they create get narrower. But because Z is area-preserving, the loop areas are the same, so the loops get longer, which leads to many intersections among them, as illustrated in the figure.

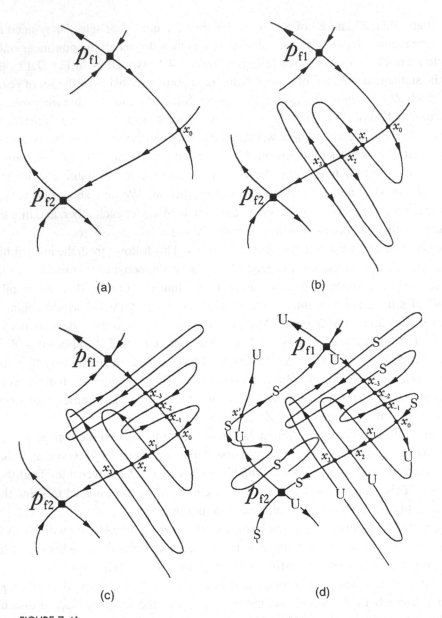

FIGURE 7.41

A heteroclinic intersection. (a) Two hyperbolic fixed points p_{f1} and p_{f2}, and an intersection x_0 of the unstable manifold of p_{f1} with the stable manifold of p_{f2}. (b) Adding the forward maps $Z^k x_0$ of the intersection. (c) Adding the backward maps $Z^{-k} x_0$ of the intersection. (d) Adding another intersection x' and some of its backward maps. U – unstable manifold; S – stable manifold.

This may seem confusing, for how does x_0 know whether to move along $\mathbb{W}^u$ or $\mathbb{W}^s$? But remember that these are not integral curves of a dynamical system: they show how integral curves of the underlying Hamiltonian system intersect the Poincaré manifold. A point x_0 does not "move along" $\mathbb{W}^s$ or $\mathbb{W}^u$ under the action of Z, but rather it jumps discontinuously to a new position. The stable (or unstable) manifolds of Fig. 7.41 are continuous because they show *all* the points that are asymptotic to the fixed point under the action of Z (or of

Z^{-1}). The action of Z (or of Z^{-1}) moves any *particular* point on the manifold in discrete jumps.

The result is that infinitely many intersection points pack into the region bounded by the stable and unstable manifolds of p_{f1} and p_{f2}. Figure 7.41(d) shows why the loops, and hence the intersections, must lie in those bounds. In that figure x' is another intersection of $\mathbb{W}^u$ and $\mathbb{W}^s$; some of the $Z^k x'$ intersections and their loops are shown for negative k and none for positive k. The $\mathbb{W}^u$ curves are labeled U, and the $\mathbb{W}^s$ curves are labeled S. Since stable manifolds cannot intersect each other, the $\mathbb{W}^s$ curve emanating from p_{f1} forms a barrier to the one emanating from p_{f2}, and vice versa. Similarly, unstable manifolds form barriers for each other (this is not illustrated in the figure). Thus intersections between the loops are all confined to a region bounded by the stable and unstable manifolds.

As the loops get longer and longer they are therefore forced to bend around, and this leads to even further complications, which we do not attempt to illustrate. For a more detailed discussion of the complex tangle of loops that results, see, for instance, Jackson (1990), Lichtenberg and Lieberman (1992) or Ott (1993). For our purposes it is enough to point out in conclusion that there is an infinite number of convolutions in a region close to hyperbolic fixed points. Although we have not discussed homoclinic intersections, the situation is quite similar. [Actually, there are homoclinic intersections even in Fig. 7.41(d) where a loop of the p_{f1} stable manifold can be seen to intersect a loop of the p_{f1} unstable manifold. Many more would be seen if the diagram were more detailed.]

Invertibility of Z implies that stable and unstable manifolds of hyperbolic points cannot cross elliptic orbits. Therefore elliptic orbits form barriers to both the $\mathbb{W}^u$ and $\mathbb{W}^s$. So far we have been discussing only the rational (periodic) orbits, ignoring the irrational (quasiperiodic) ones. Any irrational ones that have not been destroyed (and in the next subsection it will be shown that there are many of them, at least for small ϵ) consist of elliptic orbits. In two dimensions these elliptic orbits also form boundaries that cannot be crossed by the complicated network of loops: the tangles therefore lie in rings between such boundaries.

> **REMARK:** In more than two dimensions quasiperiodic invariant tori do not form such barriers because the interior and exterior of a torus are not well defined (see Fig. 6.7). Hence iteration of the map can send points far from their initial positions, even when few of the invariant tori are destroyed. Motion on such trajectories is very slow. The phenomenon is called Arnol'd diffusion (Arnol'd, 1964; see also Saslavsky et al., 1991).   □

We have seen that for small ϵ hyperbolic fixed points on periodic tori alternate with elliptic ones. The closed orbits around the elliptic points, like the quasiperiodic tori, form barriers to the $\mathbb{W}^u$ and $\mathbb{W}^s$ manifolds of the hyperbolic points. The resulting picture is roughly illustrated in Fig. 7.42, which shows two hyperbolic points and the elliptic point between them. Stable manifolds are shown intersecting with unstable ones. Actually the pattern is far more complex, as the loops get tighter and bend around. The picture of Fig. 7.42 repeats throughout each chain of islands that gets formed between pairs of remaining irrational tori, yielding the pattern indicated in Fig. 7.43.

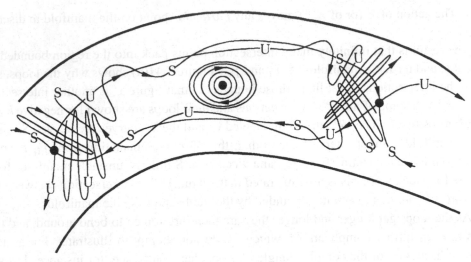

FIGURE 7.42
Two hyperbolic fixed points separated by an elliptic one, all sandwiched between two invariant tori. Arrows indicate directions of motion. U – unstable manifold; S – stable manifold.

Figure 7.43 is a schematic diagram of some of the island chains of elliptic points separated by hyperbolic points and the webs formed by their stable and unstable manifolds. Only one island chain is shown in any detail; three other smaller ones are shown in decreasing detail. Each chain is bounded by irrational invariant tori that have not yet been destroyed. This figure does not apply directly to the standard map but to perturbed integrable systems in two dimensions in general. Nevertheless it reflects the patterns that were obtained by numerical calculation in Fig. 7.35.

The picture obtained is self-similar in that each elliptic point in an island chain repeats the pattern of the whole. Indeed, it was pointed out near the end of the previous subsection that the argument for the breakup of the periodic tori can be applied also to the closed orbits about each of the small elliptic points. Therefore if the neighborhood of any one of the small elliptic points were enlarged, it would have the same general appearance as the whole diagram.

THE TRANSITION TO CHAOS

Consider an initial point of the Poincaré section lying somewhere in the tangle formed by the infinite intersections of $\mathbb{W}^u$ and $\mathbb{W}^s$. As time progresses, Z will carry this point throughout the tangle, and because $\mathbb{W}^u$ and $\mathbb{W}^s$ loop and bend in complicated ways, the time development is hard to follow. Two points that start out close together can end up far apart after a few applications of Z. This is the onset of chaos.

In the two-dimensional case, the separation of two such points cannot grow arbitrarily, because their motion is bounded by those of the irrational invariant tori that remain whole. In more than two dimensions, however, they can end up far apart. Even in two dimensions, as ϵ increases and more irrational tori are destroyed, such points can wander further and further apart. In contrast, chaos does not affect points that lie on remaining irrational tori: on them the motion develops in an orderly way.

FIGURE 7.43

The Poincaré map of a perturbed Hamiltonian system in two dimensions. The figure shows four island chains of stable fixed points separated by unstable fixed points. The largest island chain also shows some detail of the webs formed by intersecting loops of stable and unstable manifolds. Each island chain lies in a ring bounded by invariant tori that are still undestroyed. The greater the perturbation (i.e., the larger ϵ), the fewer the invariant tori that remain undestroyed (Arnol'd and Avez, 1968).

The map Z on the Poincaré section reflects the underlying Hamiltonian dynamics. That means that under some initial conditions, chaos sets in for the Hamiltonian dynamics as soon as ϵ becomes greater than zero; hence canonical perturbation theory is not applicable to those initial conditions. In Section 6.3.2 we described some problems inherent in canonical perturbation theory. The results of the Poincaré analysis show that those problems are of crucial importance: even for very small ϵ canonical perturbation theory must be used with caution. Perturbation theory is valid for points that lie on those irrational tori that remain, if there are any (even for the smallest perturbations some of the irrational tori break up), but not for points that do not. If we want to apply perturbation theory, we need to know which of the irrational tori remain. At the very least, as a gauge of the applicability of perturbation theory, we need to know what fraction of them remains. Since this fraction depends on ϵ, the degree of caution required in using perturbation theory depends on ϵ.

These matters are addressed by the KAM theorem.

7.5.4 THE KAM THEOREM

BACKGROUND

The Poincaré–Birkhoff theorem applies to the rational (resonant or periodic) tori. In this section we turn to the irrational (quasiperiodic) ones. The method for dealing with them, known as the KAM theorem (or simply KAM), was first enunciated by Kolmogorov (1957) in 1954 and later proven and extended under more general conditions by Arnold in 1963 and Moser in 1962 (see also Moser, 1973). It asserts that for *sufficiently small* perturbations of the Hamiltonian, *most* quasiperiodic orbits are only slightly changed, or to put it another way, that most of the irrational tori are stable under a sufficiently small perturbation of an integrable Hamiltonian system. We will state this more precisely in what follows. The Poincaré–Birkhoff theorem is strictly valid only for two-dimensional Poincaré sections, but KAM applies more generally to perturbations of Hamiltonian dynamical systems.

The question answered by KAM was originally posed in the nineteenth century in the context of celestial mechanics: is the solar system stable? Do the perturbations the planets impose on each other's orbits cause the orbits to break up (in the sense, for example, of Poincaré–Birkhoff), or do the perturbed orbits remain close to the unperturbed elliptical ones of Kepler theory? The first relevant nontrivial problem studied in this connection was the so-called *three-body problem* in which three bodies attract each other gravitationally. This problem can be treated by canonical perturbation theory by taking the integrable Hamiltonian of just two of the bodies as the unperturbed Hamiltonian and adding the effect of the third body, if its mass is sufficiently small, as a perturbation. Poincaré found that small denominators cause the Fourier series (Section 6.3.2) to diverge even in the lowest orders of ϵ. Is the gravitational two-body system consequently unstable against small perturbations, or is it the mathematical treatment that is at fault? The Poincaré–Birkhoff theorem would make it seem that such problems are inherently unstable, for it shows that, at least in two dimensions, the rational tori all break up under even the smallest perturbation, and the rational tori are in some sense dense in the space of all tori. But since there are many more irrational tori than rational ones in the phase manifold, just as (in fact because) there are many more irrational numbers than rational ones, the Poincaré–Birkhoff treatment just scratches the surface of the problem. The larger question is answered, reassuringly, by the KAM theorem, which proves that most irrational tori are preserved if the perturbation is sufficiently small. That means that for most initial conditions the three-body system is stable, and this has given some hope for the stability of the solar system as long as the perturbations are quite small (see Hut, 1983). It has been found recently, however, that the solar system actually does exhibit chaotic behavior (see Lasker, 1989, and Sussman and Wisdom, 1992).

Before going on to describe the KAM theorem and its proof, we make some definitions. We emphasize that the systems now under discussion have in general $n \geq 2$ freedoms. The Hamiltonian

$$H(\phi_0, J_0, \epsilon) = H_0(J_0) + \epsilon H_1(\phi_0, J_0) \tag{7.98}$$

is written from the start in the AA variables $\phi_0 = \{\phi_0^1, \phi_0^2, \ldots, \phi_0^n\}$ and $J_0 = \{J_{01}, J_{02}, \ldots,$

J_{0n}} of the completely integrable unperturbed Hamiltonian H_0. The perturbation is $\epsilon H_1(\phi_0, J_0)$. The unperturbed flow is on n-tori $\mathbb{T}_0^n$ defined by values of J_0. Recall that the unperturbed frequencies are functions of J_0:

$$\{v_0^1, v_0^2, \ldots, v_0^n\} = \left\{\frac{\partial H_0}{\partial J_{01}}, \frac{\partial H_0}{\partial J_{02}}, \ldots, \frac{\partial H_0}{\partial J_{0n}}\right\} \equiv v_0 \equiv \frac{\partial H_0}{\partial J_0}. \qquad (7.99)$$

The KAM theorem is quite intricate and is difficult to understand. Even pure mathematicians find it difficult to follow the proof, so we will describe it only in outline, together with an explanation of its assumptions and restrictions. We will first give a statement of the theorem, then proceed with a more discursive description of what it establishes, and finally give a brief outline of one method of proof (J. Bellissard, 1985 and private communication). [For the interested reader who wants more than we present, we recommend, in addition to Bellissard, the reviews by Arnold (1963), Moser (1973), and Bost (1986), as well as the original statement of the theorem given by Kolmogorov (1957).] Although the original approach to KAM used canonical perturbation theory, the Lie transformation method (Section 6.3.2) is probably better suited to it, for explicit expressions can be written to arbitrary orders in the perturbation and it does not give rise to complications having to do with inversion of functions. [The Lie method of proof is described also by Bennetin et al. (1984).] Recall that both of these methods are based on finding a canonical transformation Φ_ϵ to new AA variables $(\phi, J) = \Phi_\epsilon(\phi_0, J_0)$ such that $H(\phi_0, J_0, \epsilon) \equiv H(\Phi_\epsilon^{-1}(\phi, J), \epsilon) = H'(J, \epsilon)$, that is, such that the new Hamiltonian is a function only of the new action variables. Both perturbation-theory methods try to construct Φ_ϵ as a power series in ϵ. The problem with such an approach is that the actual perturbed system is not completely integrable and hence does not even have AA variables. Thus if the perturbative treatment leads to a solution, the power series for Φ_ϵ cannot converge to a canonical transformation. Moreover, because of small denominators the procedure can break down even at the lowest orders in ϵ. The achievement of KAM is to show that for sufficiently small ϵ and under certain conditions, there are *regions of the phase manifold* in which the perturbative program converges to *all* orders in ϵ.

TWO CONDITIONS: HESSIAN AND DIOPHANTINE

The regions of convergence in $\mathbf{T}^*\mathbb{Q}$ are specified by two conditions. The first is a Hessian condition:

$$\det\left|\frac{\partial v_0^a}{\partial J_{0\beta}}\right| \equiv \det\left|\frac{\partial^2 H_0}{\partial J_{0\alpha}\partial J_{0\beta}}\right| \neq 0. \qquad (7.100)$$

This condition guarantees that the function $v_0(J_0)$ is invertible, or that there exists a function $J_0(v_0)$, and allows the treatment to proceed equivalently in terms of the action variables or the frequencies (such frequencies are called *nondegenerate*). Recall that the unperturbed tori are defined by their J_0 values; this condition thus implies that each unperturbed frequency also uniquely defines a torus $\mathbb{T}_0^n(v_0)$. Wherever the Hessian condition is *not* satisfied, H_0 is a linear function of at least some of the action variables. The corresponding frequencies do not vary from torus to torus ($\partial H_0/\partial J_{0\alpha} \equiv v_0^\alpha$ is a constant for some α), and the frequencies cannot be used as unique labels for the tori. This restricts the $J_{0\alpha}$ to some subset $\Omega_H \subset \mathbb{R}^n$ (where the subscript H stands for Hessian), and that in turn means that the treatment is restricted to a subset $\Omega_H \times \mathbb{T}^n \subset \mathbf{T}^*\mathbb{Q}$. Often such *good* regions are traversed by *bad* ones, in which (7.100) is not satisfied, so the good region may not be connected. The treatment is then further restricted to a connected component of the good region (see Problems 14 and 15). There are some degenerate systems, however, for which (7.100) is nowhere satisfied. We do not discuss such systems, even though they include, for example, the harmonic oscillator. From now on we restrict Ω_H to sets such that $\Omega_H \times \mathbb{T}^n$ are connected regions of $\mathbf{T}^*\mathbb{Q}$ and write Ω_H^ν for the image of Ω_H under the function $\nu_0(J_0)$.

WORKED EXAMPLE 7.5

(Finding Ω_H in one freedom.) Consider the one-freedom Hamiltonian $H = aJ^3/3$ in which J takes on values in the open interval $-1 < J < 1$, where $a > 0$ is constant. Find all possible Ω_H and Ω_H^ν.

Comment: Although this is an artificial example, it helps to clarify the Hessian condition. See also Problems 14 and 15.

Solution. The Hessian condition is

$$\frac{d^2 H}{dJ^2} = 2aJ \neq 0.$$

This is satisfied everywhere except at $J = 0$, which cuts the interval into two parts: $J > 0$ and $J < 0$. Thus Ω_H is either $\Omega_{H+} \equiv [0, 1]$ or $\Omega_{H-} \equiv [-1, 0]$. Either will do; choose Ω_{H+} (Ω_{H-} is treated similarly). This is illustrated in Fig. 7.44. In Ω_{H+} the

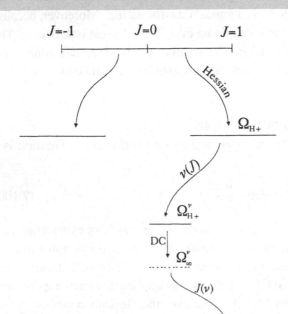

FIGURE 7.44
The transition from $[-1, 1]$ to Ω_∞. The Hessian splits the J manifold into two parts, of which we consider only one, which we call Ω_{H+}. Then $\nu(J)$ maps this to Ω_{H+}^ν. The DC eliminates many frequencies, leaving only Ω_∞^ν, indicated here by dots. Then $J(\nu)$ maps that to Ω_∞.

one-to-one relation between J and ν is $\nu(J) \equiv dH/dJ = aJ^2$ and $J(\nu) = \sqrt{\nu/a}$ [on Ω_{H-} this becomes $J(\nu) = -\sqrt{\nu/a}$]. There is no equivalent one-to-one map on the entire J manifold. Then Ω_{H+}^ν consists of all ν values that satisfy $\nu(J) = aJ^2$ with $J \in \Omega_{H+}$, that is, $\Omega_{H+}^\nu \equiv [0, a]$.

There is, in addition, a second condition restricting the frequencies. Recall that we are dealing now only with nonresonant (incommensurate, irrational) frequencies, those for which $(\nu_0 \cdot m) \equiv \sum \nu_0^\alpha m_\alpha = 0$ with integer $m = \{m_1, m_2, \ldots, m_n\}$ implies that $m_\alpha = 0$ for all α. In other words, if the frequencies are incommensurate, then $|\nu_0 \cdot m| > 0$ for any nonzero integer m; but $|\nu_0 \cdot m|$ may get arbitrarily small. KAM does not treat all incommensurate frequencies, only those for which the $|\nu_0 \cdot m|$ are bounded from below by the *weak diophantine condition*

$$|\nu_0 \cdot m| \geq \gamma |m|^{-\kappa} \quad \text{for all integer } m, \tag{7.101}$$

where $|m| = \sqrt{(m \cdot m)}$ and γ and $\kappa > n$ are positive constants. This condition is a way of dealing with the problem of small denominators. We will have much more to say about this condition later.

THE THEOREM

Let H_0 be the Hamiltonian of a completely integrable system in the sense of the Liouville integrability theorem (Section 6.2.2) and assume that $H(\phi_0, J_0, \epsilon)$ is a well-behaved (specifically, analytic) function of all of its arguments. The theorem then makes two statements: 1. For sufficiently small ϵ, frequencies in Ω_H^ν can be found that (a) satisfy (7.101) with an *appropriately chosen* γ and (b) define tori on which the perturbed flow is quasiperiodic at those frequencies. 2. The *measure* of such *quasiperiodic frequencies* (essentially the fraction they make up of all the frequencies in Ω_H^ν) increases as ϵ decreases. There are other, *aperiodic frequencies*, that do not define tori with quasiperiodic flows: their tori break up when $\epsilon > 0$. Let the measure of the aperiodic frequencies be μ. Then μ depends on ϵ, and in the unperturbed limit μ vanishes: $\mu|_{\epsilon=0} = 0$.

The diophantine condition (DC) (7.101) is a way to deal with the problem of small denominators, and the proof we describe of the KAM theorem is based on the Lie transformation method. Yet we discussed the small-denominator problem only in connection with canonical perturbation theory. We must first show, therefore, that the problem exists also in the Lie method. To do so we rewrite the Lie method's first-order equation (6.132b):

$$E_1 = H_1 - \{G_1, H_0\};$$

$\partial G_1/\partial t = 0$ because the CT from the initial variables $\xi = (\phi_0, J_0)$ to the new ones $\eta = (\phi, J)$ is time independent (see the subsection on the Lie transformation method in Section 6.3.2). For convenience, we drop the subscript 1 from G_1, as it is the only generator we'll be using for a while. The first-order Hamiltonian E_1 leads to quasiperiodic motion on a torus if $\partial E_1/\partial \phi = 0$, where ϕ will become the first-order angle variable (we will consider higher orders later). Thus

$$\{G, H_0\} = H_1 + h, \tag{7.102}$$

where h is some function, yet to be determined, independent of ϕ. Now assume that G and H_1, the only functions in Eq. (7.102) that depend on ϕ, can be expanded in Fourier series:

$$G(\phi_0, J_0) = \sum_m G_m(J_0)e^{i(\phi_0 \cdot m)}, \tag{7.103}$$

$$H_1(\phi_0, J_0) = \sum_m H_{1,m}(J_0)e^{i(\phi_0 \cdot m)}. \tag{7.104}$$

The PB is

$$\{G, H_0\} = \sum_{\alpha=1}^n \left[\frac{\partial G}{\partial \phi^\alpha} \frac{\partial H_0}{\partial J_\alpha} - \frac{\partial H_0}{\partial \phi^\alpha} \frac{\partial G}{\partial J_\alpha} \right].$$

The second term vanishes, and $\partial H_0/\partial J_\alpha = \nu^\alpha(J)$ (we drop the subscript 0 on ν, ϕ, and J), so that when expanded in a Fourier series Eq. (7.102) becomes

$$\sum i(\nu \cdot m)G_m e^{i(\phi \cdot m)} = h + \sum H_{1,m}e^{i(\phi \cdot m)},$$

where ν, G_m, and $H_{1,m}$ are all functions of J. There is no $m = \{0, 0, \ldots, 0\}$ term on the left-hand side of this equation, so h is chosen to cancel out the corresponding term in the sum on the right-hand side (this is essentially the averaging procedure of Section 6.3.2).

Hence the Fourier coefficients of G are given by

$$G_m(J) = \frac{H_{1,m}(J)}{i(\nu(J) \cdot m)}. \tag{7.105}$$

Compare this to Eq. (6.112): small denominators arise in the Lie transformation method in almost the same way as they do in canonical perturbation theory.

The diophantine condition (7.101) bypasses the problem caused by small denominators by affecting the convergence of the Fourier series (7.103). Of course (7.103) will fail to converge if $\nu \cdot m = 0$ for any integer m whose $H_{1,m} \neq 0$, which means that it will not work for the rational tori. But convergence requires more: the $|G_m|$ must decrease rapidly enough with increasing $|m|$. According to Eq. (7.105), the rate of decrease of the $|G_m|$ depends both on the $|H_{1,m}|$ and on the denominators $|\nu \cdot m|$, so even if the $|H_{1,m}|$ decrease fast enough for (7.104) to converge, (7.103) will not converge if the $|\nu \cdot m|$ decrease too rapidly. The DC, by placing a lower bound on the $|\nu \cdot m|$, controls their rate of decrease.

To see more clearly the mechanism by which the DC excludes certain frequencies, turn to the space of the ν^α (remember that the subscript 0 has been dropped). Replace the $\geq$ in the DC initially by $>$. Then if γ is set equal to zero, the DC allows all incommensurate frequencies, excluding only the commensurate ones, for which $|\nu \cdot m| = 0$ for some m. The way this happens in two dimensions is illustrated in Fig. 7.45. The lattice in the figure consists of all points with integer coordinates. Suppose m_1 is an integer vector. Then the DC with $\gamma = 0$ eliminates all frequencies on the line passing through the origin

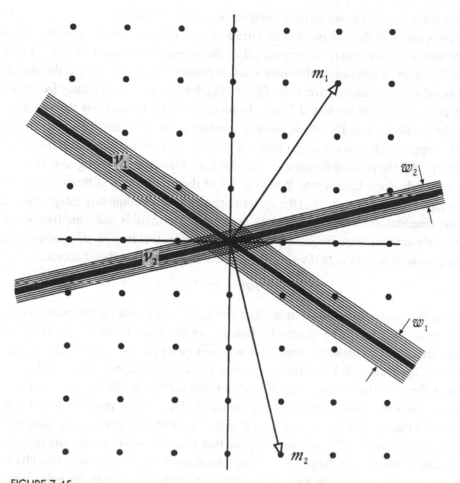

FIGURE 7.45

The v plane for two degrees of freedom. The dots are at points with integer coordinates. See the text for an explanation.

perpendicular to m_1, the single thick line labeled v_1 in the figure. Thus it eliminates all the periodic frequencies that lie on what we will call *periodic lines*, that is, those passing through the origin perpendicular to any integer vector. The line labeled v_2 is another example. The periodic lines are all those that pass through the origin and any lattice point of the figure; they form a set of measure zero in the space of lines through the origin, but they are dense in the sense that between any two periodic lines there is an infinite number of others. (In exactly the same way, the rational numbers on the unit interval form a dense set of measure zero. See the number-theory appendix to this chapter.) By eliminating the periodic frequencies, the DC with $\gamma = 0$ eliminates all rational tori. By leaving all the quasiperiodic frequencies, it leaves the irrational tori. We have illustrated this in two dimensions, but a similar result is obtained in more than two: the degenerate DC eliminates a measure-zero dense set of *periodic hyperplanes* orthogonal to integer vectors, again leaving all the quasiperiodic frequencies and their tori.

Now replace the $\geq$ in the DC and consider $\gamma > 0$. This broadens each periodic line to a strip of width $w = \gamma \, |m|^{-\kappa}$ and eliminates the union of all such strips. In Fig. 7.45 this is illustrated for m equal to m_1 and m_2: both strips are eliminated. If more integer vectors m are drawn, more of the ν plane will be eliminated. It may appear that this broadening will eliminate the entire plane, swamping all of the quasiperiodic frequencies. In fact if γ is small enough, it can fail to eliminate a set of nonzero measure. That is discussed in the number-theory appendix. We write $\Omega_\infty^\nu \subset \Omega_H^\nu$ for the set of remaining frequencies (the subscript ∞ will be explained later). In more than two dimensions the situation is similar: the nondegenerate DC eliminates a *hyperslab* about each periodic hyperplane, but for small enough γ it leaves a nonzero-measure set $\Omega_\infty^\nu \subset \Omega_H^\nu$ of quasiperiodic frequencies and their tori. All the rational frequencies are eliminated from Ω_∞^ν, leaving something that looks very much like a Cantor set. What the KAM theorem proves is that on this set of initial conditions (or rather on Ω_∞) the system behaves like a completely integrable one.

In one freedom, for which m is simply an integer and there is only one frequency ν, the analog of commensurate frequencies is rationality of ν, that is, $\nu \equiv p/q$, where p and q are integers with no common divisor. The diophantine condition then becomes

$$|\nu - p/q| > \gamma q^{-\kappa}$$

(see the number-theory appendix). In Worked Example 7.5 this eliminates many values of ν that lie in $\Omega_{H+}^\nu = [0, a]$. The values of ν that are not eliminated form $\Omega_\infty^\nu \subset \Omega_{H+}^\nu$. Then Ω_∞ is the image of Ω_∞^ν under the map $J(\nu) = \sqrt{\nu/a}$ (see Fig. 7.44). From $\Omega_\infty^\nu \subset \Omega_{H+}^\nu$ it follows that $\Omega_\infty \subset \Omega_{H+}$. KAM is proven for $\nu \in \Omega_\infty^\nu$ or, equivalently, for $J \in \Omega_\infty$.

That is the way the DC restricts the frequencies for the KAM theorem, but how ϵ enters these considerations has not yet been indicated. That will be pointed out in the next subsection. In the meantime, we state only that the "appropriate" choice of γ depends on ϵ. As ϵ increases, so does the minimum γ, and that eliminates ever increasing regions of Ω_H^ν. Eventually, when ϵ gets large enough, the minimum γ becomes so large that all of the frequencies are swamped by the hyperslabs surrounding the periodic frequencies, and the ϵ series fails to converge anywhere on $\mathbf{T}^*\mathbb{Q}$. But as ϵ gets smaller, so does the minimum γ and the measure of the admitted frequencies increases. In the limit $\gamma|_{\epsilon=0} = 0$, only the periodic frequencies (a set of measure zero) are eliminated. Moreover, KAM shows that on the tori belonging to the admitted frequencies (i.e., to frequencies in Ω_∞^ν) the motion of the perturbed dynamics is quasiperiodic and that if $J(\nu) \in \Omega_\infty$, the perturbation series converges to *infinite* orders in ϵ.

A BRIEF DESCRIPTION OF THE PROOF OF KAM

KAM is a theorem about the convergence of a perturbative procedure. But the KAM procedure differs in two important ways from other perturbative procedures we have discussed. One difference is that in KAM the frequencies remain fixed as the order of perturbation increases. The other is that the convergence is much more rapid in KAM: the perturbation parameter goes in successive steps as $\epsilon, \epsilon^2, \epsilon^4, \epsilon^8, \dots, \epsilon^{2^{l-1}}, \dots$, rather than as $\epsilon, \epsilon^2, \epsilon^3, \epsilon^4, \dots, \epsilon^l, \dots$, as in other perturbative routines. The proof we outline is based on the Lie method, which allows the calculation of G_l at each order by solving equations that do not involve G_j with $j < l$.

For reasons that will eventually become clear, we write ϵ_0 for ϵ. Both Fourier series (7.103) and (7.104) must converge, and that places conditions not only on the G_m, but also on the $H_{1,m}$.

The condition that a Fourier series converge (an analyticity condition), as applied to the $H_{1,m}$, is

$$|H_{1,m}(J)| \leq (\text{const.}) \, e^{-|m|s} \qquad (7.106)$$

for some real $s > 0$. As applied to the G_m it is

$$|G_m| \leq (\text{const.}) \, e^{-|m|r}. \qquad (7.107)$$

But the G_m and $H_{1,m}$ are related through (7.105), in which the frequencies appear in the denominator. It is the DC that keeps the denominators from getting too small and allows (7.103) to converge, and the DC involves γ. This results in a relation among s, r, and γ, which is used in the proof in such a way as to avoid divergences at each step of the perturbative procedure.

In the first step the Fourier series is truncated (this is a method introduced by Arnol'd) to a finite sum. The point of truncation is chosen so that the remaining infinite sum is small compared with the finite one. Specifically, the Fourier series of Eq. (7.104) is broken into two parts

$$H_1^{<L}(\phi, J) = \sum_{|m|<L} H_{1,m}(J)e^{i(\phi \cdot m)} \qquad (7.108)$$

and

$$H_1^{>L}(\phi, J) = \sum_{|m|\geq L} H_{1,m}(J)e^{i(\phi \cdot m)}, \qquad (7.109)$$

with L chosen to make $H_1^{>L} \sim O(\epsilon_0^2)$; then L depends on ϵ_0 (this is where the dependence on ϵ_0 enters). The sizes of the sums depend on the rates of convergence of the Fourier series, so L depends on s and r. Now $H_1^{>L}$ is of second order in ϵ_0, which means that it can be neglected in first order, and $H_1^{<L}$ is a finite sum of trigonometric functions, which means that there is no question of its convergence.

Continuing with the first step, a kind of truncated theorem is proven, one that applies only to the ms for which $|m| < L$. Only the lower ms are involved in the truncated series, and the DC can be weakened by replacing *all* integer m in Eq. (7.101) by just the lower ms. This defines a region Ω_1^ν of ν-space that is larger than Ω_∞^ν (more frequencies are allowed) and consequently a region $\Omega_1 \subset \Omega_H$ of J-space. Thus $\Omega_H \supset \Omega_1 \supset \Omega_\infty$. This achieves a solution of Eq. (7.102) to first order in ϵ_0 on $\Omega_1 \times \mathbb{T}^n$: a new Hamiltonian, independent of the new, first-order angle variables ϕ_1, is obtained. This new first-order Hamiltonian H_{01} is a function defined on $\Omega_1 \times \mathbb{T}^n$, that is, defined only for the frequencies admitted by the weakened DC. It can be written in the form

$$H_{01} \equiv H_0 + \epsilon_0 h_0, \qquad (7.110)$$

where $\epsilon_0 h_0$ is the first-order addition to H_0. Moreover, on $\Omega_1 \times \mathbb{T}^n$ this defines a completely integrable system.

We are now ready for the second step. In this step H_{01} is treated as the unperturbed Hamiltonian and the cut-off part $H_1^{>L}$ is introduced as a perturbation. But $H_1^{>L}$ is of order

ϵ_0^2, so it can be written as $\epsilon_0^2 H_2$, and the perturbation parameter becomes $\epsilon_1 \equiv \epsilon_0^2$, not ϵ_0. The Hamiltonian is then

$$H = H_{01} + \epsilon_1 H_2,$$

analogous to (7.98), and this is treated in the same way that $H_0 + \epsilon_0 H_1$ was in the first step: the Fourier series is truncated, now at a larger L, cutting off a part that is of order $\epsilon_1^2 = \epsilon_0^4$, and a procedure similar to the one that led to (7.110) now leads to

$$H_{02} = H_{01} + \epsilon_1 h_1,$$

where $\epsilon_1 h_1$ is the first-order (in ϵ_1) addition to H_{01}. The "unperturbed" Hamiltonian H_{01} is defined on Ω_1, but the truncation leads to a weaker DC, which cuts down the region further to a subset Ω_2 of Ω_1 that is also larger than Ω_∞ (again, because more frequencies are allowed). Thus $\Omega_H \supset \Omega_1 \supset \Omega_2 \supset \Omega_\infty$. What is obtained in this way is a solution of Eq. (7.102) to first order in $\epsilon_1 \equiv \epsilon_0^2$ on $\Omega_2 \times \mathbb{T}^n$.

The third step is similar. It yields a solution of order $\epsilon_1^2 \equiv \epsilon_0^4$, and so on. As the proof goes on, it constructs a sequence of Hamiltonians H_{0l} that converge to some final Hamiltonian $H_{0\infty}$. Each H_{0l} is defined on $\Omega_l \times \mathbb{T}^n$, where $\Omega_H \supset \Omega_1 \supset \Omega_2 \supset \cdots \supset \Omega_l \supset \cdots \supset \Omega_\infty$.

An important tool for proving KAM is *Newton's method* for finding the roots of functions (Press, 1992). In Newton's method an educated guess is taken for the root and then successive approximations are made that converge exponentially fast (*superconverge*) to the true root. In the KAM proof the successive approximations also superconverge: in the $(l+1)$st step a solution is obtained to order $\epsilon_{l-1}^2 \equiv \epsilon_0^{2^l}$. Originally, when the perturbative approach was first attempted, it was suspected that if perturbative expansion converges at all, it does so very slowly. Kolmogorov suggested the approach that uses a generalization of Newton's method. In fact the proof of KAM actually involves finding roots of functional equations, and it is precisely there that Newton's method enters.

What KAM shows is that for small enough ϵ_0 and for frequencies ν that satisfy the DC this process converges in the limit $l \to \infty$ to a Hamiltonian $H_{0\infty}(J_\infty)$ that is independent of the final angle variables ϕ_∞. This Hamiltonian is a function on $\Omega_\infty \times \mathbb{T}^n$, where Ω_∞ is the final subset to which the Ω_l converge. That is, $\Omega_\infty = \bigcap_{l \geq 0} \Omega_l$ (this explains the subscript ∞). If the initial conditions for the motion lie in $\Omega_\infty \times \mathbb{T}^n \subset \mathbf{T}^*\mathbb{Q}$, the motion remains on a certain torus $\mathbb{T}_\infty^n(\nu)$ defined by the initial value of J_∞, or equivalently, according to (7.100), by ν. In this way KAM tells us not only that the motion stays on $\mathbb{T}_\infty^n(\nu)$, but that its frequency is ν.

This brief description leaves out several important aspects of the proof and of the theorem itself. Among them is a description of how Newton's method is actually used. Another is that ϵ_{l+1} is not strictly proportional to ϵ_l^2. There is an l-dependent factor Γ_l that complicates matters, so that in successive orders

$$\epsilon_{l+1} \leq \Gamma_l \epsilon_l^2.$$

This factor is important because it involves γ (as well as r and s), and at each successive step a new γ_l is defined. There is a relation between γ_l and ϵ_l, and what the theorem proves is that in the limit a γ_∞ exists if ϵ_0 is small enough. Another aspect of KAM that we did not cover is that $\mathbb{T}_\infty^n(\nu)$ is close to the unperturbed torus $\mathbb{T}_0^n(\nu)$ belonging to the same frequency (closeness has to be defined in terms of some topology on $\mathbf{T}^*\mathbb{Q}$). Hence the perturbed and unperturbed motions are not only at the same frequency but also close. This is seen in the example of Figs. 7.35. For instance, the preserved tori of Fig. 7.35(b), those with the higher values of $|J|$, are close to the unperturbed ones of Fig. 7.35(a) and may even be called distorted versions of them. Yet another uncovered aspect is that Ω_∞^ν and Ω_∞ are Cantor sets.

In sum, the proof of the KAM theorem reduces the admitted (or *good*) frequencies in each step. Before the perturbation is turned on, when $\epsilon = 0$ the system is completely integrable and all frequencies are good. But all frequencies satisfy the equation $|\nu \cdot m| \geq 0$, so for the unperturbed system, an admissible DC has $\gamma = 0$. When $\epsilon > 0$, however, fewer frequencies are still good, only those that satisfy the DC with some $\gamma > 0$. As ϵ grows, so does the γ in the DC, and the measure α of good frequencies gets smaller. Normalize α so that $\alpha_{max} = 1$. Eventually γ gets so large that the DC eliminates all frequencies (α goes to zero). In the limit as $\epsilon \to 0$, all frequencies are allowed, so $\alpha \to 1$. This is the main import of KAM: *for sufficiently small ϵ almost all tori are preserved.* Is the solar system stable against small perturbations? Well, for almost all initial conditions it is!

We end the discussion of KAM by remarking that what the theorem provides is essentially an algorithm and a qualitative statement. It gives no method for calculating either the maximum value of ϵ for which the theorem holds or of γ as a function of ϵ. On the one hand some estimates for the maximum, or critical, value of ϵ are ridiculously small ($\epsilon \sim 10^{-50}$). On the other hand, we have seen that for the example of the standard map the critical value, found numerically by careful analysis, is $\epsilon_c \approx 0.9716$: up to this value some tori are preserved. To our knowledge a rigorous formal estimate of a realistic critical value for ϵ remains an open question.

PROBLEMS

1. For the model system of Section 7.1.1, draw the velocity-phase diagram for the two cases $a < l$ and $a \geq l$.

2. (a) Verify the eigenvalues and eigenvectors z_1 and z_2 of the **A** matrix of Eq. (7.28).
    (b) Do the same for Eq. (7.29). Show how the **A** matrix operates on **u** and **v**. Which is the major and which the minor axis of the ellipse?
    (c) Give a physical interpretation of the ellipses found at Eq. (7.29): show that this is the small-vibration approximation and find the frequency of small vibrations.

3. Draw Poincaré maps for the three cases near the end of the second subsection of Section 7.2.2, indicating the order in which the successive points are plotted on $\mathbb{P}$. [Partial answer: Consider Case 3 with the tangent matrix given by

$$ \mathbf{A} = \begin{bmatrix} -1 & 0 \\ \mu & -1 \end{bmatrix} \equiv -\mathbf{I} + \mathbf{M}. $$

Show that $\mathbf{A}^n = (-1)^n(\mathbf{I} - n\mathbf{M})$. The fixed point is the origin: $\mathbf{p}_f = 0$. We choose $\mu = 1$ and plot the Poincaré map in Fig. 7.46 for several different initial points, called $\mathbf{a}_0, \mathbf{b}_0, \ldots, \mathbf{f}_0$, all chosen on the 1 axis. Since the origin $\mathbf{a}_0$ is a fixed point, all the $\mathbf{a}_n \equiv \mathbf{A}^n \mathbf{a}_0$ are also at the origin. The $\mathbf{b}_n, \ldots, \mathbf{f}_n$ are shown in the figure for those n that fit on the diagram.]

4. Complete the discussion of the (T, D) plane:

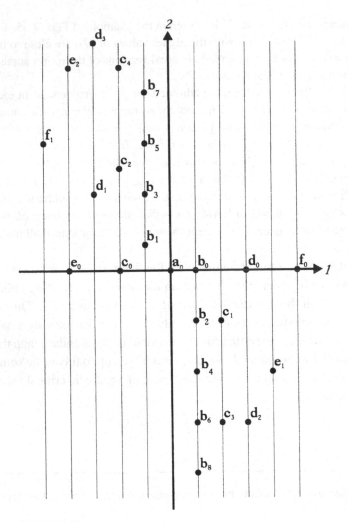

FIGURE 7.46

(a) Where are the two eigenvalues equal? If they are equal, the map may not be diagonalizable. For that case, plot the map in the same detail as in the previous problem.

(b) Discuss the various possibilities for $D = -1$ (see the discussion for $D = 1$).

5. For the linearized Hénon map, find the equations of the curves corresponding to the three
curves of Fig. 7.21 whose equations are $D = T - 1$, $D = -T - 1$, and $D = (T/2)^2$.
Find the stability region of the Hénon map, that is, the region corresponding to the stability
triangle of Fig. 7.21. [*Partial answer*: None of the three curves are straight lines. The
parabola corresponds to $a^2 + 2ab - b^2 + b(1 + b^2) = 0$. This curve, with two branches,
is tangent to both other curves, just as the parabola is tangent to the lines in Fig. 7.21.]

6. (a) Show graphically that for the logistic map the trivial fixed point at $x = 0$ is a repellor.

(b) In Fig. 7.26 pick x_0 to lie within the interval defined by the limit cycle, somewhere in the
interval $[0.49, 0.83]$. Graphically construct the x_n sequence to see how it approaches
the limit cycle.

7. For the logistic map, find the a value a_1 at which the fixed points x_+ and x_- of F_a^2 in
Eq. (7.67) become unstable. Do this without referring to ϵ of Worked Example 7.3.

8. Obtain a closed-form expression for $x_n = F_4^n(x_0)$ (see Worked Example 7.4).

9. Write out the 16th-degree polynomial $F_a^4(x)$.

10. The Ulam–von Neumann map is defined by (Ulam & von Neumann, 1947)

$$\xi_{n+1} \equiv G_r(\xi_n) = 1 - r\xi_n^2,$$

where $\xi \in [-1, 1]$ and the tuning parameter $r \in [0, 2]$. Write a computer program that will generate the bifurcation diagram for the Ulam–von Neumann map. [Note: The substitution $\xi = \cos\theta$ relates G_2 to the map of Worked Example 7.4. G_r can also be related directly to the logistic map F_a by writing $\xi = (4x-2)/(a-2)$ and $r = \frac{1}{4}a(a-2)$.]

11. Show that the two definitions of $\rho(\Theta, K)$ for the circle map are equivalent. Show that the general definition gives the correct result for $K = 0$, $\Theta/2\pi$ rational.

12. Find the value of $K > 0$ at which the fixed point of the circle map becomes unstable for $\Theta > 0$. Find the corresponding fixed point θ_f.

13. For the standard map Z_ϵ:

    (a) Show that the map is conservative (area preserving).

    (b) Show that $(0, 0)$ and $(\pi, 0)$ are fixed points. By linearizing the map about these fixed points, find their nature (i.e., whether they are elliptic, hyperbolic, etc.). Find $\mathbb{E}^u$ and $\mathbb{E}^s$ if either is hyperbolic. Discuss the axes of the ellipse if either is elliptic. Compare with Figs. 7.35.

    (c) Show that $(0, \pi)$ and (π, π) are a conjugate pair of period-two points (i.e., are mapped into each other by Z_ϵ). Find their nature as fixed points of Z_ϵ^2. [*Partial answer*: The Jacobian matrix is

$$A = \begin{bmatrix} 1+E & 1 \\ E & 1 \end{bmatrix},$$

where $E = \epsilon \cos\phi$. $\det A = 1$. Eigenvalues are: $\rho_\pm = 1 + E/2 \pm \sqrt{E + E^2/4}$. Eigenvectors: $([\rho - 1]/E, 1)$. The eigenvectors give $\mathbb{E}^{u,s}$.]

14. Consider the one-freedom Hamiltonian $H = E[1 - \exp(-J^2/2)]$, where E is a positive constant and J varies from $-\infty$ to $+\infty$. Show that the Hessian condition allows three possible Ω_H manifolds, find the map $\nu(J)$ on each, and obtain Ω_H^ν for each.

15. Consider the two-freedom Hamiltonian $H = \frac{1}{2}aJ_1^2J_2^2$, where a is a positive constant and the J manifold is $\mathbb{R}^2$. Show that the Hessian condition allows four possible Ω_H manifolds, find the maps $\nu(J)$ and $J(\nu)$ on each, and obtain Ω_H^ν for each. Write the DC $|\nu \cdot m| \geq \gamma |m|^{-\kappa}$ explicitly in terms of the J_α (these are the equations that define the four possible Ω_∞ sets).

16. Show that a necessary and sufficient condition that the Floquet exponent be real is that $|e_1(\tau) + \dot{e}_2(\tau)| < 2$.

17. Find the vectors $\tilde{e}_k(t)$ when the eigenvalues of $\mathbf{R}$ are equal, but $\mathbf{R}$ is not diagonalizable. That is, calculate their components in the standard basis $\{e_1(t), e_2(t)\}$. (This is done in the text for the case $e_2(\tau) \neq 0$. Do it for the case $e_2 = 0$.)

18. (a) Without using the equivalence principle (i.e., by imposing the appropriate constraint in Cartesian coordinates), write down the Lagrangian $L = T - V$ for the pendulum whose point of support oscillates in the y direction according to $y = f(t)$, where f is a given function. Obtain the equation of motion and show that it is (7.57) if $f(t) = A\cos\nu t$.

(b) Show that a gauge transformation (adding a function of the form $dF(\theta, t)/dt$ to the Lagrangian; Section 4.2.2) yields a new Lagrangian $L' = T' - V'$, where T' is the usual kinetic energy of a pendulum and V' is the potential of a time-varying gravitational field. Compare the two Hamiltonians.

19. A degenerate case of Hill's equation is the harmonic oscillator, in which $G = $ const. $= \omega^2$. Calculate the standard basis and, taking the periodicity τ to be undetermined, the representation of $\mathbf{R}$ in the standard basis.

(a) Show that $T \equiv \operatorname{tr} \mathbf{R} = 2\cos\omega\tau$, and discuss the case in which $\tau = 2\pi/\omega$.

(b) Calculate the values of τ that yield Cases 1 and 2.

(c) Argue that this degenerate case of Hill's equation, when applied to the inverted pendulum, is the limit as $h \to 0$. From the definition of ϵ for the inverted pendulum, calculate the values of ϵ at the points where the tongues reach the h axis.

20. Obtain Eq. (7.64) for the damped pendulum with the vibrating pivot.

21. For the inverted pendulum show that the minimum frequency $\nu_{\min}$ for stability increases as $1/A$ in the limit as $A \to 0$.

22. The equation

$$m\ddot{y} + \left(B\dot{y}^2 - A\right)\dot{y} + ky = 0$$

describes a self-sustained oscillator known as the Rayleigh model (Rayleigh, 1883). Show that by an appropriate change of variables this can be reduced to the van der Pol oscillator and analyze the properties of the motion on the basis of Section 7.1.3.

NUMBER THEORY

In this appendix we first discuss some properties of rational and irrational numbers as they apply to the nonlinear dynamical systems. In particular, we explain how in the case of the circle map it is possible that the union of finite $\Delta\Theta_r$ intervals at all rationals does not completely cover [0,1]. We also explain how the diophantine condition of the KAM theorem, which seems at first glance to eliminate all frequencies, actually fails to eliminate a substantial fraction of them. We do both of these by first establishing a relevant result on the unit interval, which can be applied directly to the circle map. We then extend that result from the unit interval to the circle and eventually to the DC on the plane. Then we turn to the cases of one and, briefly, two freedoms and show how representing real numbers by continued fractions helps us understand the transition to chaos (Khinchin, 1964).

THE UNIT INTERVAL

As is known (Cantor, 1915; James, 1966, pp. 35–36), the rational numbers (*the rationals*) on the unit interval [0, 1] form an everywhere dense set of measure zero. That they are everywhere dense means that the neighborhood of any point, no matter how small, contains rationals. That their measure is zero means roughly that in spite of being everywhere dense, they occupy no space on [0, 1]. This is because they are countable: unlike the irrationals, they can be put into one-to-one correspondence with the positive integers. Their countability is used in the following way to show that their measure is zero. Having ordered them (i.e., put them into one-to-one correspondence with the integers), construct a small open interval of length $\epsilon < 1$ about the first rational, and one of length ϵ^2 about the second, ..., and one of length ϵ^n about the nth. The sum of all these little intervals (this is a geometric series) is $\sigma = \epsilon/(1 - \epsilon)$, which can be made arbitrarily small by choosing ϵ small enough. Thus the space occupied by the rationals is less than any positive number; hence it is zero.

Countability can also be used to show that it is possible to remove a finite open interval about each rational without removing all of [0, 1]. To generalize the argument we just used to show that their measure is zero, replace each ϵ^n, $n = 1, 2, \ldots$, by $\Gamma\epsilon^n$, where Γ may be any positive number. The sum of all the little finite intervals is $\sigma = \Gamma\epsilon/(1 - \epsilon)$, which can also be made arbitrarily small by choosing ϵ small enough. In fact $\sigma < 1$ if $\epsilon < 1/(\Gamma + 1)$. That means that it is possible to remove a *finite* interval about every rational and still leave some of [0, 1], in spite of the fact that the rationals are dense! Moreover, the total length λ removed, necessarily positive, is even less than σ, since there is bound to be some overlap among the small intervals. The actual size of λ depends on how the rationals are ordered (there are many ways to order them), but if $\epsilon < 1/(\Gamma + 1)$, then $0 < \lambda < 1$. That means that there must be points that have not been removed, and these must be irrational. The remaining irrationals are totally disconnected (there are no intervals remaining), for between any two of them are rational points about which small intervals have been removed. Nevertheless, the measure of the remaining irrationals, the length they occupy, is $\mu = 1 - \lambda > 0$. We will call this the ϵ-*procedure* for removing intervals about the rationals.

Recall that the rotation number in the circle map is constant on finite widths $\Delta\Theta_{rK}$ about all of the rationals, yet is not constant everywhere. Now we can see how that is possible: just as the total length occupied by the $\Gamma\epsilon^n$ intervals is less than 1, so the total length occupied by the $\Delta\Theta_{rK}$ widths is less than the full variation 2π of Θ. Moreover, just as the length occupied by the $\Gamma\epsilon^n$ intervals is less than their sum σ, so the length occupied by the $\Delta\Theta_{rK}$ is less than $\sum \Delta\Theta_{rK}$, also as a result of overlaps. That is why we wrote $\cup\Delta\Theta_{rK}$ in Section 7.4.2 rather than $\sum \Delta\Theta_{rK}$. There are many ways to construct finite intervals about all rationals without covering all of [0, 1], so the $\Delta\Theta_{rK}$ need not be in any way equal to or proportional to the $\Gamma\epsilon^n$. Indeed, the ϵ-procedure is just a particularly simple example of how such intervals can be chosen.

A DIOPHANTINE CONDITION

We illustrate the existence of other possibilities with a different example, one that is relevant to the diophantine condition of the KAM theorem. Write each rational in [0, 1] in its lowest form p/q, and about each one construct an interval of length $1/q^3$. For each q there are at most $q - 1$ rationals. For instance if $q = 5$, there are four: $\frac{1}{5}, \frac{2}{5}, \frac{3}{5}$, and $\frac{4}{5}$. But if $q = 6$ there are only two: $\frac{1}{6}$ and $\frac{5}{6}$; for $\frac{2}{6} = \frac{1}{3}$ and $\frac{4}{6} = \frac{2}{3}$ belong to $q = 3$ and $\frac{3}{6} = \frac{1}{2}$ belongs to $q = 2$. Thus, for a given q, no more than $(q - 1)/q^3$ is covered by the intervals, and the total length Q that is covered is less (because of overlaps) than the sum of these intervals over all q:

$$Q < \sum_{q=2}^{\infty} \frac{q-1}{q^3} < \sum_{q=2}^{\infty} \frac{1}{q^2}.$$

Now, it is known [this is $\zeta(2)$, where ζ is the Riemann zeta function] that

$$\sum_{q=1}^{\infty} \frac{1}{q^2} = \frac{1}{6}\pi^2 \approx 1.64493407,$$

so that

$$Q < 0.64493407.$$

That means that the total length λ_q covered in this procedure satisfies $\lambda_q < 0.65$. In fact λ_q can be decreased if $1/q^3$ is replaced by Γ/q^3, where $\Gamma < 1$. In any case, as in the ϵ-procedure, the irrationals that remain uncovered occupy a length $\mu = 1 - \lambda_q > 0$. We will call this example, in general with $\Gamma \leq 1$, the q^3-*procedure* for covering finite intervals about the rationals. We stress again that the ϵ- and q^3-procedures are just two among many.

The advantage of the q^3-procedure over the ϵ-procedure is that the covered and uncovered irrationals are easily characterized. Every irrational ν that satisfies

$$\left| \nu - \frac{p}{q} \right| < \frac{\Gamma}{q^3} \quad \text{for some } q \tag{7.111}$$

lies in one of the covered intervals. Hence the only (disconnected) irrationals left uncovered are those that satisfy a diophantine condition, namely those for which

$$\left| \nu - \frac{p}{q} \right| \geq \frac{\Gamma}{q^3} \quad \forall \, q. \tag{7.112}$$

That plenty of irrationals satisfy (7.112) follows from the fact that their measure is $\mu > 0$. The same thing can be done with q^3 replaced by q^n, $n > 2$, leading to a different diophantine condition (Khinchin, 1964, p. 45).

This procedure also offers a way to characterize the degree of irrationality of a number: the irrationals that satisfy (7.112) are further away from all rationals than those that satisfy (7.111) and they can therefore be considered more irrational. We will return to this idea again, in discussing approximation of irrationals by rationals as applied to the DC in two dimensions (one freedom).

THE CIRCLE AND THE PLANE

We now want to demonstrate that finite strips can be removed about all periodic lines in the plane without removing all the points of the plane [see the discussion in the paragraphs following Eq. (7.105)]. At issue in this demonstration is the rationality not of points on the unit interval, but of slopes of lines, that is, tangents of certain angles: analogs of the ϵ- or q^3-procedures must be applied to tangents of angles (think of the angles as coordinates on a circle). We describe only the q^3-procedure. First consider only angles θ that are on the arc $[0, \pi/4]$; the q^3-procedures can be applied directly to their tangents, for they are all on the unit interval. Do this in the following way: for every rational value p/q of $\tan \theta$ construct a small finite arc of length $\alpha(\theta) = \frac{1}{4}\pi/q^3$ on the circle about θ. The total arc length covered in this way is $A = \frac{1}{4}\pi\lambda_q$, which is less than all of $[0, \pi/4]$, and all the points that remain uncovered have irrational values of $\tan\theta$. This takes care of one eighth of the circle. Now extend this process to the arcs $[\pi/4, \pi/2]$, $[0, -\pi/4]$, and $[-\pi/4, -\pi/2]$ by changing signs and/or using cotangents instead of tangents. This now takes care of half the circle, which proves to be sufficient for our purposes. Call θ a *rational angle* if $|\tan\theta| < 1$ or $|\cot\theta| < 1$ is rational.

Now place the circle, of some fixed radius R_0, at the origin on the plane and call $l(\theta)$ the line drawn through the origin at the angle θ. Draw all lines $l(\theta)$ for rational angles θ on the half circle. Each such $l(\theta)$ passes through a lattice point of Fig. 7.42 and hence is what we have called a periodic line. Moreover, *all* periodic lines are obtained in this way. Finally, remove a finite strip about each $l(\theta)$ so as to remove precisely the arc length $\alpha(\theta)$ on the circle. This leaves a nonzero measure of irrational points on the circle, as well as all points on the plane that radiate outward from them. That is, if there are points left on the circle of radius R_0, there will also be points left on circles of radius $R > R_0$. (The different radii correspond to different values of Γ in the analogs on the line.) This demonstrates what we set out to show: it is possible to remove a finite strip about every periodic line without removing all the points of the plane. We do not try to prove the analogous result for higher dimensions.

KAM AND CONTINUED FRACTIONS

The origin of the DC in the KAM theorem lies in the fact that even if the ν^α are an incommensurate set, $(\nu \cdot m)$ can get very small [if the ν^α are a commensurate set, there is some integer m vector for which $(\nu \cdot m) = 0$]. The DC is designed to eliminate sets of frequencies that, though incommensurate, are in some sense closer than others to commensurate ones. In two dimensions (one freedom, when there is only one ν variable) these equations cease to be meaningful. From now on we consider only systems with one freedom (it may help to refer to Figs. 7.35). This case is illustrated by the (discrete) standard map of Section 7.5.1. The one-freedom analog of a commensurate set of frequencies is a rational rotation number $\nu = p/q$ in the equation

$$\phi_{n+1} = [\phi_n + \nu(J)] \bmod (2\pi)$$

[compare the first of Eqs. (7.89)]. The one-freedom analog of a rational torus is a value of J for which ν is rational (we will call the closed-curve orbits tori, as we did for the standard map, and write $J(\nu)$ for such a torus). The danger of rational frequencies lies not in vanishing or small denominators, but in the breakup of the rational tori, as described by the Poincaré–Birkhoff theorem for one freedom: as soon as the perturbation ϵ is greater than zero, the rational tori break up and begin to swamp the irrational tori closest to them. The one-freedom analog of the DC is designed to eliminate rotation numbers that, though irrational, are close enough to rational ones to get swamped. It is consequently of the form

$$\left| \nu - \frac{p}{q} \right| \geq \frac{\gamma}{q^\kappa} \quad \forall\, q \tag{7.113}$$

for some κ. The one-freedom analog of the Hessian condition is the condition that there be a one-to-one relation between J and the rotation number ν.

When the rational tori break up, they all spread out into rings of finite thickness, swamping the irrational tori that lie in those rings. Yet, as we have learned from the previous discussion, in spite of the finite thickness of the rings there can still be irrational tori that survive, and the "farther" an irrational torus is from all of the rational ones, the more likely it is to be among the survivors. Equation (7.113) provides a way to describe how different an irrational rotation number is from any rational one, hence (by the analog of the DC) how different an irrational torus is from its rational neighbors. This idea can be made more precise in the context of *continued fractions* (Richards, 1981, is an informal complement to Khinchin).

Every real positive number α can be written as a *continued fraction* in the form

$$\alpha = a_0 + \cfrac{1}{a_1 + \cfrac{1}{a_2 + \cdots}} \equiv [a_0; a_1, a_2, \ldots], \tag{7.114}$$

where the a_j are all positive integers (except that a_0 may be zero). A continued fraction terminates (i.e., there is an integer N such that $a_j = 0$ for all $j > N$) if and only if it represents a rational number. Indeed, if it terminates it can of course be written in the form $\alpha = p/q$. Hence if α is irrational, its continued-fraction representation $[a_0; a_1, a_2, \ldots]$ does not terminate. Rational approximations for an irrational α are called its *convergents*, obtained by terminating its representation. That is, the kth convergent is the rational $\alpha_k \equiv p_k/q_k$ represented by $[a_0; a_1, a_2, \ldots a_k]$:

$$\alpha_k = a_0 + \cfrac{1}{a_1 + \cfrac{1}{a_2 + \cfrac{1}{\ddots \cfrac{}{a_{k-1} + \cfrac{1}{a_k}}}}}.$$

Then $\alpha = \lim_{k \to \infty} \alpha_k$. The kth convergent can be calculated from the recursion relation (Khinchin, 1964, p. 4)

$$\begin{aligned} p_k &= a_k p_{k-1} + p_{k-2}, \\ q_k &= a_k q_{k-1} + q_{k-2}, \end{aligned} \tag{7.115}$$

with initial conditions $p_0 = a_0, q_0 = 1, p_1 = a_1 a_0 + 1, q_1 = a_1$.

We list some examples:

$$\begin{aligned} e &= [2; 1, 2, 1, 1, 4, 1, 1, 6, 1, 1, 8, 1, \ldots], \\ \pi &= [3; 7, 15, 1, 292, 1, 1, 1, 2, 1, 3, 1, 14, 2, 1, 1, 2, 2, 2, 2, 1, 84, 2, \ldots], \\ \sqrt{2} &= [1; 2, 2, 2, \ldots], \\ \varphi &\equiv \frac{1 + \sqrt{5}}{2} = [1; 1, 1, 1, \ldots], \end{aligned} \tag{7.116}$$

There are several number-theoretic theorems about continued-fractions. One is that the continued-fraction representation of a number α is periodic iff α is a *quadratic* irrational, one that is a solution of a quadratic equation with integer coefficients ($\sqrt{2}$ and φ are particularly simple examples, both of period one). Another is that *transcendental* numbers, like e and π, which are the solutions of no algebraic equations, have aperiodic continued fractions. Of use in approximating irrationals by rationals are the following two theorems: 1. The kth convergent α_k is the best approximation to an irrational α by a rational number p/q whose denominator $q \leq q_k$. 2. If α is an irrational number whose kth convergent is p_k/q_k, then

$$\left| \alpha - \frac{p_k}{q_k} \right| < \frac{1}{q_k q_{k+1}}. \tag{7.117}$$

Notice the resemblance of this inequality to (7.111).

For reasons that date back to ancient Greece, the number $\varphi \approx 1.61803399$ of Eq. (7.116) is called the *golden mean*. The series of its convergents approach the golden mean more slowly than does the series of convergents for any other irrational. Indeed, for φ all of the $a_k = 1$, and then the second of Eqs. (7.115) implies that the q_k grow as slowly as possible (specifically, $q_k = q_{k-1} + q_{k-2}$; these turn out to be the *Fibonacci numbers* $q_k \equiv F_k$). Thus (7.116) imposes the largest possible upper bound on the rate at which the convergents approach their limit. In these terms, the golden mean is as far as an irrational can get from the rationals; it is as irrational as a number can be.

Now we relate some of these number-theoretic results to KAM.

Suppose a torus $J(\nu)$ has an irrational rotation number ν whose kth convergent is $\nu_k = p_k/q_k$. Since ν_k is the closest rational number to ν with denominator $q \leq q_k$, it follows that $J(\nu_k)$ is the closest rational torus to $J(\nu)$ with points of period $q \leq q_k$. Such considerations tell us something about which rational tori are close to irrational ones and are therefore likely to swamp them. For instance, if $\gamma/q_k^\kappa > 1/q_k q_{k+1}$, then according to (7.117) $|\nu - \nu_k| < \gamma/q_k^\kappa$ and hence $J(\nu)$ is swamped by the breakup of $J(\nu_k)$ and is therefore not a torus for which the KAM theorem holds. Notice that this depends on how large γ is, and hence on ϵ, as was explained at the end of Section 7.5.3. This is illustrated by the numerical calculations that yielded Figs. 7.35: as ϵ increases, fewer and fewer irrational tori are preserved. The torus that is farthest from the rational ones is the one whose rotation number ν is most difficult to approximate by rational numbers. For the standard map (this is not true more generally), as ϵ increases, the last torus to succumb is the one whose rotation number is the golden mean.

It is indeed remarkable that such subtle properties of numbers play such a significant role in the analysis of dynamical systems.

Although we have related continued fractions only to systems with one freedom, they are relevant also to systems with two (but not more; see the last Remark in Section 7.5.3). We will only hint at this. Recall Fig. 7.45. Commensurate frequencies are represented in that figure by rational lines, those with rational slope, so that for them $\nu^1/\nu^2 = p/q$ is a rational number. Incommensurate frequencies are represented by lines that pass through no lattice points and whose (irrational) slopes can be written as nonterminating continued fractions. The general DC can be replaced in this two-freedom case by

$$\left| \frac{\nu_1}{\nu_2} - \frac{p}{q} \right| \geq \frac{\gamma}{q^\kappa}$$

and can be treated, as was the one-freedom case, in terms of continued fractions. The geometry of the integer lattice of the plane can, in fact, be used to prove many of the theorems concerning continued fractions, so the two-freedom case lends itself naturally to such a treatment. We will not go into that, but we refer the interested reader to Stark (1970).

CHAPTER 8

RIGID BODIES

CHAPTER OVERVIEW

The motion of rigid bodies, also called *rotational dynamics*, is one of the oldest branches of classical mechanics. Interest in this field has grown recently, motivated largely by problems of stability and control of rigid body motions, for example in robotics (for manufacturing in particular) and in satellite physics. Our discussion of rigid-body dynamics will first be through Euler's equations of motion and then through the Lagrangian and Hamiltonian formalisms. The configuration manifold of rotational dynamics has properties that are different from those of the manifolds we have so far been discussing, so the analysis presents special problems, in particular in the Lagrangian and Hamiltonian formalisms.

8.1 INTRODUCTION

8.1.1 RIGIDITY AND KINEMATICS

Discussions of rigid bodies often rely on intuitive notions of rigidity. We want to define rigidity carefully, to show how the definition leads to the intuitive concept, and then to draw further inferences from it.

DEFINITION

A rigid body is an extended collection of point particles constrained so that the distance between any two of them remains constant.

To see how this leads to the intuitive idea of rigidity, consider any three points A, B, and C in the body. The definition implies that the lengths of the three lines connecting them remain constant, and then Euclidean geometry implies that so do the angles: triangle ABC moves rigidly. Since this is true for every set of three points, it holds also by triangulation for sets of more than three, so the entire set of points moves rigidly. In other words, fix the lengths and the angles will take care of themselves.

What is the configuration manifold $\mathbb{Q}$ of a rigid body? Suppose the distances between all the points of a rigid body are known. The configuration of the body can then be specified by giving 1. the position of an arbitrary point A (this involves three coordinates), 2. the direction to another point B (two more coordinates), and 3. the orientation of the plane containing A, B, and a third point C (one more coordinate, for a total of six): $\mathbb{Q}$ has dimension six. Three of the dimensions have to do with the position of the arbitrary point A, and three with the orientation about that point. The part of $\mathbb{Q}$ associated with A is easy to describe: it is simply Euclidean three-space $\mathbb{R}^3$. The part associated with orientation is more complicated and imbues $\mathbb{Q}$ with a nontrivial geometric structure, which we put off until the next section.

The motion of a rigid body through its configuration manifold, its dynamics, is determined by the forces acting on it. Generally these forces cause it to move in all of $\mathbb{Q}$, that is, to accelerate as a whole and to change its orientation simultaneously. If the sum of the forces vanishes, however, the center of mass does not accelerate [recall Eq. (1.56)] and can be taken as the origin of an inertial system, and then it is convenient to choose the center of mass to be A, the point about which the orientation is described. A common situation is one in which the body has a fixed point, a *pivot* (we will call it a pivot so as not to confuse it with the usage of Chapter 7, where we refer to a fixed point of the dynamics). Then it is convenient to choose the pivot to be A. These two cases are formally the same, as will be demonstrated later. In the rest of this introduction we assume that there is an *inertial point* A in the body, one that is known to move at constant velocity in an inertial system or to remain fixed: a pivot.

THE ANGULAR VELOCITY VECTOR ω

We will now describe the motion of a rigid body in the inertial frame and show in several steps how rigidity implies that 1. there is a line $\tilde{\omega}$ in the body that passes through the origin A and is instantaneously at rest and 2. all points in the body move at right angles to $\tilde{\omega}$ at speeds proportional to their distance from it (Fig. 8.1). That is, the condition of rigidity imposes rotation about the *instantaneous axis of rotation* $\tilde{\omega}$.

In the first step move to the inertial frame and consider any point $X \neq A$ in the body. Let the position vector of X be **x**. Since the distance $x = \sqrt{x^2}$ from the origin

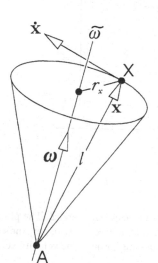

FIGURE 8.1
The relations among the position vector **x** of a point X in the body, its position and velocity vectors **x** and $\dot{\mathbf{x}}$, and the angular velocity vector ω. The axis of rotation is labeled $\tilde{\omega}$, and r_x is the distance from X to $\tilde{\omega}$. The pivot point is labeled A, and l is the line from A to X.

A to X is fixed,

$$\frac{d}{dt}x^2 = 2\dot{\mathbf{x}} \cdot \mathbf{x} = 0,$$

so the velocity of every point in the body is either perpendicular to its position vector or is zero. The line l from A to X remains straight, so l pivots about A. Therefore the velocities of all points on l are parallel (hence parallel to $\dot{\mathbf{x}}$) and proportional to their distances from A.

Second, consider the plane P containing l that is perpendicular to $\dot{\mathbf{x}}$ (Fig. 8.1). Rigidity constrains every point on P to have a velocity parallel to $\dot{\mathbf{x}}$. Indeed, let Y be any point on P with arbitrary position vector $\mathbf{y}$ in P. Because $|\mathbf{x} - \mathbf{y}|^2$ is fixed,

$$\frac{1}{2}\frac{d}{dt}|\mathbf{x} - \mathbf{y}|^2 = (\mathbf{x} - \mathbf{y}) \cdot (\dot{\mathbf{x}} - \dot{\mathbf{y}}) = -\mathbf{x} \cdot \dot{\mathbf{y}} = 0,$$

for $\dot{\mathbf{x}}$ is perpendicular to every vector in P and $\mathbf{y} \cdot \dot{\mathbf{y}} = 0$. Hence $\dot{\mathbf{y}}$, which is perpendicular to both $\mathbf{x}$ and $\mathbf{y}$, is perpendicular to P.

Third, consider three points X, Y, Z on a line l' on P, with position vectors $\mathbf{x}, \mathbf{y}, \mathbf{z}$ (Fig. 8.2; in the figure P is viewed on edge). Because they are on a line, there is some number β such that

$$\mathbf{y} - \mathbf{x} = \beta(\mathbf{z} - \mathbf{x}).$$

Each point Y on l' belongs to a unique value of β and each value of β specifies a unique point on l'. The time derivative yields

$$\dot{\mathbf{y}} = \beta\dot{\mathbf{z}} + (1 - \beta)\dot{\mathbf{x}}. \tag{8.1}$$

This means that if $\dot{\mathbf{x}}$ and $\dot{\mathbf{z}}$ are known, the velocity $\dot{\mathbf{y}}$ of every other point Y between X and Z on l' is also known. Moreover, as long as $\dot{\mathbf{z}} \neq \dot{\mathbf{x}}$ there is one point Y_0 on l' that has

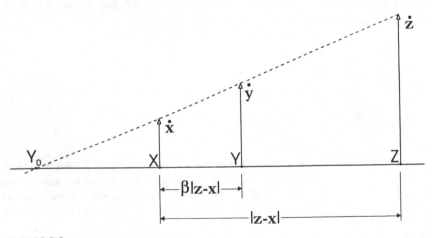

FIGURE 8.2
Two points X and Z that define a line l' on the plane P. Y is a point between X and Z. The plane is viewed edge on, so is represented by the horizontal line. The three velocities $\dot{\mathbf{x}}$, $\dot{\mathbf{y}}$, $\dot{\mathbf{z}}$ are perpendicular to the plane. There is a stationary point Y_0 where the dotted line passing through the velocity vectors meets P.

velocity zero. Indeed, $\dot{\mathbf{y}}_0 = 0$ if its β_0 satisfies $\beta_0(\dot{\mathbf{z}} - \dot{\mathbf{x}}) = \dot{\mathbf{x}}$, or if $\beta_0 = \pm\dot{x}/(\dot{z} \pm \dot{x})$ (some care is required with the signs). The special case of $\dot{\mathbf{z}} = \dot{\mathbf{x}}$ is dealt with in Problem 1.

Now, A does not necessarily lie on l', while $\mathbf{Y}_0$ does, so A and $\mathbf{Y}_0$ are in general distinct points. Both are at rest, so there is a line on P that is instantaneously stationary. We call this $\tilde{\omega}$, and assertion 1 is thereby proved. Assertion 2 follows immediately from the fact that every point on each line perpendicular to $\tilde{\omega}$ must move with a speed proportional to its distance from $\tilde{\omega}$. Hence the body is instantaneously rotating about $\tilde{\omega}$. We emphasize that $\tilde{\omega}$ is the *instantaneous* axis of rotation and may in general wander about in the body. This means that the speed of any point X in the body is $\dot{x} = \omega r_x$, where ω is a constant called the *instantaneous angular speed* or *rate of rotation*, and r_x is the distance of X from $\tilde{\omega}$ (Fig. 8.1). Define the *instantaneous angular velocity vector* $\boldsymbol{\omega}$ to be the vector of magnitude ω along $\tilde{\omega}$. Then from elementary vector algebra it follows that $|\boldsymbol{\omega} \wedge \mathbf{x}| = \omega r_x$, and the instantaneous velocity of each point $\mathbf{x}$ in the body is

$$\dot{\mathbf{x}} = \boldsymbol{\omega} \wedge \mathbf{x}. \tag{8.2}$$

As always when the cross product is involved, one has to be careful about signs (the right-hand rule).

8.1.2 KINETIC ENERGY AND ANGULAR MOMENTUM

KINETIC ENERGY

The kinetic energy of the rigid body will be calculated in an inertial system whose origin is at A. Let the mass density of the body be $\mu(\mathbf{x})$, where $\mathbf{x}$ is the position vector (see Chapter 1, in particular Problem 1.17). Then the kinetic energy is

$$T = \frac{1}{2} \int \dot{x}^2(\mathbf{x})\mu(\mathbf{x})\, d^3x \equiv \frac{1}{2} \int \dot{x}^2\, dm, \tag{8.3}$$

where $dm = \mu(\mathbf{x})\, d^3x$, the integral is taken over the entire body, and $\dot{x}^2 \equiv \dot{\mathbf{x}} \cdot \dot{\mathbf{x}}$ is the square of the speed, a function of $\mathbf{x}$.

To calculate this in some coordinate system, rewrite Eq. (8.2) in the form

$$\dot{x}_k = \epsilon_{klm}\omega_l x_m,$$

so that

$$\begin{aligned}
\dot{x}^2 \equiv \dot{x}_k \dot{x}_k &= \epsilon_{klm}\omega_l x_m \epsilon_{kij}\omega_i x_j \\
&= (\delta_{li}\delta_{mj} - \delta_{lj}\delta_{mi})\omega_l\omega_i x_m x_j \\
&= \omega_l(\delta_{li}x^2 - x_l x_i)\omega_i.
\end{aligned}$$

Hence the kinetic energy is

$$\begin{aligned}
T &= \frac{1}{2}\omega_l\left[\int (\delta_{li}x^2 - x_l x_i)\, dm\right]\omega_i \\
&= \frac{1}{2}\omega_l I_{li}\omega_i = \frac{1}{2}\boldsymbol{\omega} \cdot \mathbf{I}\boldsymbol{\omega}, \tag{8.4}
\end{aligned}$$

where the

$$I_{li} = \int (\delta_{li} x^2 - x_l x_i)\, dm \qquad (8.5)$$

are elements of the *inertia tensor* or *inertia matrix*, which depends only on the geometry of the rigid body and its mass distribution; $\mathbf{I}$ is the operator whose representation in that particular coordinate system is I; and $\mathbf{I}\omega$ is the vector obtained when $\mathbf{I}$ is applied to ω.

The inertia tensor is an important property of a rigid body. It is the analog in rotational motion of the mass in translational motion: if ω is the analog of velocity, then $T = \omega \cdot \mathbf{I}\omega/2$ can be thought of as the analog of $T = \mathbf{v} \cdot m\mathbf{v}/2$. The I_{ij} depend on the orientation of the body in the inertial system and change with time as the body changes its orientation. We will see later how time-independent matrix elements I_{jk} calculated for the rigid body at rest can be used to find the time-dependent matrix elements $I_{jk}(t)$.

It is important that $\mathbf{I}$ has an inverse $\mathbf{I}^{-1}$. This can be proven from physical considerations alone. If the body is rotating, the ω_j are not all zero, and the kinetic energy is also nonzero (in fact positive). Thus

$$I_{jk}\omega_j\omega_k > 0 \quad \forall \omega \neq 0, \qquad (8.6)$$

where ω is the column vector whose components are the ω_j. That means that the I matrix annihilates no nonzero vector and hence has an inverse (see the book's appendix). [A matrix such as I that satisfies (8.6) is called *positive definite*.]

Equation (8.5) implies that I is a symmetric matrix. Therefore (see the appendix) it can be diagonalized by an orthogonal transformation, which means that there is an orthogonal coordinate system whose basis vectors are eigenvectors of $\mathbf{I}$. A coordinate system in which I is diagonal is called a *principal-axis* system, and the eigenvalues of $\mathbf{I}$ are called the *principal moments* or the *moments of inertia* of the body, usually labeled I_1, I_2, I_3. Equation (8.6) implies that the I_k are all positive. In the principal-axis system, Eq. (8.4) becomes

$$T = \frac{1}{2}\left(I_1\omega_1^2 + I_2\omega_2^2 + I_3\omega_3^2\right), \qquad (8.7)$$

where the ω_k are the components of ω in the principal-axis system.

WORKED EXAMPLE 8.1.1

(a) Find the I_{jk} of a uniform cube of side s whose pivot A is at a corner (Fig. 8.3) and whose sides are lined up along the axes of an orthonormal coordinate system.
(b) Find the principal-axis system and the moments of inertia.

Solution. (a) In this example μ is constant. According to Eq. (8.5)

$$I_{11} = I_{22} = I_{33} = \mu \int_0^s dx_1 \int_0^s dx_2 \int_0^s dx_3 \left(x_2^2 + x_3^2\right) = \frac{2}{3}\mu s^5,$$

$$I_{12} = I_{23} = I_{31} = -\mu \int_0^s dx_1 \int_0^s dx_2 \int_0^s dx_3 (x_1 x_2) = -\frac{1}{4}\mu s^5.$$

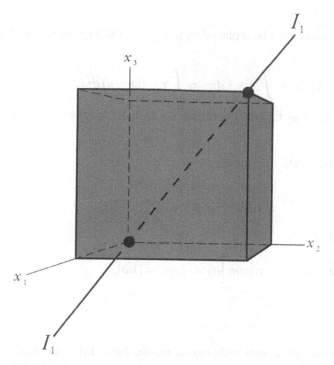

FIGURE 8.3
A uniform cube in an orthonormal system of coordinates, with its origin is at one of the corners. One of the principal axes of the inertia tensor about the origin lies along the diagonal of the cube and is labeled $I_1 I_1$ in the figure. The other two are perpendicular to it and are not shown.

Hence in this coordinate system the I matrix is

$$I = \frac{2}{3} M s^2 A \equiv \frac{2}{3} M s^2 \begin{bmatrix} 1 & \alpha & \alpha \\ \alpha & 1 & \alpha \\ \alpha & \alpha & 1 \end{bmatrix},$$

where $M \equiv \mu s^3$ is the mass of the cube, $\alpha = -3/8$, and A is the 3×3 matrix.
   (b) To find the principal axes, diagonalize A. The eigenvalue equation is

$$(1 - \lambda)^3 - 3\alpha^2(1 - \lambda) + 2\alpha^3 \equiv (1 - \lambda - \alpha)^2(1 - \lambda + 2\alpha) = 0,$$

whose solutions are $\lambda_1 = 2\alpha + 1 = 1/4$, $\lambda_2 = \lambda_3 = 1 - \alpha = 11/8$. The eigenvector belonging to λ_1 (not normalized) is $(1, 1, 1)$, and thus it lies along the diagonal of the cube: one of the principal axes of the cube pivoted at a corner is along its diagonal. The other two belong to equal eigenvalues, so they form a two-dimensional eigenspace. Any two mutually perpendicular directions perpendicular to the diagonal of the cube can serve as the other two principal axes. The moments of inertia are $I_1 = \frac{1}{6} M s^2$, $I_2 = I_3 = \frac{11}{12} M s^2$.

ANGULAR MOMENTUM

According to Section 1.4 (sums must be replaced by integrals), the angular momentum of the rigid body is

$$\mathbf{J} = \int \mu(\mathbf{x})\mathbf{x} \wedge \dot{\mathbf{x}}\, d^3x \equiv \int \mathbf{x} \wedge \dot{\mathbf{x}}\, dm = \int \mathbf{x} \wedge (\boldsymbol{\omega} \wedge \mathbf{x})\, dm,$$

(where we use $\mathbf{J}$ rather than $\mathbf{L}$ so as to avoid confusion with the Lagrangian). The ith component of $\mathbf{x} \wedge (\boldsymbol{\omega} \wedge \mathbf{x})$ is

$$[\mathbf{x} \wedge (\boldsymbol{\omega} \wedge \mathbf{x})]_i = \epsilon_{ijk}x_j\epsilon_{klm}\omega_l x_m$$
$$= (\delta_{il}\delta_{jm} - \delta_{im}\delta_{jl})x_j\omega_l x_m$$
$$= (\delta_{il}x^2 - x_i x_l)\omega_l.$$

Hence the ith component of $\mathbf{J}$ is

$$J_i = \left[\int (\delta_{il}x^2 - x_i x_l)\, dm\right]\omega_l = I_{il}\omega_l = [\mathbf{I}\boldsymbol{\omega}]_i,$$

or

$$\mathbf{J} = \mathbf{I}\boldsymbol{\omega}. \tag{8.8}$$

Note that the analogy extends to momentum and angular momentum: $\mathbf{I}\boldsymbol{\omega}$ is the analog of $m\mathbf{v}$. Equation (8.8) implies that $\mathbf{J}$ is parallel to $\boldsymbol{\omega}$ iff $\boldsymbol{\omega}$ is an eigenvector of $\mathbf{I}$.

In the principal-axis system, in which I is diagonal, the kth component of the angular momentum is $J_k = I_k\omega_k$ (no sum), where the ω_k are the components of $\boldsymbol{\omega}$ in the principal-axis system.

WORKED EXAMPLE 8.1.2

The cube in Worked Example 8.1.1, oriented as in Fig. 8.3, rotates instantaneously about the edge that is lined up along the x_1 axis. Find $\mathbf{J}$ and the angle between $\mathbf{J}$ and $\boldsymbol{\omega}$.

Solution. If $\boldsymbol{\omega}$ is along the x_1 axis, $\omega = (\Omega, 0, 0)$, where $\Omega = |\boldsymbol{\omega}|$. In that coordinate system

$$J = \frac{2}{3}Ms^2\begin{bmatrix} 1 & \alpha & \alpha \\ \alpha & 1 & \alpha \\ \alpha & \alpha & 1 \end{bmatrix}\begin{bmatrix} \Omega \\ 0 \\ 0 \end{bmatrix} = \frac{2}{3}Ms^2\Omega\begin{bmatrix} 1 \\ \alpha \\ \alpha \end{bmatrix} = Ms^2\Omega\begin{bmatrix} \frac{2}{3} \\ -\frac{1}{4} \\ -\frac{1}{4} \end{bmatrix}.$$

Clearly $\boldsymbol{\omega}$ is not an eigenvector of $\mathbf{I}$, and $\mathbf{J}$ is not parallel to $\boldsymbol{\omega}$ (see Fig. 8.4). The angle θ between $\mathbf{J}$ and $\boldsymbol{\omega}$ is given by

$$\cos\theta = \frac{\mathbf{J}\cdot\boldsymbol{\omega}}{J\Omega} = \frac{Ms^2\Omega^2}{Ms^2\Omega^2}\frac{2/3}{\sqrt{4/9 + 2/16}} = 0.8835,$$

or $\theta = \arccos 0.8835 = 0.4876$ rad $= 27.9°$.

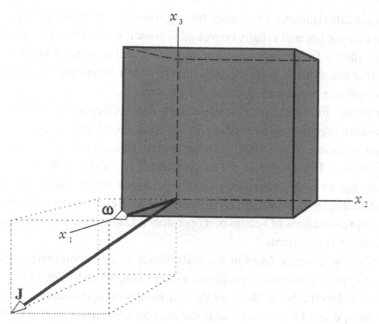

FIGURE 8.4
The cube of Fig. 8.3 with ω (instantaneously) along the x_1 axis. The sides of the dotted box are proportional to the components $(\frac{2}{3}, -\frac{1}{4}, -\frac{1}{4})$ of the angular momentum: ω and $\mathbf{J}$ are not collinear. Both ω and $\mathbf{J}$ are drawn at the origin of the body system, although they live in two different vector spaces: angular momentum and angular velocity cannot be added to each other or to position vectors.

8.1.3 DYNAMICS

The equation that specifies the dynamics of rigid body motion, the analog of $\mathbf{F} = \dot{\mathbf{p}}$, is Eq. (1.73):

$$\mathbf{N_Z} = \dot{\mathbf{J}}_Z. \tag{8.9}$$

Recall that the subscript $\mathbf{Z}$ in Chapter 1 means that the torque must be calculated about the center of mass or about an inertial point. What (8.9) requires is the ability to calculate the time derivative of $\mathbf{J} = I\omega$ about a suitable point. That in turn requires calculating how both I and ω vary in time in an inertial coordinate system, which can get quite complicated; for instance, the integral defining I in an inertial system will in general change as the body moves. In order to proceed, we must discuss suitable coordinate systems for performing the calculations.

SPACE AND BODY SYSTEMS

In this subsection we define two coordinate systems, only one of which is inertial.

The noninertial one, called the *body system* $\mathfrak{B}$, is fixed in the body and moves with it. Its origin A is by assumption an inertial point or the center of mass (perhaps even a pivot) and its orientation is chosen for convenience. The position vector $\mathbf{x}$ of any point in the body has fixed components in $\mathfrak{B}$ and is represented by a fixed (column) vector called $x_{\mathfrak{B}}$. Since $x_{\mathfrak{B}}$ is fixed, $\dot{x}_{\mathfrak{B}} = 0$ and it is impossible to describe the motion in terms of $x_{\mathfrak{B}}$. Nonetheless, the $x_{\mathfrak{B}}$ vectors can be used as permanent labels for the points of the body.

The inertial system is called the *space system* $\mathfrak{S}$. Its origin is also at A, and its orientation is also chosen for convenience but will usually be picked to coincide with that of $\mathfrak{B}$ at some particular time t. We write $x_{\mathfrak{S}}$ for the representation of $\mathbf{x}$ in $\mathfrak{S}$, and in general $\dot{x}_{\mathfrak{S}} \neq 0$, so it can be used to describe the motion. It can be used not only for a kinematic description, but also for the dynamical one because $\mathfrak{S}$ is an inertial system.

In the discussion around Eq. (8.2) the velocity vector $\dot{\mathbf{x}}$ was not defined carefully: the components of the position vector vary differently in different coordinate systems. The dynamically important variation of $\mathbf{x}$ is relative to the inertial space system $\mathfrak{S}$, so what is actually of interest is $\dot{x}_{\mathfrak{S}}$. However, it is not as convenient to calculate the I_{jk} in $\mathfrak{S}$, where they keep changing, as in $\mathfrak{B}$, where they are fixed. Many other objects are also most conveniently calculated in $\mathfrak{B}$, so what will be needed is a careful analysis of how to compare the $\mathfrak{S}$ and $\mathfrak{B}$ representations of vectors and operators and how to transform results of calculations between the two systems.

Equation (8.2) tells how a vector fixed in the body varies in *any* coordinate system whose origin is at A. Suppose some such coordinate system is specified. Then the $\mathbf{x}$ on the right-hand side is fixed in the body, the $\dot{\mathbf{x}}$ on the left-hand side is the rate of change of $\mathbf{x}$ as viewed from the specified system, and ω is the angular velocity of the body with respect to the specified system. In particular, if the system specified is the $\mathfrak{S}$ *that coincides instantaneously with* $\mathfrak{B}$, Eq. (8.2) reads

$$\dot{x}_{\mathfrak{S}} = \omega_{\mathfrak{B}} \wedge x_{\mathfrak{B}}. \tag{8.10}$$

The body components of ω are used here because $\mathfrak{S}$ and $\mathfrak{B}$ coincide instantaneously, so the body and space components are the same.

Equation (8.10) can be extended to an arbitrary vector $\mathbf{s}$ in the body, one that may be moving in the body system $\mathfrak{B}$. Its velocity $\dot{s}_{\mathfrak{S}}$ relative to an inertial system is the velocity $\omega \wedge s_{\mathfrak{B}}$ it would have if it were fixed in $\mathfrak{B}$ plus its velocity $\dot{s}_{\mathfrak{B}}$ relative to $\mathfrak{B}$:

$$\dot{s}_{\mathfrak{S}} = \omega_{\mathfrak{B}} \wedge s_{\mathfrak{B}} + \dot{s}_{\mathfrak{B}}. \tag{8.11}$$

Equation (8.11) tells how to transform velocity vectors between $\mathfrak{B}$ and $\mathfrak{S}$. In particular if $\mathbf{s}$ is the angular velocity, this equation reads

$$\dot{\omega}_{\mathfrak{S}} = \dot{\omega}_{\mathfrak{B}}, \tag{8.12}$$

so that the body and space representations not only of ω, but also of $\dot{\omega}$, are identical. From now on we leave the subscript off ω.

DYNAMICAL EQUATIONS

The next step is to apply this to the angular momentum and to Eq. (8.9). If $\mathbf{s}$ is the angular momentum $\mathbf{J}$, Eq. (8.11) reads

$$\dot{J}_{\mathfrak{S}} = \omega \wedge J_{\mathfrak{B}} + \dot{J}_{\mathfrak{B}} = \omega \wedge (I\omega) + I\dot{\omega}, \tag{8.13}$$

where I is the *time-independent* representation of $\mathbf{I}$ in $\mathfrak{B}$. This is a great simplification for it eliminates the need to calculate the inertia tensor in the space system, where its elements keep changing. According to Eq. (8.9), $\dot{J}_{\mathfrak{S}}$ has to be equated to the space-system representation $N_{\mathfrak{S}}$ of the torque **N**:

$$N_{\mathfrak{S}} = \omega \wedge (I\omega) + I\dot{\omega}. \tag{8.14}$$

The analogy with particle mechanics goes one step further. If Eq. (8.14) is written in terms of vectors and operators rather than their representations, it is

$$\dot{\mathbf{J}} \equiv \mathbf{N} = \boldsymbol{\omega} \wedge (\mathbf{I}\boldsymbol{\omega}) + \mathbf{I}\dot{\boldsymbol{\omega}},$$

which means that $\boldsymbol{\omega} \cdot \mathbf{N} = \boldsymbol{\omega} \cdot \mathbf{I}\dot{\boldsymbol{\omega}}$. The kinetic energy changes according to $\dot{T} = \frac{1}{2}(\dot{\boldsymbol{\omega}} \cdot \mathbf{I}\boldsymbol{\omega} + \boldsymbol{\omega} \cdot \mathbf{I}\dot{\boldsymbol{\omega}}) = \boldsymbol{\omega} \cdot \mathbf{I}\dot{\boldsymbol{\omega}}$, where we have used the symmetry of **I**, so $\boldsymbol{\omega} \cdot \mathbf{N} = \dot{T}$, which is the analog of $\mathbf{v} \cdot \mathbf{F} = \dot{T}$.

Before proceeding, we write out (8.14) in the principal-axis system:

$$\left.\begin{array}{l} N_1 = (I_3 - I_2)\omega_3\omega_2 + I_1\dot{\omega}_1, \\ N_2 = (I_1 - I_3)\omega_1\omega_3 + I_2\dot{\omega}_2, \\ N_3 = (I_2 - I_1)\omega_2\omega_1 + I_3\dot{\omega}_3. \end{array}\right\} \tag{8.15}$$

These are known as the *Euler equations for the motion of a rigid body*. In these equations everything is calculated in $\mathfrak{B}$.

As a special case consider torque-free motion of a rigid body with a pivot A, the common origin of $\mathfrak{S}$ and $\mathfrak{B}$. (This applies equally to a body falling freely in a uniform gravitational field: in the falling frame the equivalent gravitational field vanishes and the center of mass can be taken as A.) Then all of the left-hand sides of (8.15) vanish. If the body is instantaneously rotating about one of its principal axes, two of the ω_k also vanish, and all of the first terms on the right-hand sides vanish. The solution of the resulting equations is $\dot{\omega}_k = 0$, $k = 1, 2, 3$, so the angular velocity vector is stationary. Thus if the body starts rotating about a principal axis, $\boldsymbol{\omega}$ remains fixed, and the body continues to rotate about the same axis. If, however, the rotation is not about a principal axis, $\boldsymbol{\omega}$ varies in time and wanders through the body, which is then observed to wobble, flip, and in general to move in complicated ways. This happens in spite of the fact that the angular momentum is constant, for there is no torque on the body.

Consider ω-space, the vector space in which $\boldsymbol{\omega}$ moves as the motion proceeds. We have found three fixed points in ω-space for the torque-free motion of a rigid body: the three vectors $(\Omega, 0, 0)$, $(0, \Omega, 0)$, and $(0, 0, \Omega)$ in the principal-axis system, where Ω is a constant. Actually, these are *fixed rays* rather than fixed points, because Ω may be any constant. We will turn to their stability later. It should be borne in mind, however, that ω-space is not the carrier manifold for the dynamics, so these fixed points are not the kind we have discussed before. More about all this is discussed in later sections.

In spite of the analogs we have been pointing out, the dynamical equations themselves (i.e., Euler's equations for rotational motion) are not analogs of Newton's equations for particle motion. Newton's are second-order equations for the position vector $\mathbf{x}(t)$, whereas Euler's are first-order equations for the angular velocity vector $\boldsymbol{\omega}(t)$. The solution of Euler's equations does not describe the orientation of the body, only its instantaneous axis and rate of rotation. The best analog in particle motion would be a set of equations for the velocity vector $\mathbf{v}(t)$. This will become clearer later, when we derive Euler's equations by the Lagrangian formalism, in the process taking a careful look at the configuration and tangent manifolds $\mathbb{Q}$ and $\mathbf{T}\mathbb{Q}$ for rigid-body motion.

WORKED EXAMPLE 8.2

A sphere of mass M, radius a, and moment of inertia I about its center (the inertia matrix is $I\mathbb{I}$, a multiple of the unit matrix) rolls without slipping on a horizontal plane. Obtain the general motion.

Solution. There may be a constraint force $\mathbf{F}$ parallel to the plane, forcing the sphere to roll. Call Cartesian coordinates in the plane x_1 and x_2. Newton's equations of motion in the plane are

$$F_1 = M\ddot{x}_1, \quad F_2 = M\ddot{x}_2. \tag{8.16}$$

The torque equations about the center of mass (all the I_k are equal) are

$$N_1 \equiv F_2 a = I\dot{\omega}_1, \tag{8.17a}$$
$$N_2 \equiv -F_1 a = I\dot{\omega}_2, \tag{8.17b}$$
$$N_3 \equiv 0 = I\dot{\omega}_3. \tag{8.17c}$$

The rolling constraint equations are

$$\dot{x}_1 = a\omega_2, \quad \dot{x}_2 = -a\omega_1. \tag{8.18}$$

Equations (8.16), (8.17b), and the first of (8.18) yield

$$F_1 = -I\dot{\omega}_2/a = M\ddot{x}_1 = Ma\dot{\omega}_2,$$

which implies that $\dot{\omega}_2 = -Ma^2\dot{\omega}_2/I$. But since Ma^2/I is positive, this equality is impossible unless $\dot{\omega}_2 = 0$. Thus $F_1 = 0$. Similarly, $F_2 = 0$, so $\ddot{\mathbf{x}} = 0$ and the motion is simply in a straight line. We conclude that the sphere rolls at constant speed in a straight line and that its angular velocity ω is constant.

Several aspects of this system are interesting. First, it turns out that $\mathbf{F} = \mathbf{0}$: once the sphere is rolling, no frictional force is necessary to keep it rolling. Second, ω_3 is not necessarily zero, so ω is not necessarily horizontal, as one might have guessed. That it would have been a bad guess is clear from the fact that a ball can spin about the vertical axis (the special case $\dot{\mathbf{x}} = \mathbf{0}$, ω vertical). Also, anyone who has watched soccer matches has noticed that the rolling axis is only rarely horizontal.

Comment: This example demonstrates that some rigid body problems can be solved very simply, but even then the properties of their solutions can be surprising. We will turn to a more formal solution of this problem at the end of Section 8.3.1.

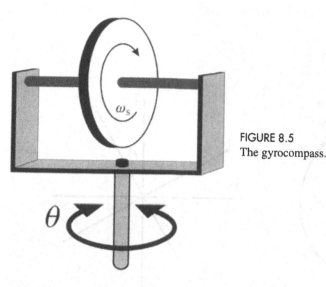

FIGURE 8.5
The gyrocompass.

EXAMPLE: THE GYROCOMPASS

A *gyrocompass* is a device used to find true North without reference to the Earth's magnetic field. It consists of a circular disk spinning about its symmetry axis, with the axis constrained to remain horizontal but free to turn in the horizontal plane (Fig. 8.5). The rate of spin ω_S is maintained constant by an external driving mechanism. Actually, gyrocompasses that are used on ships and planes are more complicated than this, but we will treat just this idealized system. Our treatment follows essentially that of Moore (1983).

Suppose the compass is set in motion at a point P on the Earth whose latitude is α and with its axis initially at some angle θ_0 with respect to true North. We will show that if $\omega_S \gg \Omega_E$ (Ω_E is the rate of rotation of the Earth), the axis of the gyrocompass oscillates about true North.

To analyze this system, we choose two local Cartesian coordinate systems at P. The first (primed) system is attached to the Earth. The $3'$ axis always points north, the $1'$ axis always points west, and the $2'$ axis is always vertical. This system rotates with an angular velocity Ω whose components are (Fig. 8.6)

$$\Omega_{1'} = 0, \quad \Omega_{2'} = \Omega_E \sin \alpha, \quad \Omega_{3'} = \Omega_E \cos \alpha.$$

The second (unprimed) system is lined up with the gyrocompass frame. The 2 axis coincides with $2'$, the 3 axis is along the gyrocompass symmetry axis, and the 1 axis is as shown in Fig. 8.6(b). The unprimed system rotates with respect to the primed one, and its total angular velocity has components

$$\omega_1 = -\Omega_E \cos \alpha \sin \theta, \quad \omega_2 = \Omega_E \sin \alpha + \dot{\theta}, \quad \omega_3 = \Omega_E \cos \alpha \cos \theta. \tag{8.19}$$

In the unprimed system the angular momentum of the gyrocompass is given by $J_k = I_{kj} \bar{\omega}_j$, where the $\bar{\omega}_j$ are obtained by adding ω_S along the 3 direction to the ω_j: then $\bar{\omega}_{1,2} = \omega_{1,2}$ and $\bar{\omega}_3 = \omega_3 + \omega_S$. In this idealized version we assume that the mounting frame itself has negligible inertia. Then symmetry implies that $I_1 = I_2$, so that

$$J_1 = -I_1 \Omega_E \cos \alpha \sin \theta,$$
$$J_2 = I_1 (\Omega_E \sin \alpha + \dot{\theta}),$$
$$J_3 = I_3 (\Omega_E \cos \alpha \cos \theta + \omega_S).$$

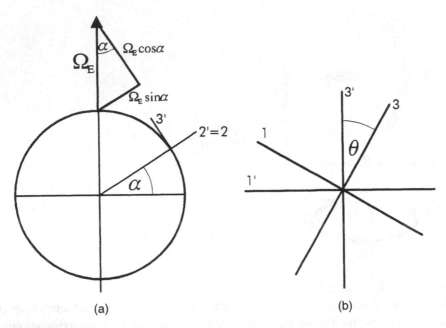

FIGURE 8.6
Coordinates in the calculations for the gyrocompass. (a) Coordinates on the Earth; α is the latitude. (b) Looking down vertically (down the 2 axis) at latitude α. North is in the $3'$ direction, and the gyrocompass axis is in the 3 direction.

This gives the J_k in the unprimed system. Although the unprimed system is not the body system (for the gyrocompass is spinning in it), what is of interest is θ, the angle between the 3 and $3'$ directions. Equation (8.11) is valid for relating the components of vectors in *any* two relatively rotating frames, so for the purposes of this discussion, the primed frame can be labeled $\mathfrak{S}$ and the unprimed one $\mathfrak{B}$. Then the rate of change of $\mathbf{J}$ in the primed system is

$$\dot{J}_{\mathfrak{S}} \equiv \dot{J}' = \omega_{\mathfrak{B}} \wedge J_{\mathfrak{B}} + \dot{J}_{\mathfrak{B}},$$

where $\omega_{\mathfrak{B}}$ is given by (8.19). The primed system is almost inertial, so $\dot{J}'$ is essentially the torque N on the gyrocompass. But the $2'$ component of the torque is clearly zero, for the gyrocompass is free to rotate about its vertical spindle. Hence (this is calculated for a fixed latitude α)

$$N_{2'} = \dot{J}_{2'} \equiv \dot{J}_2 + \omega_3 J_1 - \omega_1 J_3$$
$$= I_1 \ddot{\theta} + \Omega_E \cos\alpha \sin\theta \left[\Omega_E (I_3 - I_1)\cos\alpha\cos\theta + I_3\omega_S\right] = 0.$$

A gyrocompass runs in a regime such that

$$|\Omega_E(I_3 - I_1)\cos\alpha\cos\theta| \ll I_3\omega_S,$$

so

$$I_1\ddot{\theta} + I_3\omega_S\Omega_E\cos\alpha\sin\theta \approx 0. \tag{8.20}$$

This looks like the pendulum equation, since I_1, I_3, ω_S, Ω_E, and α are all constants. In the small-angle approximation, therefore, θ oscillates about zero with (circular) frequency

$$\omega_{\text{osc}} = \sqrt{\frac{I_3}{I_1}\omega_S \Omega_E \cos\alpha}. \tag{8.21}$$

This result shows that the gyrocompass will seek north in the same sense as a pendulum seeks the vertical. The frequency ω_{osc} with which it oscillates about north depends on the latitude α: it is a maximum at the equator, where $\cos\alpha = 1$. At the north pole, where $\cos\alpha = 0$, Eq. (8.20) reads $\ddot\theta = 0$, so θ changes at a constant rate. This is what would be expected. For instance, if $\dot\theta$ is initially set equal to $-\Omega_E$, the gyrocompass will remain stationary while the Earth rotates under it. From the rotating Earth it will look as though the gyrocompass rotates once a day – at the pole the gyrocompass is useless.

MOTION OF THE ANGULAR MOMENTUM J

We now want to discuss questions of stability for a freely rotating body (with no applied torques). We will do this by studying the motion of the angular momentum vector **J** through the body.

A freely rotating body may be falling (so that the gravitational force applies no torques) and rotating about its center of mass, or it may be pivoted at an arbitrary point with no other forces. Since $\mathbf{N} = \mathbf{0}$, the angular momentum vector **J** (in the space system) is invariant, a vector-valued constant of the motion. But its body-system representation $J \equiv J_{\mathfrak{B}}$ will in general not be invariant. Indeed, according to Eq. (8.13) $\dot J = \omega \wedge J$, which is equal to zero only if ω and J are parallel.

Much can be learned about the motion by studying the way J moves in the three-dimensional space $\mathbb{R}_J$ of all J values. But just as finding $\omega(t)$ does not solve for the motion, neither does finding $J(t)$. Indeed, although $\mathbb{R}_J$ is three dimensional, it is not the rotational part of the configuration manifold of the rigid body, for it does not specify its orientation. And of course it cannot be **TQ**, for its dimension is wrong. Nevertheless $\mathbb{R}_J$ provides insight into the motion.

In general $J = I\omega$ (these are representations in the body system). If ω is an eigenvector of I, the angular velocity ω is parallel to J, and since that means that $\dot J = 0$, if the body is spinning about one of its principal axes J is fixed: in $\mathbb{R}_J$ the principal directions define fixed rays for the motion of J. Since then ω is parallel to J, the angular velocity ω also remains fixed in these directions, so if the body is spinning along a principal axis, it will continue to do so.

A neat way see this in more detail is to make use of two important scalar constants of the motion, J^2 and the kinetic energy $T = \frac{1}{2}\omega \cdot I\omega$. Because these are scalars, they are the same in all coordinate systems, constant in all. In particular they are constants in the body system. To see this in $\mathbb{R}_J$, rewrite T in terms of J by using $\omega = I^{-1}J$ (recall that I is nonsingular): the two scalar invariants are

$$2T \equiv J \cdot I^{-1}J = \text{const.}, \quad J^2 = \text{const.} \tag{8.22}$$

The angular momentum vector J (in the body system) satisfies these two equations simultaneously.

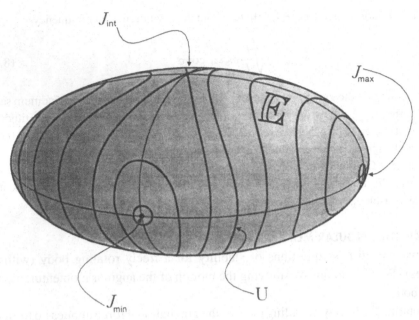

FIGURE 8.7
The energy ellipsoid $\mathbb{E}$ in $\mathbb{R}_J$ showing lines of intersection with the spheres $\mathbb{S}$ whose radii are different values of $|J|$.

The second of Eqs. (8.22) implies that J lies on a sphere $\mathbb{S} \subset \mathbb{R}_J$. The first of Eqs. (8.22), written explicitly in terms of the components of J and the matrix elements of I^{-1}, is

$$2T = I_{kl}^{-1} J_k J_l > 0. \qquad (8.23)$$

This is the equation of an ellipsoid $\mathbb{E} \subset \mathbb{R}_J$ (ignore the special case $J = 0$), the energy surface in $\mathbb{R}_J$ (Fig. 8.7): $\mathbb{E}$ is an invariant submanifold of $\mathbb{R}_J$. We now consider the motion of J on $\mathbb{E}$.

FIXED POINTS AND STABILITY

Since J satisfies both equations it lies on the intersection of $\mathbb{E}$ with $\mathbb{S}$, two concentric surfaces that can intersect only if the radius $|J|$ of $\mathbb{S}$ is larger than the minimum semiaxis of $\mathbb{E}$ and smaller than the maximum. These semiaxes are along the principal directions. Indeed, the principal-axis system diagonalizes I^{-1} as well as I, so in it Eq. (8.23) becomes the equation of an ellipsoid with its axes along the coordinate directions:

$$\frac{J_1^2}{2T I_1} + \frac{J_2^2}{2T I_2} + \frac{J_3^2}{2T I_3} = 1.$$

Assume that the I_k are unequal and that $I_1 > I_2 > I_3$. Then the semiaxes of $\mathbb{E}$ are $\sqrt{2T I_1} > \sqrt{2T I_2} > \sqrt{2T I_3}$. The condition for intersection is therefore

$$\sqrt{2T I_1} > |J| > \sqrt{2T I_3}. \qquad (8.24)$$

Hence for fixed T there is an upper and lower bound for the angular momentum.

In general a sphere intersects a concentric ellipsoid in two curves (if they intersect at all). There are two special cases. If the radius $|J|$ of $\mathbb{S}$ is minimal, $J_{\min} \equiv \sqrt{2TI_3}$ and $\mathbb{S}$ and $\mathbb{E}$ make contact only at two points, the ends of the smallest semiaxis of $\mathbb{E}$. Similarly, if $|J|$ is maximal, $J_{\max} \equiv \sqrt{2TI_1}$ and $\mathbb{S}$ and $\mathbb{E}$ make contact only at the ends of the largest semiaxis. Since J lies on an intersection throughout its motion, it cannot move from one of these points if $J = J_{\min}$ or $J = J_{\max}$. In other words, these are fixed points on $\mathbb{E}$, confirming what was said above about fixed rays lying along the principal axes. There is also a third fixed point at the intermediate axis; this will be discussed later.

Figure 8.7 shows the energy ellipsoid $\mathbb{E}$ with some of the curves along which it intersects $\mathbb{S}$. The three fixed points are at $J_{\min}$, $J_{\max}$, and $J_{\mathrm{int}} \equiv \sqrt{2TI_2}$. In general, as the body changes its orientation and J moves on the ellipsoid, it continues to satisfy both of Eqs. (8.22), so J moves on one of the intersection curves. Precisely how J moves on such a curve has yet to be established, but it is clear that $J_{\min}$ and $J_{\max}$ are stable (elliptic) points. Indeed, if J starts out close to one of these points, say one of the small circles in Fig. 8.7, it remains close to it. In contrast, J_{int} *looks* like an unstable (hyperbolic) fixed point, but that has yet to be established. Suppose, for example, that J starts out close to J_{int} on the curve marked U in Fig. 8.7. It may continue to remain close to J_{int} or it may move far away on U. To establish just how J moves on U we return to Euler's equations (8.15).

Since this is the torque-free case, the N_j are all zero. To deal with questions of stability consider motion close to the fixed points. If J is close to one of the principal axes, so is $\omega \equiv I^{-1}J$, which means that the body is rotating about a line close to one of the principal axes. Although it has already been seen that $J_{\min}$ and $J_{\max}$ are stable, it is worth seeing how this is implied by (8.15). Suppose that the rigid body is rotating about a line very close to I_1. Then in the first approximation ω_2 and ω_3 can be neglected in the first of Eqs. (8.15), so ω_1 is constant. The second and third of Eqs. (8.15) read $\dot{\omega}_2 = \omega_1\omega_3(I_3 - I_1)/I_2$ and $\dot{\omega}_3 = \omega_1\omega_2(I_1 - I_2)/I_3$, and if ω_1 is constant, these are a pair of coupled equations for ω_2 and ω_3. Taking their time derivatives uncouples them, yielding

$$\ddot{\omega}_k = \frac{\omega_1^2(I_1 - I_3)(I_2 - I_1)}{I_2 I_3}\omega_k, \quad k = 2, 3. \tag{8.25}$$

The factor multiplying ω_k on the right-hand side is negative, so this is the equation of a harmonic oscillator: to the extent that this approximation is valid, ω_2 and ω_3 oscillate harmonically about zero with frequency

$$\omega_1 \sqrt{\frac{(I_1 - I_3)(I_1 - I_2)}{I_2 I_3}},$$

remaining small. Therefore they can continue to be neglected in the first of Eqs. (8.15), and the approximation remains valid. A similar result is obtained if the body is rotating about a line very close to I_3. Then it is found that ω_3 is approximately constant and ω_1 and ω_2 oscillate about zero with frequency

$$\omega_3 \sqrt{\frac{(I_1 - I_3)(I_2 - I_3)}{I_1 I_2}}.$$

This verifies that the fixed points at J_{max} and J_{min} are stable: the body continues to rotate about a line close to either of them and J stays close to its initial value.

The situation close to J_{int} is quite different. There the equation corresponding to (8.25) is

$$\ddot{\omega}_k = \frac{\omega_2^2 (I_2 - I_3)(I_1 - I_2)}{I_1 I_3} \omega_k, \quad k = 1, 3, \tag{8.26}$$

which is similar, except that now the factor multiplying ω_k on the right-hand side is positive. Hence ω_1 and ω_3 start out to grow exponentially, and the approximation breaks down: ω_2 does not remain approximately constant. All three components of ω change significantly, so the angular velocity vector moves far away from the intermediate principal axis, and therefore so does the angular momentum.

This confirms what seemed likely from Fig. 8.7. If J starts out close to J_{int} on the orbit labeled U it moves far away; therefore J_{int} is an unstable fixed point.

A simple experiment illustrates these stability properties. Take a body with three different moments of inertia, say, a chalk eraser. Toss it lightly into the air, at the same time spinning it about its maximal principal axis, the shortest dimension. Then do the same, spinning it about its minimal principal axis, the longest dimension. In both cases the body continues to spin about that axis, perhaps wobbling slightly. This shows that these two axes are stable fixed points. Finally, do the same, spinning it about the intermediate principal axis. The object almost always (i.e., unless you happen to hit the axis exactly) flips around in a complicated way, demonstrating that this axis is unstable.

THE POINSOT CONSTRUCTION

Thus far we have been studying the motion of J in $\mathbb{R}_J$, but a similar approach, called the *Poinsot construction,* can be used to visualize the motion of the body itself in the space system $\mathfrak{S}$.

Consider the kinetic energy $T = \frac{1}{2}\omega \cdot \mathbf{I}\omega$ of a body whose inertial tensor is $\mathbf{I}$. The equation for T can be put in the form

$$\frac{\omega}{\sqrt{2T}} \cdot \mathbf{I} \frac{\omega}{\sqrt{2T}} \equiv \mathbf{s} \cdot \mathbf{I}\mathbf{s} = 1,$$

where $\mathbf{s} = \omega/\sqrt{2T}$. This is also an equation for an ellipsoid, this one called the *inertial* ellipsoid. (The inertial ellipsoid and $\mathbb{E}$ are different: the lengths of their semiaxes are inversely proportional.) Because its semiaxes are always parallel to the principal axes of the rigid body, the motion of the inertial ellipsoid mimics that of the body in $\mathfrak{S}$. The Poinsot construction studies the motion of the body by means of the equivalent motion of the inertial ellipsoid.

As the body rotates about its pivot or its center of mass, the angular velocity ω moves both in $\mathfrak{S}$ and in the body, and hence $\mathbf{s}$ moves both in $\mathfrak{S}$ and *on* (i.e., on the surface of) the inertial ellipsoid. In $\mathfrak{S}$ it moves so that its projection on the (constant) angular momentum vector remains constant. Indeed,

$$\mathbf{s} \cdot \mathbf{J} = \mathbf{s} \cdot \mathbf{I}\omega = \mathbf{s} \cdot \mathbf{I}\mathbf{s}\sqrt{2T} = \sqrt{2T}.$$

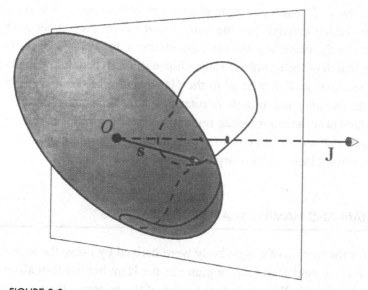

FIGURE 8.8

The inertial ellipsoid rolling on the invariable plane. The angular momentum vector **J** is perpendicular to the plane. The two curves trace out the paths of the contact point on the plane and the ellipsoid.

On the surface of the ellipsoid **s** moves so that its contact point with the surface always has a normal **n** that is parallel to **J**. This normal **n** is given by

$$\mathbf{n} = \nabla_s \left(\mathbf{s} \cdot \mathbf{Is} \right) = 2\mathbf{Is} = 2\frac{\mathbf{I}\omega}{\sqrt{2T}} = \sqrt{\frac{2}{T}}\mathbf{J}.$$

These two facts can be used to construct the following picture (Fig. 8.8): at the contact point of **s** and the surface of the ellipsoid, draw the plane tangent to the ellipsoid. This plane is perpendicular to **n** (not drawn in the figure, but **n** is parallel to **J**) and therefore also to the constant **J**, so it does not change its orientation as the body rotates. The rest of the picture is drawn with the plane's position, as well as its orientation, held fixed (it is then called the *invariable plane*). Because the projection of **s** perpendicular to the plane is constant, the distance from the **s** vector's tail to the plane is constant. The tail of **s** is then placed at a stationary point O, and **s** moves (i.e., its arrow moves) always on the invariable plane. But **s** also lies on the ellipsoid, which must therefore contact the invariable plane where **s** does. The inertial ellipsoid is tangent to the invariable plane at the contact point because its normal **n** is parallel to **J**, and it is rotating about **s** (the rigid body is rotating about ω). Hence the contact point is instantaneously stationary: the ellipsoid is *rolling* on the invariable plane while its center remains stationary at O. As it rolls, the rate at which it rotates is proportional to $s \equiv |\mathbf{s}|$ (the projection of **s** on **J** is constant, but s is not). This constitutes a complete description of the motion of the inertial ellipsoid and hence also of the rigid body in $\mathfrak{S}$.

The stability of the fixed points can also be understood in terms of the Poinsot construction. When the inertial ellipsoid contacts the plane at a point near the largest semiaxis (corresponding to the largest moment of inertia), the distance from O to the plane is close to the length of the semiaxis. If contact is to be maintained between the plane and the ellipsoid, the angle between the semiaxis and the normal to the plane cannot get too big. When the ellipsoid contacts the plane near the smallest semiaxis, the distance from O to the plane is close to the length of the semiaxis. If the ellipse is not to penetrate the plane, the angle between the semiaxis and the normal to the plane cannot get too big. In both cases that implies that the motion remains close to rotation about the semiaxis. When the ellipsoid contacts the plane near the intermediate semiaxis, these constraints on the angle disappear and the motion need not remain close to rotation about that semiaxis. In fact it can be seen that the angle must increase by a large amount.

8.2 THE LAGRANGIAN AND HAMILTONIAN FORMULATIONS

Euler's equations for the motion of a rigid body were derived by using the approach of Chapter 1 and thus involve neither the Lagrangian nor the Hamiltonian formalism on which much of this book is based. We now return to one of the principal themes of this book and derive the Euler–Lagrange equations by a systematic application of the procedure of Chapters 2 and 3 and then move on to the Hamiltonian formalism through the Legendre transform of Chapter 5. It will be seen that the Euler–Lagrange and Hamilton's canonical equations are entirely equivalent to Euler's equations of Section 8.1.

To do this requires understanding the configuration manifold of a rigid body, to which we devote Section 8.2.1.

8.2.1 THE CONFIGURATION MANIFOLD $\mathbb{Q}_R$

It was stated in Section 8.1 that the configuration manifold $\mathbb{Q}$ of a rigid body consists of two parts: $\mathbb{R}^3$, which describes the position of a point A in the body, and another part that we will call $\mathbb{Q}_R$ for the time being, which describes the orientation of the body about A. Hence $\mathbb{Q} = \mathbb{R}^3 \times \mathbb{Q}_R$. We relax the requirement that A be an inertial point: A is now either a fixed pivot or the center of mass of the body. This choice guarantees that Eq. (8.9) is still valid even if the center of mass is not itself an inertial point (torques can be taken about an inertial point or the center of mass). It was established that it takes three variables to fix the orientation of the body, so dim $\mathbb{Q}_R = 3$.

INERTIAL, SPACE, AND BODY SYSTEMS

Since a coordinate system centered at A will not in general be inertial, the definitions of $\mathfrak{S}$ and $\mathfrak{B}$ must be generalized. Let $\mathfrak{I}$ be a truly inertial system in which the center of mass A has position vector $\mathbf{r}(t)$. If $\mathbf{x}$ is the position vector, relative to A, of some point X in the body, the position vector relative to $\mathfrak{I}$ is

$$\mathbf{y} = \mathbf{r} + \mathbf{x}. \tag{8.27}$$

See Fig. 8.9. This vector equation can be written in any coordinate system. For instance,

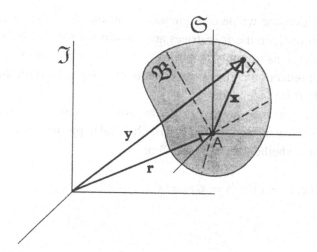

FIGURE 8.9
The inertial, space, and body systems $\mathfrak{I}$, $\mathfrak{S}$, and $\mathfrak{B}$ and the vectors $\mathbf{r}$, $\mathbf{y}$, and $\mathbf{x}$. The origins of both $\mathfrak{S}$ (solid lines) and $\mathfrak{B}$ (dotted lines) are within the rigid body, and $\mathfrak{B}$ is fixed in the body and rotates with it. The heavy dots represent the points A and X of the text.

in $\mathfrak{I}$ it becomes

$$y' = r' + x'; \tag{8.28}$$

primes will be used to indicate representations of vectors in $\mathfrak{I}$. Now we define a new *space system* $\mathfrak{S}$ whose origin is at the center of mass and whose axes remain always parallel to those of $\mathfrak{I}$. This system need not be inertial, but the components of any vector (e.g., of $\mathbf{x}$) are the same in $\mathfrak{I}$ as in $\mathfrak{S}$, so the prime in Eq. (8.28) indicates representations in $\mathfrak{S}$ as well as in $\mathfrak{I}$. The body system $\mathfrak{B}$ is the same as defined in Section 8.1.

The representations x' and $x_{\mathfrak{B}}$ in $\mathfrak{S}$ and $\mathfrak{B}$ of any vector $\mathbf{x}$ are related through the coordinate transformation connecting the two systems. Let R be the transformation matrix from $\mathfrak{B}$ to $\mathfrak{S}$, i.e., let $x' = Rx$. The R matrix preserves lengths (in fact inner products), so it is the real analog of a unitary matrix, called *orthogonal*, which means that $RR^{\mathrm{T}} = \mathbb{I}$, as will be shown below (see also the appendix). From now on we drop the subscript $\mathfrak{B}$ for representations in the body system, and then (8.28) becomes

$$y' = r' + Rx. \tag{8.29}$$

As the body moves, $\mathfrak{B}$ rotates with it, while $\mathfrak{S}$ remains always lined up with $\mathfrak{I}$. Hence R changes in time (x is the fixed body label of the point), and the time dependence on the right-hand side of (8.29) is all in r' and R. The dynamically relevant coordinates of a point X in the body, the coordinates in the inertial system $\mathfrak{I}$, are therefore contained in its label x and in r' and R.

THE DIMENSION OF $\mathbb{Q}_R$

The same r' and R work for any point in the rigid body; all that varies from point to point is x. Since, therefore, r' and R can be taken as generalized coordinates for the entire rigid body, they must range over the whole configuration manifold $\mathbb{Q} \equiv \mathbb{R}^3 \times \mathbb{Q}_R$. Clearly r' runs through the $\mathbb{R}^3$ part of $\mathbb{Q}$, so the R matrices are coordinates for $\mathbb{Q}_R$. Now, R has nine matrix elements, which seems to imply that dim $\mathbb{Q}_R = 9$, but it was shown in Section 8.1 that dim $\mathbb{Q}_R = 3$. The reduction comes from six equations that constrain the matrix elements of R, leaving only three independent.

The six relations arise from the fact that the R matrices are rotations: they preserve lengths and dot products (inner or scalar products). That is, the scalar product of two arbitrary vectors $\mathbf{x}$ and $\mathbf{z}$ is the same whether calculated in $\mathfrak{B}$ or in $\mathfrak{S}$:

$$(\mathbf{x}, \mathbf{z}) = (x, z) = (x', z') = x_k z_k = x'_k z'_k.$$

Since $x' = Rx$,

$$(x, z) = (Rx, Rz) = (R^\dagger Rx, z),$$

where we use the notation of the appendix. This shows that R must be a unitary matrix, and since it is real, it is *orthogonal* (that is, the Hermitian conjugate $R^\dagger$ should be replaced by the transpose R^T):

$$R^T R = R R^T = \mathbb{I}. \tag{8.30}$$

In terms of the matrix elements ρ_{jk} of R,

$$\rho_{jk} \rho_{lk} = \rho_{kj} \rho_{kl} = \delta_{jl}. \tag{8.31}$$

These are the equations that constrain the matrix elements. There are six of them, not nine (i and j run from 1 to 3), because δ_{jl} is symmetric. Thus only three of the ρ_{jk} are independent and each rotation matrix R can be specified by giving just three numbers: dim $\mathbb{Q}_R = 3$.

THE STRUCTURE OF $\mathbb{Q}_R$

Of course the dimension of a manifold does not characterize it completely. For example, the plane $\mathbb{R}^2$, the two-sphere $\mathbb{S}^2$, and the two-torus $\mathbb{T}^2$, all of dimension two, are very different manifolds. The next step is to exhibit the geometry (or topology) of $\mathbb{Q}_R$. Just as the geometry of $\mathbb{S}^2$ can be derived from the equation $x_k x_k = 1$ connecting its coordinates in three-space, so the geometry of $\mathbb{Q}_R$ can be derived from Eq. (8.30) or (8.31). We proceed in three steps to unfold the properties of $\mathbb{Q}_R$.

Step 1. We show that det $R = 1$; that is, every rotation is an orthogonal matrix of determinant one. The determinant of R is a polynomial of degree three in its matrix elements. According to the theory of determinants, det $R = \det R^T$ and det $\mathbb{I} = 1$. Taking the determinants of both sides of Eq. (8.30) yields $(\det R)^2 = 1$, or det $R = \pm 1$. But only the positive sign is valid for rotations. Indeed, suppose the rigid body is oriented so that $\mathfrak{S}$ and $\mathfrak{B}$ are connected by some matrix R_0 (or, as we shall say, that the orientation of the body is given by R_0). Rotate the body smoothly about the common origin of $\mathfrak{B}$ and $\mathfrak{S}$ until $\mathfrak{B}$ is brought into $\mathfrak{S}$. In the process, the transformation matrix R from $\mathfrak{B}$ to $\mathfrak{S}$ changes, starting out as R_0 and ending up as $\mathbb{I}$ when $\mathfrak{B}$ coincides with $\mathfrak{S}$, and det R changes from det R_0 to det $\mathbb{I} = 1$. This change of det R is smooth, without discontinuities, for it is a continuous function of the matrix elements. The only values it can take on are $+1$ or -1, and since it ends up at $+1$, smoothness implies that must have been $+1$ all along. Hence det $R_0 = 1$, and since R_0 is arbitrary, every rotation matrix is *special* (i.e., has determinant one).

This works as long as it is possible to rotate $\mathfrak{B}$ into $\mathfrak{S}$, which is possible if both are either right-handed or left-handed systems, but not otherwise. We therefore require all coordinate systems to be right-handed. Nevertheless, orthogonal matrices with determinant -1 also exist, for example the diagonal matrix with elements $1, 1, -1$ [*reflection* through the $(1, 2)$ plane]. We state without proof that every orthogonal matrix of determinant -1 is a rotation followed by this reflection and maps left-handed into right-handed systems and vice versa.

In short, $\mathbb{Q}_R$ is the set (actually the *group*) of rotation matrices, **SO**(3) (Special Orthogonal matrices in 3 dimensions). In Step 3 we discuss the geometry of the **SO**(3) manifold.

Step 2. We now prove a theorem due to Euler: every R is a rotation *about an axis*, that is, every rotation leaves at least one vector (and all of its multiples) invariant. This means that if a sphere in some initial orientation is rotated in an arbitrary way about its center, at least two diametrically opposed points on its surface end up exactly where they started. If the sphere is rotated further in order to move those points out of their initial positions, two other points will fall into their own initial positions. The theorem therefore implies that there is at least one vector (actually one *ray* or direction) that has the same components in $\mathfrak{S}$ as in $\mathfrak{B}$.

Hence Euler's theorem is the assertion that every R has at least one eigenvector belonging to eigenvalue $+1$. The proof goes as follows: that R is orthogonal implies that it can be diagonalized with eigenvalues $\{\lambda_1, \lambda_2, \lambda_3\}$ of unit magnitude on the main diagonal. The λ_j are the three roots of the equation $\det(R - \lambda \mathbb{I}) = 0$, a cubic equation with real coefficients, so they are either all real or occur in complex conjugate pairs. Their product is $\lambda_1 \lambda_2 \lambda_3 = \det R = 1$. Since there is an odd number of them, at least one must be real, and the other two are either (a) both real or (b) form a complex conjugate pair. In Case (a) the λ_j are all ± 1. Since their product is $+1$, two or none are negative and the other is equal to $+1$. In Case (b) let λ_2 and λ_3 be the complex eigenvalues: $\lambda_3 = \lambda_2^*$. Unit magnitude implies that $|\lambda_2|^2 \equiv \lambda_2 \lambda_3 = 1$. Hence

$$\lambda_1 \lambda_2 \lambda_3 = \lambda_1 \equiv 1,$$

and again the eigenvalue $+1$ occurs. Diagonalizability implies that this eigenvalue has an eigenvector. This proves Euler's theorem.

Case (a) includes only the unit matrix and reflections. Case (b) is the general one: in general R cannot be diagonalized over the real numbers. But physical three-space is a vector space over the reals, so for physical purposes most rotations cannot be diagonalized. Nevertheless, the real eigenvalue $+1$ *always* occurs.

Step 3. We now explain the geometry of **SO**(3). According to Euler's theorem, every rotation $R \in$ **SO**(3) can be specified by its axis and by the angle θ of rotation about that axis. Let **n** be the unit vector pointing along the axis. The angle can be restricted to $0 \le \theta \le \pi$, for a rotation by $\theta + \pi$ is the same (i.e., yields the same $\mathfrak{S}$ from a given $\mathfrak{B}$) as a rotation in the opposite sense about the same axis by $\pi - \theta$. Hence any rotation can be represented by a three-dimensional "vector" **v** (the quotation marks will be explained shortly) whose direction is along **n**, whose magnitude is θ, and whose arrow gives the sense of the rotation by the right-hand rule [we will sometimes write $\mathbf{v} \equiv (\mathbf{n}, \theta)$]. The components of the **v** "vectors" are generalized coordinates for $\mathbb{Q}_R$, as valid as three independent matrix elements. The maximum length of a **v** "vector" is π so all the **v**s lie within a sphere of radius π. These coordinates show that $\mathbb{Q}_R$, the manifold of **SO**(3), is a three-dimensional *ball* (the interior of a sphere plus its surface) of radius π whose diametrically opposed surface points are *identified* (Section 6.2.3), since they represent the same rotation.

The geometry of $\mathbb{Q}_R$ differs from the geometry of an ordinary three-ball, whose opposite sides are not identified. For instance, a straight line extended in any direction from the center of $\mathbb{Q}_R$ arrives again at the center. Another caution is that $\mathbb{Q}_R$ is not a vector space. That is, if $\mathbf{v}_1$ and $\mathbf{v}_2$ represent two rotations, $\mathbf{v}_1 + \mathbf{v}_2$ is in general not in the ball of radius π and represents nothing physical, certainly not the result of the successive rotations. That is why we place quotes on "vector." This is an important distinction from the physical point of view. Rotation is not a commutative operation, whereas vector addition is. The result of two rotations depends on the order in which they are performed (see Problem 3).

8.2.2 THE LAGRANGIAN

We now move on to the Lagrangian formalism. In this section we write down the Lagrangian, using the matrix elements of the R matrices as the generalized coordinates in $\mathbb{Q}_R$.

KINETIC ENERGY

The kinetic energy of the body consists of two parts, that of the center of mass plus the kinetic energy about the center of mass [see Eq. (1.65)]:

$$T = \frac{1}{2}M\dot{r}'^2 + T_{\text{rot}}(\mathbb{Q}_R),$$

where M is the total mass. (The prime is necessary because the kinetic energy must be written in $\mathfrak{I}$.)

The kinetic energy $T_{\text{rot}}(\mathbb{Q}_R)$ about A is given by an integral like the one in (8.3), but now it will be rewritten in the generalized coordinates of $\mathbb{Q}_R$ and their velocities. Let $\hat{\mu}(x')$ be the mass density as a function of the space system coordinates, that is,

$$M = \int \hat{\mu}(x')\, d^3x'.$$

This integral can be taken over all space, but since $\hat{\mu}$ vanishes outside the body, the integral is effectively only over the rigid body itself. The rotational kinetic energy (8.3) is

$$T_{\text{rot}} = \frac{1}{2} \int \hat{\mu}(x') \| \dot{x}' \| \, d^3 x', \tag{8.32}$$

where $\| \dot{x}' \| \equiv (\dot{x}', \dot{x}')$. To put this in terms of the generalized coordinates on $\mathbb{Q}_R$, write $x' = Rx$ and use the fact that $\dot{x} = 0$, so that

$$\dot{x}' = \dot{R}x.$$

Now let $\mu(x) = \hat{\mu}(x') \equiv \hat{\mu}(R^{-1}x)$ (the notation now agrees with the notation of Section 8.1.2) and change the variable of integration in Eq. (8.32) from x' to x (the Jacobian of the transformation is $\det R = 1$). Then (8.32) becomes

$$T_{\text{rot}} = \frac{1}{2} \int \mu(x) \| \dot{R}x \| \, d^3 x = \frac{1}{2} \int \mu(x) \dot{\rho}_{jk} x_k \dot{\rho}_{jl} x_l \, d^3 x$$
$$= \frac{1}{2} \dot{\rho}_{jk} K_{kl} \dot{\rho}_{jl} = \frac{1}{2} \text{tr}(\dot{R} K \dot{R}^{\text{T}}), \tag{8.33}$$

where K is the symmetric matrix whose elements are

$$K_{kl} \equiv K_{lk} = \int \mu(x) x_k x_l \, d^3 x. \tag{8.34}$$

The matrix elements of K are called the *second moments* of the mass distribution. K is a constant matrix which, like the I of Section 8.1, describes some of the inertial and geometric properties of the rigid body and depends only on the way the mass is distributed in it. Eventually K will lead us again to the inertia tensor I.

The time dependence of T_{rot} in Eq. (8.33) is now contained in $\dot{R}$ and $\dot{R}^{\text{T}}$: the rotational part of the kinetic energy is now stated entirely in terms of coordinates on $\mathbb{Q}_R$.

THE CONSTRAINTS

Assume that the potential energy $\hat{V}$ of the body is the sum of a term depending on the center-of-mass position and another depending on R:

$$\hat{V}(r', R) = U(r') + V(R).$$

This excludes many systems, for instance bodies in nonuniform gravitational fields, whose rotational potential energy depends on the location of the body, so $\hat{V}$ is a more general function of r' and R. It does, however, include a rigid body pivoted at a fixed point, no matter what the force field (then the common origin of $\mathfrak{S}$ and $\mathfrak{B}$ is the pivot, and $\mathfrak{S}$ is the same as $\mathfrak{J}$). This assumption will make it possible to uncouple the motion on $\mathbb{R}^3$ from the motion on $\mathbb{Q}_R$.

The Lagrangian of the system is then

$$L(r', \dot{r}', R, \dot{R}) = \frac{1}{2} M \dot{r}'^2 + \frac{1}{2} \text{tr}(\dot{R} K \dot{R}^{\text{T}}) - U(r') - V(R). \tag{8.35}$$

The condition (8.30) or (8.31) is a set of constraint equations on the ρ_{jk} and can be handled by the standard procedure for dealing with constraints in the Lagrangian formalism contained in Eq. (3.19): multiply each constraint equation $\rho_{jk}\rho_{jl} - \delta_{kl} = 0$ by a Lagrange multiplier λ_{kl}, and add the sum over k and l to the Lagrangian. The symmetry (there are only six constraint equations, not nine) is taken care of automatically by making the λ_{kl} symmetric: $\lambda_{kl} = \lambda_{lk}$; the sum over all k and l will add in each constraint equation twice, but both times it will be multiplied by the same Lagrange multiplier and will effectively enter only once. The sum is

$$\lambda_{kl}(\rho_{jk}\rho_{jl} - \delta_{kl}) = \text{tr}(\Lambda R^{\mathsf{T}} R - \Lambda), \tag{8.36}$$

where $\Lambda = \Lambda^{\mathsf{T}}$ is the matrix of the λ_{kl}. Then in accordance with Eq. (3.19) the resulting generalized Lagrangian is

$$\mathcal{L}(r', \dot{r}', R, \dot{R}, \Lambda) = \frac{1}{2}M\dot{r}'^2 + \frac{1}{2}\text{tr}(\dot{R}K\dot{R}^{\mathsf{T}}) - U(r') - V(R) + \text{tr}(\Lambda R^{\mathsf{T}} R - \Lambda). \tag{8.37}$$

If the body is rotating about a stationary pivot, $\mathcal{L}$ has no $(r', \dot{r}')$ dependence.

8.2.3 THE EULER–LAGRANGE EQUATIONS

DERIVATION

Now that the Lagrangian has been obtained, the next step is to find the EL equations. The r' equations simply restate the center-of-mass theorem, and we do not discuss them in any detail:

$$M\dot{r}' = -\nabla' U \equiv F', \tag{8.38}$$

where ∇' is the gradient with respect to r', and F' is the representation of the total external force $\mathbf{F}$ in $\mathfrak{J}$ or $\mathfrak{S}$.

We will obtain the equations on $\mathbb{Q}_R$ in matrix form by treating the R matrices themselves as generalized coordinates. That will require taking derivatives of $\mathcal{L}$ with respect to R, $\dot{R}$, and their transposes, which can be done by first taking the derivatives with respect to their matrix elements. Except for the potential energy, the only terms in which the matrices appear are traces, so what needs to be calculated are expressions like $\partial[\text{tr}(AB)]/\partial\alpha_{jk}$ where A and B are matrices and the α_{jk} are the matrix elements of A. By straightforward calculation,

$$\frac{\partial\,\text{tr}(AB)}{\partial\alpha_{jk}} \equiv \frac{\partial\alpha_{il}\beta_{li}}{\partial\alpha_{jk}} = \delta_{ij}\delta_{lk}\beta_{li} = \beta_{kj},$$

and this can be put in the matrix form

$$\frac{\partial\,\text{tr}(AB)}{\partial A} = B^{\mathsf{T}}. \tag{8.39}$$

Similarly, since $\alpha_{jk}^{\mathrm{T}} = \alpha_{kj}$,

$$\frac{\partial \, \mathrm{tr}(AB)}{\partial A^{\mathrm{T}}} = B. \tag{8.40}$$

These formulas can be used to write down the EL equations on $\mathbb{Q}_R$ in matrix form. For example,

$$\frac{\partial \mathcal{L}}{\partial \dot{R}} = \frac{1}{2}[\dot{R}K + (K\dot{R}^{\mathrm{T}})^{\mathrm{T}}] = \dot{R}K,$$

where the symmetry of K has been used. In this way the EL equation is found to be (Λ is also symmetric)

$$\ddot{R}K + \frac{\partial V}{\partial R} = 2R\Lambda. \tag{8.41}$$

The $\partial V / \partial R$ appearing here is the matrix whose elements are the $\partial V / \partial \rho_{jk}$. For a rigid body pivoted about a fixed point this is the entire equation: (8.38) does not enter. The motion in $\mathbb{Q}_R$ is governed by Eq. (8.41) and the constraint equation (8.30). Because it is a matrix equation, the order of factors needs to be carefully maintained.

The Lagrangian $\mathcal{L}$ involves the Lagrange-multiplier matrix Λ which, like R and r', gives rise to its own EL equation. But there is a way to eliminate Λ and ignore its equation (but see Problem 7). First multiply (8.41) on the left by R^{T}:

$$R^{\mathrm{T}}\ddot{R}K + R^{\mathrm{T}}\frac{\partial V}{\partial R} = 2\Lambda. \tag{8.42}$$

Because the right-hand side of this equation is manifestly symmetric, subtracting its transpose eliminates Λ. The result is

$$R^{\mathrm{T}}\ddot{R}K - (R^{\mathrm{T}}\ddot{R}K)^{\mathrm{T}} \equiv R^{\mathrm{T}}\ddot{R}K - K\ddot{R}^{\mathrm{T}}R = \frac{\partial V}{\partial R^{\mathrm{T}}}R - R^{\mathrm{T}}\frac{\partial V}{\partial R}. \tag{8.43}$$

This is the EL equation, a second-order differential equation for R. Now (8.43) and (8.30) describe the motion in $\mathbb{Q}_R$. Like all EL equations, (8.43) can be interpreted as a first-order equation for the motion on $\mathbf{T}\mathbb{Q}_R$. The dimension of $\mathbf{T}\mathbb{Q}_R$ is twice that of $\mathbb{Q}_R$, so another first-order equation is needed, the usual kind: $dR/dt = \dot{R}$. Just as R is the generic point in $\mathbb{Q}_R = \mathbf{SO}(3)$, so the combination $(R, \dot{R})$ is the generic point in $\mathbf{T}\mathbb{Q}_R = \mathbf{TSO}(3)$, with $\dot{R}$ defining the point in the fiber (Section 2.4) above R.

The Lagrangian dynamical system described by Eqs. (8.30) and (8.43) is very different from the kind we have been studying, which usually took place on $\mathbb{R}^n$ (or on $\mathbf{T}\mathbb{R}^n$). This is due in large part to the complicated geometry of $\mathbb{Q}_R$ and the even more complicated geometry of $\mathbf{T}\mathbb{Q}_R$ (which we do not attempt to describe in any detail). It is different also because $\mathbb{Q}_R$ has the additional algebraic structure of a group: this is a dynamical system on $\mathbf{SO}(3)$.

THE ANGULAR VELOCITY MATRIX Ω

In the next step the constraint condition (8.30) is combined with the EL equation (8.43) to rewrite the equation of motion. First take the derivative of (8.30) with respect to the time:

$$R^T \dot{R} + \dot{R}^T R \equiv \Omega + \Omega^T = 0,$$

where Ω is the antisymmetric matrix defined by

$$\Omega = R^T \dot{R} \quad \text{or} \quad \dot{R} = R\Omega. \tag{8.44}$$

The antisymmetry of Ω is important. It will be shown in Section 8.2.5 that every antisymmetric 3×3 matrix can be put into one-to-one correspondence with a three-vector, and the vector that corresponds to Ω will be seen to be the angular velocity vector ω of Section 8.1. This makes Ω physically significant; we will call it the *angular velocity matrix*.

Equation (8.43) will be written in terms of Ω. In fact Eq. (8.44) shows that Ω and $\dot{R}$ determine each other uniquely for a given R, so Ω can be used instead of $\dot{R}$ to specify the point on the fiber above R. Thus the generic point of $\mathbf{TQ}_R$ can be denoted (R, Ω) as well as $(R, \dot{R})$ and that is what makes it possible to write (8.43) in terms of Ω. The kinetic energy given in (8.33) can also be written in terms of Ω rather than $\dot{R}$: it is found by using $\text{tr}(AB) = \text{tr}(BA)$ and the orthogonality of R that (see Problem 6)

$$T_{\text{rot}} = \frac{1}{2}\text{tr}(\Omega K \Omega^T). \tag{8.45}$$

As the equation of motion involves $\ddot{R}$, its rewritten form will involve $\dot{\Omega}$. The relation between $\dot{\Omega}$ and $\ddot{R}$ is obtained by differentiating Eq. (8.44) with respect to t. The t derivative of the second of Eqs. (8.44), multiplied on the left by R^T, is

$$R^T \ddot{R} = R^T(\dot{R}\Omega + R\dot{\Omega}) = \Omega^2 + \dot{\Omega},$$

whose transpose is

$$\ddot{R}^T R = (\Omega^2)^T + \dot{\Omega}^T = \Omega^2 - \dot{\Omega}.$$

These equations can be inserted into Eq. (8.43) to yield

$$K\dot{\Omega} + \dot{\Omega}K + \Omega^2 K - K\Omega^2 = G, \tag{8.46}$$

where G is the antisymmetric matrix defined by

$$G = -G^T = \left(\frac{\partial V}{\partial R^T}\right)R - R^T\left(\frac{\partial V}{\partial R}\right). \tag{8.47}$$

Now Eq. (8.46) is the EL equation for the rigid body. The matrix G, since it depends on V, reflects the forces applied to the body. We will show in Section 8.2.5 that (8.46) is a matrix form of Euler's equations for the motion of a rigid body; thus we have derived those equations by the Lagrangian recipe.

The equation is of first order in (R, Ω) and, like all EL equations, it requires another first-order equation, the analog of $\dot{R} = dR/dt$. Now the other equation is (8.44) in the form $\Omega = R^T dR/dt$. A solution $(R(t), \Omega(t))$ of these equations with initial values $R(0)$ and $\Omega(0)$ represents a complete solution of the dynamical system, for when $R(t)$ is known, the orientation of the body is known as a function of t.

8.2.4 THE HAMILTONIAN FORMALISM

We deal briefly only with the rotational part of Eq. (8.35), namely

$$L(R, \dot{R}) = \frac{1}{2}\text{tr}(\dot{R}K\dot{R}^T) - V(R). \qquad (8.48)$$

To obtain the Hamiltonian through the Legendre transform, first take the derivative of L with respect to $\dot{R}$ to obtain the *momentum matrix*:

$$P \equiv \frac{\partial L}{\partial \dot{R}} = \frac{1}{2}(K\dot{R}^T)^T + \frac{1}{2}\dot{R}K = \dot{R}K. \qquad (8.49)$$

The solution $\dot{R} = PK^{-1}$ is now inserted into

$$H \equiv p_{jk}\dot{\rho}_{jk} - L \equiv \text{tr}(P\dot{R}^T) - L,$$

where the p_{jk} are the matrix elements of P, to yield (use symmetry of K^{-1})

$$H = \frac{1}{2}\text{tr}(PK^{-1}P^T) + V(R). \qquad (8.50)$$

Note how similar this is to the usual expression for a Hamiltonian.

This procedure is valid only if K has an inverse. To establish that it does, we repeat its definition of Eq. (8.34):

$$K_{jk} = \int \mu(x)\, x_j x_k \, d^3x.$$

The mass density μ is always positive. If the body is truly three dimensional (i.e., not an infinitesimally thin plane or rod), the diagonal matrix elements are all positive definite, for they are integrals over squares of variables. Because K is symmetric, it can be diagonalized in an orthogonal system, one in which $\det K$ is the product of the diagonal elements. Thus $\det K$ is positive definite, and hence K^{-1} exists.

The next step is to write Hamilton's canonical equations. They are easily found to be

$$\dot{R} = PK^{-1}, \quad \dot{P} = -\frac{\partial V}{\partial R}. \qquad (8.51)$$

The first of these merely repeats the definition of P at Eq. (8.49). The second defines the dynamics.

As for the Lagrangian formalism, we will show that the second of Eqs. (8.51) is equivalent to Euler's equations.

8.2.5 EQUIVALENCE TO EULER'S EQUATIONS

We now show that the EL equations (8.46) and Hamilton's canonical equations (8.51) are equivalent to Euler's equations (8.14). This will be done by first establishing a correspondence between antisymmetric 3×3 matrices and 3-vectors.

ANTISYMMETRIC MATRIX–VECTOR CORRESPONDENCE

An antisymmetric matrix B in three dimensions has only three independent matrix elements:

$$B \equiv \begin{bmatrix} 0 & B_{12} & B_{13} \\ B_{21} & 0 & B_{23} \\ B_{31} & B_{32} & 0 \end{bmatrix} = \begin{bmatrix} 0 & -b_3 & b_2 \\ b_3 & 0 & -b_1 \\ -b_2 & b_1 & 0 \end{bmatrix}.$$

Hence there is a one-to-one correspondence between column three-vectors $b = (b_1, b_2, b_3)$ and antisymmetric 3×3 matrices B. We write

$$B \leftrightarrow b \quad \text{or} \quad B_{ij} = -\epsilon_{ijk} b_k \quad \text{or} \quad b_i = -\frac{1}{2} \epsilon_{ijk} B_{jk}. \qquad (8.52)$$

Since G and Ω of Eq. (8.46) are antisymmetric, this correspondence can be used to define two column vectors g and ω by

$$G \leftrightarrow g \quad \text{and} \quad \Omega \leftrightarrow \omega. \qquad (8.53)$$

Both sides of Eq. (8.46) are themselves antisymmetric matrices, so they also correspond to column vectors. The first term on the left-hand side of (8.46) yields

$$-\frac{1}{2} \epsilon_{ijk} (K\dot{\Omega})_{jk} = -\frac{1}{2} \epsilon_{ijk} K_{jh} \dot{\Omega}_{hk}$$

$$= \frac{1}{2} \epsilon_{ijk} \epsilon_{hkl} \dot{\omega}_l K_{jh}$$

$$= \frac{1}{2} (\delta_{il} \delta_{jh} - \delta_{ih} \delta_{lj}) \dot{\omega}_l K_{jh}$$

$$= \frac{1}{2} (K_{jj} \delta_{il} - K_{li}) \dot{\omega}_l$$

$$\equiv \frac{1}{2} I_{il} \dot{\omega}_l = \frac{1}{2} (I\dot{\omega})_i,$$

where

$$I \equiv \mathbb{I}(\operatorname{tr} K) - K \qquad (8.54)$$

is the inertia matrix of Eq. (8.5). Moving around some indices shows that the second term on the left-hand side of (8.46) is the same as the first. In fact the two together form the antisymmetric part of $K\Omega$:

$$K\dot{\Omega} + \dot{\Omega}K \leftrightarrow I\dot{\omega}.$$

Similar calculations for the other terms yield

$$\Omega^2 K - K\Omega^2 \leftrightarrow -\omega \wedge K\omega.$$

Because $\omega \wedge \omega = 0$, any multiple of the unit matrix $\mathbb{I}$ can be added to K in the last expression, and therefore it is equally true that

$$\Omega^2 K - K\Omega^2 \leftrightarrow -\omega \wedge (K - \mathbb{I}\operatorname{tr}K)\omega = \omega \wedge (I\omega).$$

With these results, the vector form of (8.46) becomes

$$g = \omega \wedge I\omega + I\dot{\omega}. \tag{8.55}$$

This is the same as (8.14) provided that it can be shown that $g = N_{\mathfrak{B}}$ and that ω is the angular velocity vector.

THE TORQUE

That $N_{\mathfrak{B}} = g$ can be seen by calculating the components of both vectors. To avoid confusion of indices we drop the subscript $\mathfrak{B}$ on N.

First we tackle g. According to Eq. (8.52) and the definition (8.47) of g,

$$g_k = -\frac{1}{2}\epsilon_{ijk}\left[\frac{\partial V}{\partial \rho_{hi}}\rho_{hj} - \rho_{hi}\frac{\partial V}{\partial \rho_{hj}}\right] = -\epsilon_{ijk}\frac{\partial V}{\partial \rho_{hi}}\rho_{hj}. \tag{8.56}$$

Now we deal with N. This is a problem in physical understanding, because N_k is the kth component of the torque produced by V: how is the torque related to the potential that produces it? N_k is the negative rate of change of V as the body is rotated about the k axis. That is, $N_k = -\partial V/\partial\theta_k$, where θ_k is the angle of this rotation. The body starts in some initial orientation given by R_0 and is rotated further about the k axis through an angle θ_k. Its new orientation is given by

$$R = R_0 S(\theta_k),$$

where $S(\theta_k)$ is the rotation about the k axis that brings it back to the orientation given by R_0. What must be calculated is

$$N_k = -\frac{\partial V}{\partial\theta_k} = -\frac{\partial V}{\partial R}\frac{\partial R}{\partial\theta_k} = -\frac{\partial V}{\partial\rho_{hi}}\frac{\partial\rho_{hi}}{\partial\theta_k};$$

the ρ_{jk} are the matrix elements of R. The derivative of R is

$$\frac{\partial R}{\partial \theta_k} = R_0 \frac{\partial S}{\partial \theta_k} = R_0 S S^T \frac{\partial S}{\partial \theta_k} = R S^T \frac{\partial S}{\partial \theta_k} \equiv R T_k,$$

where $T_k = S^T (\partial S / \partial \theta_k)$ is a different matrix for each k. For instance, for $k = 3$

$$S(\theta_3) = \begin{bmatrix} \cos \theta_3 & \sin \theta_3 & 0 \\ -\sin \theta_3 & \cos \theta_3 & 0 \\ 0 & 0 & 1 \end{bmatrix},$$

and straightforward calculation yields

$$T_3 = \begin{bmatrix} 0 & 1 & 0 \\ -1 & 0 & 0 \\ 0 & 0 & 0 \end{bmatrix},$$

or $(T_3)_{ij} = \epsilon_{ij3}$. Similar calculations for rotations about the other two axes lead to

$$(T_k)_{ij} = -\epsilon_{ijk}.$$

Therefore

$$\frac{\partial \rho_{hi}}{\partial \theta_k} = \rho_{hj} (T_k)_{ji}$$

and

$$N_k = -\frac{\partial V}{\partial \rho_{hi}} \frac{\partial \rho_{hi}}{\partial \theta_k} = -\frac{\partial V}{\partial \rho_{hi}} \rho_{hj} (T_k)_{ji} = -\frac{\partial V}{\partial \rho_{hi}} \rho_{hj} \epsilon_{ijk}.$$

Comparison with Eq. (8.56) shows that $N_k = g_k$. Hence Eq. (8.55) reads

$$I \dot{\omega} + \omega \wedge I \omega = N. \tag{8.57}$$

THE ANGULAR VELOCITY PSEUDOVECTOR AND KINEMATICS

Equation (8.57) looks like (8.14), but it remains to be shown that the ω that appears here is the same as the one of Section 8.1. This ω is obtained from Ω through Eq. (8.52). Both ω and Ω are representations: ω is a column vector representing a vector $\boldsymbol{\omega}$, and R and $\Omega \equiv R^T \dot{R}$ are matrices representing operators $\mathbf{R}$ and $\boldsymbol{\Omega} \equiv \mathbf{R}^T \dot{\mathbf{R}}$. In a new coordinate system Ω is represented by a new matrix Ω', and this new matrix can be used in (8.52) to define a new column vector ω'. The question is whether ω' is the representation in the new coordinate system of the same $\boldsymbol{\omega}$. If it is not, the vector represented by ω in (8.57) is frame dependent and therefore cannot be the angular velocity vector.

We consider only coordinate systems related by orthogonal transformations. If W is the orthogonal transformation matrix from one such system to another, the two matrices representing Ω are related by

$$\Omega' = W\Omega W^{\mathrm{T}}.$$

Then the vector ω exists, if

$$\omega' = W\omega. \tag{8.58}$$

It is shown in Problem 13, however, that

$$\omega' = (\det W)\, W\omega. \tag{8.59}$$

Equations (8.58) and (8.59) differ, so the ω' defined by (8.52) is frame dependent. Because W is orthogonal, however, $\det W = \pm 1$. Therefore (8.58) and (8.59) differ at most in sign, and then only for orthogonal matrices with negative determinant – those that involve reflection. Provided we stick to right-handed coordinate systems W is *special* orthogonal, $\det W$ is always $+1$, and (8.58) and (8.59) are identical.

Technically, a mathematical object that transforms according to (8.59) is called a *pseudovector*, but we will continue to call ω a vector.

TRANSFORMATIONS OF VELOCITIES

It is now seen that ω is a vector, but it must still be shown that it is the angular velocity vector of Section 8.1. To do this, calculate the time derivative of a vector in $\mathfrak{S}$ in terms of its time derivative in $\mathfrak{B}$, rederiving Eq. (8.11). Let **s** be a vector with representations s and s' in the body and space systems, respectively (primes for the space system and no primes the body system). We have

$$\dot{s}' \equiv \frac{d}{dt}s' = \frac{d}{dt}(Rs) = \dot{R}s + R\dot{s} = \dot{R}R^{\mathrm{T}}s' + R\dot{s}; \tag{8.60}$$

$\dot{s}'$ is the time derivative of the $\mathfrak{S}$ representation s' of **s**, and $\dot{s}$ of its $\mathfrak{B}$ representation s. The fact that $\dot{s}' \neq R\dot{s}$ implies that $\dot{s}'$ and $\dot{s}$ are not the $\mathfrak{S}$ and $\mathfrak{B}$ representations of some vector $\dot{\mathbf{s}}$: there is no unique $\dot{\mathbf{s}}$. When coordinate systems are, like $\mathfrak{S}$ and $\mathfrak{B}$, in relative motion the time derivatives of vectors have to be treated differently from vectors.

This is done by turning to two time-derivative vectors associated with **s**. The first is $\dot{\mathbf{s}}_{(\mathfrak{B})}$, *defined* as the vector whose representation in $\mathfrak{B}$ is $\dot{s}$. According to Eq. (8.60), however, $\dot{s}'$ is not the representation $\dot{s}'_{(\mathfrak{B})}$ of $\dot{\mathbf{s}}_{(\mathfrak{B})}$ in $\mathfrak{S}$. That is given in the usual way by

$$\dot{s}'_{(\mathfrak{B})} = R\dot{s},$$

for $\dot{\mathbf{s}}_{(\mathfrak{B})}$ is *by definition* a vector. The subscript $(\mathfrak{B})$ names the vector, not its representation, and primes (or their absence) denote representations. Observe that $\dot{s}'_{(\mathfrak{B})}$ is the second term on the right-hand side of (8.60).

Now turn to the first term on the right-hand side of (8.60). The matrix appearing there is

$$\dot{R}R^{\mathrm{T}} = R(R^{\mathrm{T}}\dot{R})R^{\mathrm{T}} = R\Omega R^{\mathrm{T}} \equiv \Omega',$$

the representation of Ω in $\mathfrak{S}$. The matrix Ω' is associated with the column (pseudo)vector ω' that represents ω in $\mathfrak{S}$, and the ith component of the expression in question is

$$(\Omega' s')_i = \Omega'_{ij} s'_j = -\epsilon_{ijk} \omega'_k s'_j = (\omega' \wedge s')_i.$$

When these expressions are inserted into (8.60) it becomes

$$\dot{s}' = \omega' \wedge s' + \dot{s}'_{(\mathfrak{B})}. \tag{8.61}$$

Comparing this with Eq. (8.11) Section 8.1 now shows that ω' is the $\mathfrak{S}$ representation of the angular momentum vector ω of Section 8.1. It thereby completes the proof of equivalence of the vector and matrix forms of Euler's equations for the motion of a rigid body. In a sense this is a rederivation of (8.11).

The right-hand side of (8.61) is the $\mathfrak{S}$ representation of the vector $\omega \wedge s + \dot{s}_{(\mathfrak{B})}$. The equation can be written in $\mathfrak{B}$, or for that matter in any other coordinate system, if the left-hand side is used to define a vector. Therefore we define a second vector $\dot{s}_{(\mathfrak{S})}$, whose definition in the space system parallels the definition of $\dot{s}_{(\mathfrak{B})}$ in the body system: $\dot{s}_{(\mathfrak{S})}$ is *defined* as the vector whose representation in $\mathfrak{S}$ is $\dot{s}'$. Then (8.61) implies that

$$\dot{s}_{(\mathfrak{S})} = \omega \wedge s + \dot{s}_{(\mathfrak{B})}. \tag{8.62}$$

This is now a vector equation; it can be represented in any coordinate system. Equation (8.62) holds for any vector s. It implies in particular that $\dot{\omega}_{(\mathfrak{S})} = \dot{\omega}_{(\mathfrak{B})}$, a result obtained already in Section 8.1.

The derivative that is of dynamical interest is $\dot{s}_{(\mathfrak{S})}$, for it is the time derivative of s in an inertial system.

HAMILTON'S CANONICAL EQUATIONS

We now demonstrate that Hamilton's canonical equations (8.51) are also equivalent to Euler's equations, but in the form $\dot{J}' = N'$, essentially the $\mathfrak{S}$ representation of Eq. (8.9). Since (8.51) is a matrix equation, this will be done by finding the matrix Υ that corresponds to J' through Eq. (8.52). Then the canonical equations will be written in terms of Υ, which can be translated to equations for J'. Equations (8.51) are in terms of the momentum matrix P of (8.49), so the first step is to establish the relation between J' and P. That relation follows directly from the definition of $\mathbf{J}$ in Section 8.1, which in $\mathfrak{S}$ is

$$\begin{aligned}
J'_k &= \int \hat{\mu}(x') \epsilon_{klm} x'_l \dot{x}'_m \, d^3 x' \\
&= \int \mu(x) \epsilon_{klm} R_{lr} \dot{R}_{ms} x_r x_s \, d^3 x \\
&= \epsilon_{klm} R_{lr} K_{rs} \dot{R}_{ms} = \epsilon_{klm} R_{lr} (\dot{R} K)_{mr} \\
&= \epsilon_{klm} R_{lr} P_{mr} = \epsilon_{klm} (P R^{\mathrm{T}})_{ml}.
\end{aligned} \tag{8.63}$$

The next step is to relate P to Υ. That is left to Problem 14, where it is shown that $\Upsilon = (P R^{\mathrm{T}} - R P^{\mathrm{T}})$, which is twice the antisymmetric part of $P R^{\mathrm{T}}$. [Every matrix M

can be written as the sum of its *symmetric* and *antisymmetric parts* $M = A + S$, where $A = -A^T = \frac{1}{2}(M - M^T) \equiv \mathfrak{A}(M)$ and $S = S^T = \frac{1}{2}(M + M^T)$.] Thus $\Upsilon = 2\mathfrak{A}(PR^T)$ and

$$\dot{\Upsilon} = 2\mathfrak{A}(\dot{P}R^T + P\dot{R}^T). \tag{8.64}$$

Now the canonical equations (8.51) are inserted for $\dot{P}$. After multiplying on the left by R^T and on the right by R, (8.64) becomes

$$R^T\dot{\Upsilon}R = 2\mathfrak{A}(-R^TV' + \Omega K\Omega^T),$$

where $V' = \partial V/\partial R$. The first term here is just G of Eq. (8.47). The second term is symmetric, so its antisymmetric part vanishes. Hence

$$R^T\dot{\Upsilon}R = G. \tag{8.65}$$

This is (8.51) rewritten in terms of Υ.

The final step is to find the vector form of (8.65). By the same reasoning that was used to establish that ω is a pseudovector, it follows from $\Upsilon \leftrightarrow J'$ that $R^T\dot{\Upsilon}R \leftrightarrow R^T\dot{j}'$. It has been shown that $G \leftrightarrow N$. Thus the vector form of (8.65) is $R^T\dot{j}' = N$ or

$$\dot{j}' = RN \equiv N'. \tag{8.66}$$

That completes the demonstration.

8.2.6 DISCUSSION

Much of this section has been quite formal. It had several objectives. The first was to exhibit the configuration manifold $\mathbf{SO}(3)$ of the rigid body. The complicated geometry and group structure of $\mathbf{SO}(3)$ was an omen that the motion would be difficult to describe. Another object was to show that in spite of these complications, the equations of motion can be obtained by the usual formalisms. Yet another was to analyze how the time derivatives of vectors transform between moving frames.

Although such details may now be understood, one should not be lulled into thinking that from here on the road is an easy one. The matrix EL equations we have obtained are difficult to handle and essentially never used to calculate actual motion.

The Hamiltonian formalism, probably even more complicated, has been studied in some detail (Trofimov and Fomenko, 1984, and references therein), in particular Liouville integrability. It involves the theory of Lie groups (more correctly, of Lie algebras) and their actions on manifolds. For such reasons we have said nothing about the cotangent bundle $\mathbf{T}^*\mathbf{SO}(3)$ and have not tried to define a Poisson bracket in terms of matrices. In fact the structure of $\mathbf{SO}(3)$ and especially of $\mathbf{T}^*\mathbf{SO}(3)$ makes it pointless to try to use canonical coordinates of the (q, p) type. For example, one may think that the three components of the angular momentum would serve as ideal momentum variables and try writing the Hamiltonian of the free (zero-torque) rigid rotator in the principal-axis body system in the form

$$H = \frac{1}{2}\left(\frac{J_1^2}{I_1} + \frac{J_2^2}{I_2} + \frac{J_3^2}{I_3}\right),$$

for this is the total energy. But $\{J_i, J_j\} = \epsilon_{ijk}J_k \neq 0$, so the J_k cannot serve as canonical momenta. Nevertheless, this H can be used as the Hamiltonian, but one must bear in mind that it is not written in terms of canonical coordinates. We will not discuss these matters any further but will return essentially to the vector form of Euler's equations, applying them to specific physical systems. This will be done in terms of a standard set of coordinates on $\mathbf{SO}(3)$, called Euler's angles.

8.3 EULER ANGLES AND SPINNING TOPS

8.3.1 EULER ANGLES

Since $\dim \mathbb{Q}_R = 3$ there should be a way to specify a rotation R from $\mathfrak{B}$ to $\mathfrak{S}$ by naming just three rotations through angles θ_1, θ_2, and θ_3 about specified axes in a specified order. For example, the three rotations could be about the 1, 2, and 3 axes of $\mathfrak{S}$ in that order. A way to calculate $(\theta_1, \theta_2, \theta_3)$ from a knowledge of R, and vice versa, would then have to be found. Such systems are used in practice to define rotations: the heading, pitch, and banking angle of an airplane are given by successive rotations about the vertical, transverse, and longitudinal axes of the aircraft (in $\mathfrak{B}$). As in all such systems, the order matters, but if the angles are small enough the result is nearly independent of order. Why this is so will be explained later.

DEFINITION
One such system, of great usefulness in dealing with rigid body motion, consists of the *Euler angles*. We define them on the example of a solid sphere whose center is fixed at the common origin of $\mathfrak{S}$ and $\mathfrak{B}$ (Fig. 8.10). Arrange the sphere so that $\mathfrak{S}$ and $\mathfrak{B}$ coincide initially, and mark the point on the surface where it is pierced by the body 3-axis; call this point the pole. At the pole draw a short arrow pointing to the place where the negative 2-axis pierces the surface, as shown in the figure. The final orientation of the sphere can be specified by giving the final position of the pole and the final direction of the arrow. This is done by rotating the sphere in three steps from the initial orientation of Fig. 8.10(a) to the final one of Fig. 8.10(d). First, rotate through an angle ϕ about the pole (the common 3-axis) until the arrow points to the final position of the pole, as in Fig. 8.10(b); ϕ is one of the Euler angles. Second, rotate about the body 1-axis through an angle θ, moving the pole in the direction of the arrow to its final position, as in Fig. 8.10(c); θ is another Euler angle. Third, rotate about the body 3-axis through an angle ψ until the arrow is pointing in its final direction, as in Fig. 8.10(d); ψ is the third Euler angle. The final orientation of $\mathfrak{B}$ and $\mathfrak{S}$, with the Euler angles indicated, is shown in Fig. 8.11.

It is clear that ϕ, θ, ψ specify R uniquely. Conversely, R specifies ϕ, θ, ψ by the reverse procedure that rotates the sphere back from its final to its initial orientation. It is

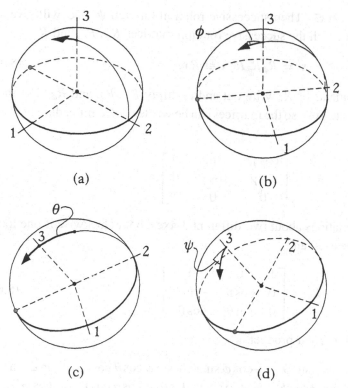

FIGURE 8.10

The Euler angles. In all these pictures the arrow points at the gray dot located on the negative 2 axis. The 1 axis in (b) and (c) lies along the *line of nodes*.

this reverse procedure that yields the transformation R from $\mathfrak{B}$ (the dotted axes in Fig. 8.11) to the fixed $\mathfrak{S}$ (the solid-line axes in Fig. 8.11). The one-to-one relation between R and the Euler angles means that ϕ, θ, ψ form a set of generalized coordinates on $\mathbb{Q}_R$. Most important, since there are just three of them, the rigid-body constraints are inherently included in the Euler angles; no additional constraints are needed. We return to this point when we return to the Lagrangian and the Euler–Lagrange equations.

R IN TERMS OF THE EULER ANGLES

If the Euler angles determine R, they also determine its matrix elements. To write the ρ_{jk} in terms of ϕ, θ, ψ, rotate an arbitrary position vector $\mathbf{x}$ in the rigid body, with

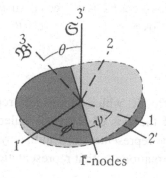

FIGURE 8.11

The relation between the body and space systems. The axes of the space system $\mathfrak{S}$ are solid lines and are labeled with primes. The $(1', 2')$ plane of the space system has a solid border. The axes of the body system $\mathfrak{B}$ are dashed lines and are labeled without primes. The $(1, 2)$ plane of the body system has a dashed border. The intermediate position of the 1 axis, called the line of nodes (see Fig. 8.10), is labeled $1''$-nodes. The space system is that of Fig. 8.10(a) and the body system is that of Fig. 8.10(d).

components x in $\mathfrak{B}$ and x' in $\mathfrak{S}$. Then successive rotation through ψ, θ, ϕ will give the transformation from x to x'. Call the successive rotation matrices R_ψ, R_θ, and R_ϕ:

$$x' = R_\phi R_\theta R_\psi x \equiv Rx. \tag{8.67}$$

To find the matrix elements of R, write out and multiply R_ψ, R_θ, and R_ϕ. The three rotations are about coordinate axes, so the matrices can be written down immediately. They are

$$R_\gamma = \begin{bmatrix} \cos\gamma & -\sin\gamma & 0 \\ \sin\gamma & \cos\gamma & 0 \\ 0 & 0 & 1 \end{bmatrix} \tag{8.68}$$

for $\gamma = \phi, \psi$ (these are rotations about two different 3-axes, both clockwise if one looks toward the origin along the axis) and

$$R_\theta = \begin{bmatrix} 1 & 0 & 0 \\ 0 & \cos\theta & -\sin\theta \\ 0 & \sin\theta & \cos\theta \end{bmatrix} \tag{8.69}$$

(a rotation about the 1-axis). Their product is

$$R = \begin{bmatrix} \cos\phi\cos\psi - \sin\phi\cos\theta\sin\psi & -\cos\phi\sin\psi - \sin\phi\cos\theta\cos\psi & \sin\phi\sin\theta \\ \sin\phi\cos\psi + \cos\phi\cos\theta\sin\psi & -\sin\phi\sin\psi + \cos\phi\cos\theta\cos\psi & -\cos\phi\sin\theta \\ \sin\theta\sin\psi & \sin\theta\cos\psi & \cos\theta \end{bmatrix}. \tag{8.70}$$

This expression can be used not only to find the ρ_{jk} from the Euler angles, but also to find the Euler angles from the ρ_{jk}. Indeed, if R is any rotation matrix, $\theta = \cos^{-1}\rho_{33}$, and ϕ and ψ (actually their sines and cosines) can be obtained from elements in the last row and column. Thus all of the elements of R are determined (up to sign) by ρ_{23}, ρ_{32}, and ρ_{33} (Problem 5), reflecting the three-dimensionality of $\mathbb{Q}_R$.

The inverse of R gives the transformation from x' to x, that is, from $\mathfrak{S}$ to $\mathfrak{B}$. It can be obtained from R by interchanging ϕ and ψ and changing the signs of all the angles:

$$R^{-1}$$
$$= \begin{bmatrix} \cos\phi\cos\psi - \sin\phi\cos\theta\sin\psi & \cos\psi\sin\phi + \sin\psi\cos\theta\cos\phi & \sin\psi\sin\theta \\ -\sin\psi\cos\phi - \cos\psi\cos\theta\sin\phi & -\sin\phi\sin\psi + \cos\phi\cos\theta\cos\psi & \cos\psi\sin\theta \\ \sin\theta\sin\phi & -\sin\theta\cos\phi & \cos\theta \end{bmatrix}. \tag{8.71}$$

As a check, R^{-1} is R^{T}.

The matrix R is the representation of a rotation operator $\mathbf{R}$, which can also be represented in any other (orthogonal) coordinate system. If the axis of $\mathbf{R}$ (recall Euler's theorem) is chosen as the 3-axis of a coordinate system, the representation R in that system takes the form of Eq. (8.68), where γ is the angle of rotation. In that representation

$\operatorname{tr} R = 1 + 2\cos\gamma$, and because the trace of a matrix is independent of coordinates (see appendix), that is, because $\operatorname{tr} R = \operatorname{tr} \mathbf{R}$, this equation holds in any system. Hence the angle of rotation γ can always be found in any coordinate system from the equation

$$\cos\gamma = \frac{\operatorname{tr} R - 1}{2}. \tag{8.72}$$

ANGULAR VELOCITIES

Rotations through finite angles are not commutative (Problem 3), but angular velocities (and hence infinitesimal rotations) are. More specifically, they can be added as vectors (i.e., angular-velocity pseudovectors form a vector space), so the angular velocity ω can be written as the sum of the Euler angular velocities $\dot{\theta} + \dot{\phi} + \dot{\psi}$.

To be precise, let a time-dependent rotation $R(t)$ from $\mathfrak{B}$ to $\mathfrak{S}$ be made up of two component rotations, each with its own time dependence:

$$R(t) = R_c(t)R_b(t). \tag{8.73}$$

All three rotations in this equation have angular velocities associated with them. For instance the angular velocity matrix associated with R_c is $\Omega_c = R_c^T \dot{R}_c$, and the corresponding angular velocity vector ω_c can be obtained from Ω_c. That the angular velocity vectors form a vector space means that

$$\omega = \omega_c + \omega_b, \tag{8.74}$$

where ω is the angular velocity vector associated with $R(t)$. Commutativity of angular velocities then follows immediately from commutativity of vector addition.

The proof of (8.74) depends on keeping careful track of which coordinate system is being used to represent the ω vectors and Ω operators. Suppose that R is a rotation from $\mathfrak{B}$ to $\mathfrak{S}$. Then R_b transforms part of the way, to an intermediate system that we label $\mathfrak{C}$, and R_c transforms the rest of the way, from $\mathfrak{C}$ to $\mathfrak{S}$. According to (8.44),

$$\Omega \equiv R^T \dot{R} = R_b^T R_c^T \left(\dot{R}_c R_b + R_c \dot{R}_b \right) = R_b^T \Omega_c R_b + \Omega_b. \tag{8.75}$$

The definition $\Omega = R^T \dot{R}$ always gives the representation of Ω in the initial coordinate system, the one that R maps *from*, and up to now that has always been $\mathfrak{B}$. In Eq. (8.75) Ω and Ω_b are again the representation of operators Ω and Ω_b in $\mathfrak{B}$, for both R and R_b are transformations from $\mathfrak{B}$ to other systems. However, R_c is a transformation from $\mathfrak{C}$, so Ω_c is the representation of Ω_c in $\mathfrak{C}$. To represent Ω_c in $\mathfrak{B}$, its $\mathfrak{C}$ representation must be transformed back to $\mathfrak{B}$ by applying the inverse of R_b, namely R_b^T. That is already done by the first term on the right-hand side of (8.75), which is therefore the $\mathfrak{B}$ representation of the operator equation

$$\Omega = \Omega_b + \Omega_c$$

(addition of operators is commutative). But each Ω operator corresponds to its angular velocity (pseudo-)vector ω, so this equation leads directly to (8.74).

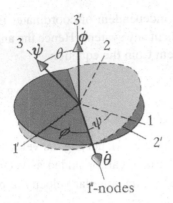

FIGURE 8.12
Same as Fig. 8.11, but with the angular velocity vectors $\dot{\phi}$, $\dot{\theta}$, $\dot{\psi}$ drawn in. Each angular velocity vector is along the axis of its corresponding rotation.

This analysis can be applied immediately to the equation $R = R_\phi R_\theta R_\psi$. It follows from (8.75) that

$$\Omega = R_\psi^{\mathrm{T}} R_\theta^{\mathrm{T}} \Omega_\phi R_\theta R_\psi + R_\psi^{\mathrm{T}} \Omega_\theta R_\psi + \Omega_\psi$$

and hence that

$$\omega = \dot{\theta} + \dot{\phi} + \dot{\psi}, \tag{8.76}$$

where the three angular velocities on the right-hand side are the angular velocities of the rotations through θ, ϕ, and ψ, respectively. Each of these angular velocity vectors is directed along its instantaneous axis of rotation. For example, R_ϕ is a rotation about the original body 3-axis, so $\dot{\phi}$ is directed along the body 3-axis, as illustrated in Fig. 8.12. The directions of $\dot{\phi}$ and $\dot{\psi}$, also shown in the figure, are determined similarly.

Equation (8.76) can be represented in any coordinate system. The components of $\dot{\theta}$, $\dot{\phi}$, and $\dot{\psi}$ in either $\mathfrak{B}$ or $\mathfrak{S}$ can be read off Fig. 8.12 and used with (8.76) to construct the components of ω in those systems. For example, the 1 and 2 components of the vector $\dot{\phi}$ are obtained by first multiplying its magnitude $\dot{\phi}$ by $\sin\theta$ to project it onto the $(1, 2)$ plane and then multiplying the projection by $\sin\psi$ and $\cos\psi$ to obtain the 1 and 2 components: $\dot{\phi}_1 = \dot{\phi}\sin\theta\sin\psi$ and $\dot{\phi}_2 = \dot{\phi}\sin\theta\cos\psi$. When this is done for $\dot{\theta}$ and $\dot{\psi}$, the body-system components of ω are found to be

$$\left.\begin{array}{l} \omega_1 = \dot{\phi}\sin\theta\sin\psi + \dot{\theta}\cos\psi, \\ \omega_2 = \dot{\phi}\sin\theta\cos\psi - \dot{\theta}\sin\psi, \\ \omega_3 = \dot{\phi}\cos\theta + \dot{\psi}. \end{array}\right\} \tag{8.77}$$

Similarly, the space-system components of ω are

$$\left.\begin{array}{l} \omega_1' = \dot{\psi}\sin\theta\sin\phi + \dot{\theta}\cos\phi, \\ \omega_2' = -\dot{\psi}\sin\theta\cos\phi + \dot{\theta}\sin\phi, \\ \omega_3' = \dot{\psi}\cos\theta + \dot{\phi}. \end{array}\right\} \tag{8.78}$$

These two representations can be obtained from each other by applying R or R^{T}; they can also be obtained by interchanging ϕ and ψ and making some sign changes, which reflects a sort of mathematical symmetry between $\mathfrak{B}$ and $\mathfrak{S}$.

DISCUSSION

If the body rotates at a constant ω, the angular velocity vector is $\omega = d\rho/dt$, where $\rho(t)$ is a vector along ω whose magnitude represents the total angle of rotation. This can be put in the form $d\rho = \omega\, dt$, where $d\rho$ is an infinitesimal rotation. In these terms Eq. (8.74) and similar ones can be written as $d\rho \equiv \omega\, dt = d\rho_b + d\rho_c$, which can then be interpreted as showing that infinitesimal rotations, unlike finite ones, are commutative. For finite rotations, the smaller their angles (i.e., the more closely they approach infinitesimal ones), the more nearly they commute, which is why the result of successive rotations through very small angles is nearly independent of order.

We have seen that (R, Ω) label the generic point in $\mathbf{T}\mathbb{Q}_R$. So do $(\phi, \theta, \psi; \omega)$, for (ϕ, θ, ψ) is equivalent to R and ω to Ω. But just as Ω is not itself the time derivative of any generalized coordinate on $\mathbb{Q}_R$, neither is ω. That is, there is no set of generalized coordinates $q = \{q_1, q_2, q_3\}$ on $\mathbb{Q}_R$ such that ω (or ω') is dq/dt, and hence the components of the angular velocity are not generalized velocities in the usual sense.

The right-hand sides of Eqs. (8.77) and (8.78) reflect that fact, but they also yield an important practical consequence by relating ω (or rather its representation in $\mathfrak{B}$ or $\mathfrak{S}$) to time derivatives of the Euler angles. Through them Euler's equations for the motion of a rigid body, of first order in ω, become second-order equations in the Euler angles. The Euler angles with their derivatives, namely $(\phi, \theta, \psi, \dot{\phi}, \dot{\theta}, \dot{\psi})$, form the usual kind of coordinates for $\mathbf{T}\mathbb{Q}_R$ and in terms of them Euler's equations are the usual kind of dynamical equations: the Lagrangian can be written in terms of the Euler angles and can then be handled in the usual way to yield a set of second-order differential equations. As has been mentioned, no additional constraint equations are needed. When writing the Lagrangian in terms of the rotation matrices, we had to include the constraints in the $\operatorname{tr}(\Lambda R^T R - \Lambda)$ term. By passing to the Euler angles, the need for such a term is obviated. These are the principal advantage of the Euler angles.

Lagrange multipliers, however, are related to the forces of constraint. Here this means that the elements of the Λ matrix are related to the internal forces of the rigid body. In the Euler-angle approach, since the Λ matrix does not appear, the internal forces cannot be calculated. See Problem 7.

Results similar to the ones we have obtained for the Euler angles can be obtained with other sets of coordinates, for example rotations θ_1, θ_2, and θ_3 about the 1, 2, and 3 axes mentioned at the start of this section.

WORKED EXAMPLE 8.3

We return to the rolling sphere of Worked Example 8.2, a sphere of mass M, radius a, and moment of inertia I rolling on a horizontal plane.

Solution. This time we solve the problem more formally, by using Euler angles invoking the nonholonomic constraints. The constraints in this problem are those that cause rolling, not the ones that cause rigidity mentioned earlier in this section.

As before, take Cartesian coordinates (x_1, x_2) on the plane. Before invoking constraints the Lagrangian is the kinetic energy (in $\mathfrak{S}$)

$$L = T = \frac{1}{2}M\dot{x}'^2 + \frac{1}{2}I\omega'^2,$$

where $x' = \{x'_1, x'_2\}$ represents the center-of-mass position vector **x**. In terms of the Euler angles this equation becomes

$$L = \frac{1}{2}M(\dot{x}_1'^2 + \dot{x}_2'^2) + \frac{1}{2}I(\dot{\phi}^2 + \dot{\psi}^2 + \dot{\theta}^2 + 2\dot{\phi}\dot{\psi}\cos\theta).$$

The rolling constraints are given by Eq. (8.18) of Worked Example 8.2 (there they are written in $\mathcal{B}$) and can also be rewritten in terms of the Euler angles:

$$\left.\begin{aligned} f_1 &\equiv -\dot{\psi}\sin\theta\cos\phi + \dot{\theta}\sin\phi - \dot{x}_1'/a = 0, \\ f_2 &\equiv \dot{\psi}\sin\theta\sin\phi + \dot{\theta}\cos\phi + \dot{x}_2'/a = 0. \end{aligned}\right\} \tag{8.79}$$

These constraints are nonholonomic, so the equations of motion cannot be obtained from Eq. (3.19) and the Lagrangian $\mathcal{L}$, but from Eq. (3.21): they are given by

$$\frac{d}{dt}\frac{\partial L}{\partial \dot{q}^\alpha} - \frac{\partial L}{\partial q^\alpha} = \sum_l \lambda_l \frac{\partial f_l}{\partial \dot{q}^\alpha},$$

where the λ_l are the Lagrange multipliers. Hence the equations of motion are

$$M\ddot{x}_1' = -\frac{\lambda_1}{a}, \tag{8.80a}$$

$$M\ddot{x}_2' = \frac{\lambda_2}{a}, \tag{8.80b}$$

$$I\frac{d}{dt}[\dot{\phi} + \dot{\psi}\cos\theta] \equiv I\dot{\omega}_3' = 0, \tag{8.80c}$$

$$I[\ddot{\theta} + \dot{\phi}\dot{\psi}\sin\theta] = \lambda_1\sin\phi + \lambda_2\cos\phi, \tag{8.80d}$$

$$I\frac{d}{dt}[\dot{\psi} + \dot{\phi}\cos\theta] = \sin\theta\,(-\lambda_1\cos\phi + \lambda_2\sin\phi). \tag{8.80e}$$

These five equations and Eqs. (8.79) are seven equations for seven variables: two Lagrange multipliers plus two Cartesian coordinates plus three Euler angles.

So far two constants of the motion have been obtained: the total energy T and the vertical component ω_3' of the angular velocity. To go further, multiply (8.80d) by $\cos\phi$ and (8.80e) by $\sin\phi/\sin\theta$ and add the results. This yields

$$\lambda_2 = I\{\ddot{\theta}\cos\phi + \dot{\phi}\dot{\psi}\sin\theta\cos\phi + \ddot{\psi}\sin\phi/\sin\theta + \ddot{\phi}\sin\phi\cos\theta/\sin\theta - \dot{\theta}\dot{\phi}\sin\phi\}.$$

But from (8.80c), $\ddot{\phi} = -\ddot{\psi}\cos\theta + \dot{\theta}\dot{\psi}\sin\theta$, so

$$\begin{aligned} \lambda_2 &= I\{\ddot{\theta}\cos\phi - \dot{\theta}\dot{\phi}\sin\phi + \ddot{\psi}\sin\phi(1 - \cos^2\theta)/\sin\theta \\ &\quad + \dot{\phi}\dot{\psi}\sin\theta\cos\phi + \dot{\theta}\dot{\psi}\cos\theta\sin\phi\} \\ &= I\frac{d}{dt}[\dot{\theta}\cos\phi + \dot{\psi}\sin\theta\sin\phi] = I\dot{\omega}_1' = Ma\ddot{x}_2' = -Ma^2\dot{\omega}_1'; \end{aligned}$$

the last two equalities use Eq. (8.80b) and the second of Eqs. (8.79). As in Worked Example 8.2, I, M, and a^2 are all positive, so $\lambda_2 = I\dot{\omega}'_1 = -Ma^2\dot{\omega}'_1$ implies that $\lambda_2 = 0$, which in turn implies that $\dot{\omega}'_1 = 0$ and $\ddot{x}'_2 = 0$. Without performing a similar calculation for λ_1, simply from the fact that there is nothing special about the x'_2 direction, it follows that $\dot{\omega}'_2 = 0$ and $\ddot{x}'_1 = 0$.

Thus we conclude again that the sphere rolls at constant speed in a straight line and that its angular velocity ω is constant.

8.3.2 GEOMETRIC PHASE FOR A RIGID BODY

Associated with the free rotation of a rigid body is a *geometric phase*, which is analogous to the geometric part of the Hannay angle discussed in Section 6.4.3. It is derived from studying a one-form on the phase manifold. Here the derivation does not depend on the adiabatic assumption or on any other approximations. Our treatment is based on a paper by Montgomery (1991).

Although the angular momentum **J** of a free rigid body is fixed in the space frame $\mathfrak{S}$, it is not fixed in the body frame $\mathfrak{B}$, where it moves along one of the curves on the ellipsoid of Fig. 8.7. If it lies on a typical curve, not the separatrix passing through $\mathbf{J}_{\text{int}}$, the body-system angular momentum $J_{\mathfrak{B}}$ returns periodically to its initial position. Suppose that $\mathfrak{B}$ and $\mathfrak{S}$ coincide initially. Then after one period τ of the $J_{\mathfrak{B}}$ motion, $\mathfrak{B}$ and $\mathfrak{S}$ will not coincide in general. That is, when $J_{\mathfrak{B}}$ has made one circuit of a curve on the ellipsoid, the body ends up in a new orientation ΔR. We want to find the rotation ΔR.

In $\mathfrak{S}$ the angular momentum $J_{\mathfrak{S}} \equiv \mathbf{J}$ docs not change. The fact that its representation $J_{\mathfrak{B}}$ is the same at $t = 0$ and $t = \tau$ means that the body moves so that **J** is directed along the same line in the body at both times. Therefore the axis of ΔR is along the fixed vector **J**. All that remains to be found is $\Delta\alpha$, which is the *geometric phase*.

The carrier manifold $\mathbb{M}$ of the system, whether it be $\mathbf{T}\mathbb{Q}_R$, $\mathbf{T}^*\mathbb{Q}_R$, or any other convenient manifold, has dimension six. The state of the system can be specified by naming the three components of $J_{\mathfrak{B}}$ and the three coordinates of R, the rotation that brings $\mathfrak{B}$ into $\mathfrak{S}$. Indeed, R gives the orientation of the body, and $J_{\mathfrak{B}}$ can be used together with R to find the angular momentum $J_{\mathfrak{S}}$ (or **J**) and the kinetic energy T, or for that matter, any other dynamical variable. We will write ξ for the phase point in $\mathbb{M}$: $\xi(t) \equiv [R(t), J_{\mathfrak{B}}(t)]$ (both $J_{\mathfrak{B}}$ and R vary in time). At time $t = 0$ the two frames coincide, so $R = \mathbb{I}$ and $J_{\mathfrak{B}} = J_{\mathfrak{S}}$, or $\xi(0) = [\mathbb{I}, J_{\mathfrak{S}}]$. At time $t = \tau$ the body angular momentum $J_{\mathfrak{B}}$ is again equal to $J_{\mathfrak{S}}$, but in general $R(\tau) \equiv \Delta R \neq \mathbb{I}$, or $\xi(\tau) = [\Delta R, J_{\mathfrak{S}}] \neq \xi(0)$.

In the time from $t = 0$ to τ the system moves along the physical trajectory C_P consisting of the points $\xi(t)$, $0 < t < \tau$. Although after one circuit of C_P the axis of $R(\tau) \equiv \Delta R$ is along **J**, the axis of $R(t)$ varies on C_P. Consider also a second curve C_R in $\mathbb{M}$ swept out if $J_{\mathfrak{B}}$ is held *fixed* at $J_{\mathfrak{S}}$ and the body is rotated along the fixed **J** axis from $\xi(0)$ to $\xi(\tau)$. The curve C_R is not a physical trajectory: along it the body rotates about a fixed axis through an angle α that changes from $\alpha = 0$ to $\alpha = \Delta\alpha$. The geometric phase angle $\Delta\alpha$ will be found by considering both C_P and C_R.

Both curves start at $\xi(0)$ and end at $\xi(\tau)$, and on both of them the body ends up rotated by the angle $\Delta\alpha$. We will show that $\Delta\alpha$ can be calculated by integrating the canonical one-form $\vartheta_0 = p_\alpha \, dq^\alpha$, introduced in Section 5.2, around the closed curve $C = C_P - C_R$ (in Chapter 5 we wrote θ_0 for the canonical one-form, but now we write ϑ_0 to avoid confusion with the Euler angle). The integral turns out to be nonzero, reflecting the fact that ϑ_0 is not exact. The geometric part of the Hannay angle also depended on integrating an inexact one-form (Section 6.4.3) around a closed curve.

In Cartesian coordinates in the space system

$$\vartheta_0 = \sum_j \mathbf{p}_j \cdot d\mathbf{x}_j,$$

where the sum is taken over the particles of the rigid body. The body is rotating with instantaneous angular velocity ω, so $d\mathbf{x}_j = d\omega \wedge \mathbf{x}_j$, and

$$\vartheta_0 = \sum_j \mathbf{p}_j \cdot d\omega \wedge \mathbf{x}_j = \left(\sum_j \mathbf{x}_j \wedge \mathbf{p}_j \right) \cdot d\omega = \mathbf{J} \cdot d\omega.$$

Then the integral to be calculated is

$$W \equiv \oint_C \vartheta_0 = \oint_C \mathbf{J} \cdot d\omega.$$

This integral will be calculated in two ways, first by calculating the integral on C_P and C_R separately and then by using Stokes's theorem around C. These two methods give two expressions for W and hence for $\Delta\alpha$, and a formula for $\Delta\alpha$ is obtained when the two expressions are equated.

First consider C_P. This curve represents the motion, so on it $\mathbf{J} = \mathbf{I}\omega$ and $d\omega = \omega \, dt$. Hence

$$\int_{C_P} \mathbf{J} \cdot d\omega = \int_{C_P} \omega \cdot \mathbf{I}\omega \, dt = \int_{C_P} 2T \, dt = 2T\tau.$$

Next consider C_R. On C_R the rotation is about the fixed $\mathbf{J}$ direction. Parametrize C_R by the angle α, which changes from $\alpha = 0$ to $\alpha = \Delta\alpha$ on the curve. Then $d\omega$ represents the differential rate of rotation of the body with respect to α; hence it is a vector of magnitude $d\omega = d\alpha$ pointing along $\mathbf{J}$: we write $d\omega = \omega \, d\alpha/\omega$ and note that $\mathbf{J} \cdot \omega/\omega = J$ is the (fixed) magnitude of $\mathbf{J}$. Hence

$$\int_{C_R} \mathbf{J} \cdot d\omega = J \int_{C_R} d\alpha = J\Delta\alpha.$$

Therefore

$$W = 2T\tau - J\Delta\alpha. \tag{8.81}$$

The next step is to calculate W on C by using Stokes's theorem, according to which

$$W = \oint_C \vartheta_0 = \int_S d\vartheta_0, \tag{8.82}$$

where d is the exterior derivative and S is a surface capping C (i.e., whose boundary is C). Because C is a one-dimensional curve, S is two dimensional. These geometric objects are defined in $\mathbb{M}$, but because $\mathbf{J}$ is constant along C and S is *any* surface capping C, they can both be taken to lie in the three-dimensional submanifold of $\mathbb{M}$ for which $\mathbf{J} = \text{const}$. Applying the theorem requires calculating ϑ_0 on S and taking its exterior derivative.

This can be done in terms of the Euler angles. Choose coordinates such that $\mathbf{J}$ is pointing in the space 3-direction, which implies that $\mathbf{J} \cdot d\omega = J_3' \, d\omega_3' \equiv J \, d\omega_3'$, where the primes denote components in $\mathfrak{S}$. Then interpreting Eq. (8.78) as a statement about infinitesimal angles, as in the Discussion above Worked Example 8.3, leads to $\mathbf{J} \cdot d\omega = J(d\psi \cos\theta + d\phi)$, so that according to (8.82)

$$
\begin{aligned}
W &= \int_S d\{J(d\psi \cos\theta + d\phi)\} = J \int_S \sin\theta \, d\psi \wedge d\theta \\
&= J \int_S d\Omega = J\Omega,
\end{aligned}
$$

where Ω (not to be confused with the $\Omega = R^{\mathrm{T}} R$ of Section 8.2) is the solid angle subtended by S on the Euler-angle sphere. We state without proof (see Montgomery's paper) that Ω is also the solid angle subtended by the cone formed by $J_\mathfrak{B}$ as it goes once around its circuit on the energy ellipsoid of Fig. 8.7

This result and (8.81) yield

$$
\Delta\alpha = \frac{2T\tau}{J} - \Omega.
$$

Although our discussion has been abstract, geometric phase has practical applications, in particular in the modern field of robotics. It is relevant also in explaining the cat's famous ability to land upright when dropped, no matter what its initial orientation. Both of these applications, however, extend the treatment from rigid bodies to systems whose moments of inertia depend on time. For a survey on this subject see Marsden et al. (1991).

8.3.3 SPINNING TOPS

This section concerns an axially symmetric top spinning without friction about a fixed point at its tip (Fig. 8.13). The pivot point about which it spins is taken as the common origin of $\mathfrak{S}$ and $\mathfrak{B}$ and the symmetry axis is taken as the body 3 axis.

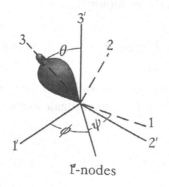

1'-nodes

FIGURE 8.13

The top is spinning on the $(1', 2')$ plane, with gravity pointing straight down in the negative $3'$ direction. The Euler angles and the orientations of the body (unprimed) and space (primed) axes are the same as in Figs. 8.11 and 8.12.

THE LAGRANGIAN AND HAMILTONIAN

We concentrate mostly on a top spinning under the influence of gravity. The kinetic energy $T = \frac{1}{2}\omega \cdot \mathbf{I}\omega$ can be represented either in $\mathfrak{S}$ or in $\mathfrak{B}$, depending on where $\mathbf{I}$ and ω are represented. It is always more convenient to represent $\mathbf{I}$ in the body system, where its matrix elements are constants, so $\mathfrak{B}$ is chosen. Because the body is axially symmetric about the 3 axis, $I_1 = I_2$ and

$$
\begin{aligned}
T &= \frac{1}{2}\left\{I_1\left(\omega_1^2 + \omega_2^2\right) + I_3\omega_3^2\right\} \\
&= \frac{1}{2}I_1(\dot{\theta}^2 + \dot{\phi}^2 \sin^2\theta) + \frac{1}{2}I_3(\dot{\psi} + \dot{\phi}\cos\theta)^2.
\end{aligned}
$$

The potential energy is $Mgh = Mgl\cos\theta$, where M is the mass of the top, g is the acceleration of gravity, and l is the distance (along the 3 axis) from the fixed point to the center of mass.

Thus the Lagrangian is

$$
L = \frac{1}{2}I_1(\dot{\theta}^2 + \dot{\phi}^2 \sin^2\theta) + \frac{1}{2}I_3(\dot{\psi} + \dot{\phi}\cos\theta)^2 - Mgl\cos\theta. \tag{8.83}
$$

Since ϕ and ψ are ignorable coordinates, it follows immediately that

$$
p_\phi \equiv \frac{\partial L}{\partial \dot{\phi}} = I_1\dot{\phi}\sin^2\theta + I_3(\dot{\psi} + \dot{\phi}\cos\theta)\cos\theta \tag{8.84}
$$

and

$$
p_\psi \equiv \frac{\partial L}{\partial \dot{\psi}} = I_3(\dot{\psi} + \dot{\phi}\cos\theta) \tag{8.85}
$$

are conserved angular momenta. We will return to them again later.

If the equation

$$
p_\theta \equiv \frac{\partial L}{\partial \dot{\theta}} = I_1\dot{\theta} \tag{8.86}
$$

is added to Eqs. (8.84) and (8.85), the three equations for the momenta can be inverted (Problem 8), and then the usual Legendre transform yields the Hamiltonian

$$
H = \frac{p_\theta^2}{2I_1} + \frac{(p_\phi - p_\psi\cos\theta)^2}{2I_1\sin^2\theta} + \frac{p_\psi^2}{2I_3} + Mgl\cos\theta. \tag{8.87}
$$

The first three of Hamilton's canonical equations, those for the time derivatives of the Euler angles, are obtained in Problem 9. The only nontrivial equation for the conjugate momenta is

$$
\dot{p}_\theta = \frac{(p_\phi - p_\psi\cos\theta)(p_\phi\cos\theta - p_\psi)}{I_1\sin^3\theta} + Mgl\sin\theta. \tag{8.88}
$$

The momenta p_ϕ and p_ψ are by definition in involution with each other, and being constants of the motion they are in involution with H. There exist three constants of the motion in involution; thus this system is completely integrable in the sense of the Liouville integrability theorem, as was first noted by Lagrange (1889) in 1788.

To go further we use a method due to Piña (1993) based on a time-dependent canonical transformation. The Type 3 generating function of this transformation is

$$F^3(p_\phi, p_\theta, p_\psi; \Phi, \Theta, \Psi; t) \equiv -\frac{1}{2}tp_\psi^2\left(\frac{1}{I_3} - \frac{1}{I_1}\right) - p_\psi\Psi - p_\theta\Theta - p_\phi\Phi. \qquad (8.89)$$

According to Problem 5.9, $q^\alpha = -\partial F^3/\partial p_\alpha$, $P_\alpha = -\partial F^3/\partial Q^\alpha$, and $K = H + \partial F^3/\partial t$. This implies that

$$\Phi = \phi, \quad \Theta = \theta, \quad P_\Phi = p_\phi, \quad P_\Theta = p_\theta, \quad P_\Psi = p_\psi,$$

but that

$$\Psi = \psi - tp_\psi\left(\frac{1}{I_3} - \frac{1}{I_1}\right). \qquad (8.90)$$

Although Ψ differs from ψ by a constantly increasing angle (modulo 2π), P_Ψ remains a constant of the motion (so does P_Φ). The new Hamiltonian is

$$K = \frac{P_\Theta^2}{2I_1} + \frac{P_\Phi^2 - 2P_\Phi P_\Psi \cos\Theta + P_\Psi^2}{2I_1\sin^2\Theta} + Mgl\cos\Theta. \qquad (8.91)$$

Because it is time independent, K is also a constant of the motion (Problem 10). Two nice features of K are that it is symmetric under interchange of P_Φ and P_Ψ, which will be relevant later, and that it is independent of I_3 [I_3 is buried in the definition of Ψ in Eq. (8.90)].

THE MOTION OF THE TOP

The variable Θ appears in K only in $\cos\Theta$ and in $\sin^2\Theta = 1 - \cos^2\Theta$, so it is convenient to write the equations in terms of the new variable

$$z \equiv \cos\Theta. \qquad (8.92)$$

In addition to Θ, only P_Θ need be written in terms of z, for P_Φ and P_Ψ are constants:

$$P_\Theta = I_1\dot\Theta = -I_1\frac{d(\cos\Theta)/dt}{\sin\Theta} = -I_1\frac{\dot z}{\sqrt{1-z^2}}.$$

When this and Eq. (8.92) are inserted into Eq. (8.91) and $\dot z$ is separated out, the equation becomes

$$\frac{1}{2}\dot z^2 = \frac{K}{I_1}(1-z^2) - \frac{Mgl}{I_1}(1-z^2)z - \frac{1}{2I_1^2}\left(P_\Phi^2 - 2P_\Phi P_\Psi z + P_\Psi^2\right). \qquad (8.93)$$

This equation can be interpreted as the kinetic energy of a unit-mass particle moving in a cubic potential (the negative of the right-hand side) with total energy zero. This will be called the *equivalent z system*. Its $\mathbb{Q}$ manifold is the z interval $[-1, 1]$. In order to make the analysis even more transparent, the time t is replaced by the dimensionless time parameter $\tau = t/T$, where $T = \sqrt{I_1/2Mgl}$. Then Eq. (8.93) can be written in the form

$$\frac{1}{2}\left(\frac{dz}{d\tau}\right)^2 + V(z) = 0, \tag{8.94}$$

with

$$V(z) = -\frac{1}{2}\left[(1 - z^2)(a_0 - z) + 2a_1 a_2 z - a_1^2 - a_2^2\right], \tag{8.95}$$

where

$$a_0 = \frac{K}{Mgl}, \quad a_1 = \frac{TP_\Phi}{I_1}, \quad a_2 = \frac{TP_\Psi}{I_1}.$$

In Problem 11 it is found that

$$\left.\begin{aligned}
\dot{\Phi} &= \frac{P_\Phi - P_\Psi z}{I_1(1 - z^2)}, \\
\dot{\Psi} &= \frac{P_\Psi - P_\Phi z}{I_1(1 - z^2)}.
\end{aligned}\right\} \tag{8.96}$$

Note the symmetry between Φ and Ψ.

Equation (8.94) is the energy first integral of the equivalent z system (the total energy is zero) and can be used to find $z(\tau)$ and hence $z(t)$. The solution depends on the three a_i parameters and hence on the constants of the motion K, P_Φ, and P_Ψ, on the energy and on two angular momenta. Then $\Phi(t)$, $\Theta(t)$, and $\Psi(t)$, and hence $\phi(t)$, $\theta(t)$, and $\psi(t)$, can be found from $z(\tau)$. That yields a complete solution of the problem.

Before describing $z(t)$ in detail, we describe some general properties of the motion. Write the cubic potential $V(z)$ in the form

$$V(z) = -\frac{1}{2}(z - z_1)(z - z_2)(z - z_3), \tag{8.97}$$

with $z_1 < z_2 < z_3$. The relation between the three roots z_k and the a_j parameters is obtained by multiplying out both (8.95) and (8.97):

$$\begin{aligned}
z_1 + z_2 + z_3 &= a_0, \\
z_1 z_2 + z_2 z_3 + z_3 z_1 &= 2a_1 a_2 - 1, \\
z_1 z_2 z_3 &= a_1^2 + a_2^2 - a_0.
\end{aligned} \tag{8.98}$$

According to (8.94) the total energy is zero. Therefore the equivalent z system moves only where $V(z)$ is negative. Equation (8.97) shows that $V \to +\infty$ as $z \to -\infty$ and $V \to -\infty$ as $z \to +\infty$. The graph of $V(z)$ is shown in Fig. 8.14: it goes from positive to negative at

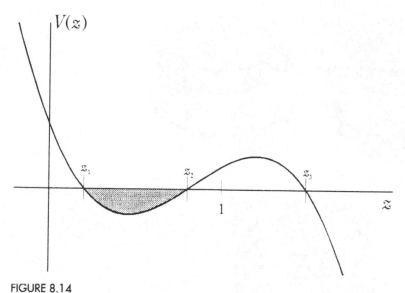

FIGURE 8.14

The potential of the equivalent z system. The shaded region is the only one available for the motion. In general $-1 \le z \le 1$. In the graph $z = -1$ is beyond the left side of the diagram.

the first root z_1, goes back to positive at the middle root z_2, and finally becomes negative at the third root z_3. Figure 8.14 and Problem 12 show that $-1 < z_1 < z_2 < 1 < z_3$ (recall that $\mathbb{Q}$ is $z \in [-1, 1]$). The only region where $V(z)$ is negative is between z_1 and z_2, so the equivalent z system moves between these two roots, in the shaded region on Fig. 8.14.

Different initial conditions for the top do not change the total energy of the equivalent z system; they change $V(z)$. Under some initial conditions z_1 and even z_2 will be negative. There are also special initial conditions under which $z_1 = z_2$. Since the total energy of the equivalent z system is always zero, if $z_1 = z_2$ the z system is stationary.

NUTATION AND PRECESSION

In general the equivalent z system oscillates in the potential well between z_1 and z_2, and the motion of the top can then be analyzed in terms of $z(\tau)$. Probably the easiest aspect to understand is the motion of the angle θ, which gives the tilt of the top (Fig. 8.13). From (8.92) it follows that $\dot\theta \equiv \Theta = -\dot z/\sqrt{1 - z^2}$ (the dot on z stands for the derivative with respect to τ). Hence the turning points of z, the points where $\dot z = 0$, correspond to turning points of θ, which therefore oscillates between $\theta_1 = \arccos z_1$ and $\theta_2 = \arccos z_2$. The resulting up-and-down motion of the upper pole of the top is called *nutation*. Note that if z_1 and z_2 coalesce, there is no nutation: $\theta = \text{const}$.

Now consider the angle ϕ, the azimuth angle of the upper pole about the fixed vertical space axis (the 3' axis in Fig. 8.13), whose time derivative $\dot\phi \equiv \Phi$ is given in terms of z by Eq. (8.96). Since $P_\Phi \equiv p_\phi$ and $P_\Psi \equiv p_\psi$ are constant, $\dot\phi$ oscillates with z. Whether $\dot\phi$ changes sign depends on the relative magnitudes of p_ϕ and p_ψ and on the minimum and maximum z values z_1 and z_2 (which themselves depend, among other things, on p_ϕ and p_ψ). Three possibilities are illustrated in Fig. 8.15. The upper pole of the top moves on a sphere (nutates) between the parallels of latitude at θ_1 and θ_2, and ϕ is the longitude on

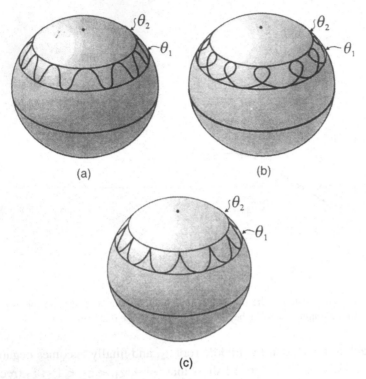

FIGURE 8.15
Some possible motions of the upper pole of the spinning top.

the sphere. The meandering curve between the two parallels traces out the motion of the pole. In Fig. 8.15(a) $\dot\phi$ is either positive or negative throughout the motion: the top is said to *precess* as well as nutate. (In the special case of constant z, the limit latitudes θ_1 and θ_2 coalesce and the motion of the top is pure precession.) In Fig. 8.15(b) $\dot\phi$ changes sign, and the pole precesses in opposite directions at θ_1 and θ_2. In Fig. 8.15(c) $\dot\phi$ is zero at θ_2 (the precession vanishes), which means that $P_\Phi - P_\Psi z_1 = 0$. We call these Cases (a), (b), and (c), respectively.

Case (c) may seem exceptional, and in a way it is: it represents the separatrix between Cases (a) and (b). Nevertheless this precessionless motion, or motion very close to it, is often observed, for when a top is set into motion tilted at some angle to the vertical, both θ and ϕ are usually at rest or almost at rest. That corresponds to one of the cusps at θ_2 in Case (c). We do not prove that the initial θ is θ_2 and not θ_1, but that results from the pull of gravity, the positivity of K, and the initial stationary values of $\dot\theta$ and $\dot\phi$.

The equivalent z system can be solved explicitly in terms of elliptic integrals. According to Eqs. (8.94) and (8.97),

$$\tau = \int \frac{dz}{\sqrt{(z - z_1)(z - z_2)(z - z_3)}}.$$

This is an *elliptic integral of Jacobi type* from which it follows that

$$z(\tau) = z_1 + (z_2 - z_1)\operatorname{sn}^2(\alpha\tau, k), \tag{8.99}$$

where

$$\alpha = \frac{1}{2}\sqrt{z_3 - z_1}, \quad k^2 = \frac{z_2 - z_1}{z_3 - z_1},$$

and $\mathrm{sn}(x, k)$ is tabulated (Abramowitz and Stegun, 1972) as the *Jacobian elliptic function* (it is periodic in both variables). Solutions for Φ and Ψ can now be obtained by using the results of Problem 11.

According to Problem 11,

$$\left. \begin{aligned} \frac{d(\Psi + \Phi)}{d\tau} &= T(\dot{\Psi} + \dot{\Phi}) = \frac{a_1 + a_2}{1 + z}, \\ \frac{d(\Psi - \Phi)}{d\tau} &= T(\dot{\Psi} - \dot{\Phi}) = \frac{a_1 - a_2}{1 - z}. \end{aligned} \right\}$$

On insertion of (8.99), these equations become

$$\left. \begin{aligned} \frac{d(\Psi + \Phi)}{d\tau} &= \frac{a_1 + a_2}{1 + \{z_1 + (z_2 - z_1)\, \mathrm{sn}^2(\alpha\tau, k)\}}, \\ \frac{d(\Psi - \Phi)}{d\tau} &= \frac{a_1 - a_2}{1 - \{z_1 + (z_2 - z_1)\, \mathrm{sn}^2(\alpha\tau, k)\}}. \end{aligned} \right\} \tag{8.100}$$

Now solutions for $\Psi(\tau)$ and $\Phi(\tau)$ can be formally written down as integrals over τ involving elliptic functions and they can then be evaluated numerically. The special case in which $z_1 = z_2$, when $\theta = \mathrm{const.}$, is particularly simple to analyze. Then $\mathrm{sn}(\alpha\tau, k)$ drops out of both terms in (8.100) and both $d\Psi/d\tau$ and $d\Phi/d\tau$ are constants. Some algebra leads to

$$\left(\frac{d\Psi}{d\tau} \right)^2 + \left(\frac{d\Phi}{d\tau} \right)^2 = z_3$$

and

$$\frac{d\Psi}{d\tau} \frac{d\Phi}{d\tau} = \frac{1}{2}.$$

Thus these angular velocities lie on one of the four intersections of a circle of radius $\sqrt{z_3}$ and an equilateral hyperbola (Fig. 8.16).

Most tops that children play with differ from the kind of top we have been discussing in two significant ways. First, the point on the ground is not fixed, but is free to move on the plane on which the top is spinning, usually a horizontal one. Second, a real top is acted upon by frictional forces which slow it down. That causes the nature of its motion to change during the course of its spin. For example, it may start without nutation initially, but before it comes to rest it will assuredly pass through a regime in which it does nutate.

We have been discussing only a symmetric top, a rare example of a completely integrable system in rigid body motion. See Liemanis (1965) for some other examples. If the top is asymmetric (i.e., if $I_1 \neq I_2 \neq I_3$), the system is not integrable and may become chaotic (Galgani et al., 1981). We will not pursue that particular discussion but mention only that whether a system is integrable or not depends on its symmetry (i.e., on its inertia tensor) and on the potential $V(R)$ in Eq. (8.50).

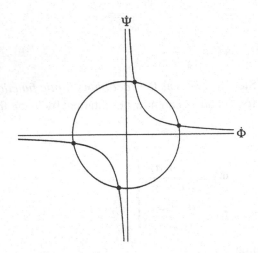

QUADRATIC POTENTIAL; THE NEUMANN PROBLEM

Another system that is integrable in the sense of Liouville is a symmetric top (as before, with $I_1 = I_2$) rotating about a fixed point in the presence not of a uniform gravitational field, but of an *anisotropic* quadratic potential (Bogoyavlensky, 1986). This dynamical system is mathematically equivalent to the Neumann system discussed in Worked Example 2.1 and again in Section 6.2.2: a mass point constrained to the surface of a sphere $\mathbb{S}^2$ and subjected to a quadratic potential. To prove integrability, we exhibit the equivalence by showing that the EL equations are the same as the equations of Worked Example 2.1.

Except for the potential, the Lagrangian is the same as (8.83):

$$L = \frac{1}{2}I_1(\dot\theta^2 + \dot\phi^2 \sin^2\theta) + \frac{1}{2}I_3(\dot\psi + \dot\phi\cos\theta)^2 - V.$$

Now the potential V will be allowed to depend not only on θ, as in the previous example, but also on ϕ, that is, generally on the direction of the top's axis. Such a V is conveniently given in terms of the *Poisson vector*

$$\mathbf{x} \equiv (x_1, x_2, x_3) = (\sin\theta\sin\phi, \sin\theta\cos\phi, \cos\theta), \qquad (8.101)$$

which is the Cartesian position vector of a point on a unit sphere $\mathbb{S}^2$, whose components are redundant, for $\mathbf{x}\cdot\mathbf{x} \equiv |x|^2 = 1$. The redundancy is handled as a constraint in the Lagrangian treatment of the system.

To proceed, calculate the $\dot{x}_\alpha$:

$$\dot{x}_1 = \dot\theta\cos\theta\sin\phi + \dot\phi\sin\theta\cos\phi,$$
$$\dot{x}_2 = \dot\theta\cos\theta\cos\phi - \dot\phi\sin\theta\sin\phi,$$
$$\dot{x}_3 = -\dot\theta\sin\theta,$$

and note that $|\dot{x}|^2 = (\dot\theta^2 + \dot\phi^2\sin^2\theta)$. Since V does not depend on ψ, the conjugate momentum $p_\psi = I_3(\dot\psi + \dot\phi \cos\theta)$, the same as in Eq. (8.85), is conserved. Therefore the Lagrangian is

$$L = \frac{1}{2}I_1|\dot x|^2 + \frac{1}{2}I_3^{-1}p_\psi^2 - V.$$

We now specify V more precisely:

$$V(\mathbf{x}) = k_1 x_1^2 + k_2 x_2^2 + k_3 x_3^2,$$

with three different values of the k_α. Finally, then, the Lagrangian is

$$L = \frac{1}{2}I_1|\dot x|^2 + K_\psi - \sum k_\alpha x_\alpha^2, \tag{8.102}$$

where $K_\psi = \frac{1}{2}I_3^{-1}p_\psi$ is a constant that depends on the initial conditions; it can be thought of as setting the zero of the potential and will not contribute to the equations of motion.

Before concluding, we count the number of freedoms. Originally the problem has the three rotational freedoms of a rigid body, but since ψ is a cyclic variable, the freedoms are reduced to two for each p_ψ, namely θ and ϕ. Equation (8.101) increases this to three, but with one constraint, which reduces it again to two.

In conclusion, the EL equations obtained from (8.102) are those of the Neumann problem of Worked Example 2.1, which establishes the equivalence of the two dynamical systems. Complete integrability was established in Section 6.2.2. Bogoyavlensky (1986) obtains explicit solutions of this rigid-body system in terms of the Riemann theta function of two variables.

8.4 CAYLEY–KLEIN PARAMETERS

Cayley–Klein (CK) parameters are another coordinate system on $\mathbb{Q}_R$, that is, on **SO**(3). As will be seen, they offer a direct way to specify a rotation R in terms of its axis and angle $(\mathbf{n}, \theta)$ (Section 8.2.1). They also have important applications in quantum mechanics.

8.4.1 2 × 2 MATRIX REPRESENTATION OF 3-VECTORS AND ROTATIONS

3-VECTORS

The CK parameters are defined from the way vectors in 3-space are transformed by rotations. Three-vectors can be represented not only as column vectors, but also as 2×2 complex, Hermitian, zero-trace matrices. Specifically, if $\mathbf{v} = \{v_1, v_2, v_3\}$ is a real 3-vector, it can be represented by

$$V \equiv A(\mathbf{v}) = \begin{bmatrix} v_3 & v_1 - iv_2 \\ v_1 + iv_2 & -v_3 \end{bmatrix}. \tag{8.103}$$

Every vector is represented in this way by a unique matrix, and every zero-trace Hermitian two-by-two matrix V, necessarily of this form, uniquely defines a vector $\mathbf{v}$: if $\mathbf{v}$ is given, the matrix $V = A(\mathbf{v})$ is unique, and if V is given, the vector $\mathbf{v} = A^{-1}(V)$ is unique. Moreover, $\det V = -(v_1^2 + v_2^2 + v_3^2) \equiv -\|\mathbf{v}\| \equiv -|\mathbf{v}|^2$, so the magnitude of $\mathbf{v}$ can be calculated directly from V.

Rotating $\mathbf{v}$ to a new vector $\mathbf{v}'$ induces a matrix transformation from $V \equiv A(\mathbf{v})$ to a new matrix $V' = A(\mathbf{v}')$. Similarly, transforming V somehow to a new Hermitian traceless matrix V' induces a vector transformation from $\mathbf{v} = A^{-1}(V)$ to $\mathbf{v}' = A^{-1}(V')$.

ROTATIONS

First consider the matrix transformations. If T is any nonsingular 2×2 matrix, the *similarity transformation* $V' = TVT^{-1}$ leaves both the trace and the determinant invariant: $\det V' = \det V = -\|\mathbf{v}\|$ and $\operatorname{tr} V' = \operatorname{tr} V = 0$. Then if V' is Hermitian, it can represent a vector. In general V' will not be Hermitian and will therefore not represent a vector, but if T is unitary, V' is Hermitian. Indeed, if $T^{-1} = T^{\dagger}$, then

$$(V')^{\dagger} = (TVT^{\dagger})^{\dagger} = (T^{\dagger})^{\dagger} V^{\dagger} T^{\dagger} = TVT^{\dagger} \equiv V'.$$

To emphasize that T is unitary we will now call it U. Thus a similarity transformation on V by a unitary U matrix induces a transformation from $\mathbf{v} = A^{-1}(V)$ to $\mathbf{v}' = A^{-1}(UVU^{-1})$ such that $\|\mathbf{v}'\| = \|\mathbf{v}\|$. Symbolically,

$$\mathbf{v} \xrightarrow{A} V \xrightarrow{U} V' \xrightarrow{A^{-1}} \mathbf{v}' \equiv \tilde{U}\mathbf{v}, \tag{8.104}$$

defining an operator $\tilde{U}$ on three-space. Since $\tilde{U}$ preserves all lengths, it is a rotation. (To be perfectly rigorous, we would have to show also that $\tilde{U}$ is linear.) The U matrices will be restricted even further to those for which $\det U = 1$. This is a small restriction, since $UU^{\dagger} = \mathbb{I}$ implies that $|\det U| = 1$ and $\det U$ has no effect on the transformation from V to V'. The U matrices we will now be discussing form the set (actually the group) called $\mathbf{SU}(2)$ [Special (determinant 1) Unitary matrices in 2 dimensions].

In summary, one way to construct a rotation R in 3-space is by the scheme of (8.104) with $U \in \mathbf{SU}(2)$. Since R is defined by U, we will write $\tilde{U} \equiv R_U$:

$$\mathbf{v} \xrightarrow{A} V \xrightarrow{U} V' \xrightarrow{A^{-1}} \mathbf{v}' \equiv R_U\mathbf{v}. \tag{8.105}$$

The reverse operation, that is, whether every rotation $R \in \mathbf{SO}(3)$ defines a $U \in \mathbf{SU}(2)$, has yet to be established.

8.4.2 THE PAULI MATRICES AND CK PARAMETERS

DEFINITIONS

Because R_U is a rotation, it has an axis and angle $(\mathbf{n}, \theta)$ that form the direction and magnitude of the "vector" $\mathbf{r}_U$ in the 3-ball $\mathbb{Q}_R$, defined in Section 8.2.1. This "vector" can be obtained explicitly from U, as will now be shown.

Obtaining $\mathbf{r}_U$ is made easier by writing the U matrices, the elements of $\mathbf{SU}(2)$, in a convenient way. Every U is of the form

$$U = \begin{bmatrix} \mu & \nu \\ -\nu^* & \mu^* \end{bmatrix}, \quad |\mu|^2 + |\nu|^2 = 1 \tag{8.106}$$

(see Problem 21). A standard way of writing these is in terms of the three *Pauli spin matrices* (used in quantum mechanics to describe the phenomenon of half-integral spin)

$$\sigma_1 = \begin{bmatrix} 0 & 1 \\ 1 & 0 \end{bmatrix}, \quad \sigma_2 = \begin{bmatrix} 0 & -i \\ i & 0 \end{bmatrix}, \quad \sigma_3 = \begin{bmatrix} 1 & 0 \\ 0 & -1 \end{bmatrix}. \tag{8.107}$$

The Pauli matrices are Hermitian, unitary, and of zero trace. Together with the 2×2 unit matrix $\mathbb{I}$ they form a basis for all 2×2 matrices. To put the elements of $\mathbf{SU}(2)$ into *standard form* is to write them in this basis. Writing μ and ν in terms of their real and imaginary parts, put $\mu = \alpha_0 + i\alpha_3$ and $\nu = \alpha_2 + i\alpha_1$ with all α_ρ real, $\rho = \{0, 1, 2, 3\}$. Then every element of $\mathbf{SU}(2)$ is of the form

$$\begin{aligned} U &= \begin{bmatrix} \alpha_0 + i\alpha_3 & \alpha_2 + i\alpha_1 \\ -\alpha_2 + i\alpha_1 & \alpha_0 - i\alpha_3 \end{bmatrix} \\ &= \alpha_0 \mathbb{I} + i\alpha_1\sigma_1 + i\alpha_2\sigma_2 + i\alpha_3\sigma_3 \\ &\equiv \alpha_0 \mathbb{I} + i\mathbf{a} \cdot \mathbf{s}, \end{aligned} \right\} \tag{8.108}$$

where $\mathbf{a}$ is the real 3-vector with components $\{\alpha_1, \alpha_2, \alpha_3\}$ and $\mathbf{s}$ is a new kind of 3-vector whose three components are the three Pauli matrices, so $\mathbf{a} \cdot \mathbf{s} \equiv \sum_1^3 \alpha_k\sigma_k$ is a 2×2 matrix. The determinant condition still has to be satisfied. According to Eq. (8.106) this condition translates to a condition on the α_ρ:

$$\sum_0^3 \alpha_\rho^2 \equiv \alpha_0^2 + |\mathbf{a}|^2 = 1. \tag{8.109}$$

In Problem 21 two useful formulas are derived: 1. The Pauli matrices satisfy the multiplication rule

$$\sigma_j\sigma_k = i\epsilon_{jkl}\sigma_l + \delta_{jk}\mathbb{I} \tag{8.110}$$

and 2. if $\mathbf{v}, \mathbf{u}$ are two real 3-vectors, then

$$(\mathbf{v} \cdot \mathbf{s})(\mathbf{u} \cdot \mathbf{s}) = (\mathbf{v} \cdot \mathbf{u})\mathbb{I} + i(\mathbf{v} \wedge \mathbf{u}) \cdot \mathbf{s}, \tag{8.111}$$

where $\mathbf{v} \wedge \mathbf{u}$ is the usual cross product in 3-space.

FINDING R_U

This standard form for U will now be used to obtain R_U explicitly in terms of U. Since R_U transforms every 3-vector $\mathbf{v}$ to another 3-vector $\mathbf{v}'$, the action of R_U will be found if $\mathbf{v}'$ is written explicitly in terms of $\mathbf{v}$ and U. But U determines the α_ρ and vice versa, so it will be found also by writing $\mathbf{v}'$ explicitly in terms of $\mathbf{v}$ and the α_ρ. We proceed to do just that.

According to Eq. (8.103) and the definition of the Pauli matrices,

$$V \equiv A(\mathbf{v}) = \mathbf{v} \cdot \mathbf{s}. \tag{8.112}$$

Then from (8.108),

$$\begin{aligned}
V' \equiv UVU^\dagger &= (\alpha_0\mathbb{I} + i\mathbf{a} \cdot \mathbf{s})(\mathbf{v} \cdot \mathbf{s})(\alpha_0\mathbb{I} - i\mathbf{a} \cdot \mathbf{s}) \\
&= \left[\alpha_0^2\mathbf{v} + \mathbf{a}(\mathbf{v} \cdot \mathbf{a}) + (\mathbf{v} \wedge \mathbf{a}) \wedge \mathbf{a} + 2\alpha_0\mathbf{v} \wedge \mathbf{a}\right] \cdot \mathbf{s} \\
&\equiv \mathbf{v}' \cdot \mathbf{s},
\end{aligned}$$

from which

$$\mathbf{v}' = \alpha_0^2\mathbf{v} + \mathbf{a}(\mathbf{v} \cdot \mathbf{a}) + (\mathbf{v} \wedge \mathbf{a}) \wedge \mathbf{a} + 2\alpha_0\mathbf{v} \wedge \mathbf{a} \equiv R_U\mathbf{v}. \tag{8.113}$$

This equation exhibits how R_U acts on any 3-vector $\mathbf{v}$: given $\mathbf{v}$ and the matrix U in standard form, it shows the vector $\mathbf{v}'$ into which $\mathbf{v}$ is rotated. It demonstrates decisively that R_U as defined by Eq. (8.105) is linear [see the words of caution after (8.104)]. The α_ρ are the *Cayley–Klein parameters* of R_U. (These CK parameters and the Pauli matrices form a representation of the *quaternions*, first introduced by Hamilton, 1969.)

AXIS AND ANGLE IN TERMS OF THE CK PARAMETERS

The "vector" $\mathbf{r}_U \equiv (\mathbf{n}, \theta)$ associated with R_U has yet to be exhibited. This will now be done in two steps.

First, it is easily shown that $\mathbf{n}$ is parallel to $\mathbf{a}$ by showing that a $\mathbf{v}$ that is a multiple of $\mathbf{a}$ is not changed by R_U. Indeed, if they are parallel, $\mathbf{v} \wedge \mathbf{a} = 0$ and (8.113) and (8.109) imply that $\mathbf{v}' = \mathbf{v}$. Hence

$$\mathbf{n} = \frac{\mathbf{a}}{|\mathbf{a}|}. \tag{8.114}$$

Second, θ is found by choosing a $\mathbf{v}$ that is perpendicular to $\mathbf{n}$ and hence also to $\mathbf{a}$. The angle between $\mathbf{v}'$ and $\mathbf{v}$ is then the angle θ of rotation. If they are perpendicular, $\mathbf{v} \cdot \mathbf{a} = 0$ and

$$\mathbf{v}' = \mathbf{v}(\alpha_0^2 - |\mathbf{a}|^2) + 2\alpha_0\mathbf{v} \wedge \mathbf{a}. \tag{8.115}$$

The dot product $\mathbf{v} \cdot \mathbf{v}' = |\mathbf{v}||\mathbf{v}'|\cos\theta = \|\mathbf{v}\|\cos\theta$, from which it follows that

$$\cos\theta = \alpha_0^2 - |\mathbf{a}|^2. \tag{8.116}$$

Equations (8.114) and (8.116) define $\mathbf{r}_U$ explicitly, and hence R_U, in terms of the CK parameters.

The inverse procedure is to find the CK parameters from the axis and angle. As a first step note that Eq. (8.109) implies that there exists a ϕ such that $\alpha_0 = \cos\phi$ and $|\mathbf{a}| = \sin\phi$, or $\mathbf{a} = \mathbf{n}\sin\phi$. Equation (8.116) shows that $\theta = 2\phi$, so

$$\alpha_0 = \cos\frac{1}{2}\theta, \quad \mathbf{a} = \mathbf{n}\sin\frac{1}{2}\theta. \tag{8.117}$$

The U matrix belonging to $(\mathbf{n}, \theta)$ is

$$U = \mathbb{I}\cos\frac{1}{2}\theta + i\mathbf{n}\cdot\mathbf{s}\sin\frac{1}{2}\theta. \qquad (8.118)$$

This means that the axis and angle of a rotation can be directly read off the U matrix. According to (8.116) and (8.117) R_U rotates any vector $\mathbf{v}$ into

$$\mathbf{v}' = \mathbf{v}\cos\theta + \mathbf{v}\wedge\mathbf{n}\sin\theta + \mathbf{n}(\mathbf{v}\cdot\mathbf{n})(1 - \cos\theta). \qquad (8.119)$$

This closes the logical circle, for (8.119) can be reobtained from geometrical considerations alone. Problem 22 shows how to find the CK parameters of two successive rotations.

8.4.3 RELATION BETWEEN SU(2) AND SO(3)

The CK parameters define a rotation $R \in \mathbf{SO}(3) \equiv \mathbb{Q}_R$ uniquely, so it may seem strange that although $\dim\mathbb{Q}_R = 3$, there are four CK parameters. But because the CK parameters are related by Eq. (8.109), only three of them are independent. However, (8.109) is quadratic, and that leads to a sign redundancy, which is made explicit by (8.118). If θ is changed by 2π while $\mathbf{n}$ is fixed, U changes sign, whereas R_U does not change at all. To put it differently, although U defines a unique rotation, U and $-U$ define the same rotation: $R_U = R_{-U}$. As a result, R does not define its U uniquely [see the sentence after Eq. (8.105)].

This can be understood in terms of the structures of $\mathbf{SU}(2)$ and $\mathbf{SO}(3)$. Both $\mathbf{SU}(2)$ and $\mathbf{SO}(3)$ are groups of linear transformations, which means essentially that each group contains an identity element (namely $\mathbb{I}$), that every group element has an inverse in the group, and that the products of group elements remain in the group. Hence R_U is a map from one group to another, from $\mathbf{SU}(2)$ to $\mathbf{SO}(3)$.

This map has the important property that it preserves products. Indeed,

$$U_1 U_2 V(U_1 U_2)^\dagger = U_1\big(U_2 V U_2^\dagger\big)U_1^\dagger$$

implies that

$$R_{U_1 U_2} = R_{U_1} R_{U_2}.$$

(The map is then called a *group homomorphism*.) $\mathbf{SU}(2)$ is called a *representation* of $\mathbf{SO}(3)$ in which U represents R_U. Although related in this way, the two groups are nevertheless really different because their elements depend so differently on θ.

This can be seen also in terms of the CK parameters. To each set of CK parameters there is one and only one $U \in \mathbf{SU}(2)$ and vice versa. But since U and $-U$ represent the same element of $\mathbf{SO}(3)$, there are two sets of CK parameters for every rotation matrix. Equation (8.109) shows that the CK parameters lie on $\mathbb{S}^3$ (the surface of a sphere imbedded in Euclidean 4-space $\mathbb{R}^4$), so the manifold of $\mathbf{SU}(2)$ is $\mathbb{S}^3$. The representation R_U of $\mathbf{SO}(3)$ by $\mathbf{SU}(2)$ thus yields a map from $\mathbb{S}^3$ to the 3-ball $\mathbb{Q}_R$, a map in which pairs of points in $\mathbb{S}^3$ are mapped to single points in $\mathbb{Q}_R$. It is said that $\mathbb{S}^3$ *covers* $\mathbb{Q}_R$ [or $\mathbf{SU}(2)$ covers $\mathbf{SO}(3)$] twice, once for each of the two possible

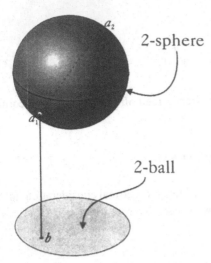

2-sphere

2-ball

FIGURE 8.17

Illustration, in one less dimension, of the double covering of the 3-ball $\mathbf{SO}(3)$ by the 3-sphere $\mathbf{SU}(2)$. Here the 2-sphere covers the 2-ball (disk). There are two points, a_1 and a_2, on the diameter of the sphere. The entire diameter, both points on the sphere, projects down to the single point b in the ball, just as U and $-U$ in $\mathbf{SU}(2)$ represent the same $R \in \mathbf{SO}(3)$. As the diameter rotates and a_1 approaches the equator of the sphere, b moves to the edge of the ball. If the rotation is continued, a_2 moves onto the lower hemisphere, and the diameter projects onto the diametrically opposite point of the ball. This corresponds to the identification of diametrically opposite points in the ball of $\mathbf{SO}(3)$.

signs of U for a given R. Figure 8.17 attempts to illustrate this double covering by means of the analogous double covering of the 2-ball (the circular disk) by $\mathbb{S}^2$.

The $\mathbf{SU}(2)$ representation of the rotations is important in quantum mechanics. It is used to describe the half-integral spin of many particles (e.g., the electron, the neutrino, the proton); the half integer is closely associated with the factor of $\frac{1}{2}$ in the relation between θ and ϕ at Eq. (8.117).

WORKED EXAMPLE 8.4

Find the CK parameters of a general rotation in terms of its Euler angles.

Solution. We use Eq. (8.70) to write the U matrices that represent the R matrices of Eqs. (8.68) and (8.69). Then we will multiply them out, rather than use the results of Problem 22.

Since the axis of the R_γ of (8.68) is the 3 axis, it follows from (8.118) that (the signs of the angles must be handled carefully)

$$U_\gamma = \mathbb{I}\cos\frac{1}{2}\gamma - i\sigma_3 \sin\frac{1}{2}\gamma = \begin{bmatrix} e^{-i\gamma/2} & 0 \\ 0 & e^{i\gamma/2} \end{bmatrix}.$$

Since the axis of the R_θ of (8.69) is the 1 axis, it follows from (8.118) that

$$U_\theta = \mathbb{I}\cos\frac{1}{2}\theta - i\sigma_1 \sin\frac{1}{2}\theta = \begin{bmatrix} \cos\theta/2 & -i\sin\theta/2 \\ -i\sin\theta/2 & \cos\theta/2 \end{bmatrix}.$$

Then matrix multiplication leads to

$$U \equiv U_\phi U_\theta U_\psi = \begin{bmatrix} e^{-i(\phi+\psi)/2}\cos\theta/2 & -ie^{-i(\phi-\psi)/2}\sin\theta/2 \\ -ie^{i(\phi-\psi)/2}\sin\theta/2 & e^{i(\phi+\psi)/2}\cos\theta/2 \end{bmatrix}. \qquad (8.120)$$

The CK parameters can now be read off this equation:

$$\alpha_0 = \cos\frac{1}{2}\theta \cdot \cos\frac{1}{2}(\phi + \psi),\ \alpha_1 = -\sin\frac{1}{2}\theta \cdot \cos\frac{1}{2}(\phi - \psi),$$

$$\left.\alpha_3 = \sin\frac{1}{2}\theta \cdot \sin\frac{1}{2}(\phi - \psi),\ \alpha_2 = -\cos\frac{1}{2}\theta \cdot \sin\frac{1}{2}(\phi + \psi).\right\}$$

PROBLEMS

1. Suppose $\dot{z} = \dot{x}$ in Fig. 8.2. Show that either (a) the body is not rotating or (b) l' is parallel to $\tilde{\omega}$.

2. Prove that every rotation R can be *block diagonalized* to the form

$$R \longrightarrow \begin{bmatrix} 1 & 0 & 0 \\ 0 & \cos\theta & \sin\theta \\ 0 & -\sin\theta & \cos\theta \end{bmatrix}$$

by a real orthogonal transformation.

3. Calculate the rotation matrix $R(\theta)$ that results from a rotation of θ about the 3 axis followed by another of θ about the 2 axis. Now calculate the result of applying them in the opposite order. Note the difference. Specialize to $\theta = \pi/2$. Find the axes and angles of the $\pi/2$ rotations.

4. (a) Calculate the rotation matrix $R(\alpha)$ that results from a rotation of α about the 3 axis followed by another of α about the 2 axis followed by a third of α about the 1 axis.

   (b) Find the angle and axis of $R(\pi/2)$. Then find the angle and axis of $R(\pi/4)$. [Comment: Part (b) shows that even if the component rotations about the three axes are equal, the axis of the resultant rotation depends on the magnitude of the component rotations.]

5. Use Eq. (8.70) to show explicitly that three elements of R determine the rest. [Suggestion: Use ρ_{23}, ρ_{32}, ρ_{33}.]

6. Show that $\text{tr}(\Omega K \Omega^T) = \omega \cdot \mathbf{I}\omega$.

7. Let $\mathbf{f}$ be the internal force density in the rigid body. That is, the total force $\mathbf{F}_v$ on a volume v of the body is given by $\mathbf{F}_v = \int_v \mathbf{f}\, d^3x$. Show that in the absence of external torques the elements of Λ are given by

$$\lambda_{jk} = \frac{1}{2}\int f_j x_k\, d^3x,$$

where the integral is taken over the whole body [see Eq. (8.42)]. Show that the symmetry of Λ implies the absence of total torque due to internal forces.

Problems 8–12 involve some calculations for the spinning top system.

8. Invert the equations for the momenta conjugate to the Euler angles to obtain $\dot{\theta}$, $\dot{\phi}$, $\dot{\psi}$ in terms of the momenta and the angles and obtain the given expression for H.

9. Derive the canonical equation (8.88) for p_θ.

10. Prove that the conserved quantity K (the new Hamiltonian) is positive as long as the center of mass of the top is higher than the pivot. Write it in terms of the total energy and conserved angular momenta of the top.

11. Use Eq. (8.92) to write the Eqs. (8.96) for $\dot{\Phi}$ and $\dot{\Psi}$ in terms of z.

12. Prove that $V(-1)$ and $V(1)$ are both nonnegative, as shown in Fig. 8.14, and that at least one is positive unless $p_\phi = p_\psi = 0$.

13. Let B be a 3×3 antisymmetric matrix and $B' = WBW^T$, where W is an orthogonal matrix. Let $b \leftrightarrow B$ and $b' \leftrightarrow B'$ in accordance with Eq. (8.52). Find the components of b' in terms of those of b and the matrix elements of W and show that the result can be put in the form $b' = (\det W) Wb$, as in Eq. (8.59).

14. Show that $J'_k = \epsilon_{klm}(PR^T)_{ml}$ and $\Upsilon \leftrightarrow J'$ imply that $\Upsilon = (PR^T - RP^T)$.

15. The mass quadrupole tensor of a body is defined as

$$Q_{ij} = \int \mu(\mathbf{x})(3x_i x_j - x^2 \delta_{ij})\, d^3x,$$

where $\mu(\mathbf{x})$ is the mass density and the integral is taken over the volume of the body. Express Q_{ij} in terms of the inertia tensor I_{kl}.

16. Find the principal moments of inertia of a sphere of radius R that has a spherical cavity of radius $r < R$ tangent to its surface (i.e., the center of the cavity is at a distance $R - r$ from the center of the large sphere).

17. (a) Find the angular velocity ω, the kinetic energy T, and the angular momentum $\mathbf{J}$ of a right circular cone of mass M, height h, and radius R at the base, rolling without slipping on its side on a horizontal plane. Assume that the cone returns to its original position with period P.

    (b) Find the total torque $\mathbf{N}$ on the cone and analyze how gravitation and constraint forces provide this torque.

    (c) Find the minimum coefficient of friction $\mu_{\min}$ for which the cone will not slip.

    (d) Assuming that $\mu > \mu_{\min}$, find the minimum value of P for which the cone will not tilt.

18. A door of mass M, uniform density, height H, width W, and thickness T is swinging shut with angular velocity ω.

    (a) Find the torque $\mathbf{N}$ about the point P (Fig. 8.18).

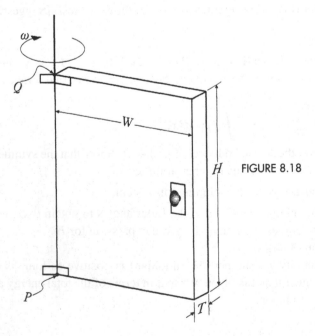

FIGURE 8.18

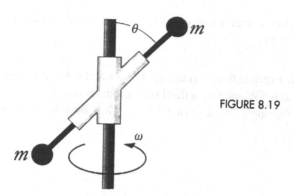

FIGURE 8.19

(b) In the limit $T = 0$ determine what you can about the forces the hinges apply to the door (assume that the hinges are at the very corners of the door).

19. Figure 8.19 shows two point masses m on a light rod of length b. The rod rotates without energy loss with angular velocity ω about a vertical axis. The angle θ it makes with the vertical is constrained to remain constant (the midpoint of the rod is fixed).

(a) Calculate representations of $\mathbf{I}$ and ω in both $\mathfrak{S}$ and $\mathfrak{B}$. Calculate (in $\mathfrak{S}$) $J' \equiv \sum x'_j \wedge p'_j$ and show explicitly that $\mathbf{J} = \mathbf{I}\omega$.

(b) Calculate the torque of constraint on the system. Discuss physically how this torque can be provided by the mechanism of the figure.

(c) Describe the motion that would result if the constraining torque were suddenly turned off. (The subsequent motion may be counterintuitive.)

20. This problem is based on an example given by Feynman (1985); see also Mehra (1994). A free circular disk, say an idealized flat pie plate or Frisbee, spins about an axis inclined at a small angle to its symmetry axis. While spinning, the plane of the of the disk wobbles.

(a) Show that there is no nutation, only precession.

(b) Show that the rates of wobble and spin are both constant.

(c) Show that if the spin axis is close to the symmetry axis, the rates of wobble and spin are approximately in the ratio $2:1$. (Note: In the 1985 reference, Feynman mistakenly stated the ratio to be $1:2$. He was corrected by Chao in 1989, which is acknowledged in the Mehra reference.)

21. (a) Show that every element $U \in \mathbf{SU}(2)$ is of the form

$$U = \begin{bmatrix} \mu & \nu \\ -\nu^* & \mu^* \end{bmatrix}, \quad |\mu|^2 + |\nu|^2 = 1.$$

(b) Derive the multiplication rule $\sigma_j \sigma_k = i\epsilon_{jkl}\sigma_l + \delta_{jk}\mathbf{I}$.

(c) Use part (b) to derive $(\mathbf{v} \cdot \mathbf{s})(\mathbf{u} \cdot \mathbf{s}) = (\mathbf{v} \cdot \mathbf{u})\mathbb{I} + i(\mathbf{v} \wedge \mathbf{u}) \cdot \mathbf{s}$.

22. Given the Cayley–Klein parameters of two matrices in $\mathbf{SU}(2)$, find the Cayley–Klein parameters of their product. [Note: This gives a way to find the axis and angle of the product of two rotations whose axes and angles are known.] [Answer: If the Cayley–Klein parameters of the component matrices are $\{\alpha_0, \mathbf{a}\}$ and $\{\beta_0, \mathbf{b}\}$, those of their product are $\{\alpha_0\beta_0 - \mathbf{a} \cdot \mathbf{b}, \ \alpha_0\mathbf{b} + \beta_0\mathbf{a} - \mathbf{a} \wedge \mathbf{b}\}$. The only part that depends on the order of multiplication is the cross product term at the end.]

23. Verify explicitly in terms of the Euler angles that $\cos(\Theta/2) = (\operatorname{tr} U)/2$ and $\cos \Theta = (\operatorname{tr} R - 1)/2$ give the same result for Θ.

24. See Problem 4.

    (a) Calculate the $U(\alpha)$ matrix that results from a rotation of α about the 3 axis followed by another of α about the 2 axis followed by a third of α about the 1 axis.

    (b) Use CK parameters to find the angle and axis of $U(\pi/2)$. Then find the angle and axis of $U(\pi/4)$.

CHAPTER 9

CONTINUUM DYNAMICS

CHAPTER OVERVIEW

The Lagrangian and Hamiltonian formalisms of particle dynamics can be generalized and extended to describe continuous systems such as a vibrating rod or a fluid. In such systems each point $\mathbf{x}$, influenced by both external and internal forces, moves independently. In the rod, points that start out very close together always remain close together, whereas in the fluid they may end up far apart. The displacement and velocity of the points of the system are described by functions $\psi(\mathbf{x}, t)$ called fields.

In this chapter we describe the classical (nonquantum) theory of fields. Particle dynamics will be transformed to field theory by allowing the number of particles to increase without bound while their masses and the distance between them go to zero in such a way that a meaningful limit exists. This is called passing to the continuum limit.

9.1 LAGRANGIAN FORMULATION OF CONTINUUM DYNAMICS

9.1.1 PASSING TO THE CONTINUUM LIMIT

THE SINE–GORDON EQUATION

In this section we present an example to show how to pass to the continuum limit. Consider the system illustrated in Fig. 9.1, consisting of many simple plane pendula of length l and mass m, all suspended from a horizontal rod with a screw thread cut into it. The planes of the pendula are perpendicular to the rod (only two of the planes are drawn in the figure), and the pendula are attached to (massless) nuts that move along the rod as the pendula swing in their planes [for more details, see Scott (1969)]. The pitch of the screw is β; that is, when a pendulum swings through an angle ϕ, its nut, together with the plane in which the pendulum swings, moves a distance $x = \beta\phi$ along the rod from its equilibrium position. The nuts, in turn, are coupled through springs, and as the pendula swing back and forth, the movement of the nuts compresses and stretches the springs by amounts proportional to the angles ϕ_j through which the pendula swing.

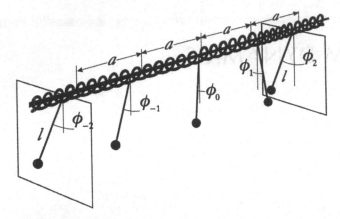

FIGURE 9.1
A chain of pendula coupled through springs. a is the separation between the equilibrium positions of
the pendula. The deviation of the nth suspension point from its equilibrium positon is x_n.

At equilibrium, that is, when the pendula hang straight down (all the $\phi_j = 0$), the
distance between the points of suspension is a. In the figure the points of suspension have
moved in proportion to the angles ϕ_j and the springs are correspondingly stretched and
compressed.

The kinetic energy of the jth pendulum is

$$T_j = \frac{1}{2}m\left(l^2\dot{\phi}_j^2 + \dot{x}_j^2\right) = \frac{1}{2}m(l^2 + \beta^2)\dot{\phi}_j^2 \equiv \frac{1}{2}m\lambda^2\dot{\phi}_j^2.$$

Assume that there are $2n + 1$ pendula numbered from $-n$ to n, as in Section 4.2.3. Then
the Lagrangian of the system is

$$L = \frac{1}{2}m\lambda^2 \sum_{j=-n}^{n} \dot{\phi}_j^2 - \left\{ mgl \sum_{j=-n}^{n}(1 - \cos\phi_j) + \frac{1}{2}k\beta^2 \sum_{j=-n}^{n-1}(\phi_{j+1} - \phi_j)^2 \right\}, \qquad (9.1)$$

where the first term is the kinetic energy and the second, in braces, is the potential, consisting
of a gravitational and an elastic term. The constant k in the elastic term is the force
constant of the springs (assumed to be the same all along the rod). The gravitational
term is an *external* force on the system, whereas the elastic term is an *internal* force,
the interaction between the particles. If $g = 0$, there is no external force and the system
becomes essentially the one in Section 4.2.3. If $k = 0$, there is no elastic term, and the
system becomes a chain of noninteracting pendula.

We write the Euler–Lagrange equations only for particles numbered $-(n - 1) \le j \le$
$n - 1$. In the final analysis n will go to infinity, so neglecting the two end pendula will
make no difference. After factoring out $m\lambda^2$, the EL equations are

$$\ddot{\phi}_j - \omega^2[(\phi_{j+1} - \phi_j) - (\phi_j - \phi_{j-1})] + \Omega^2 \sin\phi_j = 0, \quad -(n - 1) \le j \le n - 1,$$
$$(9.2)$$

where $\omega^2 = k\beta^2/m\lambda^2$ and $\Omega^2 = gl/\lambda^2$. This is a set of coupled second-order ordinary differential equations for the ϕ_j, something like those of Section 4.2.3 but now nonlinear (because of the sine term).

The continuum limit is obtained by replacing each pendulum of Eq. (9.1) by s others of the same length l and passing to the limit $s \to \infty$. The s new pendula have masses $\Delta m = m/s$ and they are distributed at displacements $\Delta x = a/s$ along the rod, so that the mass per unit length along the rod, that is, the linear mass density $\rho \equiv \Delta m/\Delta x = m/a$, is independent of s. Each of the reduced-mass pendula is labeled by the coordinate x along the rod of its point of suspension at equilibrium: ϕ_j is rewritten as $\phi(x)$. The spring remains the same, so if the constant is k for a length a, it is sk for a length a/s. Then Y, defined by $Y\lambda^2 \equiv ak\beta^2 = (a/s)(sk\beta^2)$, is also independent of s. The Lagrangian for the reduced-mass pendula is then

$$L = \frac{m}{2s}\lambda^2 \sum \dot{\phi}(x)^2 - \left\{ \frac{m}{s}gl \sum [1 - \cos\phi(x)] + \frac{1}{2}sk\beta^2 \sum [\phi(x + \Delta x) - \phi(x)]^2 \right\}$$

$$= \sum \Delta x \left\{ \frac{1}{2}\rho\lambda^2\dot{\phi}^2(x) - \rho\Omega^2\lambda^2[1 - \cos\phi(x)] - \frac{1}{2}Y\lambda^2 \left(\frac{\phi(x + \Delta x) - \phi(x)}{\Delta x} \right)^2 \right\}.$$

$$(9.3)$$

After factoring out $\rho\lambda^2\Delta x \equiv (m/s)\lambda^2$, the corresponding EL equation becomes

$$\ddot{\phi}(x) - \frac{Y}{\rho}\frac{[\phi(x + \Delta x) - 2\phi(x) + \phi(x - \Delta x)]}{(\Delta x)^2} + \Omega^2 \sin\phi(x) = 0. \qquad (9.4)$$

Because this result is independent of s, it is possible to pass to the $s \to \infty$ limit with no difficulty. In that limit $\Delta x \to 0$, and

$$\lim_{\Delta x \to 0} \left\{ \frac{[\phi(x + \Delta x) - 2\phi(x) + \phi(x - \Delta x)]}{(\Delta x)^2} \right\}$$

$$= \lim_{\Delta x \to 0} \left\{ \frac{1}{\Delta x}\left[\frac{\phi(x + \Delta x) - \phi(x)}{\Delta x} - \frac{\phi(x) - \phi(x - \Delta x)}{\Delta x} \right] \right\}$$

$$= \frac{\partial^2\phi(x, t)}{\partial x^2},$$

where we write $\phi(x, t)$ since ϕ for each x depends on t. Hence the EL equation reads

$$\frac{\partial^2\phi}{\partial t^2} - v^2\frac{\partial^2\phi}{\partial x^2} + \Omega^2 \sin\phi(x) = 0, \qquad (9.5)$$

with $v^2 = Y/\rho$. This is the *one-dimensional real sine–Gordon equation* (we will call it simply the sine–Gordon, or sG, equation). The function $\phi(x, t)$ is called the *wave function* or *field function* or just the *field*.

In the $s \to \infty$ limit the equation (9.3) for the Lagrangian becomes an integral:

$$L = \int \rho\lambda^2 \left\{ \frac{1}{2}\left(\frac{\partial\phi}{\partial t} \right)^2 - \frac{1}{2}v^2\left(\frac{\partial\phi}{\partial x} \right)^2 - \Omega^2(1 - \cos\phi) \right\}dx. \qquad (9.6)$$

The integrand

$$\mathcal{L}\left(\phi, \frac{\partial \phi}{\partial x}, \frac{\partial \phi}{\partial t}\right) \equiv \rho \lambda^2 \left\{ \frac{1}{2}\left(\frac{\partial \phi}{\partial t}\right)^2 - \frac{1}{2}v^2\left(\frac{\partial \phi}{\partial x}\right)^2 - \Omega^2(1 - \cos \phi) \right\} \qquad (9.7)$$

is called the *Lagrangian density* for the sG equation. Then $L = \int \mathcal{L}\, dx$.

Solutions of the sG equation will be discussed in Section 9.4.

THE WAVE AND KLEIN–GORDON EQUATIONS

In the sG system the *external* force is provided by the pendula and is reflected by the third term in $\mathcal{L}$. If some other external force binds the particles to their equilibrium positions, the last term is different. We give two examples, both linear.

First, suppose $g = 0$, so that there is no external force. Then the pendula do not contribute to the potential energy and the discrete system becomes essentially a chain of interacting oscillators, as in Section 4.2.3. It is then simplest to replace $\phi(x, t)$ by $\eta(x, t) \equiv \lambda \phi(x, t)$ as the generalized coordinate of the particle whose equilibrium position is x. The Lagrangian density becomes

$$\mathcal{L}\left(\frac{\partial \eta}{\partial x}, \frac{\partial \eta}{\partial t}\right) \equiv \rho \left\{ \frac{1}{2}\left(\frac{\partial \eta}{\partial t}\right)^2 - \frac{1}{2}v^2\left(\frac{\partial \eta}{\partial x}\right)^2 \right\} \qquad (9.8)$$

(this $\mathcal{L}$ is independent of η itself) and the EL equation reads

$$\frac{\partial^2 \eta}{\partial t^2} - v^2\frac{\partial^2 \eta}{\partial x^2} = 0, \qquad (9.9)$$

which is the one-dimensional wave equation with wave velocity v. We found in Section 4.2.3 that the solutions of the discrete system have wavelike properties, and it is now seen that in the continuum limit the solutions become simple one-dimensional waves.

> **REMARK:** In the discrete system, the η function must be restricted. For example, in the sine–Gordon equation, ϕ varies from $-\pi$ to π, but in addition ϕ may need to be restricted further to prevent adjacent planes from overlapping. Similarly, in the wave equation the proximity of the masses restricts η. However, after going to the continuum limit such restrictions are dropped: ϕ can vary from $-\pi$ to π, and η in the wave equation can vary from $-\infty$ to ∞. $\qquad \square$

For the second example, suppose the external force binding each particle to its equilibrium position is provided not by a pendulum but by an elastic spring with spring constant K: the force is $-K\eta$ (the generalized coordinate is again called η). It is shown in Problem 1 that in the $s \to \infty$ limit the Lagrangian density is

$$\mathcal{L}\left(\eta, \frac{\partial \eta}{\partial x}, \frac{\partial \eta}{\partial t}\right) \equiv \rho \left\{ \frac{1}{2}\left(\frac{\partial \eta}{\partial t}\right)^2 - \frac{1}{2}v^2\left(\frac{\partial \eta}{\partial x}\right)^2 - \frac{1}{2}\Omega^2\eta^2 \right\} \qquad (9.10)$$

(this $\mathcal{L}$ depends explicitly on η) and the EL equation is

$$\frac{\partial^2 \eta}{\partial t^2} - v^2\frac{\partial^2 \eta}{\partial x^2} + \Omega^2\eta = 0, \qquad (9.11)$$

where $\Omega^2 = K/m$. Equation (9.11) is the small-ϕ limit (in fact the linearization) of the sine–Gordon equation; it is known as the one-dimensional (real) *Klein–Gordon* equation. Its complex three-dimensional analog plays an important role in relativistic quantum mechanics (which will be discussed in Section 9.2.2). (Although it is the small-ϕ limit of the sine–Gordon equation, historically Klein–Gordon came first. In fact the name sine–Gordon was invented to rhyme with the other.)

9.1.2 THE VARIATIONAL PRINCIPLE

INTRODUCTION

The three one-dimensional examples of Section 9.1.1 began our discussion of the Lagrangian formulation of continuum dynamics. Now we reformulate continuum dynamics without reference to discrete systems and show how the EL equations of motion can be obtained directly from the Lagrangian densities.

The simplest system to start with is the wave equation, for which it is easily verified that Eq. (9.9) is given in terms of the Lagrangian density of (9.8) by

$$\frac{\partial}{\partial t}\frac{\partial \mathcal{L}}{\partial(\partial\eta/\partial t)} + \frac{\partial}{\partial x}\frac{\partial \mathcal{L}}{\partial(\partial\eta/\partial x)} = 0.$$

Let us call the wave function ψ for both ϕ and η. Then in the other two examples, Klein–Gordon and sine–Gordon, in which $\mathcal{L}$ depends on ψ as well as on its derivatives, the EL equations are obtained from

$$\frac{\partial}{\partial t}\frac{\partial \mathcal{L}}{\partial(\partial\psi/\partial t)} + \frac{\partial}{\partial x}\frac{\partial \mathcal{L}}{\partial(\partial\psi/\partial x)} - \frac{\partial \mathcal{L}}{\partial\psi} \equiv -\frac{\delta L}{\delta\psi} = 0, \tag{9.12}$$

defining $\delta L/\delta\psi$, which is called the *functional* or *variational derivative* of L with respect to ψ. We will now obtain these equations from a variational principle that is an extension of the one used for discrete (particle) systems.

VARIATIONAL DERIVATION OF THE EL EQUATIONS

The extension from discrete to continuum systems will be obtained by taking notice of the similarity between their EL equations: L is replaced by $\mathcal{L}$, the coordinate variable q is replaced by the field variable ψ, and $(d/dt)\partial L/\partial\dot{q}$ is replaced by the first two terms in (9.12). To understand the third replacement, recall that q depends only on the parameter t, whereas ψ depends (in the one-dimensional examples) on two parameters, x and t. With those changes, the continuum case is analogous to the discrete case. This analogy suggests that the equations of motion (9.12) can be derived without any reference to discrete systems, by an extension of the variational derivation of Section 3.1. This we now proceed to show.

Start by changing from a one-dimensional x to a three-dimensional $\mathbf{x}$. Then (9.9), with $\partial^2/\partial x^2$ replaced by ∇^2, describes longitudinal (one-component) waves in 3-space. Transverse waves, as is well known, have two components, both perpendicular to the direction of propagation, so the wave function has two components. More generally, waves in 3-space may be both transverse and longitudinal, and then the wave function is a three-component vector. We deal from now on, even more generally, with N component wave functions ψ_I, where I takes on $N \geq 1$ values. The ψ_I will be functions of four parameters, $\mathbf{x} \equiv (x^1, x^2, x^3)$ and the time t.

As in Section 5.1.1, we will write $x \equiv (t, x^1, x^2, x^3) \equiv (x^0, x^1, x^2, x^3)$ for the 4-space coordinates, so the wave functions will appear in the form $\psi_I(x)$ (Greek indices run from 0 to 3 and lower case Latin ones from 1 to 3). The calculations will take place in 4-space and we will write the Lagrangian density in the form $\mathcal{L}(\psi(x), \nabla\psi(x), x)$, where ψ stands for the collection of the N functions ψ_I, and $\nabla\psi$ stands for the collection of their $4N$ derivatives $\partial_\alpha \psi_I \equiv \psi_{I,\alpha}$.

The four x^μ now play the role of t in Chapter 3, so the action is an integral of the form

$$S = \int \mathcal{L}(\psi, \nabla\psi, x)\, d^4x, \tag{9.13}$$

where $d^4x = dx^0 dx^1 dx^2 dx^3$. In particle dynamics the action was $S = \int L\, dt$. In continuum dynamics the Lagrangian is $L = \int \mathcal{L}\, d^3x$, so (9.13) also reads $S = \int L\, dt$.

The problem in particle dynamics was to find the $q(t)$ functions in a fixed interval of time, given their values at the end points (on the boundary) of the interval. The problem in continuum dynamics is to find the $\psi_I(x)$ functions in a given region of 4-space, fixing their values on the boundary of the region. Let R be a region of 4-space with a three-dimensional boundary ∂R (Fig. 9.2). We proceed more or less paraphrasing Section 3.1.

Suppose the $\psi_I(x)$ are given on ∂R. The problem is to find them inside R. According to Eq. (9.13) each choice of $\psi_I(x)$ inside R yields a value for S. The variational principle states that of all the possible $\psi_I(x)$ functions in R, the physical ones are those that minimize S. More precisely, let $\psi_I(x; \epsilon)$ be a one-parameter family of functions inside R, differentiable in ϵ, and let the physical ψ_I functions be in that family. Choose the parametrization so that the $\psi_I(x; 0)$ are the physical functions. That the functions take on the given values on ∂R

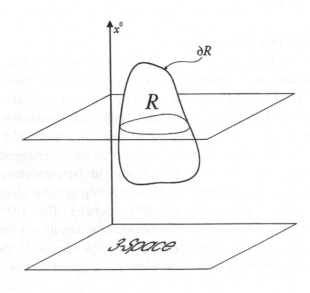

FIGURE 9.2
A region in four-space. Three-dimensional configuration space is represented by a two-dimensional plane.

can be put in the form

$$\psi_I(x;\epsilon) = \psi_I(x;0) \equiv \psi_I(x), \quad \forall x \in \partial R,$$

from which

$$\frac{\partial \psi_I}{\partial \epsilon} = 0, \quad \forall x \in \partial R. \tag{9.14}$$

When such a family is inserted into the definition of the action and the integral is taken over R, the action becomes a function of ϵ. The variational principle states that for every ϵ-family satisfying (9.14),

$$\frac{dS}{d\epsilon}\bigg|_{\epsilon=0} \equiv \left[\frac{d}{d\epsilon}\int_R \mathcal{L}\, d^4x\right]_{\epsilon=0} = 0.$$

As in the discrete case, we abbreviate $d/d\epsilon|_{\epsilon=0}$ by δ:

$$\delta S \equiv \delta \int_R \mathcal{L}\, d^4x = 0. \tag{9.15}$$

The next step, also as in the particle case, is to calculate the derivative of the integral with respect to ϵ. Because R is independent of ϵ, all that need be found is (summing over I from 1 to N and over α from 0 to 3)

$$\delta \mathcal{L} = \frac{\partial \mathcal{L}}{\partial \psi_I}\delta\psi_I + \frac{\partial \mathcal{L}}{\partial \psi_{I,\alpha}}\delta\psi_{I,\alpha}.$$

Now use

$$\frac{\partial \psi_{,\alpha}}{\partial \epsilon} = \frac{\partial}{\partial \epsilon}\frac{\partial \psi}{\partial x^\alpha} = \frac{\partial}{\partial x^\alpha}\frac{\partial \psi}{\partial \epsilon},$$

which means that $\delta\psi_{I,\alpha} = \partial_\alpha(\delta\psi_I)$. Then after some manipulations similar to those of the particle case, we get

$$\delta \mathcal{L} = \left[\frac{\partial \mathcal{L}}{\partial \psi_I} - \frac{\partial}{\partial x^\alpha}\frac{\partial \mathcal{L}}{\partial \psi_{I,\alpha}}\right]\delta\psi_I + \frac{\partial}{\partial x^\alpha}\left[\frac{\partial \mathcal{L}}{\partial \psi_{I,\alpha}}\delta\psi_I\right]. \tag{9.16}$$

For later use note that so far we have not used (9.14). Now Eq. (9.15) becomes

$$0 = \int_R \left[\frac{\partial \mathcal{L}}{\partial \psi_I} - \frac{\partial}{\partial x^\alpha}\frac{\partial \mathcal{L}}{\partial \psi_{I,\alpha}}\right]\delta\psi_I\, d^4x + \int_R \frac{\partial}{\partial x^\alpha}\left[\frac{\partial \mathcal{L}}{\partial \psi_{I,\alpha}}\delta\psi_I\right]d^4x. \tag{9.17}$$

The second integral can be converted to a surface integral by Stokes's theorem:

$$\int_R \frac{\partial}{\partial x^\alpha}\left[\frac{\partial \mathcal{L}}{\partial \psi_{I,\alpha}}\delta\psi_I\right]d^4x = \oint_{\partial R}\left[\frac{\partial \mathcal{L}}{\partial \psi_{I,\alpha}}\delta\psi_I\right]dn_\alpha,$$

where dn_α is now the normal surface element on ∂R. Since R is a four-dimensional region and ∂R is a three-dimensional hypersurface, dn_α is a four-component vector. Equation (9.14) implies that the integral over ∂R vanishes, which means that the second term in (9.17) vanishes.

Only the first integral remains in (9.17). Because the equation must hold for arbitrary ϵ-family, each expression (i.e., for each I) in square brackets in the integrand must vanish. We write its negative:

$$\frac{\partial}{\partial x^\alpha}\frac{\partial \mathcal{L}}{\partial \psi_{I,\alpha}} - \frac{\partial \mathcal{L}}{\partial \psi_I} \equiv -\frac{\delta L}{\delta \psi_I} = 0. \tag{9.18}$$

This extends to three space dimensions the definition of the functional derivative given in Eq. (9.12). Equations (9.18) are the EL equations of continuum dynamics.

The definition of the Lagrangian is then $L = \int \mathcal{L}(\psi, \nabla\psi, x)d^3x$, and its value depends on the ψ functions inserted into this definition. In other words L is a *functional* of ψ, a map from functions ψ to $\mathbb{R}$. The x dependence is integrated out, but the time dependence is not, so L is a time-dependent functional of ψ.

THE FUNCTIONAL DERIVATIVE

The $\delta L/\delta\psi_I$ defined in (9.18) is called the functional derivative because it is in a sense the derivative of L with respect to $\psi_I(x)$. It tells how L changes in the limit of *small variations of* ψ. To calculate this, let ψ depend on a parameter ϵ; then the field functions become $\psi_I(x, \epsilon)$. Repeating the calculation that led to Eq. (9.16) yields

$$\frac{dL}{d\epsilon} = \int \left[\frac{\partial \mathcal{L}}{\partial \psi_I} - \frac{\partial}{\partial x^\alpha}\frac{\partial \mathcal{L}}{\partial \psi_{I,\alpha}} \right] \frac{\partial \psi_I}{\partial \epsilon} d^3x + \sigma,$$

where σ is a surface integral that is assumed to vanish for large enough regions of integration. Small variation of ψ means that $\partial\psi/\partial\epsilon$ is zero everywhere except in a small region R, of volume V_R, about some point $\mathbf{x}$ of 3-space. If R is very small, the integral is nearly V_R times the integrand. Then in the limit of small V_R, as R closes down on $\mathbf{x}$

$$\lim_{R \to 0} \frac{1}{V_R}\frac{dL/d\epsilon}{d\psi_I/d\epsilon} \equiv \frac{\delta L}{\delta \psi_I} = \frac{\partial \mathcal{L}}{\partial \psi_I} - \frac{\partial}{\partial x^\alpha}\frac{\partial \mathcal{L}}{\partial \psi_{I,\alpha}}.$$

The functional derivative depends on the point $\mathbf{x}$ at which it is calculated and on t because L is a function of t; hence $\delta L/\delta\psi$ is a function of $(\mathbf{x}, t)$.

DISCUSSION

The variational derivation shows that if the Lagrangian is known, the dynamical EL equations, the *field equations*, can be obtained without reference to any discrete system. How to obtain the Lagrangian is another matter. In particle dynamics L was often, but not always, $T - V$. In field theory it is not as simple, but some examples will be given in later sections. As in particle dynamics, the Lagrangian of two *noninteracting* fields ψ_I and ϕ_B is the sum of their two Lagrangians, for the set of equations it gives is simply the union of the two separate sets. If the fields interact, however, a term describing the interaction must be added to the Lagrangian. More about that will also be presented later.

Recall that in particle dynamics if two Lagrangians differ by the total time derivative $d\Phi(q, t)/dt$ of a function only of the qs and t, their EL equations are the same. The analog of this theorem in continuum dynamics is that if two Lagrangian *densities* differ by the divergence $\partial_\alpha \Phi^\alpha(\psi, x)$ of a vector function only of the field variables ψ_I and the coordinates x^μ (but not of the derivatives $\psi_{I,\mu}$ of the field variables), their EL equations are the same. This is proved in Problem 2.

The indices on our equations are beginning to proliferate, and in order to neaten the equations we will often abbreviate the notation in the way we did in the previous subsection. When it proves convenient, we drop all indices, both what we will call the *spin indices* $I, J, \ldots$ and the 4-space indices $\mu, \nu, \ldots$. An expression of the form $\partial_\alpha V^\alpha$ will be written $\nabla \cdot V$, and one of the form $\partial_\alpha S$ will be written ∇S, reminiscent of 3-vector notation, but without the boldface. The fields will then be written ψ instead of ψ_I. Although such abbreviated notation hides some details, they are recoverable, and the final results of calculations will be written with the indices restored. In abbreviated form Eq. (9.18) becomes

$$\nabla \cdot \frac{\partial \mathcal{L}}{\partial \nabla \psi} - \frac{\partial \mathcal{L}}{\partial \psi} \equiv -\frac{\delta L}{\delta \psi} = 0 \tag{9.19}$$

and the second term of (9.16) becomes

$$\nabla \cdot \left[\frac{\partial \mathcal{L}}{\partial \nabla \psi} \right] \delta \psi. \tag{9.20}$$

9.1.3 MAXWELL'S EQUATIONS

Maxwell's equations are an important example of field equations derived from a Lagrangian density. Their discussion requires using and extending the notation and the ideas of the special theory of relativity [Section 5.1.1; we continue to write (x^0, x^1, x^2, x^3) for the time and the three-space coordinates].

SOME SPECIAL RELATIVITY

Relativistic equations involve things like four-vectors and tensors, referred to as *covariant objects*. Examples are the four-velocity $u^\mu = dx^\mu/d\tau$ of Eq. (5.23) (where τ is the invariant *proper time*) and the relativistic momentum p^μ at Eq. (5.26). If the definition $u^\mu = dx^\mu/d\tau$ is to hold for all observers, the components of u^μ must transform under Lorentz transformations (5.17) the same way as the coordinates. Suppose that the coordinates x'^μ of an event in one reference frame are given in terms of the coordinates x^μ of the same event in another by

$$x'^\mu = \Lambda^\mu{}_\nu x^\nu, \tag{9.21}$$

where the $\Lambda^\mu{}_\nu$ are the elements of the Lorentz transformation. Then if the definition of the u^μ is to be frame independent, its components must transform according to

$$u'^\mu = \Lambda^\mu{}_\nu u^\nu.$$

Other covariant objects include 16-component *four-tensors*, like the $R^{\mu\nu}$ in the equation $a^\mu b^\nu = R^{\mu\nu}$. If such equations are to be frame independent, the $R^{\mu\nu}$ must transform according to

$$R'^{\alpha\beta} = \Lambda^\alpha{}_\mu \Lambda^\beta{}_\nu R^{\mu\nu}.$$

Vectors and tensors that are functions of x, the coordinates in *Minkowski space*, are called vector and tensor *fields*. Recall the Minkowski metric $g_{\mu\nu}$ (and its inverse $g^{\mu\nu}$) whose components are all zero if $\mu \neq \nu$ and whose nonzero components are

$$g_{11} = g_{22} = g_{33} = -g_{00} = 1 = g^{11} = g^{22} = g^{33} = -g^{00}.$$

These are used to lower and raise indices:

$$a_\mu = g_{\mu\nu} a^\nu, \quad R_\mu{}^\nu = g_{\mu\lambda} R^{\lambda\nu}, \quad R^\mu{}_\nu = g_{\nu\lambda} R^{\mu\lambda}.$$

In a similar way, tensor fields of higher *rank* can be defined, for example, objects like $R^{\kappa\lambda\nu}$.

ELECTROMAGNETIC FIELDS

This covariant approach is used in the relativistic description of the electromagnetic field. We do not intend here to describe the logic by which Maxwell's equations are assimilated into special relativity but give only the results. The electric and magnetic fields are identified with an antisymmetric tensor $f_{\mu\nu} = -f_{\nu\mu}$ according to

$$\begin{aligned}
\mathbf{E} &\equiv (E_1, E_2, E_3) = (f_{10}, f_{20}, f_{30}), \\
\mathbf{B} &\equiv (B_1, B_2, B_3) = (f_{23}, f_{31}, f_{12}),
\end{aligned} \tag{9.22}$$

or (the columns and rows are numbered from 0 to 3)

$$[f_{\mu\nu}] = \begin{bmatrix} 0 & -E_1 & -E_2 & -E_3 \\ E_1 & 0 & B_3 & -B_2 \\ E_2 & -B_3 & 0 & B_1 \\ E_3 & B_2 & -B_1 & 0 \end{bmatrix}. \tag{9.23}$$

Then Maxwell's inhomogeneous equations (i.e., $\nabla \cdot \mathbf{E} = \rho$ and $\nabla \wedge \mathbf{B} - \partial \mathbf{E}/\partial t = \mathbf{j}$) are

$$\partial_\nu f^{\mu\nu} = j^\mu, \tag{9.24}$$

where $j = (\rho, \mathbf{j})$ is the *four-vector electric current*: ρ is the charge density and $\mathbf{j}$ is the current density. We write one of these in three-vector notation. Let $\mu = 1$. Then (9.24) reads

$$\partial_0 f^{10} + \partial_1 f^{11} + \partial_2 f^{12} + \partial_3 f^{13} = -\frac{\partial E_1}{\partial t} + 0 + \frac{\partial B_3}{\partial x^2} + \frac{\partial(-B_2)}{\partial x^3}$$

$$= (\nabla \wedge \mathbf{B})_1 - \frac{\partial E_1}{\partial t} = j_1.$$

The antisymmetry of $f^{\mu\nu}$ implies that

$$\partial_\mu \partial_\nu f^{\mu\nu} \equiv \partial_\mu j^\mu = 0. \tag{9.25}$$

This is the *continuity equation* $\nabla \cdot \mathbf{j} + \partial\rho/\partial t = 0$, the equation of charge conservation. Maxwell's homogenous equations (i.e., $\nabla \cdot \mathbf{B} = 0$ and $\nabla \wedge \mathbf{E} + \partial\mathbf{B}/\partial t = \mathbf{0}$) are

$$\partial_\lambda f_{\mu\nu} + \partial_\mu f_{\nu\lambda} + \partial_\nu f_{\lambda\mu} = 0 \tag{9.26}$$

(use the metric to lower the indices). We write one of these in three-vector notation:

$$\partial_0 f_{12} + \partial_1 f_{20} + \partial_2 f_{01} = \frac{\partial B_3}{\partial t} + \frac{\partial E_2}{\partial x^1} + \frac{\partial(-E_1)}{\partial x^2}$$

$$= \frac{\partial B_3}{\partial t} + (\nabla \wedge \mathbf{E})_3 = 0.$$

It is important to understand the relation between (9.25) and Maxwell's equations. When j is a *given external current*, it must satisfy (9.25) or Maxwell's equations would be inconsistent: Equation (9.25) is a condition on admissible four-currents. This formulation assumes that the current is known a priori. In reality the electromagnetic field affects the motion of whatever charges are present and therefore reacts back on j, and that reaction is described by other equations. A complete theory derives not only (9.24), equations that describe how the j^μ affect the $f^{\mu\nu}$, but also other equations that describe how the $f^{\mu\nu}$ affect the j^μ, together forming a set of equations for both the $f_{\mu\nu}$ and the j^μ. Such a complete theory will be described in Section 9.2.2. The resulting j^μ will always satisfy (9.25).

It is shown in standard works on electromagnetism (e.g., Jackson, 1975) that Eq. (9.26) imply that there exist *four-vector potentials* A^μ satisfying the differential equations (again, use the metric to lower the indices)

$$\partial_\mu A_\nu - \partial_\nu A_\mu = f_{\mu\nu}. \tag{9.27}$$

It is easy to show that (9.26) follows directly from (9.27), but the converse requires some proof, which is given in standard works. For a given tensor field $f_{\mu\nu}$ the solution of (9.27) for the A^μ is not unique. Any two solutions A^μ and $\bar{A}^\mu$ differ by a *gauge transformation* (first mentioned in Section 2.2.4)

$$\bar{A}^\mu = A^\mu + g^{\mu\nu}\partial_\nu\chi, \tag{9.28}$$

where χ is any (scalar) function of the x^λ.

THE LAGRANGIAN AND THE EL EQUATIONS

We now show that Maxwell's equations for the electromagnetic field are the EL equations obtained from the Lagrangian density

$$\mathcal{L}(A, \nabla A) = -\frac{1}{4} f_{\mu\nu} f^{\mu\nu} - A_\mu j^\mu, \tag{9.29}$$

where the A_μ are taken to be the field variables ψ_I of Section 9.1.2. The spin index I on ψ_I becomes μ on A_μ and the $f^{\mu\nu}$ must be written in accordance with (9.27) as functions of the A_λ and their derivatives $\partial_\alpha A_\lambda \equiv A_{\lambda,\alpha}$.

The EL equation (9.19) is now

$$\nabla \cdot \frac{\partial \mathcal{L}}{\partial \nabla A} - \frac{\partial \mathcal{L}}{\partial A} = 0.$$

With the indices restored, this is

$$\frac{\partial}{\partial x^\alpha} \frac{\partial \mathcal{L}}{\partial A_{\lambda,\alpha}} - \frac{\partial \mathcal{L}}{\partial A_\lambda} = 0.$$

The derivative $\partial \mathcal{L}/\partial \nabla A$ is obtained from (9.27), which can now be written in the form $f_{\mu\nu} = A_{\nu,\mu} - A_{\mu,\nu}$, and from (9.29):

$$\left(\frac{\partial \mathcal{L}}{\partial \nabla A} \right)^{\lambda\alpha} = -\frac{1}{4} \left(\frac{\partial f_{\mu\nu}}{\partial A_{\lambda,\alpha}} f^{\mu\nu} + f_{\mu\nu} \frac{\partial f^{\mu\nu}}{\partial A_{\lambda,\alpha}} \right).$$

The second term on the right-hand side is the same as the first, as can be seen by raising and lowering indices according to $f_{\mu\nu} = g_{\mu\rho} g_{\nu\sigma} f^{\rho\sigma}$ and $f^{\mu\nu} = g^{\mu\rho} g^{\nu\sigma} f_{\rho\sigma}$ and by using the fact that the $g_{\mu\nu}$ form a constant matrix whose inverse is $g^{\mu\nu}$. Thus

$$\left(\frac{\partial \mathcal{L}}{\partial \nabla A} \right)^{\lambda\alpha} = -\frac{1}{2} \frac{\partial f_{\mu\nu}}{\partial A_{\lambda,\alpha}} f^{\mu\nu} = -\frac{1}{2} \left(\delta^\lambda_\mu \delta^\alpha_\nu - \delta^\lambda_\nu \delta^\alpha_\mu \right) f^{\mu\nu}$$

$$= -\frac{1}{2} (f^{\lambda\alpha} - f^{\alpha\lambda}) = -f^{\lambda\alpha}. \tag{9.30}$$

The $\partial \mathcal{L}/\partial A$ term is just the four-vector j, so the EL equations are

$$-\partial_\alpha f^{\lambda\alpha} + j^\lambda = 0, \tag{9.31}$$

in agreement with (9.24). We will return to these results, in particular to (9.29), when we discuss the energy content of the field in the next section.

Although the EL equations (9.24) are gauge invariant, the Lagrangian of (9.29) from which they are derived is not, for it depends on the A_μ. We came across a similar situation when discussing the dynamics of a charged particle in an electromagnetic field (Section 2.2.4). If $j^\mu = 0$ in Eq. (9.29), however, the resulting *free-field* Lagrangian is gauge invariant.

9.2 NOETHER'S THEOREM AND RELATIVISTIC FIELDS

9.2.1 NOETHER'S THEOREM

THE THEOREM

Constants of the motion, or conserved quantities, can be found for continuum dynamics as they can be found in particle dynamics, by applying Noether's theorem. In fact Noether (1918) originally proved her theorem for fields, not particles. Our treatment, analogous here to the one in Section 3.2.2, is based on invariance of the Lagrangian, but many authors (e.g., Bogoliubov and Shirkov, 1958) base it on invariance of the action.

In Section 3.2.2 it was shown that to every q-symmetry of the Lagrangian L corresponds a constant of the motion $K(q, \dot{q})$. A q-symmetry is a q-transformation, one of the form $q(t) \mapsto q(t; \epsilon)$, that leaves L invariant. That K is a constant of the motion means that $dK(q, \dot{q})/dt = 0$. The field-theory analogs of q-transformations are transformations of the field variables ψ_I. The symmetries are of the Lagrangian density $\mathcal{L}$ rather than of L itself. The analog of t is the set of variables x^0, x^1, x^2, x^3, so the conclusion of the theorem will not involve just the time derivative of some function but derivatives with respect to all of the x^α.

In analogy with Section 3.2.2, consider an ϵ-family $\psi_I(x; \epsilon)$ of transformations of the field variables. At $\epsilon = 0$ the fields $\psi_I(x; 0)$ are solutions of the EL equations (9.18) (i.e., of $\delta L/\delta \psi_I = 0$). The transformations are not restricted to some region R of four-space and they do not vanish on any particular submanifold. According to Eqs. (9.16) and (9.20), when the field equations are satisfied (δ is $d/d\epsilon$ at $\epsilon = 0$)

$$\delta\mathcal{L} = \nabla \cdot \left[\frac{\partial \mathcal{L}}{\partial \nabla \psi} \delta\psi \right]. \tag{9.32}$$

Therefore, invariance of $\mathcal{L}$ under the ϵ-family (i.e., $\delta\mathcal{L} = 0$) implies that the right-hand side vanishes. As in the particle case [see Eq. (3.49)], $\mathcal{L}$ does not have to be strictly invariant but can change by a four-divergence, as discussed at the end of Section 9.1.2 and in Problem 2 (then $\mathcal{L}$ is *quasisymmetric*; the analog in Chapter 3 was called gauge invariance). Hence if there exist four functions $\Phi^\alpha(\psi, x)$ such that

$$\delta\mathcal{L} = \frac{\partial \Phi^\alpha}{\partial x^\alpha} \equiv \nabla \cdot \Phi,$$

then

$$\nabla \cdot \left[\frac{\partial \mathcal{L}}{\partial \nabla \psi} \delta\psi - \Phi \right] \equiv \nabla \cdot G \equiv \partial_\alpha G^\alpha = 0, \tag{9.33}$$

where

$$G^\alpha = \frac{\partial \mathcal{L}}{\partial \psi_{I,\alpha}} \delta\psi_I - \Phi^\alpha. \tag{9.34}$$

This is the analog of the result at Eq. (3.50): the four functions G are not constants of the motion, but $\partial_\alpha G^\alpha = 0$. The theorem of Chapter 3 required the existence of just one function Φ that depended on the qs and t but not on the $\dot{q}$s. The new theorem requires four functions Φ^α that depend on the ψ_I and the x^α but not on the $\psi_{I,\alpha}$. In Chapter 3, d/dt was the derivative of both the explicit t dependence and the t dependence through q. Here the ∂_α are derivatives of both the explicit x^α dependence of the Φ^α and the x^α dependence through the ψ_I (in Chapter 3 the total time derivative d/dt was used, but here there are four x^α, so it can't be written that way).

The result is Noether's theorem for field theory. It tells us that a quasisymmetry of the Lagrangian density corresponds to a *conserved current*, that is, to a set G of four functions G^α that satisfy Eq. (9.33): the divergence of G vanishes (we have not yet used Minkowskian geometry).

CONSERVED CURRENTS

The four G^α form a conserved current, and that leads to a constant of the motion. We show this for the familiar example of the four-vector electric current. The continuity equation Eq. (9.25) can be written $\nabla \cdot j = 0$. Now write $G = j$, or $G^0 = \rho$ and $\{G^{1,2,3}\} \equiv$ $\mathbf{G} = \mathbf{j}$. We show explicitly how, with one reasonable assumption, this equation leads to charge conservation.

What needs to be shown is that the integral of G^0 over all 3-space does not vary in time. Consider a fixed region V of three-space. For simplicity make V a sphere. Let R be the region that V sweeps out in four-space in a finite time $t_1 - t_0$ (Fig. 9.3). Then according to the four-dimensional Stokes's (divergence) theorem

$$0 = \int_R \partial_\alpha G^\alpha = \oint_{\partial R} G^\alpha dn_\alpha,$$

where dn_α is the normal hypersurface element on the boundary ∂R of R. The components of dn_α can be read off Fig. 9.3. On the upper sphere (disk) dn_α has only a zero component

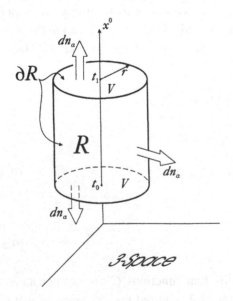

FIGURE 9.3
A region R of four-space swept out by a fixed region V of three-space. The dn_α represent normals.

equal to the hypersurface (volume) element of three-space, which is represented in the figure by the plane in which the disk lies: $dn_\alpha = (d^3x, 0, 0, 0)$. On the lower sphere $dn_\alpha = (-d^3x, 0, 0, 0)$. On the other side dn_α has no zero component. A small square on the side of the cylinder is of height $dx_0 \equiv dt$ and width $d\Sigma$, where $d\Sigma$ is the surface element on the boundary ∂V of V, so $dn_\alpha = (0, dx_0 d\Sigma)$. Hence the integral over ∂R is

$$\int_V G^0(t_1)d^3x - \int_V G^0(t_0)d^3x + \int_{t_0}^{t_1} dt \oint_{\partial V} \mathbf{G} \cdot d\Sigma = 0. \tag{9.35}$$

If the last term here can be shown to vanish, the other two integrals are equal for arbitrary t_0 and t_1 and therefore the integral of G^0 does not vary in time.

Therefore consider the last term. Its integral is over the surface ∂V, which grows as r^2 with increasing r. Assume that the magnitude of $\mathbf{G}$ decreases faster than r^{-2}. Then the integral can be made arbitrarily small by making r sufficiently large. If the integral is taken over all of three-space it then vanishes, and (9.35) reads

$$\int G^0(t_1)d^3x - \int G^0(t_0) d^3x = 0,$$

or

$$\frac{d}{dt} \int G^0 d^3x \equiv \frac{dQ}{dt} = 0. \tag{9.36}$$

Thus if the magnitude of the current density $|\mathbf{G}|$ decreases sufficiently rapidly, in particular if it vanishes outside some finite region W, the integral of G^0 over all space, the total charge, is conserved (it is conserved even over any region containing W). More spectacularly, $\nabla \cdot G \equiv \nabla \cdot j = 0$ implies that the total charge in the universe is constant.

All that was used in this derivation was $\nabla \cdot G = 0$ and the rapid decrease of $|\mathbf{G}|$. The identification of G with j is irrelevant, so what has been proven is that if G is a rapidly decreasing conserved current, $\int G^0 d^3x$ is a conserved quantity.

ENERGY AND MOMENTUM IN THE FIELD

The Lagrangian of a given physical system should not depend on where or when it is written: it should be independent of the time at which clocks are started and of the origin in three-space – independent, that is, of the origin in four-space. (Since we will be dealing almost entirely with Lagrangian densities, we call them simply Lagrangians when there is no confusion.) This is true, for example, of the Lagrangian (9.29) for the electromagnetic field. Such an $\mathcal{L}$ is symmetric, or at least quasisymmetric, under displacements of the origin, and we now show that this leads to conservation laws for both momentum and energy.

Under the displacement of the origin by a four-vector h, the field functions $\psi(x)$ are replaced by $\psi(x + h)$. Consider one-parameter families of such displacements and write $\psi(x, \epsilon) = \psi(x + \epsilon h)$ for fixed h. Then $\delta\psi = h \cdot \nabla\psi$, or $\delta\psi_I = h^\alpha \partial_\alpha \psi_I \equiv h^\alpha \psi_{I,\alpha}$. If $\mathcal{L}$ is origin independent, it does not depend explicitly on the coordinates x. Nevertheless, the change in $\psi(x) \to \psi(x, \epsilon)$ induces a change in $\mathcal{L}$, whose variation with ϵ is then given by

$$\delta\mathcal{L} = h^\alpha \partial_\alpha \mathcal{L} = \partial_\alpha(h^\alpha \mathcal{L}) \equiv \nabla \cdot (h\mathcal{L});$$

the change turns out to be the four-divergence of the vector function $\Phi = h\mathcal{L}$. Hence Eq. (9.33) can be applied to yield [using (9.32) and writing out the indices for the final expression]

$$0 = \nabla \cdot \left[\frac{\partial \mathcal{L}}{\partial \nabla \psi} \delta \psi - \Phi \right] = \nabla \cdot \left[\frac{\partial \mathcal{L}}{\partial \nabla \psi} h \cdot \nabla \psi - h\mathcal{L} \right]$$

$$= h^\mu \partial_\alpha T_\mu{}^\alpha, \tag{9.37}$$

where

$$T_\mu{}^\alpha \equiv \frac{\partial \mathcal{L}}{\partial \psi_{I,\alpha}} \psi_{I,\mu} - \delta_\mu^\alpha \mathcal{L}. \tag{9.38}$$

Because h is arbitrary (9.37) implies that

$$\partial_\alpha T_\mu{}^\alpha = 0. \tag{9.39}$$

Origin independence of the Lagrangian therefore leads to not one, but four conserved currents, one for each μ, each of which has its conserved quantity as established at Eq. (9.36). The four conserved quantities, indexed by μ, are

$$P_\mu \equiv \int T_\mu{}^0 \, d^3x. \tag{9.40}$$

The μ comes from the displacement h^μ, so if the displacement is in time alone, $\mu = 0$ and the conserved current associated with invariance under time displacement is $T_0{}^\alpha$. Similarly, displacement in the k direction leads to the conserved current $T_k{}^\alpha$.

Recall that in particle dynamics $\partial L / \partial t = 0$ implies conservation of energy, and $\partial L / \partial q^\alpha = 0$ implies conservation of the conjugate momentum p_α. Partial derivatives can be interpreted as translation properties: for example, $\partial L / \partial q^\alpha = 0$ means that the Lagrangian is invariant under translation in the q^α direction. Hence invariance under time displacement implies energy conservation and invariance under space displacement implies momentum conservation. Therefore P_0 is taken to be the total energy contained in the field and P_k to be the total kth component of the momentum. In analogy with the four-vector electric current density, the $T_\mu{}^0$ are the momentum and energy densities, and the $T_\mu{}^k$ are the energy and momentum current densities. For example $T_0{}^k$ is the energy flowing per unit time per unit area in the k direction, and $T_2{}^1$ is the 2 component of momentum flowing per unit time per unit area in the 1 direction. The $T_\mu{}^\alpha$ themselves are the elements of the *energy–momentum tensor*.

WORKED EXAMPLE 9.1

The Schrödinger field. Consider the (nonrelativistic) Lagrangian density

$$\mathcal{L} = \frac{i\hbar}{2}\{\psi^*\dot{\psi} - \dot{\psi}^*\psi\} - \frac{\hbar^2}{2m}\nabla\psi^* \cdot \nabla\psi - V(r)\psi^*\psi, \tag{9.41}$$

where $\psi(\mathbf{x}, t)$ is complex. **(a)** Find the EL equations for ψ. **(b)** $\mathcal{L}$ is invariant under the gauge transformation $\psi \to \psi \exp(i\epsilon)$, $\psi^* \to \psi^* \exp(-i\epsilon)$. Find the corresponding Noether conserved current. **(c)** Find the momentum density in the field.

Solution. **(a)** The negative of the functional derivative is

$$-\frac{\delta \mathcal{L}}{\delta \psi^*} \equiv \left[\nabla \cdot \frac{\partial \mathcal{L}}{\partial \nabla \psi^*}\right] + \partial_t \frac{\partial \mathcal{L}}{\partial \dot{\psi}^*} - \frac{\partial \mathcal{L}}{\partial \psi^*}$$

$$= \left[-\frac{\hbar^2}{2m}\nabla^2 \psi - \frac{i\hbar}{2}\dot{\psi}\right] - \frac{i\hbar}{2}\dot{\psi} + V(r)\psi.$$

Set this equal to zero to obtain the EL equation:

$$i\hbar\dot{\psi} = \left\{-\frac{\hbar^2}{2m}\nabla^2 + V(r)\right\}\psi.$$

This is the nonrelativistic (linear) Schrödinger equation. The $\delta\mathcal{L}/\delta\psi$ EL equation is the complex conjugate equation for ψ^*.

(b) According to (9.34) the space part of the conserved current is

$$j^k = \frac{\partial \mathcal{L}}{\partial(\partial_k \psi)}i\psi + \frac{\partial \mathcal{L}}{\partial(\partial_k \psi^*)}(-i\psi^*) = -\frac{i\hbar^2}{2m}(\psi \partial_k \psi^* - \psi^* \partial_k \psi)$$

or

$$\mathbf{j} = -\frac{i\hbar^2}{2m}(\psi\nabla\psi^* - \psi^*\nabla\psi).$$

The time part is

$$j^0 = -\hbar\psi^*\psi.$$

(c) According to (9.38) the momentum density is

$$T_k{}^0 = \frac{\partial \mathcal{L}}{\partial \dot{\psi}}\partial_k \psi + \frac{\partial \mathcal{L}}{\partial \dot{\psi}^*}\partial_k \psi^* = \frac{i\hbar}{2}\{\psi^*\partial_k\psi - \psi\partial_k\psi^*\}$$

or

$$\mathbf{T}^0 = \frac{i\hbar}{2}\{\psi^*\nabla\psi - \psi\nabla\psi^*\}.$$

Note the similarity to $\mathbf{j}$ of Part (b).

EXAMPLE: THE ELECTROMAGNETIC ENERGY–MOMENTUM TENSOR

An important example is the energy–momentum tensor of the free electromagnetic field, which we calculate in this subsection and compare with the usual expressions.

That the field is free means that there are no charges or currents present. Then the Lagrangian is just $\mathcal{L} = -\frac{1}{4} f^{\mu\nu} f_{\mu\nu}$ and $\partial \mathcal{L}/\partial \nabla \psi \equiv \partial \mathcal{L}/\partial \nabla A$ is given by Eq. (9.30). Hence the energy–momentum tensor is

$$T_\mu{}^\alpha = -f^{\lambda\alpha} A_{\lambda,\mu} + \frac{1}{4} \delta_\mu^\alpha f^{\beta\lambda} f_{\beta\lambda}. \tag{9.42}$$

There is a problem with this $T_\mu{}^\alpha$: the presence of $A_{\lambda,\mu}$ makes it clear that the tensor is not gauge invariant. To be gauge invariant it would have to contain the A_μ only in antisymmetric combinations such as $A_{\lambda,\mu} - A_{\mu,\lambda} = f_{\mu\lambda}$. There is a well-developed theory of dealing with such matters by *symmetrizing* energy–momentum tensors, but this is not the place to go into it (see Aharoni, 1959; Corson, 1955). What it boils down to in this case is subtracting the term $f^{\lambda\alpha} A_{\mu,\lambda}$, which makes the tensor gauge invariant [we will return to this point again in Section 9.2.2 when discussing the angular momentum in the field]. The additional term does not affect conservation. To see this, note first that

$$f^{\lambda\alpha} A_{\mu,\lambda} = \partial_\lambda(f^{\lambda\alpha} A_\mu) - A_\mu \partial_\lambda f^{\lambda\alpha} = \partial_\lambda(f^{\lambda\alpha} A_\mu), \tag{9.43}$$

since according to (9.24) $\partial_\lambda f^{\lambda\alpha} = 0$ if $j = 0$. Therefore, subtracting the term yields

$$\partial_\alpha \left[T_\mu{}^\alpha - f^{\lambda\alpha} A_{\mu,\lambda} \right] = \partial_\alpha \left[T_\mu{}^\alpha - \partial_\lambda(f^{\lambda\alpha} A_\mu) \right] = \partial_\alpha T_\mu{}^\alpha$$

because $\partial_\alpha \partial_\lambda(f^{\lambda\alpha} A_\mu) = 0$ from symmetry considerations alone. From now on

$$\begin{aligned} \bar{T}_\mu{}^\alpha &\equiv T_\mu{}^\alpha - f^{\lambda\alpha} A_{\mu,\lambda} \\ &= -f^{\lambda\alpha} f_{\lambda\mu} + \frac{1}{4} \delta_\mu^\alpha f^{\beta\lambda} f_{\beta\lambda} \end{aligned} \tag{9.44}$$

will be taken as the energy–momentum tensor of the electromagnetic field (see Problem 3).

The tensor $\bar{T}_\mu{}^\alpha$ is called the *symmetrized* form of $T_\mu{}^\alpha$ because when $g^{\kappa\mu}$ is used to raise the first index, the tensor is symmetric:

$$\bar{T}^{\kappa\alpha} = \frac{1}{4} g^{\kappa\alpha} f^{\beta\lambda} f_{\beta\lambda} - g^{\kappa\nu} f^{\lambda\alpha} f_{\lambda\nu}. \tag{9.45}$$

The first term is obviously symmetric in κ and α. The second term may be written in the form $f^{\lambda\alpha} f_\lambda{}^\kappa = f_\lambda{}^\alpha f^{\lambda\kappa} \equiv f^{\lambda\kappa} f_\lambda{}^\alpha$, which demonstrates its symmetry. The symmetry of the energy–momentum tensor implies, for example, that the k component of momentum flowing in the j direction is equal to the j component of momentum flowing in the k direction.

WORKED EXAMPLE 9.2

Check that $\bar{T}_\mu{}^\alpha$ yields the correct expression for the Poynting vector.

Solution. The Poynting vector $\mathbf{\Pi} = \mathbf{E} \wedge \mathbf{B}$ gives the energy current density (flux), so according to the last paragraph of the previous subsection $\bar{T}_0{}^k$ should be Π_k. From (9.44) it follows that $\bar{T}_0{}^k = -f^{\lambda k} f_{\lambda 0}$. We check this only for $k = 1$:

$$\bar{T}_0{}^1 = -f^{21} f_{20} - f^{31} f_{30} = B_3 E_2 - B_2 E_3 = (\mathbf{E} \wedge \mathbf{B})_1 \equiv \Pi_1.$$

More generally, $\bar{T}_\mu{}^\alpha$ yields the correct expressions for energy and momenta and their fluxes in terms of $\mathbf{E}$ and $\mathbf{B}$ (Jackson, 1975). We leave the rest of the proof to Problem 4.

9.2.2 RELATIVISTIC FIELDS

Relativistic fields are those whose Lagrangians are invariant under Lorentz transformations (i.e., are *Lorentz scalars*). In this section Minkowskian geometry will be used to show how Lorentz invariance leads to the conservation of angular momentum and to other conservation laws.

LORENTZ TRANSFORMATIONS

In Eq. (9.21) the Lorentz transformations were written as

$$x'^\mu = \Lambda^\mu{}_\nu x^\nu. \tag{9.46}$$

In abbreviated notation this is $x' = \Lambda x$. All *contravariant* vectors, those whose components are labeled by upper indices, transform like the x^μ in (9.46). The *norm* of a vector v^μ is the scalar $|v|^2 \equiv v^\alpha v_\alpha = g_{\alpha\beta} v^\alpha v^\beta = g^{\alpha\beta} v_\alpha v_\beta$, which does not change under a Lorentz transformations. That means that it can be written in any frame:

$$|v|^2 = g_{\alpha\beta} v'^\alpha v'^\beta = g_{\alpha\beta} \Lambda^\alpha{}_\nu \Lambda^\beta{}_\mu v^\nu v^\mu = g_{\nu\mu} v^\nu v^\mu,$$

and this implies that

$$g_{\alpha\beta} \Lambda^\alpha{}_\nu \Lambda^\beta{}_\mu = g_{\nu\mu}.$$

If both sides are multiplied by $g^{\nu\lambda}$ (and summed) and the gs are used to raise and lower indices, this becomes

$$\Lambda_\beta{}^\lambda \Lambda^\beta{}_\mu = \delta^\lambda_\mu,$$

or

$$\hat{\Lambda}^\mathrm{T} \Lambda = \mathbf{1},$$

where the transpose Λ^T is defined by $(\Lambda^\mathrm{T})^\lambda{}_\beta \equiv \Lambda^\beta{}_\lambda$ (changes the order of indices) and the $\hat{}$ operation is defined by $(\hat{\Lambda})^\beta{}_\lambda \equiv \Lambda_\beta{}^\lambda$ (raises and lowers the indices), so that $(\hat{\Lambda}^\mathrm{T})^\lambda{}_\beta = \Lambda_\beta{}^\lambda$. In other words, $\hat{\Lambda}^\mathrm{T} = \Lambda^{-1}$ and

$$x^\mu = (\Lambda^{-1})^\mu{}_\nu x'^\nu = \Lambda_\nu{}^\mu x'^\nu. \tag{9.47}$$

To see how *covariant* vectors, those with lower indices, transform, write $v_\mu = g_{\mu\nu}v^\nu$. Then

$$v'_\mu \equiv g_{\mu\nu}v'^\nu = g_{\mu\nu}\Lambda^\nu{}_\lambda g^{\lambda\kappa}v_\kappa \equiv \Lambda_\mu{}^\kappa v_\kappa. \qquad (9.48)$$

LORENTZ INVARIANT $\mathcal{L}$ AND CONSERVATION

If four functions Φ^α can be found such that $\delta\mathcal{L} = \nabla \cdot \Phi$ (i.e., if $\delta\mathcal{L}$ is a divergence), then according to (9.34) $G = (\partial\mathcal{L}/\partial\nabla\psi)\delta\psi - \Phi$ is a conserved current. We will proceed in two steps: in Step 1 we will show that $\delta\mathcal{L}$ is a divergence and find Φ; in Step 2 we will calculate G from Φ.

Finding $\delta\mathcal{L}$ in Step 1 requires understanding how $\mathcal{L}$ changes under Lorentz transformations, and that in turn requires understanding how the field functions transform. There are two aspects to the transformation of the fields. First, in the transformed frame the fields are rewritten as functions of the new x'^μ coordinates instead of the original x^μ. For example, in Section 9.2.1 the fields were rewritten as functions of the $x^\mu + h^\mu$.

Under a Lorentz transformation there is an additional change. Consider, for instance, the electromagnetic field A_μ. The new fields are not simply the A_μ written out as functions of the x'^μ, for they are themselves transformed in accordance with (9.48) to new functions $A'_\mu = \Lambda_\mu{}^\nu A_\nu$, linear combinations of the old ones. Thus the full transformation is $A'_\mu(x') = \Lambda_\mu{}^\nu A_\nu(x)$; this gives the new field functions in terms of the old ones at a point in four-space whose old coordinates are x and whose new ones are x'. For fields other than the electromagnetic the components of the field may be geometric objects other than vectors (e.g., they may be tensors), and then the linear combinations that give the new field in terms of the old can be more complicated. Being linear combinations of the old ones, they are obtained from some transformation matrix D. That is, the new field functions are given in terms of the old by an equation of the form

$$\psi'_J(x') = D^I_J\psi_I(x).$$

The D matrix depends on Λ and on the *geometry* of the ψ_I (whether the ψ_I are vectors, tensors, etc.). In the electromagnetic case, in which the ψ_I are the covariant A_μ vectors, D is the Lorentz transformation of Eq. (9.48). More generally, the D matrix is called a *representation* of Λ; it should properly be written $D^I_J(\Lambda)$, but in the interests of simplicity we will usually drop the Λ dependence. Because $x = \Lambda^{-1}x'$, the full transformation can be written in the form (dropping the prime from x')

$$\psi'_J(x) = D^I_J\psi_I(\Lambda^{-1}x) \qquad (9.49)$$

or, in abbreviated notation,

$$\psi'(x) = D\psi(\Lambda^{-1}x).$$

This is the transformation equation for the field functions in two coordinate systems x and x' related by a Lorentz transformation.

Assume as before that $\mathcal{L}$ is a Lorentz scalar. This means that the expression for $\mathcal{L}$ is of the same form in all coordinate systems; hence the field functions and their derivatives occur in $\mathcal{L}$ only in scalar combinations, so that the D part of their transformation drops out. The only change in $\mathcal{L}$ under a Lorentz transformation is due to the change from x to x' in the arguments of the field functions ψ and their derivatives $\nabla\psi$ (assume also, as in the previous subsection, that $\mathcal{L}$ has no explicit x dependence). For example, in the free-field Maxwell Lagrangian the A_μ and their derivatives appear only in the combination $f^{\mu\nu}f_{\mu\nu}$, which is manifestly a scalar. Such combinations have exactly the same form in the new coordinate system as they have in the old. As a result, only the change in x need be considered in calculating $\delta\mathcal{L}$.

Step 1. Consider an ϵ-family of Lorentz transformations $\Lambda(\epsilon)$ starting at $\Lambda(0) = \mathbb{I}$ (recall that δ is the derivative at $\epsilon = 0$). Then

$$\delta\mathcal{L} = (\partial_\alpha\mathcal{L})\delta\left(\Lambda_\mu{}^\alpha x^\mu\right) = (\nabla\mathcal{L})\delta(\Lambda^{-1}x)$$
$$\equiv \Omega x\nabla\mathcal{L} = \nabla\cdot(\Omega x\mathcal{L}) - \mathcal{L}\nabla\cdot(\Omega x),$$

defining

$$\Omega = \delta\Lambda^{-1}.$$

Then $\delta\mathcal{L}$ is a divergence if the second term vanishes or if $\nabla\cdot(\Omega x) = 0$. It vanishes owing to a symmetry property of Ω: because $\Lambda^{-1} = \hat{\Lambda}^{\mathrm{T}}$ (remember that $\Lambda(0) = \mathbb{I}$),

$$0 = \delta(\Lambda\hat{\Lambda}^{\mathrm{T}}) = \hat{\Omega}^{\mathrm{T}} + \Omega.$$

Write this out in index form

$$\omega_\mu{}^\nu + \omega^\nu{}_\mu = 0,$$

where $\omega_\mu{}^\nu = \delta\Lambda_\mu{}^\nu \equiv (\Omega)_\mu{}^\nu$. Now use the gs to raise and lower indices, arriving at

$$\omega^{\mu\nu} + \omega^{\nu\mu} = \omega_{\mu\nu} + \omega_{\nu\mu} = 0,$$

which shows that $\omega_{\mu\nu}$ is an antisymmetric tensor. It follows that $\omega_{\mu\mu} = 0$ and then, by raising one index, that $\omega_\mu{}^\mu \equiv \mathrm{tr}\,\Omega = 0$. But then

$$\nabla\cdot(\Omega x) \equiv \partial_\alpha\left(\omega_\mu{}^\alpha x^\mu\right) = \omega_\mu{}^\alpha\delta_\alpha^\mu = \omega_\mu{}^\mu = 0.$$

Thus $\delta\mathcal{L}$ is a divergence:

$$\delta\mathcal{L} = \nabla\cdot(\Omega x\mathcal{L}) \equiv \nabla\cdot\Phi,$$

defining

$$\Phi \equiv \Omega x\mathcal{L}. \tag{9.50}$$

In component form, $\Phi^\alpha = \omega_\mu{}^\alpha x^\mu\mathcal{L}$.

Step 2. Now that Φ has been found, $\delta\psi$ in the first term of (9.33) must be calculated. For ϵ-families, (9.49) becomes $\psi(x, \epsilon) = D(\epsilon)\psi(\Lambda^{-1}(\epsilon)x)$. Because the Lorentz transformation is the identity at $\epsilon = 0$, so is D, and then the derivative at $\epsilon = 0$ is

$$\delta\psi = c\psi + (\nabla\psi)\Omega x \tag{9.51}$$

(the last term comes from $\delta x^\mu \partial_\mu \psi_I$), where

$$c_J^I = \frac{dD_J^I}{d\epsilon}\bigg|_{\epsilon=0} = \frac{dD_J^I}{d\Lambda_\nu^{\ \mu}} \delta\Lambda_\nu^{\ \mu} \equiv \frac{1}{2}\Gamma_J^{I\nu}{}_\mu \omega_\nu^{\ \mu}, \tag{9.52}$$

defining $\Gamma \equiv dD/d\Lambda$. Raising and lowering indices implies that $\Gamma_J^{I\nu}{}_\mu \omega_\nu^{\ \mu} = \Gamma_J^{I\nu\mu} \omega_{\nu\mu}$, and because $\omega_{\nu\mu}$ is antisymmetric, $\Gamma_J^{I\nu\mu}$ can also be chosen antisymmetric in μ and ν. In abbreviated notation this reads

$$c = \frac{1}{2}\Gamma\delta\Lambda^{-1} \equiv \frac{1}{2}\Gamma\Omega,$$

and then

$$\delta\psi = \frac{1}{2}(\Gamma\Omega)\psi + (\nabla\psi)\Omega x. \tag{9.53}$$

Now insert (9.53) into (9.34) and combine it with (9.50) to arrive at the expression

$$G = \frac{\partial\mathcal{L}}{\partial\nabla\psi}\left[\frac{1}{2}(\Gamma\Omega)\psi + (\nabla\psi)\Omega x\right] - \Omega x\mathcal{L}$$

for the conserved current. The Ω can be factored out and this can then be written in the form

$$G = \Omega\left\{\left(\frac{1}{2}\Gamma\psi + x\nabla\psi\right)\frac{\partial\mathcal{L}}{\partial\nabla\psi} - x\mathcal{L}\right\}. \tag{9.54}$$

Much is hidden here by the abbreviated notation, in particular the factoring out of Ω. When it is unraveled, the expanded equation becomes (see Problem 5)

$$G^\alpha = \frac{1}{2}\omega_{\mu\nu}\left\{\left(\Gamma_I^{J\mu\nu}\psi_J + x^\mu g^{\nu\beta}\psi_{I,\beta} - x^\nu g^{\mu\beta}\psi_{I,\beta}\right)\frac{\partial\mathcal{L}}{\partial\psi_{I,\alpha}} - g^{\alpha\nu}\mathcal{L}x^\mu + g^{\alpha\mu}\mathcal{L}x^\nu\right\}$$

$$\equiv \frac{1}{2}\omega_{\mu\nu}M^{\mu\nu\alpha}, \tag{9.55}$$

defining the $M^{\mu\nu\alpha}$. Since the $\omega_{\mu\nu}$ are constants, the $M^{\mu\nu\alpha}$ themselves constitute six conserved currents, one for each combination of μ and ν (six, not sixteen, because they are antisymmetric in μ and ν):

$$\partial_\alpha M^{\mu\nu\alpha} = 0. \tag{9.56}$$

The $M^{\mu\nu\alpha}$ are the conserved currents derived from the Lorentz invariance of $\mathcal{L}$.

Some of the terms in the expression for $M^{\mu\nu\alpha}$ can be grouped in pairs and written in terms of the energy–momentum tensor $T_\nu{}^\mu$ of Eq. (9.38):

$$\frac{\partial\mathcal{L}}{\partial\psi_{I,\alpha}}x^\mu g^{\nu\beta}\psi_{I,\beta} - x^\mu g^{\alpha\nu}\mathcal{L} = x^\mu g^{\nu\beta}\left(\frac{\partial\mathcal{L}}{\partial\psi_{I,\alpha}}\psi_{I,\beta} - \delta_\beta^\alpha\mathcal{L}\right)$$

$$= x^\mu g^{\nu\beta}T_\beta{}^\alpha = x^\mu T^{\nu\alpha}.$$

Hence

$$M^{\mu\nu\alpha} = \Gamma_I^{J\mu\nu}\psi_J\frac{\partial\mathcal{L}}{\partial\psi_{I,\alpha}} + (x^\mu T^{\nu\alpha} - x^\nu T^{\mu\alpha}). \tag{9.57}$$

We concentrate on the space components of this tensor, the $M^{jk\alpha}$. It is seen from (9.52) that in the first term they are obtained from the jk elements of the Lorentz transformation, namely from $\Lambda_j{}^k$. These are rotations in three-space, which are included among the Lorentz transformations. Just as invariance of $\mathcal{L}$ under space translations leads to momentum conservation, so invariance under rotations leads to conservation of angular momentum. The conserved-current equation that expresses this is $\partial_\alpha M^{jk\alpha} = 0$, so the angular-momentum density in the field is M^{jk0}. In fact the jk components of the last term in (9.57) actually look like the angular momentum. To see this, consider for instance

$$l^{12} \equiv x^1 T^{20} - x^2 T^{10}.$$

The T^{k0} are the components of the momentum density, so think of them as the momenta Δp_k in small regions of space. Then

$$l^{12} \sim x^1\Delta p_2 - x^2\Delta p_1$$

is the 3 component of the angular momentum in small regions of space or is the density of the 3 component of angular momentum.

The angular momentum about the axis perpendicular to the (j, k) plane is given by

$$J^{jk} = \int M^{jk0}\,d^3x. \tag{9.58}$$

More generally, the six quantities

$$J^{\mu\nu} = \int M^{\mu\nu0}\,d^3x \tag{9.59}$$

are conserved. The first term on the right-hand side of (9.57) depends on Γ, and hence on the transformation properties of ψ. If the field is a scalar, $\Gamma = 0$, and all that remains of $M^{\mu\nu\alpha}$ is $(x^\mu T^{\nu\alpha} - x^\nu T^{\mu\alpha})$. This part, through its dependence on the x^μ, depends on the choice of origin, and for that reason its contribution to M^{jk0} is interpreted as *orbital* angular momentum. The Γ-dependent part is interpreted as *intrinsic* angular momentum or *spin*. This is best understood from the quantum point of view, and we will not go into it more deeply.

Recall that the energy–momentum tensor of the electromagnetic field had to be symmetrized by adding the term of Eq. (9.43). Symmetrizing is also more generally necessary, and we state without proof (but see Aharoni, 1959, or Corson, 1955) that it is always possible to add terms so as to replace $T^{\mu\nu}$ by its symmetrized version $\bar{T}^{\mu\nu}$ without changing the values of the P^μ, and that then

$$\bar{M}^{\mu\nu\alpha} = x^\mu \bar{T}^{\nu\alpha} - x^\nu \bar{T}^{\mu\alpha} \tag{9.60}$$

will yield the same values of the $J^{\mu\nu}$ as does Eq. (9.59). In Problem 6 it is shown that the angular momentum density in the electromagnetic field is $(\mathbf{E} \wedge \mathbf{B}) \wedge \mathbf{x}$.

Although that does not exhaust the constants of the motion obtained from the $M^{\mu\nu\alpha}$ tensor, we will not try to describe the other three.

FREE KLEIN–GORDON FIELDS

So far the only relativistic field discussed in this section has been the Maxwell field. We now turn to two others, the *scalar and vector Klein–Gordon (K–G) fields* [see Eq. (9.11)]. All three are *classical fields*. Their dynamical equations are also *quantum particle* wave equations, whose solutions are the quantum wave functions for certain particles. For example, the solution of the scalar K–G equation is the free pion wave function. Because quantum particle theories are classical field theories, the physics of some of the results to be obtained can be understood quantum mechanically, so occasionally that point of view will be taken. It will in fact be necessary for discussing the interaction between the electromagnetic (Maxwell) field and the relativistic particles of the K–G field.

Both the scalar and vector K–G fields are described by the Klein–Gordon equation, which will be given below (recall also Section 9.1). The terms *scalar* and *vector* refer to the field functions, but since both fields are relativistic, both Lagrangian densities are scalars.

The Lagrangian (density) of scalar K–G field is

$$\mathcal{L} = \frac{1}{2}[g^{\alpha\beta}(\partial_\alpha\psi)(\partial_\beta\psi) + \Omega^2\psi^2], \tag{9.61}$$

where Ω^2 is a (positive) constant. Compare this with Eq. (9.10). The only significant differences are that now $v = 1$ and that there are three space dimensions instead of one. In Section 9.1 the dynamical equation was obtained as the limit of a one-dimensional system of particles interacting through springs (giving rise to v) and bound by other springs to their equilibrium positions (giving rise to Ω^2). If they were not bound to their equilibrium positions, the limit would represent longitudinal vibrations of an elastic rod.

The EL equation obtained directly from the $\mathcal{L}$ of (9.61) is

$$(\Box - \Omega^2)\psi = 0, \tag{9.62}$$

where

$$\Box \equiv g^{\alpha\beta}\partial_\alpha\partial_\beta = \nabla^2 - \partial_t^2 \tag{9.63}$$

is called the *d'Alembertian operator*. If Ω were zero this would simply be the scalar wave equation for wave velocity $v = 1$ (this is connected with the units in which the velocity of light is $c = 1$). Equation (9.62) is the quantum mechanical equation for a *scalar particle* of mass Ω. In quantum mechanics the momentum **p** of a particle becomes the operator $-i\nabla$ and its energy becomes $i\partial_t$ (in units in which $\hbar = 1$). In special relativity the mass and energy are related by $g^{\alpha\beta}p_\alpha p_\beta + m^2 = 0$, and in terms of operators this becomes $\Box - m^2 = 0$. Except for the lack of a field function for the operator to act on, this is the K–G equation with $\Omega = m$, which shows that Ω can be identified with the mass.

The Lagrangian of the vector K–G field is similar to the scalar one:

$$\mathcal{L} = \frac{1}{2}[g^{\alpha\beta}(\partial_\alpha \psi^\mu)(\partial_\beta \psi_\mu) + \Omega^2 \psi^\mu \psi_\mu], \tag{9.64}$$

where, as always, $\psi^\mu = g^{\mu\nu}\psi_\nu$. The EL equation for this field is

$$(\Box - \Omega^2)\psi_\mu = 0, \tag{9.65}$$

which can again be interpreted as the quantum mechanical equation for a *vector particle* of mass Ω. If $\Omega = 0$ this is simply the wave equation, which is satisfied by the vector potential A_μ of the Maxwell field in the *Lorentz gauge*, which is given by $\partial_\mu A^\mu = 0$.

The energy–momentum tensor of the K–G field is found by applying the general formula of (9.38):

$$T_\mu{}^\nu = \frac{1}{2}g^{\alpha\beta}\left(\delta_\alpha^\nu \partial_\beta \psi + \delta_\beta^\nu \partial_\alpha \psi\right)\partial_\mu \psi - \delta_\mu^\nu \mathcal{L}$$
$$= g^{\nu\alpha}(\partial_\alpha \psi)(\partial_\mu \psi) - \frac{1}{2}\delta_\mu^\nu[g^{\alpha\beta}(\partial_\alpha \psi)(\partial_\beta \psi) + \Omega^2 \psi^2].$$

When $g^{\rho\mu}$ is used to raise the index μ, this becomes

$$T^{\rho\nu} = \psi^{,\rho}\psi^{,\nu} - \frac{1}{2}g^{\rho\nu}[g^{\alpha\beta}(\partial_\alpha \psi)(\partial_\beta \psi) + \Omega^2 \psi^2],$$

where we have written $\psi^{,\rho} \equiv g^{\rho\nu}\partial_\nu \psi$ to show more clearly that the tensor obtained is already symmetric. We do not calculate the energy–momentum tensor for the vector field.

COMPLEX K–G FIELD AND INTERACTION WITH THE MAXWELL FIELD

So far we have been discussing free fields, both electromagnetic and K–G. Now we consider interacting fields. The K–G and Maxwell fields interact through the electric charge of the K–G particle, as yet unmentioned. It will be seen that in order for a particle to have a charge, its field must be complex; we thus assume it to be of the form $\psi_I = \varphi_I + i\chi_I$, where both φ_I and χ_I are real. There are then $2N$ real fields, rather than N, and the resulting $2N$ EL equations can be found by treating either $\{\psi_I, \psi_I^*\}$ or $\{\varphi_I, \chi_I\}$ as independent. Choose the former. According to (9.18) the $2N$ EL equations are then

$$\frac{\delta L}{\delta \psi} = 0 \quad \text{and} \quad \frac{\delta L}{\delta \psi^*} = 0. \tag{9.66}$$

Take the Lagrangian to be real, that is, to depend on the fields only in real combinations such as $\psi^*\psi$ or $\psi^*\partial_\alpha\psi$, etc. That means that $\mathcal{L}$ is invariant under transformations of the form

$$\psi_I(x, \epsilon) = \psi_I(x)e^{i\epsilon}, \quad \psi_I^*(x, \epsilon) = \psi_I^*(x)e^{-i\epsilon}$$

(called a *global gauge transformations of the first kind*), and this invariance of $\mathcal{L}$ leads to a new conserved current. Under such a gauge transformation $\delta\psi_I = i\psi_I$ and $\delta\psi_I^* = -i\psi_I^*$, so according to Eq. (9.34) (in this case $\Phi = 0$) the conserved current is

$$j^\mu = i\left(\frac{\partial\mathcal{L}}{\partial\psi_{I,\mu}}\psi_I - \frac{\partial\mathcal{L}}{\partial\psi_{I,\mu}^*}\psi_I^*\right). \tag{9.67}$$

Now apply this to the scalar K–G field. The field is complexified by replacing the Lagrangian of (9.61) by

$$\mathcal{L}_{\text{KG}} = g^{\alpha\beta}(\partial_\alpha\psi)(\partial_\beta\psi^*) + \Omega^2\psi\psi^*, \tag{9.68}$$

and then the two EL equations repeat (9.62):

$$(\square - \Omega^2)\psi = 0,$$
$$(\square - \Omega^2)\psi^* = 0.$$

These are two complex conjugate equations, and if one is satisfied, so is the other. The conserved current is (we lower the index)

$$j_\alpha = i(\psi\partial_\alpha\psi^* - \psi^*\partial_\alpha\psi). \tag{9.69}$$

In quantum mechanics this is the usual expression for the electric current four-vector (actually, a multiple of it), and it explains why the particle belonging to the complex field is said to be charged. The gauge transformation of the first kind and the entire derivation of (9.69) depends on the field being complex. In fact for a real field, j^μ vanishes. (Compare with Worked Example 9.1.)

Having obtained an expression for the electric current density, we can discuss the coupling between the K–G and Maxwell fields. The Lagrangian for both fields without any interactions is simply $\mathcal{L}_{\text{KG}} + \mathcal{L}_{\text{EM}}$, the sum of (9.68) and the free-field Maxwell Lagrangian $\mathcal{L}_{\text{EM}} \equiv -\frac{1}{4}f^{\mu\nu}f_{\mu\nu}$. Since ψ and ψ^* appear only in $\mathcal{L}_{\text{KG}}$ and the A_μ only in $\mathcal{L}_{\text{EM}}$, the resulting field equations for the K–G field will not involve the Maxwell field, and vice versa. In other words, the charged particle's behavior with this Lagrangian is the same with or without the electromagnetic field, and the electromagnetic field, in turn, is not affected by the particle. The problem is what to add to the Lagrangian, how to alter it, so that the two fields interact correctly.

Recall that in classical particle dynamics the interaction with the electromagnetic field is obtained by replacing $\mathbf{p}$ by $\mathbf{p} - e\mathbf{A}$, where e is the charge of the particle. This is extended to the four-momentum by replacing p_μ by $p_\mu - eA_\mu$. That is how the K–G and electromagnetic fields will be coupled, simply by making this replacement in $\mathcal{L}_{KG}$, but with the momentum p_μ understood quantum mechanically as $-i\partial_\mu$. Hence ∂_μ is replaced by $\partial_\mu + ieA_\mu$. Such *minimal coupling* then has to be tested for consistency (and of course compared with experiment) when the coupled equations are obtained.

To proceed, therefore, replace the Lagrangian of (9.68) by

$$\mathcal{L}'_{KG} = g^{\alpha\beta}(\partial_\alpha\psi + ieA_\alpha\psi)(\partial_\beta\psi^* - ieA_\beta\psi^*) + \Omega^2\psi\psi^* \tag{9.70}$$

(note that it is still real), so that the total Lagrangian becomes

$$\mathcal{L}_{tot} = g^{\alpha\beta}(\partial_\alpha\psi + ieA_\alpha\psi)(\partial_\beta\psi^* - ieA_\beta\psi^*) + \Omega^2\psi\psi^* - \frac{1}{4}f^{\mu\nu}f_{\mu\nu}.$$

The EL equation obtained from ψ^* is

$$-\frac{\delta L_{tot}}{\delta\psi^*} \equiv [g^{\alpha\beta}(\partial_\alpha + ieA_\alpha)(\partial_\beta + ieA_\beta) - \Omega^2]\psi = 0 \tag{9.71}$$

(the EL equation obtained from ψ is simply the complex conjugate). The EL equation obtained from A_μ is

$$\frac{\delta L_{tot}}{\delta A_\mu} \equiv \frac{\delta L_{EM}}{\delta A_\mu} + \frac{\delta L'_{KG}}{\delta A_\mu} = 0.$$

The first term here is the same as the free-field equation, that is, Eq. (9.31) with $j^\mu = 0$. The second term is

$$J^\mu \equiv -ieg^{\mu\beta}[\psi(\partial_\beta - ieA_\beta)\psi^* - \psi^*(\partial_\beta + ieA_\beta)\psi]. \tag{9.72}$$

Hence the EL equation obtained from A_μ now reads

$$\partial_\nu f^{\mu\nu} = J^\mu. \tag{9.73}$$

Equations (9.71) and (9.73), with the definition of (9.72), are now the coupled field equations.

Coupling the K–G and Maxwell fields has yielded Eqs. (9.72) and (9.73), a set of equations for both the $f^{\mu\nu}$ and the J^μ. This is therefore what was called a complete theory in the second subsection of Section 9.1.3: the current is not an external one inserted by hand but is an integral part of the theory. We do not check explicitly that $\partial_\mu J^\mu = 0$, but that follows from (9.73).

DISCUSSION OF THE COUPLED FIELD EQUATIONS

Equation (9.73) is reasonable if J^μ can be interpreted as the electric current four-vector. Comparison with (9.69) shows that J^μ is j^μ except for the factor $-e$ and the replacement of ∂_α by $\partial_\alpha + ieA_\alpha$ when acting on ψ and by its complex conjugate when acting on ψ^*, as in the Lagrangian. Hence if (9.72) is taken as the definition of the electric current four-vector, minimal coupling modifies the electromagnetic field precisely as desired. Nevertheless there is a problem: Equation (9.72), and hence the right-hand side of (9.73), is not invariant under gauge transformations of the A_μ, for when A_μ is replaced by $\bar{A}_\mu = A_\mu + \partial_\mu \chi$, the term $(\partial_\mu + ieA_\mu)\psi$ changes to

$$(\partial_\mu + ie\bar{A}_\mu)\psi = (\partial_\mu + ieA_\mu + ie\partial_\mu \chi)\psi = e^{-ie\chi}(\partial_\mu + ieA_\mu)e^{ie\chi}\psi.$$

For later consideration it is convenient to write this as

$$(\partial_\mu + ie\bar{A}_\mu)e^{-ie\chi}\psi = e^{-ie\chi}(\partial_\mu + ieA_\mu)\psi.$$

A second question relating to (9.72) is whether the J^μ it defines is the conserved current obtained from invariance of the Lagrangian under gauge transformations of the first kind (like $\mathcal{L}_{KG}$, the new Lagrangian $\mathcal{L}_{tot}$ is invariant under these transformations).

These two questions are related. They are handled by defining the following more general transformation on all the fields ψ, ψ^*, and the A_μ:

$$\begin{aligned} A_\mu &\to \bar{A}_\mu = A_\mu + \partial_\mu \chi, \\ \psi &\to \bar{\psi} = e^{-ie\chi}\psi, \\ \psi^* &\to \bar{\psi}^* = e^{ie\chi}\psi^*, \end{aligned} \tag{9.74}$$

so that

$$(\partial_\mu + ie\bar{A}_\mu)\bar{\psi} = e^{-ie\chi}(\partial_\mu + ieA_\mu)\psi. \tag{9.75}$$

Under this transformation both (9.71) and (9.73) are valid for the barred as well as the unbarred functions. Equation (9.74) is called an *extended gauge transformation* or a *gauge transformation of the second kind*. Problem 8 shows that $\mathcal{L}_{tot}$ is invariant under extended gauge transformations and the extent to which the associated Noether conserved current is J^μ (this turns out not to be as simple as one might have guessed; see Karatas and Kowalski, 1990).

9.2.3 SPINORS

SPINOR FIELDS

Whereas relativistic field theories must be Lorentz invariant, nonrelativistic linear field theories under isotropic conditions must be invariant under rotations. A common example of a nonrelativistic equation is the wave equation

$$v^2 \nabla^2 \psi_I - \frac{\partial^2 \psi_I}{\partial t^2} = 0, \tag{9.76}$$

where ψ may be a scalar or a component of a vector. The operator $\nabla^2 \equiv \sum_j \partial_j^2$ is invariant under rotations, but whether or not the entire equation is invariant depends on the transformation properties of the wave function ψ_l. When ψ is a scalar, the equation is clearly rotationally invariant. When the field is a vector function ψ with components ψ_1, ψ_2, ψ_3 in a Cartesian coordinate system, Eq. (9.76) is satisfied by each of its components separately. It can then be shown in the usual way that if the ψ_l satisfy the equation in one orthonormal coordinate system, then in any system obtained from the first by a rotation the new components will also satisfy it.

This property is called rotational invariance (a more exact term would be rotational *covariance*) of the equation. The situation for tensor fields is similar. Tensors transform like products of vectors: if the nine elements of a second-rank tensor are labeled T_{jk}, they transform among themselves like the nine products $v_k w_j$ of two vectors $\mathbf{v}$ and $\mathbf{w}$. Tensors of higher rank transform like products of more than two vectors, and in this way classical, rotationally invariant field theories can be constructed for tensors of any rank. (More significantly, similar results are obtained for relativistic theories, but we will not go into that.)

This subsection discusses another kind of rotationally invariant equation, a nonrelativistic *spinor* wave equation modeled on Dirac's relativistic spinor wave equation for the electron. (The equation we present is not the *Pauli equation* for the nonrelativistic electron. Ours is of first order in the space variables, whereas the Pauli equation is of second order, and ours applies to mass zero, whereas the mass of the electron enters into the Pauli equation – as well as the Dirac equation.) The rotational properties of spinors depend on the **SU**(2) representation of the rotation group **SO**(3), discussed in Section 8.4. Unlike scalar fields, spinor fields can carry angular momentum, and their importance in quantum mechanics is that they describe *fermions*, particles of half-integral spin.

A spinor field Ψ is a two-component object represented by a column vector (or *column spinor*) of the form

$$\Psi = \begin{pmatrix} \psi_1(x, t) \\ \psi_2(x, t) \end{pmatrix}, \tag{9.77}$$

and under rotations it is transformed by the two-dimensional U matrices of Section 8.4.2. That is, under a rotation $R \in \mathbf{SO}(3)$ a spinor transforms according to

$$\Psi(x, t) \rightarrow \Psi'(x, t) = U_R \Psi(R^{-1}x, t), \tag{9.78}$$

where U_R is the **SU**(2) representation of R. Compare with Eq. (9.49). Because of the factor of $\frac{1}{2}$ in the connection between **SO**(3) and **SU**(2), a continuous change from $R = \mathbb{I}$ to a rotation through an angle of 2π will change Ψ to $-\Psi$.

The Lagrangian of the spinor field theory is written in terms of Ψ and its *Hermitian conjugate* (or *Hermitian adjoint*), the *row spinor*

$$\Psi^\dagger = (\psi_1^\dagger, \psi_2^\dagger) = (\psi_1^*, \psi_2^*)$$

(it is as though Ψ were flipped across a nonexistent main diagonal and the complex conjugate were taken). Under a rotation

$$\Psi^\dagger(x, t) \rightarrow \Psi^{\dagger'}(x, t) = \Psi^\dagger(R^{-1}x, t)U_R^\dagger.$$

This and Eq. (9.78) will be written in the abbreviated form

$$\Psi' = U_R \Psi \quad \text{and} \quad \Psi^{\dagger\prime} = \Psi^\dagger U_R^\dagger.$$

The unitarity of the U_R matrices implies that the *norm*

$$\|\Psi\| \equiv \Psi^\dagger \Psi \equiv \sum_{k=1}^{2} |\psi_k|^2$$

is invariant under rotation.

A SPINOR FIELD EQUATION

Our spinor example is modeled on the Dirac equation: the Lagrangian density is a nonrelativistic analog, namely (see Section 8.4)

$$\mathcal{L} = \frac{i}{2}\{\Psi^\dagger \partial_t \Psi + v\Psi^\dagger \mathbf{s} \cdot \nabla\Psi\} - \frac{i}{2}\{(\partial_t \Psi^\dagger)\Psi + v(\nabla\Psi^\dagger) \cdot \mathbf{s}\Psi\}, \qquad (9.79)$$

where v is a velocity (which will turn out to be the velocity of propagation). This looks also a bit like the Schrödinger Lagrangian of Worked Example 9.1. In the Schrödinger Lagrangian each term appears with its complex conjugate, and in our spinor Lagrangian each term appears with its Hermitian adjoint (the Pauli matrices are all Hermitian). But the Schrödinger Lagrangian is quadratic in ∇, whereas our spinor Lagrangian is linear in ∇. Rotational invariance follows from the fact that the field appears in rotationally invariant combinations of Ψ and $\Psi^\dagger$ and that $\mathbf{s} \cdot \nabla$ is a scalar operator, so $\mathcal{L}$ is a rotational scalar.

The field equations are obtained by treating Ψ and $\Psi^\dagger$ as independent, as were ψ and ψ^*. They are

$$-i\frac{\delta\mathcal{L}}{\delta\psi} = \partial_t \Psi^\dagger + v\nabla\Psi^\dagger \cdot \mathbf{s} = 0$$

and its Hermitian adjoint

$$i\frac{\delta\mathcal{L}}{\delta\Psi^\dagger} = \partial_t \Psi + v\mathbf{s} \cdot \nabla\Psi = 0. \qquad (9.80)$$

As is the case for complex fields, each of these two equations implies the other, so it is enough to consider only (9.80).

In solving Eq. (9.80) it is first shown that any solution Ψ is also a solution of (9.76). Indeed, apply ∂_t to (9.80):

$$\partial_t^2 \Psi + v\mathbf{s} \cdot \nabla(\partial_t \Psi) = 0.$$

If Ψ satisfies (9.80), the second term becomes [use (8.110) and (8.111)]

$$v^2(\mathbf{s} \cdot \nabla)^2 \Psi \equiv v^2 \nabla^2 \Psi,$$

so Ψ satisfies (9.76). Thus each component of the spinor field satisfies the scalar wave equation. But in addition the spinor field satisfies Eq. (9.80), and that leads to a relation between ψ_1 and ψ_2.

Consider a spinor *plane wave solution* of (9.76) propagating in the positive 3 direction, that is, a solution of the form $\Psi(x_3 - vt) \equiv \Psi(\xi)$, where $\xi = x_3 - vt$. Then only the 3-derivative in $\nabla \Psi$ is nonzero, equal to $\partial_3 \Psi = \partial_\xi \Psi$; also, $\partial_t \Psi = -v \partial_\xi \Psi$. Then (9.80) implies that

$$\partial_\xi \Psi = \sigma_3 \partial_\xi \Psi.$$

Thus $\partial_\xi \Psi$ is an *eigenspinor* of σ_3 belonging to eigenvalue 1, and integration shows that Ψ itself is an eigenspinor of σ_3 belonging to eigenvalue 1. It follows from (8.107) that $\psi_2 = 0$. There is nothing special about the 3 direction, so this result can be generalized: for a plane wave propagating in an arbitrary direction $\mathbf{n}$, the spinor wave is an eigenvector of $\mathbf{s} \cdot \mathbf{n}$ belonging to eigenvalue 1.

In classical field theory this result has no particular meaning, but in quantum mechanics the $\mathbf{s}$ operator determines the spin of the particle associated with the field. The quantum interpretation is that a particle that is in the state represented by a plane wave has its spin lined up with the direction of propagation. As was mentioned in Chapter 8, $\mathbf{SU}(2)$ is associated with spin $\frac{1}{2}$, so what this means is that the spin of a massless spin-$\frac{1}{2}$ particle is along the direction of its motion. We do not proceed any further with this discussion.

9.3 THE HAMILTONIAN FORMALISM

9.3.1 THE HAMILTONIAN FORMALISM FOR FIELDS

DEFINITIONS

If the Lagrangian density $\mathcal{L}$ is independent of the Ath field function ψ_A, the Ath EL equation reads

$$\partial_\mu \frac{\partial \mathcal{L}}{\partial \psi_{A,\mu}} = 0, \tag{9.81}$$

which means that the $\partial \mathcal{L} / \partial \psi_{A,\mu}$ form a conserved current. The particle analog of this is that if q^α does not appear in L, the conjugate momentum $p_\alpha \equiv \partial L / \partial \dot{q}^\alpha$ is conserved. This expression was taken as the definition of p_α and used in the transition from the Lagrangian to the Hamiltonian formalism in particle dynamics. The analogous transition from the Lagrangian to the Hamiltonian formalism in continuum dynamics must be done carefully, for the analogs of $\dot{q}^\alpha$ are the $\psi_{I,\mu}$, derivatives of the field function with respect to *four* parameters. If this analogy were used for the transition to the Hamiltonian formalism in continuum dynamics, there would be *four* momenta $\partial \mathcal{L} / \partial \psi_{I,\mu}$ conjugate to each field function.

Therefore the *momentum function* conjugate to $\psi_{I,\mu}$ is defined differently, as

$$\pi^I(x) = \frac{\partial \mathcal{L}}{\partial \psi_{I,0}} \tag{9.82}$$

(recall that $\psi_{I,0}$ is $\partial_t \psi_I$). If (9.81) is satisfied, then from the discussion of conserved currents in Section 9.2.1 it follows that

$$\frac{d}{dt}\int \pi^I d^3x \equiv \frac{d}{dt}\int \frac{\partial L}{\partial \psi_{I,0}} d^3x = 0.$$

In the Hamiltonian formalism, therefore, the role of $t \equiv x^0$ is different from that of the three other x^k. In spite of this, at least for a while we will use the notation developed for the discussion of relativistic fields, which treats all the x^μ on a more equal footing.

With the definition of the conjugate momentum function, the Hamiltonian is constructed in analogy with the particle equation $H = \dot{q}^\alpha p_\alpha - L$:

$$H = \int \psi_{I,0} \pi^I d^3x - L \equiv \int (\psi_{I,0}\pi^I - L)d^3x \equiv \int \mathcal{H}d^3x,$$

where

$$\mathcal{H} = \psi_{I,0}\pi^I - L. \tag{9.83}$$

This defines the Hamiltonian H and the *Hamiltonian density* $\mathcal{H}$. Notice that according to Eq. (9.38), $\mathcal{H} = T^0_0$, so the Hamiltonian density is the energy density.

The Hamiltonian formulation proceeds also in analogy with particle dynamics. Equation (9.82) is solved for the $\psi_{I,0}$ as functions of π^I, $\psi_{I,k}$, x^μ, and these solutions are inserted into (9.83) to yield the function $\mathcal{H}(\psi_I, \pi^I, \psi_{I,k}, x^\mu)$. The analogs of Hamilton's canonical equations are then obtained as described next.

THE CANONICAL EQUATIONS

Equation (9.83) is used to write

$$\frac{\partial \mathcal{H}}{\partial \pi^I} = \psi_{I,0} + \frac{\partial \psi_{J,0}}{\partial \pi^I}\pi^J - \frac{\partial L}{\partial \psi_{J,0}}\frac{\partial \psi_{J,0}}{\partial \pi^I}.$$

The last two terms cancel according to (9.82), so that

$$\psi_{I,0} = \frac{\partial \mathcal{H}}{\partial \pi^I},$$

which is a close analog of the particle equation $\dot{q}^\alpha = \partial H/\partial p_\alpha$. This represents half of the canonical equations.

Obtaining the other half starts from

$$\frac{\partial \mathcal{H}}{\partial \psi_J} = \pi^I \frac{\partial \psi_{I,0}}{\partial \psi_J} - \frac{\partial L}{\partial \psi_{I,0}}\frac{\partial \psi_{I,0}}{\partial \psi_J} - \frac{\partial L}{\partial \psi_J}.$$

Again the first two terms cancel according to (9.82). The field equations (9.18) are used to rewrite the last term:

$$\frac{\partial L}{\partial \psi_J} = \frac{\partial}{\partial x^\alpha}\frac{\partial L}{\partial \psi_{J,\alpha}} = \frac{\partial}{\partial x^k}\frac{\partial L}{\partial \psi_{J,k}} + \frac{\partial}{\partial x^0}\frac{\partial L}{\partial \psi_{J,0}}$$

$$= \frac{\partial}{\partial x^k}\frac{\partial L}{\partial \psi_{J,k}} + \frac{\partial \pi^J}{\partial x^0}.$$

According to Eq. (9.83), the first term is

$$\frac{\partial \mathcal{L}}{\partial \psi_{J,k}} = -\frac{\partial \mathcal{H}}{\partial \psi_{J,k}} + \pi^I \frac{\partial \psi_{I,0}}{\partial \psi_{J,k}} - \frac{\partial \mathcal{L}}{\partial \psi_{I,k}} \frac{\partial \psi_{I,0}}{\partial \psi_{J,k}}.$$

Again the last two terms cancel, so

$$\frac{\partial \pi^I}{\partial x^0} \equiv \pi^I_{,0} = -\frac{\partial \mathcal{H}}{\partial \psi_I} + \frac{\partial}{\partial x^k} \frac{\partial \mathcal{H}}{\partial \psi_{I,k}}.$$

The derivative with respect to x^0 is the time derivative, so this last equation is the analog of $\dot{p}_\alpha = -\partial H / \partial q^\alpha$. It can be made to look more like the analog if it is restated in terms of the functional derivative defined at Eq. (9.12) or (9.18). Since $\mathcal{H}$ does not depend on the $\pi^I_{,\mu}$ or on $\psi_{I,0}$, both sets of equations can be written as

$$\psi_{I,0} = \frac{\delta H}{\delta \pi^I}, \quad \pi^I_{,0} = -\frac{\delta H}{\delta \psi_I}. \tag{9.84}$$

In spite of the similarity of these equations to the canonical equations of particle dynamics, the analogy is rather weak (but see Section 9.6): the argument of $\mathcal{H}$ includes the space derivatives of the ψ_I but not of the π^I. As a result, the ψ_I and π^I are different kinds of variables, and there are no evident analogs of $\mathbf{T}^*\mathbb{Q}$ or canonical transformations.

WORKED EXAMPLE 9.3

Consider a scalar field whose Lagrangian density is of the form

$$\mathcal{L} \equiv \frac{1}{2}\left\{(\partial_t \psi)^2 - v^2(\partial_x \psi)^2\right\} - U(\psi).$$

(a) Apply the Hamiltonian formalism to this system: obtain $\mathcal{H}$ and the field equation.
(b) Specialize to the sG and K–G fields.

Solution. (a) According to (9.82)

$$\pi = \frac{\partial \mathcal{L}}{\partial(\partial_t \psi)} = \partial_t \psi,$$

and then from (9.83)

$$\mathcal{H} = \pi^2 - \mathcal{L} = \frac{1}{2}\{\pi^2 + v^2(\partial_x \psi)^2\} + U(\psi).$$

The canonical equations are

$$\partial_t \psi = \pi \quad \text{and} \quad \partial_t \pi = -U'(\psi) + v^2 \partial_x^2 \psi.$$

As usual, the first of these merely repeats the definition of π. It can be combined with the second to yield the field equation

$$(\partial_t^2 - v^2\partial_x^2)\psi + U'(\psi) = 0.$$

(b) The sG Lagrangian density is given by (9.7):

$$\mathcal{L}_{sG} \equiv \frac{1}{2}\{(\partial_t\psi)^2 - v^2(\partial_x\psi)^2 - 2\Omega^2(1 - \cos\psi)\}.$$

Hence $U_{sG}(\psi) = \Omega^2(1 - \cos\psi)$ and the field equation is

$$(\partial_t^2 - v^2\partial_x^2)\psi + \Omega^2\sin\psi = 0,$$

which agrees with (9.5). The Hamiltonian is

$$\mathcal{H}_{sG} = \frac{1}{2}\{\pi^2 + v^2(\partial_x\psi)^2\} + \Omega^2(1 - \cos\psi).$$

In the linearized limit the sG equation becomes the K–G equation for a general speed v, not necessarily the speed of light c. The Lagrangian density becomes

$$\mathcal{L}_{K-G} \equiv \frac{1}{2}[(\partial_t\psi)^2 - v^2(\partial_x\psi)^2 - \Omega^2\psi^2]. \tag{9.85}$$

Hence $U_{K-G}(\psi) = \Omega^2\psi^2$ and the equation of motion is

$$(\partial_t^2 - v^2\partial_x^2)\psi + \Omega^2\psi = 0,$$

which agrees with (9.11). The Hamiltonian is

$$\mathcal{H}_{K-G} = \frac{1}{2}\{\pi^2 + v^2(\partial_x\psi)^2\} + \Omega^2\psi^2.$$

POISSON BRACKETS

Although canonical transformations have no evident analog in this formalism, Poisson brackets of dynamical variables do. Dynamical variables are now defined to be functions of the ψ_I, π^I, $\psi_{I,k}$, and perhaps the x^μ. Suppose $\mathcal{F}$ is a dynamical variable, and write

$$F = \int \mathcal{F} d^3x.$$

Then

$$\frac{dF}{dt} = \int \left(\partial_t\mathcal{F} + \frac{\partial\mathcal{F}}{\partial\psi_I}\psi_{I,0} + \frac{\partial\mathcal{F}}{\partial\pi^I}\pi^I_{,0} + \frac{\partial\mathcal{F}}{\partial\psi_{I,k}}\psi_{I,k0}\right) d^3x,$$

where $\psi_{I,k0} = \partial_t \psi_{I,k} \equiv \partial_0 \psi_{I,k}$. The last term in the integral can be written in the form

$$\frac{\partial \mathcal{F}}{\partial \psi_{I,k}} \psi_{I,k0} = \partial_k \left(\frac{\partial \mathcal{F}}{\partial \psi_{I,k}} \psi_{I,0} \right) - \partial_k \left(\frac{\partial \mathcal{F}}{\partial \psi_{I,k}} \right) \psi_{I,0}.$$

The first term here leads to a surface integral. Assume, as we've done several times before, that surface integrals of field variables vanish in the limit as regions of integration grow to cover all of space. Then

$$\frac{dF}{dt} = \int \partial_t \mathcal{F} d^3x + \int \left[\frac{\partial \mathcal{F}}{\partial \psi_I} \psi_{I,0} - \partial_k \left(\frac{\partial \mathcal{F}}{\partial \psi_{I,k}} \right) \psi_{I,0} + \frac{\partial \mathcal{F}}{\partial \pi^I} \pi^I_{,0} \right] d^3x$$

$$= \int \partial_t \mathcal{F} d^3x + \int \left[\frac{\delta F}{\delta \psi_I} \psi_{I,0} + \frac{\delta F}{\delta \pi^I} \pi^I_{,0} \right] d^3x$$

$$= \int \partial_t \mathcal{F} d^3x + \int \left[\frac{\delta F}{\delta \psi_I} \frac{\delta H}{\delta \pi^I} - \frac{\delta F}{\delta \pi^I} \frac{\delta H}{\delta \psi_I} \right] d^3x$$

$$\equiv \int \partial_t \mathcal{F} d^3x + \{F, H\}^f . \tag{9.86}$$

This defines $\{F, H\}^f$, the *Poisson bracket* (or PB) *for fields*.

As in particle dynamics, PBs provide a way of obtaining the time variation of dynamical variables without first solving the equations of motion. For instance, PBs can be used to show that the integral over all space of $\mathcal{G} = \pi \partial_x \psi$ is a constant of the motion in the one-dimensional scalar K–G field. To do so, first calculate the functional derivatives of H and G. The canonical equations give (see the end of Worked Example 9.3; $\mathcal{H}_{K-G}$ is now simply $\mathcal{H}$)

$$\frac{\delta H}{\delta \psi} = \Omega^2 \psi - v^2 \partial_x^2 \psi, \qquad \frac{\delta H}{\delta \pi} = \pi;$$

for $\mathcal{G}$ they are

$$\frac{\delta G}{\delta \psi} = -\partial_x \pi, \qquad \frac{\delta G}{\delta \pi} = \partial_x \psi.$$

Then

$$\{G, H\}^f = \int \left[-\pi \partial_x \pi - \Omega^2 \psi \partial_x \psi + v^2 (\partial_x \psi) \partial_x^2 \psi \right] dx$$

$$= \frac{1}{2} \int \partial_x [-\pi^2 - \Omega^2 \psi^2 + v^2 (\partial_x \psi)^2] \, dx.$$

This can be converted to a surface integral (in this one-dimensional case it is merely the difference between end-point values), which vanishes for large enough regions by the usual assumption. Hence $\{G, H\}^f = 0$, and

$$\frac{dG}{dt} = \int \partial_t \mathcal{G} d^3x = 0.$$

Although the Hamiltonian formalism is not very useful in classical field theory, it is important in making the transition to quantum field theory; the PBs of interest are analogs of $\{\xi^\alpha, \xi^\beta\}$. Because of the special role time plays in the Hamiltonian formalism, the needed field PBs turn out to be of the form $\{\psi_I(x^0, \mathbf{y}), \pi^J(x^0, \mathbf{z})\}^f$, PBs of field functions at different points $\mathbf{y}$ and $\mathbf{z}$ in space and at the same time x^0. In quantum field theory they appear as *equal time commutators*. But field PBs are defined for the *integrals* of functions over 3-space, so in order to calculate these, the field functions must be written as such integrals. This is done with the aid of the delta function:

$$\psi_I(x^0, \mathbf{y}) = \int \delta(\mathbf{x} - \mathbf{y})\psi_I(x)\, d^3x,$$

$$\pi^J(x^0, \mathbf{z}) = \int \delta(\mathbf{x} - \mathbf{z})\pi^J(x)\, d^3x.$$

Then (leaving out the argument x^0)

$$\left.\begin{aligned}
\frac{\delta\psi_I(\mathbf{y})}{\delta\psi_K} &= \frac{\partial}{\partial\psi_K}\left[\delta(\mathbf{x} - \mathbf{y})\psi_I(x)\right] = \delta(\mathbf{x} - \mathbf{y})\delta_I^K, \\
\frac{\delta\pi^J(\mathbf{z})}{\delta\psi_K} &= \frac{\partial}{\partial\psi_K}\left[\delta(\mathbf{x} - \mathbf{z})\pi^J(x)\right] = 0.
\end{aligned}\right\} \tag{9.87}$$

Similar results can be obtained for the functional derivatives with respect to π^K. The result is

$$\{\psi_I(\mathbf{y}), \pi^J(\mathbf{z})\}^f = \int \delta(\mathbf{x} - \mathbf{y})\delta_I^K \delta(\mathbf{x} - \mathbf{z})\delta_K^J\, d^3x = \delta(\mathbf{y} - \mathbf{z})\delta_I^J. \tag{9.88}$$

These equations are obvious analogs of the $\{q, p\}$ brackets, except that for the continuous variables $\mathbf{y}$ and $\mathbf{z}$ the Kronecker delta is replaced by the delta function. The field PBs of the ψ_I with each other and of the π^I with each other all vanish, also in analogy with particle dynamics. Field theory involves the $\psi_{I,k}$ as well, and they have no particle analogs. A calculation for them shows that

$$\{\psi_{I,k}(\mathbf{y}), \psi_J(\mathbf{z})\}^f = 0,$$
$$\{\psi_{I,k}(\mathbf{y}), \pi^J(\mathbf{z})\}^f = \delta_I^J \partial_k \delta(\mathbf{y} - \mathbf{z}).$$

See Problem 9. (We have considered here only the nonrelativistic case.)

9.3.2 EXPANSION IN ORTHONORMAL FUNCTIONS

Field theory was obtained as the $N \to \infty$ limit of systems of N interacting particles. Such systems are fruitfully understood in terms of normal modes (see the Chapter 4). This section will describe a similar approach in classical field theory, one based on expanding the fields in sets of functions that define the equivalent of normal modes. This leads to a more canonical Hamiltonian formalism than the direct analog approach of the previous section and it separates out the time in a more reasonable way.

ORTHONORMAL FUNCTIONS

Suppose the field functions are defined over a manifold $M = \mathbb{R} \times M_S$, where $\mathbb{R}$ is the range of t and M_S is the space manifold. Usually M_S is three-space, but it could be the line, as in the one-dimensional wave equation, or the unit circle in the complex plane, or some other manifold. For the present we take it to be ordinary three-space. Many of the ideas that follow are obtained by extending the theory of vector spaces (see the appendix) to infinite dimensions.

Let $\varphi_n(\mathbf{x})$, $n = \ldots, -2, -1, 0, 1, 2, \ldots$ be an infinite *complete set of orthonormal functions* over three-space. *Completeness* means that any sufficiently smooth function $\psi(\mathbf{x}) \in \mathcal{F}(M_S)$ that goes to zero rapidly enough as $|\mathbf{x}| \to \infty$ can be *expanded* in a series of these functions with unique coefficients: the series *converges to ψ in the mean* (Byron and Fuller, 1969). We add $t = x^0$ as a parameter to the argument of ψ, and then

$$\psi(x^0, \mathbf{x}) \equiv \psi(x) = \sum_n a_n(x^0)\varphi_n(\mathbf{x}). \tag{9.89}$$

Derivatives of the ψ functions can also be expanded. The time derivative of ψ is

$$\psi_{,0}(x) = \sum_n \dot{a}_n(x^0)\varphi_n(\mathbf{x}), \tag{9.90}$$

and the kth space derivative is

$$\partial_k \psi(x) = \sum_n a_n(x^0)\partial_k \varphi_n(\mathbf{x}). \tag{9.91}$$

Orthonormality means that

$$(\varphi_n, \varphi_m) \equiv \int \varphi_n^*(\mathbf{x})\varphi_m(\mathbf{x})\, d^3x = \delta_{nm}, \tag{9.92}$$

which also defines a *Hermitian inner product* on $\mathcal{F}$ (actually not on all of $\mathcal{F}$, but only on the acceptable – sufficiently smooth and rapidly decreasing – functions):

$$(\psi, \chi) = \int \psi^*(\mathbf{x})\chi(\mathbf{x})d^3x. \tag{9.93}$$

Orthonormality is used to obtain explicit expressions for the expansion coefficients of ψ in terms of ψ itself:

$$(\varphi_n, \psi) = \left(\varphi_n, \sum a_m\varphi_m\right) = \sum a_m(\varphi_n, \varphi_m) = a_n. \tag{9.94}$$

That is,

$$a_n = \int \varphi_n^*(\mathbf{x})\psi(\mathbf{x})\, d^3x. \tag{9.95}$$

If ψ_I is a field function, it determines and is determined uniquely by its expansion coefficients $a_{In}(t)$, which must carry the same index I as the functions.

The most familiar orthonormal set is $\varphi_n = (2\pi)^{-1/2}e^{in\theta}$ for which M_S is the interval $0 \le \theta \le 2\pi$ (or the unit circle in the complex plane), and the expansion of a function is called its Fourier series. We will use Fourier series in some examples.

PARTICLE-LIKE EQUATIONS

Since the $\psi_I(t, \mathbf{x})$ and the $a_{In}(t)$ are equivalent, the field theory can be rewritten in terms of the a_{In} in both its Lagrangian and Hamiltonian formalisms. It will now be shown that when rewritten, the theory looks like particle dynamics for an infinite number of particles. Specifically, in terms of the a_{In} the EL equations are

$$\frac{d}{dt}\frac{\partial L}{\partial \dot{a}_{In}} - \frac{\partial L}{\partial a_{In}} = 0, \tag{9.96}$$

where $L = \int \mathcal{L}\, d^3x$ is the Lagrangian as previously defined.

Equation (9.96) can be verified by calculating the derivatives that appear in it. First,

$$\frac{\partial L}{\partial a_{In}} = \frac{\partial}{\partial a_{In}}\int \mathcal{L}\, d^3x = \int \left[\frac{\partial \mathcal{L}}{\partial \psi_J}\frac{\partial \psi_J}{\partial a_{In}} + \frac{\partial \mathcal{L}}{\partial \psi_{J,k}}\frac{\partial \psi_{J,k}}{\partial a_{In}}\right] d^3x.$$

According to (9.89), $\partial \psi_J/\partial a_{In} = \delta_J^I \varphi_n$, and according to (9.91), $\partial \psi_{J,k}/\partial a_{In} = \delta_J^I \partial_k \varphi_n$. Thus

$$\frac{\partial L}{\partial a_{In}} = \int \left[\frac{\partial \mathcal{L}}{\partial \psi_I}\varphi_n + \frac{\partial \mathcal{L}}{\partial \psi_{I,k}}\partial_k \varphi_n\right] d^3x.$$

If the second term is integrated by parts and a surface integral thrown away, the result is

$$\frac{\partial L}{\partial a_{In}} = \int \left[\frac{\partial \mathcal{L}}{\partial \psi_I} - \frac{\partial}{\partial x_k}\frac{\partial \mathcal{L}}{\partial \psi_{I,k}}\right] \varphi_n\, d^3x.$$

Next, a similar procedure yields

$$\frac{\partial L}{\partial \dot{a}_{In}} = \int \frac{\partial \mathcal{L}}{\partial \psi_{I,0}}\varphi_n d^3x. \tag{9.97}$$

From the last two equations it follows that

$$\frac{d}{dt}\frac{\partial L}{\partial \dot{a}_{In}} - \frac{\partial L}{\partial a_{In}} = \int \left[\frac{\partial}{\partial x^0}\frac{\partial \mathcal{L}}{\partial \psi_{I,0}} - \frac{\partial \mathcal{L}}{\partial \psi_I} + \frac{\partial}{\partial x_k}\frac{\partial \mathcal{L}}{\partial \psi_{I,k}}\right] \varphi_n d^3x$$

$$= \int \left[\frac{\partial}{\partial x^\mu}\frac{\partial \mathcal{L}}{\partial \psi_{I,\mu}} - \frac{\partial \mathcal{L}}{\partial \psi_I}\right] \varphi_n d^3x.$$

If the EL equations for the field are satisfied, the integrand vanishes and Eqs. (9.96) follow. The converse is also true. Indeed, if Eqs. (9.96) are satisfied, the last integral vanishes. Completeness then implies that the bracketed term in its integrand is zero (the EL equations for the field are satisfied), so Eqs. (9.96) are equivalent to the field equations.

Since (9.96) looks like the usual EL equations (for an infinite number of particles whose generalized coordinates are the a_{In}), the corresponding Hamiltonian formalism is obtained by the usual Legendre transform for particle dynamics. The momentum conjugate to a_{In} is defined by

$$b_n^I = \frac{\partial L}{\partial \dot{a}_{In}}. \tag{9.98}$$

(We raise only the field indices I, etc. because the summation convention will be used only for them. We will use summation signs on the expansion indices n, etc.)

The π^J and b_n^J are related; their relation is obtained by comparing Eqs. (9.97) and (9.98) [using Hermiticity and (9.82)]:

$$b_n^I = \int \pi^I \varphi_n \, d^3x \equiv (\pi^{I*}, \varphi_n) = (\varphi_n, \pi^{I*})^*,$$

or $b_n^{I*} = (\varphi_n, \pi^{I*})$. This means, according to Eq. (9.94), that the b_n^{I*} are the expansion coefficients of π^{I*}, that is, $\pi^{I*} = \sum b_n^{I*} \varphi_n$. Taking complex conjugates yields

$$\pi^I = \sum b_n^I \varphi_n^*. \tag{9.99}$$

As usual, the Hamiltonian is defined by

$$H = \sum b_n^I \dot{a}_{In} - L,$$

and it can then be shown (Problem 10) that $H = \int \mathcal{H} \, d^3x$. The canonical equations are

$$\dot{a}_{In} = \frac{\partial H}{\partial b_n^I} \equiv \{a_{In}, H\},$$

$$\dot{b}_n^I = -\frac{\partial H}{\partial a_{In}} = \{b_n^I, H\}.$$

The Poisson brackets here are the usual ones (not the field PBs of Section 9.3.1), except that the sums are infinite. The fundamental PBs are

$$\{a_{In}, b_m^J\} = \delta_I^J \delta_{nm},$$

and all others vanish. We do not go further with the general description of the Hamiltonian formalism, but it is evident that it arises naturally in this treatment.

EXAMPLE: KLEIN–GORDON

Usually $\mathbb{M}_S$ is unbounded, for example the infinite line or all of 3-space. It is often easiest, however, to write the theory on a finite region and pass to the infinite limit at the end. As an illustration, consider the one-dimensional scalar K–G field of Worked Example 9.3, now on a finite interval: $0 \le x \le X$. Take the $\varphi_n(x)$ to be of the form Ne^{ik_nx}, where N is a normalizing constant and $k_n = 2\pi n/X = -k_{-n}$, which leads to a Fourier decomposition of the field. Normalization requires that $N = X^{-1/2}$. Write

$$\psi(x, t) = X^{-1/2} \sum_{n=-\infty}^{\infty} a_n(t) e^{ik_nx},$$

$$\partial_t \psi = X^{-1/2} \sum_{n=-\infty}^{\infty} \dot{a}_n(t) e^{ik_nx},$$

$$\partial_x \psi = iX^{-1/2} \sum_{n=-\infty}^{\infty} k_n a_n(t) e^{ik_nx}.$$

If ψ is to be real, $k_n = -k_{-n}$ implies that $a_{-n} = a_n^*$.

The Lagrangian is obtained by first writing $\mathcal{L}$ in accordance with (9.85):

$$\mathcal{L} = \frac{1}{2X} \sum_{n,m} \{\dot{a}_n \dot{a}_m + (v^2 k_n k_m - \Omega^2) a_n a_m\} e^{i(k_n + k_m)x}.$$

The integral to obtain L involves only the x-dependent part, outside the brackets, for which

$$\int_0^X e^{2\pi i(n+m)/X} dx = X \delta_{n,-m},$$

so that

$$L = \frac{1}{2} \sum_n [\dot{a}_n \dot{a}_{-n} + (v^2 k_n k_{-n} - \Omega^2) a_n a_{-n}].$$

Finally, use the condition that $a_n^* = a_{-n}$ to write this in the form

$$L = \frac{1}{2} \sum_n \dot{a}_n^* \dot{a}_n - \frac{1}{2} \sum_n (v^2 k_n^2 + \Omega^2) a_n^* a_n.$$

If one ignores for the moment the fact that the a_n are complex, the first term looks like the kinetic energy for an infinite collection of particles, and the second term, except for the Ω^2, looks like the potential energy of harmonic oscillators.

The EL equations taken with respect to the a_n^* are the complex conjugates of the ones taken with respect to the a_n, so one set is sufficient:

$$\frac{1}{2} \ddot{a}_n + \frac{1}{2} (v^2 k_n^2 + \Omega^2) a_n = 0.$$

The equations are uncoupled (as is characteristic of free fields). This is what makes the decomposition in terms of orthonormal functions look like normal modes. In the special case $\Omega = 0$, for which this becomes the scalar wave equation, these are the equations for a set of uncoupled harmonic oscillators, each at its own frequency $\omega_n = vk_n = 2\pi nv/X$. Their solutions are $a_n(t) = \alpha_n e^{i\omega_n t}$ (we leave out the complex conjugate) and the general solution is then

$$\psi(x, t) = \sum A_n e^{i(k_n x - \omega_n t)} = \sum A_n e^{ik_n(x - vt)}.$$

This is the usual Fourier decomposition for waves on a finite string; it is a superposition of independent running waves, all at the same velocity v.

The situation is somewhat different for $\Omega^2 \neq 0$. That still yields a set of uncoupled Harmonic oscillator equations, but now the frequencies (we will now call them v_n) are given by

$$v_n = \sqrt{v^2 k_n^2 + \Omega^2}. \tag{9.100}$$

This is a dispersion relation in the same sense as Eq. (4.54): it gives the relation between the frequency ν_n and wave vector k_n (or equivalently the wavelength $\lambda_n = 2\pi/k_n$). There is now a minimum *cutoff* frequency $\nu_0 = \Omega$ corresponding to $n = 0$, and the exponential in the solution is now not $k_n(x - vt)$ but

$$k_n x - t\sqrt{v^2 k_n^2 + \Omega^2}.$$

This means that different frequencies propagate at different velocities

$$\nu_n = \sqrt{v^2 + \frac{\Omega^2}{k_n^2}} = \sqrt{v^2 + \frac{\Omega^2 X^2}{4\pi^2 n^2}}.$$

The lower n, the more ν_n deviates from v, and $\lim_{n\to 0} \nu_n$ is infinite. We mention without proof that the frequency dependence of the rate of propagation implies that wave packets will not keep their shape, spreading as time increases.

The general solution of the K–G equation is

$$\psi(x, t) = \sum_n A_n e^{i(k_n x - \nu_n t)} \equiv \sum_n \psi_n(x, t)$$

with ν_n given by (9.100). The K–G equation is the quantum field theoretic equation for a relativistic particle of mass Ω (in a two-dimensional space–time in this example). This is not the place to describe in any depth the quantum properties of the K–G equation, but we make the following reminders. In quantum mechanics (with $\hbar = 1$) the momentum operator is $\hat{p} = -i\partial/\partial x$ and the momentum of a Fourier component of ψ_n of ψ is obtained by calculating

$$\hat{p}\psi_n \equiv \hat{p} A_n e^{i(k_n x - \nu_n t)} = -i\frac{\partial}{\partial x} A_n e^{i(k_n x - \nu_n t)}$$
$$= k_n A_n e^{i(k_n x - \nu_n t)} = k_n \psi_n.$$

The equation $\hat{p}\psi_n = k_n \psi_n$ shows that ψ_n is an *eigenfunction* of the momentum operator $\hat{p}$ *belonging to eigenvalue* $p_n \equiv k_n$ (we emphasize that p_n is not an operator but a number). The energy is given by the operator $i\partial/\partial t$, and a similar calculation shows that ψ_n is an eigenfunction also of the energy operator $\hat{E}$ belonging to the eigenvalue $E_n \equiv \nu_n$ (also a number). This is interpreted as saying that ψ_n is the wave function of a particle of definite momentum p_n and definite energy E_n, and then what Eq. (9.100) says (set $v = c = 1$) is that

$$E_n^2 - p_n^2 = \Omega^2.$$

That is, the energy and momentum of such a particle satisfy the equation for a relativistic free particle of mass Ω. This demonstrates again, as at Eq. (9.62), that the K–G equation describes particles of mass Ω.

9.4 NONLINEAR FIELD THEORY

The very first field equation obtained in this chapter was the sine–Gordon equation, an important example of a *nonlinear* field theory. Yet so far we have discussed only linear equations, such as Maxwell's and the K–G equation (as has been mentioned, the latter is a linearization of the sG equation). In this section we discuss some aspects of two nonlinear equations, singling out certain particular solutions called *solitons* and *kinks*.

One of the principal differences between linear and nonlinear equations is that solutions to the linear ones satisfy a superposition principle: the sum of solutions is also a solution. That is what makes it possible, for instance, to obtain plane-wave decompositions of general solutions to Maxwell's equation through Fourier analysis. Superposition lies at the basis of the orthogonal function expansion of Section 9.3.2. Solutions of nonlinear equations cannot in general be superposed to yield other solutions and they differ in significant other ways from solutions of linear ones.

The nonlinear equations we will discuss are the sG and the cubic K–G equations. In Section 9.5, in treating hydrodynamics, we will discuss also the Korteweg–de Vries equation.

There are, of course, other equations, among them those for which no solutions have been found. In particle dynamics (i.e., when the number of freedoms is finite) there is much that can be said about nonintegrable systems (Chapter 7). There is no equivalent way to understand nonintegrable wave equations, when the number of freedoms becomes infinite.

9.4.1 THE SINE–GORDON EQUATION

So far solutions of the sine–Gordon equation (9.5) have been discussed only for its linearized Klein–Gordon form. We now turn to solutions of the full nonlinear sG equation itself. Bäcklund (1873) was the first to discuss this equation, in connection with the differential geometry of surfaces with constant negative curvature. Later we will say a few words about one of its contemporary applications, the theory of long Josephson junctions.

To start the discussion, rewrite Eq. (9.5) in terms of the dimensionless coordinates

$$t' = \Omega t, \quad x' = \frac{\Omega}{v}x, \tag{9.101}$$

which transforms the equation into (dropping the primes)

$$\phi_{tt} - \phi_{xx} + \sin \phi = 0, \tag{9.102}$$

where $\phi_t = \partial_t \phi$, $\phi_{tt} = \partial_t \phi_t$, etc. In this form the sG equation is manifestly Lorentz invariant (in one space and one time dimension) as long as ϕ is a relativistic scalar. Indeed, it can be written in the form

$$\Box \phi - \sin \phi = 0,$$

where $\square$ is the d'Alembertian defined at Eq. (9.63). Thus the sG equation is an example of a nonlinear relativistic field theory. The speed of light is $v \equiv c$, which is equal to 1 in the dimensionless coordinates.

SOLITON SOLUTIONS

It was mentioned near the end of Section 9.3 that dispersion (different speeds for different frequencies) causes wave packets of linear equations to spread. A way to see this is to construct a packet at time $t = 0$ by Fourier analysis, so that the initial packet is the sum of trigonometric waves with amplitudes and phases that depend on their frequencies. As the trigonometric waves propagate at different velocities, however, their phase relations change: they no longer interfere in the way that shaped the original packet, so the shape changes. It turns out that such packets always spread as $t \to \infty$. The total energy is conserved in a packet of the linear equations studied so far, but as the packet spreads the energy becomes dispersed over larger and larger regions. (If in addition there are dissipative terms in the wave equation, however, the total energy is not conserved.)

In nonlinear equations the situation is different. Among the solutions that have been found to some nonlinear equations are those called *solitons*. These are localized traveling waves, somewhat like wave packets of linear equations, but with two remarkable properties: A) Unlike the wave packets of linear equations they do not change shape as they move and B) when two of them meet, they emerge from the scattering region unscathed. In this section we discuss some soliton solutions of the sG equation.

To find the first such solution, try to solve (9.102) with a function of the form (this is called making an *ansatz*)

$$\phi(x, t) = \psi(x - ut) = \psi(\xi), \tag{9.103}$$

where u is some constant velocity (because x and t are dimensionless, so is u, which is now measured in units of v). The first question is whether such solutions can be found and, if they can, for what values of u.

With this ansatz (9.102) becomes

$$(1 - u^2)\psi''(\xi) = \sin \psi(\xi), \tag{9.104}$$

where the prime denotes differentiation with respect to ξ. Before obtaining the soliton solutions, we explain how they arise. Think of ψ as a generalized coordinate q in a one-freedom particle dynamical system and of ξ as the time. We will call this the *auxiliary* dynamical system. Then (9.104) can be obtained from the one-freedom auxiliary Lagrangian

$$L_\psi = \frac{1}{2}\psi'^2 - \gamma^2 \cos \psi,$$

where $\gamma = (1 - u^2)^{-1/2}$ is a Lorentz contraction factor that reflects the relativistic nature of the sG equation. The potential energy of the auxiliary system is $V_\psi(\psi) = \gamma^2 \cos \psi$, discussed in Chapter 1 [around Eq. (1.49) and again in Section 1.5.1; Fig. 1.5 is essentially a graph of $-V_\psi$]. Recall that originally the sG equation was derived as the continuum limit of a one-dimensional chain of pendula connected by springs; $\psi \equiv \varphi = 0$ represents the pendula hanging straight down, and $\psi = 2\pi$ represents them also hanging straight down but having undergone one revolution about their support. If only one revolution is under consideration, ψ varies only from 0 to 2π. Within this range, the auxiliary system has unstable fixed points at $\psi = 0$ and $\psi = 2\pi$ and a stable one at $\psi = \pi$.

The energy first integral of the auxiliary system (not to be confused with the actual energy in the sG field) is

$$E_\psi \equiv \gamma^2 C = \frac{1}{2}\psi'^2 + \gamma^2 \cos \psi, \tag{9.105}$$

and the methods of Chapter 1 lead to

$$\xi = \frac{\pm 1}{\sqrt{2}\gamma} \int_{\psi(0)}^{\psi(\xi)} \frac{d\psi}{\sqrt{C - \cos \psi}}. \tag{9.106}$$

If Equation (9.105) is rewritten as $2(C - \cos \psi) = (1 - u^2)\psi'^2$, it is seen that a real solution requires that the sign of $1 - u^2$ be the same as that of $C - \cos \psi$. Choose $u^2 < 1$ and, for the time being, $C = 1$. The latter choice means that the solution is the one lying on the separatrix between bound and unbound solutions (see Fig. 1.5), which connects the unstable fixed points of the auxiliary system. It takes an infinite time ξ to reach either of these fixed points, so the solutions look like those of Fig. 9.4. An explicit solution is obtained by performing the integration in Eq. (9.106):

$$\int_\pi^{\psi(\xi)} \frac{d\psi}{\sin \psi/2} \equiv 2 \ln(\tan \psi/4) = \pm 2\gamma\xi,$$

or

$$\psi(\xi) = 4 \arctan\{e^{\pm \gamma \xi}\}, \tag{9.107}$$

which is what is actually plotted in Fig. 9.4. The solution is called a soliton if the positive sign is chosen in the exponent and an *antisoliton* if the negative sign is chosen. Solitons can be obtained for all values of u such that $u^2 < 1$.

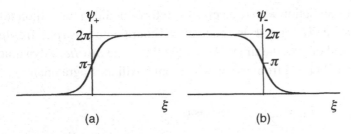

FIGURE 9.4
(a) A soliton. As time progresses the soliton moves either to the right or left, depending on the sign of u. (b) An antisoliton. As time progresses the antisoliton moves either to the right or left, depending on the sign of u.

Rewritten in terms of x and t the solitons are

$$\phi(x, t) = 4 \arctan\left\{e^{\pm\gamma(x-ut)}\right\},\tag{9.108}$$

which means that the graph of Fig. 9.4(a) or (b) moves in the positive or negative x direction, depending on whether u is positive or negative. We emphasize that nonlinearity implies that multiples of these solutions are not themselves solutions; even $\arctan\{e^{\pm\gamma(x-ut)}\}$ is not a solution of the sG equation.

PROPERTIES OF sG SOLITONS

Winding Number. Since ψ is the pendulum angle, the soliton represents one complete twist of the chain of pendula about the line joining their supports. According to (9.108) the twists, whether positive or negative, propagate up or down the chain depending on the sign of u. The *rotation* or *winding* number ρ of the soliton, the number of times it twists around the chain, is $\rho = 1$, and the winding number of an antisoliton is $\rho = -1$ (its twist is negative, i.e., in the opposite sense).

More generally, the total winding number ρ_T for any solution of the sG equation is defined as

$$\rho_T = \frac{1}{2\pi}[\psi(\infty) - \psi(-\infty)] = \frac{1}{2\pi}\int_{-\infty}^{\infty}\psi'\,d\xi.\tag{9.109}$$

This can also be stated in terms of the Lagrangian density $\mathcal{L}$ of the sG equation, Eq. (9.7). In the dimensionless variables and with irrelevant constant factors dropped,

$$\mathcal{L} = \frac{1}{2}\phi_t^2 - \frac{1}{2}\phi_x^2 - (1 - \cos\phi).$$

But according to (9.103) $\phi_x = \psi'$; the integral over ξ can be replaced by an integral over x, so

$$\rho_T = \frac{1}{2\pi}\int_{-\infty}^{\infty}\frac{\partial\mathcal{L}}{\partial\phi_x}dx \equiv \frac{1}{2\pi}\int_{-\infty}^{\infty}\rho(x)\,dx,\tag{9.110}$$

where $\rho(x) = \partial\mathcal{L}/\partial\phi_x$ is the *winding-number density*. If a solution has n solitons and m antisolitons (it will soon be seen that such solutions exist), the total number of twists in the chain is $\rho_T = n - m$, which is therefore also the *soliton number*. The relativistic nature of the sG equation means that any local disturbance can not reach $x = \pm\infty$ in a finite time, and that means that the total twist of the chain is the same at all times. Hence soliton number is conserved.

Energy. We treat the energy in part by a worked example.

WORKED EXAMPLE 9.4

(a) Find the Hamiltonian density $\mathcal{H}$ of the sine–Gordon equation [use the dimensionless parameters of Eq. (9.101)]. Write the energy E in the field in terms of ϕ and its derivatives. **(b)** For a solitonlike solution satisfying (9.103) rewrite E in terms of ψ and its derivatives. Find the energy density $\varepsilon(\xi)$ (energy per unit length) in terms of ψ' alone. **(c)** Specialize to a $C = 1$ soliton of Eq. (9.105). Write down $\varepsilon(\xi)$ explicitly and draw a graph of it. Find the total energy E carried by such a soliton.

Solution. **(a)** The Lagrangian density is given by Eq. (9.7):

$$\mathcal{L} = \left\{ \frac{1}{2} \left(\frac{\partial \phi}{\partial t} \right)^2 - \frac{1}{2} v^2 \left(\frac{\partial \phi}{\partial x} \right)^2 - \Omega^2 (1 - \cos \phi) \right\}.$$

In terms of the dimensionless parameters of Eq. (9.101)

$$\mathcal{L} = \Omega^2 \left\{ \frac{1}{2} \left(\frac{\partial \phi}{\partial t'} \right)^2 - \frac{1}{2} \left(\frac{\partial \phi}{\partial x'} \right)^2 - (1 - \cos \phi) \right\}.$$

Define a new Lagrangian density $\mathcal{L}' = \mathcal{L}/\Omega^2$. Then dropping all primes,

$$\mathcal{L} = \frac{1}{2} [\phi_t^2 - \phi_x^2 - 2(1 - \cos \phi)].$$

Henceforth this new $\mathcal{L}$ will be used as the Lagrangian density; then the new $\mathcal{H}$ and subsequently the energy E will be off by a factor of Ω^2.

According to Eq. (9.82) the conjugate momentum is

$$\pi = \frac{\partial \mathcal{L}}{\partial \phi_t} = \phi_t,$$

and

$$\mathcal{H} = \phi_t \pi - \mathcal{L} = \frac{1}{2} [\pi^2 + \phi_x^2 + 2(1 - \cos \phi)].$$

The energy in the field is given by

$$E \equiv H = \frac{1}{2} \int [\pi^2 + \phi_x^2 + 2(1 - \cos \phi)] \, dx \tag{9.111}$$

$$= \frac{1}{2} \int [\phi_t^2 + \phi_x^2 + 2(1 - \cos \phi)] \, dx.$$

More precisely, the energy in an interval $a \leq x \leq b$ is given by taking the integral in (9.111) between the limits a and b.

(b) According to (9.103) $\phi_t = -u\psi'$, $\phi_x = \psi'$ and $dx = d\xi$, so

$$E = \int \left[\frac{1}{2} (u^2 + 1)\psi'^2 + (1 - \cos \psi) \right] d\xi.$$

According to Eq. (9.105)

$$C - \cos \psi = \frac{1}{2} (1 - u^2)\psi'^2,$$

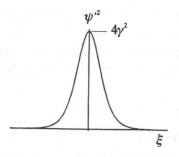

FIGURE 9.5
The energy distribution in a soliton or antisoliton.

so

$$E = \int [\psi'^2 + (1 - C)]\,d\xi.$$

The energy density is $\varepsilon = \psi'^2 + (1 - C)$. But the $1 - C$ term is a constant "background" density, so the energy density is renormalized to $\varepsilon(\xi) = \{\psi'(\xi)\}^2$. Then

$$E = \int_{-\infty}^{\infty} \psi'^2 \, d\xi. \tag{9.112}$$

(c) If $C = 1$ (then renormalization is unnecessary), ψ is given by (9.107) and

$$\varepsilon(\xi) \equiv \psi'^2(\xi) = 4\gamma^2 (\operatorname{sech} \gamma\xi)^2. \tag{9.113}$$

Figure 9.5 is a graph of $\varepsilon(\xi)$. It shows that the energy of the soliton is localized. The total energy carried by the soliton is (9.112):

$$E = 4\gamma \int_{-\infty}^{\infty} \{\operatorname{sech}\gamma\xi\}^2 d(\gamma\xi) = 4\gamma \int_{-\infty}^{\infty} \frac{4}{(e^z + e^{-z})^2} dz$$

$$= 8\gamma \int_{-\infty}^{\infty} \frac{2e^{2z} dz}{(e^{2z} + 1)^2} = 8\gamma = \frac{8}{\sqrt{1 - u^2}}. \tag{9.114}$$

With the results of Worked Example 9.4, solitons begin to look a lot like relativistic particles, whose energy is given by the equation $E = m/\sqrt{1 - u^2}$, similar to (9.114). Soliton-number conservation looks like particle-number conservation (see Perring and Skyrme, 1962). Moreover, Figs. 9.4 and 9.5 demonstrate that the energy is essentially localized. The contribution to E is almost entirely from the ξ interval in which ψ undergoes most of its variation. The half-width of the packet in Fig. 9.5 is $\Delta\xi = 1.7627/\gamma$ (Problem 11). This also looks relativistic, for it is proportional to $1/\gamma = \sqrt{1 - u^2}$: the faster the soliton moves, the narrower it becomes by exactly the Lorentz contraction factor.

MULTIPLE-SOLITON SOLUTIONS

Solutions also exist with multiple solitons, but since the sG equation is nonlinear, they can't be obtained simply by adding up single-soliton solutions: they comprise another kind of solution entirely. (It will be shown later, however, that there is a way to use the single soliton as a starting point for generating other solutions.) In principle such solutions can be found from appropriate initial conditions, but we will not do that. Instead, we simply present here some other solutions and leave it up to the reader to show that they do indeed satisfy the sG equation (Problem 12).

Soliton–Soliton. For a soliton interacting with another soliton the solution is

$$\phi_{ss}(x, t) = 4 \arctan[u \sinh(\gamma x) \operatorname{sech}(u\gamma t)]. \tag{9.115}$$

Figure 9.6 is a graph of this solution. As in the single-soliton solution, the two x intervals where ϕ_{ss} is not constant represent twists of the pendulum chain, both in the same sense. The two solitons approach each other as t increases from negative values to zero and then pass through each other, emerging undistorted and moving apart as t increases through positive values. The total winding number of this solution is $\rho_{ss} = 2$.

As was mentioned, this two-soliton solution cannot be obtained by summing two one-soliton solutions or by any other simple manipulation of them. Nevertheless it is seen from the figure that ϕ_{ss} looks like two solitons moving in opposite directions. This can also be seen analytically by studying the behavior at large t. Indeed, rewrite (9.115) in the form

$$\phi_{ss} = 4 \arctan\left[u \frac{e^{\gamma x} - e^{-\gamma x}}{e^{\gamma ut} + e^{-\gamma ut}} \right].$$

At large values of t the second term in the denominator of the bracketed fraction may be neglected, and then the fraction itself becomes nonnegligible only when $|\gamma x| \geq \gamma ut$. Between $x = -ut$ and $x = ut$, therefore, $\lim_{t \to \infty} \phi_{ss} \approx 0$. The nonzero regions move away from the origin with speed u: they look like two solitons moving apart, in agreement with the picture in Fig. 9.6. A similar result is obtained in the limit $t \to -\infty$.

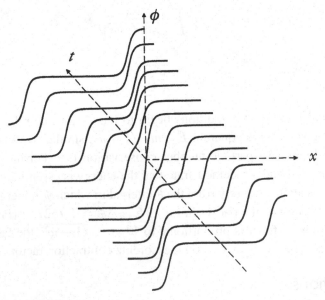

FIGURE 9.6
A soliton–soliton solution. The time axis points into the paper.

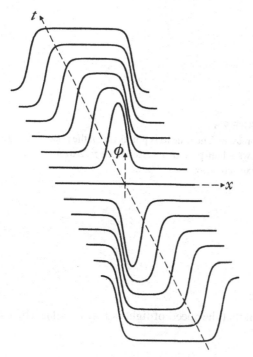

FIGURE 9.7
A soliton–antisoliton solution. The time axis points into the paper.

Soliton–Antisoliton. For a soliton interacting with an antisoliton the solution is

$$\phi_{sa}(x, t) = 4 \arctan [u^{-1} \operatorname{sech}(\gamma x) \sinh(u\gamma t)]. \tag{9.116}$$

Fig. 9.7 is a graph of this solution. The x intervals in which ϕ_{sa} increases represent a soliton and the intervals in which ϕ_{sa} decreases represent an antisoliton. Again, as time increases from negative values to zero the soliton and antisoliton approach each other. At $t = 0$ they cancel (annihilate), but at positive values of t they move apart undistorted. The twists are in opposite senses, so the total winding number is $\rho_{sa} = 0$.

This solution can be analyzed in the same way as ϕ_{ss} in the $t \to \pm\infty$ limits.

The Breather. Another solution is

$$\phi_b(x, t) = 4 \arctan[(u\gamma)^{-1} \operatorname{sech}(\gamma^{-1}x) \sin(ut)].$$

Figure 9.8 is a graph of this solution. Since $\sin(ut)$ varies only between -1 and 1, the pattern of this figure repeats over and over, and it is clear why it is called a breather. This solution is notable in that the pendula near the center of the chain (near $x = 0$) oscillate back and forth, but the disturbance is localized: it does not propagate in either direction along the chain. The total winding number is $\rho_b = 0$.

GENERATING SOLITON SOLUTIONS

Bäcklund (1873) developed a method for using solutions of the sG equation to generate new ones. This is best explained through another change of coordinates. Let $w = \frac{1}{2}(x - t)$ and

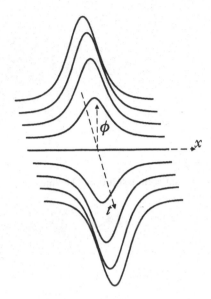

FIGURE 9.8
A breather. The time axis points out of the paper. The graphs shows a half-period: t varies from $\pi/2$ to $3\pi/2$ (the x and ϕ axes are drawn at $t = \pi$).

$z = \frac{1}{2}(x + t)$ be *light-cone* coordinates (recall that the speed of light is $v \equiv c = 1$). Then with $\chi(w, z) = \phi(x, t)$ the sG equation becomes

$$\partial_w \partial_z \chi \equiv \chi_{wz} = \sin \chi.$$

Any function χ satisfying this equation, when rewritten in the (x, t) coordinates, is a solution of the sG equation. Now consider two functions χ and $\bar\chi$ that satisfy the coupled equations

$$\frac{1}{2} \frac{\partial}{\partial w}(\bar\chi + \chi) = \lambda^{-1} \sin \frac{1}{2}(\bar\chi - \chi),$$

$$\frac{1}{2} \frac{\partial}{\partial z}(\bar\chi - \chi) = \lambda \sin \frac{1}{2}(\bar\chi + \chi),$$

(9.117)

where $\lambda \neq 0$ is an arbitrary positive or negative parameter. The z derivative of the first of these minus the w derivative of the second yields

$$\chi_{wz} = \lambda^{-1} \cos \frac{1}{2}(\bar\chi - \chi) \cdot \lambda \sin \frac{1}{2}(\bar\chi + \chi) - \lambda \cos \frac{1}{2}(\bar\chi + \chi) \cdot \lambda^{-1} \sin \frac{1}{2}(\bar\chi - \chi)$$

$$= \sin \chi,$$

so χ is a solution of the sG equation. Similarly, the z derivative of the first plus the w derivative of the second yields

$$\bar\chi_{wz} = \sin \bar\chi.$$

Hence $\bar\chi$ is also a solution.

This procedure can be used to construct new solutions from old ones. Given a solution χ of the sG equation, any solution $\bar\chi$ of Eqs. (9.117) is another solution of the sG equation. It is not obvious that it is easier to solve (9.117) than it is to solve the sG equation, but solving it can be avoided in the following way.

To start with, take χ to be identically zero (this is the trivial, or null, solution). Then Eqs. (9.117) become

$$\frac{1}{2}\bar{\chi}_w = \lambda^{-1} \sin \frac{1}{2}\bar{\chi},$$

$$\frac{1}{2}\bar{\chi}_z = \lambda \sin \frac{1}{2}\bar{\chi}. \tag{9.118}$$

Each of these is an ordinary differential equation of the form

$$\frac{dF}{dy} = a \sin F$$

and can be integrated to obtain an expression similar to (9.107). This gives two equations:

$$\bar{\chi}(w, z) = 4 \arctan\{\exp \lambda^{-1}[w + f(z)]\},$$
$$\bar{\chi}(w, z) = 4 \arctan\{\exp \lambda[z + g(w)]\},$$

where the functions f and g must be found by setting the two expressions equal. When this is done the result is

$$\bar{\chi}(w, z) = 4 \arctan\{\exp(\lambda^{-1}w + \lambda z)\}.$$

Finally, writing

$$\lambda = \pm\sqrt{(1 - u)/(1 + u)} \tag{9.119}$$

and changing back to the (x, t) variables leads to

$$\bar{\chi}(x, t) = 4 \arctan\{e^{\gamma(x+ut)}\},$$

which is a special case of (9.108). In this example the initial χ from which the procedure started was the null solution, but it works with other initial solutions, as will be seen below.

The $\bar{\chi}$ of the example depends on λ through u, and hence applying this procedure with different λs yields different $\bar{\chi}$s. This is true even with initial solutions other than the null solution. When this procedure is applied to a given initial solution χ_0 with a certain value λ_a of λ we will write $\chi_0 \overset{\lambda_a}{\to} \chi_{1a}$. Then χ_{1a} can be used as a new initial solution and the procedure can be applied again with a different value of λ. Let

$$\chi_0 \overset{\lambda_a}{\to} \chi_{1a} \overset{\lambda_b}{\to} \chi_{2ab} \quad \text{and} \quad \chi_0 \overset{\lambda_b}{\to} \chi_{1b} \overset{\lambda_a}{\to} \chi_{2ba}.$$

We state without proof (but see Bianchi, 1879) that $\chi_{2ab} = \chi_{2ba} \equiv \chi_2$. This is illustrated diagrammatically in Fig. 9.9. Problem 13 then shows that this leads to the following *algebraic* equation for χ_2:

$$\tan \frac{1}{4}(\chi_2 - \chi_0) = \frac{(\lambda_a + \lambda_b)}{(\lambda_b - \lambda_a)} \tan \frac{1}{4}(\chi_{1a} - \chi_{1b}). \tag{9.120}$$

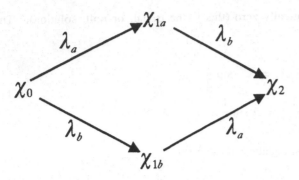

FIGURE 9.9
Generating sG solutions from sG solutions.

WORKED EXAMPLE 9.5

Use Eq. (9.120) to obtain the soliton–soliton solution by taking χ_0 to be the null solution and $\chi_{1a,b}$ to be the one-soliton solutions

$$\chi_{1a}(x, t) = 4 \arctan\{e^{\gamma(x-u_a t)}\},$$
$$\chi_{1b}(x, t) = 4 \arctan\{e^{-\gamma(x-u_b t)}\},$$

where $u_{a,b}$ is related to $\lambda_{a,b}$ according to (9.119).

Solution. With $\chi_0 = 0$ Eq. (9.120) becomes

$$\chi_2 = 4 \arctan\left[\frac{\lambda_a + \lambda_b}{\lambda_b - \lambda_a} \tan Z\right],$$

where

$$Z = \arctan\{e^{\gamma(x-u_a t)}\} - \arctan\{e^{-\gamma(x-u_b t)}\}.$$

The trigonometric identity

$$\tan(A + B) = \frac{\tan A + \tan B}{1 - \tan A \tan B}$$

then implies that

$$\tan Z = \frac{e^{\gamma(x-u_a t)} - e^{-\gamma(x-u_b t)}}{1 + e^{-\gamma\{u_a - u_b\}t}}.$$

Now choose $u_a \equiv u = -u_b$, so that

$$\tan Z = \frac{e^{\gamma x} - e^{-\gamma x}}{1 + e^{-2u\gamma t}} e^{-u\gamma t} = \frac{e^{\gamma x} - e^{-\gamma x}}{e^{u\gamma t} + e^{-u\gamma t}}$$
$$= \sinh(\gamma x)\operatorname{sech}(u\gamma t).$$

From the choice of u_a and u_b it follows that $\lambda_a = \pm\sqrt{(1-u)/(1+u)} = \pm 1/\lambda_b \equiv \lambda$. Since in general λ may be positive or negative, choose $\lambda_a = \lambda$ and $\lambda_b = -1/\lambda$.

Then

$$\frac{\lambda_a + \lambda_b}{\lambda_b - \lambda_a} = -\frac{\lambda^2 - 1}{\lambda^2 + 1} = u.$$

Hence

$$\chi_2 = 4 \arctan[u \sinh(\gamma x) \operatorname{sech}(u\gamma t)],$$

which is exactly the solution of Eq. (9.115).

NONSOLITON SOLUTIONS

The solutions discussed so far are solitons according to the definition at the beginning of this section: A) They do not spread in time and B) when they collide (i.e., the two-soliton and the soliton–antisoliton solutions) they emerge unscathed. Solutions exist, however, that satisfy Property A but not B. Property B can be checked only for a solution that in the $t \to -\infty$ limit looks like solitons approaching each other. The three solutions we will now exhibit involve multiple twists or kinks all moving in the same direction at the same speed. For them Property B can be checked only if they can be imbedded in other solutions, more or less the way the single soliton was imbedded in the soliton–soliton solution. That we will not attempt to do. The three solutions have been studied numerically, however, and have been found not to satisfy Property B. Analytic methods exist for discussing such matters in nonlinear field theories, in particular with localized solutions (inverse scattering transform; see Gardner et al., 1967), but these are beyond the scope of this book.

To obtain these solutions, return to the approach used for obtaining single-soliton solutions. The soliton solution was obtained from Eq. (9.106), which was derived directly from the sG equation with the ansatz of (9.103), by choosing $C = 1$ and $u^2 < 1$. Now choose $1 < C < \infty$ and $u^2 < 1$, corresponding to unbound solutions of the auxiliary system. Then (9.106) can be rewritten as

$$\xi = \frac{\pm 1}{2\gamma} \int_\pi^\psi \frac{d\psi}{\sqrt{C' + \sin^2(\psi/2)}}, \tag{9.121}$$

where $C' = \frac{1}{2}(C - 1)$ so $0 < C' < \infty$. Change variables from ψ to $l = \cos(\psi/2)$ and define the parameter k by

$$k = \frac{1}{\sqrt{C' + 1}} \quad \text{or} \quad C' = \frac{1 - k^2}{k^2}.$$

Then $0 \le k < 1$ and Eq. (9.121) becomes

$$\xi = \frac{\pm 1}{\gamma} \int_0^{\cos(\psi/2)} \frac{k \, dl}{[(1 - l^2)(1 - k^2 l^2)]^{1/2}}. \tag{9.122}$$

This integral can be related to the *doubly periodic Jacobian elliptic function of modulus k,* denoted $\mathrm{sn}(y, k)$, itself defined (Abramowitz and Stegun, 1972) by the integral

$$y = \int_0^{\mathrm{sn}(y,k)} \frac{k\,dl}{[(1 - l^2)(1 - k^2 l^2)]^{1/2}}.$$

The two integrals can be identified by setting

$$\cos(\psi/2) = \pm\mathrm{sn}(y, k) = \pm\mathrm{sn}(\gamma\xi/k, k),$$

and then

$$\psi_\pm(\xi) = 2\arcsin[\pm\mathrm{sn}(\gamma\xi/k, k)] + \pi. \tag{9.123}$$

For a fixed value of k the function $\mathrm{sn}(y, k)$ satisfies the periodicity conditions

$$\mathrm{sn}(y + 4K(k), k) = -\mathrm{sn}(y + 2K(k), k) = \mathrm{sn}(y, k),$$

where K is the complete elliptic function of the first kind

$$K(k) = \int_0^1 \frac{dl}{[(1 - l^2)(1 - k^2 l^2)]^{1/2}}$$

(between $k = 0$ and 1, the function $K(k)$ varies from $\pi/2$ to $\ln[4(1 - k^2)^{-1/2}]$). Thus sn is an odd periodic function with period $4K(k)$, and hence $\psi(\xi)$ has period P given by

$$P = \frac{4kK(k)}{\gamma}. \tag{9.124}$$

To understand the periodicity of ψ, recall that $\psi(\xi) = \phi(x - ut)$ is the angle the pendula make with the vertical. When ψ goes from 0 to 2π, the pendulum angle has gone once around and is back where it started: the ψ manifold is a circle. Just as the torus $\mathbb{T}^2$ can be represented by a plane with periodic cells (Chapter 6), the circle $\mathbb{S}$ can be represented by a line with periodic intervals. This is what is shown in Fig. 9.10, a graph of ψ in which the vertical axis is such a line. Because the axis represents $\mathbb{S}$, the graph is not on a plane but on a cylinder, and in one period P the pendula go once around. The curve is a helix whose pitch (the slope of the curve) is periodic with period P, and the solution ϕ corresponds to a distribution of the pendula in such a helix. The helix moves along the cylinder with velocity u.

The exact shape of the helix depends on k. In the $k \to 0$ limit $\mathrm{sn}(y, k) \to \sin y$, and Eq. (9.123) becomes

$$\psi_\pm(\xi) = \pi \pm 2\gamma\xi/k,$$

which is a straight line of slope $\pm 2\gamma/k$, – a helix of constant pitch. In the opposite limit, when $k \to 1$, the period of $\mathrm{sn}(y, k)$ tends to infinity and $\mathrm{sn}(y, k) \to \tanh y$. This leads

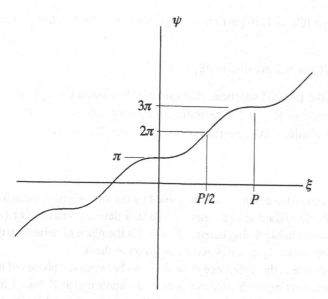

FIGURE 9.10

Graph of the sG solution of Eq. (9.123) for small k. The period P is defined in Eq. (9.124).

to the soliton and antisoliton solutions, which is to be expected, as in this limit $C' = 0$ ($C = 1$).

Other solutions are obtained with other choices of C and u. With the choice $|C| < 1$ (bound solutions of the auxiliary system) and $u^2 < 1$, for example, Eq. (9.106) can then be put in the form

$$\xi = \frac{\pm 1}{2\gamma} \int_\pi^{\psi(\xi)} \frac{d\psi}{\sqrt{\sin^2(\psi/2) - k_1^2}},$$

where $0 < k_1^2 \equiv \frac{1}{2}(1 - C) < \frac{1}{2}$. An additional change of variable, this time to $l = \sin(\psi/2)$, transforms the integral to

$$\xi = \frac{\pm 1}{\gamma} \int_1^{\sin(\psi/2)} \frac{dl}{[(1 - l^2)(l^2 - k_1^2)]}.$$

From this it follows that $\psi(\xi)$ can be written in terms of the Jacobian elliptic function $dn(y, k)$:

$$\psi(\xi) = 2 \arcsin[\pm dn(\gamma\xi, k_1)]. \tag{9.125}$$

This is an even function and periodic, but with period $2K(k)$:

$$dn(y + 2K(k), k) = dn(y, k).$$

Its maximum value is $dn(0, k) = 1$ and its minimum is k, so ψ oscillates between π and $\arcsin(k_1)$, not winding around the cylinder.

Finally, consider $u^2 > 1$. For $|C| > 1$, in particular, logic similar to that of the preceding two cases leads to

$$\psi(\xi) = \pm 2 \arcsin[\text{sn}(\tilde{\gamma}\xi/k, k)],$$

where $\tilde{\gamma} = [u^2 - 1]^{-1/2}$. Like the (9.123) solution, this one winds around the cylinder in a sense that depends on the sign. For $|C| < 1$ a periodic oscillatory solution is obtained, similar to (9.125). For $|C| = 1$ a solitonlike solution is obtained, but for inverted pendula and therefore unstable.

JOSEPHSON JUNCTIONS

There exist real physical systems whose behavior is described by the sine–Gordon equation. One of these is the Josephson junction (Barone and Paterno, 1982), a device consisting of two superconductors separated by a very thin insulating barrier. This is not the place to delve into the theory of Josephson junctions, but we briefly describe what can go on in them.

The sG field variable in this device is the difference $\psi = \psi_1 - \psi_2$ between the phases of the *macroscopic wave functions* characterizing each superconductor. Josephson (1962) found that the dynamical equation for ψ is exactly Eq. (9.102), so all the solutions that we have described in this section apply in principle to Josephson junctions. A soliton corresponds physically to a jump by 2π in the phase difference across the insulating barrier, and this is associated with a change in the superconducting current flowing through the barrier. Josephson showed that the current through the barrier depends on $\sin \psi(x, t)$ and is constant only if the phase difference is constant. But ψ satisfies the sG equation, whose solutions include solitons. The localized nature of the solitons causes the magnetic field resulting from the corresponding superconducting currents, which is proportional to the spatial derivative of ψ, to be localized. These localized packets of magnetic flux are themselves called solitons (also fluxons or vortices) and must all be multiples of the fundamental quantum of flux $\Phi_0 = h/2e$, where h is Planck's constant and e is the electron charge.

Of course in a real Josephson junction there is also some dissipation, so the sG equation does not describe it in detail. But the dissipative effects can be minimized and controlled experimentally and the system brought so close to the theoretical sG system that even Lorentz contraction of the fluxons has been measured experimentally (Laub et al., 1995).

9.4.2 THE NONLINEAR K–G EQUATION

THE LAGRANGIAN AND THE EL EQUATION

A nonlinear equation that occurs in particle physics and in the theory of phase transitions is the cubic Klein–Gordon equation for a scalar field φ. Here it will be treated in one-plus-one dimensions (one space and one time dimension). The Lagrangian density is

$$\mathcal{L} = \frac{1}{2}[\varphi_t^2 - v^2\varphi_x^2] + \lambda(\varphi^2 - 1)^2.$$

In the theory of phase transitions this is known as the Ginzburg–Landau Lagrangian, and in particle physics, with $v = c$, as the φ^4 field Lagrangian.

The EL equation is

$$\varphi_{tt} - v^2\varphi_{xx} - 4\lambda(\varphi^2 - 1)\varphi = 0.$$

As in the sG case, a change of variables to

$$t' = \sqrt{2\lambda}t \quad \text{and} \quad x' = \frac{\sqrt{2\lambda}}{v}x \qquad (9.126)$$

brings this to a simpler form (dropping the primes):

$$\varphi_{tt} - \varphi_{xx} = 2(\varphi^2 - 1)\varphi. \qquad (9.127)$$

This is the *cubic* K–G equation because φ^3 is the highest power of the field that appears in it.

KINKS

As in the case of the sG equation, nondispersive solutions will be obtained from the substitution $\varphi(x, t) = \psi(x - ut) \equiv \psi(\xi)$ of Eq. (9.103). In terms of ξ Eq. (9.127) becomes

$$(1 - u^2)\psi'' = 2(\psi^2 - 1)\psi. \qquad (9.128)$$

Again, this is treated as an equation for an auxiliary one-freedom dynamical system in which ψ is the generalized coordinate and ξ is the time. The solutions obtained in this way, called *kinks*, resemble solitons in that they propagate without changing shape (what we called Property A), but kinks differ from solitons in that they become distorted when they collide.

Equation (9.128) can be obtained from the one-freedom auxiliary Lagrangian

$$L_\psi = \frac{1}{2}\psi'^2 + \frac{1}{2}\gamma^2(\psi^2 - 1)^2,$$

where $\gamma = (1 - u^2)^{-1/2}$, as in the sG equation. The potential V_ψ of the auxiliary system is plotted in Fig. 9.11.

There are three fixed points, two of them unstable, at $\psi = \pm 1$.

The energy first integral of the auxiliary system (not to be confused with the actual energy in the K–G field; see Problem 14) is

$$E_\psi \equiv \gamma^2 C = \frac{1}{2}\psi'^2 - \frac{1}{2}\gamma^2(\psi^2 - 1)^2, \qquad (9.129)$$

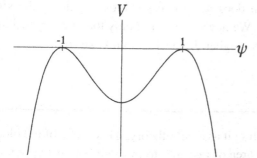

FIGURE 9.11

The potential $V(\psi)$ of the auxiliary system. The zero energy level is tangent to the two unstable fixed points, ψ_- and ψ_+.

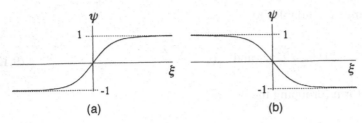

FIGURE 9.12
Two solutions of the cubic K–G equation: (a) a kink; (b) an antikink.

and the methods of Chapter 1 lead to

$$\xi = \pm \frac{1}{\gamma\sqrt{2}} \int \frac{d\psi}{\sqrt{C + (\psi^2 - 1)^2}}. \tag{9.130}$$

Again, we choose $u^2 < 1$ and, in order to obtain the auxiliary solution that connects the two unstable fixed point, $C = 0$. It takes an infinite time ξ to reach these fixed points and thus the solutions look very much like those of Fig. 9.4. They differ from those in that they go from $\psi = -1$ at $\xi = -\infty$ to $\psi = 1$ at $\xi = +\infty$. Put $C = 0$ into (9.130) and use $\psi^2 \leq 1$ (for the solution that lies *between* the two fixed points). Then

$$\xi = \pm \frac{1}{\gamma\sqrt{2}} \int \frac{d\psi}{1 - \psi^2} = \pm \frac{1}{\gamma\sqrt{2}} \operatorname{arctanh} \psi$$

or

$$\psi(\xi) \equiv \varphi(x - ut) = \pm \tanh \sqrt{2}\gamma\xi = \pm \tanh \sqrt{2}\gamma(x - ut). \tag{9.131}$$

These solutions are plotted in Fig. 9.12. The solution of Figs. 9.12(a) and (b) are a kink and an antikink, respectively.

Like the solitons of the sG theory, these propagate with velocity u, but numerical calculations (Makhankov, 1980) show that when a kink and an antikink collide they do not behave like solitons. The exact nature of such a collision depends on the initial kink–antikink approach velocity: the larger the velocity, the more elastic the collision. If the velocity of approach is smaller than a certain critical value, however, the kink and antikink may form a rather long-lived bound state, called a *bion*. Eventually the bion decays, "radiating" its energy in the form of small-amplitude oscillations, similar to normal modes of a linear system, which are superposed on the interacting kinks.

Although the nonlinear K–G equation does not possess soliton solutions, the kink solutions are important to its applications. We have considered only the one-dimensional case, but in higher dimensions there are also higher dimensional kinklike solutions.

9.5 FLUID DYNAMICS

This section deals with fluid flow, treating it as a field theory. This is one of the oldest fields of physics. It remains a very active area of research to this day both because there remain fundamental questions to be answered and because of its obvious technological importance.

Fluid dynamics is just one of two nonlinear classical field theories concerned with continuous media; the other is elasticity theory. The former has been studied more extensively recently because of the important phenomenon of turbulence, which we will discuss briefly below, and because the discovery of completely integrable equations has served as a paradigm for further developments. Elasticity theory, in spite of its importance, will not be treated in this book.

Our description of fluid dynamics will ignore molecular structure: we think of the fluid as composed of *fluid cells*, each of which contains a large number of molecules. Yet each cell is assumed so small that the system composed of these cells can be treated as a continuum.

9.5.1 THE EULER AND NAVIER–STOKES EQUATIONS

SUBSTANTIAL DERIVATIVE AND MASS CONSERVATION

A fluid, say water, can be described by its mass density $\rho(\mathbf{x}, t)$, velocity $\mathbf{v}(\mathbf{x}, t)$, and the local pressure $p(\mathbf{x}, t)$. These make up a set of five field variables, for the velocity vector has three components. They are field variables in the same sense as those of the preceding sections, and $\mathbf{x}$ and t play the same role as in other field theories. The $\mathbf{x}$ in their arguments describes fixed points in three-space, and as the fluid flows past each point $\mathbf{x}$, the five field variables change in time t. The cells of the flowing fluid, which have no analog in the other fields, move through space among the fixed position vectors $\mathbf{x}$, so the position vector $\mathbf{x}$ of each cell keeps changing.

In what follows we will need to know how the properties of a fluid cell (e.g., its momentum or the values of the field variables in it) change as the fluid flows and the cell moves through space. Let $\psi(\mathbf{x}, t)$ be some such property, for the time being a scalar. Then even if ψ were $\psi_S(\mathbf{x})$, static (independent of t), its value in each cell would change as the cell moved through space. Its rate of change would be simply

$$\frac{d}{dt}\psi_S = \lim_{\Delta t \to 0} \frac{1}{\Delta t} \{\psi_S(\mathbf{x} + \mathbf{v}\Delta t) - \psi_S(\mathbf{x})\} = \mathbf{v} \cdot \nabla \psi_S.$$

But ordinarily ψ is not static, depending on both $\mathbf{x}$ and t, so in general as the cell moves with the flow its rate of change is

$$\frac{D\psi}{Dt} = \partial_t \psi + \mathbf{v} \cdot \nabla \psi \equiv D_t \psi. \tag{9.132}$$

In fluid dynamics this is called the *substantial derivative* of ψ. [Compare with the discussion around Eq. (5.161).] If ψ is a vector, Eq. (9.132) applies to each component of ψ.

The mass M_Ω in any region Ω of the fluid is

$$M_\Omega = \int_\Omega \rho \, d^3 x. \tag{9.133}$$

Conservation of mass is analogous to conservation of charge and is given by a continuity equation such as (9.25): $\partial_t \rho + \nabla \cdot \mathbf{j} = 0$, where $\mathbf{j}$ is the mass current. We do not prove this: it can be derived by applying the divergence theorem to Eq. (9.133). In Problem 15 it is shown that in a fluid $\mathbf{j} \equiv \mathbf{v}\rho$, so the differential equation of mass conservation is

$$\partial_t \rho + \nabla \cdot (\rho \mathbf{v}) = 0. \tag{9.134}$$

EULER'S EQUATION

A basic equation for an ideal fluid, called Euler's equation, is obtained when Newton's law $\mathbf{F} = \dot{\mathbf{P}}$ is applied to a *fixed* mass M_Ω of the fluid occupying a *moving* volume $\Omega(t)$ (an ideal fluid is one without dissipation and at absolute temperature $T = 0$). The $\mathbf{F}$ here is the total force on $\Omega(t)$, the sum of the forces on its cells, and $\mathbf{P}$ is the total momentum in $\Omega(t)$. These will be called $\mathbf{F}_\Omega$ and $\mathbf{P}_\Omega$, so Newton's law reads $\mathbf{F}_\Omega = \dot{\mathbf{P}}_\Omega$. Both of these must be calculated in terms of the five field functions and then inserted into the equation.

The total force is the sum of the forces on the cells, and the force on each cell is obtained from $p(\mathbf{x}, t)$, the force per unit area applied to the fluid at point $\mathbf{x}$ and time t, plus any external *body forces* acting on the fluid, such as gravity. We will ignore body forces, so the total force on the fluid in any volume $\Omega(t)$ is due to the pressure on its surface $\partial\Omega$ (the internal pressure forces cancel out) and is given by

$$\mathbf{F}_\Omega = -\oint_{\partial\Omega} p \, d\mathbf{\Sigma} \equiv -\int_\Omega \nabla p \, d^3x, \tag{9.135}$$

where $d\mathbf{\Sigma}$ is the outward normal surface element on $\partial\Omega$. The second equality follows from Stokes's (or the divergence) theorem. The total momentum is the integral of the momentum density $\rho\mathbf{v}$:

$$\mathbf{P}_\Omega = \int_{\Omega(t)} \rho\mathbf{v} \, d^3x. \tag{9.136}$$

Finding $\dot{\mathbf{P}}_\Omega$ is made complicated by having to take into account both the change of momentum within $\Omega(t)$ and the change resulting from the motion of $\Omega(t)$ itself (Fig. 9.13). We perform this calculation for an arbitrary well-behaved scalar function $W(\mathbf{x}, t)$, rather than for $\rho\mathbf{v}$. For such a function

$$\frac{d}{dt} \int_{\Omega(t)} W(\mathbf{x}, t) \, d^3x = \int_{\Omega(t)} \partial_t W \, d^3x + I,$$

where I is the contribution from the motion of $\Omega(t)$. Figure 9.14 is an enlargement of part of Fig. 9.13. In a time dt the infinitesimal patch dA on the surface $\partial\Omega(t)$ of $\Omega(t)$, whose normal is tilted at an angle θ to the velocity, moves through a distance $dl = v \, dt$, sweeping out the volume $dV = dA \cos\theta \, dl \equiv \mathbf{v} \cdot d\mathbf{A} \, dt$. This volume contributes $W\mathbf{v} \cdot d\mathbf{A} \, dt$ to

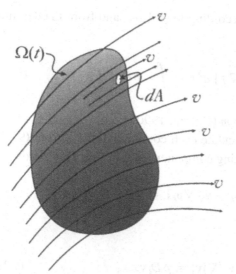

FIGURE 9.13
A moving region $\Omega(t)$ of the fluid, showing the velocity field.

$\int W d^3x$ in the time dt, that is, contributes at the rate $W\mathbf{v} \cdot d\mathbf{A}$. Then I is the integral of this rate over the entire surface:

$$I = \int_{\partial\Omega(t)} W\mathbf{v} \cdot d\mathbf{A} \equiv \int_{\Omega(t)} \nabla \cdot (W\mathbf{v}) \, d^3x.$$

Finally,

$$\frac{d}{dt} \int_{\Omega(t)} W(\mathbf{x}, t) \, d^3x = \int_{\Omega(t)} [\partial_t W + \nabla \cdot (W\mathbf{v})] \, d^3x. \tag{9.137}$$

Now, every function W can be written in the form $W = \rho f$, where $f(\mathbf{x}, t)$ is some other function, and then (9.137) becomes

$$\frac{d}{dt} \int_{\Omega(t)} \rho f \, d^3x = \int_{\Omega(t)} \{\partial_t(\rho f) + \nabla \cdot (\rho f \mathbf{v})\} \, d^3x$$

$$= \int_{\Omega(t)} \{\rho \partial_t f + f[\partial_t \rho + \nabla \cdot (\rho \mathbf{v})] + \rho \mathbf{v} \cdot \nabla f\} \, d^3x.$$

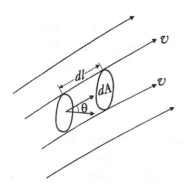

FIGURE 9.14
An enlargement of part of Fig. 9.13. $d\mathbf{A}$ is an element of area on the surface $\partial\Omega$, and θ is the angle between the velocity and the normal to $d\mathbf{A}$.

The expression in square brackets vanishes according to (9.134), and from the definition (9.132) of D_t it follows that

$$\frac{d}{dt} \int_{\Omega(t)} \rho f \, d^3x = \int_{\Omega(t)} \rho \{\partial_t f + \mathbf{v} \cdot \nabla f\} \, d^3x \equiv \int_{\Omega(t)} \rho D_t f \, d^3x. \qquad (9.138)$$

Equation (9.138) is called the *transport theorem* (Chorin, 1990).

The transport theorem can be used to calculate each component of $\dot{\mathbf{P}}_\Omega$ from (9.136), which can then be set equal to the corresponding component of (9.135). This yields

$$-\partial_k p = \rho[\partial_t v_k + \mathbf{v} \cdot \nabla v_k],$$

so that

$$-\nabla p = \rho \{\partial_t \mathbf{v} + (\mathbf{v} \cdot \nabla)\mathbf{v}\} \equiv \rho D_t \mathbf{v}. \qquad (9.139)$$

This is *Euler's equation for an ideal fluid*. Because the field functions are ρ, p, and the three components of $\mathbf{v}$, the expression $\rho \{\partial_t \mathbf{v} + (\mathbf{v} \cdot \nabla)\mathbf{v}\}$ makes the equation highly non-linear. It is therefore difficult to solve, even though it is an equation for ideal fluids.

In spite of such difficulties, Euler's equation can be used to establish momentum conservation in general for an ideal fluid. The rate of change of momentum density is $\partial_t(\rho \mathbf{v})$ or, in component form,

$$\partial_t(\rho v_i) = (\partial_t \rho)v_i + \rho(\partial_t v_i).$$

By using the continuity equation to rewrite the first term on the right-hand side and Euler's to rewrite the second, this can be put in the form

$$\partial_t(\rho v_i) + \partial_k \Pi_{ik} = 0, \qquad (9.140)$$

where

$$\Pi_{ik} = p\delta_{ik} + \rho v_i v_k \qquad (9.141)$$

is called the *momentum-stress tensor* of the fluid. Equation (9.140) defines a conserved current density: it is the equation of momentum conservation. The space part of the conserved current density, the symmetric Π_{ik} tensor, gives the flux of the ith component of momentum in the k direction.

VISCOSITY AND INCOMPRESSIBILITY

Euler's equation and mass conservation imply momentum conservation. Conversely, Eqs. (9.134), (9.140), and (9.141) (i.e., mass and momentum conservation and the definition of the momentum-stress tensor) together are equivalent to Euler's equation. This is a fruitful way of stating it because it allows extending the treatment to dissipative fluids. All real fluids (except superfluid helium at $T = 0$) lose momentum and energy as they flow: they are dissipative. So far, however, we have ignored that, and what must now be done is to extend Euler's equation to less idealized situations.

Viscosity is a property of a fluid that causes relatively moving adjacent parts of the fluid to exert forces on each other, forces that vanish in the limit when the relative velocity goes to zero. One of its effects is to destroy momentum conservation, which is accounted for by adding to Π_{ik} a term (see Landau & Lifshitz, 1987) that destroys the equality of (9.140):

$$\Pi_{ik} = p\delta_{ik} + \rho v_i v_k - \sigma'_{ik} \equiv \sigma_{ik} + \rho v_i v_k, \tag{9.142}$$

where the σ'_{ik} term (called the *viscosity stress tensor*) of the *stress tensor* σ_{ik} contains the information on viscosity.

Assume that the v_k are slowly varying functions of the position, so that σ'_{ik} can be approximated by the first-degree terms of a Taylor expansion, which we write in the form

$$\sigma'_{ik} = \mu \left(\partial_k v_i + \partial_i v_k - \frac{2}{3} \delta_{ik} \partial_j v_j \right) + \xi \delta_{ik} \partial_j v_j, \tag{9.143}$$

where μ and ξ arc called the *viscosity coefficients*. These parameters are usually determined experimentally, but they can also be found from an approximate microscopic theory. At room temperature $\nu \equiv \mu/\rho$, called the *kinematic viscosity*, is about 10^{-2} cm²/s for water and 0.15 cm²/s for air. The $\partial_k v_i$ and $\partial_i v_k$ terms are multiplied by the same coefficient because the viscous forces vanish if the fluid is rotating as a whole.

Another property of fluids is their *compressibility*: in general the density ρ depends on the pressure p. An *incompressible* fluid is one whose density is constant, that is, one for which $\partial_t \rho = 0$ and $\nabla \rho = 0$. Under these conditions the mass-conservation equation (9.134) becomes $\nabla \cdot \mathbf{v} = 0$, the left-hand side of (9.140) becomes $\rho \partial_t v_i$, and the $\partial_j v_j$ terms drop out of Eq. (9.143). Then if (9.142) is used to write Π_{ik}, Eq. (9.140) becomes

$$\begin{aligned} \rho \partial_t v_i &= -\partial_k \left[p\delta_{ik} + \rho v_i v_k - \nu\rho(\partial_k v_i + \partial_i v_k) \right] \\ &= -\partial_i p - \rho v_k \partial_k v_i + \nu\rho \partial_k \partial_k v_i. \end{aligned} \tag{9.144}$$

THE NAVIER–STOKES EQUATIONS

The famous *Navier–Stokes equations* for incompressible fluid flow consist of (9.144) (divided by ρ and put in vector form) and the incompressibility equation:

$$\partial_t \mathbf{v}(\mathbf{x}, t) + \{ \mathbf{v}(\mathbf{x}, t) \cdot \nabla \} \mathbf{v}(\mathbf{x}, t) = -\frac{1}{\rho} \nabla p(\mathbf{x}, t) + \nu \nabla^2 \mathbf{v}(\mathbf{x}, t), \tag{9.145}$$

$$\nabla \cdot \mathbf{v} = 0. \tag{9.146}$$

These equations have defied full analytic solution. Although numerical calculations with modern computers give results that compare favorably with experiment, the results obtained remain incomplete.

REMARK: If $v = 0$, Eq. (9.145) becomes Euler's equation (9.139). In fact v is the one parameter that remains from σ'_{ik}, so if $v = 0$ there is no dissipation and energy is conserved. The velocities calculated with Euler's equation are then generally much greater than those that may be expected in a real fluid. □

The Navier–Stokes equation is best treated in dimensionless variables, for then the results apply to many situations, independent of the particular fluid. To rewrite (9.145) in dimensionless terms, find the dimensions of v. According to (9.145) they are $l^2 t^{-1}$, so the magnitude of v depends on the units in which it is calculated. The unit λ of length may be chosen as some characteristic length for the particular flow under consideration. Flow usually takes place in some container (e.g., a pipe), so a characteristic dimension of the container (e.g., the diameter of the pipe) will do for λ. Similarly, suppose that the flow has some average or characteristic velocity $\bar{v}$. Then v is replaced by a dimensionless parameter R_e, called the *Reynolds number*, defined in terms of λ and $\bar{v}$ as

$$R_e = \frac{\lambda \bar{v}}{v}.$$

To make the entire equation dimensionless, the units of time, distance and velocity are scaled in accordance with

$$t \to t' = \bar{v}t/\lambda,$$
$$\mathbf{x} \to \mathbf{x}' = \mathbf{x}/\lambda,$$
$$\mathbf{v} \to \mathbf{v}' = \mathbf{v}/\bar{v}.$$

That is, distance is measured in multiples of λ and time in multiples of the time it takes the average flow to cover λ. The Navier–Stokes equation can then be written in the form (dropping the primes)

$$\partial_t \mathbf{v}(\mathbf{x}, t) + \{\mathbf{v}(\mathbf{x}, t) \cdot \nabla\} \mathbf{v}(\mathbf{x}, t) = -\frac{1}{\rho}\nabla p(\mathbf{x}, t) + \frac{1}{R_e}\nabla^2 \mathbf{v}(\mathbf{x}, t), \qquad (9.147)$$

where ρ and p must also be rescaled. Once the Navier–Stokes equation is written in this dimensionless form all that need be specified to describe the motion of a fluid is its Reynolds number. Different fluids in different containers but with the same R_e flow the same way in the rescaled units, and scaling arguments can be used to obtain important results. Small v corresponds to large R_e and vice versa. The Reynolds number is discussed further in the next section.

TURBULENCE

The Navier–Stokes equations are particularly interesting (and very difficult to solve) in *turbulent* regimes. Describing turbulence is an important unsolved problem of physics; its history goes back more than a hundred years (see Lamb, 1879). This book will touch on it only briefly, without a detailed discussion.

Turbulence has no simple definition but manifests itself in irregularities of the velocity field so random (or almost random) that they must be dealt with statistically. It is a dissipative process in which kinetic energy is transferred from large eddies to small ones, from them to smaller ones, then to even smaller ones, and eventually is dissipated to internal energy of the fluid. To make up for such dissipation, energy must constantly be supplied to a turbulent fluid to keep it flowing. Turbulent flow is ubiquitous in nature, for example in the uppermost layers of the atmosphere, in most rivers, and in currents below oceanic surfaces. It is more common than regular (*laminar*) flow. Turbulence always arises at low nonzero viscosities, when the viscosity ν in Eq. (9.145) is a small parameter, that is, when R_e is large.

Problems that arise in (9.145) at small ν occur in other differential equations, often called singular, when the highest derivative is multiplied by a small parameter. Small-ν singularities in such *differential* equations are analogous to the singularity in the following *algebraic* example. Consider the quadratic equation

$$\nu x^2 + x - 1 = 0.$$

When $\nu = 0$ the equation becomes linear and has the obvious solution $x = 1$. For other values of ν the two solutions of the equation are

$$x_\pm = \frac{1 \pm \sqrt{1 + 4\nu}}{2\nu},$$

but only one of the solutions converges to the $\nu = 0$ solution as $\nu \to 0$. Indeed, if we expand the two solutions in powers of ν about the $\nu = 0$ solution we get

$$x_- = 1 - \nu + 2\nu^2 - O(\nu^3),$$
$$x_+ = -1/\nu - 1 + \nu - 2\nu^2 + O(\nu^3).$$

Clearly x_- converges to the $\nu = 0$ solution, but x_+ blows up: $x_+ \to -\infty$ (it is instructive to graph the equation for various values of ν). This is not an artifact of the expansion: x_+ is a legitimate solution of the quadratic equation for all $\nu \neq 0$, but for small ν it would not be found by a search near the $\nu = 0$ solution. Similarly, solutions of the Navier–Stokes equation for small ν can be missed if they are sought near some $\nu = 0$ solution, for instance by a perturbation procedure. Such *singular perturbation* problems are characteristic of singular differential equations.

To discuss this in terms of R_e, turn to Eq. (9.147). The Reynolds number is a measure of the relative importance of the nonlinear and viscous terms in (9.147) – the smaller R_e, the more dominant the viscous term. When $R_e \ll 1$, the nonlinearity can be neglected and some solutions of the equation can be found in closed form. In most real cases, however, when $R_e \gg 1$, the nonlinearity dominates and there are no stationary solutions. The velocity flow is then convoluted, consisting of vortices at different length scales: turbulence sets in. Laminar flows have R_e values from about 1 to 10^2, and turbulence may start at around $R_e = 10^3$ with strong turbulence up to about 10^{14} (Batchelor, 1973). For turbulent flow it is futile to try to find $\mathbf{v}(\mathbf{x}, t)$ explicitly. The goal is then to obtain a fully developed statistical theory of turbulence. For more on turbulence, the interested reader can consult, for instance, Monin and Yaglom (1973).

Because the Navier–Stokes equation presents such difficulties, it is of significant interest to find simplifications or approximations that lend themselves to relatively full solutions. We now turn to such simplifications.

9.5.2 THE BURGERS EQUATION

THE EQUATION

A one-dimensional approximation of the Navier–Stokes equation (9.145) was introduced by Burgers (1948, also 1974). The Burgers equation omits the pressure term; it is

$$v_t + vv_x = \nu v_{xx}. \tag{9.148}$$

Except for the missing pressure term, this is a faithful one-dimensional model of the Navier–Stokes equation: it is of first order in t, is of second order in x, and has a quadratic term in the velocity field and a viscous dissipative term.

The Burgers equation is not adapted to perturbative methods in ν for the same reason as the Navier–Stokes equation. Nor is it adapted to perturbation in the velocity field, say by writing $v = v_0 + \epsilon v_1 + O(\epsilon^2)$ in the nonlinear term (i.e., replacing vv_x by $v_0 v_x$ to first order in the perturbation parameter ϵ, solving, and going to $\epsilon = 1$). This would hide some important physical phenomena such as turbulence. It turns out, however, that an exact formal solution of (9.148) can be found (Hopf, 1950; Cole, 1951; see also Whitham, 1974) through a sequence of transformations that convert it to a *diffusion equation*. The diffusion equation, a type of field equation studied in many other contexts, is what would remain if the quadratic vv_x term were removed from (9.148). But the strength of the Hopf–Cole method lies in obtaining a diffusion equation without dropping any part of (9.148), just by redefining the dependent variable and transforming the entire equation.

The equation is first rewritten in terms of what could be called the potential $f(x, t)$ of the velocity field $v(x, t)$, defined by

$$v = f_x. \tag{9.149}$$

In terms of f, Eq. (9.148) becomes

$$\left(f_t + \frac{1}{2} f_x^2 \right)_x = \nu (f_{xx})_x,$$

whose integral is

$$f_t + \frac{1}{2}(f_x)^2 = \nu f_{xx} \tag{9.150}$$

(the integration constant can be set equal to zero without loss of generality). The second change is nonlinear, from $f(x, t)$ to a new function $\phi(x, t)$ defined by

$$\phi = \exp\left(-\frac{f}{2\nu} \right) \quad \text{or} \quad f = -2\nu \log \phi. \tag{9.151}$$

Then Eq. (9.150) becomes

$$\phi_t = \nu\phi_{xx}.\tag{9.152}$$

This is a diffusion equation with the viscosity ν playing the role of the *diffusion constant*. To solve (9.152) take the Fourier transform in x of ϕ:

$$\phi(x, t) = \int_{-\infty}^{\infty} e^{ikx}\tilde\phi(k, t)\,dk.$$

Insert this into (9.152) and write the equation for the Fourier transform:

$$\tilde\phi_t(k, t) = -k^2\nu\tilde\phi(k, t).$$

The solution is

$$\tilde\phi(k, t) = \tilde\phi(k, 0)e^{-\nu k^2 t}.$$

Since $\tilde\phi(k, 0)$ is the inverse Fourier transform of $\phi(x, 0)$, that is, since (we leave off factors of 2π; see Hopf, 1950)

$$\tilde\phi(k, 0) = \int_{-\infty}^{\infty} e^{-ikx}\phi(x, 0)\,dx,$$

the Fourier integral for ϕ becomes

$$\phi(x, t) = \int_{-\infty}^{\infty} dy \int_{-\infty}^{\infty} e^{-\nu k^2 t + ik(x-y)}\phi(y, 0)\,dk.$$

The integral over k can be performed by using the identity

$$\int_{-\infty}^{\infty} e^{-ay^2 + iby}\,dy = \sqrt{\frac{\pi}{a}}e^{-b^2/4a},\tag{9.153}$$

so the general solution of the diffusion equation is

$$\phi(x, t) = \sqrt{\frac{\pi}{\nu t}}\int_{-\infty}^{\infty} e^{-(x-y)^2/4\nu t}\phi(y, 0)\,dy.\tag{9.154}$$

Finally, this solution has to be transformed back in terms of $v(x, t)$. According to (9.149) and (9.151)

$$v = -2\nu\frac{\phi_x}{\phi} \quad\text{or}\quad \phi(x, t) = \exp\left\{-\frac{1}{2\nu}\int^x v(y, t)\,dy\right\}.$$

If $v(x, 0) = R(x)$ is the initial velocity field, the final solution is

$$v(x, t) = \frac{1}{t}\frac{\int_{-\infty}^{\infty} dy\,(x-y)e^{-(x-y)^2/4\nu t}\exp\left\{-\frac{1}{2\nu}\int^y R(z)\,dz\right\}}{\int_{-\infty}^{\infty} dy\,e^{-(x-y)^2/4\nu t}\exp\left\{-\frac{1}{2\nu}\int^y R(z)\,dz\right\}}.\tag{9.155}$$

This is the general solution of the Burgers equation in terms of the initial condition $v(x, 0) = R(x)$. The solution we discuss for $v(x, t)$ will look like a wave packet, and we will sometimes refer to its leading edge of the packet as the *wave front*.

The viscosity enters the solution through exponentials of $-1/v$, which are not analytic functions, so the solution can not be expanded in a Taylor series about $v = 0$. This is not too surprising: it is another manifestation of the general problem of a small parameter multiplying the highest derivative in a differential equation. Equation (9.155) is nevertheless a solution in closed form, one that can be used to study the small-v regime simply by evaluating the integrals, as will now be shown.

One advantage of the Burgers equation is that its exact solvability in one dimension allows analysis of its solutions with different initial data. Although it was introduced as a simple approximation to the more general Navier–Stokes equation, the fact that it is among the simplest nonlinear diffusion equations lends it more general applicability in real physical problems. It is used to represent fluid turbulence in more than one dimension (Funaki and Woyczynski, 1996) and for other problems, such as interfacial growth (Godreche, 1990). Its extension to higher dimensions have been extensively studied but no exact or approximate analytic formal solutions have been found.

ASYMPTOTIC SOLUTION

When $v = 0$, solutions to the Burgers equation may exhibit a type of discontinuity called a *shock wave* and may even become multivalued. (When $0 < v \ll 1$, these pathologies get smoothed out and eliminated.) This can be demonstrated by using the *method of steepest descent* (Morse and Feshbach, 1953) to evaluate the small-v limit of the integrals in (9.155).

We give a brief outline of the method for evaluating integrals of the form

$$g(\lambda) = \int_{-\infty}^{\infty} \varphi(y)e^{-\lambda f(y)}dy$$

in the limit of large λ. Assume that both f and φ are continuous functions and that f has one maximum, occurring at $y = y_0$ (this treatment can be generalized to a finite number of local maxima and we will apply it later to two). Expand the integrand in a Taylor series about y_0 (use the fact that $f'(y_0) = 0$):

$$g(\lambda) \approx \int_{-\infty}^{\infty} \{\varphi(y_0) + \varphi'(y_0)(y - y_0) + O((y - y_0)^2)\}$$
$$\times e^{-\lambda[f(y_0)+\frac{1}{2}f''(y_0)(y-y_0)^2+O((y-y_0)^3)]}\, dy.$$

To lowest order in $y - y_0$ this yields

$$g(\lambda) \approx \varphi(y_0)e^{-\lambda f(y_0)}\int_{-\infty}^{\infty} e^{-\frac{1}{2}\lambda f''(y_0)(y-y_0)^2}dy + O((y - y_0)^4)$$

(since the integral of the Gaussian times the odd function $y - y_0$ vanishes).

Now use (9.153). Then

$$g(\lambda) \approx \varphi(y_0)e^{-\lambda f(y_0)}\sqrt{\frac{2\pi}{\lambda|f''(y_0)|}}. \qquad (9.156)$$

This result can be shown (Bruijn, 1958) to approach the exact result asymptotically in the limit as $\lambda \to \infty$, so this method works only for large λ. In our application λ is replaced by $1/\nu$, so it may be used for small ν. The resulting expression for $v(x, t)$ will be called the *asymptotic solution*.

To proceed requires finding the extrema of the function in the exponentials of the integrands of (9.155). They are obtained from the equation

$$0 = \frac{df(y)}{dy} \equiv -\frac{d}{dy}\left\{\frac{1}{2}\int^y R(z)\,dz + \frac{(x-y)^2}{4t}\right\} = -\frac{1}{2}R(y) + \frac{x-y}{2t}. \tag{9.157}$$

In general this equation has more than one solution $y_c(x, t)$. Assume, however, that there is just one. This assumption restricts the form of R, but even with this restriction solving (9.157) for y_c can be quite complicated. When it is solved, however, the asymptotic solution for v is

$$v(x, t) \approx \frac{x - y_c(x, t)}{t} = R(y_c(x, t)) = R(x - R(y_c)t) = R(x - vt) \tag{9.158}$$

[the rest of the integrals in (9.155) cancel out in the numerator and denominator].

Equation (9.158) is an implicit solution of (9.155). Note first that it implies that

$$v_t = \frac{-vR'}{1 + R't}, \qquad v_x = \frac{R'}{1 + R't}, \tag{9.159}$$

so that $v_t + vv_x = 0$; thus the $v(x, t)$ of Eq. (9.158) satisfies the Burgers equation (9.148) for $\nu = 0$. The second of Eqs. (9.159) shows that v_x becomes infinite if $1 + R't = 0$. This is at the critical time t_{cr} when, for some value of x, the equation

$$t_{cr} = -\frac{1}{R'} \tag{9.160}$$

is first satisfied. An infinite value for v_x means that the wave front becomes vertical at that value of x (the wave at time t_1 in Fig. 9.15). A physical explanation of how this happens is that different parts of the wave move at different speeds, and the fast-moving parts eventually catch up with the slower moving ones. Such discontinuities are called

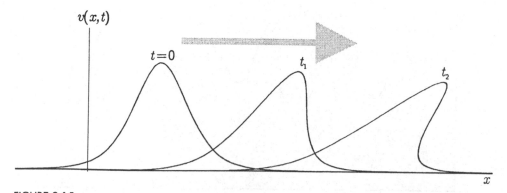

FIGURE 9.15

Formation of a shock wave. The wave moves to the right: $0 < t_1 < t_2$. The wave front at t_1 is just about vertical.

shock waves or simply *shocks*. As time goes on, v_x can become negative and the wave front can take on multiple values (the wave at time t_2 in Fig. 9.15).

As mentioned, these pathologies appear only in the $\nu = 0$ Burgers equation, but shock waves can occur also for $\nu \neq 0$ if Eq. (9.157) for the extremum y_c has more than one solution. For example, suppose that there are two solutions y_c^1 and y_c^2 of the form given by Eq. (9.158). Then at different times one or the other of these will dominate: at one time $f(y_c^1) > f(y_c^2)$ and at others $f(y_c^2) > f(y_c^1)$. The condition for a shock is that $f(y_c^1) = f(y_c^2)$. We do not go further into this, but see Whitham (1974).

9.5.3 SURFACE WAVES

A nontrivial application of Euler's equation is to the propagation of surface waves of a fluid, say water, in a gravitational field **g**. These waves, frequently seen on bodies of water such as ponds or the ocean, result from the action of **g** tending to force the fluid surface to its unperturbed equilibrium level. Their wavelengths vary from fractions of a meter to several hundred meters. Although they propagate on the surface of the fluid, their properties depend on its depth.

In addition to the gravitational force, surface tension and viscosity provide restoring forces, but they are important only for short wavelengths, on the order of a few centimeters. We will be interested, however, only in longer wavelength gravitational waves and will neglect surface and viscosity effects. For an overview of this subject, see Lamb (1879) and Debnath (1994).

EQUATIONS FOR THE WAVES

Consider the situation shown in Fig. 9.16. A volume of fluid, say water, in a uniform gravitational field of strength g is in a long narrow channel. At equilibrium the water lies to a depth h in the channel and is bounded below by a fixed hard horizontal floor and above by its surface in contact with the air. The surface is free to oscillate and the

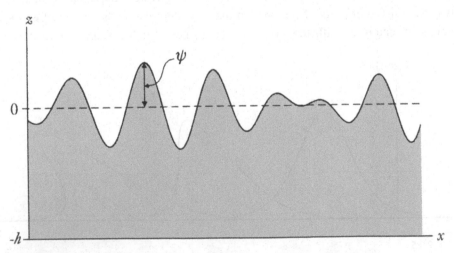

FIGURE 9.16
Water waves in a channel. The wave amplitude is exaggerated.

present treatment will concentrate on long-wavelength *gravitational* waves. Assume that the channel is so narrow that the y direction can be neglected. Choose the origin at the equilibrium surface, so that at equilibrium $z = 0$ everywhere. In general, in the presence of waves, the surface is given by a function of x and t, that is, $z = \psi(x, t)$. This will lead to a one–plus–one-dimensional field theory in which the field variable is ψ.

The treatment will be based on Euler's equation (9.139) with the appropriate boundary conditions. In the long-wavelength limit, turbulence, a short-range viscosity-dependent phenomenon, can be neglected. It will be assumed that the flow in the channel is *irrotational*, that is, that $\nabla \wedge \mathbf{v} = 0$ (when y is neglected this becomes $\partial_z v_x = \partial_x v_z$). This implies that there is a scalar function ϕ, the *velocity potential*, such that $\mathbf{v} = \nabla \phi$. Incompressibility (i.e., $\nabla \cdot \mathbf{v} = 0$) then means that ϕ satisfies Laplace's equation

$$\partial_x^2 \phi + \partial_z^2 \phi = 0 \tag{9.161}$$

in the water, that is, for $-h < z < \psi(x, t)$.

The left-hand side $-\nabla p$ of Euler's equation, essentially the force per unit volume, was the only force on the fluid, but in the present situation the gravitational force $-\rho g \hat{\mathbf{k}}$ has to be added ($\hat{\mathbf{k}}$ is the unit vector in the z direction). Also, on the right-hand side the $(\mathbf{v} \cdot \nabla)\mathbf{v}$ term can be rewritten by using the vector identity

$$\mathbf{v} \wedge (\nabla \wedge \mathbf{v}) = \frac{1}{2} \nabla v^2 - (\mathbf{v} \cdot \nabla)\mathbf{v}.$$

Since we are assuming that $\nabla \wedge \mathbf{v} = 0$, Euler's equation reduces to

$$-\nabla p - \rho g \hat{\mathbf{k}} = \rho \left\{ \partial_t \mathbf{v} + \frac{1}{2} \nabla v^2 \right\}.$$

In terms of ϕ this is (remember that ρ is constant)

$$\nabla \left[\partial_t \phi + \frac{1}{2} (\nabla \phi)^2 + \frac{p}{\rho} + gz \right] = 0. \tag{9.162}$$

Equation (9.162) can be integrated to yield

$$\partial_t \phi + \frac{1}{2} (\nabla \phi)^2 + \frac{p}{\rho} + gz = C(t), \tag{9.163}$$

where $C(t)$ is an arbitrary function of t alone. Both $C(t)$ and the pressure term can be eliminated by making a kind of gauge transformation from ϕ to

$$\phi' = \phi + \int \left[\frac{p}{\rho} - C(t) \right] dt,$$

and then

$$\partial_t \phi' + \frac{1}{2} (\nabla \phi')^2 + gz = 0.$$

The difference between $\nabla\phi$ and $\nabla\phi'$ is essentially $t\nabla p/\rho$. At the surface, which is what will be under consideration much of the time, the pressure is atmospheric and constant, so there $\nabla p = 0$ and $\nabla\phi$ is exactly equal to $\nabla\phi'$. Assume further that the channel is so shallow that the pressure variation in the water is small compared to atmospheric, so that the ∇p contribution to $\nabla\phi'$ can be neglected throughout the liquid, and $\mathbf{v}$ can be taken as the gradient of ϕ' as well as ϕ (that is the sense in which this is a gauge transformation: it does not affect the velocity). We drop the prime and rewrite (9.163) as

$$\partial_t\phi + \frac{1}{2}(\nabla\phi)^2 + gz = 0, \tag{9.164}$$

which is known as *Bernoulli's equation*. Equations (9.161) and (9.164) are the basic equation for the velocity field.

Boundary conditions must now be added to these equations. At the floor the boundary condition requires that no flow occur across the boundary: $v_z = 0$. In terms of ϕ this is

$$\partial_z\phi = 0 \quad \text{at} \quad z = -h. \tag{9.165}$$

At the upper surface the boundary condition is more complicated. Consider a fluid cell in the liquid with coordinates (x, z). Within the fluid x and z are independent variables, but at the surface the value of x and the shape of the wave determine z. If the boundary condition at the surface, like the one at floor, is that there is no flow across the boundary, the cell stays on the surface. Then since $z = \psi(x, t)$,

$$\frac{dz}{dt} = \frac{dx}{dt}\partial_x\psi + \partial_t\psi.$$

Now, $dz/dt = v_z \equiv \partial_z\phi$ and $dx/dt = v_x \equiv \partial_x\phi$, so the boundary condition on ϕ is

$$\partial_z\phi = (\partial_x\phi)\partial_x\psi + \partial_t\psi \quad \text{at the upper surface.} \tag{9.166}$$

Equations (9.161) and (9.164)–(9.166) define the problem completely: they are equations for both ϕ and ψ. They are nonlinear and not easy to solve. In fact a complete solution has not been found, and one turns to reasonable physical approximations. We will first discuss solutions to the linearization of these equations, for the linear solutions are interesting in themselves. Later we will obtain a nonlinear approximation called the *Korteweg–de Vries (KdV) equation*, that can sometimes be solved exactly.

LINEAR GRAVITY WAVES

Consider very small amplitude waves produced, for example, by a small stone dropped onto the initially flat surface of a relatively deep channel. Suppose that the amplitudes are so small that it is sufficient to treat the linearized limit. In this limit Eqs. (9.161), (9.165), (9.166), and (9.164) become (drop quadratic terms and set $z \sim 0$ at the surface)

$$\partial_x^2\phi + \partial_z^2\phi = 0, \tag{9.167}$$

$$\partial_z\phi = 0 \quad \text{at } z = -h, \tag{9.168}$$

$$\partial_t\psi - \partial_z\phi = 0 \quad \text{at the surface,} \tag{9.169}$$

$$\partial_t\phi + g\psi = 0 \quad \text{at the surface.} \tag{9.170}$$

Differentiate (9.170) with respect to t and use (9.169) to obtain

$$\phi_{tt} + g\phi_z = 0 \quad \text{at the surface.} \tag{9.171}$$

The next step is to solve Eqs. (9.171), (9.167), and (9.168) for ϕ, and then (9.170) can be used to find ψ.

Wavelike solutions for ϕ throughout the fluid, propagating in the x direction with wavelength λ and (circular) frequency ω, will be assumed of the form

$$\phi = Z(z)\sin(kx - \omega t), \tag{9.172}$$

where $k = 2\pi/\lambda$. With this ansatz Eq. (9.167) becomes

$$\frac{d^2 Z(z)}{dz^2} - k^2 Z(z) = 0,$$

whose solution is

$$Z(z) = Ae^{kz} + Be^{-kz}, \tag{9.173}$$

where A and B depend on the boundary conditions. When (9.173) is inserted into Eq. (9.168), it is found that $B = Ae^{-2hk}$, which leads to

$$\phi = 2Ae^{-kh}\cosh k(z+h)\sin(kx - \omega t).$$

This can be inserted into the other boundary condition (9.169) and the result integrated over time to yield (the z dependence will be taken care of shortly)

$$\psi = \frac{2kA}{\omega}e^{-kh}\sinh k(z+h)\cos(kx - \omega t) \equiv a(z)\cos(kx - \omega t), \tag{9.174}$$

where the wave amplitude $a(z)$ is

$$a(z) = \frac{2kA}{\omega}e^{-kh}\sinh k(z+h).$$

This amplitude can be inserted into the expression for ϕ:

$$\phi = \frac{a(z)\omega\cosh k(h+z)}{k\sinh kh}\sin(kx - \omega t). \tag{9.175}$$

The linearized surface waves are given by (9.174) only when z is set equal to zero:

$$\psi = \frac{2kA}{\omega}e^{-kh}\sinh kh\cos(kx - \omega t) \equiv a_0\cos(kx - \omega t),$$

where $a_0 = a(0)$ depends on k and h. Because of this dependence, the linearized waves satisfy a dispersion relation (Section 4.2.3), which can be obtained by applying Eq. (9.171) to the ϕ of (9.175):

$$\omega^2 = gk\tanh kh. \tag{9.176}$$

Accordingly, the frequency depends strongly on $kh \equiv 2\pi h/\lambda$ and hence on the ratio of the depth of the channel to the wavelength. This is a manifestation of competition between inertial and gravitational forces. The phase velocity is

$$c \equiv \frac{\omega}{k} = \left(\frac{g}{k} \tanh kh \right)^{1/2}.$$

In the limits of small and large values of kh linear waves behave differently. They are called shallow and deep-water waves, but note that the channel depth h is compared to the wavelength, not to the amplitude.

Shallow Linear Water Waves: $kh \ll 1$ or $\lambda \gg h$. Use

$$\tanh kh \approx kh - \frac{(kh)^3}{3} + \cdots$$

to expand Eq. (9.176) in powers of kh. Then to lowest order,

$$\omega = c_0 k \left[1 - \frac{(kh)^2}{6} + O(k^4 h^4) \right], \tag{9.177}$$

where $c_0 \equiv \sqrt{gh}$. This shows that when the wavelength is much larger than the depth, there is essentially no dispersion. Also, the phase velocity c_0 increases as the square root of the channel depth.

Deep-water Linear Waves: $kh \gg 1$ or $\lambda \ll h$. In this limit $\tanh kh \approx 1$ and

$$\omega = \sqrt{gk}. \tag{9.178}$$

In this case there is dispersion, and the dispersion relation is independent of the channel depth. The phase velocity is $c_\infty \equiv \sqrt{g/k}$, so long wavelength waves travel faster than short ones.

NONLINEAR SHALLOW WATER WAVES: THE KdV EQUATION

In the linearized approximation surface waves propagate in general dispersively: a wave packet made up of such waves broadens and loses amplitude as it moves along. It was then a surprise, first recorded in a famous recollection by Russell (1844), that nonlinear effects could stabilize the shape of such a wave at increased amplitudes. Russell gave an account of a single lump of surface water that detached itself from a boat that stopped suddenly after being pulled through a canal by horses; the wave moved at about 8–9 miles/hour, maintaining its shape for about a couple of miles. What he saw was the first recorded experimental example of a propagating soliton.

In this subsection we derive an equation for such waves, called the KdV equation (Korteweg–de Vries, 1895). This is a nonlinear, dispersive, nondissipative equation that is completely integrable and has soliton solutions.

Deriving the KdV equation is a nontrivial procedure (in what is called *asymptotics*) requiring several approximations and involving a perturbation expansion in the two physical parameters

$$\epsilon \equiv \frac{a}{h} \quad \text{and} \quad \mu \equiv \left(\frac{h}{\lambda}\right)^2. \tag{9.179}$$

Of these parameters, ϵ measures the ratio of the wave amplitude to the channel depth, assumed small in the previous subsection, and μ is proportional to $(hk)^2$, which was seen at Eq. (9.177) to be a measure of dispersive effects. The double perturbative expansion assumes that ϵ and μ are of the same order, which means that the low-amplitude and dispersive effects balance each other, and that is what allows soliton solutions.

For the derivation it is convenient to rewrite the surface wave equations in terms of ϵ and μ and to reset the z origin to the bottom of the channel and to transform to dimensionless coordinates. These coordinates are defined by $x' \equiv x/\lambda$, $z' \equiv z/h$, $t' \equiv tc_0/\lambda$, $\psi' \equiv \psi/a$, and $\phi' \equiv \phi h/(ac_0)$, where $c_0 = \sqrt{gh}$ as above. Then Eqs. (9.161), (9.165), (9.166), and (9.164) become (we drop the primes and use subscript notation for derivatives)

$$\mu\phi_{xx} + \phi_{zz} = 0, \tag{9.180}$$

$$\phi_z = 0 \quad \text{at } z = 0, \tag{9.181}$$

$$\mu[\psi_t + \epsilon\psi_x\phi_x] - \phi_z = 0 \quad \text{at the surface,} \tag{9.182}$$

$$\phi_t + \frac{\epsilon}{2}\phi_x^2 + \frac{\epsilon}{2\mu}\phi_z^2 + \psi = 0 \quad \text{at the surface.} \tag{9.183}$$

The surface is now defined by $z = 1 + \epsilon\psi$.

For the perturbative treatment, the velocity potential is expanded in powers of μ:

$$\phi = \phi_0 + \mu\phi_1 + \mu^2\phi_2 + \cdots. \tag{9.184}$$

To zeroth order in μ Eq. (9.180) becomes $\phi_{0zz} = 0$, and then from (9.181), $\phi_{0z} = 0$. Thus $\phi_0 \equiv \phi_0(x, t)$, a function of only x and t. The first- and second-order equations are

$$\phi_{0xx} + \phi_{1zz} = 0, \tag{9.185a}$$

$$\phi_{1xx} + \phi_{2zz} = 0. \tag{9.185b}$$

Equation (9.185a) implies that

$$\phi_1 = -\frac{z^2}{2}u_x,$$

where $u \equiv \phi_{0x}$, and then (9.185b) yields

$$\phi_2 = \frac{z^4}{24}u_{xxx}.$$

Next, these results are inserted into Eq. (9.183) to lowest order in ϵ and μ (use $u_z = 0$ and $z = 1 + \epsilon\psi \approx 1$):

$$\phi_{0t} - \frac{\mu}{2}u_{xt} + \frac{\epsilon}{2}u_x^2 + \psi = 0.$$

The x derivative of this equation is

$$u_t - \frac{\mu}{2}u_{xxt} + \epsilon u u_x + \psi_x = 0. \tag{9.186}$$

Equation (9.182) is now written to second order in μ and ϵ:

$$\psi_t + \epsilon u \psi_x + (1 + \epsilon \psi)u_x - \frac{\mu}{6}u_{xxx} = 0. \tag{9.187}$$

Equations (9.186) and (9.187), a pair of coupled partial differential equations for u and ψ, are the lowest order corrections to the linearized equations for shallow water. In zeroth order they read

$$\psi_t + u_x = 0 \quad \text{and} \quad \psi_x + u_t = 0, \tag{9.188}$$

which lead to

$$\psi_{tt} - \psi_{xx} = 0 \quad \text{and} \quad u_{tt} - u_{xx} = 0.$$

Thus to this order both ψ and u satisfy the wave equation, so $u(x, t) = u(x \pm t)$ and $\psi(x, t) = \psi(x \pm t)$. Thus in zeroth order $\psi_x \mp \psi_t = 0$.

To obtain the KdV equation, an ansatz is made for a solution to the next order in ϵ and μ:

$$u = \psi + \epsilon P(x, t) + \mu Q(x, t). \tag{9.189}$$

The functions P and Q here must be determined from the consistency condition that the ansatz work in both (9.186) and (9.187). To first order, when (9.189) is inserted into those equation, they become

$$\psi_t + \psi_x + \epsilon(P_x + 2\psi \psi_x) + \mu \left(Q_x - \frac{1}{6}\psi_{xxx} \right) = 0, \tag{9.190a}$$

$$\psi_x + \psi_t + \epsilon(P_t + \psi \psi_x) + \mu \left(Q_t - \frac{1}{2}\psi_{txx} \right) = 0. \tag{9.190b}$$

In first order ψ is no longer a function simply of $x \pm t$ but depends more generally on x and t. It will be seen in the next subsection, however, that the soliton solutions are functions of $x \pm wt$ for some other velocity w.

Now use (9.188) together with (9.189) in these equations and subtract the first from the second. The result is

$$\epsilon(2P_x + \psi \psi_x) + \mu \left(2Q_x - \frac{2}{3}\psi_{xxx} \right) = 0.$$

Since the ϵ and μ terms must be independent, it follows that

$$P = -\frac{1}{4}\psi^2 \quad \text{and} \quad Q = \frac{1}{3}\psi_{xx}.$$

When these results are put back into Eq. (9.190a), the result is

$$\psi_t + \psi_x + \frac{3}{2}\epsilon\psi\psi_x + \frac{1}{6}\mu\psi_{xxx} = 0. \tag{9.191}$$

This is one form of the desired KdV equation, but we will rewrite it in a more convenient form.

First we return to dimensional variables. Then Eq. (9.191) becomes

$$(\psi_t + c_0\psi_x) + \frac{3c_0}{2h}\psi\psi_x + \frac{c_0h^2}{6}\psi_{xxx} = 0. \tag{9.192}$$

Each term here represents an aspect of surface waves. The term in parentheses describes a surface wave propagating at the constant speed c_0. The next term is quadratic and represents the nonlinearity of the equation. The last term is related to dispersion. As a KdV wave evolves, the nonlinear and dispersive terms compete, and it is this competition that gives rise to soliton solutions.

We discuss the soliton solutions in the next subsection, but before doing so we redefine the wave function and recast the equation in a new dimensionless form. The new wave function is

$$\xi \equiv \frac{\psi}{4h} + \frac{1}{6},$$

and the new dimensionless variables are

$$t' = \frac{t}{6}\sqrt{\frac{g}{h}}, \quad x' = -\frac{x}{h}.$$

The sign change of x is for convenience; it is permissible because the argument of the wave solutions can be $x + c_0t$ or $x - c_0t$, depending on whether the wave is traveling in the negative of positive x direction. Then (9.192) reads (again we drop the primes)

$$\xi_t = \xi\xi_x + \xi_{xxx}, \tag{9.193}$$

which is the form of the KdV equation that we will mostly consider from now on.

SINGLE KdV SOLITONS

Soliton solutions are obtained by making a further ansatz, similar to the ones made to obtain them in the sG and K–G cases. Assume that ξ is of the form

$$\xi(x, t) = \eta(x + wt) \equiv \eta(\zeta), \tag{9.194}$$

where w is some (dimensionless) velocity. (The positive sign is chosen for convenience.) With this ansatz, Eq. (9.193) becomes

$$w\eta' - \eta\eta' - \eta''' = 0.$$

A first integral is

$$w\eta - \frac{1}{2}\eta^2 - \eta'' = A,$$

(9.195)

where A is a constant. Now use

$$\frac{1}{2}\frac{d\eta'^2}{d\eta} = \eta'\frac{d\eta'}{d\zeta}\frac{d\zeta}{d\eta} = \eta'\eta''\frac{1}{\eta'} = \eta''$$

to rewrite (9.195) in the form

$$\frac{1}{2}\frac{d\eta'^2}{d\eta} = w\eta - \frac{1}{2}\eta^2 - A.$$

A second integration, this time with respect to η, yields

$$\frac{1}{2}\eta'^2 = \frac{1}{2}w\eta^2 - \frac{1}{6}\eta^3 - A\eta + B,$$

(9.196)

where B is also a constant. Imposing the boundary condition $(\eta, \eta', \eta'') \to 0$ as $\zeta \to \infty$ forces $A = 0$ in Eq. (9.195) and $B = 0$ in (9.196), so that with this condition,

$$\eta'^2 = \eta^2\left(w - \frac{1}{3}\eta\right).$$

Then

$$\zeta = \int \frac{d\eta}{\eta'} = \sqrt{3}\int \frac{d\eta}{\eta\sqrt{3w - \eta}}.$$

This can be integrated by making the substitution $\eta = 3w\,\text{sech}^2\,\vartheta$. Some algebra leads to

$$\zeta - \zeta_0 = \frac{2}{\sqrt{w}}\int d\vartheta = \frac{2}{\sqrt{w}}\vartheta,$$

and so

$$\eta(\zeta) = 3w\,\text{sech}^2\left\{\frac{1}{2}\sqrt{w}(\zeta - \zeta_0)\right\}.$$

(9.197)

Equation (9.197) is an exact solution of the KdV equation representing a nondispersive traveling wave packet (remember, $\zeta = x + wt$). Note that the speed w is proportional to the amplitude: the taller the packet, the faster it moves. To determine whether it is in fact a soliton requires studying the interaction of more than one such packet. That has been done numerically and will be discussed in the next subsection. There are also other exact solutions of the KdV equation that can be obtained analytically. They obey different

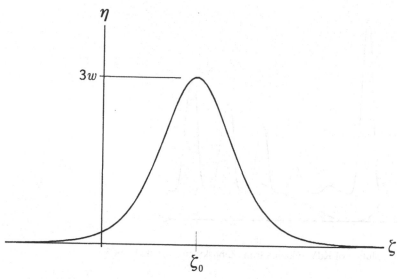

FIGURE 9.17
The KdV soliton of Eq. (9.197).

boundary conditions, so $A \neq 0$ and $B \neq 0$, and we will not address them. Figure 9.17 is a graph of the $\eta(\zeta)$ of Eq. (9.197).

MULTIPLE KdV SOLITONS

The revival of interest in the KdV equation dates to a numerical study by Zabusky and Kruskal in 1965. In this subsection we briefly describe their analysis. It is interesting that they were not trying to model surface waves on a fluid. Rather, they were trying to understand one of the earliest discoveries made by numerical simulation – some work by Fermi, Pasta, and Ulam (1955) on ergodicity – when they modeled a solid as a lattice of nonlinearly coupled oscillators.

In the Zabusky–Kruskal calculation the boundary conditions are chosen for convenience to be periodic, which effectively replaces the infinitely long channel by an annular one: $x \in [0, 2]$. Specifically, their initial condition is

$$\xi(x, 0) = \cos \pi x. \tag{9.198}$$

The KdV equation is written in the form

$$\xi_t + \xi \xi_x + \delta^2 \xi_{xxx} = 0, \tag{9.199}$$

where the factor δ^2 is added to control the competition between the nonlinear and dispersive terms. Zabusky and Kruskal choose $\delta = 0.022$, which makes the dispersion small compared to the nonlinearity. (The difference in sign is of no consequence: it involves replacing x by $-x$.)

What is then found is shown in Fig. 9.18. The dotted curve is the initial condition $\xi(x, 0)$. The broken curve is the wave at $t = 1/\pi$ and the solid curve is at $t = 3.6/\pi$. It is seen that as time flows the initial shape becomes deformed. There is almost a shock appearing at $t = 1/\pi$, which arises because this is too early for the low dispersion to have much effect. Up to this time the dominating $\xi_t + \xi \xi_x$ leads to a shock like those of the zero-viscosity Burgers equation [in fact for very small δ Eq. (9.199) looks very much like (9.148) with $\nu = 0$]. The shock does not

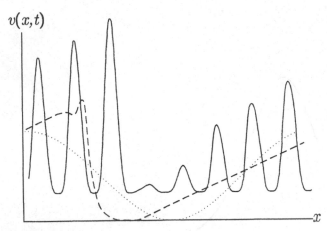

FIGURE 9.18
Multiple-soliton solution of KdV. (Copied from Zabusky and Kruskal, 1965.)

develop into a singularity, however, because for later t the dispersion begins to take effect. At $t = 3.6/\pi$ there are what look like eight solitary wave packets, each of which separately can almost be fitted to the single soliton of the previous subsection.

Because the equation from which these waves are derived is so much like the zero-ν Burgers equation, the time at which the shock wave forms can be obtained from Eq. (9.160). The initial $R \equiv \xi(x, 0) = \cos \pi x$, so the initial velocity profile yields

$$t_{cr} = -\frac{1}{R'} = 1/\pi.$$

This is the value obtained in the numerical simulation. The location of the discontinuity can be obtained similarly by taking the x derivative of the velocity for the $\nu = 0$ equation. We state without proof that this gives $x = 1/2$, which also agrees approximately with the numerical result.

As described above, the taller packets move faster than the shorter ones. They catch up and interact, losing their individual character for a short time. But they emerge from the interaction region with almost the same shape they had on entering it, except for a possible phase change. That is what makes this a soliton solution. As time runs on dispersion continues to alter the size and shape of the wave packets, and eventually the entire wave returns almost exactly to the initial cosine shape. As time continues to run on this scenario repeats many times. After a very long time round-off error begins to affect the numerical calculation, and it ceases to be reliable.

Two years after the Zabusky–Kruskal work, Gardner et al. (1967) used an analytic method called *the inverse scattering transform* to show that the KdV equation is fully integrable for localized initial conditions and that its remarkable properties were related to this integrability. Still later Hirota (1971) gave an algorithm for constructing an N-soliton solution for the KdV equation analytically.

9.6 HAMILTONIAN FORMALISM FOR NONLINEAR FIELD THEORY

Our discussion of nonlinear field theory has been restricted to a few special cases, and it may seem that the properties of their solutions are also quite special. But there are many other nonlinear differential equations whose solutions share the same properties. In this last section of the book we close the discussion with a true Hamiltonian treatment of nonlinear field equations, reflecting Chapters 6 and 7 and providing a framework for more general treatment of such systems. This treatment will be applied particularly to the KdV equation and briefly to the sG equation.

9.6.1 THE FIELD THEORY ANALOG OF PARTICLE DYNAMICS

FROM PARTICLES TO FIELDS

Section 9.3.1 extended the Hamiltonian formalism of particle dynamics, with a finite number of freedoms, to field theory, with an infinite number of freedoms. The extension was imperfect, however, for several reasons (see, for instance, the comments just above Worked Example 9.3). In this section we discuss a different extension of Hamiltonian dynamics, one that is particularly applicable to nonlinear field theory. Our treatment will be largely by analogy, necessarily imperfect (we will point out the imperfections as we go along).

The infinite-freedom analog of the set of coordinates $\xi_k(t)$ on $\mathbf{T}^*\mathbb{Q}$ are now the field functions, which will be called $\xi(x, t)$ (we treat just one x dimension): the analog of k is x and the ξs are assumed to be C^∞ (possessing continuous derivatives to all orders) real functions of x. Two kinds of systems will be considered. In one kind x lies on the infinite line $(-\infty < x < \infty)$ and it is assumed that $\lim_{|x| \to \infty} \xi(x) = 0$. In the other x lies on a circle $(0 \leq x < 2\pi)$ and it is assumed that $\lim_{x \to 2\pi} \xi(x) = \xi(0)$ (the field functions are periodic on the circle). For both kinds it is assumed that ξ is square integrable:

$$\int \xi^2 \, dx < \infty,$$

where the integral is taken over the entire infinite line or once over the circle, respectively.

In both cases the "phase manifold" is thus the space of all square integrable functions, called L_2. But L_2 is not a finite-dimensional manifold in the sense of earlier chapters (which is why we write it in quotes the first time). In fact L_2 is not obtained by going to the $n \to \infty$ limit of some finite-dimensional $\mathbf{T}^*\mathbb{Q}$, so it is not really analogous to $\mathbf{T}^*\mathbb{Q}$. The number of freedoms, mirrored in the dimension of L_2, is uncountably infinite.

In finite freedoms vector fields are indexed by k, like the coordinates themselves. In fact the ξ_k themselves are a special vector field: at each point $\xi \in \mathbf{T}^*\mathbb{Q}$ the kth *component* of the vector field is equal to the *coordinate* ξ_k. The infinite-freedom analogs of vector fields are functions of x, indexed by the analog x of k. In the analogy, $\xi(x)$ itself is a special kind of function: at each $\xi \in L_2$ the *value* of the function for each x is equal to the *value* $\xi(x)$ (the analog of both "component" and "coordinate" is "value").

An inner product is defined on L_2: if $\xi, \eta \in L_2$, their inner product is

$$\int \xi(x)\eta(x) \, dx \equiv (\xi, \eta), \tag{9.200}$$

analogous to the sum $\sum_k \xi_k \eta_k \equiv (\xi, \eta)$ in particle mechanics. More generally, analogs of sums in particle mechanics are integrals in field theory.

DYNAMICAL VARIABLES AND EQUATIONS OF MOTION

The next step is to define the analogs of dynamical variables. In particle systems dynamical variables, which are functions of the phase variables, are maps from $\mathbf{T}^*\mathbb{Q}$ to $\mathbb{R}$: each set of coordinates ξ_k (each point in $\mathbf{T}^*\mathbb{Q}$) is mapped to a number in $\mathbb{R}$. In field theory dynamical variables are maps from the phase manifold L_2 to $\mathbb{R}$: each function $\xi \in L_2$ is mapped to a number in $\mathbb{R}$. A map from functions to numbers is a *functional*, so dynamical variables are functionals. An example of a functional is

$$F(\xi) = \int \left(\frac{1}{6}\xi^3 - \frac{1}{2}\xi_x^2 \right) dx \equiv H^K(\xi), \tag{9.201}$$

where $\xi_x = \partial \xi / \partial x$. This functional is given the special name $H^K(\xi)$ because it will prove important later, when the technique is applied to the KdV equation.

In particle dynamics the equations of motion are of the form

$$\dot{\xi}_k = F_k(\xi). \tag{9.202}$$

Their analogs in field theory are the field equations (assumed, at first, to be of first order in t), of the form

$$\xi_t \equiv \partial_t \xi = \tilde{K}\xi, \tag{9.203}$$

where $\tilde{K}$ is an operator, in general nonlinear. (An operator maps functions to functions.) For example, in the KdV equation

$$\tilde{K}\xi = \xi\xi_x + \xi_{xxx}. \tag{9.204}$$

9.6.2 THE HAMILTONIAN FORMALISM

Equation (9.202) represents a Hamiltonian system only if F_k can be written as $\sum \omega_{kj}\partial_j H$. That is the condition for *canonicity* in particle dynamics. Analogously, (9.203) represents a Hamiltonian field theory only if $\tilde{K}\xi$ can be written as the analog of $\sum \omega_{jk}\partial_k H$. Thus to obtain the condition for canonicity in field theory requires defining the analogs of the symplectic form ω and of the gradient $\partial_k H$ of a dynamical variable H.

THE GRADIENT

We start by redefining the gradient in finite dimensions in a way that lends itself to generalization to L_2. Consider a dynamical variable $f(\xi)$ in finite dimensions. Let η be any vector (η has $2n$ components); η defines a direction in $\mathbf{T}^*\mathbb{Q}$, and the *directional derivative of f along η* at any point $\xi \in \mathbf{T}^*\mathbb{Q}$ is

$$\lim_{\epsilon \to 0} \frac{1}{\epsilon}(f(\xi + \epsilon\eta) - f(\xi)) \equiv \frac{d}{d\epsilon}f(\xi + \epsilon\eta)\Big|_{\epsilon=0} = \sum\{\partial_k f(\xi)\}\eta_k = (\nabla f, \eta). \tag{9.205}$$

Note that ∇f is a *vector field*: its components at each $\xi \in \mathbf{T}^*\mathbb{Q}$ are the $\partial_k f(\xi)$. Equation (9.205) can be used as a definition: the gradient ∇f of f at $\xi \in \mathbf{T}^*\mathbb{Q}$ is the vector whose inner product with any vector η gives the directional derivative of f along η.

This definition is suitable for extension to infinite dimensions. Consider a dynamical variable $F(\xi)$ in infinite dimensions (a functional). Let $\eta \in L_2$ be arbitrary. The *directional derivative of F along η* at any function $\xi \in L_2$ is

$$\lim_{\epsilon \to 0} \frac{1}{\epsilon}(F(\xi + \epsilon\eta) - F(\xi)) \equiv \frac{d}{d\epsilon}F(\xi + \epsilon\eta)\Big|_{\epsilon=0} = (F_\xi, \eta), \qquad (9.206)$$

where the inner product is defined by (9.200) (sums become integrals). This defines the gradient F_ξ of F: it is the *function* whose inner product with any $\eta \in L_2$ gives the directional derivative along η.

WORKED EXAMPLE 9.6

Use (9.206) to find the gradient H_ξ^K of the functional $H^K(\xi)$ of Eq. (9.201).

Solution. First calculate $H^K(\xi + \epsilon\eta)$:

$$H^K(\xi + \epsilon\eta) = \int \left[\frac{1}{6}(\xi + \epsilon\eta)^3 - \frac{1}{2}(\xi_x + \epsilon\eta)^2\right]dx$$

$$= H^K(\xi) + \epsilon \int \left(\frac{1}{2}\xi^2\eta - \xi_x\eta_x + O(\epsilon)\right)dx,$$

so

$$\lim_{\epsilon \to 0} \frac{1}{\epsilon}[H^K(\xi + \epsilon\eta) - H^K(\xi)] = \int \left(\frac{1}{2}\xi^2\eta - \xi_x\eta_x\right)dx.$$

The second term can be integrated by parts:

$$\int \xi_x\eta_x\,dx = [\xi_x\eta] - \int \xi_{xx}\eta\,dx.$$

The term in square brackets is the integrated part, evaluated at the end points, so it vanishes. (If x lies on the infinite line, the functions vanish at $\pm\infty$, and if on the circle, the two end points are the same.) As a result, the integral is

$$\int \left(\frac{1}{2}\xi^2 + \xi_{xx}\right)\eta\,dx \equiv (H_\xi^K, \eta),$$

so according to (9.206) the gradient is

$$H_\xi^K = \frac{1}{2}\xi^2 + \xi_{xx}. \qquad (9.207)$$

Comment: The procedure used here will be used again: the x derivative is shifted from one factor to the other in an integral and the sign is changed.

Notice that H_ξ^K is a function: Equation (9.207) should properly be written $H_\xi^K(x) = \frac{1}{2}[\xi(x)]^2 + \xi_{xx}(x)$. Since it is indexed by x, it is the analog of a vector field in finite dimensions, and indeed, a gradient ∇f in finite dimensions is a vector field. Omitting the x in (9.207) is equivalent to writing ∇f rather than $\partial_k f$.

THE SYMPLECTIC FORM

The analog of $\sum \omega_{kj} \partial_j H$ requires also the analog Ω of the symplectic form ω (Section 5.2.3). In finite freedoms ω can map vectors to vectors: $\sum \omega_{jk} \xi_k$ is a vector. Hence in infinite freedoms Ω must map functions to functions: Ω is a linear operator. In addition, an essential property of ω is its antisymmetry: $\omega_{ij} = -\omega_{ji}$. This lends itself to generalization if stated in terms of the finite-dimensional inner product:

$$(\xi, \omega \eta) \equiv \sum \xi_j \omega_{jk} \eta_k = -\sum \omega_{kj} \xi_j \eta_k \equiv -(\omega \xi, \eta).$$

Analogously, Ω is antisymmetric with respect to the inner product in L_2:

$$(\xi, \Omega \eta) = -(\Omega \xi, \eta). \tag{9.208}$$

There are two important differences between ω and Ω. In particle dynamics ω was a nonsingular matrix, but Ω will not always be nonsingular. Also, ω was the same for all dynamical systems, but Ω will be different for different systems.

THE CONDITION FOR CANONICITY

We now use the definition of the gradient and of Ω to state the condition that Eq. (9.203) represent the analog of a Hamiltonian particle dynamical system: it is that $\tilde{K} \xi$ on the right-hand side of Eq. (9.203) be of the form $\tilde{K} \xi = \Omega H_\xi$, where Ω is a linear antisymmetric operator on L_2 and H_ξ is the gradient of a functional $H(\xi)$. In other words the system is Hamiltonian if it can be written in the form

$$\xi_t = \Omega H_\xi. \tag{9.209}$$

Then $H(\xi)$ is called the Hamiltonian (functional) of the system. Because ω is the same for all systems in particle dynamics, a system is characterized by its Hamiltonian H. Because Ω is different for different systems in field theory, a system is characterized by both H and Ω.

POISSON BRACKETS

In particle dynamics the Poisson bracket of two dynamical variables is defined as

$$\{f, g\} = \sum (\partial_k f) \omega_{kj} (\partial_j g) = (\nabla f, \omega \nabla g).$$

In complete analogy, the PB of two functionals $F(\xi)$ and $G(\xi)$ is defined as

$$\{F, G\} \equiv (F_\xi, \Omega G_\xi). \tag{9.210}$$

We state without proof that the PB so defined has the three properties of the classical PB: it is bilinear, is antisymmetric, and satisfies the Jacobi identity. As in particle mechanics, the PB with the Hamiltonian gives the time variation. The analogs of sums are integrals, so the analog of

$$\frac{df}{dt} = \sum \frac{\partial f}{\partial \xi_k} \dot{\xi}_k$$

is

$$\frac{d}{dt} F(\xi) = \int F_\xi \xi_t \, dx$$

(we will deal only with functionals that do not depend explicitly on t). Inserting (9.209) into the right-hand side yields

$$\frac{d}{dt} F(\xi) = \int F_\xi \Omega H_\xi \, dx = (F_\xi, \Omega H_\xi) = \{F, H\}.$$

Therefore $F(\xi)$ is a constant of the motion if and only if $\{F, H\} = 0$, as in particle dynamics. Of course antisymmetry implies that $\{H, H\} = 0$, so (a time independent) $H(\xi)$ itself is always a constant of the motion. Any dynamical variable can be taken as the Hamiltonian of a dynamical system, and then it is clear that if F is a constant for the H system, H is a constant for the F system. This is the same as in particle dynamics.

9.6.3 THE KdV EQUATION

We now apply these considerations to the KdV equation of Section 9.5.3.

KdV AS A HAMILTONIAN FIELD

The KdV equation (9.193) is a field equation, but not obviously a Hamiltonian one. In particle dynamics the treatment of a Hamiltonian system generally started with the Hamiltonian function, and the equations of motion were obtained from it. In this case, however, the equation of motion was derived first, and now the problem is to show that it can be put into the canonical form of the Hamiltonian formalism.

The right-hand side of (9.193) is the x derivative of the gradient H_ξ^K found in Worked Example 9.6; therefore (9.193) can be written in the form

$$\xi_t = \Delta H_\xi^K, \tag{9.211}$$

where $\Delta \equiv \partial/\partial x$. This is in canonical form provided that Δ is an antisymmetric operator on L_2. It is clearly an operator because it maps functions to functions. Antisymmetry follows from

$$(\xi, \Delta\eta) = \int \xi \eta_x \, dx = [\xi\eta] - \int \xi_x \eta \, dx = -(\Delta\xi, \eta),$$

where the integrated term vanishes as usual.

This shows that (9.211) is in canonical form: the KdV field is Hamiltonian with $H \equiv H^K$ given by (9.201) and with $\Omega = \Delta$.

CONSTANTS OF THE MOTION

A constant of the motion for the KdV field is a functional for which $\{F, H^K\} = 0$.

WORKED EXAMPLE 9.7

Show that

$$F_1(\xi) = \int \frac{1}{2}\xi^2 \, dx \tag{9.212}$$

is a constant of the motion.

Solution. First calculate $F_{1\xi}$ using (9.206):

$$\left.\frac{dF_1}{d\epsilon}\right|_{\epsilon=0} = \int \xi\eta \, dx = (F_{1\xi}, \eta),$$

which implies that $F_{1\xi} = \xi$. Then according to (9.210)

$$\{F_1, H^K\} = \left(F_{1\xi}, \Delta H_\xi^K\right) = \int \xi(\xi\xi_x + \xi_{xxx}) \, dx.$$

Integrating the first term by parts yields

$$\int \xi^2 \xi_x \, dx = -2 \int \xi_x \xi^2 \, dx \equiv 0.$$

The second term, involving three transpositions of the subscript x, is the negative of itself (see the comment at the end of Worked Example 9.6). Thus $F_1(\xi)$ is a constant of the motion for the KdV equation.

Since H^K itself is a constant of the motion, F_1 and H^K are two constants of the motion in involution. It turns out that these are two of an infinite set of independent constants of the motion F_k, $k = 0, 1, 2, \ldots$, all in involution with each other (Miura et al., 1968),

$$\{F_j, F_k\} = 0.$$

This begins to look like complete integrability in the sense of the LI theorem (Section 6.2.2), and the KdV equation is sometimes called integrable. Yet the situation in infinite dimensions is infinitely more complicated than in finite dimensions, and the LI theorem does not apply. (Neither, incidentally, does the Liouville volume theorem.) The existence of this infinite set of constants of the motion for a given field equation does not imply that the flow is on an infinite-dimensional torus, even if the Σ_j they define are compact (Schwarz, 1991 and private communication). Nonetheless, the infinite set of constants of the motion provides insight into the dynamical system. Moreover, each of them can itself be taken as the Hamiltonian of a new dynamical system and inserted into (9.211) in place of H^K:

$$\xi_t = \Delta F_{k\xi}. \tag{9.213}$$

Involution implies that each such system has the same infinite set of constants of the motion and is as "integrable" as the KdV equation.

The F_k can be constructed explicitly by a recursion procedure. We will describe the procedure in the next subsection, but first we write down four of the F_k (see Problem 17):

$$\left.\begin{aligned}
F_0(\xi) &= \int 3\xi \, dx, \\
F_1(\xi) &= \int \frac{1}{2}\xi^2 \, dx, \\
F_2(\xi) &= \int \left(\frac{1}{6}\xi^3 - \frac{1}{2}\xi_x^2 \right) dx, \\
F_3(\xi) &= \int \frac{1}{2} \left(\frac{5}{36}\xi^4 - \frac{5}{3}\xi\xi_x^2 + \xi_{xx}^2 \right) dx.
\end{aligned}\right\} \tag{9.214}$$

(Note that $F_2 = H^K$.) Because their gradients play a role in the recursion procedure, we write them down as well:

$$\left.\begin{aligned}
F_{0\xi} &= 3, \\
F_{1\xi} &= \xi, \\
F_{2\xi} &= \frac{1}{2}\xi^2 + \xi_{xx}, \\
F_{3\xi} &= \frac{1}{2}\left(\frac{5}{9}\xi^3 + \frac{5}{3}\xi_x^2 + \frac{10}{3}\xi\xi_{xx} + 2\xi_{xxxx} \right).
\end{aligned}\right\} \tag{9.215}$$

GENERATING THE CONSTANTS OF THE MOTION

The recursion procedure for generating the constants of the motion is not a formula for the F_j themselves but a differential equation for their gradients $F_{j\xi}$. Specifically, it is

$$LF_{j\xi} = \Delta F_{(j+1)\xi}, \tag{9.216}$$

where L, called the Lenard operator, is

$$L = \Delta^3 + \frac{2}{3}\xi\Delta + \frac{1}{3}\xi_x. \tag{9.217}$$

(See Problem 18.) Note that the gradients are all that are needed for equations such as (9.213).

If (9.216) can be solved successively for the $F_{(j+1)\xi}$ and if the solutions can then be integrated to yield the F_{j+1} themselves, this procedure will generate all of the constants of the motion from F_0. Three things have to be proven: 1. that this differential equation has a solution $F_{(j+1)\xi}$, 2. that the solution is indeed the gradient of some functional $F_{j+1}(\xi)$, and 3. that the F_{j+1} so obtained is a constant of the motion. Points 1 and 2 are more involved, and together they imply 3. We will first assume that 1 and 2 are true and then prove 3. We put off the discussion of 1 and 2 to the next subsection.

Assume, therefore, that Points 1 and 2 have been established: solutions to $F_{j\xi}$ (9.216) exist and are gradients. Then the proof that the F_j are constants of the motion requires showing that they are all in involution with $H^K \equiv F_2$. In fact we will show that they are all in involution with each other, that is, $\{F_j, F_k\} = 0 \forall j, k$. According to Eqs. (9.210) and (9.216)

$$\{F_j, F_k\} = (F_{j\xi}, \Delta F_{k\xi}) = (F_{j\xi}, L F_{(k-1)\xi}).$$

Since L involves only odd orders of differentiation, it is an antisymmetric operator on L_2. Now use antisymmetry of L and Δ on the last term to get

$$(F_{j\xi}, L F_{(k-1)\xi}) = -(L F_{j\xi}, F_{(k-1)\xi}) = -(\Delta F_{(j+1)\xi}, F_{(k-1)\xi})$$
$$= (F_{(j+1)\xi}, \Delta F_{(k-1)\xi}) = \{F_{(j+1)}, F_{(k-1)}\}.$$

Thus

$$\{F_j, F_k\} = \{F_{(j+1)}, F_{(k-1)}\};$$

the index j has increased by 1 and k has decreased by 1. Continue this process until you arrive at

$$\{F_j, F_k\} = \{F_{j+k}, F_0\} = (F_{(j+k)\xi}, \Delta F_{0\xi}) \equiv 0.$$

Thus the F_j are in involution with each other. In particular, they are in involution with H^K and are therefore constants of the motion. This shows that Points 1 and 2 imply Point 3.

MORE ON CONSTANTS OF THE MOTION

Point 1: Until Point 2 has been discussed, we write G_k for $F_{k\xi}$ (the notation $F_{k\xi}$ assumes prematurely that it is a gradient) and rewrite (9.216) as

$$LG_j = \Delta G_{j+1}. \tag{9.218}$$

What must be shown is that these differential equations have solutions G_{j+1}. The analogous problem in finite dimensions would be to show that the differential equation $g_k = \partial_k f$ has a solution f. But it does not always have a solution: it can have one only if the g_k satisfy the integrability conditions $\partial_k g_j = \partial_j g_k$, which state that derivatives commute, that is, that $\partial_k \partial_j f = \partial_j \partial_k f$. In infinite dimensions there is an analogous integrability condition. Consider a simplified version of (9.218), the differential equation $\gamma = \Delta\phi$, where $\gamma \in L_2$ is given and $\phi \in L_2$ is to be found. If it has a solution ϕ, then

$$\int \gamma \, dx \equiv \int \Delta\phi \, dx = [\phi],$$

where the integrated part, in square brackets, vanishes for all $\phi \in L_2$. Hence the existence of a solution ϕ implies that

$$\int \gamma \, dx = 0. \tag{9.219}$$

This is the integrability condition in infinite dimensions.

Note, however, that in both the finite and infinite dimensions these are only necessary conditions. The sufficient condition is what we really need, the converse: if (9.219) is satisfied, a solution ϕ exists. In finite dimensions the converse is in general true only locally, and we did not prove even that. Similarly, we will not attempt to prove the converse in infinite dimensions.

What we will show, however, is the necessity, namely that (9.219) is satisfied by the left-hand side of (9.218). The proof (by induction) is similar to the one that established involution. It starts from the general formula $\int 3\phi \, dx \equiv (\phi, F_{0\xi}) \equiv (\phi, G_0)$ as applied to $\phi = LG_j$:

$$3 \int LG_j \, dx = (LG_j, G_0) = -(G_j, LG_0) = -(G_j, \Delta G_1)$$
$$= (\Delta G_j, G_1) = \left(LG_{(j-1)}, G_1\right).$$

Again one index is raised and the other lowered. The process continues until the two indices are the same on one of the two terms of the second line. Then antisymmetry of L or Δ establishes that the inner product is equal to its negative and hence vanishes. Thus the left-hand side of (9.218) satisfies (9.219).

Point 2: The proof that each of the $G_j(\xi)$ is the gradient $F_{j\xi}(\xi)$ of a functional $F_j(\xi)$ is beyond the scope of this book (for a proof, see Lax, 1976). However, we will explain what is at stake. The finite-dimensional analog states that a set of functions g_k form the gradient of some function f, that is, that the equation $g_k = \partial_k f$ can be solved for f. That again requires the integrability condition $\partial_j g_k = \partial_k g_j$, but now it is carried over to L_2 in a different way, by invoking the second derivatives of f. Independence of the order of differentiation can be restated in terms of inner products and directional derivatives. The derivative of the gradient $\partial_k f$ of f is the matrix $\partial_j \partial_k f = \partial_j g_k \equiv g_{kj}$, and independence of the order of differentiation implies that the g_{kj} matrix is symmetric. Symmetry is defined in terms of the inner product, and the derivatives of the g_k can also be defined in terms of the inner product. Write

$$\frac{d}{d\epsilon} g_k(\xi + \epsilon\eta)\bigg|_{\epsilon=0} = \sum \eta_j \partial_j g_k \equiv \sum g_{kj}\eta_j. \tag{9.220}$$

The matrix g of the g_{kj} is symmetric, that is, the g_k form the gradient of some function, only if

$$(\xi, g\eta) \equiv \sum \xi_k g_{kj}\eta_j = \sum \xi_k g_{jk}\eta_j \equiv (g\xi, \eta) \quad \forall \xi, \eta \in L_2.$$

That is what will be generalized to infinite dimensions. Suppose that $G \in L_2$. Then the analog of the g_{kj} matrix is the ξ-dependent *operator* $\Gamma(\xi)$ defined by the analog of (9.220):

$$\lim_{\epsilon \to 0} \frac{1}{\epsilon}[G(\xi + \epsilon\eta) - G(\xi)] = \Gamma(\xi)\eta. \tag{9.221}$$

This equation, showing the action of $\Gamma(\xi)$, establishes that it acts linearly on any $\eta \in L_2$, although its ξ dependence is not linear (see Worked Example 9.8 below and Problem 19). In terms of the inner product, symmetry of $\Gamma(\xi)$ means that

$$(\Gamma(\xi)\eta, \psi) = (\Gamma(\xi)\psi, \eta);$$

G is the gradient of a functional $F(\xi)$ only if its $\Gamma(\xi)$ operator is symmetric. This is a necessary (but not a sufficient) condition for G to be a gradient.

WORKED EXAMPLE 9.8

Find the $\Gamma(\xi)$ operator of $G(\xi) \equiv F_{2\xi}(\xi) = \frac{1}{2}\xi^2 + \xi_{xx}$.

Solution. Use (9.221):

$$\Gamma(\xi)\eta = \lim_{\epsilon \to 0} \frac{1}{\epsilon}\left(\frac{1}{2}[\xi^2 + 2\epsilon\xi\eta + \epsilon^2\eta^2] - \frac{1}{2}\xi^2 + \epsilon\eta_{xx}\right)$$

$$= \lim_{\epsilon \to 0}\left(\xi\eta + \frac{1}{2}\epsilon\eta^2 + \eta_{xx}\right) = \xi\eta + \eta_{xx}.$$

Note that $\lim_{\epsilon \to 0}(1/\epsilon)$ is essentially the derivative at $\epsilon = 0$. It follows that $\Gamma(\xi) = \xi + \partial_x^2$.

Lax (1976) shows that if the G_k satisfy (9.218), their $\Gamma_k(\xi)$ operators are symmetric (the index on Γ comes from G) and hence the G_k are gradients and can be written $F_{k\xi}$. He even proves the sufficiency, that is, that the $F_{k\xi}$ can be integrated to define functionals $F_k(\xi)$. In keeping with the Greek–Latin juxtaposition (the operator corresponding to G_k is $\Gamma_k(\xi)$), we write $\Phi_{k\xi}(\xi)$ for the operator corresponding to $F_{k\xi}$ (so the Γ operator of Worked Example 9.8 is $\Phi_{2\xi}$). Problem 19 explicitly demonstrates the symmetry of $\Phi_{2\xi}(\xi)$ and $\Phi_{3\xi}(\xi)$.

> **REMARK:** Many authors write G_ξ for what we have called $\Gamma(\xi)$ and $F_{\xi\xi}(\xi)$ for what we are calling $\Phi_\xi(\xi)$. We do not use that notation because it masks the different nature of operators and functions.   □

9.6.4 THE SINE–GORDON EQUATION

Because the sine–Gordon equation is of second order in t, the way it is included in this scheme is a little different. This section discusses this very briefly.

TWO-COMPONENT FIELD VARIABLES

We repeat (9.102), the sG equation in dimensionless variables:

$$\phi_{tt} - \phi_{xx} + \sin\phi = 0,$$

where $\phi(x)$ is the angle of the pendulum at position x. This equation is of second order in the time derivative, so is not in the form of (9.203), which is a prerequisite for the

Hamiltonian formalism. The way to get around this is to make ϕ_t itself a field variable by defining the two-component field

$$\boldsymbol{\xi} = \begin{bmatrix} \xi_1 \\ \xi_2 \end{bmatrix} \equiv \begin{bmatrix} \phi \\ \phi_t \end{bmatrix}.$$

The phase manifold is now $L_2 \times L_2$ and its elements, the field variables, are now two-component $\boldsymbol{\xi}$ vectors. The inner product on $L_2 \times L_2$ is defined as follows: let $\boldsymbol{\xi}, \boldsymbol{\eta}$ be two elements of $L_2 \times L_2$. Their inner product is

$$(\boldsymbol{\xi}, \boldsymbol{\eta}) \equiv \sum_{\alpha=1}^{2} \int \xi_\alpha \eta_\alpha \, dx. \tag{9.222}$$

In terms of $\boldsymbol{\xi}$ the sG equation is

$$\boldsymbol{\xi}_t \equiv \begin{bmatrix} \phi_t \\ \phi_{tt} \end{bmatrix} = \begin{bmatrix} \phi_t \\ \phi_{xx} - \sin \phi \end{bmatrix} = \begin{bmatrix} 0 & 1 \\ -1 & 0 \end{bmatrix} \begin{bmatrix} \sin \phi - \phi_{xx} \\ \phi_t \end{bmatrix}$$

$$\equiv JB\boldsymbol{\xi}, \tag{9.223}$$

where

$$J = \begin{bmatrix} 0 & 1 \\ -1 & 0 \end{bmatrix}$$

and B is the (nonlinear) operator on $L_2 \times L_2$ defined by

$$B\boldsymbol{\xi} \equiv B \begin{bmatrix} \xi_1 \\ \xi_2 \end{bmatrix} \equiv \begin{bmatrix} \sin \xi_1 - \xi_{1xx} \\ \xi_2 \end{bmatrix}. \tag{9.224}$$

Equation (9.223) is of first order in the time derivative. It can be put into Hamiltonian form if $JB\boldsymbol{\xi}$ can be written as $\Omega\mathbf{H}$, where Ω is an antisymmetric operator and $\mathbf{H}$ is the gradient of some functional $H(\boldsymbol{\xi})$. This requires extending the definition of the gradient to $L_2 \times L_2$, which is done simply by taking the gradient separately in each component of $L_2 \times L_2$. That is, if $F(\boldsymbol{\xi})$ is a functional (a map from $L_2 \times L_2$ to $\mathbb{R}$), then its gradient $\mathbf{F}$ is defined as the two-component vector

$$\mathbf{F} = \begin{bmatrix} F_{\xi_1} \\ F_{\xi_2} \end{bmatrix}.$$

This is similar to the usual definition of the gradient in a finite-dimensional vector space: the nth component of the gradient is the derivative with respect to the nth coordinate.

sG AS A HAMILTONIAN FIELD

The energy in the sG field was found in Worked Example 9.4 to be

$$E = \frac{1}{2} \int \left[\phi_t^2 + \phi_x^2 + 2(1 - \cos \phi) \right] dx.$$

It will now be shown that this expression for the energy, written out in terms of $\boldsymbol{\xi}$, can serve as the Hamiltonian H^s for the sG equation (dropping the constant term, which does not contribute to the equations of motion):

$$H^s(\boldsymbol{\xi}) \equiv H^s(\xi_1, \xi_2) = \int \left(\frac{1}{2}\xi_2^2 + \frac{1}{2}\xi_{1x}^2 - \cos \xi_1 \right) dx.$$

Proving this requires finding the gradient of H^s and showing that the sG equation can be written in canonical form with the resulting gradient.

WORKED EXAMPLE 9.9

Find the gradient $\mathbf{H}^s$ of H^s.

Solution. According to (9.206) the first component $H_{\xi_1}^s$ of $\mathbf{H}^s$ is given by

$$\lim_{\epsilon \to 0} \frac{1}{\epsilon} (H^s(\xi_1 + \epsilon\eta, \xi_2) - H^s(\xi_1, \xi_2))$$

$$\equiv \lim_{\epsilon \to 0} \frac{1}{\epsilon} \int \epsilon[\xi_{1x}\eta_x + \eta \sin \xi_1 + O(\epsilon^2)] \, dx$$

$$= \int [-\xi_{1xx} + \sin \xi_1]\eta \, dx \equiv \int H_{\xi_1}^s \eta \, dx.$$

The $\sin \xi_1$ term comes from

$$\cos(\xi_1 + \epsilon\eta_1) - \cos \xi_1 = -\epsilon\eta_1 \sin \xi_1 + O(\epsilon^2).$$

The ξ_{1xx} term comes from integrating $\xi_{1x}\eta_x$ (move the subscript x from η to ξ and change the sign). Thus

$$H_{\xi_1}^s = -\xi_{1xx} + \sin \xi_1.$$

The second component $H_{\xi_2}^s$ of $\mathbf{H}^s$ is given by

$$\lim_{\epsilon \to 0} \frac{1}{\epsilon} (H^s(\xi_1, \xi_2 + \epsilon\eta) - H^s(\xi_1, \xi_2))$$

$$\equiv \lim_{\epsilon \to 0} \frac{1}{\epsilon} \int \epsilon[\xi_2\eta + O(\epsilon^2)] \, dx$$

$$= \int \xi_2\eta \, dx.$$

Thus

$$H_{\xi_2}^s = \xi_2.$$

The result is

$$\mathbf{H}^s = \begin{bmatrix} -\xi_{1xx} + \sin \xi_1 \\ \xi_2 \end{bmatrix}. \tag{9.225}$$

Comparison of Eqs. (9.225) and (9.224) shows that $\mathbf{H}^s = B\boldsymbol{\xi}$. Therefore, with ξ_1 identified as ϕ and ξ_2 as ϕ_t, the sG equation (9.223) becomes

$$\boldsymbol{\xi}_t = J\mathbf{H}^s. \tag{9.226}$$

This is in the required Hamiltonian form with $J = \Omega$ if J is antisymmetric with respect to the inner product of (9.222). But antisymmetry follows immediately from the fact that $J_{ik} = -J_{ki}$, so (9.226) is the Hamiltonian form of the sG equation. In fact J is, up to sign, just the symplectic form $-\omega$ of two-freedom Hamiltonian particle dynamics.

The operator $\Omega \equiv J$ of the sG equation is simpler than the operator $\Omega \equiv \Delta$ of the KdV equation, for it involves no derivatives. In part this is because of the third derivative that appears in the KdV equation and in part because of the way L_2 is doubled in the sG equation.

The Poisson bracket is still defined by Eq. (9.210), but now the inner product is given by (9.222) and $\Omega = J$. As before, the PB yields time derivatives: if $F(\boldsymbol{\xi})$ is a functional, then

$$\frac{d}{dt} F(\boldsymbol{\xi}) = \int \left(F_{\xi_1} \xi_{1t} + F_{\xi_2} \xi_{2t} \right) dx = \sum_\alpha \int F_{\xi_\alpha} \xi_{\alpha t} \, dx$$

$$\equiv (\mathbf{F}, \boldsymbol{\xi}_t) = (\mathbf{F}, J\mathbf{H}^s) \equiv \{F, H^s\}.$$

We go no further in discussing the Hamiltonian treatment of the sG equation. However, as for the KdV equation, there also exists an infinite set of constants of the motion in involution for the sG equation (Faddeev et al., 1975).

PROBLEMS

1. Obtain the one-dimensional Klein–Gordon equation and its Lagrangian density by going to the $s \to \infty$ limit for a system of n particles interacting through a spring force, as in the sine–Gordon equation, and whose external force is provided not by a pendulum but by a spring force. Take the spring constant of the external force to be K, as in the text. [Hint: in order to obtain the text's $s \to \infty$ limit the external force on each length Δx of the distributed system (or on each of the divided masses Δm) must be chosen proportional to Δx (or to Δm).]

2. Show that if two Lagrangian densities differ by a four-divergence of the form $\partial_\alpha \Phi^\alpha(\psi, x)$, their Euler-Lagrange equations are the same.

3. The energy tensor $T_\mu{}^\alpha$ of the electromagnetic field was replaced by its symmetrized form $\bar{T}_\mu{}^\alpha$ on the strength of the fact that $\partial_\alpha T_\mu{}^\alpha = \partial_\alpha \bar{T}_\mu{}^\alpha$. Does it follow that the P_μ defined using $\bar{T}_\mu{}^\alpha$ is the same as that defined using $T_\mu{}^\alpha$? Prove that it does.

4. Find the energy and momentum densities in the electromagnetic field in terms of **E** and **B** from the expression for $\bar{T}_\mu{}^\alpha$. Find the flux of the kth component of momentum in the jth direction.

5. Obtain the expanded form (9.55) of (9.54).

6. From the expression for $\bar{T}^{\mu\nu}$ and Eq. (9.60) show that the angular momentum density in the electromagnetic field is $(\mathbf{E} \wedge \mathbf{B}) \wedge \mathbf{x}$.

7. The free-field Maxwell Lagrangian $\mathcal{L} = -\frac{1}{4} f^{\mu\nu} f_{\mu\nu}$ is gauge invariant. Gauge transformations, because they involve χ functions, do not form ϵ-families in the sense of the Noether theorem. (The analogs of such transformations in particle dynamics would be ϵ-families in which ϵ is a function of t.) Nevertheless if an ϵ-family of gauge transformations of the form $\bar{A}_\mu = A_\mu + \epsilon \partial_\mu \chi$ is constructed with an arbitrary *fixed* χ function, it can be used to try to find an associated Noether constant. Show that what is obtained in this way is a tautology when the free-field equations are taken into account. See Karatas and Kowalski (1990).

8. (a) Show that $\mathcal{L}_{\text{tot}}$ is invariant under extended gauge transformations.

   (b) As discussed in Problem 7, gauge invariance with arbitrary χ functions can not be used to find Noether constants. But if $\chi \equiv K$ is a real constant, the resulting ϵ-family of extended gauge transformations leads to a conserved current. Show that the conserved current is J^μ as given by (9.72). [Comments: 1. If $\chi = K$, the gauge transformation becomes almost trivial. 2. If you try to push through the gauge transformation as in the previous problem, you get the spurious result that χJ^μ is conserved for an arbitrary function χ. See Karatas and Kowalski (1990).]

9. (a) Calculate the functional derivatives of the ψ_I and π^I with respect to the π^J, similar to Eqs. (9.87), and derive Eq. (9.88).

   (b) Calculate the functional derivatives of the $\psi_{I,k}$ with respect to the ψ_J and the π^J, also similar to Eqs. (9.87).

   (c) Calculate the field PBs $\{\psi_I(\mathbf{y}), \psi_J(\mathbf{y})\}^f$, $\{\pi^I(\mathbf{z}), \pi^J(\mathbf{z})\}^f$, $\{\psi_{I,k}(\mathbf{y}), \psi_J(\mathbf{z})\}^f$, $\{\psi_{I,k}(\mathbf{y}), \pi^J(\mathbf{z})\}^f$, and $\{\psi_{I,k}(\mathbf{y}), \psi_{J,l}(\mathbf{z})\}^f$.

10. Show the consistency of $H = \sum b_n^I \dot{a}_{In} - L$ and $H = \int \mathcal{H} d^3x$.

11. Calculate the half-width of the energy packet of the soliton of Eq. (9.113).

12. Show explicitly that the three functions

   (a) soliton–soliton:

$$\psi_{ss}(x, t) = 4 \arctan\left[\frac{u \sinh \gamma x}{\cosh u\gamma t}\right],$$

   (b) soliton–antisoliton:

$$\psi_{sa}(x, t) = 4 \arctan\left[\frac{\sinh u\gamma t}{u \cosh \gamma x}\right], \text{ and}$$

   (c) breather:

$$\psi_b(x, t) = 4 \arctan\left[(u\gamma)^{-1} \frac{\sin ut}{\cosh(\gamma^{-1}x)}\right]$$

   satisfy the sine–Gordon equation.

13. Use the second of Eqs. (9.117) and the diagram of Fig. 9.9 to derive (9.120). [Hint: Prove that $\sin x - \sin y = 2 \sin \frac{1}{2}(x - y) \cos \frac{1}{2}(x + y)$.]

14. (a) Write out the Lagrangian $\mathcal{L}$ of the cubic K–G equation in the transformed variables of Eq. (9.126) and redefine a new $\mathcal{L}$ by dividing out the λ dependence (see Worked Example 9.4). Complete this problem by using this new $\mathcal{L}$.

    (b) Find the Hamiltonian $\mathcal{H}$. Write the energy E in the field in terms of φ and its derivatives.

    (c) For a kink-like solution satisfying (9.103) rewrite E in terms of ψ and its derivatives. Find the energy density $\varepsilon(\xi)$ (energy per unit length) in terms of ψ' alone.

    (d) Specialize to a $C = 0$ kink of Eq. (9.131). Find the energy carried in a finite ξ interval by a kink of Eq. (9.131). Write down $\varepsilon(\xi)$ explicitly and find the total energy E carried by such a kink. Optional: plot a graph of $\varepsilon(\xi)$. [Answer to Part (a): $\mathcal{L} = \frac{1}{2}\{\varphi_t^2 - \varphi_x^2 + (\varphi^2 - 1)^2\}$.]

15. By considering a small (infinitesimal) volume in a fluid show that the mass current is $\mathbf{j} = \mathbf{v}\rho$.

16. Show that the Burgers equation (9.148) admits a solution of the form

$$v(x, t) = c[1 - \tanh\{c(x - ct)/2\nu\}].$$

17. (a) Find the gradient $F_{3\xi}$ of the functional

$$F_3(\xi) = \frac{1}{2} \int \left(\frac{5}{36}\xi^4 - \frac{5}{3}\xi\xi_x^2 + \xi_{xx}^2 \right) d\xi.$$

    (b) Show explicitly that $\{F_3, F_1\} = 0$, where F_1 is given by Eq. (9.212).

    (c) Show explicitly that F_3 is a constant of the motion for the KdV equation. [Answer to Part (a): $F_{3\xi} = \frac{1}{2}(\frac{5}{9}\xi^3 + \frac{5}{3}\xi_x^2 + \frac{10}{3}\xi\xi_{xx} + 2\xi_{xxxx})$.]

18. Show explicitly that $LF_{j\xi} = \Delta F_{(j+1)\xi}$ for the KdV equation with $j = 0, 1, 2$.

19. For the KdV equation: (a) Find $\Phi_{3\xi}\eta$. (b) Show that $\Phi_{2\xi}$ as found in Worked Example 9.8 and $\Phi_{3\xi}$ as found in (a) are symmetric operators.

EPILOGUE

In its treatment of dynamical systems, this book has emphasized several themes, among them nonlinearity, Hamiltonian dynamics, symplectic geometry, and field equations. The last section on the Hamiltonian treatment of nonlinear field equations unites these themes in bringing the book to an end. To arrive at this point we have traveled along the roads laid out by the Newtonian, Lagrangian, Hamiltonian, and Hamilton–Jacobi formalisms, eventually replacing the traditional view by a more modern one. On the way we have roamed through many byways, old and new, achieving a broad understanding of the terrain.

The end of the book, however, is not the end of the journey. The road goes on: there is much more terrain to cover. In particular, statistical physics and quantum mechanics, for both of which the Hamiltonian formalism is crucial, lie close to the road we have taken. These fields contain still unsolved problems bordering on the matters we have discussed, in particular questions about ergodic theory in statistical physics and about chaos in quantum mechanics. As usual in physics, the progress being made in such fields builds on previous work. In this case it depends on subject matter of the kind treated in this book.

Although they are deterministic, classical dynamical systems are generically so sensitive to initial conditions that their motion is all but unpredictable. Even inherently, therefore, classical mechanics still offers a rich potential for future discoveries and applications.

APPENDIX

VECTOR SPACES

APPENDIX OVERVIEW

Vector space concepts are important in much of physics. They are used throughout this book and are essential to its understanding. In particular, the concept of a manifold relies strongly on them, for locally a manifold looks very much like a vector space, and the fibers of tangent bundles and cotangent bundles are vector spaces. Vector spaces arise in rotational dynamics, in the discussion of oscillations, in linearizations (maps and chaos), and in our treatment of constraints. For many such reasons, the reader should thoroughly understand vector spaces and linear algebra and be able to manipulate them with confidence.

This appendix is included as a brief review of mostly finite-dimensional vector spaces. It contains definitions and results only – no proofs.

GENERAL VECTOR SPACES

A *vector space* $\mathbb{V}$ *over the real numbers* $\mathbb{R}$ or *complex numbers* $\mathbb{C}$ consists of a set of elements called *vectors* $\mathbf{x}, \mathbf{y}, \ldots$, for which there is a commutative operation called *addition*, designated by the usual symbol $+$. Addition has the following properties:

1. The space $\mathbb{V}$ is *closed* under addition: if $\mathbf{x}, \mathbf{y} \in \mathbb{V}$, then there is a $\mathbf{z} \in \mathbb{V}$ such that $\mathbf{z} = \mathbf{x} + \mathbf{y} = \mathbf{y} + \mathbf{x}$.
2. Addition is *associative*: $\mathbf{x} + (\mathbf{y} + \mathbf{z}) = (\mathbf{x} + \mathbf{y}) + \mathbf{z}$.
3. There exists a *null vector* $\mathbf{0} \in \mathbb{V}$: $\mathbf{x} + \mathbf{0} = \mathbf{x}$ for any $\mathbf{x} \in \mathbb{V}$.
4. Each $\mathbf{x}$ has its *negative*, $-\mathbf{x}$, whose sum with $\mathbf{x}$ yields the null vector: $\mathbf{x} + (-\mathbf{x}) \equiv \mathbf{x} - \mathbf{x} = \mathbf{0}$.

There is a second operation on $\mathbb{V}$, called *multiplication* by numbers. Multiplication by numbers has the following properties ($\mathbb{R}$ should be replaced by $\mathbb{C}$ in what follows if $\mathbb{V}$ is over the complex numbers):

5. The space is *closed* under multiplication: if $\alpha \in \mathbb{R}$ and $\mathbf{x} \in \mathbb{V}$, then $\alpha\mathbf{x} \in \mathbb{V}$.
6. Multiplication is *associative*: $\alpha(\beta\mathbf{x}) = (\alpha\beta)\mathbf{x}$.
7. Multiplication is *linear*: $\alpha(\mathbf{x} + \mathbf{y}) = \alpha\mathbf{x} + \alpha\mathbf{y}$.

8. Multiplication is *distributive*: $(\alpha + \beta)\mathbf{x} = \alpha\mathbf{x} + \beta\mathbf{x}$.

9. Finally, $1\mathbf{x} = \mathbf{x} \; \forall \mathbf{x} \in \mathbb{V}$.

This completes the definition of a vector space.

A *subspace* of $\mathbb{V}$ is a subset of $\mathbb{V}$ that is itself a vector space.

Vectors $\mathbf{x}_1, \mathbf{x}_2, \ldots, \mathbf{x}_k$ are *linearly independent* if the equation $\alpha_i\mathbf{x}_i = 0$ can be satisfied only if $\alpha_i = 0$ for $i = 1, 2, \ldots, k$. Otherwise the $\mathbf{x}_i$ are *linearly dependent*.

A *basis* (or *coordinate system*) in $\mathbb{V}$ is a set of vectors $\Xi = \{\mathbf{x}_1, \mathbf{x}_2, \ldots\}$ such that every vector in $\mathbb{V}$ can be written *uniquely* as a *linear combination* of them, that is, such that for each $\mathbf{y} \in \mathbb{V}$ there is one and only one set of numbers $\eta_1, \eta_2, \ldots$ satisfying

$$\mathbf{y} = \eta_i\mathbf{x}_i. \tag{A.1}$$

If $\mathbb{V}$ has any basis containing a finite number n of vectors $\mathbf{x}_i$, then 1. every other basis also contains n vectors, 2. every set of n linearly independent vectors is a basis, and 3. there can be no more than n vectors in any linearly independent set. This unique number n associated with $\mathbb{V}$ is the *dimension* of $\mathbb{V}$. (A vector space with no finite basis is *infinite dimensional*. We deal almost exclusively with finite-dimensional spaces.)

The coefficients η_i are the *components* of $\mathbf{y}$ in the basis Ξ. The set of components of a vector is called the *representation* y of the vector $\mathbf{y}$ in the basis. They are often arranged in a column which is identified with y:

$$y = \begin{bmatrix} \eta_1 \\ \eta_2 \\ \vdots \\ \eta_n \end{bmatrix}. \tag{A.2}$$

The result is the *column vector* representation of $\mathbf{y}$ in Ξ. There are many bases in a vector space and therefore many representations of each vector.

Let $\Omega = \{\mathbf{a}_1, \mathbf{a}_2, \ldots, \mathbf{a}_n\}$ and $\Xi = \{\mathbf{b}_1, \mathbf{b}_2, \ldots, \mathbf{b}_n\}$ be two bases in an n-dimensional vector space $\mathbb{V}$, and let the components of $\mathbf{y} \in \mathbb{V}$ be η_i' in Ω and η_i in Ξ:

$$\mathbf{y} = \eta_i\mathbf{b}_i = \eta_i'\mathbf{a}_i. \tag{A.3}$$

Since every vector can be represented in a basis, so can basis vectors. Let

$$\mathbf{b}_i = \tau_{ki}\mathbf{a}_k :$$

the τ_{ki} are the components of $\mathbf{b}_i$ in the Ω basis. Then it follows that

$$\eta_k' = \tau_{ki}\eta_i. \tag{A.4}$$

It is important to realize that this is the relation between the two representations (sets of components) of *a single vector* $\mathbf{y}$, not the relation between two vectors in $\mathbb{V}$.

The τ_{kj} can be arranged in a square array called an n-by-n *matrix* which we designate by the letter T:

$$T = \begin{bmatrix} \tau_{11} & \tau_{12} & \cdots & \tau_{1n} \\ \tau_{21} & \tau_{22} & \cdots & \tau_{2n} \\ \vdots & \vdots & \ddots & \vdots \\ \tau_{n1} & \tau_{n2} & \cdots & \tau_{nn} \end{bmatrix}. \tag{A.5}$$

T is the *transformation matrix* from Ξ to Ω, and the τ_{ij} are its *matrix elements*. Let y be the column vector whose components are the η_j and y' be the column vector whose components are the η'_j. Then Eq. (A.4) is written in *matrix notation* in the form

$$y' = Ty. \tag{A.6}$$

This says that the matrix T is applied to the column vector y to yield the column vector y'. Again we emphasize that this equation connects two representations of one and the same vector, not two different vectors.

LINEAR OPERATORS

A *linear operator* **A** on $\mathbb{V}$ is a map that carries each vector **x** into another **y** (one writes **y** = **Ax**) so that for any **x**, **y** $\in \mathbb{V}$ and any $\alpha, \beta \in \mathbb{R}$

$$\mathbf{A}(\alpha\mathbf{x} + \beta\mathbf{y}) = \alpha(\mathbf{Ax}) + \beta(\mathbf{Ay}).$$

Linear operators can be added and multiplied by numbers:

$$(\alpha\mathbf{A} + \beta\mathbf{B})\mathbf{x} = \alpha(\mathbf{Ax}) + \beta(\mathbf{Bx}).$$

Linear operators can also be multiplied by each other:

$$(\mathbf{AB})\mathbf{x} \equiv \mathbf{A}(\mathbf{Bx}) \equiv \mathbf{ABx}.$$

Operator multiplication so defined is associative, that is, $\mathbf{A}(\mathbf{BC}) = (\mathbf{AB})\mathbf{C} \equiv \mathbf{ABC}$. In general, however, $\mathbf{AB} \neq \mathbf{BA}$, that is, operator multiplication is not *commutative*. In those cases when $\mathbf{AB} = \mathbf{BA}$, the operators **A** and **B** are said to *commute*. The operator $[\mathbf{A}, \mathbf{B}] \equiv \mathbf{AB} - \mathbf{BA}$ is called the *commutator* of **A** and **B**.

Let **y** = **Ax**, where **x** has components ξ_j and **y** has components η_j in some basis $\Omega = \{\mathbf{a}_1, \ldots, \mathbf{a}_n\}$. If

$$\mathbf{Aa}_k = \alpha_{jk}\mathbf{a}_k, \tag{A.7}$$

the α_{jk} are the components of the vector $\mathbf{Aa}_k$ in the Ω basis. Then

$$\eta_j = \alpha_{jk}\xi_k. \tag{A.8}$$

Let A be the matrix of the α_{jk}. Then in matrix notation Eq. (A.8) can be written

$$y = Ax.$$

It is important to realize that this is an equation connecting the representations of *two different vectors* [cf. Eq. (A.4)]. The matrix A is the representation of the operator **A** in the Ω basis. If **A** is represented by A with elements α_{jk} and **B** by B with elements β_{jk}, then **AB** is represented by the matrix whose elements are $\alpha_{jh}\beta_{hk}$. This matrix is called the (*matrix*) *product* AB of A and B: in other words the operator product is represented by the matrix product.

The representation of an operator, like that of a vector, changes from basis to basis: if $y = Ax$, then $y' = A'x'$. Analogous to (A.4) and (A.6) are

$$\left.\begin{aligned} \alpha'_{ij} &= \tau_{ik}\alpha_{kh}\tau^{-1}_{hj}, \\ A' &= TAT^{-1}. \end{aligned}\right\} \tag{A.9}$$

(See the next section for the definition of T^{-1}.)

INVERSES AND EIGENVALUES

The *inverse* of a linear operator $\mathbf{A}$ is an operator $\mathbf{A}^{-1}$, if such exists, which undoes the action of $\mathbf{A}$: if $\mathbf{A}\mathbf{x} = \mathbf{y}$, then $\mathbf{A}^{-1}\mathbf{y} = \mathbf{x}$, and vice versa. The operator $\mathbf{A}\mathbf{A}^{-1} = \mathbf{A}^{-1}\mathbf{A} \equiv \mathbf{1}$ is the *unit operator* or *identity*, which maps every vector $\mathbf{x}$ into itself: $\mathbf{1}\mathbf{x} = \mathbf{x}$. The operator $\mathbf{1}$ is represented in all coordinate systems by the *unit matrix* $\mathbb{I}$, whose components are all zeros except on the main diagonal (the one running from upper left to lower right), which are all 1s. An example is the 4-dimensional (i.e., 4-by-4) unit matrix:

$$\mathbb{I}_4 = \begin{bmatrix} 1 & 0 & 0 & 0 \\ 0 & 1 & 0 & 0 \\ 0 & 0 & 1 & 0 \\ 0 & 0 & 0 & 1 \end{bmatrix}.$$

The matrix elements of $\mathbb{I}$ are generally denoted δ_{jk}, called the Kronecker delta symbol: $\delta_{jk} = 1$ if $j = k$, and $\delta_{jk} = 0$ if $j \neq k$. If $\mathbf{B}$ is any operator, then $\mathbf{1}\mathbf{B} = \mathbf{B}\mathbf{1} = \mathbf{B}$. If B is any matrix, then $\mathbb{I}B = B\mathbb{I} = B$.

Not all operators have inverses. If $\mathbf{A}^{-1}$ exists, $\mathbf{A}$ is called *nonsingular*. A necessary and (in finite dimensions) sufficient condition for the existence of $\mathbf{A}^{-1}$ is that the equation $\mathbf{A}\mathbf{x} = \mathbf{0}$ have only the solution $\mathbf{x} = \mathbf{0}$. Represented in a basis, this condition requires that the equations $\alpha_{jk}\xi_k = 0$ have only the trivial solution $\xi_k = 0 \, \forall k$. As is well known from elementary algebra, this means that the *determinant* of the α_{jk}, written det A, must be nonzero. Thus the condition that $\mathbf{A}$ be nonsingular is that det $A \neq 0$.

If a nonsingular operator $\mathbf{A}$ is represented in some basis by A, the representation of $\mathbf{A}^{-1}$ in that basis is written A^{-1}, and it follows that $AA^{-1} = \mathbb{I}$. From elementary algebra it is known that the matrix elements of A^{-1} are given by

$$(A^{-1})_{jk} = \frac{\gamma_{jk}}{\det A},$$

where γ_{jk} is the *cofactor* of α_{kj} (note the inversion of the indices). The matrix A^{-1} exists iff det $A \neq 0$, and then A is also called nonsingular. The inverse of an operator $\mathbf{A}$ can be obtained by finding its representation A in some basis, calculating A^{-1} in the same basis, and identifying $\mathbf{A}^{-1}$ with the operator whose representation in that basis is A^{-1}.

Recall the definition of a transformation matrix. If T is the transformation matrix from Ξ to Ω, then T^{-1} is the transformation matrix from Ω to Ξ. A transformation matrix does not represent an operator.

The determinant of the representation of an operator is the same in all bases: one can thus write det **A** instead of det A. The *rank* of **A** is the size of the largest nonzero determinant in A.

If **A** is singular and **Ax** = **b**, then **A**(**x** + **n**) = **b**, where **n** is any vector in the *null space* or *kernel* of **A** (the subspace of $\mathbb{V}$ consisting of all vectors such that **An** = **0**). If **b** is some nonzero vector, the equation **Ax** = **b** has a unique solution **x** iff **A** is nonsingular.

The *trace* of a matrix A is the sum of the elements on its main diagonal:

$$\text{tr } A = \alpha_{kk}.$$

The trace of an operator is the trace of its representation and, like the determinant, is basis independent; that is, one can write tr **A** instead of tr A.

Given an operator **A**, one often needs to find vectors **x** and numbers λ that satisfy the *eigenvalue equation*

$$\mathbf{Ax} = \lambda\mathbf{x}. \tag{A.10}$$

Solutions exist when $\det(\mathbf{A} - \lambda\mathbf{1}) = 0$. In a given basis this equation is represented by an nth degree polynomial equation of the form

$$\det(A - \lambda\mathbb{I}) = \sum_{k=0}^{n} s_k \lambda^k = 0, \tag{A.11}$$

which in general has n solutions (some may be complex even if the vector space is real), called the *eigenvalues* of **A**. Briefly, the method of solving an *eigenvalue problem* is to find the roots of (A.11) and then to insert them one at a time into (A.10) to find the *eigenvector* **x** that *belongs* to each eigenvalue [i.e., that satisfies (A.10) with each particular value λ_k of λ]. There is at least one eigenvector **x** belonging to each eigenvalue. If **x** is an eigenvector belonging to eigenvalue λ, so is $\alpha\mathbf{x}$ for any number α. Eigenvectors belonging to different eigenvalues are linearly independent. If a root λ of Eq. (A.11) has multiplicity greater than one, there may be (but not necessarily) more than one linearly independent eigenvector belonging to λ.

If **A** has n linearly independent eigenvectors (always true if the eigenvalues are all different) they form a basis for $\mathbb{V}$. In such a basis $\Xi = \{\mathbf{x}_1, \mathbf{x}_1, \ldots, \mathbf{x}_n\}$

$$\mathbf{Ax}_j = \lambda_j \mathbf{x}_j$$

(No sum on j). A is a *diagonal matrix* (i.e., one with elements only on the main diagonal) whose matrix elements are the eigenvalues of **A**. Finding the basis Ξ is called *diagonalizing* **A**.

INNER PRODUCTS AND HERMITIAN OPERATORS

The *inner* or *scalar product* on an n-dimensional space $\mathbb{V}$ generalizes the three-dimensional dot product.

With every pair of vectors **x**, **y** $\in \mathbb{V}$ the inner or scalar product associates a number in $\mathbb{R}$ (or in $\mathbb{C}$), written (**x**, **y**). This association has three properties:

1. It is *linear*: $(\mathbf{x}, \alpha\mathbf{y} + \beta\mathbf{z}) = \alpha(\mathbf{x}, \mathbf{y}) + \beta(\mathbf{x}, \mathbf{z})$

2. It is *Hermitian*: $(\mathbf{x}, \mathbf{y}) = (\mathbf{y}, \mathbf{x})^*$, where the asterisk indicates the complex conjugate (on a real space the inner product is symmetric: the order of the vectors within the parentheses is irrelevant).

3. It is *positive-definite*: $(\mathbf{x}, \mathbf{x}) \geq 0$ and $(\mathbf{x}, \mathbf{x}) = 0$ iff $\mathbf{x} = 0$.

Not every vector space has an inner product defined on it. A complex vector space with an inner product on it is called a *unitary vector space*. A real one is called *Euclidean* or *orthogonal*. The *norm* of a vector $\mathbf{x}$ is $(\mathbf{x}, \mathbf{x}) \equiv \|\mathbf{x}\| \equiv |\mathbf{x}|^2$. Two vectors $\mathbf{x}, \mathbf{y}$ are *orthogonal* iff $(\mathbf{x}, \mathbf{y}) = 0$. The *orthogonal complement* of $\mathbf{x} \in \mathbb{V}$ is the subspace of all vectors in $\mathbb{V}$ orthogonal to $\mathbf{x}$.

A *unit* or *normal vector* $\mathbf{e}$ is one for which $\|\mathbf{e}\| = 1$. An *orthonormal basis* $\mathbf{E} = \{\mathbf{e}_1, \mathbf{e}_2, \ldots, \mathbf{e}_n\}$ is composed entirely of vectors that are normal and orthogonal to each other:

$$(\mathbf{e}_j, \mathbf{e}_k) = \delta_{jk}.$$

The components of a vector $\mathbf{x} = \xi_j \mathbf{e}_j$ are given in an orthonormal basis by

$$\xi_j = (\mathbf{e}_j, \mathbf{x})$$

and the matrix elements of an operator by

$$\alpha_{jk} = (\mathbf{e}_j, \mathbf{A}\mathbf{e}_k).$$

The inner product of two vectors $\mathbf{x} = \xi_j \mathbf{e}_j$ and $\mathbf{y} = \eta_j \mathbf{e}_j$ is

$$(\mathbf{x}, \mathbf{y}) = \xi_j^* \eta_j.$$

In a unitary space every operator $\mathbf{A}$ can be associated with another, called its *Hermitian conjugate* or *adjoint* $\mathbf{A}^\dagger$, defined by

$$(\mathbf{x}, \mathbf{A}\mathbf{y}) = (\mathbf{A}^\dagger \mathbf{x}, \mathbf{y}) \quad \forall\, \mathbf{x}, \mathbf{y} \in \mathbb{V}.$$

The matrix elements of $\mathbf{A}^\dagger$ are related to those of $\mathbf{A}$ by

$$\alpha_{kj}^\dagger = \alpha_{jk}^*.$$

If $\mathbf{A} = \mathbf{A}^\dagger$, then $\mathbf{A}$ is a *Hermitian* operator. If $\mathbf{U}^\dagger = \mathbf{U}^{-1}$, then $\mathbf{U}$ is a *unitary* operator. The matrix elements α_{jk} (in an orthonormal basis) of a Hermitian operator satisfy the equations

$$\alpha_{jk} = \alpha_{kj}^*$$

and the matrix elements ω_{jk} of a unitary operator satisfy the equations

$$\omega_{kj}\omega_{ij}^* \equiv \omega_{jk}\omega_{ji}^* = \delta_{ki}.$$

A real Hermitian operator is *symmetric* ($\alpha_{jk} = \alpha_{kj}$), and a real unitary one is called *orthogonal*. The matrix of a unitary (orthogonal) operator is a *unitary (orthogonal) matrix*, and that of a

Hermitian (symmetric) operator is a *Hermitian (symmetric) matrix*. The transformation matrix between two (orthonormal) bases is always unitary.

The eigenvalues of a Hermitian operator are always real, while those of a unitary operator always lie on the unit circle. Eigenvectors belonging to different eigenvalues of both types of operators are orthogonal to each other. Both types of operators can be diagonalized in orthonormal bases.

Many of the statements made about operators can be reinterpreted as statements about matrices. For example, the last statement about diagonalization implies that if A is a Hermitian or unitary matrix, there is at least one unitary matrix U such that UAU^{-1} is diagonal.

BIBLIOGRAPHY

Abraham, R., and Marsden, J. E., *Foundations of Mechanics*, 2nd edition, Addison-Wesley, Reading (1985).

Abramowitz, M., and Stegun, I. A., *Handbook of Mathematical Functions with Formulas, Graphs, and Mathematical Tables*, U.S. Govt. Printing Office, Washington (1972).

Acheson, D. J., "A Pendulum Theorem," *Proc. Roy. Soc. A* **443**, 239 (1993).

Acheson, D. J., and Mulllin, T., "Upside-Down Pendulums," *Nature* **366**, 215 (1993).

Adler, R. L., and Rivlin, T. J., "Ergodic and Mixing Properties of the Chevyshev Polynomials," *Proc. Am. Math. Soc.* **15**, 794 (1964).

Aharoni, J., *The Special Theory of Relativity*, Oxford University Press, London (1959).

Aharonov, Y., and Bohm, D., "Significance of the Eletromagnetic Potentials in the Quantum Theory," *Phys. Rev.* **115**, 485 (1959).

Alfvén, H., *Cosmical Electrodynamics*, Oxford University Press, London (1950).

Arnold, T. W., and Case, W., "Nonlinear Effects in a Simple Mechanical System," *Am. J. Phys.* **50**, 220 (1982).

Arnol'd, V. I., "Small Denominators, II: Proof of a Theorem by A. N. Kolmogorov on the Preservation of Conditionally-Periodic Motion Under a Small Perturbation of the Hamiltonian," *Uspekhi Mat. Nauk* **18**, 13 (1963). [*Russ. Math. Surveys* **18**, 9 (1963).]

Arnol'd, V. I., "Small Denominators and Problems of Stability of Motion in Classical and Celestial Mechanics," *Uspekhi Mat. Nauk* **18**, 91 (1963). [*Russ. Math. Surveys* **18**, 85 (1963).]

Arnold, V. I., "Instability of Dynamical Systems with Many Degrees of Freedom," *Dokl. Akad. Nauk SSSR* **156**, 9 (1964). [*Sov. Math. Dokl.* **5**, 581 (1964).]

Arnol'd, V. I., "Small Denominators, I: Mappings of the Circumference into Itself," *AMS Transl. Series 2* **46**, 213 (1965).

Arnol'd, V. I., *Mathematical Methods of Classical Mechanics*, 2nd edition, Springer-Verlag, Berlin (1988).

Arnol'd, V. I., *Huygens and Barrow, Newton and Hoyle*, Birkäuser, Basel (1990).

Arnol'd, V. I., and Avez, A., *Ergodic Problems of Classical Mechanics*, Benjamin, New York (1968).

Bäcklund, A. V., "Einiges über Curven- und Flächen-Transformationen," Lunds Univ. rs-skrift, **10**, Für Ar 1873. II. *Afdelningen für Mathematic och Naturenskap*, 1 (1873).

Bak, P., "The Devil's Staircase," *Physics Today* **39**, No. 12, 39 (1986).

Barnsley, M., *Fractals Everywhere*, Academic Press, San Diego (1990).

Barone, A., and Paterno, G., *Physics and Applications of Josephson Junctions*, Wiley, New York (1982).

Batchelor, G. K., *An Introduction to Fluid Dynamics*, Cambridge University Press, Cambridge (1973).

Bellissard, J., "Stability and Instability in Quantum Mechanics," in *Trends and Developments in the Eighties*, Albeverio and Blanchard, eds., World Scientific, Singapore (1985).

Bennetin, G., Galgani, L., Giorigilli, A., and Strelcyn, J. M., "A Proof of Kolmogorov's Theorem on Invariant Tori Using Canonical Transformations Defined by the Lie Method," *Nuovo Cimento* **79B**, 201 (1984).

Bercovich, C., Smilansky, U., and Farmelo, G., "Demonstration of Classical Chaotic Scattering," *Eur. J. Phys.* **12**, 122 (1991).

Bergmann, P. G., *Introduction to the Theory of Relativity*, Dover, New York (1976).

Berry, M. V., "Quantal Phase Factors Accompanying Adiabatic Changes," *Proc. Roy. Soc. A* **392**, 45 (1984); "Classical Adiabatic Angles and Quantal Adiabatic Phase," *J. Phys. A* **18**, 15 (1985).

Berry, M. V., "The Geometric Phase," *Scientific American* **259**, 46 (1988).

Birkhoff, G. D., *Dynamical Systems*, Am. Math, Soc. Colloquium publications Vol. IX, New York (1927).

Bianchi, L., "Ricerche sulle superficie a curvatura costante e sulle elicoidi," *Annali della R. Scuola Normale Superiore di Pisa* **2**, 285 (1897).

Bishop, R. L., and Goldberg, S. I., *Tensor Analysis on Manifolds,* Dover, New York (1980).

Blackburn, J. A., Smith, H. J. T., and Jensen, N. G., "Stability and Hopf Bifurcations in the Inverted Pendulum," *Am. J. Phys.* **60**, 903 (1992).

Bleher, S., Grebogi, C., and Ott, E., "Bifurcation to Chaotic Scattering," *Physica D* **46**, 87 (1990).

Bogoliubov, N. N., and Shirkov, D. V., *Introduction to the Theory of Quantized Fields*, Wiley, New York (1958).

Bogoyavlensky, O. I., "Integrable Cases of Rigid Body Dynamics and Integrable Systems on the Ellipsoids," *Comm. Math. Phys.* **103**, 305 (1986).

Born, M., *The Mechanics of the Atom*, F. Ungar, New York (1960).

Bost, J. B., "Tores Invariants des Systèmes Dynamiques Hamiltoniens," *Asterisque* **639**, 133 (1986).

Bruijn, N. G., *Asymptotic Methods in Analysis*, North Holland, Amsterdam (1958).

Burgers, J. M., "A Mathematical Model Illustrating the Theory of Turbulence," *Adv. Appl. Mech.* **1**, 171 (1948).

Burgers, J. M., *Nonlinear Diffusion Equation. Asymptotic Solutions and Statistical Problems,* D. Reidel, Dortrecht-Holland/Boston (1974).

Byron, F. W., and Fuller, R. W., *Mathematics of Classical and Quantum Physics*, Addison-Wesley, Reading (1969).

Cantor, G., *Contributions to the Founding of the Theory of Transfinite Numbers*, Dover, New York (1915).

Cary, J. R., "Lie Transform Perturbation Theory for Hamiltonian Systems," *Phys. Rep.* **79**, 129 (1981).

Chao, B. F., "Feyman's Dining Hall Dynamics," *Physics Today*, February, p. 15 (1989).

Chirikov, R. V., "A Universal Instability of Many-Dimensional Oscillator Systems," *Phys. Rep.* **52**, 263 (1979).

Chorin, A. J., and Marsden, J. E., *A Mathematical Introduction to Fluid Mechanics*, 2nd edition, Springer-Verlag, Berlin (1990).

Cole, J. D., "On a Quasilinear Parabolic Equation Occurring in Aerodynamics," *Q. J. Appl. Math.* **9**, 225 (1951).

Corson, E. M., *Introduction to Tensors, Spinors, and Relativistic Wave-Equations*, Blackie & Sons, London (1955).

Crampin, M., and Pirani, F., *Applicable Differential Geometry*, Cambridge University Press, Cambridge (1986).

Cromer, A. H., and Saletan, E. J., "A Variational Principle for Nonholonomic Constraints," *Am. J. Phys.* **38**, 892 (1970).

Crutchfield, J. P., and Huberman, B. A., "Chaotic States of Anharmonic Systems in Periodic Fields," *Phys. Rev. Lett.* **43**, 1743 (1979).

Currie, D. G., and Saletan, E. J., "Canonical Transformations and Quadratic Hamiltonians," *Nuovo Cimento* **9B**, 143 (1972).

De Almeida, M., Moreira, I. C., and Yoshida, H., "On the Non-Integrability of the Störmer Problem," *J. Phys. A.* **25**, L227 (1992).

Debnath, L., *Nonlinear Water Waves*, Academic Press, New York (1994).

Delaunay, C. "Théorie du mouvement de la lune," *Mém. Acad. Sci. Paris*, XXVIII (1860); XXIX (1867).

Deprit, A., "Canonical Transformations Depending on a Small Parameter," *Cel. Mech.* **1**, 12 (1969).

Devaney, R. L., *An Introduction to Chaotic Dynamical Systems*, Benjamin/Cummings, Menlo Park (1986).

Dhar, A., "Non-Uniqueness in the Solutions of Newton's Equations of Motion," *Am. J. Phys.* **61**, 58 (1993).

Dirac, P. A. M., *The Principles of Quantum Mechanics*, 4th edition, Oxford University Press, London (1958).

Dirac, P. A. M., *Lectures in Quantum Field Theory*, Academic Press, New York (1966).

Doubrovine, B., Novikov, S., and Fomenko, A., *Géométrie Contemporaine*, première partie, Éditions MIR, Moscow (1982).

Dragt, A. J., "Trapped Orbits in a Magnetic Dipole Field," *Rev. Geophys.* **3**, 225 (1965).

Duffing, G., *Erzwungene Schwingungen bei Veränderlicher Eigenfrequenz*, Vieweg, Braunschweig (1918).

Eckhardt, B., "Irregular Scattering," *Physica* **D33**, 89 (1988).

Eisenbud, L., "On the Classical Laws of Motion," *Am. J. Phys.* **26**, 144 (1958).

Euler, L., "Problème: un corps étant attiré en raison réciproque carrée de distance vers deux points fixes donnés. Trouver le mouvement du corps en sens algébraique," *Mem. de Berlin für* **1760**, 228 (1767).

Faddeev, L., Takhatadzhyan, L., and Zakharov, V., "Complete Description of Solutions of the 'sine-Gordon' Equation," *Sov. Phys. Dokl.* **19**, 824 (1975). [Russian original in *Dokl. Akad. Nauk SSSR* **219**, 1334 (1974).]

Feigenbaum, M. J., "Quantitative Universality for a Class of Nonlinear Transformations," *J. Stat. Phys.* **19**, 25 (1978); "The Universal Metric Properties of Nonlinear Transformations," *J. Stat. Phys.* **21**, 669 (1979).

Feit, S. D., "Characteristic Exponents and Strange Attractors," *Commun. Math. Phys.* **61**, 249 (1978).

Fermi, E., Pasta, J., and Ulam, S. M., "Studies on Nonlinear Problems, Los Alamos Lab Report LA-1940, May 1955," in Fermi, E., *Collected Papers*, Vol. 2, p. 978, Univ. Chicago Press, Chicago (1965).

Feynman, R. P., *Surely You're Joking, Mr. Feynman!*, Norton, New York (1985).

Feynman, R. P., and Hibbs, A. R., *Quantum Mechanics and Path Integrals*, McGraw-Hill, New York (1965).

Feynman, R. P., Leighton, R. B., and Sands, M., *The Feynman Freshman Lectures on Physics*, Addison-Wesley, Reading (1963).

Floquet, G., "Sur les Equations Différentielles Linéaires," *Ann. de L'Ecole Normale Supérieure* **12**, 47 (1883).

Forbat, N., *Analytische Mechanik der Schwingungen*, VEB Deutscher Verlag der Wissenschaften, Berlin (1966).

Funaki, T., and Woyczynski, W. A., eds. *Nonlinear Stochastic PDE's: Hydrodynamic Limit of Burgers' Turbulence*, The IMA Volumes in Mathematics and Its Applications, Vol. 77, Springer-Verlag, New York (1996).

Galgani, L., Giorgilli, A., and Strelcyn, J. M., "Chaotic Motions and Transition to Stochasticity in the Classical Problem of the Heavy Rigid Body with a Fixed Point, II," *Nuovo Cimento* **61B**, 1 (1981).

Gardner, C. S., Green, J. M., Kruskal, D. M., and Miura, B. M., "Method for Solving the Korteweg de Vries Equation," *Phys Rev. Lett.* **19**, 1095 (1967).

Gaspard, P., and Rice, S. A., "Scattering from a Classically Chaotic Repellor [sic]," *J. Chem. Phys.* **90**, 2225 (1989).

Gel'fand, I. M., and Fomin, S. V., *Calculus of Variations*, Prentice-Hall, Englewood Cliffs (1963).

Gel'fand, I. M., and Shilov, G. E., *Generalized Functions. Vol. 1, Properties and Operations*, Academic, New York (1964).

Giacaglia, G. E. O., *Perturbation Methods in Non-linear Systems*, Applied Math. Sci., Vol. 8, Springer-Verlag, Berlin (1972).

Godreche, C., ed. *Solids Far from Equilibrium: Growth, Morphology and Defects*, Cambridge Univ. Press, Cambridge (1990).

Goldman, T., Hughes, R. J, and Nieto, M. M., "Gravity and Antimatter," *Scientific American*, p. 48, March (1988).

Goldstein, H., *Classical Mechanics*, 2nd edition, Addison-Wesley, Reading (1981).

Halliday, D., Resnick, R., and Walker, J., *Fundamentals of Physics*, 4th edition, Wiley, New York (1993).

Halmos, P. R., *Measure Theory*, Van Nostrand, New York (1950).

Hamilton, W. R., in *Elements of Quaternions*, C.J. Joly, ed., Chelsea, New York (1969).

Hannay, J. H., "Angle Variable Holonomy in Excursion of an Integrable Hamiltonian," *J. Phys. A* **18**, 221 (1985).

Hao, B.-L., *Chaos*, World Scientific, Singapore (1984).

Hartman, P., *Ordinary Differential Equations*, Birkhäuser, Boston (1982).

Heagy, J. F., "A Physical Interpretation of the Hénon Map," *Physica* **D57**, 436 (1992).

Hecht, E., and Zajac, A., *Optics*, Addison-Wesley, Reading (1974).

Hénon, M., "A Two-Dimensional Mapping with a Strange Attractor," *Commun. Math. Phys.* **50**, 69 (1976).

Herman, M. R., In *Geometry and Topology*, ed. J. Palais, Lecture Notes in Mathematics, Vol. 597, 271, Springer-Verlag, Berlin (1977).

Hill, H. G., "Mean Motion of the Lunar Perigee," *Acta. Math.* **8**, 1 (1886).

Hirota, R., "Exact Solution of the Korteweg-de Vries Equation for Multiple Collisions of Solitons," *Phys. Rev. Lett.* **27**, 1192 (1971).

Hopf, E., "The Partial Differential Equation $v_x + vv_x = vv_{xx}$," *Comm. Pure Appl. Math.* **3**, 201 (1950).

Hori, G., "Theory of General Perturbations with Unspecified Canonical Variables," *Publ. Astron. Soc. Japan* **18**, 287 (1966).

Hut, P., "The Topology of Three Body Scattering," *Astron. J.* **88**, 1549 (1983).

Jackson, E. A., *Perspectives of Nonlinear Dynamics*, Cambridge University Press, Cambridge (1990).

Jackson, J. D., *Classical Electrodynamics*, Wiley, New York (1975).

Jacobson, M. V., "Absolute Continuous Measures for One-Parameter Families of One-Dimensional Maps," *Commun. Math. Phys.* **81**, 39 (1981).

James, R. C., *Advanced Calculus*, Wadsworth, Belmont, CA (1966).

Jensen, M. H., Bak, P., and Bohr, T., "Transition to Chaos by Interaction of Resonances in Dissipative Systems. I. Circle Maps," *Phys. Rev.* **A30**, 1960 (1984).

José, J., Rojas, C., and Saletan, E. J., "Elastic Particle Scattering from Two Hard Disks," *Am. J. Phys.* **60**, 587 (1992).

Josephson, B. D., "Possible New Effects in Superconducting Tunneling," *Phys. Lett.* **A1**, 1 (1962).

Jung, C., and Scholz, H., "Chaotic Scattering off the Magnetic Dipole," *J. Phys.* **A21**, 2301 (1988).

Karatas, D. L., and Kowalski, K. L., "Noether's Theorem for Local Gauge Transformations," *Am. J. Phys.* **58**, 123 (1990).

Karplus, M., and Porter, R. N., *Atoms and Molecules*, Benjamin/Cummings, Boston (1970).

Keller, J. B., Kay, I., and Shmoys, J. "Determination of the Potential from Scattering Data," *Phys. Rev.* **102**, 557 (1956).

Khinchin, A. Ya., *Continued Fractions*, University of Chicago, Chicago (1964).

Kolmogorov, A. N., "General Theory of Dynamical Systems in Classical Mechanics," in *Proceedings of the 1954 International Congress of Mathematics*, p. 315. North Holland, Amsterdam (1957). [Translated as an appendix in Abraham and Marsden (1985).]

Korteweg, D. J., and de Vries, G., "On the Change of Form of Long Waves Advancing in a Rectangular Canal, and a New Type of Long Standing Waves," *Phil. Mag.* **39**, 422 (1895).

Krim, J., "Friction at the Atomic Scale," *Scientific American*, p. 74, October (1996).

Lagrange, J. L., "Recherches sur le mouvement d'un corps qui est attiré vers deux centres fixes," *Auc. Mem. de Turin* **4**, 118 (1766–69).

Lagrange, J. L., "Mécanique Analytique," in *Oeuvres*, Vol. XII, p. 25, Gauthier Villars, Paris (1889).

Lamb, H., *Hydrodynamics*, Cambridge University Press, Cambridge (1879). (Dover Reprint, 1945.)

Landau, L., and Lifshitz, E., *The Classical Theory of Fields*, 4th edition, Pergamon, Bristol (1975).

Landau, L., and Lifshitz, E., *Mechanics*, 3rd edition, Pergamon, Oxford (1976).

Landau, L., and Lifshitz, E., *Fluid Dynamics*, 2nd edition, Pergamon, Oxford (1987).

Lasker, J., "A Numerical Experiment on the Chaotic Behavior of the Solar System," *Nature* **338**, 237 (1989).

Lax, P. D., "Almost Periodic Solutions of the KdV Equation," *SIAM Review* **18**, 351 (1976).

Laub, A., Doderer, T., Lachenmann, S. G., Heubener, R. P., and Oboznov, V. A., "Lorentz Contraction of Flux Quanta Observed in Experiments with Annular Josephson Tunnel Junction," *Phys. Rev. Lett.* **75**, 1372 (1995).

Liapounoff, M. A., *Problème Général du Mouvement*, Princeton University Press, Princeton (1949).

Liapunov: see Liapounoff.

Lichtenberg, A. J., *Phase-Space Dynamics of Particles*, Wiley, New York (1969).

Lichtenberg, A. J., and Lieberman, M. A., *Regular and Chaotic Dynamics*, 2nd ed., Applied Math. Sci., Vol. 38, Springer-Verlag, Berlin (1992).

Liemanis, E., *The General Problem of the Motion of Coupled Rigid Bodies about a Fixed Point*, Springer-Verlag, Berlin (1965).

Linsay, P., "Period Doubling and Chaotic Behavior in a Driven Anharmonic Oscillator," *Phys. Rev. Lett.* **47**, 1349 (1981).

Liouville, J., "Note sur la Théorie de la Variation de Constantes Arbitraires," *J. Mathémathique Appliqûee* **3**, 342 (1838).

Lyapunov: see Liapounoff.

Mach, E., *Science of Mechanics* (5th Edition), Open Court, New York (1942).

Magnus, W., and Winkler, S., *Hill's Equation*, Wiley, New York (1966).

Makhankov, V. G., "Computer Experiments in Soliton Theory," *Comp. Phys. Comm.* **21**, 1 (1980).

Mandelbrot, B. B., *The Fractal Geometry of Nature*, Freeman, New York (1982).

Marmo, G., Saletan, E. J., Simoni, A., and Vitale, B., *Dynamical Systems; a Differential Geometric Approach to Symmetry and Reduction*, Wiley, Chichester (1985).

Marmo, G., Saletan, E. J., Schmid, R., and Simoni, A., "Bi-Hamiltonian Dynamical Systems and the Quadratic Hamiltonian Theorem," *Nuovo Cimento* **100B**, 297 (1987).

Marsden, J. E., Reilly, O. M., Wicklin, F. J., and Zombro, B. W., "Symmetry, Stability, Geometric Phases, and Mechanical Integrators," *Nonlinear Science Today* **1**, 4 (1991).

Maslov, V. P., and Fedoriuk, M. V., *Semi-Classical Approximations in Quantum Mechanics*, Reidel, Boston (1981).

Mattis, D. C., in *The Many-body Problem: an Encyclopedia of Exactly Solved Models in One Dimension*, Daniel C. Mattis, ed., World Scientific, Singapore (1993).

May, R. M., "Simple Mathematical Models with Very Complicated Dynamics," *Nature* **261**, 459 (1976).

McLachlan., N. W., *Theory and Applications of Mathieu Functions*, Clarendon Press, Oxford (1947).

Mehra, J., *The Beat of a Different Drummer: the Life and Science of Richard Feynman*, Clarendon Press, Oxford (1994).

Miller, W. M., "WKB Solution of Inversion Problems for Potential Scattering," *J. Chem. Phys.* **51**, 3631 (1969).

Miura, R. M., Gardner, C. S., and Kruskal, M. D., "Korteweg-de Vries Equations and Generalizations. II Existence of Conservation Laws and Constants of Motion," *J. Math. Phys.* **9**, 1204 (1968).

Monin, A. S., and Yaglom, A. M., *Statistical Fluid Mechanics: Mechanics of Turbulence*, Vols. I, II, MIT Press, Cambridge, MA (1973).

Montgomery, R., "How Much Does the Rigid Body Rotate? A Berry's Phase from the 18th Century," *Am. J. Phys.* **59**, 394 (1991).

Moore, E. N., *Theoretical Mechanics*, Wiley, New York (1983).

Morandi, G., Ferrario, C., LoVecchio, G., Marmo, G., and Rubano, C., "The Inverse Problem of the Calculus of Variations and the Geometry of the Tangent Bundle," *Phys. Rep.* **188**, 149 (1990).

Morse, P. M., and Feshbach, H., *Methods of Theoretical Physics*, International Series in Pure and Applied Physics, McGraw-Hill, New York (1953).

Moser, J., "On Invariant Curves of Area Preserving Mappings of an Annulus," *Nachr. Akad. Wiss. Göttingen Math. Phys.*, K1.II 1 (1962).

Moser, J., *Stable and Random Motion in Dynamical Systems*, Princeton University Press (Ann. Math. Studies), Princeton (1973).

Nayfeh, A. H., and Mook, D. T., *Nonlinear Oscillations*, Wiley-Interscience, New York (1979).

Neumann, C., "Deproblemate quodam mechanico, quod ad primam integralium ultraellipticorum classem revocatur," *Journal für die Reine und Angewandte Mathematik*, **56**, 46–63 (1859).

Noether, E., "Invariante Variationsprobleme," *Nachr. d. König. Ges. d. Wiss., Göttingen*, 235 (1918).

Ohanian, H. C., *Physics*, 2nd edition, Norton, New York (1989).

Ott, E., *Chaos in Dynamical Systems*, Cambridge University Press, Cambridge (1993).

Ott, E., and Tél, T., "Chaotic Scattering: an Introduction," *Chaos* **3**, 417 (1993).

Ozorio de Almeida, A. M., *Hamiltonian Systems: Chaos and Quantization*, Cambridge University Press, Cambridge (1990).

Perelomov, A. M., *Integrable Systems of Classical Mechanics and Lie Algebras*, Birkhäuser, Boston (1990).

Perring, J. K., and Skyrme, T. H. R., "A Model of Unified Field Equations," *Nucl. Phys.* **31**, 550 (1962).

Piña, E., "On the Motion of the Symmetric Lagrange Top," *Rev. Mex. Fis.* **39**, 10 (1993).

Poincaré, H., *Leçons de Mécanique Céleste*, Vol. 2, Gauthiers-Villars, Paris (1909–12).

Poincaré, H., *New Methods in Celestial Mechanics*, Vol. 2, ed. D.L. Goroff, AIP (1993). Original: *Les Méthodes Nouvelles de la Mécanique Celeste* (1892–9).

Press, W. H. et al., *Numerical Recipes in FORTRAN, 2nd edition*, Cambridge University Press, Cambridge (1992). [Note: Other versions exist for Basic, C, Pascal, etc.]

Pullen R. A., and Edmonds A. R., "Comparison of Classical and Quantal Spectra for a Totally Bound Potential," *J. Phys. A, Math. Phys. Gen.* **14**, L477–L485 (1981).

Reitz, J. R., and Milford, F. J., *Foundations of Electromagnetic Theory*, Addison-Wesley, Reading (1964).

Richards, I., "Continued Fractions Without Tears," *Mathematics Magazine* **54**, 163 (1981).

Rivlin, T. J., *The Chebyshev Polynomials*, Wiley, New York (1974).

Russell, J. S., "Report on Waves," *14th Meeting of the British Association Report*, York (1844).

Russell, D. A., Hanson, J. D., and Ott, E., "Dimension of Strange Attractors," *Phys. Rev. Lett.* **15**, 1175 (1980).

Saletan, E. J., and Cromer, A. H., *Theoretical Mechanics*, Wiley, New York (1971).

Sandoval-Vallarta, S. M., "Theory of the Geomagnetic Effect of Cosmic Radiation," *Handbuch Der Physik* **XLVI/I**, 88 (1961).

Saslavsky, G. M., Sagdeev, R. Z., Usikov, D. A., and Chernikov, A. A., *Weak Chaos and Quasi-Regular Patterns*, Cambridge Univ. Press, Cambdridge (1991).

Schutz, B., *Geometrical Methods of Mathematical Physics*, Cambridge Univ. Press, Cambridge (1980).

Schwartz, L., *Théorie des Distributions*, Hermann, Paris (1966).

Schwarz, M., "Commuting Flows and Invariant Tori: Korteweg-de Vries," *Adv. Math.* **89**, 192 (1991).

Scott, A. C., "A Nonlinear Klein–Gordon Equation," *Am. J. Phys.* **37**, 52 (1969).

Shaw, R., "Strange Attractors, Chaotic Behavior, and Information Flow," *Z. Naturforsch* **36a**, 80 (1981).

Shilov, G. E., *An Introduction to the Theory of Linear Spaces*, Dover, New York (1974).

Sommerfeld, A., "Zur Quantentheorie der Spektrallinien," *Ann. Physik* **51**, 1 (1916).

Spivak, M., *Calculus on Manifolds; A Modern Approach to Classical Theorems of Advanced Calculus*, Benjamin, New York (1968).

Stark, H., *An Introduction to Number Theory*, Markham, Chicago (1970).

Störmer, C., *The Polar Aurora*, Oxford at the Clarendon Press, London (1955).

Strand, M. P., and Reinhardt, W. P., "Semiclassical Quantization of the Low-Lying Electronic States of H_2^+," *J. Chem. Phys.* **70**, 3812 (1979).

Strogatz, S. H., *Nonlinear Dynamics and Chaos*, Addison-Wesley, Reading (1994).

Sudarshan, E. C. G., and Mukunda, N., *Classical Dynamics: A Modern Perspective*, Wiley, New York (1974).

Sussman, G. J., and Wisdom, J., "Chaotic Evolution of the Solar System," *Science* **257**, 56 (1992).

Tabor, M., *Chaos and Integrability in Nonlinear Dynamics*, Wiley, New York (1989).

Taylor, E., and Wheeler, J. A., *Spacetime Physics*, Freeman, San Francisco (1964).

Testa, J., Pérez, J., and Jeffries, C., "Evidence for Universal Chaotic Behavior of a Driven Nonlinear Oscillator," *Phys. Rev. Lett.* **48**, 714 (1982).

Toda, M., *Theory of Nonlinear Lattices*, Springer-Verlag, Berlin (1987).

Tricomi, F. G., *Integral Equations*, Wiley, New York (1957).

Trofimov, V. V., and Fomenko, A. T., "Liouville Integrability of Hamiltonian Systems on Lie Algebras," *Uspekhi Mat. Nauk* **39**, 3 (1984) [in Russian; transl: *Russian Math. Surveys* **39**, 1 (1984)].

Uhlenbeck, K., "Equivariant Harmonic Maps into Spheres," in "Harmonic Maps," *Proceedings*, New Orleans, 1980, edited by R. J. Knill, M. Kalka and H. C. Sealy. *Springer Lecture Notes in Mathematics*, Vol. 949, 146 (1982).

Ulam, S. M., and von Neumann, J., "On Combinations of Stochastic and Deterministic Processes," *Bull. Am. Math. Soc.* **53**, 1120 (1947).

van der Pol, B., "On Relaxation-Oscillations," *Phil. Mag.*, **2**, 987 (1926).

von Zeipel, H., "Recherches sur le Mouvement des Petits Planets," *Arkiv. Astron. Mat. Phys.*, **11, 12, 13** (1916–17).

Whitham, G. B., *Linear and Nonlinear Waves*, Pure and Applied Mathematics, Wiley, New York (1974).

Whittaker, E. T., *A Treatise on the Analytical Dynamics of Particles and Rigid Bodies*, Dover, New York (1944).

Whittaker, E. T., and Watson, G. N., *A Course of Modern Analysis*, Cambridge University Press, Cambridge (1943).

Wiggins, S., *Introduction to Applied Nonlinear Dynamical Systems and Chaos*, Springer-Verlag, Berlin (1990).

Wisdom, J., see Sussman, G. J.

Zabusky, N. J., and Kruskal, M. D., "Interaction of Solitons in a Collisionless Plasma and the Recurrence of Initial States," *Phys. Rev. Lett.* **15**, 240 (1965).

INDEX

Printed in the United States
By Bookmasters